38th Riso International Symposium on Materials Science 2017

IOP Conference Series: Materials Science and Engineering
Volume 219

Riso, Denmark
4 – 8 September 2017

ISBN: 978-1-5108-4557-2
ISSN: 1755-1307

Printed from e-media with permission by:

Curran Associates, Inc.
57 Morehouse Lane
Red Hook, NY 12571

CURRAN ASSOCIATES INC. proceedings.com

Some format issues inherent in the e-media version may also appear in this print version.

Printed by Curran Associates, Inc. (2017)

For permission requests, please contact the Institute of Physics
at the address below.

Institute of Physics
Dirac House, Temple Back
Bristol BS1 6BE UK

Phone: 44 1 17 929 7481
Fax: 44 1 17 920 0979

techtracking@iop.org

Additional copies of this publication are available from:

Curran Associates, Inc.
57 Morehouse Lane
Red Hook, NY 12571 USA
Phone: 845-758-0400
Fax: 845-758-2633
Email: curran@proceedings.com
Web: www.proceedings.com

Table of contents

Volume 219

38th Risø International Symposium on Materials Science
4–8 September 2017, Risø, Denmark

Accepted papers received: 6 July 2017
Published online: 1 August 2017

Preface

Invited papers

Contributed papers

38th Risø International Symposium on Materials Science - 4-8 September 2017

Advanced Metallic Materials by Microstructural Design

Materials design is a multidisciplinary theme which combines processing, microstructure, properties and performance of materials. The symposium will focus on design of advanced metallic materials with a structural scale from the micrometer to the nanometer dimension in the bulk and at the surface/subsurface region. Material groups will reach from conventional polycrystalline metals to multilayer composites, gradient metals and super strong steels. Processing will cover a variety of plastic and thermal treatments including syntheses by deposition techniques and 3D printing. Characterization techniques will be ex-situ and in-situ by electron, x-ray and neutron diffraction methods.

A goal of the symposium is to extend our knowledge of physical mechanisms, which control metallurgical reactions and microstructure. Experimental characterization will form the basis of analytical and numerical modelling on multiple length scales, also including materials performance under a diversity of mechanical and chemical loading conditions. An expected outcome of the symposium is guidelines and models with predictive capability of scientific and industrial interest and of relevance to the use of metallic materials in important sectors in society as energy, transport and defence.

The Proceedings contain 9 invited papers, and 46 contributed papers.

The 38th Risø International Symposium is organized by the Section of Materials Science and Advanced Characterization, Department of Wind Energy, Technical University of Denmark (DTU), at the Risø Campus. We would like to thank all those at DTU Wind Energy who assisted in the preparation of the Symposium and we gratefully acknowledge financial support from Civilingeniør Frederik Leth Christiansens Almennyttige Fond and Fabrikant Mads Clausens Fond.

Peer review statement

All papers published in this volume of *IOP Conference Series: Materials Science and Engineering* have been peer reviewed through processes administered by the proceedings Editors. Reviews were conducted by expert referees to the professional and scientific standards expected of a proceedings journal published by IOP Publishing.

Degradation of metallic materials studied by correlative tomography

T L Burnett[1]*, N J H Holroyd[2], J J Lewandowski[2], M Ogurreck[3], C Rau[3], R Kelley[4], E J Pickering[1], M Daly[1], A H Sherry[4], S Pawar[1], T J A Slater[1], P J Withers[1]

[1]School of Materials, The University of Manchester, Oxford Road, Manchester, M13 9PL, UK

[2]Department of Materials Science and Engineering, Case Western Reserve University, Cleveland, OH, USA

[3]Diamond Light Source, Harwell Science and Innovation Campus, Fermi Ave, Didcot OX11 0DE

[4]Thermo Fisher (previously FEI Company), 5350 NE Dawson Creek Drive, Hillsboro, OR 97124, USA

[5]National Nuclear Laboratory, Chadwick House, Birchwood Park, Warrington, WA3 6AE, UK

*timothy.burnett@manchester.ac.uk

Abstract. There are a huge array of characterization techniques available today and increasingly powerful computing resources allowing for the effective analysis and modelling of large datasets. However, each experimental and modelling tool only spans limited time and length scales. Correlative tomography can be thought of as the extension of correlative microscopy into three dimensions connecting different techniques, each providing different types of information, or covering different time or length scales. Here the focus is on the linking of time lapse X-ray computed tomography (CT) and serial section electron tomography using the focussed ion beam (FIB)-scanning electron microscope to study the degradation of metals. Correlative tomography can provide new levels of detail by delivering a multiscale 3D picture of key regions of interest. Specifically, the Xe^+ Plasma FIB is used as an enabling tool for large-volume high-resolution serial sectioning of materials, and also as a tool for preparation of microscale test samples and samples for nanoscale X-ray CT imaging. The exemplars presented illustrate general aspects relating to correlative workflows, as well as to the time-lapse characterisation of metal microstructures during various failure mechanisms, including ductile fracture of steel and the corrosion of aluminium and magnesium alloys. Correlative tomography is already providing significant insights into materials behaviour, linking together information from different instruments across different scales. Multiscale and multifaceted work flows will become increasingly routine, providing a feed into multiscale materials models as well as illuminating other areas, particularly where hierarchical structures are of interest.

1. Introduction to correlative tomography

In this paper correlative tomography [1] will be introduced and in particular how plasma focused ion beam microscopy (PFIB) [2] has been used to extend this.

Much can be learnt from observations of metallographic sections, to better understand many metal systems. However 3D characterization is also sometimes required, for example to understand complex interfaces, porous networks and phase distributions. Fortunately, there are an increasing number of powerful characterisation techniques available, for example, X-ray computed tomography (CT) [3], serial section tomography (SST) using either optical or electron imaging [4-5], transmission electron tomography [6], and atom probe tomography [7], with each covering different length scales.

Traditionally, data collected at different scales are not spatially correlated. Instead, features at the different scales are captured by various different techniques in parallel on different samples and the data brought together where necessary in a statistical manner. This approach is limited when the observed behaviour is rare or relies on extremes, for example the initiation and growth of fatigue cracks. In such cases it is sometimes highly desirable to track the evolution of specific sites (e.g. locations where microcracks begin to form) and to study these unusual features and understand their behaviour across a range of length scales. In many cases, including the fatigue example, the process evolves over time to cover different length scales. For example, a fatigue crack may initiate due to a persistent slip band in a single grain of a metal microstructure at the surface of a specimen, the crack then starts to form and starts to propagate across the grains and soon will cover many grains. Final failure will typically occur once the fatigue crack reaches a critical length and there is fast catastrophic failure of the entire sample or component. As such the key length scales evolve from microns (e.g. a single grain or smaller), to hundreds of microns (e.g. multiple grains), to the tens of centimetre (entire component) scale. Moreover, with each different length scale there is also a different timescale that should be used to effectively observe the process.

Figure 1 illustrates a multiscale and multi-instrument scheme. The collection of chemical information as well as morphological information also provides a multi-faceted picture of the sample. The example shows a sample imaged in 3D non-destructively (e.g. by X-ray CT) under a mechanical or environmental load (1). A region of interest (RoI) is identified from this sample and this sub-region is prepared for subsequent imaging (2), which could be laser machining or electro-discharge machining (EDM) out the RoI, for example. The sample is then ready to be imaged destructively in 3D (3). Several higher resolution 3D imaging techniques rely on the scanning electron microscope (SEM), for imaging but use different methods of slicing to prepare the serial sections, for example, mechanical polishing [8], broad ion beam milling [9], and focussed ion beam milling [10], or ultramicrotomy for softer materials [11]. During this high resolution 3D imaging a further region of interest is identified, which could, for example, be a buried inclusion or the tip of a crack or a particular grain boundary. A sample including the region of interest is then prepared (4) at this scale usually by FIB machining, but possibly also by etching/electropolishing/microtomy, although site specific targeting is more difficult with these latter techniques. Lastly, the nanoscale chemistry and morphology of this region of the sample could be characterized (5), for example by atom probe tomography or STEM-EDS tomography.

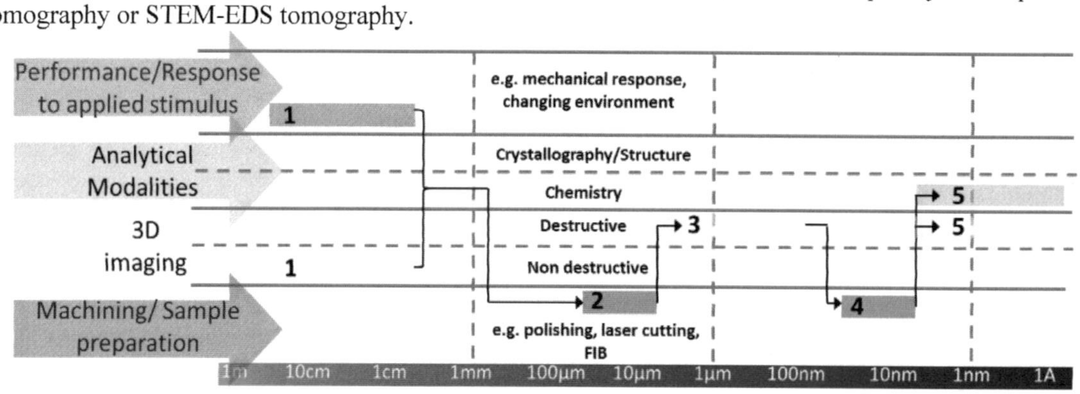

Figure 1. Diagram showing a workflow for a multiscale materials analysis. The horizontal axis is lengthscale and the large coloured arrows show different aspects related to understanding material

properties. The workflow, connected by the black arrows, gives an example combination of techniques and scales.

It is becoming increasingly important to be able to efficiently carry out such workflows in order to build up information for multiscale models or to identify the critical length scales that govern a damage mechanism. It is the ability to effectively couple existing and highly specialized pieces of equipment to follow the same RoI, often with demanding sample requirements, that will allow the fullest exploitation of the currently available technology. There are several major benefits to this approach, such as being able to find rare regions of interest through coarse scanning for subsequent detailed analysis and also by providing context and proof of the representativeness of a minute region chosen for more detailed analysis. One of the major challenges is that many instruments and techniques have been refined and developed regardless of whether this makes it more difficult to effectively link it to other techniques. There is also benefit in targeting specific areas for detailed analysis relating to managing the data generated. Without proper targeting it is easy to collect very detailed information for large volumes and create unwieldy datasets which are costly and time-consuming to analyse. If it is possible to positively identify smaller regions which require the detailed analysis, this not only limits the quantity of data collected it may also be more practical to collect further facets of information, for example the chemistry, crystallography, mechanical response etc. which would be practically prohibitive for larger volumes at high resolution.

The correlative aspect of such workflows comes in two ways: coordinate-correlated, and feature-correlated. Coordinate-correlated volumes use dimensional measurements to provide the spatial registry through a coordinate system, for example the coordinates of the sample stage in the microscope. Alternatively it is possible to register to sub-volumes by identifying specific features common to both the original data volume and the sub-volume. However in order for a method to be correlative it is necessary to be able to register the sample features from one level to another. The resolution of the two techniques means that the same features can be seen in both images and confidently registered with complementary information coming from both results [12-14]. Multiple layers of different information can be built up and correlated based on common features between many different techniques e.g. within the AFM [15].

In this paper a number of characterization approaches will be presented. Each of these will be discussed individually but then also with respect to integration into more extensive correlative tomography workflows.

2. Time-lapse imaging by X-ray CT

Time-lapse imaging using X-ray CT, also sometimes loosely referred to as 4D imaging, is an extremely powerful way of understanding behaviour and performance, as it is possible to observe the evolution of the phenomena [16]. It is often not possible to identify post-mortem the sequence of events that has led to the final condition of the specimen. As such, time-lapse imaging can be invaluable, however it is often experimentally challenging, creates enormous amounts of data and may demand the use of national facilities such as synchrotron sources, which can lead to high costs and/or long lead times. The big-data challenge is particularly formidable as detailed post-processing and analysis is often required to make sense of this data.

Time-lapse imaging has been used to investigate many processes including fatigue in metals and composites [17-18], ductile fracture [19], corrosion [20], battery performance [21], geological processes [22] and even insect life-cycles [23] and flight [24]. It is possible to push the temporal resolution to exceptionally fast times, down to <<1 second, for a full tomogram [3], but equally, long duration experiments are possible, with samples imaged once every few days or weeks so that long duration processes may also be observed. The intention in each case to observe the phenomena at realistic time scales. This leads to one of the major considerations in carrying out time-lapse experiments, that of matching the timescales of the process to the requirements of the imaging.

Taking the example of a growing corrosion pit and the fact that the pit takes 12 hours to grow to full size of 100 μm across (the appearance of the pit is a hemispherical surface cavity, with a diameter of 100 μm), then assuming a constant growth rate we would expect a growth of just over 8 μm per hour. From ex situ experimentation it was discovered that during the total growth there are periods of faster and slower growth such that it is estimated the growth rate fluctuates between 4 and 12 μm growth per hour. Here the growth rate determines the time within which a tomography scan must be recorded. For example, it is clear that if the 3D imaging takes 6 hours, only one tomogram can be recorded within the growth period of the pit. Furthermore, as the pit will have grown significantly in the time between the first projection and the last the image will be blurred, leading to poor quality data. The spatial resolution of the imaging is also intimately linked to the temporal resolution. Taking the average growth rate of 8 μm/h and assuming one tomogram takes 1 hour to record, this will only be worthwhile if the spatial resolution is better than 8 μm in order to resolve the changes. This is a careful balance however as too high resolution would again lead to blurring in the reconstructed images as small changes are occurring during the collection of the raw projection data. Another consideration is the field of view, which must be sufficient to capture the features of interest at the start and finish of the time period where the sample is observed. This may not be the case if the resolution is too high (current detectors mean that the field of view is generally ~1000x the spatial resolution [3]).

2.1 Exemplar: environmentally induced cracking of AA5083-H131 aluminium alloy

In this exemplar the environmentally induced cracking of a 5xxx alloy is investigated. Ex-situ analysis via X-ray CT had revealed a previously un-reported mode of environmentally assisted fracture [25]. Time-lapse X-ray CT was carried out to observe the crack propagation in situ to help elucidate the nature of propagation and mechanisms of this fracture mode. In this exemplar mm-sized specimens are investigated with microscale resolution with a temporal resolution of the order of minutes.

Round bar tensile specimens of 3.2 mm diameter and 12.7 mm gauge length were prepared from a 29 mm thick rolled plate of Al-Mg-Mn alloy, AA5083-H131, in the short transverse direction. The alloy had a typical 'pancake-like' grain structure with a grain size of 150 μm (L), 80 μm (LT), and 35 μm (ST) directions respectively [26].

The samples were then sensitized at 80 °C for 250 hours to create a susceptible microstructure. Following sensitization, the tensile samples were soaked in a 0.6M NaCl solution for 250 hours. This achieved two things: 1) The creation of sites of intergranular corrosion on the sample surface, and 2) charging of the samples with hydrogen. The experiment is described in further detail by Holroyd et al [25].

Immediately following the sensitization/pre-exposure treatments, the tensile specimens were subjected to slow strain rate testing (SSRT), typically employing a nominal strain rate of $\sim 10^{-5}$ /s, while being exposed to humid laboratory air (relative humidity, 50%).

The sample was interrupted just before the final failure. This 'intact' but heavily cracked sample was then analyzed using a Zeiss Xradia Versa 520 X-ray CT system operating at 70 kV and using the 4x coupled optical magnification and an exposure time of 30 s and 2001 projections, providing a voxel size of $(1.7\ \mu m)^3$. The time-lapse data were collected on samples prepared in the same way and tested under the same conditions, but this time on the I13 imaging beamline at Diamond Light Source. A pink beam was used to maximize flux around a nominal energy of 35 keV. The detector provided a 4.5 x 5.6 mm field of view with a 2.6 μm voxel size. The exposure time was 200 ms for each of the 1200 projections. The tensile testing was conducted using a Deben CT5000 in-situ rig with local sample containment creating high local humidity.

Figure 2 shows multiple views of the tested and interrupted test specimens. The virtual slices (figure 2b) and 3D rendering of the cracking in the sample (figures 2c, d and e) show the morphology of the environmentally induced cracks that resulted. As shown in figure 3(a), and in comparison to figure 2, it is possible to see three different stages of crack propagation. The first stage is intergranular corrosion (IGC). These sites are formed in the pre-exposure step and act as stress raisers to initiate the 'Type 1' cracking, which is intergranular stress corrosion cracking (IGSCC). Type 1 cracking is characterized by its very flat, smooth and direct morphology. After a period of propagation of the

Type 1 crack the cracking changes to Type 2 cracking. Type 2 cracking is easily identified from the virtual slices of the X-ray CT data as being highly topographical, composed of steps linked by almost vertical tears. This mode of fracture has previously been overlooked and is often difficult to see from the fracture surface, partly due to the great depth of field in the scanning electron microscope not highlighting the topographical changes in the fracture surface. If taken to failure there is also a final fast fracture region consisting of a 45° shear lip.

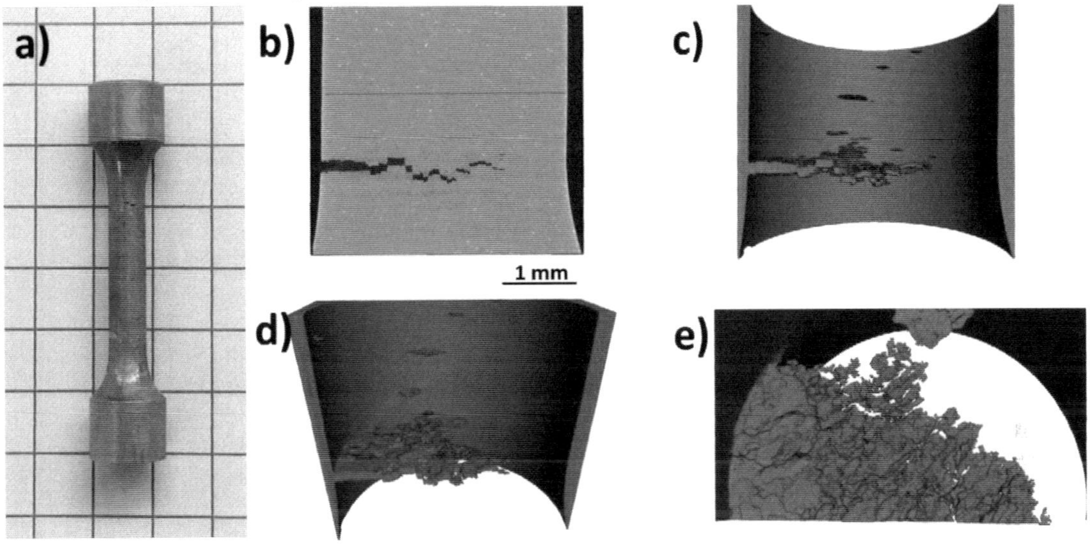

Figure 2 a) A photograph of the 3.2 mm diameter AA5083-H131 round bar tensile specimen prepared in the short transverse direction. b) virtual slice of the reconstructed X-ray CT data showing the large crack. c), d) and e) Volume Renderings of the crack shown from different directions with the crack visualized as if it were solid and the metal rendered transparent.

The other curious feature of the Type 2 cracking is the fact that surrounding this crack region there are multiple regions of isolated internal damage (figure 3a). Such regions do not occur around the Type 1 crack. The regions are associated with second phase particles (see black arrows). This observation, coupled with the knowledge of the mechanical characteristics of the failure, has led to a proposed mechanism for the propagation of the Type 2 cracking illustrated in figure 3(b). The diagram shows how the process zone ahead of the crack tip creates damage in the material which is isolated. These isolated regions of damage are then linked through regions of microvoid coalescence and the crack jumps forward. The crack halts as damage has not formed further ahead than the process zone and as such the crack arrests when there is sufficient intervening metal, i.e., no route to link isolated damage regions. The sample 're-loads' and the isolated damage is formed ahead of the new crack tip position before the next jump. This mechanism was proposed by Holroyd et al [26] from ex-situ observations of X-ray CT and SEM fractography data.

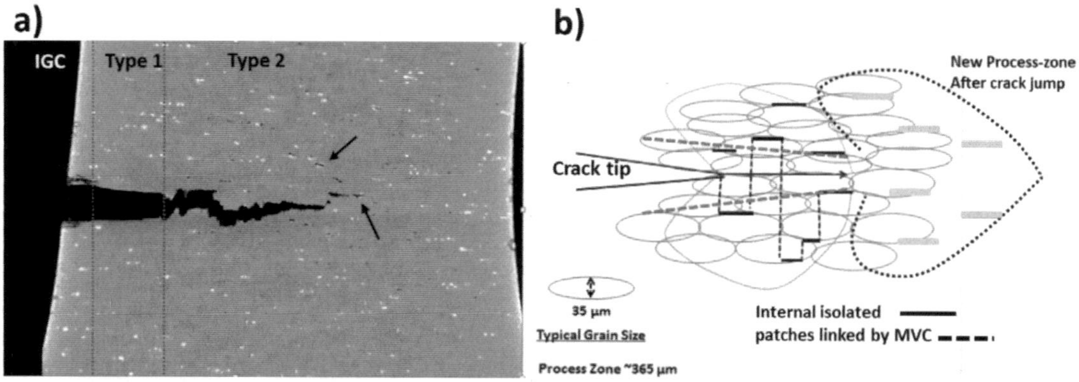

Figure 3 a) Virtual slice of the reconstructed data showing in detail the different stages of cracking and also regions of isolated damage (black arrows) ahead of the Type 2 crack. b) Diagram detailing the proposed mechanism of propagation for the Type 2 cracking.

Figure 4 shows results from the time-lapse data recorded at I13 Diamond Light Source. The data is taken from a sample tested under the same conditions and shows two examples of two different locations, where a crack tip can be seen, as well as a region of isolated damage ahead of the crack tip (time=t_1). Four minutes elapsed during the collection of the 3D tomograms, of which these are virtual slices, and in that time the crack can be seen to have linked to the isolated regions via a near-vertical steps. Once at the new position at time=$t_{1+4mins}$ new regions of isolated damage have formed ahead of the crack tip. The regions ahead of the crack shown here were confirmed to be isolated looking through multiple slices around these regions. It has been confirmed that the crack propagation stops and starts and that the crack tip only has influence a limited distance ahead of it. This confirms the theory proposed from *ex situ* testing and adds clarity to the process for this newly reported mode of environmentally induced cracking. A number of further questions have also been raised such as whether the sudden spurt in growth, also associated with a jump in crack opening displacement, leads to blunting of the crack tip.

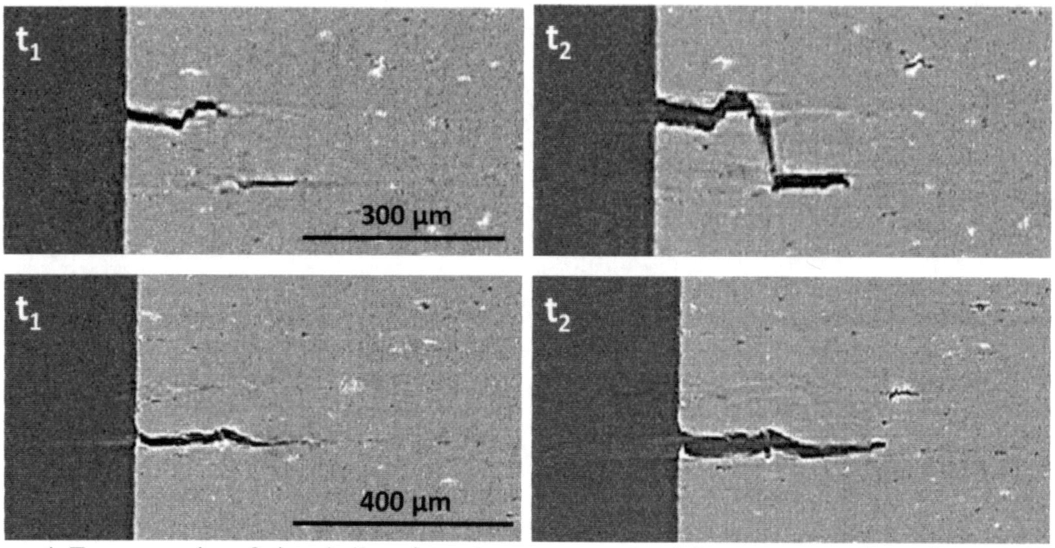

Figure 4. Two examples of virtual slices from the same sample of the reconstructed synchrotron CT data of the same position at two different time steps, t_1 and $t_{1+4mins}$. The crack tip progresses and when at the new position new regions of isolated damage ahead of the crack tip emerge which were not there previously.

3. Xe^+ ion plasma FIB: a tool for enabling multiscale analysis

Often there is the desire to investigate a specimen at multiple scales. In the example given in 2.1 it may still be desirable to investigate finer scale features, such as the grain boundary chemistry at the nanoscale. This may be an important aspect to in ultimately understanding the crack paths in the material. Recent work has shown how PFIB serial sectioning has been used for the collection of large, yet high resolution, and how datasets can be effectively used in conjunction with X-ray CT imaging (and indeed time-lapse X-ray CT) in a correlative manner [27].

The PFIB is an exciting tool that opens up the range of scales accessible with FIB technology in to the 100's of microns range whilst retaining the same nanoscale resolution [2, 28, 29]. In a 'traditional' Ga^+ FIB a volume of $(50\ \mu m)^3$ would be considered large, and whilst the milling of finer slices, down to ~10 nm thick, is possible, the practically achievable volume remains a limitation. Using a PFIB instrument volumes $(300\ \mu m)^3$ with 500 nm slice thickness have recently been demonstrated. It is also possible to combine the imaging of each slice with analytical data such as energy dispersive X-ray spectroscopy (EDS) and electron backscatter diffraction (EBSD). The desire to collect large 3D EBSD volumes is currently practically limited, with the bottleneck related to the EBSD map collection time, which typically accounts for 60-80% of the total time period for the completion of one 'slice' of data. However, recent developments in both the hardware [30] and software [31, 32] show great promise for limiting this bottleneck in the future. Whilst the nature of the damage is still being investigated, it appears that the Xe^+ is less damaging for equivalent milling parameters and, due to the inert nature of xenon, chemical incompatibility issues with elements such as Al and Cu are avoided.

3.1 Exemplar: ductile failure of SA508

In this example a volume of interest identified from X-ray CT data is analyzed in detail using serial sectioning tomography using the PFIB in order to understand the ductile failure mechanisms of a reactor pressure vessel SA508 Grade 3 steel. In this exemplar specimens are investigated from the cm-scale down the nanometer scale on a post-mortem basis.

The material investigated originated from a large SA508 Grade 3 [33] steel ring forging, in the quenched-and-tempered condition. Fracture toughness tests on the material were performed according to the European Structural Integrity Society (ESIS) P2-92 [34] standard, with tests performed on a sample of thickness B = 25 mm and width, W = 50mm using both the unloading compliance and the multi-specimen methods at room temperature. In order to probe the distribution of voids with depth using X-ray CT, cylindrical samples of ~0.5 mm diameter were prepared using electrical discharge machining (EDM) with the top of each sample incorporating the original fracture surface. A core was extracted from a region in the centre of the ductile fracture region ~2 mm ahead of the fatigue pre-crack.

X-ray CT was conducted using a Zeiss Xradia Versa 500 X-ray CT system. The voxel size was $(0.8\ \mu m)^3$ giving a resolution of 2-3 µm. The scan was performed at 100 kV, collecting 1,601 projections with an exposure time of 5 s per projection at 4x magnification. A FEI Helios PFIB was used for high-resolution investigations of the microstructure of the sample. A volume 150 x 115 x 18 µm in size was prepared with a slice thickness of 100 nm and a pixel size of 26 nm. Electron images were collected at 2 kV, 800 pA using the through-lens detector. The slices were milled using an accelerating voltage of 30 kV and current of 59 nA, providing the best balance of speed of milling and quality. Data analysis and visualization for all datasets was performed with the Avizo 9.0.0 software.

Figure 5a shows a surface rendering of the EDM core from the X-ray CT data, giving an appearance of the sample similar to that observed within the SEM. A sub-surface volume of interest was identified from the X-ray CT data including two voids of different geometries located ~600 µm below the fracture surface (see orange box in figure 5b)). The 3D rendering of the sample surface figure 5a) was used to relate the location of the voids (figure 5b)) to surface features, which were used as points of registration to plan the preparation of the volume for serial sectioning using the PFIB. The 3D rendering of the surface were directly compared to SEM images of the sample (see figure 5c)) and features registered to enable the accurate machining of the specimen for serial sectioning analysis through the identified voids. Figure 5(d) shows the PFIB prepared sample with a block created ready for serial sectioning.

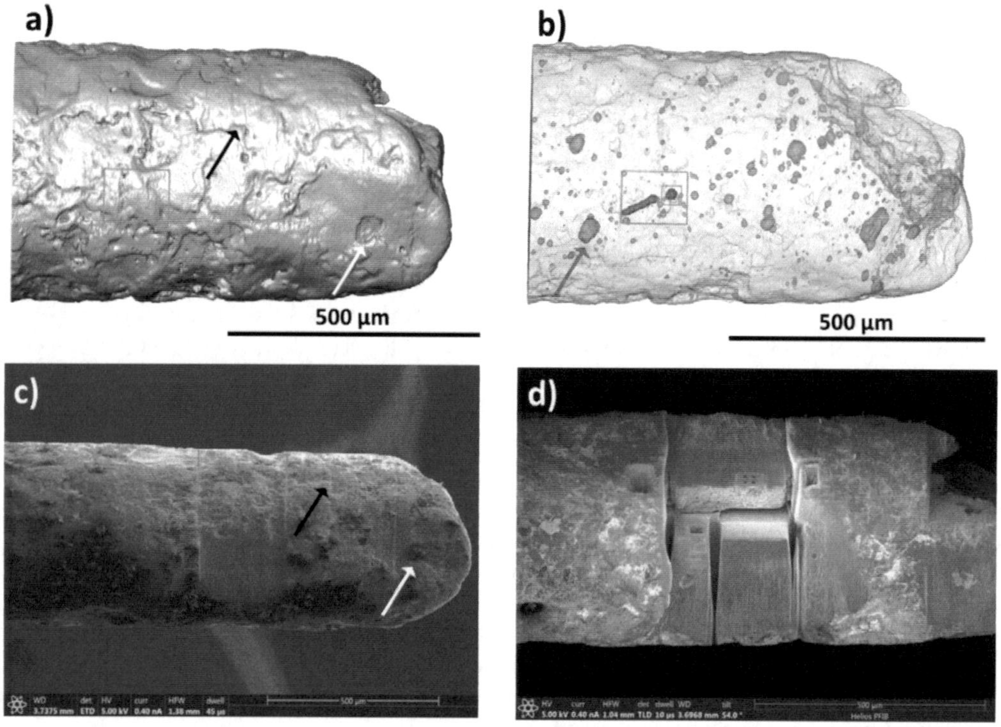

Figure 5 a) Volume Rendering of the reconstructed X-ray CT data showing the surface of the core specimen. The orange box shows the region of interest determined through comparison with b). The black and white arrows show two of the features used for registering the X-ray CT data to the SEM image of the sample shown in c). b) 3D rendering of the sample showing the internal voids. The orange box highlights two distinctive voids for further analysis. The red arrow shows another large void which can also be seen in the PFIB-prepared volumes in d) and used a reference point during machining. c) Secondary electron SEM image of the core, the contrast is dominated by some fine scale roughness but two feature used for correlation to the X-ray CT data in a) are shown with black and white arrows. d) Secondary electron image of the sample of preparation of the volume of interest for serial sectioning. The red arrow highlights a large void also visible in b) indicating the success of the correlation.

Figure 6(a) shows again the X-ray CT data highlighting the internal voids and the volume of interest. Figure 6(b) shows the region of interest as captured using the PFIB. The detailed microstructure of the steel is clear and the voids have been highlighted in red. Figure 6(d) shows the same region but this time with the inclusions captured highlighted in blue. Out of 9 of the voids captured within the PFIB sectioned volume 8 were associated with a manganese sulphide inclusion as determined by EDX (not shown). Figure 6(c) shows a magnified region of the PFIB data immediately adjacent to the large spherical void. This highlights the detail of the microstructure captured and shows an example of small voids that are associated with the carbides in the microstructure. These voids are associated with carbides and form close to the large voids and may be important in linking the large voids together in the final stages of fracture just prior to failure.

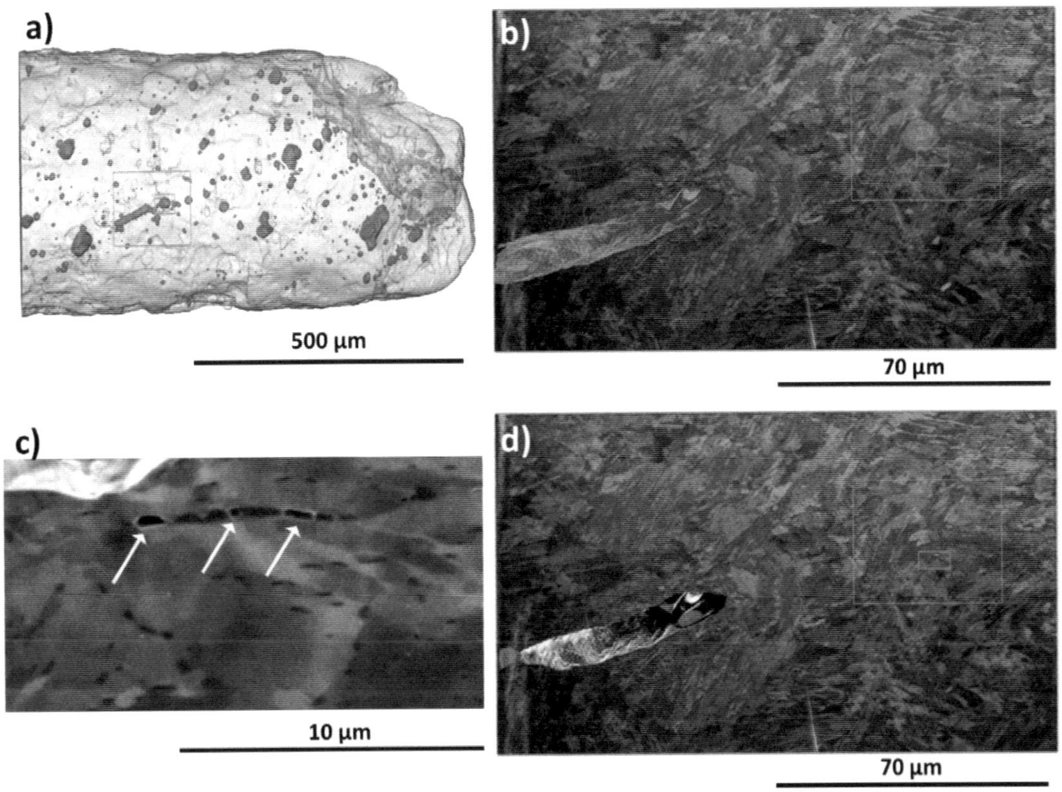

Figure 6 a) 3D rendering of the sample showing the internal voids. The orange box highlights two distinctive voids for further analysis using the PFIB. c) Though Lens Detector SEM image of one of the PFIB slices. Voids are highlighted in red, the two largest voids are also visible in a) but several other smaller voids are also visible. The large orange box highlights the large spherical void and the smaller orange box highlights the sub-region shown in c). c) shows a magnified region of b) showing the location of sub-micron voids (white arrows) associated with carbides which are found around the large voids. d) same as b) but highlighting in blue the inclusions associated with the voids.

Use of the PFIB made it possible to extend the range of scales investigated and also add additional information regarding the microstructural context of the voids. The X-ray CT imaging allowed a sufficient volume of material to be characterized to produce statistically significant information on the size distribution, void volume fraction and inter-particle distances to understand trends in the voids as a function of depth beneath the fracture surface. The PFIB serial section data is not able to extend to mm-sized volumes like the X-ray CT, but it is a multiscale dataset in its own right and effectively overlaps with the scales captured in X-ray CT. The X-ray CT volume captured effectively resolves the voids 10's of microns in size whilst at the same time capturing carbides 100's of nanometers in size. The PFIB has provided microstructural context of the large voids and made possible the critical observation that each of the voids, including many voids unresolvable from the X-ray CT data, is associated with an inclusion and that the volume of the void scales with the volume of the inclusion raised to the power 3/2, thus emphasizing the importance of the inclusion size distribution.

4. Lab-based nanoscale X-ray CT

As shown in the previous example it is often important to investigate the material at the nanoscale and as discussed in section 2 there are situations where time lapse imaging can be used to help understand processes. One of the major downsides of serial sectioning is that it is destructive and cannot be utilized for time-lapse studies. However the PFIB can be used to prepare site specific samples suitable for nanoscale X-ray CT (nano-CT) imaging, which is capable of investigating processes over time, in 3D, at the nanoscale. In this exemplar micron-sized specimens are investigated with nanoscale resolution with a temporal resolution of the order of hours.

An experimental procedure to optimize time-lapse imaging using nanoscale X-ray CT has been developed [35]. Bradley et al. explain in detail how the acquisition of both absorption and phase contrast data followed by the effective registration and combination allows for optimum phase identification. In this approach the contrast available through absorption contrast and the edge resolution possible through the phase contrast results is combined to identify the phases with precise interfaces. Here the PFIB has been used for the preparation of the site-specific NanoCT specimens,

4.1 Exemplar: corrosion of AZ31 magnesium alloy

Here, an exemplar of corrosion damage of twin roll cast AZ31 magnesium alloy is presented. There is still a real need to better understand the corrosion of magnesium alloys [36] as this is one of the principal obstacles to its uptake in many engineering and aesthetic applications. Twin-roll casting offers a route to directly produce magnesium sheet in a cost effective way. In this study the propagation of the corrosion in this alloy is investigated to better understand it's interaction with the microstructure.

A pillar of magnesium with dimensions 48 × 50 × 70 µm, was prepared from the middle of a twin roll cast section using mechanical grinding to produce a matchstick of material with dimensions of 2 x 2 x 10 mm^3, but with a sharp point < 500 µm across at one end. The sharpened point was prepared to its final dimensions using a FEI Helios PFIB. The microstructure consists of large mm-sized dendrites. The sample was subsequently immersed in 0.1 M NaCl solution, with the sample removed from the solution for X-ray imaging after cumulative exposure times of 0, 0.5 and 2 hours. Nano-CT was carried out using a Zeiss Xradia Ultra 810 system operated at 5.5 kV. 701 projections were collected with an exposure time of 60 s. Data was collected for a 65 µm field of view with a voxel size of 60 nm. An absorption contrast and phase contrast scan was acquired at each time step. The orientation of the samples was tracked between subsequent scans, and care was taken to ensure registration of the data recorded at the different time steps.

Figure 7(a) shows the time lapse sequence as a 3D rendering, while figures 7(b) and 7(c) show virtual cross sections through the volume at the positions shown by the red and blue lines respectively. The development of corrosion product is clear (seen as darker surface regions) as the surface has developed significant topography; this is quantified by the colour map on the volume surface where the depth of corrosion is shown as the different colours. Here, the depth of corrosion is presented as the difference with respect to the previous time step and not the cumulative total. In this way it is possible to more easily see if corrosion of certain regions has stopped, slowed or accelerated between each immersion step. The dendritic structure of the alloy is clearly visible from the virtual cross sections and bright, high atomic number, particles can be seen decorating the inter-dendritic regions.

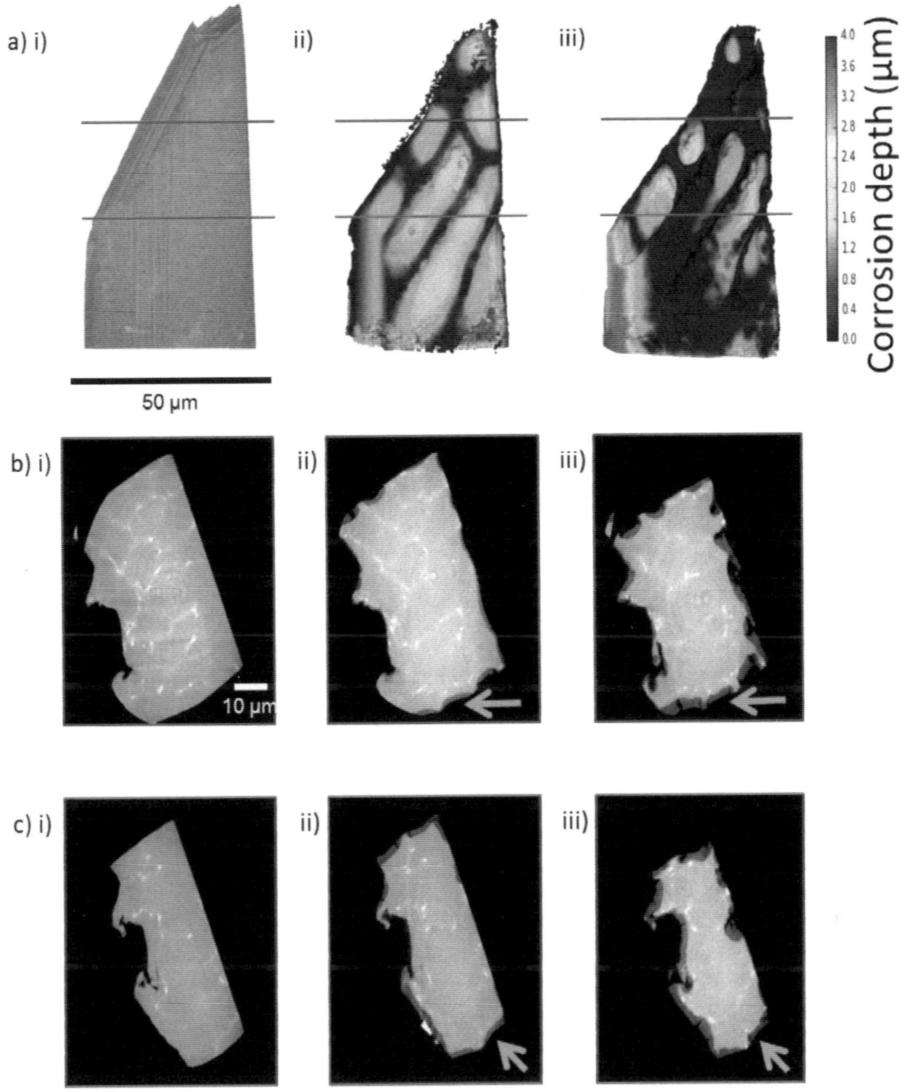

Figure 7 a) shows a volume rendering of the surface of the reconstructed nanoCT data, the red and blue lines indicate the position of virtual slices shown in b) and c). i), ii) and iii) present the same sample after two sequential immersion steps. In a) ii) and iii) the surface is coloured to show the depth of material lost since the previous scan. In b) and c) the same virtual slice is shown for the different time steps, Considerable corrosive attack can be seen, the orange arrow acts as a point of reference.

After 2 hours of exposure the corroded sample was transferred to the FIB-SEM for further analysis. Figure 8(a) shows the surface of the specimen, after being minimally polished using the PFIB (at 30 kV and with a current of 15 nA) to remove the surface topography and corrosion product to allow chemical mapping using EDX (see figures 8b) and 8c). Comparing figure 8(a) with (b) and (c) it is clear to see the correlation and confirm that the interdendritic regions have segregated particles that are Al and Mn rich, although it does appear that some of these are different particles, as the chemical distributions are slightly different. One of these interdendritic regions (from the same sample but without correlation) was analyzed in more detail using STEM-EDS (see figure 8d and 8e). From this it was possible to further understand the complexity of these sites and confirm that the interdendritic regions are rich in Al and Zn, as expected. In addition, different phase particles are also revealed in these regions as Al and Mn rich particles (which appear bright red and have a feather-like morphology) and also Mg and Si rich particles (which appear bright blue and have a cubic morphology).

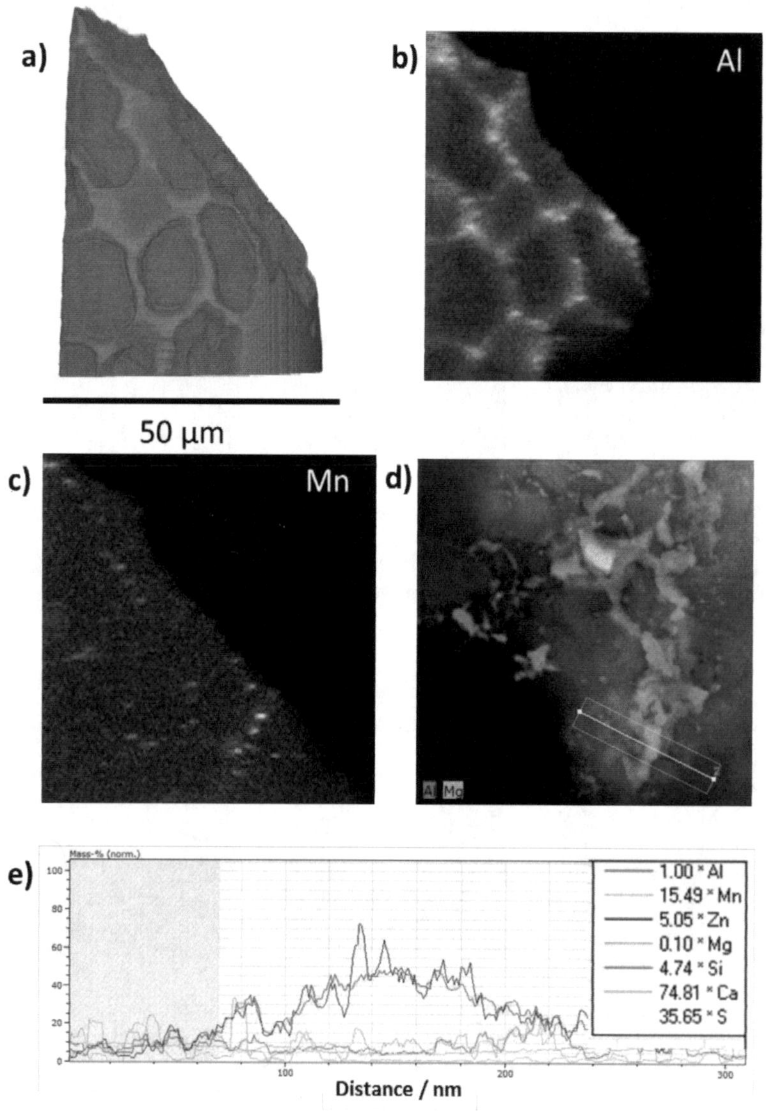

Figure 8 a) shows the surface rendering of the reconstructed nanoCT data after the final corrosion step. b) and c) show EDS maps highlighting Al and Mn respectively. In both cases enrichment appears along the interdendritic regions but there is not a one-to-one correlation of the enriched Al and Mn regions. d) a STEM-EDS spectrum map from one of the interdendritic regions highlighting Al in red and Mg in Cyan the position of a line trace is shown in yellow with the elements present along the length presented in e).

It is clear that the corrosion progresses along the dendrite arms, which are relatively Mg rich [37]. Progress is arrested when the corrosion front reaches the interdendritic regions and is due to the enrichment of the Al and Zn in these regions, although STEM-EDS data shows that the interdendritic region is quite complex with multiple phases all in close proximity. Work is continuing to understand the role of the multiple phases on the initiation and propagation of the corrosion, which is currently believed to be controlled by small impurity particles.

5. Conclusions

There is a clear desire to investigate and understand many materials systems across multiple time and length scales. To achieve this multiple instruments need to be effectively linked together as no single instrument can provide all of the required functionality. Correlative tomography describes a workflow where regions of interest are successfully targeted to build up a multiscale and multi-layered picture of a sample. The acquired information also has to be effectively connected for an analysis that makes best use of the multiple datasets, typically involving the accurate registration of the datasets. The techniques and exemplars presented in this paper outline a number of complementary approaches, each including different aspects of correlation and different scales of information, which could also be connected in a single study if desirable. In each case the intention is to identify the key time and/or length scales of the process/feature/property so that it can be better understood.

6. Acknowledgements

PJW is grateful for funding through a European Research Council Advanced Grant 695638 (CORREL-CT). The authors would like to acknowledge EPSRC funding for the Henry Moseley X-ray Imaging Facility under grants EP/M010619, EP/K004530, EP/F007906, EP/F028431.**7.** HEFCE funding established the Multidisciplinary Characterization Facility and we are grateful for funding from UK Research Partnership Investment Funding (UKRPIF) Manchester RPIF Round 2. Research support for the work on AA5083-H131 has been provided by the Office of Naval Research thorough ONR-N00014-11-1-0406 (William Mullins), ONR-N00012-14-1-0593 (Airan Perez) and ONR-N00014-15-1-2445 (William Mullins). Additional support has been provided by the Arthur P. Armington Professorship (JJL). Supply of materials was from Alcoa Technical Center.

References

[1] Burnett T L, McDonald S A, Gholinia A, Geurts R, Janus M, Slater T, Haigh S J, Ornek C, Almuaili F, Engelberg D L, Thompson G E and Withers P J 2014 Correlative Tomography *Scientific Reports* **4** 4711

[2] Burnett T L, Kelley R,Winiarski B, Contreras L, Daly M, Gholinia A, Burke M G andWithers P J 2016 Large Volume Serial Sectioning Tomography by Xe Plasma FIB Dual Beam Microscopy *Ultramicroscopy* **161** 119-129

[3] Maire E and Withers P J 2014 Quantitative X-ray tomography *Int. Mater. Rev* **59** 1 1–43

[4] Uchic M D, Holzer L, Inkson B J, Principe E L and Munroe P 2007 Three-Dimensional Microstructural Characterization Using Focused Ion Beam Tomography *MRS Bulletin* **32** (05) 408-416

[5] Thompson G, Hashimoto T, Zhong XL, Curioni M, Zhou X, Skeldon P, Withers P J, Carr J A and Monteith A G 2013 Revealing the three dimensional internal structure of aluminium alloys *Surface and Interface Analysis* **45**(10) 1536–1542

[6] Kübel C,Voigt A, Schoenmakers R, Otten M, Su D, Lee T C, Carlsson A, Bradley J 'Recent Advances in Electron Tomography: TEM and HAADF-STEM Tomography for Materials Science and Semiconductor Applications' Microscopy and Microanalysis, 11, Issue 05, 2005, pp 378-400

[7] Blavettea, D., Duguay, S., Atom probe tomography in nanoelectronics. European Physical Journal - Applied Physics, 2014. 68(1): p. 10101-p1-12

[8] Spowart J 2006 Automated serial sectioning for 3-D analysis of microstructures *Scripta Mater.* **55**(1) 5-10

[9] Winiarski B, Gholinia A, Mingard K, Gee M, Thompson G E and Withers P J 2017 Broad ion beam serial section tomography *Ultramicroscopy* **172** 52–64

[10] Groeber M, Haley B K, Uchic M D, Dimiduk D M, Ghosh S 2006 3D reconstruction and

characterization of polycrystalline microstructures using a FIB-SEM system *Mater Charact* **57**(4-5) 259-273

[11] Hashimoto T, Thompson G E, Zhou X and Withers P J 2016 3D imaging by serial block face scanning electron microscopy for materials science using ultramicrotomy *Ultramicroscopy* **163** 6-18

[12] van Driel L F, Valentijn J A, Valentijn K M, Koning R I and Koster A J 2009 Tools for correlative cryo-fluorescence microscopy and cryo-electron tomography applied to whole mitochondria in human endothelial cells *Eur. J. Cell Biol.* **88** 669–684

[13] Handschuh S, Baeumler N, Schwaha T and Ruthensteiner B A 2013 Correlative approach for combining microCT, light and transmission electron microscopy in a single 3D scenario *Front. Zoo.* **10** 44

[14] Caplan J, Niethammer M, Taylor R M and Czymmek K J 2011 The power of correlative microscopy: multi-modal, multi-scale, multi-dimensional *Curr. Op. Struc. Biol.* **21** 686–69

[15] Robinson B J, Giusca C E, Gonzalez Y T, Kay N D, Kazakova O and Kolosov O V 2015 Structural, optical and electrostatic properties of single and fewlayers MoS2: effect of substrate 2D Mater. 2 015005

[16] Stock S R 2009 MicroComputed Tomography: Methodology and Applications, (CRC Press, Boca Raton, FL, USA)

[17] Herbig M. et al. 2011 3-D growth of a short fatigue crack within a polycrystalline microstructure studied using combined diffraction and phase-contrast X-ray tomography *Acta Mat.* **59** 590–601

[18] Maire E, Owen A, Buffiere J Y and Withers P J 2001 A synchrotron X-ray study of a Ti/SiCf composite during in situ straining *Acta Mater.* **49**(1) 153-163

[19] Qian L, Toda H, Uesugi K, Kobayashi T, Ohgaki T and M. Kobayashi 2005 Application of synchrotron x-ray microtomography to investigate ductile fracture in Al alloys *Appl Phys Lett* **87** 241907

[20] King A, Johnson G, Engelberg D L, Marrow T J and Ludwig W 2008 Observation of Intergranular Stress Corrosion Cracking in a Grain-Mapped Polysrystal *Science* **321** 382–5

[21] Robinson J B, Darr J A, Eastwood D S, Hinds G, Lee P D, Shearing P R, Taiwo O O and Brett D J L 2014 Non-uniform temperature distribution in Li-ion batteries during discharge–a combined thermal imaging, X-ray micro-tomography and electrochemical impedance approach Journal of Power Sources Volume 252, 15 April 2014, pp 51-57

[22] J R. A. Godinho, K M. Gerke, A G. Stack, and P D. Lee The dynamic nature of crystal growth in pores, Sci Rep. 2016; 6: 33086

[23] Lowe, T., Garwood, RJ, Simonsen, TJ, Bradley, RS, Withers, PJ, Metamorphosis revealed time-lapse three dimensional imaging inside a living chrysalis. Journal of the Royal Society Interface, 2013. 10(84)

[24] R Mokso, D A. Schwyn, S M. Walker, M Doube, M Wicklein, T Müller, M Stampanoni, G K. Taylor & H G. Krapp, Four-dimensional in vivo X-ray microscopy with projection-guided gating, Scientific Reports 5, Article number: 8727 (2015)

[25] NJH Holroyd, TL Burnett, M Seifi, JJ Lewandowski, Improved understanding of environment-induced cracking (EIC) of sensitized 5XXX series aluminium alloys, 2017, Materials Science and Engineering: A, Volume 682, Pp. 613-62

[26] M. Seifi, N.J.H. Holroyd, J.J. Lewandowski, Deformation rate and sensitization effects on environmentally assisted cracking Al-Mg naval alloys, Corrosion 72 (2016) 264–283

[27] M Daly, TL Burnett, EJ Pickering, OCG Tuck, F Léonard, R Kelley, PJ Withers, AH Sherry, A multi-scale correlative investigation of ductile fracture, 2017, Acta Materialia Volume 130, Pp. 56-68

[28] T L Burnett, B Winiarski, R Kelley, X L Zhong, I N Boona, D W McComb, K Mani, M G Burke, P J Withers, Xe+ Plasma FIB: 3D Microstructures from Nanometers to Hundreds of Micrometers, 2016, Microscopy Today Volume 24 Issue 03 Pp. 32-39

[29] Jiruše, J., Hrnčiř, T, Lopour, F, Zadražil, M, Delobbe, A, and Salord, O, Combined plasma FIB-SEM. Microsc. Microanal., 2012. 18 ((Suppl 2)): p. 652-653

[30] **http://symmetry.oxford-instruments.com/**

[31] Yu-Hui Chen, S Un Park, D Wei, G Newstadt, M Jacksona, J P. Simmons, M De Graef A O. Hero, A Dictionary Approach to EBSD Indexing, 2015, Microscopy and Microanalysis Volume 21 Iss: 03 Pp. 739-752

[32] P G Callahan, M De Graef, Dynamical electron backscatter diffraction patterns. Part I: Pattern simulations, 2013, Microscopy and Microanalysis Volume 19 Issue 05 Pp 1255-1265

[33] ASTM Standard Specication for Quenched and Tempered Vacuum-treated Cabon and Alloy Steel Forgings for Pressure Vessel ComponentsA508A/A508M (2004)

[34] ESIS TC1 Subcommittee on Fracture Mechanics Testing Standards, ESIS Procedure for Determining the Fracture Behaviour of Materials ESIS P2-92, Tech. rep., European Structural Integrity Society (1992)

[35] R.S. Bradley, Y. Liu, T.L. Burnett, X. Zhou, S.B. Lyon, P.J. Withers, A. Gholinia, T. Hashimoto, D. Graham, S.R. Gibbon, B. Hornberger, Time-lapse lab-based x-ray nano-CT study of corrosion damage, 2017, Volume 267, Issue 1, Pp. 98–106

[36] Esmaily M, Svensson J E, Fajardo S, Birbilis N, Frankel G S, Virtanen S, Arrabal R, Thomas S and Johansson L G 2017 Fundamentals and advances in magnesium alloy corrosion *Prog Mater Sci* **89** (2017) 92–193

[37] Pawar S, Slater T J A, Burnett T L, Zhou X, Scamans G M, Fan Z, Thompson G E and Withers P J 2017 Crystallographic effects on the corrosion of twin roll cast AZ31 Mg alloy sheet *Acta Mater.* **133** 90-99

IOP Conf. Series: Materials Science and Engineering **219** (2017) 012002 doi:10.1088/1757-899X/219/1/012002

Application of characterization, modelling, and analytics towards understanding process-structure linkages in metallic 3D printing

M A Groeber, E Schwalbach, S Donegan, K Chaput, T Butler and J Miller

Air Force Research Laboratory, Materials and Manufacturing Directorate, 2230 Tenth St., Wright-Patterson AFB, OH 45419

Corresponding Author: michael.groeber@us.af.mil

Abstract. This paper presents methods for combining process monitoring, thermal modelling and microstructure characterization together to draw process-to-structure relationships in metal additive manufacturing. The paper discusses heterogeneities in the local processing conditions within additively manufactured components and how they affect the resulting material structure. Methods for registering and fusing disparate data sources are presented, and some effort is made to discuss the utility of different data sources for specific microstructural features of interest. It is the intent that this paper will highlight the need for improved understanding of metallic additive manufacturing processes and show that combining experimental data with modelling and advanced data processing and analytics methods will accelerate that understanding.

1. Introduction and background

Additive manufacturing presents a potential opportunity to introduce a paradigm shift in component design. The ability to locally tailor processing paths allows for new sophisticated designs that incorporate complex geometries with heterogeneous processing, leading to spatially varying material properties. Properly accounting for, and ultimately exploiting, this processing heterogeneity requires the ability to link the local processing state to properties and performance of the local material. Current geometry-based design approaches, such as topology optimization, do not directly account for material property evolution as the geometry is built. Most commercial additive manufacturing systems utilize fixed scanning strategies that are highly sensitive to the build geometries, resulting in potentially unavoidable changes in the local processing state. Understanding and accounting for the interaction of the component geometry with the predetermined machine scan path is critical to predicting material structure, and thus material properties. This understanding is necessary to ensure tools such as topology optimization properly account for heterogeneous material properties. An understanding of the interactions between component geometry, the processing path, and material structure is an enabling factor in the development of scan path optimization for tailored, spatially varying materials properties.

The complexity of machine scan paths and the control provided to the manufacturing engineer varies among the different printing systems. In general, most commercial machines divide the build plate into subdomains and then process them in some global order. Figure 1 shows an example of the scanning strategy employed by an EOS M280 laser powder-bed fusion machine. Two consecutive layers are colored by the time at which the laser was at each given point in space. The solid bands of apparent common color highlight a macroscopic serpentine pattern that divides the build plate into lanes termed 'stripes'. The user has the ability to change the width of these stripes at build time. Within the stripes, a finer-scale serpentine pattern exists in which the laser moves perpendicular to the long axis of the stripe. When the beam reaches one side of the stripe, it may, under certain user conditions, shut off, turn around, turn back on again, and then scan to the other side of the stripe. These individual scan vectors within a given stripe are termed 'hatches'. Similar to stripes, the user may change the spacing between neighboring hatches at build time. The global scan path is predetermined by the machine and covers the entire build plate; the only influence of the intended build geometry occurs when the hatch vectors intersect the triangulated surface mesh that defines the build, which determine when the laser is on. Figure 1 clearly shows artificial interfaces inserted within the build geometry between neighboring stripes, termed 'stripe boundaries'. The machine offers parameters to control the overlap of neighboring stripes, affecting the structure of these interfaces. The global scan path is rotated for each successive layer by a user-prescribed angle to 'randomize' the volumetric scan path and prevent alignment of the stripe boundaries throughout the build. A key observation from figure 1 is that the local processing path of any given point in each part is a convolution of the global scan path, the part geometry, and the geometry and relative placement of other parts in the bed.

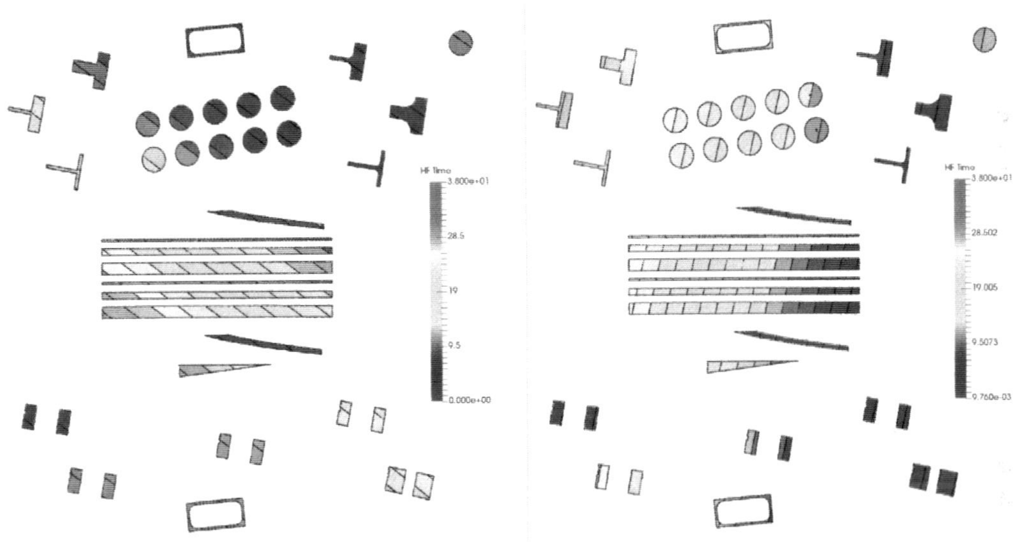

Figure 1. Two consecutive layers in a laser powder-bed fusion build displaying the time at which the laser was present at each given point in space. The macroscopic serpentine pattern of the laser raster is seen and the rotation between consecutive layers is apparent. The influence of scan path and part layout on local processing path is evident in the part at the lower right of the build plate, where it is entirely printed within a single second on one layer and takes approximately five seconds on the other layer.

In this paper, we discuss multiple characterization methods applied to samples produced by two different metallic additive manufacturing processes, namely laser powder-bed fusion (LPBF) and electron beam melting (EBM). Given the complexity of scan paths, particularly the variability between systems and the interaction of component geometries, we develop a methodology that couples experimental data and modelling to convert the scan paths into spatially resolved local thermal histories. Finally, we present initial findings on the correlations between local processing history microstructure, and defects in the two different printing processes.

2. Materials, Processes and Features of Interest

In this work, we have selected two sets of samples from two different processing techniques, LPBF and EBM. Each set of samples originates from a single build in the given process. A titanium alloy, Ti-6Al-4V, was used in both processes. Figure 1 shows the layout of specimens in the LPBF build. The samples are mostly 2.5-dimensional, extended in the build direction, which is normal to the plane of figure 1. Two components from the LPBF build are showcased in this paper: the cylindrical object in the upper right corner of the build layout, and an arch, shown as two rectangles in cross-section at the lower right corner of the build layout. The LPBF samples were investigated in an attempt to correlate the presence of voids to the local processing conditions.

Figure 2 shows the layout of the EBM build, with arrows denoting the samples showcased in this work. These two samples were taken from groups of plates arranged in an attempt to force a modulation in local processing condition. The EBM process was performed on an ARCAM A2, in which the power and speed of the beam is determined based on line length to be melted on any given scan vector. The bulk scan vectors always point in the positive or negative x or y directions and span the run from one edge of the build plate to the other. The layout of the build shown in figure 2 suggests that the group of larger plates will have a longer melted line length on every layer regardless of the scan vector directions, thus experience processing with consistently higher power and speed compared to the smaller plates. The microstructural feature of interest in these samples is the macroscopic texture and its dependence on local processing history.

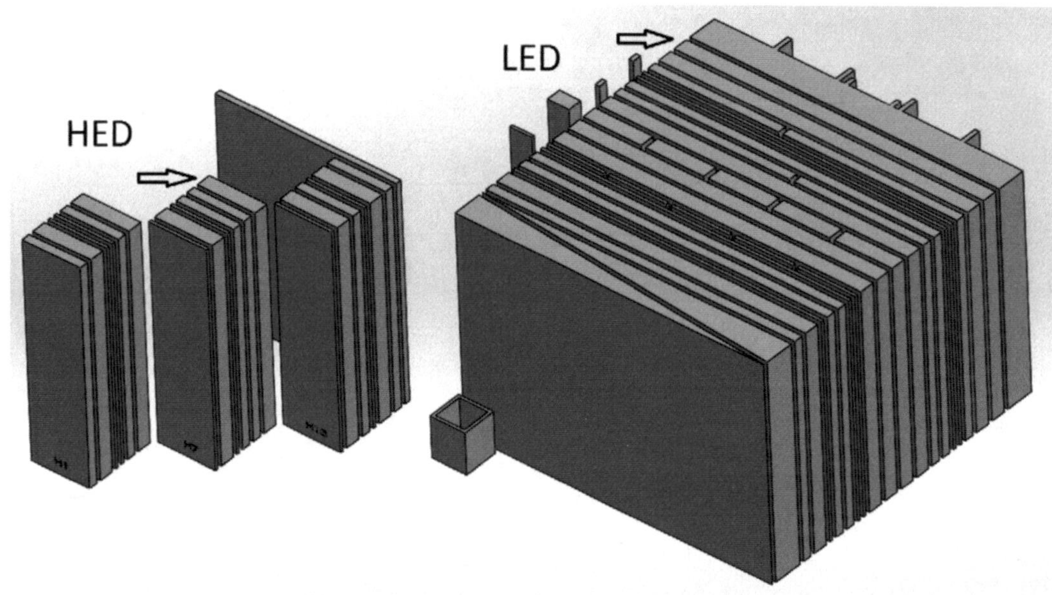

Figure 2. Layout of Ti-6Al-4V EBM build with investigated samples noted by arrows. High energy density (HED) and low energy density (LED) regions described in Section 6.2 are indicated.

3. Characterization techniques and data streams for analysis of additive manufacturing

A distinct advantage of additive manufacturing over many traditional manufacturing processes (i.e., forging or casting) is the abundant amount of data available from the process. In this work, we have data characterizing the prebuild intent, in-situ processing, and post-build structure. This section introduces the data streams available and explains why they were included in this analysis.

3.1. Build geometry

The intended build geometry is represented as a triangulated surface mesh of the part(s) to be built, typically in the form of a Stereolithography file (.stl). The surface mesh is 'sliced' by the software that controls the additive manufacturing system at the build layer thickness, resulting in 2D polygons on each print layer; the beam will be on inside of the polygons and off outside of them. The surface mesh is a discrete representation of a possibly smooth design, presenting a potential trade-off between fidelity to the original design, the size of the mesh, and the resolution that the printing system can practically achieve. It is important to note that the build geometry represents the intended geometry of the part(s); the actual printed part(s) may not correspond exactly to this geometry due to numerous factors, including residual stresses and distortions, build errors, and machine calibration errors. In this work, the intended build geometry is available for both builds.

3.2. Process intent and monitoring

During the printing process, the beam is rastered across the build plate in a potentially proprietary manner. If the machine does not supply the scan path of the beam directly, prior knowledge of the scan strategy or a third-party monitoring system is required to obtain the full path history. In addition to tracking the path of the beam, some monitoring systems also utilize additional sensors attached to the system to monitor temperature, oxygen levels, sound, beam power and voltage, and emitted light from the molten pool. In this work, the scan path is obtained via different routes for the two printing systems. In the EBM process carried out on an ARCAM A2, the scan path with voltage and speed levels was determined by internal knowledge of the scan strategy through collaboration with Oak Ridge National Laboratory's Manufacturing Demonstration Facility (ORNL MDF) and the system manufacturer. The scan path is represented as a set of scan vectors with known power and speed. In the LPBF process, a third-party monitoring system (PrintRite3D from SigmaLabs) was used to track the position of the beam, along with its power and drive, sampled at 50 kHz. In addition to tracking the beam and its properties, the PrintRite3D system measures the amount of emitted light through a photodiode on axis of the laser, the temperature of the chamber, the oxygen level of the chamber, and an audio signal within the chamber. The PrintRite3D data is represented as a point cloud of locations of the beam at discrete points in time. These data streams from the two processes were used to track the local processing history within the build chamber.

3.3. Computed tomography (CT)

After printing, it is often desired to inspect both for dimensional accuracy and for defects (i.e., voids or lack-of-fusion). Computed tomography (CT) is a common tool for non-destructively inspecting metallic components. CT is capable of quantifying measureable density gradients within parts, but has some limitations with respect to spatial resolution. It can be difficult to investigate large volumes of material with high spatial resolution using laboratory-scale CT, since the volume of inspected material is coupled inversely to the spatial resolution. Due to this limitation, surface roughness is difficult to quantify using CT, and small defects (< 50μm) are often not detected in samples larger than a few cubic centimeters. Furthermore, the complex geometries that can be printed with additive manufacturing present significant difficulties for CT reconstruction. The most common techniques for CT reconstruction, such as filtered back projection, often exhibit noticeable artifacts when applied to complex or highly anisotropic parts[1]. For this work, the relatively small LPBF samples were inspected using CT: the simple geometric form of the samples and the overall small volume allows for a reasonable spatial resolution (~23 μm) for inspecting the defects of interest.

3.4. Mechanical serial-sectioning with optical microscopy

Volumetric inspection via serial-sectioning decouples spatial resolution from the volume of inspected material, offering a distinct resolution advantage over CT while also being free from reconstruction artifacts. However, the technique is destructive and time intensive. For the investigation in this work, the imaging mode employed on each exposed section was optical microscopy (OM). OM is well suited for clearly differentiating voids in metallic samples. The LPBF sample was serial-sectioned for a relatively small volume to compare with CT. The sample inspected with serial-sectioning was imaged at approximately 10x the resolution of CT, using 1 μm section thickness with 2 μm in-plane OM pixel size. Single 2D optical images were collected for the EBM samples, but no serial-sectioning was performed on those samples.

3.5. Electron backscatter diffraction (EBSD)

EBSD is a scanning electron microscope (SEM) technique that is capable of deducing the crystallographic orientation of locations within a sample [2]. If performed at sufficient spatial resolution, individual grains can be resolved and their morphologies characterized. If EBSD is performed with a coarser spatial resolution, individual grains may be difficult to characterize, but the meso-to-macroscopic texture of the sample can be obtained. In this work, the EBM samples were inspected with EBSD. The resolution of the inspection was selected to allow for large area coverage in an effort to quantify macroscopic texture, but within reasonable time constraints. Some selected areas were inspected using a finer resolution to determine the general morphological characteristics of the grains. The prior beta grains, whose orientation distribution was the focus of this work, were relatively large and well-sampled by the chosen resolution.

4. Registration and fusion of data streams

The various data streams must be registered, or brought into spatial coincidence, to assess correlations between process and structure at specific locations. Registration of the data streams is complicated by two factors. First, the data element dimensionality varies from 0D (point clouds) to 2D (surface meshes) to 3D (image grids). Second, the data streams contain different information content that does not necessarily correlate. Thus, utilizing correlated information approaches, such as mutual information, to register the data sets could bias subsequent analyses. Ideally, the samples would contain some independent reference marks detectable in each of the data sets, which could aid registration without introducing bias. For this work, these 'reference marks' were selected to be the sample surfaces themselves. Since each data stream provides an inspection of the part geometry at a different processing step, the sample surfaces are not guaranteed to all match exactly. The surface mesh is the intended part geometry, the in situ monitoring data is where the beam rastered (which may have offsets to account for distortions and melt pool size estimates), and the CT data captures a measurement of the physical build (which could be affected by residual stresses and distortions, artifacts in the CT reconstruction, and from cutting the sample off from the build plate). Therefore, an iterative registration procedure was used to find the transformation that best brings the data streams into spatial coincidence.

The first step in the registration process is to identify 'key points' in each dataset that correspond spatially; this step involves extracting out points that represent the part surface. The second step is to use some algorithm to determine the translations, rotations and distortions necessary to bring the key points into coincidence. In the LPBF work, the iterative closest point (ICP) algorithm, implemented in the open source Point Cloud Library (PCL), was chosen to determine the necessary transformation [3,4]. The iterative closest point algorithm takes as its input the set of key points that represent the part surface. The build geometry surface mesh directly defines the sample surfaces, so each triangle was point sampled uniformly within its area. To generate an unbiased surface sampling, triangles to sample are selected randomly from the surface mesh weighted by their area. The CT data sets were thresheld using a local gradient approach to determine the voxel faces that were at the surface of the sample [5]. Those voxel faces were then triangulated and smoothed using a Laplacian surface smoothing algorithm [6] to yield a surface mesh of the printed sample. This surface mesh was then sampled in the same manner as the build geometry surface mesh. Finally, the point cloud from the PrintRite3D monitoring system was uniformly dense throughout the sample, but only the sample surface was desired for the registration. In order to identify the sample surface, a Delaunay tetrahedralization was performed using PCL and the concave hull defined by the surface triangles of the tetrahedralization was identified [7]. The points that defined the concave hull were retained as the surface of the sample. Figure 3 shows the extracted surface points and the resultant registration for an arch-shaped object from the LPBF build (lower right corner of figure 1). In the EBM work, the process intent data were directly obtained from the build geometry, and thus already shared a common reference frame. The texture information extracted from the EBSD scans of the EBM builds was associated with each sample at the macroscale, so no registration was performed to correlate with local scan path. Rather, the processing of the two samples was generalized by investigating a single point in the center of each sample. The scan strategy of the EBM process combined with the build layout makes this approach reasonable for the present analysis.

Figure 3. Data streams from CAD, CT and PrintRite3D, with point representation of sample surface. Larger inset shows final registry of full PrintRite3D data (green and yellow points) overlaid with thresheld CT data (maroon surface) and CAD surface mesh (green triangles with blue edges on top surface). Note the point representations of the sample surface in the individual datasets are in their own reference frame and do not illustrate the magnitude of initial disregistry.

5. Modeling material thermal histories from complex moving heat sources

Local thermal history is an important quantity to consider during development of additive manufacturing processes for metallic materials as it affects microstructure, residual stresses, and defects. Depending upon the specific problem, the range of physical phenomena and energy transport mechanisms involved can be extensive: conduction, buoyancy and surface-tension-driven fluid flow, the associated convective thermal transport, beam-material interactions, radiative thermal transport, and phase changes including evaporation. While they can generate considerable insight, high-fidelity solutions to such multi-physics computational problems are often expensive to compute, and thus fast-acting physics-based models have been developed that make simplifying assumptions. An example of such an analytical model is the so-called Rosenthal's solution [8]. This analytic solution describes the temperature field induced by a point source moving uniformly across a semi-infinite body, allowing only thermal conduction as the transport mechanism, and assuming temperature-independent thermo-physical parameters. This model is computationally inexpensive and the solutions exhibit many qualitative features that are also seen empirically. However, the Rosenthal resolution presents several issues. First, the energy input is a point source, so the energy input flux is effectively infinite at that point location, which then induces a temperature discontinuity. Secondly, the energy source moves uniformly in space, so the model effectively describes steady-state deposition of a single pass, and cannot describe transient effects such as beam direction changes, or modulations of power or speed. The substrate is semi-infinite, which is not unreasonable for short times, but may be unreasonable for long times compared to the thermal diffusion time scale. Finally, the assumption that thermal conduction is the only transport mechanism is flawed. Again, despite these limitations, this model is qualitatively useful and predicts many trends that are observed empirically (e.g. melt pool width scales like power over velocity, aspect ratio scales with power times velocity, etc.).

A number of other researchers expanded the model of Rosenthal by replacing the point source with a distributed energy source [10-14]. This effectively regularizes the temperature discontinuity in the Rosenthal solution immediately under the energy source. The solutions given by the two approaches converge for locations sufficiently far from the source (typically several beam diameters). While these expanded solutions admit a field maximum temperature that circumvents the temperature discontinuity, the remaining issues are still present. Furthermore, while the temperature does remain finite, the peak value is often unrealistically high when practical deposition parameters are considered, on the order of 10,000 K or higher.

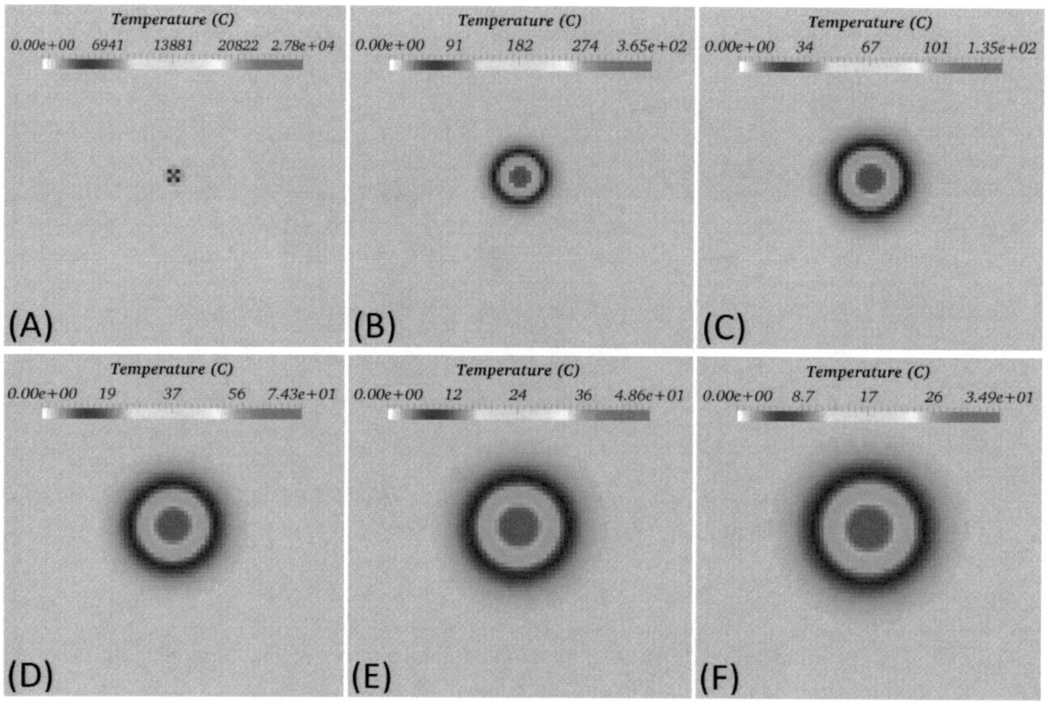

Figure 4. Sequence of time steps from the solution of a single discrete point source (A) 0.00002s, (B) 0.0002s, (C) 0.0004s, (D) 0.0006s, (E) 0.0008s, (F) 0.001s. Note the change in the temperature scale between time steps. (A) Illustrates the unrealistically high temperatures observed instantaneously after the source becomes active. Each panel has a 1mm x 1mm field-of-view.

The main goal of this work is to develop a computationally inexpensive thermal model that relaxes constraints on beam path and allows for transient conditions, including arbitrary beam motion and modulation of deposition parameters. Secondarily, we wish to estimate and include these in a 'global' sense. The first major simplifying step is to represent the continuously moving source as a series of discrete stationary energy pulses. The pulses are positioned spatially by sampling the location of the continuously moving beam at a fixed rate, placing sources at these locations, and activating them at the time the beam passes. This approach implicitly captures velocity and sequencing, while reducing the complexity of the modeling approach. We choose individual source terms to be Gaussian in space, centered at locations $r = (x_i, y_i, z_i)$ which the beam visits at times τ_i:

$$s_i(\vec{r}, t) = \frac{A_i \delta(t - \tau_i)}{(2\pi\sigma^2)^{3/2}} \exp(\frac{(\vec{r} - \vec{r}_i)^2}{2\sigma^2})$$

Here $A_i\delta(t-\tau_i)$ is the power of the source with the Dirac delta function having units of inverse time, σ controls the physical size over which the energy is deposited, and r_i and τ_i are the position and time of the ith discrete source. All the energy of the ith discrete source is deposited at time $t = \tau_i$ and then made inactive, though its deposited energy continues to influence the system for all times $t > \tau_i$. If we integrate each point source over the full time and space we get:

$$\int_t \int_{\bar{V}} s_i(\vec{r},t)dtd\vec{r} = \int_{\tau_i}^{\tau_i+\Delta t} P_i dt = P_i \Delta t$$

where we consider the half space $\bar{V}$ that includes only points where $z \leq 0$ if the source is at $z = 0$ and Δt is the time elapsed between activation of sources. Substituting into the transport equation gives:

$$\frac{\partial T}{\partial t} - \alpha\nabla^2 T = f(\vec{r},t) = \sum_i^N \frac{2P_i\Delta t}{\rho c_p} \frac{1}{(2\pi\sigma^2)^{\frac{3}{2}}} exp\left[-\frac{(\vec{r}-\vec{r}_i)^2}{2\sigma^2}\right]\delta(t-\tau_i)$$

Note the factor of two that comes from the fact that half of the energy contributed by the source is deposited above the $z = 0$ plane and the energy must be effectively doubled to ensure the energy distributed below the plane of symmetry is equivalent to the true continuous source.

To solve the transport equation, we use a Green's function approach with linear operator:

$$\mathcal{G}(\vec{r},t|\vec{q},s) = (4\pi\alpha(t-s))^{-3/2} exp\left[\frac{-(\vec{r}-\vec{q})^2}{4\alpha(t-s)}\right]$$

and the solution is given by:

$$T(\vec{r},t) = \int_V \int_t \mathcal{G}(\vec{r},t|\vec{q},s) f(\vec{q},s) d\vec{q}ds$$

Finally, executing the integrals, the temperature is given by:

$$T(\vec{r},t) = \sum_i^N \frac{P_i\Delta t\Theta(t-\tau_i)}{\rho c_p\sqrt{2}[\pi(\sigma^2+2\alpha(t-\tau_i))]^{3/2}} exp\left[-\frac{(\vec{r}-\vec{r}_i)^2}{2\sigma^2+4\alpha(t-\tau_i)}\right]$$

where Θ is the standard Heavyside step function. Figure 4 shows the solution of a single discrete point source while figure 5 shows the solution of a series of discrete point sources approximating a moving source.

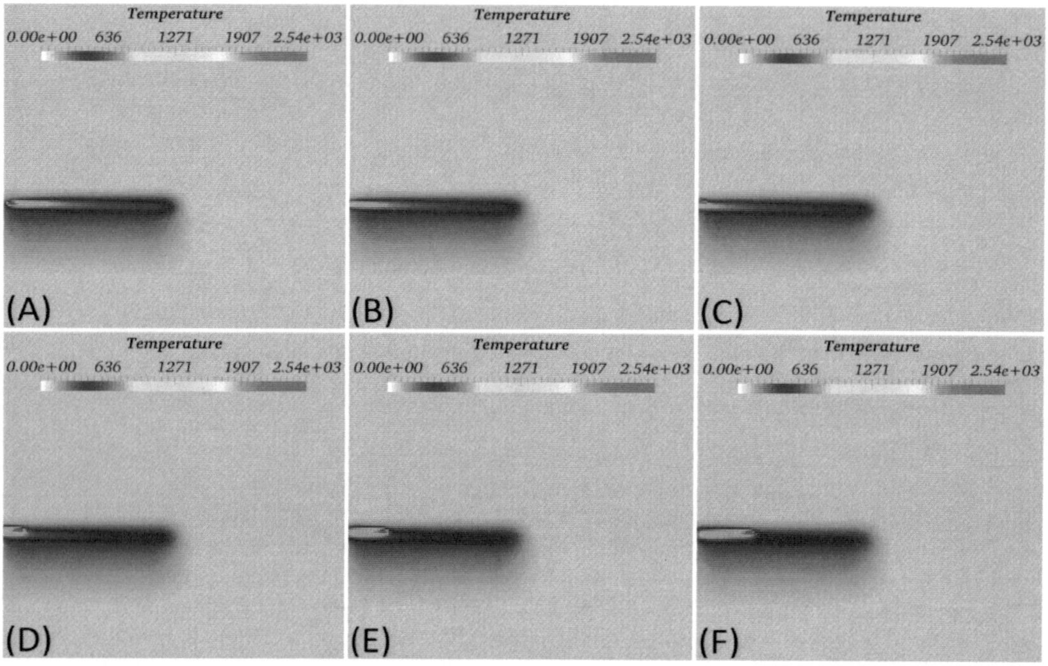

Figure 5. Sequence of time steps from the solution of a set of discrete sources turning on sequentially to represent a moving source. In the sequence, the beam approaches the edge of the region to melt (A), shuts off (B), turns around and turns back on (C), then moves away from the edge towards the other side of the region (D)-(F). The field-of-view of each panel is 14mm x 14mm. The model captures the shape of the melt pool and its dynamic size as the beam moves in a representative scan path. Note the disappearance of the melt pool immediately after the beam shuts off in (B).

6. Results and analytics

6.1. Voids in LPBF of Ti-64

CT investigation of the 10 mm cylindrical sample printed by LPBF (upper right corner of figure 1) is shown in figure 6a. The CT scan shows a significant amount of porosity that appears irregularly dispersed when looking at the entire sample. However, when a subsection of the sample is viewed (figure 6b), it is clear that the voids are correlated and often form linear arrays. The locations of the internal stripe boundaries are added in figure 6c, by displaying the points where the beam is turning around. It is clear that the linear arrays of voids coincide with these stripe boundaries.

The resolution of the CT reconstruction shown in figure 6 was 23 µm, so voids smaller than ~50 µm are generally undetected and voids smaller than ~200 µm do not have their shape well represented. The voids appear nearly equiaxed, unlike 'lack-of-fusion' voids that tend to be planar in nature. Higher resolution imaging was needed to inspect the nature of the voids. Serial sectioning using a 1 µm slice spacing was performed for a sub-region very close to that shown in figure 6b. Optical microscopy montage images were taken on each exposed section, using an in-plane resolution of 2 µm. A single optical image montage is shown in figure 7a, and the same region from the CT image is shown in figure 7b. The higher resolution image montage shows the same linear arrays as the CT image, as well as a secondary population of smaller, elongated voids. The void shapes are more clearly discernible in figures 7c and 7d. The second, smaller void population, which was generally below the detectability of the CT reconstructions, is aligned perpendicular to the arrays of larger voids. These smaller voids correspond with the individual hatches oriented perpendicular to the stripe boundaries. The morphology of the second population is indicative of lack-of-fusion porosity resulting from insufficient melt pool size, which prevents the fusion of neighboring scan melt pools. Figure 7 also shows a denuded zone of relatively dense material separating the smaller lack-of-fusion voids from both the external surface of the cylinder and the larger stripe boundary defects. Finally, in figure 7c, the larger voids appear to have a characteristic "D" shape, where the left side of the void is flatter than the right side. This characteristic is seen on many of the linear arrays.

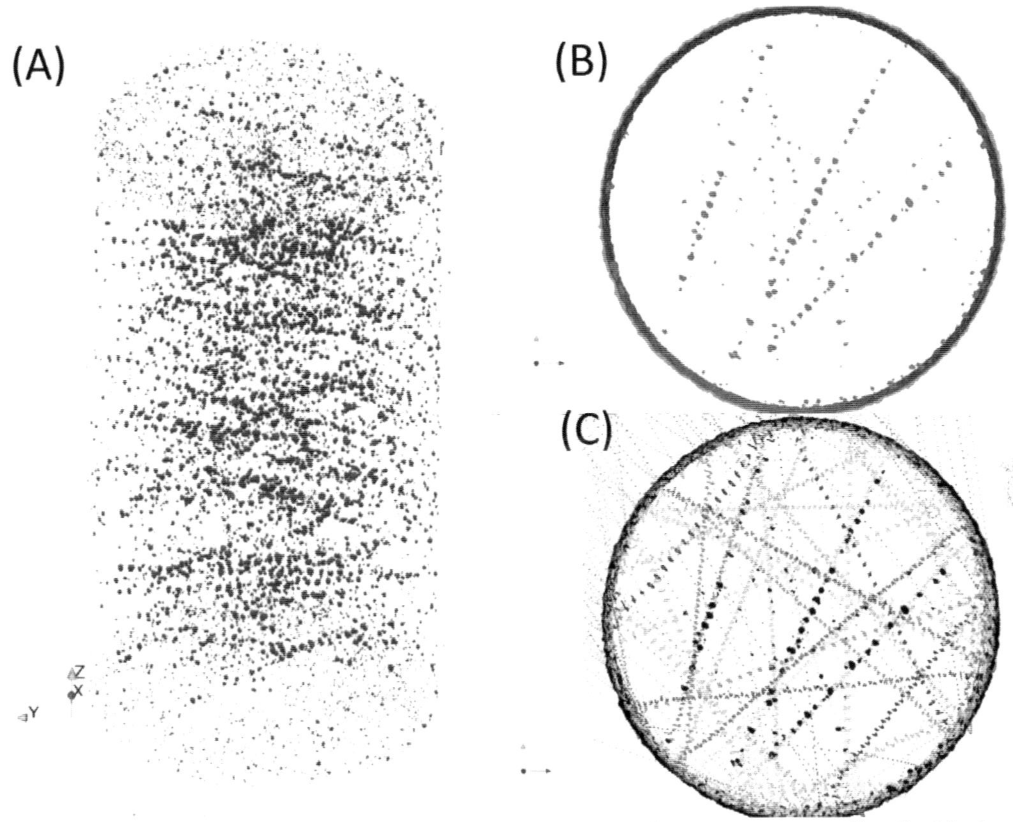

Figure 6. (A) Thresheld reconstruction of CT scan of 10mm diameter, 20 mm tall cylindrical sample showing the internal porosity in the sample, (B) 1mm thick sub-region taken from the central portion of the cylinder showing the voids arranged in linear arrays, and (C) Overlay of the voids from (B) and the internal 'stripe boundaries' on the printed layers in that sub-region. Note the linear arrays lay on boundaries, but not all boundaries align with arrays of voids.

To understand the "D" shaped voids in figure 7c, the raw monitoring data were overlaid with the optical montage. Examination of the combined data revealed that the material viewed at the left side of image was printed after the material view on the right side. Further, the voids line up with the end of scan vectors associated with printing the left side material. Because this image is from a section through the bulk of the cylinder, additional material would have been built on top of the defects. The top surface of the part can be investigated to observe a material layer that experienced only one laser pass. Figure 8a shows an image of the top surface of a part from this build. The image clearly shows a repeating pattern of "D" shaped features. Figures 8b-d show a single scan vector from above and sections through the track at the middle and end, showing that the divot is present at the end of the track.

End of track defects have been seen in higher-fidelity models that account for evaporation, convection, and fluid flow within the melt pool [15]. Using our model, we are able to determine the amount of time any point in space spends within the melt pool. Figure 9 shows discrete points along a series of scan vectors, colored by the amount of time that the material was molten at each position.

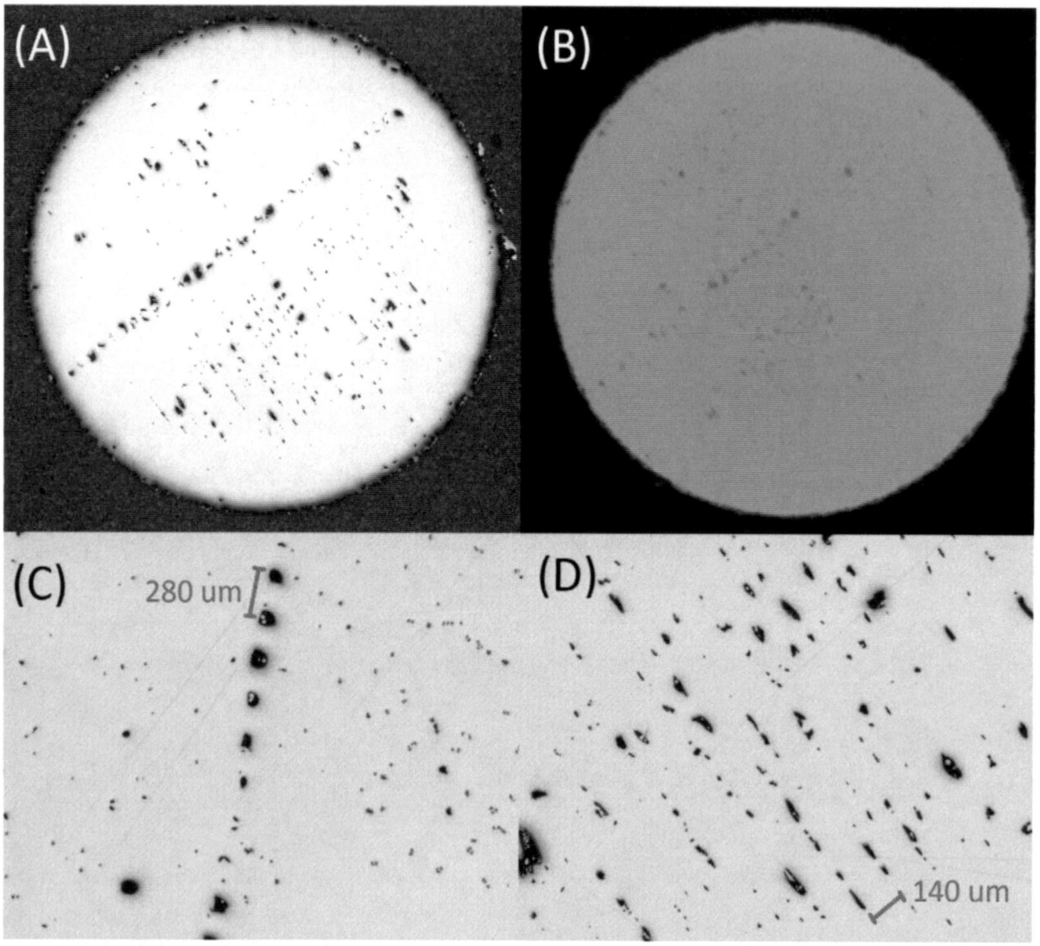

Figure 7. (A) High resolution optical microscopy montage showing the morphology of two populations of voids – montage covers the entire 10mm diameter of the part, (B) Section from CT reconstruction at same position as montage in (A), (C) Sub-image showing morphology of linear void arrays, and (D) Sub-image showing morphology of smaller, more elongated void population.

Figure 8. Images showing end-of-vector defects. (A) Top surface of sample from LPBF build, (B) Top view of single laser scan pass, (C) perpendicular section through track in (B) at midpoint along track, and (D) perpendicular section through track in (B) at endpoint where divot is present.

Points along the scan vector spend progressively less time being molten as the laser approaches the turning position. At stripe boundaries, or at the perimeter of the part, where the beam also turns off and back around, there is a periodic pattern that mirrors the pattern seen in figure 8a. As explained in [15], the molten pool solidifies too quickly and is not able to fill in the divot at the end of the track, leaving a defect. The time to solidify is material specific, depending on its thermal properties and freezing range, but for the present alloy at the power and speed values used during printing, the model appears to predict times in reasonable agreement with freezing in an end-of-track defect.

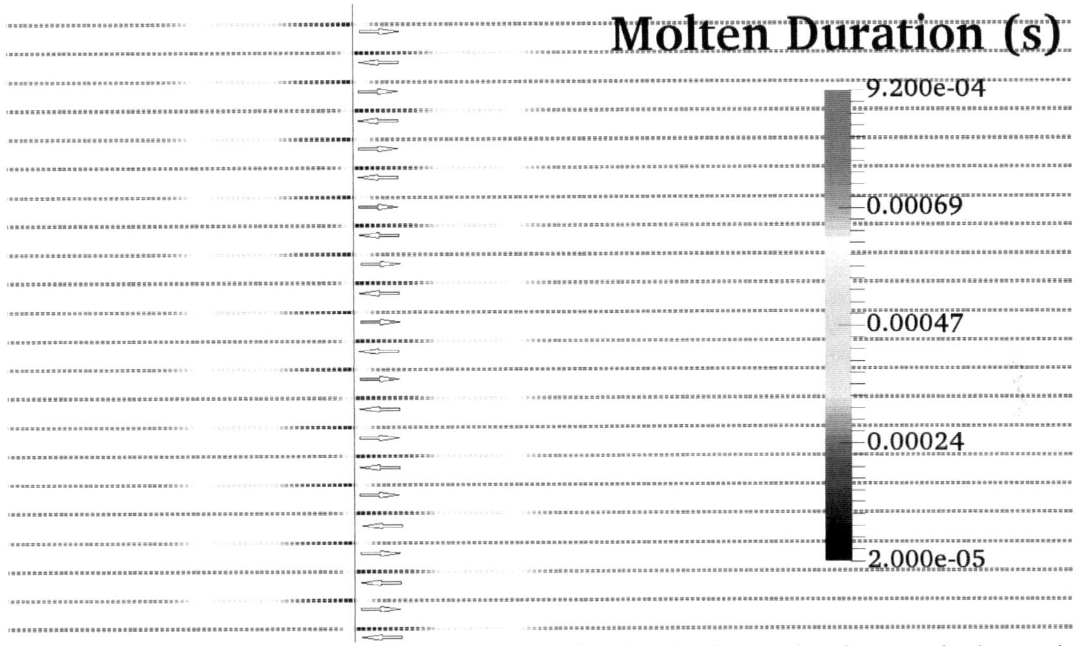

Figure 9. Discrete points along a set of scan vectors showing the time each point spent in the moving molten pool. The thin black line shows the position of the stripe boundary and the arrows show the direction of the beam into and out of the turn-around at the stripe boundary. The time spent molten decreases as the beam approaches the end of the track, leading to the 'freezing in' of a defect.

The smaller voids along the hatch vectors that appear to be lack of fusion defects were also correlated to phenomena seen in the thermal model. Figure 10 shows points along scan vectors, similar to figure 9, but with the points colored by the temperature of the point just prior to the laser passing through it (i.e., the pre-heat level prior to melting). The offset in time was selected to be the time it takes the beam to move two melt pool diameters, or about 1/5000 seconds. The figure shows that the preheat temperature is highest coming out of a turn-around, but inset from the stripe boundary or perimeter of the part. This offset results from the beam traveling faster around the turn than the heat diffusion from the previous hatch vector. The preheat temperature is directly related to the maximum width that the melt pool will achieve at that point, albeit not in a linear fashion. As a result, the individual scans do not produce a uniform width track, but rather one that narrows as the track progresses. The middle of the stripe is where the two neighboring tracks would both have narrowed and potentially created the possibility of a lack of fusion void, if the energy density were not sufficient. Figure 10 shows a region in the middle of the stripe that exhibits significantly lower preheat compared to the areas near turn-arounds, resulting in smaller melt pool widths that mirror the voids shown in figure 7d.

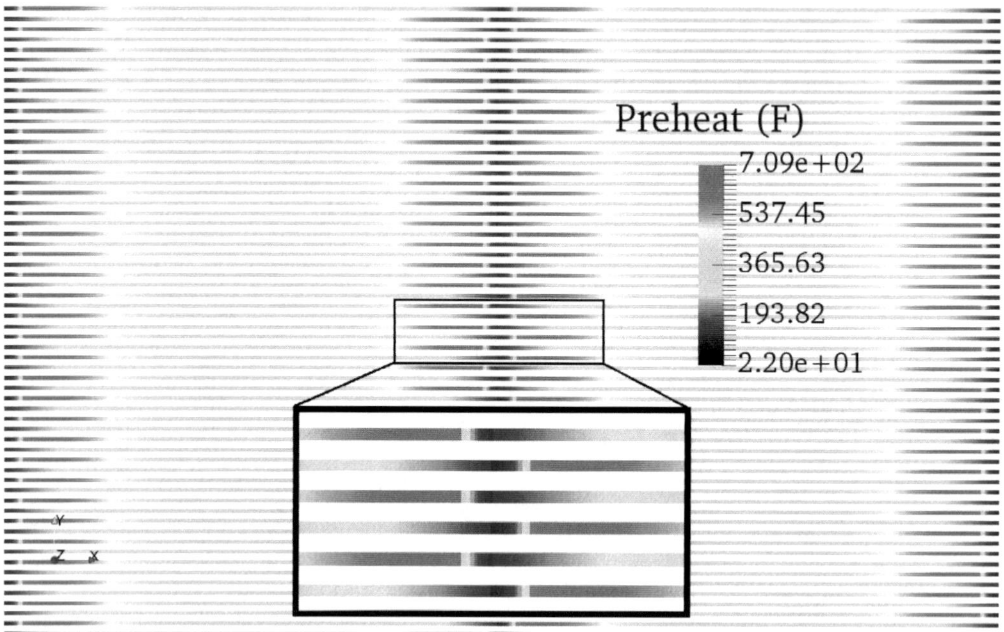

Figure 10. Discrete points along a set of scan vectors showing the temperature just prior to the laser passing through the point. The edges of the image are the perimeter of a square object and the region in the exact middle of the object is a vertical stripe boundary. The 'preheat' temperature is at a maximum for any given scan vector just after (approximately 1-2 mm) it turns on following completing a turn-around at a stripe boundary or the outer boundary of the part. The middle of the stripe is uniformly preheated relative to the regions near turn-arounds, which have both high and low preheats.

6.2. *Microstructure and texture of EBM Ti-64*

The layout of the EBM build investigated in this work was designed to push the processing conditions into two extremes. The group of smaller plates at the left of the layout in figure 2 were processed with a lower power and velocity, but a higher energy density than the larger plates at the right of the layout. For this section, the two samples will be referred to as the high energy density (HED) and low energy density (LED) samples. EBSD scans of the HED and LED samples are shown in figure 11 along with the computed pole figures. There is a noticeable change in texture between the two samples, with the HED sample exhibiting a typical texture [16] and the LED sample showing an 'abnormal' texture for the EBM process.

The scan vectors for the EBM build were input into our thermal model: the result of a representative layer is shown in figure 12. The layout of the build was such that the melt pool size and shape were relatively consistent on all layers, so for simplicity of viewing, we show a single 2D layer; note that the thermal solution is fully 3D. In the figure, a spatially-resolved temperature map is shown for a snapshot in time for the two samples. Additionally, a time-temperature plot is shown for a single point near the middle of the two samples. The spatial maps show that the melt pool and thermal field have a much higher aspect ratio in the LED sample. The time-temperature plot shows that the time above the melting temperature and cooling rate through the beta transus is longer for the HED sample. Additionally, the gradients in the thermal fields, normal of the melt pool interface, and velocity of the melt pool interface differ in the two samples. The curvature of the melt pool is larger in the LED sample, with smaller re-melt depth. The velocity of the melt pools, which affects the dominant grain growth direction, differs between the two samples. For both samples, the grains appear to grow in a columnar fashion with the long axis parallel to the build direction. However, upon higher resolution investigation of the grains in the LED sample, it is seen that the grains grow in a serrated manner, exhibiting periodic patterns associated with multiples of the layer thickness and the scan rotation strategy of the system. Figure 13 shows an EBSD map and a reconstruction of the prior beta grains from the alpha orientations [17]. The EBSD scan is perpendicular to the build direction and the grains have been annotated to show the dominant growth direction deviating from the build direction by approximately 30°.

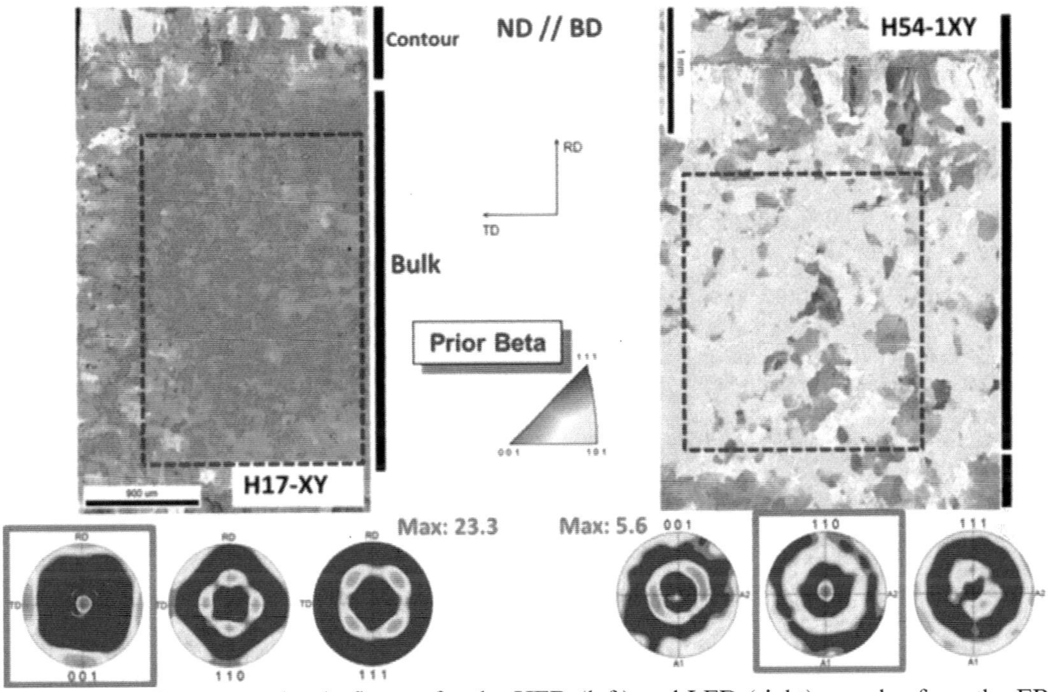

Figure 11. EBSD scans and pole figures for the HED (left) and LED (right) samples from the EBM build. The texture of the HED sample is 'typical' for EBM Ti-64, while the LED texture is 'abnormal'. Note the build direction is normal to the plane of the scan.

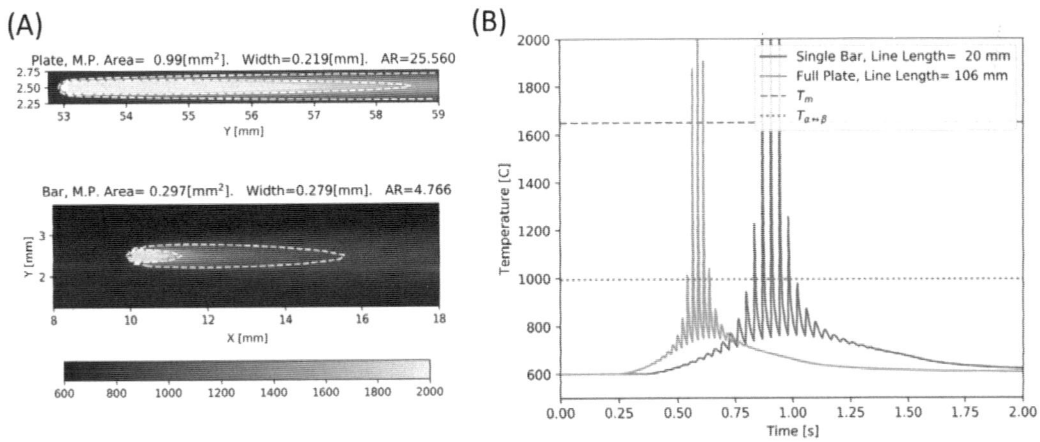

Figure 12. (A) Spatially-resolved thermal fields for a representative layer in the EBM build for the HED (bottom) and LED (top) regions, as calculated by the thermal model and (B) Time-temperature plots for a single point in the HED (blue) and LED (orange) regions. Note the differences in time above the melting temperature and the cooling rate across the beta transus.

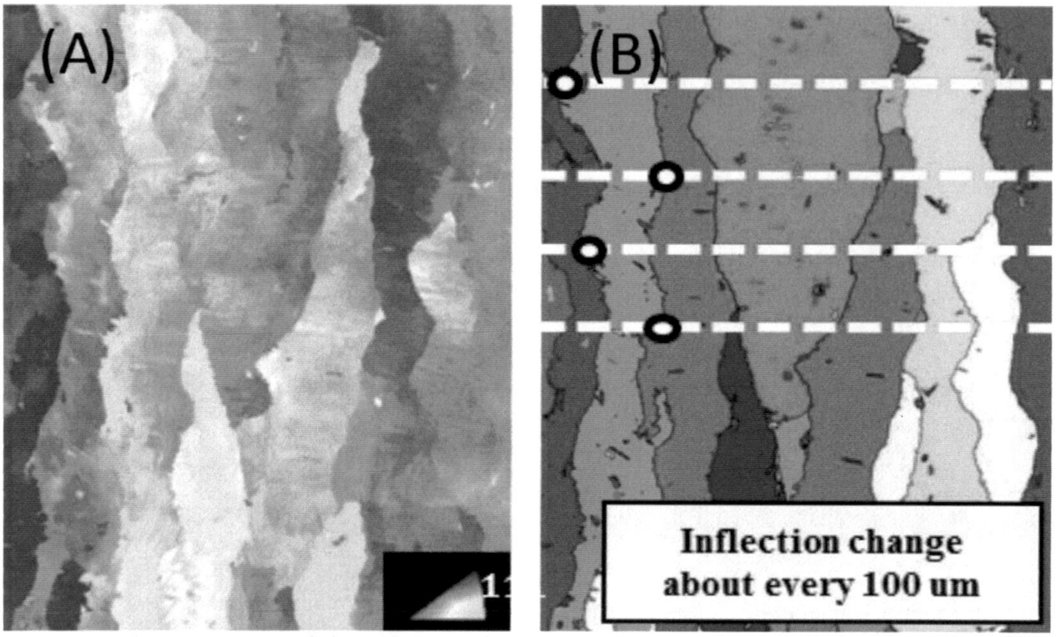

Figure 13. (A) EBSD scan of the LED sample perpendicular to build direction showing columnar features and (B) Prior beta grain reconstruction with annotations showing the serrated growth at roughly 30° to the build direction and inflections at multiples of the layer thickness (50μm), which is associated with the rotation strategy of the printing system.

7. Concluding Remarks

The printing process of metallic additively manufactured components is complex and leads to heterogeneous local processing conditions for the same global printing parameters. The geometry of individual samples, as well as their collective arrangement within the build domain, influences the processing history they will experience. This work shows that processing history affects defect and microstructure morphology, as well as crystallography. We have developed a simple yet effective model for determining spatio-temporal thermal histories within the build volume, and have shown that the histories correlate with observed microstructures. This model can be used to pre-calculate local processing conditions if the scan strategy of the system is known *a priori*. Potentially, this model could also be used to modify the build layout if it was desirable to homogenize thermal history.

References

[1] Hsieh J, Nett B, Yu Z, Sauer K, Thibault J-B and Bouman C A 2013 Recent Advances in CT Image Reconstruction. *Current Radiology Reports*

[2] Wilkinson A J and Hirsch P B. Electron diffraction based techniques in scanning electron microscopy of bulk materials. *Micron,* **28**(4) 279

[3] Besl P J and McKay N D. 1992 A method for registration of 3-D shapes. *IEEE Transactions on Pattern Analysis and Machine Intelligence,* **14**(2) 239

[4] Rusu R B and Cousins S 2011 3D is here: Point Cloud Library (PCL). *IEEE International Conference on Robotics and Automation (ICRA)*

[5] Lehmann G. 2006 Robust Automatic Threshold Selection. *The Insight Journal*

[6] Zhong X, Rowenhorst D J, Beladi H and Rohrer G S 2017 The five-parameter grain boundary curvature distribution in an autenitic and ferritic steel. *Acta Materialia,* **123** 136

[7] Barbe C B, Dobkin D P, and Huhdanpaa H T 1996 The Quickhull algorithm for convex hulls. *ACM Transactions on Mathematical Software,* **22**(4) 4693

[8] Rosenthal D. The theory of moving sources of heat and its application to metal treatments. ASME, 1946.

[9] Eagar T W and Tsai N S 1983 Temperature-Fields Produced by Traveling Distributed Heat- Sources. *Welding Journal*, **62**(12) S346

[10] Nguyen N T, Ohta A, Matsuoka K, Suzuki N, and Maeda Y 1999 Analytical solutions for transient temperature of semi-infinite body subjected to 3-D moving heat sources. *Welding Journal*, **78**(8) 265S

[11] Nguyen N T, Mai Y W, Simpson S and Ohta A 2004 Analytical approximate solution for double ellipsoidal heat source in finite thick plate. *Welding Journal*, **83**(3) 82S

[12] Elsen M V, Baelmans M, Mercelis P, and Kruth J P 2007 Solutions for modelling moving heat sources in a semi-infinite medium and applications to laser material processing. *International Journal of Heat and Mass Transfer*, **50** 4872

[13] Winczek J. Analytical solution to transient temperature field in a half-infinite body caused by moving volumetric heat source 2010. *International Journal of Heat and Mass Transfer*, **53**(25- 26) 5774

[14] Khairalluh S, Anderson A T, Rubenchik A and King W E 2016 Laser powder-bed fusion additive manufacturing: Physics of complex melt flow and formation mechanisms of pores, spatter, and denudation zones. *Acta Materialia*, **108** 36.

Strength and ductility of heavily deformed pearlitic microstructures

A. Hohenwarter[1], M. W. Kapp[2], B. Völker[1], O. Renk[2] and R. Pippan[2]

[1] Montanuniversität Leoben, Departement Materials Physics, Jahnstrasse 12, 8700 Leoben, Austria
[2] Erich-Schmid-Institute of Materials Science, Austrian Academy of Sciences, Jahnstrasse 12, 8700 Leoben, Austria

Abstract. The fracture toughness and deformation behavior of heavily deformed pearlitic steels have been investigated. A strong anisotropy of the fracture toughness and the plastic deformation behavior with respect to the lamellar orientation is observed. The consequences of this anisotropy both for processing and application, as well as for the limits in strengthening, are discussed.

1. Introduction

Eutectoid and hypereutectoid pearlitic steels are commonly used in industry. For example, nearly all rail steels have a fully pearlitic microstructure, and heavily cold drawn pearlitic wires are the standard materials for wire cables for suspension bridges, steel cord wires for tires, and the steel strings for musical instruments. In the last decades a significant increase in strength of such cold drawn pearlitic wires has been obtained [1–9]. With strengths exceeding 6 GPa in tension, such wires have further enhanced the interest in this classical structural material. In these wires the tensile strength has now reached about 1/3 of the theoretical limit, and is significantly higher than of all other currently used structural materials. Additionally, these ultra-high strength wires provide a model material for material scientists to clarify how the theoretical limit of strength of other metallic materials can be obtained.

Our interest in the last few years has been devoted to the analysis of the evolution of fracture toughness as a function of cold deformation, and to the study of the anisotropy in the deformation behavior and ductility of the nanolamellar arrangements of ferrite and cementite. The goal of the present paper is to summarize these new findings and to discuss the impact of these results with respect to the limits for processing and strengthening.

2. Fracture toughness as a function of cold working

For the processing, as well as for the in-service use of pearlitic steels, the evolution of ductility and fracture toughness as a function of strain are essential parameters. The special importance of fracture toughness and its anisotropy for ultra-strong materials has been discussed recently in [10–12]. In this section the evolution of the fracture toughness as a function of strain of pearlitic steels deformed by high pressure torsion (HPT) and of cold drawn wires are summarized. For more details, see [10,12].

The HPT processed pearlitic steel was an R 260 alloy, with a chemical composition of 0.76 wt% C, 0.35 wt% Si, 1 wt% Mn, 0.017 wt% P and 0.014 wt% S. Figure 1 contains scanning electron microscope (SEM) and transmission electron microscope (TEM) images of the undeformed and the HPT deformed microstructure for different equivalent strains viewed along the radial direction of the HPT sample. The size of the pearlite colonies in the undeformed material is in the range of 10-20 μm and the colonies are randomly oriented (figure 1a). Within the colonies the cementite lamellae spacing in the undeformed state is about 200 nm. With increasing strain the lamellae align parallel to the shear plane and the lamellar spacing decreases. As a result of the imposed strain path, unfavorably aligned lamellae become heavily bent and broken into smaller pieces, which align with further shearing (figure 1b). At equivalent stains larger than 8 the lamellae are almost fully aligned (figure 1c).

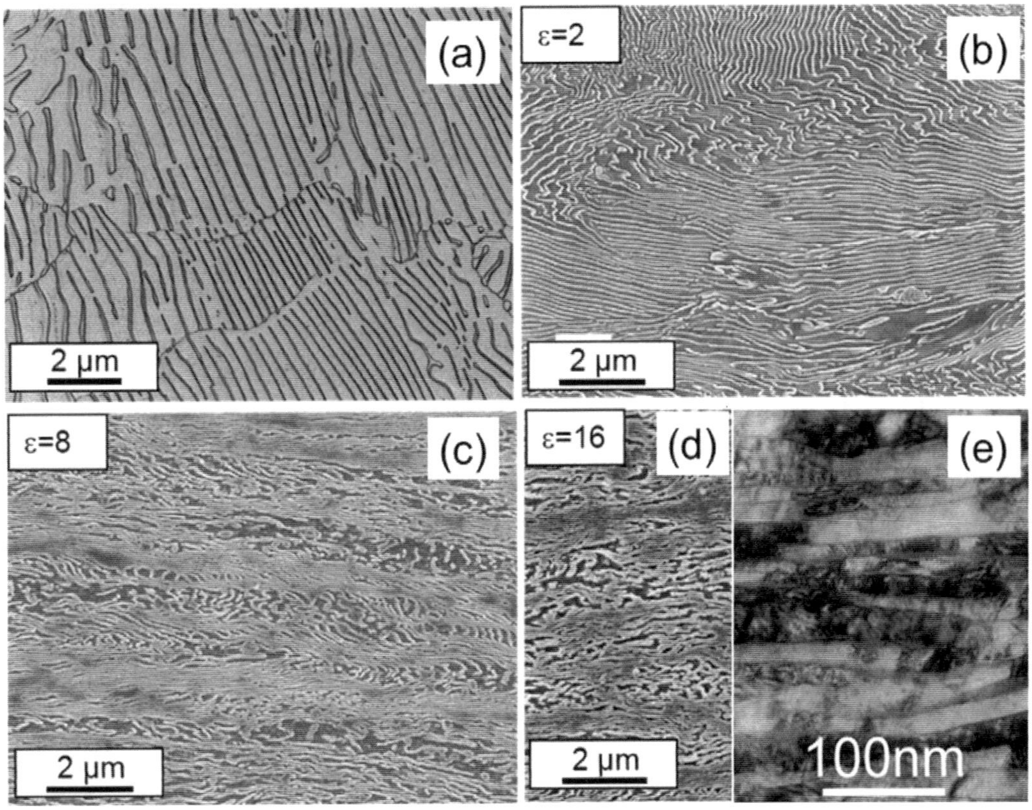

Figure 1. Scanning electron microscope images of (a) the undeformed state, (b) at an equivalent strain of ε = 2, (c) ε = 8, (d) ε = 16 and (e) transmission electron microscope image at ε=16.

At an equivalent strain of ε = 16 the lamellar spacing is reduced to 15-20 nm, as observed in the SEM (figure 1d) and TEM micrograph in figure 1e. In figure 2 the corresponding increase of hardness as a function of strain is plotted. The refinement of the microstructure is reflected in the change of the hardness, which increases strongly from 270 HV in the undeformed state to 770 HV at an equivalent strain of ~16. Along with the microstructural changes several studies have proven that a distinctive change in the chemistry of the cementite phase takes place, leading to a non-stochiometric composition and a supersaturation of the ferrite phase. Therefore, the classical description of heavily deformed pearlite as "ferrite-cementite" composite is not fully straightforward [13–15].

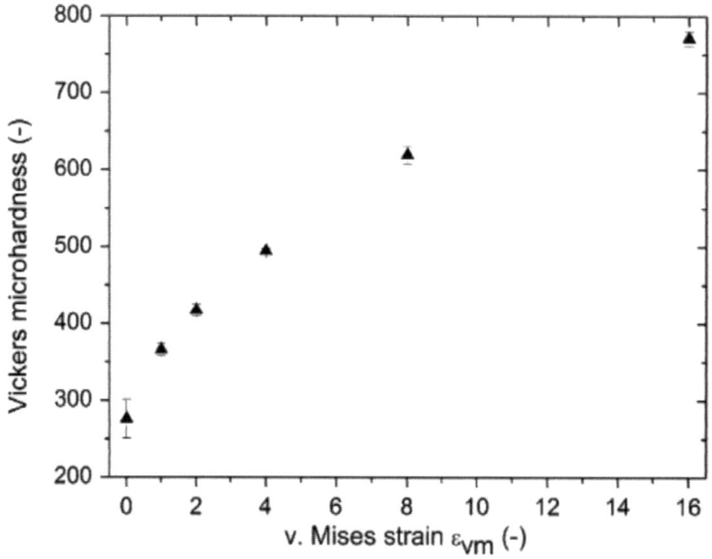

Figure 2. Evolution of Vickers microhardness with increasing equivalent strain (data from [10]).

Compact tension (CT) specimens for fracture toughness measurements were machined from the HPT samples deformed to different numbers of rotations. Depending on the number of rotations and the sample extraction radius the fracture toughness can be measured as a function of equivalent strain. The miniaturized CT specimens had a width of 5.2 mm and a thickness of 2 mm. The fracture toughness for the three different crack plane orientations, labelled as A, B and C as described in the following, were investigated [10]. In the samples with orientation A, the crack plane is parallel to the shear plane and the propagation direction is in the tangential direction of the HPT sample or parallel to the shear direction. For orientation B (data not presented here) the crack plane is perpendicular to the shear plane and the desired crack growth direction is along the axial direction of the HPT sample and perpendicular to the shear plane. For orientation C the crack plane is perpendicular to the shear plane as in orientation B, but the crack propagation direction is in the radial direction of the HPT sample.

The fracture toughness results for orientations A and C are summarized in figure 3. With increasing equivalent strain a significant anisotropy of the fracture toughness develops. For orientation A the fracture toughness decreases at smaller shear strains very quickly and reaches a nearly constant low value at higher strain levels, which is only slightly larger than the fracture toughness of a ceramic. For crack orientation C the fracture toughness increases at lower strains, but then decreases at higher strains to values similar to the undeformed material (about 40 MPa $m^{1/2}$). Fracture toughness values of orientation B, not shown in figure 3, lie between those of orientation A and C. However, the crack immediately deflects into the shear plane (perpendicular to the initial crack plane), and follows the direction of lowest toughness. Therefore, these values do not characterize the mode-I fracture toughness and correspond only to a lower limit for this loading case.

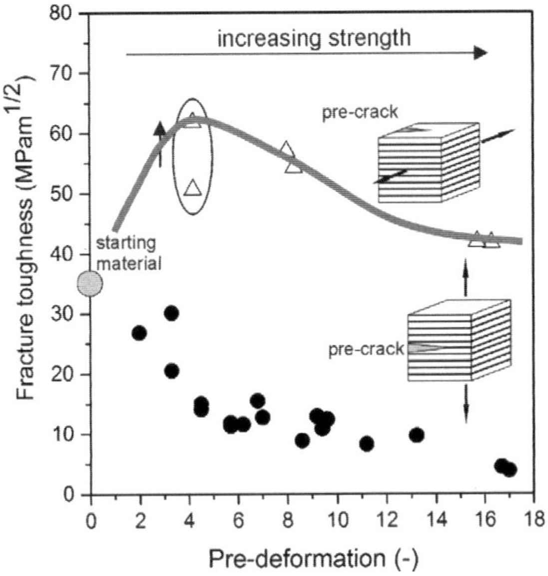

Figure 3. Evolution of fracture toughness with increasing pre-deformation, thus, equivalent strain, for two different loading directions: pre-crack parallel to (orientation A, black dots) and perpendicular to (orientation C, open triangles) the lamellar alignment. Data taken from [10].

Similar fracture toughness experiments have been performed on heavily cold drawn pearlitic wires [12]. The chemical composition of the hypereutectoid steel used was Fe, 0.98 wt.% C, 0.31 % wt. Mn, 0.2 wt.% Cr, 0.006 % wt. P and 0.007 % wt. S. Two wires cold drawn to true strains of 3.1 and 6.52, with corresponding ultimate tensile strengths of about 4 and 6.5 GPa, respectively, were investigated. The diameters of the wires were 120 µm and 24 µm, respectively. Again the anisotropy of these wires has been investigated. The microstructure in the view perpendicular to the drawing direction is very similar to the HPT microstructure perpendicular to shear plane. The thicker wire (ε = 3.1) has a lamellar spacing similar to the HPT sample with ε = 16 (figure 1d-e). In the thinner wire (ε = 6.52) the microstructure is significantly finer. The microstructure viewed along the drawing direction is very different to the HPT deformed microstructure, showing a characteristically curled microstructure. For more details regarding the microstructure and tensile properties of these wires see [9].

In order to measure the fracture toughness micrometer-sized cantilevers with the crack propagation direction parallel to the drawing direction, and single-edge-notched tension samples for crack propagation perpendicular to the drawing direction were tested. The measured anisotropy of the fracture toughness values of the thicker wire (ε = 3.1) are comparable to the HPT samples deformed to high strains (compare table 1 and figure 3). The low fracture toughness in the drawing direction (parallel to the lamellae) is comparable to, or somewhat larger than, the lowest values measured parallel to the shear plane of the HPT sample (orientation A). In both cases this is the brittle loading direction. The measured fracture toughness for orientation C of the HPT-processed material is similar to the crack plane perpendicular to the drawing direction in the wire and is in both cases distinctively higher. For both processing techniques this is the ductile loading direction. For the thinner wire (ε = 6.52), the material with the highest tensile strength, the fracture toughness decreases somewhat for the crack propagation direction in the drawing direction, but more significantly in the crack propagation direction perpendicular to the drawing direction. Thus, a pronounced anisotropy of the fracture toughness remains. The reason for the pronounced anisotropy in the HPT-processed material and the drawn wires are clearly visible from the fractographs in figures 4 and 5, for the brittle and the ductile loading direction, respectively. In all cases the transition from the pre-crack to the overload fracture is depicted. For the case of crack propagation parallel to the shear plane and the drawing direction, which is in both cases the brittle direction, a debonding of the aligned lamellar structure is evident (figure 4). In addition, the difference in the lamellar arrangement between HPT and wire drawing, namely the planar or curled microstructural arrangement, is reflected by the topography of the fractographs.

In contrast, for orientation C in the HPT sample, and for the crack orientation perpendicular to the drawing direction of the wire (figure 5), the occurrence of delamination results in a reduction of the triaxiality in front of the crack and permits a ductile failure. For further details see [10,12].

Table 1. Fracture toughness values for a pearlitic wire drawn to ε = 3.1 (low deformed) and ε = 6.52 (high deformed). Pre-cracks were introduced parallel and perpendicular to the lamellae orientation [12].

Specimen	$K_{IC,\ parallel}$ (MPa m$^{1/2}$)	$K_{IC,\ perpendicular}$ (MPa m$^{1/2}$)
low deformed-1	5.1	40.1
low deformed-2	4.9	42.5
high deformed-1	3.7	(19.7)
high deformed-2	3.8	(21.1)

Figure 4. SEM images of the fracture surfaces with the pre-crack parallel to the lamellae, thus parallel to (a) the HPT shear plane (orientation A) [10] and (b) the wire axis [12].

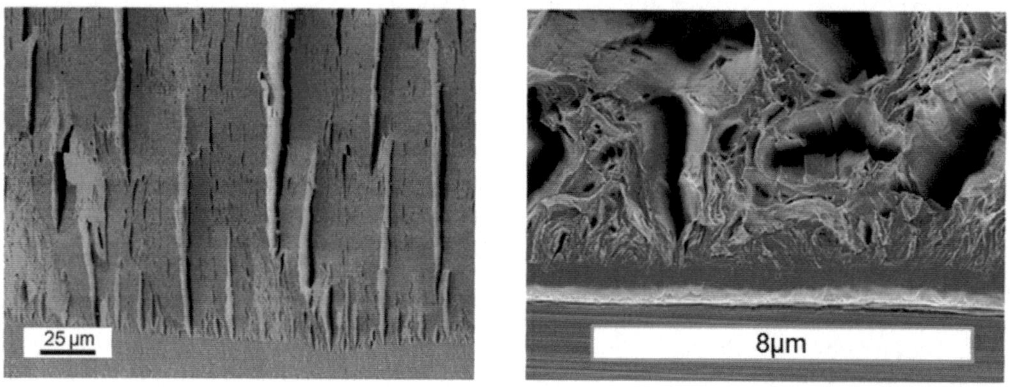

Figure 5. SEM images of the fracture surfaces with the pre-crack perpendicular to the lamellae, thus (a) along the radial direction of the HPT disc (orientation C) [10] and (b) perpendicular to the wire axis [12].

3. Effect of the lamellar arrangement during compression loading

In order to compare the mechanical behavior of an undeformed and heavily cold worked lamellar "ferrite-cementite" arrangement, micromechanical experiments on micro pillars of size about 3 x 3 x 6 μm^3 have been performed. The details of this study are described in [16]. A fully pearlitic rail steel (R260), with a chemical composition as described in section 1, was used. As already mentioned, the lamellar spacing in the undeformed pearlite is about 200 nm (see figure 1) with a colony size of 10-20 μm. To study the orientation dependency of such lamellar composites on the plastic deformation, the sample size for the undeformed pearlite has to be significantly smaller than the colony size. The heavily deformed samples were taken from a HPT sample in a region with an applied strain of 14.8; the lamellar spacing in this case was about 15 nm. For better comparison, both the undeformed and the heavily deformed samples had approximately the same dimensions. All samples were machined by focused ion beam (FIB) milling. For each material, three loading conditions, schematically depicted in figure 6, have been investigated to study potential plastic anisotropy.

The loading directions are denoted as parallel (parallel to the aligned lamellae), normal (perpendicular to the ferrite and cementite lamellae) and inclined (a loading direction between these two extrema at an inclination angle of about 45°). Measured load-displacement curves, i.e. technical stress-strain curves, are presented in [16]. In figure 7 the characteristic features are replotted in a schematic diagram to point out more clearly the differences in the deformation behavior, as a function of the loading direction for the undeformed and heavily deformed pearlite.

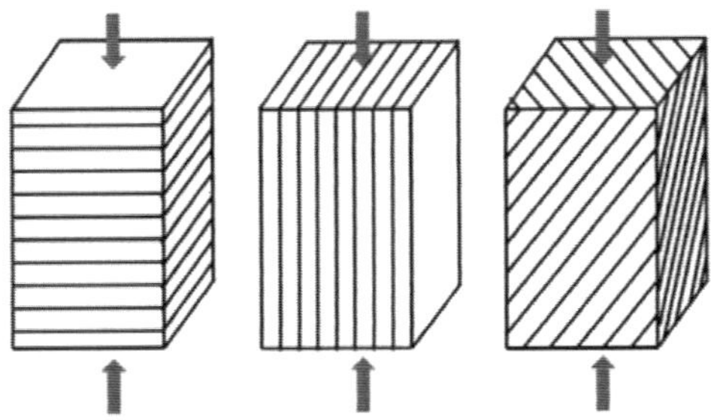

Figure 6. Schematic diagrams of the principal compression directions (normal, parallel, inclined) with respect to the ferrite and cementite lamellae alignment.

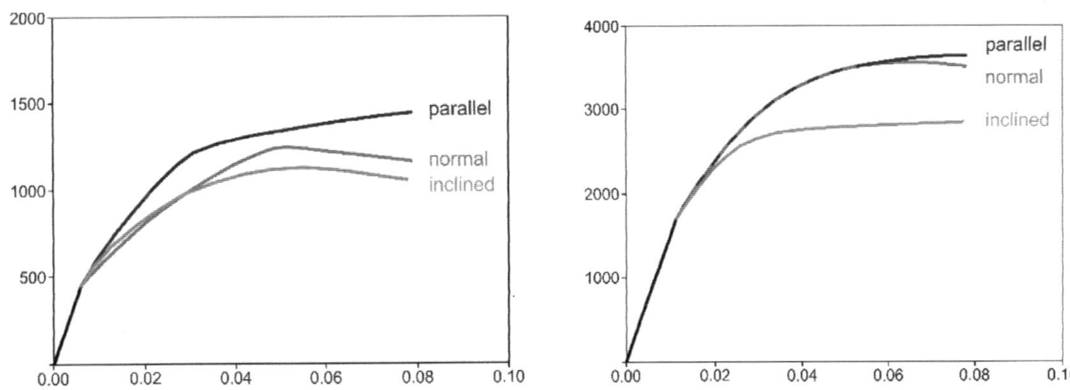

Figure 7. Schematic stress-strain curves based on data [16] (a) for the HPT processed state after an equivalent strain of ε = 16 and (b) for the non-HPT processed state, in each case for the lamellae aligned parallel, normal and inclined with respect to the loading direction.

Clearly evident is the enormous effect of HPT deformation, or in other words, the reduction of lamellar spacing from about 200 nm to 15 nm, on the mechanical behavior. Despite the significant differences in the flow stress of the two materials, there are several similarities, but also some differences with respect to the orientation dependence. The inclined direction for both microstructures is always the softest. A further pronounced similarity is the formation of only one (HPT pre-deformed to $\varepsilon = 14.8$) or two (not pre-deformed) distinct and narrow shear bands for both lamellar spacings in the normal loading direction, which are clearly visible in the SEM images of the deformed samples in figure 8. The formation of these shear bands is associated with a load drop. It seems that the realignment of the lamellar arrangement into the shear direction, compare [16], causes initially some softening and when the lamellar spacing is again sufficiently reduced, or the formed substructure is sufficiently refined (hardened), again a stable deformation takes place, presumably by a growing of the size of the shear band. Common for all three orientations is the start of plastic yielding, which is for the undeformed state at somewhat below 500 MPa and for the nanolamellar ($\varepsilon = 14.8$) at about 2 GPa. However, it should be noted that the exact determination of the onset of plastic yielding cannot be determined very precisely in such micro-compression experiments.

For the parallel and the normal lamellae arrangements with respect to the loading direction, in the undeformed material the early pronounced hardening can be explained easily by a composite model. The cementite behaves elastic and the ferrite is already nearly ideally plastic during this first few percent of plastic deformation. When the plastic deformation of the cementite starts, pronounced hardening diminishes, or is even displaced by softening, for the normal loading direction due to the formation of a shear band.

In the heavily deformed material the parallel and normal orientations behave in a very similar manner during hardening, which indicates that the composite model cannot be applied anymore. After the initial pronounced hardening regime at a flow stress of about 3.5 GPa the parallel orientation exhibits a nearly ideal plastic behavior, whereas the normal orientation shows softening due to the pronounced shear band.

The shape of the pillars after the compression experiment is compared in figure 8 by one representative pillar for each loading direction and lamellar spacing. All samples deform at higher strains on this scale in the form of shear or kink bands, but only the normally oriented pillar deforms primarily in a single band.

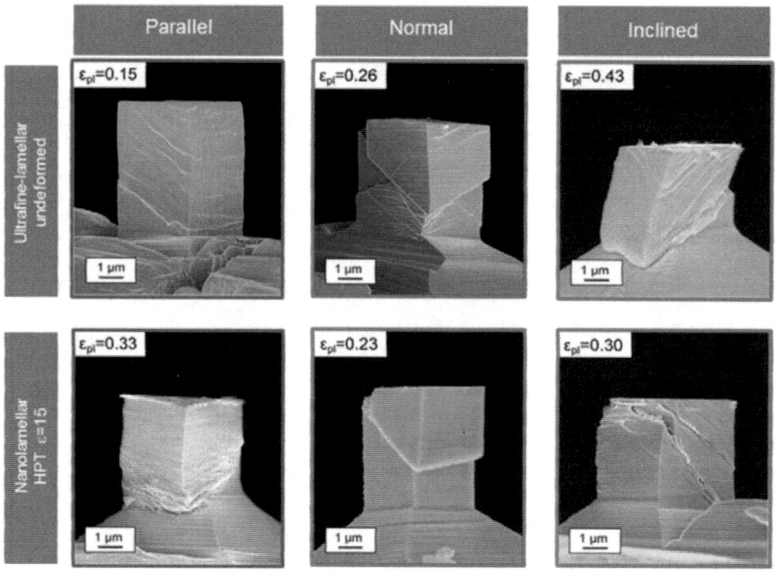

Figure 8. SEM images of the undeformed (ultrafine-lamellar) and HPT pre-deformed (nanolamellar) compressed micro pillars, with the lamellae aligned parallel, normal and inclined with respect to the loading direction.

4. Concluding remarks with respect to processing and applications

The presented results have important consequences not only for the processing and application of cold worked pearlitic steels, but also for the limits of obtainable strength and ductility in such materials. There are a large number of papers dealing with the strength of cold worked fully pearlitic steels, see for example [1–9]. In most cases it is assumed that the strength is a sum of different contributions, where the most important assumed mechanisms are the Hall-Petch or Orowan mechanisms, dislocation hardening, and solution hardening by carbon dissolution [8]. In the authors' opinion, at the limit of strength of this type of material not all of these mechanisms are of similar importance. Although the micro-compression experiments provide some hints to the hardening phenomena, in the present paper only the consequences for the cold working and application will be briefly discussed.

Anisotropy of fracture toughness is a common feature of many highly stressed biomaterials like wood or bone. The exceptional properties in certain loading directions cannot be explained without this anisotropy. This is also the case for cold drawn pearlitic wires, where the anisotropy in fracture toughness is a pre-requisite for its exceptional combination of properties.

In the typical loading direction of the wires and sheets for technical applications, or along the directions where the highest tensile deformation during the processing occurs, the fracture resistance is high. It is exceptionally high if one takes into account the strength of these materials [12]. In the directions where the stresses during application are low, or where during processing compressive stresses are present, the fracture toughness is very low. Cracks on the microscale are unavoidable during processing and in service. In the case of tensile loading of sheets or wires such cracks always experience a high fracture resistance due to crack deflection or crack delamination. The decrease in fracture toughness and the decrease in the anisotropy for cold drawn hypereutectoid steel wires with strengths above 6.5 GPa might be one of the main reasons why it becomes more and more difficult to further deform such wires. An important feature of the micro-compression experiments was the pronounced formation of a single "macro" shear band for the loading of the lamellar microstructure in normal direction, i.e. compression loading perpendicular to the lamellar plane. A consequence of this anisotropy in the shear localization seems to be that during rolling the occurrence of macro shear bands at higher strains is much more pronounced than during wire drawing.

Figure 9. Optical micrograph of a pearlitic steel (R260) cold rolled at 298 K to a logarithmic thickness reduction of $\varphi = 1.5$. Macroscopic shear bands are clearly visible in the transverse section.

From simple geometric considerations, in rolling the lamellae become nearly perfectly aligned to the rolling plane at strains of about 1.5. Such an arrangement is very sensitive to the formation of localized macro shear bands, see figure 9, whereas during wire drawing a curled arrangement of the lamellae develops, and thus is less sensitive to the formation of macro shear bands. Hence, it seems to be easier to reduce the lamellar spacing by wire drawing than by sheet rolling without failure.

The shear deformation in HPT, or on the surface of rails, is realized by a relatively homogeneous shearing. This shear deformation corresponds to deformation of the inclined micro pillar, which has the lowest flow stress. Thus, in HPT or at the surface of a rail, the macro shear direction matches well with the shear direction and with the lowest flow stress, and shearing in any other direction would require higher shear stresses. Hence, this anisotropy helps to stabilize homogenous shear deformation such as that during HPT or shear deformation near the surface of a rail.

The present results demonstrate clearly the importance of the development of an orientation dependence of the fracture toughness, and the plastic deformation both during processing and in-service, of these ultra-strong materials. Whether an extension of this combination of properties to even finer and hence stronger materials is possible, remains an open question.

Acknowledgements
Financial support by the European Research Council under ERC Grant Agreement No. 340185 USMS is gratefully acknowledged.

References

[1] Langford G 1977 *Metall. Trans.* A **8** 861–75
[2] Porter DA, Easterling KE and Smith G 1978 *Acta Metall.* **26** 1405–22
[3] Gil Sevillano J 1991 *J. Phys. III France* **1 6** 967–88
[4] Toribio J 2004 *Mater. Sci. Eng.* A **387-389** 227–30
[5] Wetscher F, Vorhauer A, Stock R and Pippan R 2004 *Mater. Sci. Eng.* A **387-389** 809–16
[6] Wetscher F, Tian BH, Stock R and Pippan R 2006 *MSF* **503-504** 455–60
[7] Wetscher F, Stock R and Pippan R 2007 *Mater. Sci. Eng.* A **445-446** 237–43
[8] Zhang X, Godfrey A, Huang X, Hansen N and Liu Q 2011 *Acta Mater.* **59** 3422–30
[9] Li Y, Raabe D, Herbig M, Choi P, Goto S, Kostka A, Yarita H, Borchers C and Kirchheim R 2014 *Phys. Rev. Lett.* **113** 106104
[10] Hohenwarter A, Taylor A, Stock R and Pippan R 2011 *Metall. Mater. Trans.* A **42** 1609–18
[11] Kammerhofer C, Hohenwarter A, Scheriau S, Brantner HP and Pippan R 2013 *Mater. Sci. Eng.* A **585** 190–6
[12] Hohenwarter A, Völker B, Kapp MW, Li Y, Goto S, Raabe D and Pippan R 2016 *Sci. rep.* **6** 33228
[13] Wetscher F, Pippan R, Sturm S, Kauffmann F, Scheu C and Dehm G 2006 *Metall. Mater. Trans.* A **37** 1963–8
[14] Sauvage X and Ivanisenko Y 2007 *J Mater Sci* **42** 1615–21
[15] Li YJ, Choi P, Borchers C, Westerkamp S, Goto S, Raabe D and Kirchheim R 2011 *Acta Mater.* **59** 3965–77
[16] Kapp MW, Hohenwarter A, Wurster S, Yang B and Pippan R 2016 *Acta Mater.* **106** 239–48

Advanced High Strength Steels: Improved Properties by Design of Textures and Microstructures

Leo A.I. Kestens[1,2], Dorien De Knijf[1,3], Felipe Castro Cerda[1,4] and Roumen H. Petrov[1,2]

[1]Ghent University, EEMMeCS Dept, Technologiepark 903, 9052 Gent, Belgium
[2]Delft University of Technology, Materials Science and Engineering Dept, Mekelweg 2, 2628 CD Delft, The Netherlands
[3]ArcelorMittal – Gent, John F. Kennedylaan 51, 9042 Gent, Belgium
[4]Universidad de Santiago de Chile, Department of Metallurgical Engineering, Av. Lib. Bdo. O'Higgins 3363, Estación Central, Santiago de Chile, Chile

leo.kestens@ugent.be

Abstract. Texture and microstructure formation during annealing of a 75% cold rolled AHS steel, reheated at a rate of 10°C/s to ~3000°C/s were studied with the purpose to investigate the interaction between the recrystallization and the austenite formation as well as the possibility of grain refinement. It was found that at ultra-high heating rates, the bcc-fcc phase transformation starts before the completion of recrystallization and the deformation texture is partially retained in the samples after annealing. The crystallographic texture observed in the martensitic phase, which is the product of intercritical austenite transformation in as-quenched samples, i.e., after the double α-γ-α' transformation, is similar to the initial cold rolling texture, which is an indication for the texture memory effect. After ultra-fast reheating with average reheating rates of 1000°C/s and higher, a significant grain refinement was observed with an average calculated ferrite grain diameter of ~2 μm. The obtained final ferrite grain size depends significantly on both the reheating temperature and the heating rate. The same effect is observed in 75% cold rolled quenching and partitioning steels for heating rates of 400°C/s, 800°C/s and 1000°C/s. In both cases the combination of fine grains, and weak RD/ND-fiber textures produce an increase of strength and ductility compared to material annealed using a conventional heating rate of ~10°C/s.

1. Introduction

The development of advanced high strength steels (AHSSs) was a reaction of the steel industry to societal requests to decrease the CO_2 emissions of the cars without sacrificing passenger safety [1]. The combination of strength and toughness is a consequence of the unique design of the microstructure in AHSS, which combines hard, ductile and metastable phases. The development of AHSS grades is one of the enduring priorities of the steel manufacturing industry today and requires often non-conventional approaches in development of stronger and tougher steel grades. The use of high strength steels in the automotive industry appears to be an effective way to decrease the weight of vehicles, to increase the safety of the passengers and to decrease the fuel consumption.

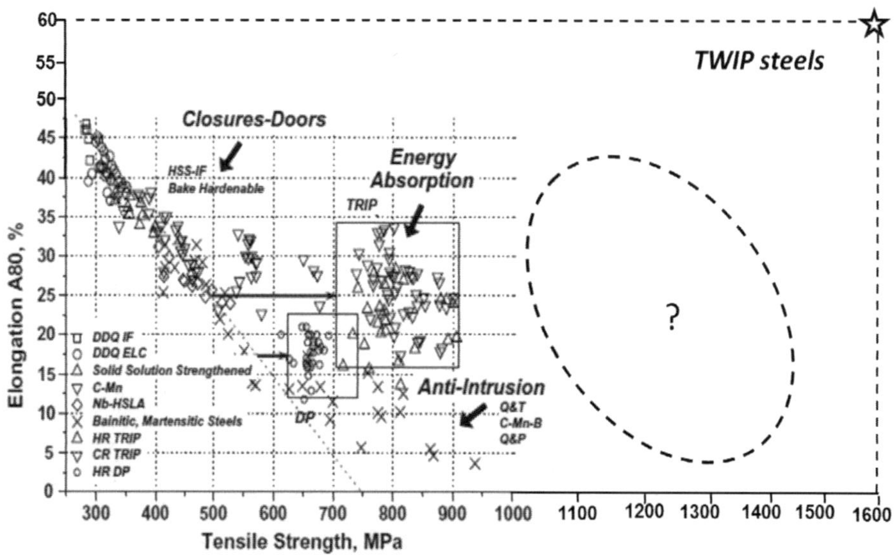

Figure 1. Distribution of the existing steel grades according their ultimate tensile strength (UTS) and tensile elongation (TE). The zone marked with a dashed ellipse is still without commercial representatives. (Adapted from ref. [2] with the data for TWIP steels).

Such steels are known as 3rd generation AHSSs and they are still under development by steel producers. The classical approach for the development of AHSS grades is two-fold, namely *(i)* the use of a specific combinations of alloying elements and *(ii)* control of the thermo-mechanical manufacturing process, which determines the formation of the microstructure. This approach appears to be very successful and has given rise to well-known steel grades such as dual phase (DP) steels, transformation induced plasticity (TRIP) steels, multiphase or complex phase steels (steels with dislocation rich microstructures such as martensite and bainite) [2] and, more recently, twin induced plasticity (TWIP) steels. The well-known strength vs ductility chart for the most popular automotive steel grades is shown schematically in Figure 1.

It is clear that there are no representatives at the moment which can cover the gap between the heavy alloyed TWIP steels and low alloyed TRIP steels. Adding of alloying elements, as done in medium Mn steels, is an easy solution, but problems with metal scarcity will restrict their extensive use more and more in the future because of cost limitations. The control of cooling rates during production is already very well studied; *i.e.*, the limits to improvement in modifying this parameter are known and significant new improvements in this direction cannot be expected, although new heat treatments like quenching and partitioning (Q&P) are now under development in order to populate the region marked with a question mark in Figure 1 [1].

A non-conventional approach is to explore the potential of the heat treatment as a tool for microstructure-properties control by using well known standard compositions and to design and produce fine grained steel grades by employing very high heating rates. This approach leads to extreme grain refinement [3–6] and to a corresponding increase of the strength of the material above the *normal* well known limits. In previous works [3–7] a number of microstructural phenomena, like remarkable grain refining and overlapping of the recrystallization with the α-γ phase transformation, were reported for reheating rates of 3000°C/s and higher (up to 7000°C/s). Although the results were very promising, it is very difficult to reproduce such heating rates in industrial lines.

The aim of the present work is to obtain a better understanding of the sensitivity of the phase transformation and recrystallization, and the corresponding changes in microstructure, texture and mechanical properties to the variation of the heating rates and initial microstructure. To study the possibilities for an industrial implementation all processing parameters are varied in a range which is considered as realistic for industrial application.

2. Experimental

Cold rolled steel with a composition shown in Table 1 is the object of study in this work. Two sets of dual phase (DP) steel, produced as hot rolled plates with a microstructure of either (i) ferrite and pearlite (FP) or (ii) 50% ferrite and 50% tempered martensite (FM) were subsequently cold rolled with a reduction of 75%, to a final thickness of 1 mm, and heat treated according the temperatures and times shown in tables 2 and 3. Samples of size 80 x 30 x 1 mm^3 were cut parallel to the rolling direction of the cold rolled sheet. The samples were reheated with heating rates of 10°C/s, ~400°C/s and ~1500°C/s and subsequently water quenched without isothermal holding (a minimum isothermal holding of 0.2s could not be avoided). The data for the heat treatment parameters are given in Table 2 and Table 3, whereby the samples with initial microstructure of ferrite and tempered martensite are designated as samples A and the samples with initial microstructure of ferrite and pearlite are designates as samples B. A record of the real "time-temperature" curves is shown as an example in Figure 2.

Q&P samples (100 x 10 x 0.5 mm^3) were 80% cold rolled and subsequently subjected to ultra-fast heating (UFH) Q&P cycles in the Gleeble™ thermo-mechanical simulator according to the cycles shown in Figure 3. Two types of heat treatments were applied: one with isothermal soaking and the other without isothermal soaking, (cf. Figure 3).

Table 1. Chemical composition of studied steels, in mass. %.

	%C	%Si	%Mn	%P	%S
DP (FM and FP)	0.114	1.263	2.072	0.0149	<0.001
Q&P	0.25	1.5	3.02	0.015	<0.001

Table 2. Heat treatment parameters and hardness of samples with microstructure 75% cold rolled F+TM.

Sample ID	Heating temperature Ta, °C	Heating rate, Vh °C/s	Quench rate, V_q, °C/s	HV3
A0	-	-	-	440
A1	674	430.2	-1281.7	405
A2	723	397.8	-821.9	360
A3	756	397.5	-1069.3	359
A4	780	328.7	-568.2	341
A6	807	357.5	-735.8	354
A7	846	323.4	-836.0	402
A8	879	317.3	-854.2	510
A9	949	352.9	-1064.4	561
A10	968	359.9	-1068.9	512
AVERAGE		362.8	-922.3	

Table 3. Heat treatment parameters and hardness of samples with microstructure 75% cold rolled F+P.

Sample ID	Heating temperature Ta, °C	Heating rate, V_h °C/s	Quench rate, V_q, °C/s	HV 3
B0	-	-	-	329
B1	749	1567.6	-803.2	329
B2	841	1445.5	-1390.2	442
B3	862	1327.5	-1539.7	421
B4	1000	1419.0	-966.9	549
B5	1053	1538.1	-1058.7	545
B6	1200	1532.2	-685.4	507
AVERAGE		1471.7	-1074.0	

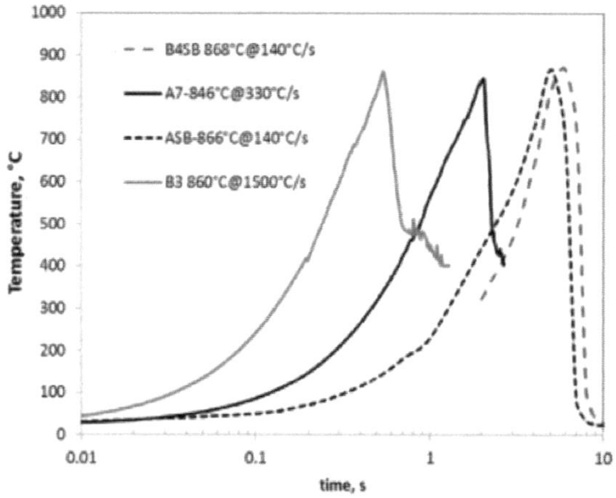

Figure 2. Record of time-temperature curves for samples with initial microstructures ferrite and tempered martensite (A) and ferrite and pearlite (B), heated in the salt-bath furnace (140°C/s) and at ulta-fast heating rates of 330°C/s and 1,500°C/s.

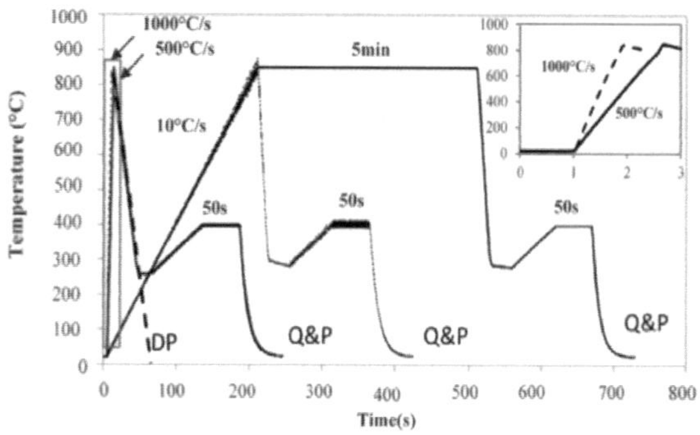

Figure 3. Heat treatment cycles performed in Gleeble on the cold rolled material, upper right corner insert is a zoom of the first 3s of heating time. The cooling rates for DP steel and for Q&P steel are marked with DP and Q&P correspondingly. Adapted from [8].

Samples for tensile tests, hardness tests and microstructure characterization were taken from the middle part of the heat-treated sheets from the zones with guaranteed homogeneous temperature distribution. Sub-size tensile test samples with a gauge length of 4 mm, width of 1mm, thickness of 1mm and total length of 10 mm were tested with a cross head velocity of 0.5 mm/min at room temperature in a micro-tensile test machine (type *Deben MICROTEST 5000N Tensile stage®*). The sample elongation was detected by displacement measurement of the machine grips. Tensile tests on sub-size samples always overestimates the strength and elongation in comparison to the standard size (A80 and A50) tensile test samples.

Similar tensile tests were applied to specimens that were water quenched from different temperatures after reheating with fast but still conventional heating rates of 140°C/s in a salt bath (SB), in order to compare the results with the data obtained from ultra-fast heating, and to correctly evaluate the contribution of ultra-fast heating to the mechanical properties. The microstructural characterization was carried out after classical metallographic sample preparation, which included mechanical grinding and polishing up to 1 μm diamond suspension and etching with 2% HNO_3 in ethanol (Nital 2%).

Samples for electron backscatter diffraction (EBSD) characterization were additionally electrolytically polished using a Struers® Lectropol 5 device with A2 type electrolyte at 22°C, 40V and electrolyte flow rate of 7. A FEI-Nova 600 scanning electron microscope was employed for the scanning electron microscope (SEM) and EBSD data acquisition on selected samples. The EBSD data were acquired at 15 kV, working distance 7 mm, sample tilt 70°, on a square scan grid with a step size of 50 nm. The data were post-processed with EDAX-TSL OIM® 5.1 data analysis software. The variations in mechanical properties as a function of quench temperature were documented by Vickers hardness measurements with a load of 3 kgf (HV3).

3. Results and Discussion

3.1. DP steel samples: microstructures

Variations in the hardness of the cold rolled and water quenched samples with different initial microstructures after heating to different temperatures with various heating rates are shown in Figure 4. The initial hardness after cold rolling depends on the microstructure and is 329HV3 for the samples

IOP Conf. Series: Materials Science and Engineering **219** (2017) 012004 doi:10.1088/1757-899X/219/1/012004

from group B with ferrite-pearlite microstructure, compared to 440HV3 for the cold rolled microstructure of ferrite and tempered martensite. The hardness does not change significantly after quenching from a temperature up to 500°C (Figure 4, zone I). An increase of the temperature with heating rates of 140°C/s (heating in a salt bath) triggers initially recovery and recrystallization in the cold rolled steel, which is associated with a decrease in the hardness (cf. Figure 4, Zone II). After quenching from temperatures situated in the dual phase ferrite-austenite or the single phase austenite regions, the hardness increases due to martensite formation from the parent austenite phase (cf. Figure 4, zone III). When the austenitisation temperature is very high the hardness after quenching slightly decreases, which can be associated with the grain growth of the parent austenite, which forms coarse and low carbon martensite (Figure 4, zone IV).

Two main tendencies are clearly demonstrated in the hardness variations – apart from the heating temperature the hardness variations are dependent on: *(i)* the initial microstructure, and *(ii)* on the heating rate. The increase of the heating rate from 140°C/s to ~360°C/s and 1500°C/s shows a clear tendency to increase the hardness after quenching from both the intercritical ($\alpha+\gamma$) and the single phase γ regions. The hardness decrease in the temperature interval 500-750°C (zone II in Figure 4), which is associated with recovery and recrystallization, becomes smaller when the heating rate is increased, which indicates that recovery and recrystallization are suppressed. No changes in the hardness are observed in samples B that were heated to the austenitic region with ~1500°C/s (cf. zone II in Figure 4- the grey line for heating rate of 1500°C/s). This is an indication that both recovery and recrystallization are completely suppressed and the α-to-γ phase transformation starts before the onset of recrystallization. This effect was previously observed after heating and quenching of cold rolled high-strength low-alloy (HSLA) steels with heating rates of ~3000°C/s and ~7000°C/s [4–7].

The tensile test results for selected samples are shown in Figure 5. The samples were water quenched from the temperatures of the two-phase ($\alpha+\gamma$) region in order to create a microstructure of ferrite and martensite, which is typical for DP steels [1,2]. It is clearly demonstrated that for intercritical annealing, an increase of the heating rate from 140°C/s to 400°C/s and 1500°C/s causes a significant increase in the ultimate tensile strength (Rm). This phenomenon could be associated with the significant grain refining effect observed in the fast-heated samples in comparison to the samples heated in the salt bath. A second important observation is that the short soaking times, practically ~1-2 s in both cases, do not allow homogenization of the carbon content of the intercritical austenite before quenching. This means that the martensite "islands" are structurally heterogeneous — i.e. they may contain zones with high carbon martensite, low carbon martensite or even retained austenite with high carbon content and correspondingly with different hardness [9].

Another relevant observation is that the initial microstructure has significant effect on the mechanical properties (type of the tensile test curve), which is more clearly pronounced at low heating rates. The tensile test curves of samples with initial ferrite and pearlite microstructures, quenched after heating with 140°C/s exhibit a well-defined yield point, and display very low work hardening, which is different in the samples with initial microstructure of ferrite and tempered martensite. It is also important to mention that the stress-strain curves reveal identical characteristics after ultra-fast heating (compare the black thick and dashed lines in Figure 5).

Figure 6 shows the microstructural evolution in samples with initial microstructure of ferrite and tempered martensite after water quenching from various temperatures in $\alpha+\gamma$ temperature interval reached with heating rates of ~140°C/s (heating in a salt bath) and fast heating with ~400°C/s. The micrographs display martensite microstructures in both group of samples, which in the case of fast heating (Figure 6b,c) appear in the recovered ferrite matrix, and the new martensite is aligned along the pancaked recovered (or partially recrystallized) ferrite. In the case of heating in the salt bath, the ferrite microstructure is fully recrystallized with well-developed almost equiaxed grains shape (cf. Figure 6d, e). The microstructures shown in Figure 7(a-e) display the corresponding changes in the steel with 75% cold rolled ferrite-pearlite microstructure after heating with ~1500°C/s and 140°C/s. The same tendency is clearly demonstrated, namely the α-to-γ phase transformation during heating starts before the onset of the recrystallization in samples heated with 1500°C/s, resulting in a fine-

IOP Conf. Series: Materials Science and Engineering **219** (2017) 012004 doi:10.1088/1757-899X/219/1/012004

grained microstructure with elongated partially recovered ferrite grains, whereas after reheating in a salt bath the phase transformation develops in the recrystallized ferrite–pearlite matrix.

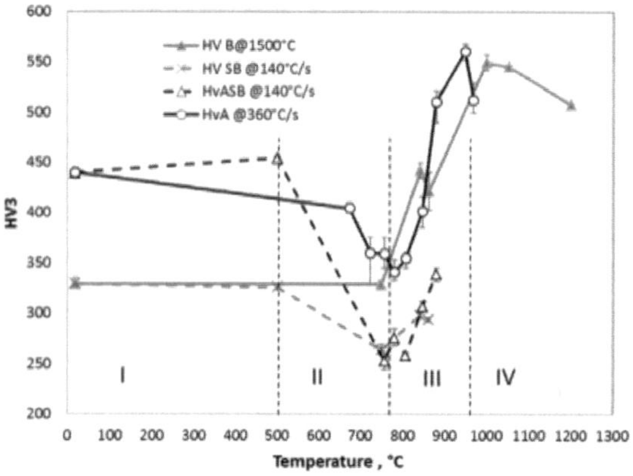

Figure 4. Hardness vs. quench temperature for different initial microstructures and heating rates. "Quench temperature" is the temperature to which cold rolled samples were heated before water quenching.

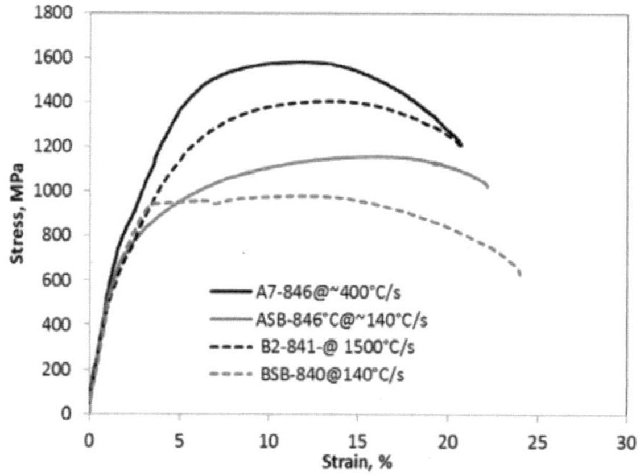

Figure 5. Tensile test results from water quenched samples after ultra-fast heating with rates of ~400°C/s and 1500°C/s (A7 and B2) and after conventional fast heating in a salt bath with ~140°C/s (samples ASB and BSB). All samples are heated for quenching to similar temperatures (~840°C) in the dual phase $\alpha+\gamma$ temperature region.

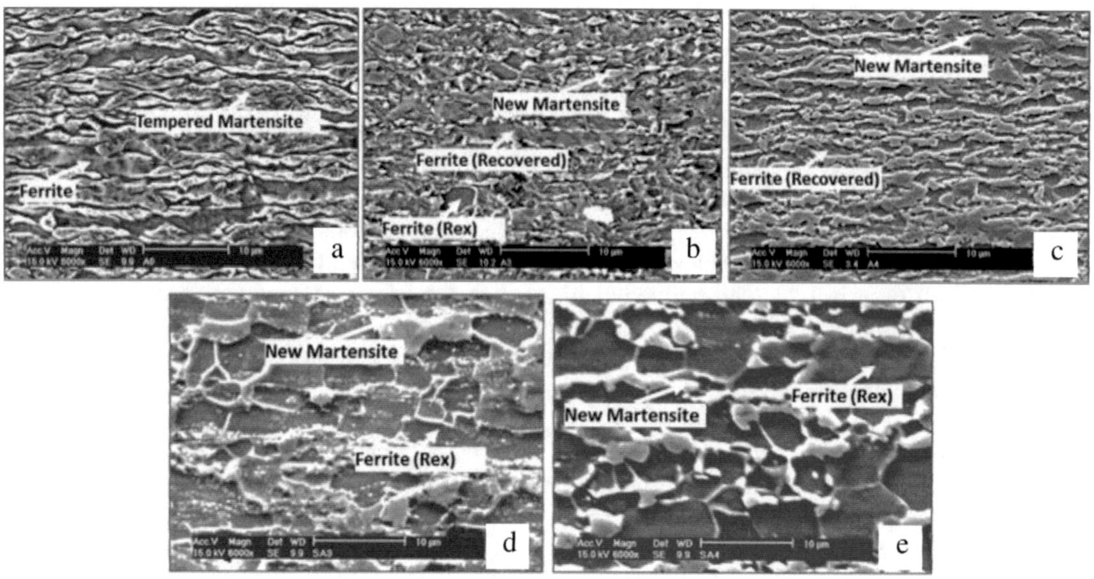

Figure 6. Microstructures of samples A and ASB (initial microstructure ferrite and tempered martensite) after: 75% cold rolling (a); reheating with ~400°C/s to 756°C (b); to 780°C (c); reheating with 140°C/s in salt bath to 756°C (d) and780°C(e). All samples are subsequently water quenched without isothermal soaking. Scale bar in all micrographs is 10μm.

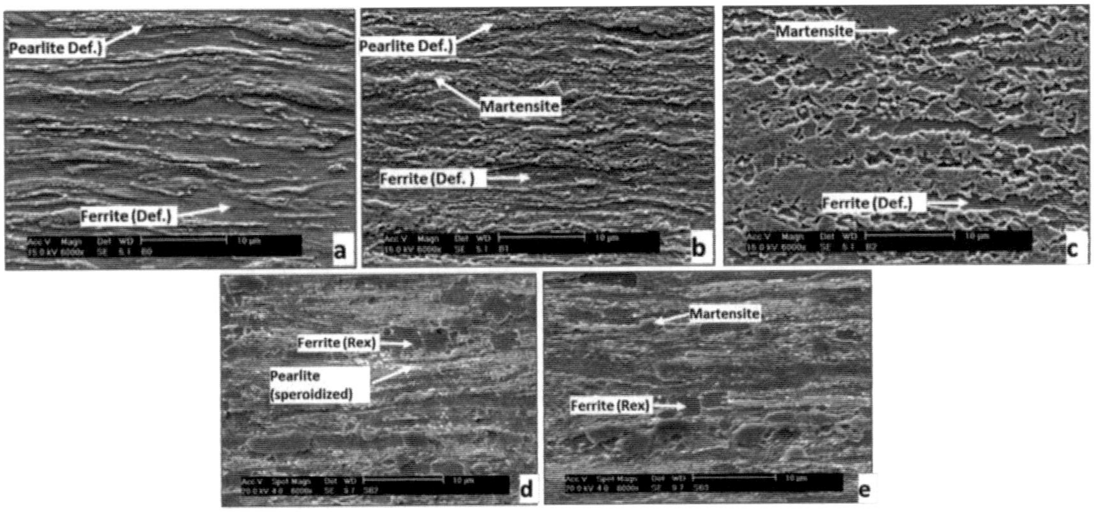

Figure 7. Microstructures of samples B and BSB (initial microstructure ferrite and pearlite) after: 75% cold rolling (a); reheating with ~1500°C/s to 750°C (b); to 840°C (c); reheating with 140°C/s in salt bath to 756°C (d) and 840°C (e). All samples are subsequently water quenched without isothermal soaking. Scale bar in all micrographs is 10 μm.

The evolution of the average ferrite grain size as a function of the heating rate was investigated on optical and SEM micrographs by means of the linear intercept method and a cross check of the result was made by EBSD measurement on samples A3 (cf. Figure 8-black triangle). The results of the measurement, which are shown in Figure 8, display that the average grain size drops from 5 μm to 1 μm with an increase of the heating rate from 140°C/s to 1500°C/s reaching a steady state. The last point in the figure represents data adopted from reference [6] for steel with the same chemical composition heated to the same temperatures with 3000°C/s.

3.2. DP steel samples: textures

Figure 9a shows the key for the main BCC texture components in the $\phi 2 = 45°$ section of Euler space and Figure 9b displays the orientation distribution function (ODF) of the 70% cold rolled initial material. Figure 9b shows a typical rolling texture with ND {111}<uvw> and RD {hkl}<110> fibers with an intensity of ~4.5 multiples of random density (mrd) [10].

Figure 10 illustrates only the ferrite textures derived from the EBSD scans using grain average image quality partitioning criteria [22]. Figure 10a, 10c, 10e, and 10g show the ODFs of the full ferrite phase (including both recrystallized and non-recrystallized grains), whereas Figure 10b, 10d, 10f, and 10h show the ODFs of recrystallized ferrite grains only. The recrystallized ferritic grains were extracted using the grain average misorientation (GAM) criterion [18, 23]. The recrystallized ferrite grains in samples heated at 150°C/s (cf. Figure 10b, 10d) show orientations close to {111}<uvw> ND fiber components. In samples heated at 150°C/s to 704 °C, the recrystallized grains have an orientation in the vicinity of the {111}<112> texture components (Figure 10b), whereas the strongest intensity after heating to 838°C at 150°C/s is 7.1 mrd and belongs mainly to the {001}<110> components.

The ODF of ferrite in samples heated at 1500°C/s to 749°C (Figure 10e) is very similar to the ODF of the cold-rolled initial material (Figure 9b). Figure 10f shows the ODF of ferrite at a very early stage of recrystallization in a sample heated at 1500°C/s. The maximum intensities of 3.1 mrd are close to {112}<110> texture component. The ODF of samples heated at 1500°C/s to 862°C (cf. Figure 10g) shows that the ferrite texture intensities are similar to the cold-rolled texture in the main components of ND {111}<uvw> and RD {hkl}<110> fibers. Recrystallized grains in the same sample show strong intensities on the {112}<110> and {111}<112> components (cf. Figure 10h).

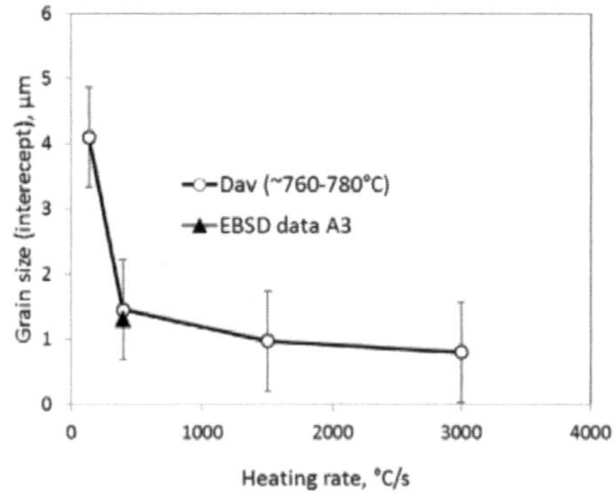

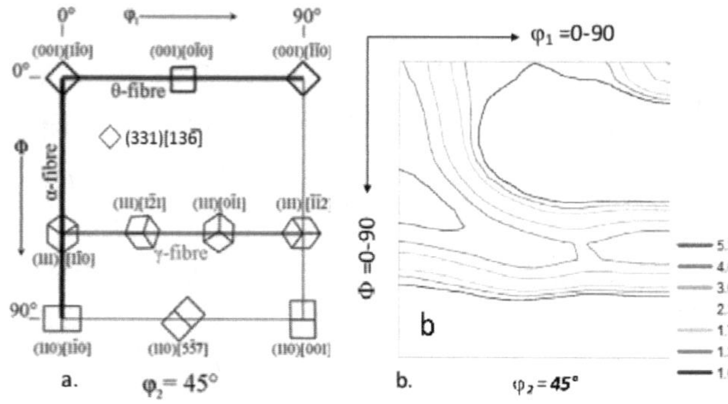

Figure 8. Average grain size (average intercept length) after water quenching as a function of heating rate. Black triangle shows average grain diameter measured from EBSD data.

Figure 9. (a) Ideal positions of the most important BCC texture components in the $\phi2 = 45°$ section of Euler space and (b) plot of ODF of the cold rolled steel used in this work.

The appearance of the Goss {110}<001> texture component with intensity of ~1.7 mrd is observed after heat treatments at both heating rates (cf. Figure 10b, 10d, and 10h). The Goss component is associated with oriented nucleation at shear bands in deformed ferrite [11]. The rotated Goss {110}<110> component, which is a high stored energy component for metals with BCC crystal structure after plane-strain compression, appears in the recrystallized ferrite after UHF at 1500°C/s (cf. Figure 10f and Figure **10**h). These components are claimed to originate along shear bands in deformed ferrite [11–13]. In some recrystallized ferrite textures also the {331}<136> component can be observed (cf. Figure 10b, Figure **10**d, Figure **10**f, and Figure **10**h). This component has been described already before [14] and is supposed to originate from specific deformation features, like intra-grain deformation bands in deformed ferrite {hkl}<110> fiber grains.

Texture measurements have shown that the first recrystallized grains of ferrite in samples treated at 150°C/s to 704°C are mainly oriented around the {111}<112> component (cf. Figure 10b). It is well known that such orientations are the most frequently occurring recrystallization component in cold-rolled low carbon steel [12,15]. Grains oriented close to {001}<110> are also present, which might be due to nucleation at low stored energy grains [16] at the beginning of recrystallization. Grains with Goss {110}<001> texture component have nucleated and grown in samples heated at 150°C/s (cf. Figure 10b).

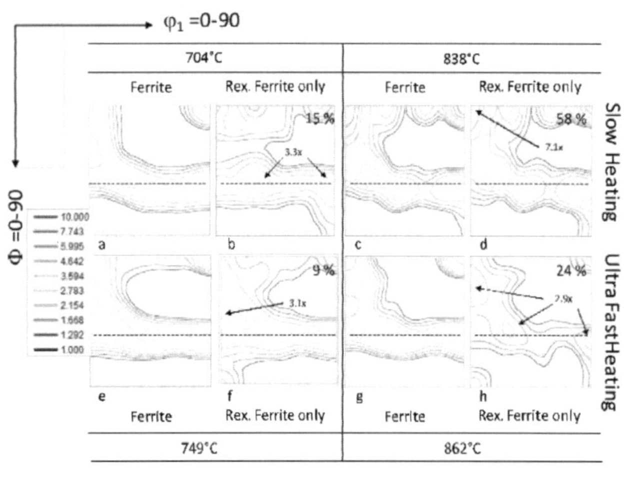

Figure 10. ODF of ferrite at $\phi2 = 45°$ from samples heated at 150°C/s (a–d) and 1500°C/s (e–h). (a) and (b) correspond to peak temperature of 704°C; (c,d) to 838°C; (e) and (f) to 749°C and (g) and (h) to 862°C. (a,c,e,g) correspond to recrystallized and non-recrystallized ferrite, whereas (b,d,f,h) correspond to recrystallized ferrite. Numbers in (b,d,f,h) indicate the recrystallized ferrite fraction. Dashed line indicates the ideal position of {111}<uvw> ND fiber.

Figure 11 and Figure **12** display EBSD maps of the samples heated at 150°C/s and 1500°C/s, respectively, to different temperatures. RD-Inverse Pole Figure (IPF) maps are plotted on top of IQ maps in Figure 11a, 11c and Figure **12**a, Figure **12**c. In these maps only the grains oriented close to the Goss {110}<001> (Figure 11a, 11c and Figure **12**c) and rotated Goss {110}<110> (Figure 12a) components with a tolerance of 10° from the ideal orientations are highlighted on the IQ maps. Figure 11b, Figure **11**d and 12b, 12d show the corresponding ND-IPF maps for each IQ map on the left-hand side. With help from the color coded RD-IPF maps, grains oriented close to the Goss texture component can be identified in the microstructure as the reddish grains in Figure 11a, Figure **11**c and 12c and greenish in Figure 11b, Figure **11**d and 12d. Grains oriented close to the rotated Goss component can be identified in the microstructure as greenish grains in Figure 12a and 12c and light blue grains in Figure 12b and 12d. Arrows indicate grains with Goss orientation and dashed circles the grains with Rotated Goss orientation in Figure 11 and Figure **12**.

Figure 11a and 11b show the recrystallized ferritic grains oriented close to the Goss {110}<001> component in the sample heated at 150°C/s to 704°C. The grains are aligned with other recrystallized grains at an angle of ~45° with RD (arrows). One could thus infer that grains oriented close to the Goss {110}<001> component in samples heated at 150°C/s are likely to originate from shear bands. Recrystallized ferritic grains oriented close to the Goss {110}<001> component are not observed in samples heated at 1500°C/s to 749°C. However, Figure 12c and 12d display grains oriented close to the Goss {110}<001> orientation with similar alignments as in Figure 11a and 11b. It can, therefore, be suggested that nucleation of ferrite at shear bands is delayed to higher temperatures in samples heated to 1500°C/s, compared to samples heated at 150°C/s.

The diffusion of carbon might also play a role in the nucleation of grains oriented in the vicinity of the Goss {110}<001> component in samples heated at 150°C/s (cf. Figure 10b) and at higher temperatures in samples heated at 1500°C/s (cf. Figure 10h). It has been claimed [17] that carbon dissolved in the cold-rolled ferrite strengthens the Goss {110}<001> component in recrystallized ferrite. The longer annealing times involved in samples heated at 150°C/s, compared to samples heated at 1500°C/s, can produce significant diffusion of carbon in ferrite and thus enhance the early appearance of Goss orientation in samples heated at 150°C/s. Consequently, the fact that grains with Goss orientation nucleate at higher temperatures in samples heated at 1500°C/s, is possibly affected by the carbon diffusion into ferrite.

In the sample heated at 1500°C/s to 749°C (cf. Figure 10f), the highest intensity in the ODF is located in the vicinity of the {112}<110> component. However, considerable intensities are present for Φ-values in the range from 0° to 90° for $\varphi_1 = 0$, and for the {111}<uvw> ND fiber components. It is suggested that the nucleation of the previously mentioned components is related to the stored energy in cold rolled ferrite. Under ultra-fast heating, the time for recovery in the cold rolled material is drastically reduced. The release of energy during the recovery is also reduced, and thus high stored energy components can nucleate preferentially. In cold rolled ferrite, the amount of energy that a component can store increases as Φ increases from 0° to 90° for $\varphi_1 = 0$ [17]. The maximum strain energy is accommodated by the rotated Goss {110}<110> component, which is present in the recrystallization textures of samples heated at 1500°C/s (cf. Figure 10f). The recrystallized ferrite from the sample heated at 150°C/s to 704°C (cf. Figure 10b) does not exhibit high stored energy components ($\Phi \rightarrow 90°$ for $\varphi_1 = 0$). This is believed to be a consequence of the larger time for recovery, compared to samples heated at 1500°C/s. The observed texture changes at 1500°C/s are consistent with data on ultra low carbon (ULC) steel under ultra-fast heating rates [18].

Figure 10h shows that grains oriented close to the rotated Goss component are still present in samples heated at 1500°C/s to 704°C and 862°C. The circles in Figure 12a and Figure **12**c indicate that such grains nucleated at grain boundaries. Moderated intensities are also observed in the sample heated at 1500°C/s to 862°C (cf. Figure 10f) in components close to {001}<110> and {113}<110>. These components were very weak at 749°C, which indicates that nucleation in some low stored energy {hkl}<001> RD fiber components is taking place at high temperatures. The appearance of similar low stored energy {hkl}<001> RD fiber components of recrystallized ferrite are observed in samples heated at 150°C/s to 838°C (cf. Figure 10d). These observations are consistent with the structures, shown in figure 7, which show that nucleation accounts for a considerable fraction of the recrystallized ferrite in samples heated at 150°C/s and 1500°C/s.

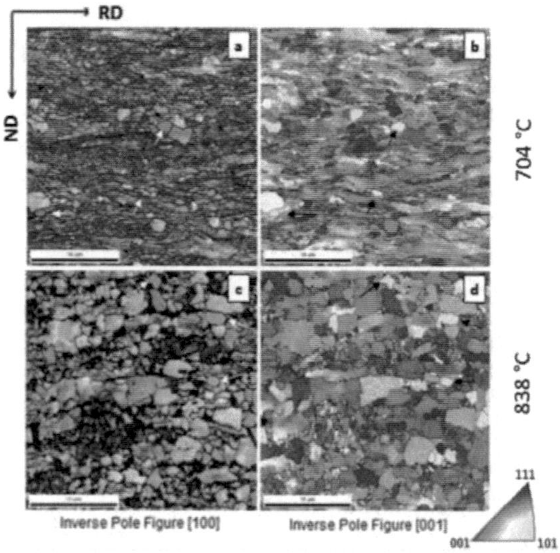

Figure 11. Microstructure of samples heated at 150°C/s to 704°C (a,b); and to 838°C (c,d). The microstructure consists of a mixture of ferrite and martensite. Arrows indicate grains oriented close to the Goss {110}<001> (with a tolerance of 10°). (c,d). (a,c) correspond to IQ maps in which the recrystallized grains of ferrite are highlighted by a [100] RD Inverse Pole Figure (IPF) map. (b,d) are [001] ND IPF maps [9].

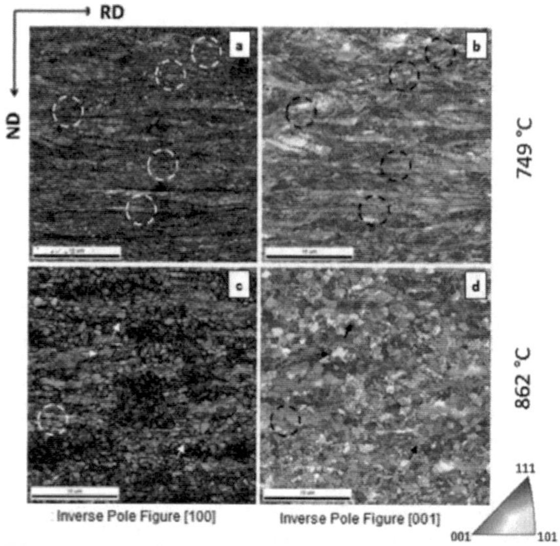

Figure 12. Microstructure of samples heated at 1500°C/s to 749°C (a,b), and to 862°C (c,d). The microstructure consists of a mixture of ferrite and martensite. Arrows indicate grains oriented close to the Goss {110}<001> (c,d), and circles the rotated Goss {110}<011> (a,b) components. (a,c) correspond to IQ maps in which the recrystallized grains of ferrite are highlighted by a [100] RD Inverse Pole Figure (IPF) map. (b,d) are [001] ND IPF maps [9].

3.3. Q&P steel samples: microstructures

Figure 13 a-d shows EBSD image quality maps (top) and phase maps (bottom) of samples heat treated according to the schedules of figure 3. The heating rate is mentioned under each map. The martensite grains in the conventionally Q&P heat treated samples are rather large (average grain diameter of 2.5 µm) with 14.1 % fresh martensite blocks, cf. Figure 13a. If a flash annealing process is conducted instead of an isothermal holding at 850°C, the microstructure is refined and the average grain diameter of the martensite reduces to 2 µm, (cf. Figure 13b). In this flash heat treatment, the austenite formation started in a fully recrystallized bcc matrix, but it can be seen that there was no time for all of the austenite grains to grow, resulting in a bimodal grain size distribution. Similar conclusions can be drawn after analyzing the microstructures obtained with a heating rate of 500°C/s, cf. Figure 13c. When the heating rate was increased from 500°C/s to 1000°C/s, the fresh martensite fraction decreased significantly due to the presence of 25% ferrite, cf. Figure 13d. The average bcc grain diameter reduced to 1.4 µm and the retained austenite grains became smaller and more homogenously distributed.

From the EBSD data, the approximate prior austenite grain size (PAGS) can be determined by plotting the rotation angles between 21.1° and 47.1° [19]. The linear interception method was used to determine the average PAGS for the Q&P samples as illustrated in Figure 13 (a-d bottom row). The average prior austenite grain diameter in the material after heating at 10°C/s and isothermal soaking is 3.8 µm and decreases to 2.9 µm without isothermal soaking. With a higher heating rate, the PAGS diameter decreased to 1.7 μm for 1000°C/s. After the Q&P heat treatment cycle without soaking, the austenite and ferrite grain sizes decreased from 0.6 μm to 0.3 μm and from 2.0 μm to 1.4 μm, respectively, as the heating rate was increased from 10°C/s to 1000°C/s. Figure 14 shows the results of quantification of the PAGS, average BCC grain diameter and the grain size of retained austenite.

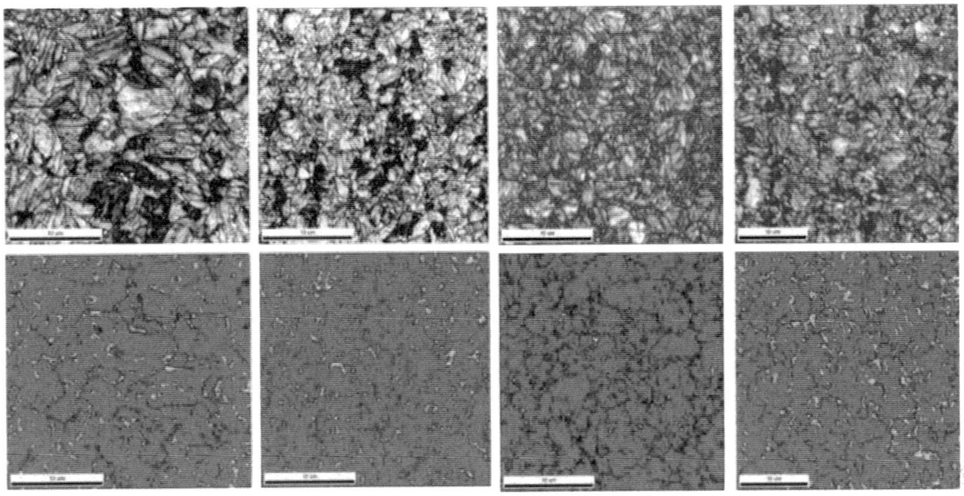

Figure 13. Image quality (IQ) maps top and phase maps (bottom) for samples Q&P with different heating rates and with and without isothermal soaking (Scale bars equal 10 μm). The black lines delineate the prior austenite grain boundaries derived according the work of Bernier et al. [19].

3.4. Q&P steel samples: textures

The textures of the samples were calculated from the EBSD data. Figure 14b displays the changes in the crystallographic textures as a function of the reheating rates. The texture of the cold rolled sample is characterized by an RD-ND fibre texture with a maximum of 8 mrd (multiples of a random distribution) on the {111}<112> texture component and a weaker rotated cube {100}<110> with an intensity of 4 mrd. The texture shows a convex curvature of the ND fibers, which is typical for cold rolled low carbon steels.

After reheating the cold rolled sample with 10°C/s and isothermal soaking, the cold rolling texture changes to a recrystallization type texture, which is characterized by a strong ND-fiber texture with a concave curvature and the disappearance of the RD-fiber component [20], cf. Figure 14b. Flash processing (i.e. without soaking) at 10°C/s changes the curvature of the ND-fiber from concave to straight as shown in Figure 14b. After increasing the heating rate to 500°C/s and 1000°C/s (cf. Figure 14b), the texture starts to resemble the original cold rolled texture with a convex curvature of the ND-fibers but with much lower intensity of ~4 mrd. Due to a texture memory effect, the martensite phase inherits the texture of the untransformed ferrite: we can conclude that this ferrite was not recrystallized but possibly only recovered [17]. Similar effects were reported by Senuma *et al.* [21] and Petrov *et al.* [6] in rapidly annealed cold-rolled extra low carbon and advanced high strength steel sheets, respectively.

3.5. Q&P steel samples: mechanical properties

Representative examples of the stress strain curves of the conventionally heat treated Q&P sample, and of samples heated at 10°C/s, 500°C/s and 1000°C/s without soaking are given in Figure 15a. The corresponding characteristic values are given in [8].

The conventionally Q&P heat treated sample has a uniform elongation of 6.1% with a yield strength and ultimate tensile strength of 782 MPa and 1050 MPa respectively. If the isothermal soaking is removed after heating at 10°C/s, the uniform elongation decreases to 2.6% with higher yield and ultimate tensile strengths of 837 MPa and 1097 MPa, respectively. Increasing the heating rate to 500°C/s results in a uniform elongation of 8.6% with an ultimate tensile strength of 1,138 MPa and a heating rate of 1,000°C/s results in a uniform elongation of 12.8% with an increased yield and ultimate tensile strengths of 811 MPa and 1,318 MPa, respectively. The ratio of yield strength to ultimate tensile strength decreases with higher heating rates without soaking, cf. Table 1. The n-values were also calculated as 0.241 for a heating rate of 10°C/s, 0.167 for 500°C/s and 0.245 for 1000°C/s.

The fracture surface of the conventionally Q&P treated and the sample heated at 10°C/s without soaking is quasi-cleavage. Figure 15b (a-b) reveals some dimples but fracture propagated mostly by cleavage resulting in a mixed ductile-brittle fracture mode. After ultra-fast heating, fracture resulted in a ductile fracture surface with large dimples, cf. Figure 15b (c-d).

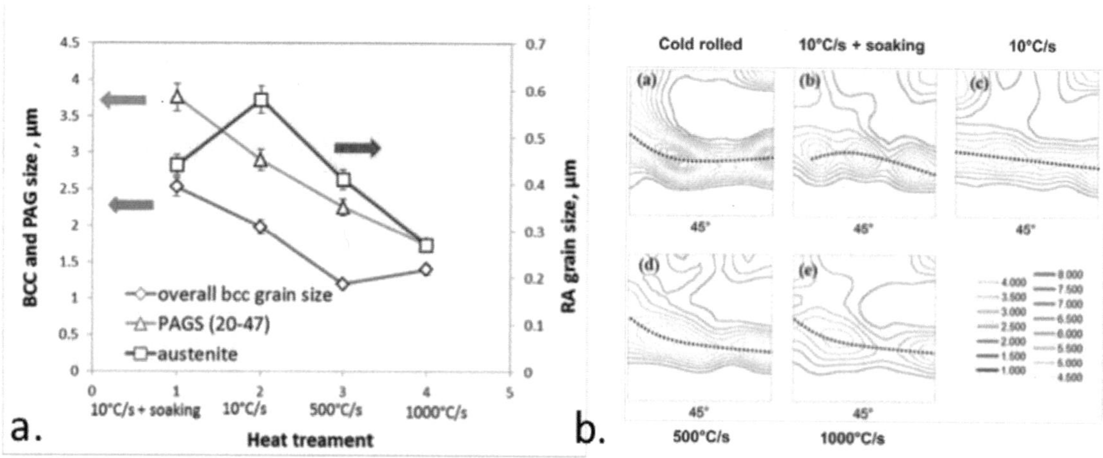

Figure 14 (a) Dependence of the average grain size from the heating rate and the isothermal soaking; (b) ODFs showing the texture evolution as a function of the heating rate and soaking time. The dotted line displays typical curvature of the ND fibres observed in cold rolled and recrystallized steels.

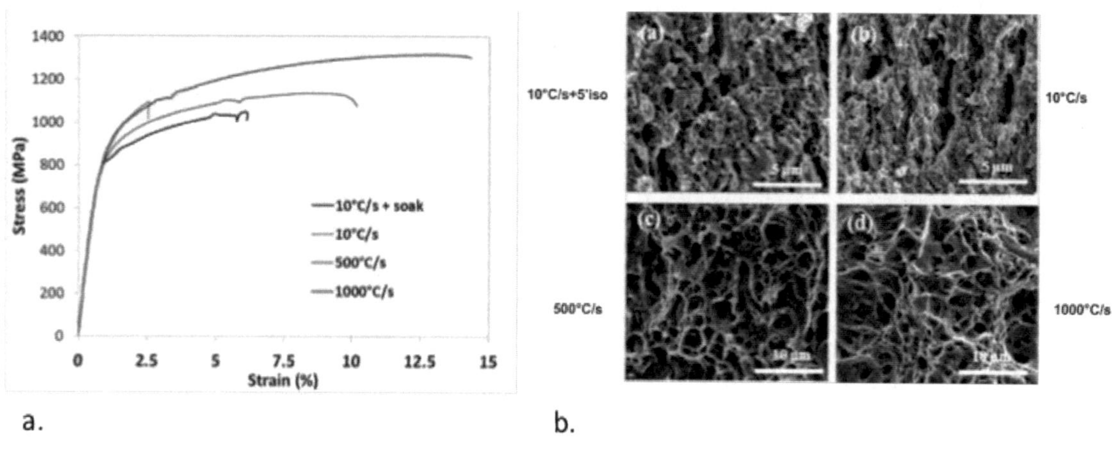

Figure 15. (a) Tensile test curves of samples after Q&P with different heating rates with and without isothermal soaking. (b) fracture surface of the samples (adapted from [8]).

4. Conclusions

The results in this study show that the microstructure and mechanical properties of existing AHSS compositions can be improved without addition of expensive alloying elements if fast heating without isothermal soaking is used. The most important findings in this study can be summarized as follows:

1. In both the DP and Q&P steels presently under investigation, the recrystallization after ultra-fast heating was not 100% complete before the onset of the α-to-γ transformation. The ultra-fast heating and the absence of the isothermal soaking produce a strong grain refinement effect in which the detailed features critically depend on the initial microstructure.
2. Ultra-fast annealing may induce an important texture shift as (i) it affects the volume fraction of ferrite that is recrystallized at the onset of α-to-γ transformation and (ii) it generally enhances oriented nucleation effects such as preferred low or high-stored energy nucleation because of an effect on the balance of driving force for recovery and nucleation.
3. The mechanical properties of the ultra-fast annealed DP and Q&P samples, with heating rates of 400°C/s, 500°C/s and 1000°C/s without isothermal soaking, are better than for the conventionally processed Q&P samples, combining an increase in strength up to 26% and ductility up to 134% (for Q&P) as compared to the conventional equivalents.
4. The fracture surface changes from quasi-cleavage in conventionally heat treated samples to ductile dimples in the ultra-fast annealed ones.

The results of the current study undoubtedly show that ultra-fast reheating of HSLA steels is a potential way to achieve significant strengthening with a minimum loss of ductility, and under certain conditions to achieve even an increase in ductility. The observed increases in strength were associated with a strong grain refinement caused by the overlapping of the α-to-γ phase transformation and ferrite recrystallization. These effects are observed also with heating rates of ~400°C/s, which can be obtained in an industrial processing line. The results also show that the control of the initial microstructures before cold rolling together with the heating rates is of crucial importance for designing HSLA steels with improved mechanical properties, i.e. to effectively fill the gap in the strength vs. ductility chart, in the domain of strengths above 1,200 MPa and elongations of ~20%, without the use of expensive alloying additions.

References

[1] De Moor E, Gibbs P J, Speer J G, Matlock D K and Schroth J G 2010 Strategies for third-generation advanced high-strength steel development Iron and Steel Technology vol 7 pp 133–44

[2] De Cooman B C 2004 Structure-properties relationship in TRIP steels containing carbide-free bainite Curr. Opin. Solid State Mater. Sci. 8 285–303

[3] Senuma T 2001 Influence of the heating rate in an annealing process on the recrystallization behavior and texture formation of a cold rolled extralow carbon steel sheet Recrystallization and grain growth, vols. 1 and 2 ed D Gottstein, G and Molodov (Heidelberger platz 3, D-14197 Berlin, Germany: Springer-Verlag Berlin) pp 1125–30

[4] Reis A C da C, Bracke L, Petrov R, Kaluba W J and Kestens L 2003 Grain Refinement and Texture Change in Interstitial Free Steels after Severe Rolling and Ultra-short Annealing ISIJ Int. 43 1260–7

[5] Petrov R, Kestens L, Kaluba W and Houbaert Y 2003 Steel grips : journal of steel and related materials. vol 1(GRIPS media)

[6] Petrov R H, Sidor J and Kestens L A I 2012 Texture formation in high strength low alloy steel reheated with ultrafast heating rates Mater. Sci. Forum 702–703 798–801

[7] Petrov R, Kestens L and Houbaert Y 2001 Recrystallization of a Cold Rolled Trip-assisted Steel

during Reheating for Intercritical Annealling. Isij 41 883–90

[8] De Knijf D, Puype A, Föjer C and Petrov R 2015 The influence of ultra-fast annealing prior to quenching and partitioning on the microstructure and mechanical properties Mater. Sci. Eng. A 627 182–90

[9] Castro Cerda F, Kestens L, Monsalve A and Petrov R 2016 The Effect of Ultrafast Heating in Cold-Rolled Low Carbon Steel: Recrystallization and Texture Evolution Metals (Basel). 6 288

[10] Kang J Y, Bacroix B, Réglé H, Oh K H and Lee H C 2007 Effect of deformation mode and grain orientation on misorientation development in a body-centered cubic steel Acta Mater. 55 4935–46

[11] Ray R K, Jonas J J and Hook R E 1994 Cold rolling and annealing textures in low carbon and extra low carbon steels Int. Mater. Rev. 39 129–72

[12] Hutchinson W B 1984 Development and control of annealing textures in low-carbon steels Int. Met. Rev. 29 25–42

[13] Nguyen Minh T, Sidor J, Petrov R and Kestens L 2012 Occurrence of shear bands in rotated Goss ({110}<110>) orientations of metals with bcc crystal structure Scr. Mater. 67 935–8

[14] Gobernado P, Petrov R H and Kestens L A I 2012 Recrystallized {311}〈136〉 orientation in ferrite steels Scr. Mater. 66 623–6

[15] Kang J-Y, Kim D-I and Lee H-C 2009 Texture Development in Low Carbon Sheet Steels for Automotive Application Microstructure and Texture in Steels (London: Springer London) pp 85–101

[16] Dillamore I L, Morris P L, Smith C J E and Hutchinson W B 1972 Transition Bands and Recrystallization in Metals Proc. R. Soc. London. Ser. A, Math. Phys. Sci. 329 405–20

[17] Ushioda K, Hutchinson W B, Ågren J and von Schlippenbach U 1986 Investigation of structure and texture development during annealing of low-carbon steel Mater. Sci. Technol. 2 807–15

[18] Castro Cerda F M, Vercruysse F, Minh T N, Kestens L, Monsalve A and Petrov R 2017 The Effect of Heating Rate on the Recrystallization Behavior in Cold Rolled Ultra Low Carbon Steel Steel Res. Int. 88

[19] Bernier N, Bracke L, Malet L and Godet S 2014 An alternative to the crystallographic reconstruction of austenite in steels Mater. Charact. 89 23–32

[20] Farideh HajyAkbary 2015 Optimising mechanical behaviour of new advanced steels based on fine non-equilibrium microstructures (TU Delft)

[21] Senuma T, Kawasaki K and Takemoto Y 2006 Recrystallization Behavior and Texture Formation of Rapidly Annealed Cold-Rolled Extralow Carbon Steel Sheets Mater. Trans. 47 1769–75

[22] Petrov R.H., Kestens L.A.I., Colás R., Totten G.E. (Eds.), Advanced High-strength Steels: Electron Backscatter Diffraction (EBSD), CRC Press, 2016, pp. 46–69 Encycl. Iron, Steel, and Their Alloys. http://dx.doi.org/10.1081/E- EISA-120050786

[23] Castro Cerda F.M., Goulas C., Sabirov I., Kestens L.A.I. and Petrov R.H. 2017, The effect of the pre-heating stage on the microstructure and texture of a cold rolled FeCMnAlSi steel under conventional and ultrafast heating, Mater. Charact. doi:10.1016/j.matchar.2017.06.010, in press

IOP Conf. Series: Materials Science and Engineering **219** (2017) 012005 doi:10.1088/1757-899X/219/1/012005

Applications of atomic-resolution HAADF-STEM and EDS-STEM characterization of light alloys

Jian-Feng Nie

Department of Materials Science and Engineering, Monash University, Victoria 3800, Australia

Jianfeng.nie@monash.edu

Abstract. Precipitation-hardened aluminium and magnesium alloys often contain a uniform distribution of plate-shaped precipitates of intermediate or equilibrium phases that form on rational planes of the matrix phase. Micro-alloying additions can change the identity and/or distribution of these precipitate plates, but the precise roles of the micro-alloying elements in the formation of such precipitate plates are still a subject of debate. A key reason for this is the lack of direct experimental observation of the details of the distribution/segregation of these elements in and surrounding precipitates from conventional transmission electron microscopy. While the advent of high-angle annular dark-field scanning transmission electron microscopy (HAADF-STEM) provides much more information at the atomic-scale, there is still a need for performing atomic-resolution chemical mapping using advanced STEM techniques, such as energy-dispersive X-ray spectroscopy (EDS-STEM) and/or electron energy-loss spectroscopy (EELS-STEM), if the precise roles of the micro-alloying elements are to be revealed.

1. Introduction

Age hardening forms the backbone of light alloy development. The age hardening phenomenon in aluminium alloys was discovered in 1906. Since then, considerable progress has been made in the improvement of the age hardening response in aluminium alloys via the addition of micro-alloying elements. By the end of 1980s, the yield strength of precipitation-hardened aluminium alloys exceeded 700 MPa, already 2–3 times higher than obtained in 1906. For the development of age-hardenable magnesium alloys, remarkable progress was made between 2000 and 2015. In this period, the yield strength of magnesium casting alloys was raised from less than 200 MPa to over 300 MPa, and that of magnesium wrought alloys from typically 300 MPa to above 475 MPa. It is foreseeable that it will require a significantly shorter period, i.e. much less than 80 years, for magnesium alloys to reach the strength level currently enjoyed by ultrahigh strength aluminium alloys. One of the most important microstructural characterization facilities that have been used in the development of age hardenable light alloys is transmission electron microscopy (TEM). The imaging, diffraction and micro-analysis techniques of TEM have played a crucial role in revealing processing-microstructure-property relationships in these alloys. In conventional transmission electron microscopy, whilst information on the identity, spatial distribution and chemistry of precipitates could be readily obtained, it was difficult to reveal, at the atomic scale, the distribution of solute atoms or atoms of micro-alloying elements. It was for this reason that the roles of micro-alloying elements in precipitation of key strengthening intermetallic phases remained unresolved. Advances in scanning transmission electron microscopy (STEM) in the past 15 years, especially the aberration-corrected STEM techniques including high-angle annular dark-field (HAADF), energy-dispersive X-ray spectroscopy (EDS) and electron energy-loss spectroscopy (EELS), provide opportunities for detecting segregated solute atoms at the atomic scale. The purpose of this article is to provide a review of the progress made in the understanding of precipitation and the role of micro-alloying elements in precipitation in selected aluminium and magnesium alloys, highlighting the importance of modern imaging and spectroscopy techniques of STEM in the characterization and development of these light alloys.

2. Aluminium alloys

2.1 Al–Cu–Sn alloys

In early 1950s, Hardy [1-2] discovered that additions of trace amounts (~0.1 at.%) of Sn to binary Al–Cu alloys could lead to a significantly enhanced age hardening response. The maximum hardness values achievable in the resultant Al–Cu–Sn alloy in the temperature range 100–200°C were substantially higher than those in the counterpart binary alloy, and this hardness increment is due to a refined distribution of θ' (Al_2Cu) precipitates. At that time, X-ray diffraction was the technique used to collect indirect information that could be used to speculate on the role of Sn in enhancing the age hardening response. Based on the streaked diffraction maxima (so-called P diffraction spots) in the X-ray diffraction patterns obtained from samples in the early stages of ageing, Silcock [3] suggested that Sn atoms segregated to the θ'/α-Al interfaces, and that this segregation reduced the interfacial energy involved in θ' nucleation and thus increased the nucleation rate of θ' precipitates. Subsequent observations made by conventional electron microscopy in the 1970s led to different views. Sankaran and Laird [4] reported that their TEM observations supported the interpretation of Silcock, even though they did not provide any direct experimental evidence. On the other hand, based on observations made by TEM and energy-dispersive X-ray spectroscopy, Kanno and coworkers [5-7] reported the formation of a Sn-containing phase, which acts as heterogeneous nucleation sites for θ'. It became possible in the 1990s to use combined techniques of imaging and microdiffraction of TEM, and one-dimensional atom probe, to examine the origin of enhanced nucleation of θ' in Al–Cu–Sn alloys. With the help of these techniques, Ringer and co-workers [8] showed the occurrence of clustering of Sn atoms in the very early stages of ageing and precipitation of incoherent particles of the equilibrium phase β-Sn at later stage of ageing. The θ' precipitates that form at the later stage of ageing were frequently found to be associated with β-Sn particles, with one β-Sn particle almost invariably located at one end facet of individual θ' plates. They further reported that Sn atoms did not segregate to the broad faces of the θ' plates. These observations have since then been widely used as direct experimental evidence for heterogeneous nucleation of θ' on β-Sn particles. Such interpretation is essentially based on the minimization of interfacial energy or volumetric misfit strain energy, and it supports the hypothesis made by Hardy [2] that Sn atoms precipitated first in the form of coherent Sn particles and subsequently act as heterogeneous nucleation sites for θ'.

Based on observations made by high-resolution TEM and microdiffraction, Bourgeois and co-workers [9] reported that the β-Sn particles associated with θ' platelets, which formed at early stage of ageing, exhibited an orientation relationship that was distinctly different from those reported previously for the Al–Cu–Sn and Al–Sn alloys. This observation implies that, while tin may assist θ' nucleation, the formation of θ' also influences the crystallography of β-Sn that forms in association with θ'. The continued ageing leads to coarsening of those β-Sn particles that are in contact with θ' plates, and eventual replacement of this orientation relationship by the one that is originally observed in the Al–Cu–Sn and Al–Sn alloys. More recent experimental observations using atomic-resolution Z-contrast HAADF-STEM [10] revealed that Sn atoms segregate at the end facets of θ' precipitate plates, as shown in figure 1. This observation indicates that the Sn particles formed in the early stages of ageing are not heterogeneous nucleation sites for θ', and that the Sn particle that is in contact with a θ' plate forms as the consequence of segregation of Sn atoms to the end facet of the θ' nucleus. The solid solubility of Sn in the aluminium matrix is very low, and any excessive segregation of Sn atoms would lead to precipitation of β-Sn. The latest observation is consistent with the model that involves shear strain energy minimization [11-12]. The formation of θ' precipitates generates a shear strain in the aluminium matrix phase. Since Sn has a large atomic size than Al, its segregation to the extension region of the matrix adjacent to the end facet of the θ' nucleus allows a reduction in the shear strain energy in the region.

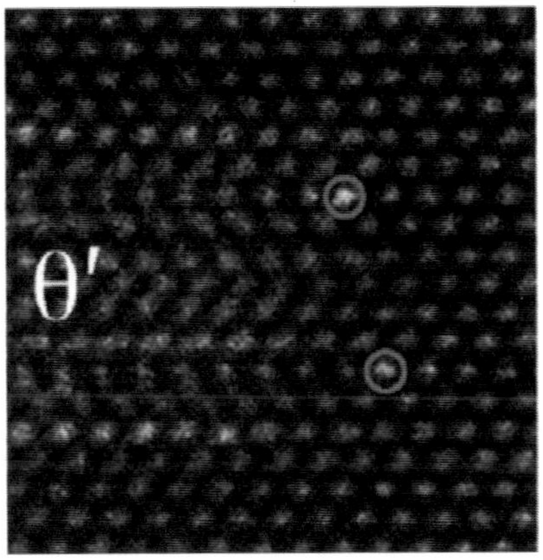

Figure 1. <011>$_{\alpha}$ HAADF-STEM image showing Sn-rich columns (circled) attached to the end facet of θ' plate in an Al–Cu–Sn alloy. The thickness of the θ' plate is 2 nm [10].

2.2 Al–Cu–Mg–Ag alloys

Combined additions of small amounts of Ag and Mg, typically in the range 0.3–0.5 wt%, to Al–Cu alloys result in almost complete replacement of metastable θ' phase by another metastable phase designated Ω (Al_2Cu). The addition of Ag alone to binary Al–Cu alloys does not facilitate nucleation of Ω precipitates. The uniform, dense and relatively thermally stable distribution of the Ω precipitates enables the alloy to exhibit a tensile yield strength above 500 MPa at ambient temperatures and improved creep resistance at temperatures up to 180–200°C.

The Ω precipitates form as thin plates on $\{111\}_{\alpha}$ planes. Although the number density and volume fraction of Ω plates is similar to those of θ' in Al–Cu binary alloys, the aspect ratio of Ω plates is much larger than that of θ'. The formation of such $\{111\}_{\alpha}$ precipitate plates, and the increased aspect ratio of these plates, reduce the effective inter-particle spacing and thus increase the alloy tensile strength. The improved creep strength is attributable to the thermal stability of the Ω phase, which is associated with a low interfacial energy between Ω and matrix phases on the habit plane.

The precise roles of Ag and Mg in stimulating the nucleation and thermal stability of the Ω phase have received considerable interest in the past, but are still not fully understood. Studies made in late 1980s and early 1990s using X-ray microanalysis [13] and one-dimensional atom probe [14] revealed the strong segregation of Ag and Mg to the broad Ω/α-Al interface and the absence of these two elements in the interior part of the Ω phase. In the following years, the development of the three-dimensional atom probe technique allowed the detection of Ag–Mg co-clusters that form in the very early stages of decomposition of the Al–Cu–Mg–Ag alloys. The Ag–Mg clusters do not have any well-defined shape at the start. During further ageing, these clusters absorb Cu atoms and gradually evolve into discs on $\{111\}_\alpha$ planes. Once Ω nuclei have developed from these clusters, Ag and Mg atoms are rejected to the broad surfaces of the Ω plates. There were controversial reports on the number of the Ag/Mg segregation layers. Early studies made by atom probe field ion microscopy claimed that Ag and Mg were segregated in a single atomic layer [15]. But subsequent work made in 2001 by atomic-resolution HAADF-STEM concluded that the Ag and Mg atoms form a double-layer in the broad surface of the Ω plate [16]. Observations made in more recent years using HAADF-STEM with better spatial resolution confirmed the existence of the double layer in the Ω/matrix interface [17]. However, in this double-layer, the Ag atoms are all distributed in a single layer adjacent to the aluminium matrix, and Cu atoms are in the next layer that is close to the precipitate, as shown in figure 2. The Ag atoms in the monolayer have a graphene-like hexagonal arrangement, with Mg atoms in the centre of individual hexagons. First-principles calculations based on these observations indicated that the Mg atoms in the layer stabilize the interfacial structure and consequently the thermal stability of the Ω phase. The segregation of Ag–Mg atoms was observed at all thicknesses of the Ω plates. It is the presence of the layer of the Ag and Mg atoms that facilitates the nucleation of Ω precipitates by reducing the interfacial energy. The presence of this layer at the Ω plate surfaces is also considered to restrict ledge formation, thereby minimizing thickening at temperatures up to 200 °C.

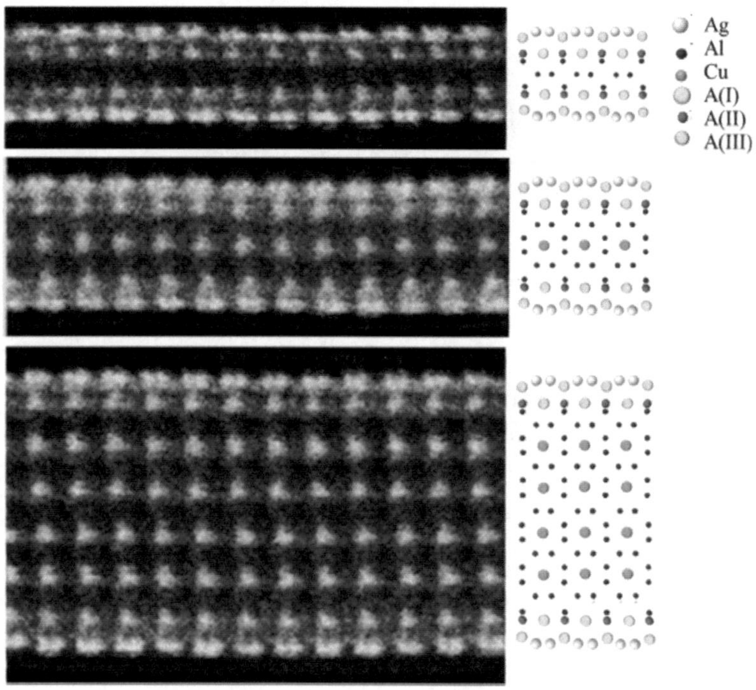

Figure 2. <112>$_\alpha$ HAADF images showing Ag segregation at the Ω/matrix interface in an Al–Cu–Mg–Ag alloy [17].

While the latest interpretation of the roles of Ag and Mg in Ω precipitation seems to be rational, it is still based on incomplete STEM observations. Since the intensity in the HAADF-STEM image is approximately proportional to the square of atomic number, it is difficult to unambiguously distinguish Mg from Al, and Ag from Cu, in HAADF-STEM images. It is for this reason that the latest interpretation remains to be validated. To resolve such issues, it will be necessary to use EDS-STEM or EELS-STEM (EELS: electron energy loss spectroscopy) in the near future to perform atomic-scale chemical mapping of the Ω/matrix interface. The latest STEM instruments are equipped with innovative technologies that significantly enhance X-ray signal collection. For example, the ChemiSTEM technology of the FEI Titan G2 has four windowless silicon drift detectors arranged symmetrically around the specimen and the objective lens pole pieces, figure 3, which greatly facilitates the chemical mapping along some critical directions of the precipitate-matrix interface. While it is relatively easy to obtain atomic-resolution chemical maps from ceramics and high-temperature metallic materials, it is still a real challenge to obtain such maps from aluminium alloys and magnesium alloys.

Figure 3. Schematic diagram of the ChemiSTEM™ design, showing 4 silicon drift detectors that are arranged symmetrically around the sample and the objective lens pole pieces [Brochure of FEI Titan™ G2 with ChemiSTEM™ Technology].

2.3 Al–Cu–Li–Mg–Ag alloys

The key strengthening precipitate phase in Al–Cu–Li alloys is the T_1 phase (Al_6CuLi). The T_1 precipitates form as plates on $\{111\}_\alpha$ planes, with an aspect ratio typically larger than 50:1. Although the T_1 phase is almost fully coherent with the aluminium matrix phase within its habit plane, it is difficult to nucleate during conventional isothermal ageing treatments. Consequently, it is common practice to cold deform the aluminium matrix phase prior to ageing, or to add micro-alloying additions of Ag and Mg to such alloys, to facilitate nucleation of the T_1 phase. The dislocations introduced by the cold-work act as heterogeneous nucleation sites for the T_1 phase and hence result in an increased number density of T_1 precipitates. The combined additions of trace amounts of Ag and Mg (typically 0.3–0.5 wt%) to Al–Cu–Li alloys remarkably enhance the nucleation and formation of T_1 phase [18]. The resultant Al–Cu–Li–Mg–Ag alloys, such as the Weldalite 049™ series of ultra-high strength aluminium alloys, can exhibit a tensile yield strength exceeding 700MPa. While the number density and volume fraction of T_1 plates are both comparable to those of Ω in Li-free Al–Cu–Mg–Ag alloys, the T_1 plates are much thinner (typically less than 2 nm), and their aspect ratio is significantly larger than that of Ω. The ultra-high strength achieved in the Al–Cu–Li–Mg–Ag alloys is attributable to the uniform distribution and an almost continuous network of T_1 plates. The Al–Cu–Li–Mg–Ag alloys also have an increased creep strength. The improved creep resistance is attributable to higher thermal stability of the T_1 phase.

Similar to the case of Al–Cu–Mg–Ag alloys, the precise roles of Ag and Mg in the precipitation of the T_1 phase have received considerable interest in the past, but are still far from being well understood. While the roles of Ag and Mg are considered to be similar to those in the precipitation of Ω phase, there have been conflicting reports on whether Ag and Mg segregate into the T_1 precipitate or just to the T_1/matrix interface. The presence of Li in the T_1 precipitate as well as its thinness make it extraordinarily difficult to accurately detect the distribution of Ag and Mg atoms within the region containing the T_1 precipitate, when conventional TEM is used alone. Even studies made by atom probe tomography gave controversial reports. Murayama and Hono [19] and Araullo-Peters and co-workers [20] reported that both Ag and Mg segregate into the T_1/matrix interface, whereas Gault *et al.* [21] reported that Ag and Mg segregate to the internal part of the T_1 precipitate rather than the T_1/matrix interface. It is unclear whether such controversial reports are due to the variation of the spatial resolution of the atom probes used. However, recent studies made by atomic-resolution HAADF-STEM [22-23] suggest that Ag atoms segregate into the T_1/matrix interface, certainly at the later stage of T_1 growth but probably also at the nucleation stage of the T_1 precipitate. The segregated Ag atoms are distributed in a single atomic layer that is adjacent to the aluminium matrix phase, similar to that observed in the Ω/matrix interface. Again, due to the limitations of the imaging technique, it is still difficult to accurately detect the distribution of Mg atoms in the T_1/matrix interface. Therefore, the precise roles of Ag and Mg in stimulating the nucleation and thermal stability of T_1 precipitate remain vague. It again seems necessary to employ atomic-resolution EDS-STEM or EELS-STEM in the following years to determine the distribution of Ag and Mg atoms in the T_1/matrix interface.

3. Magnesium alloys

3.1 Mg–RE–Ag and Mg–RE–Zn alloys

Binary Mg–Gd and Mg–Y alloys with Gd or Y less than 6 wt% are not age hardenable. Ternary additions of 2 wt% or more Ag to these alloys can generate a remarkable age hardening response in the temperature range 200–250 °C [24-25]. Ternary additions of 1–2 wt% Zn to Mg–6wt%Gd alloys can also induce a remarkable age hardening response, which leads to a significant improvement in tensile and creep properties [26]. This enhancement in mechanical properties is associated with a uniform and dense distribution of basal precipitate plates that do not form in binary Mg–Gd or Mg–Y alloys. These basal plates look remarkably similar in the Mg–Gd–Ag, Mg–Y–Ag, Mg–Gd–Zn and Mg–Y–Ag–Zn alloys. These precipitate plates are very thin and their characterization and analysis using conventional TEM is difficult. However, it is now possible to obtain atomic-resolution images of these precipitates using HAADF-STEM, as shown in in figure 4, although it remains difficult to distinguish columns rich in RE atoms from columns rich in Zn or Ag in these images. Therefore, it is again necessary to use atomic-resolution EDS-STEM or EELS-STEM in order to identify the factors that are critical in controlling the thickness of the plates.

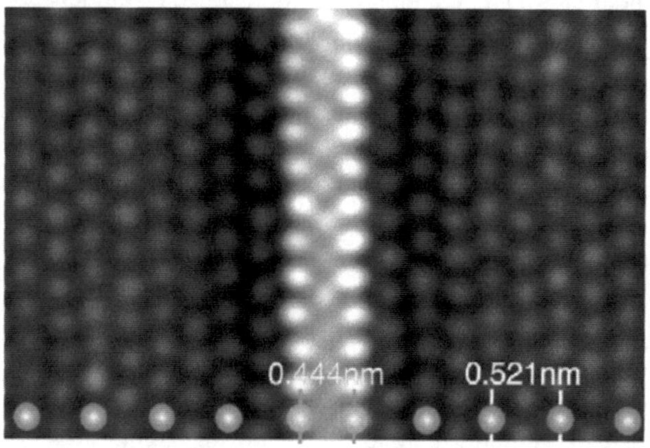

Figure 4. $<1\bar{2}10>_\alpha$ HAADF image showing a basal precipitate plate comprising Mg, Gd and Zn in a Mg–Gd–Zn alloy [27].

3.2 Mg–Sn–Zn alloys

Ternary additions of 1wt% Zn to binary Mg–Sn alloys remarkably enhance the age hardening response. Microstructural observations made by conventional TEM [28-29] indicated that the enhanced age hardening response was due to a finer distribution of β precipitates (Mg_2Sn). These observations suggested that the Zn addition stimulate the nucleation of the β precipitate, but the precise role of Zn in the precipitation process was unclear. A key reason was that it was difficult to establish the distribution of Zn element in the Mg–Sn–Zn alloys from conventional TEM images. Given that both Sn and Zn have much higher atomic numbers than Mg, it is also difficult to distinguish Zn from Sn in HAADF-STEM images. A recent study made by EDS-STEM mapping [30] indicated unambiguously that Zn segregates strongly to the β/matrix interface, for a variety of boundaries associated with different orientation relationships of the β phase. One such example is shown in figure 5. These segregation layers are quite thin, and it has proven difficult to accurately establish their thickness and structure. For this reason, it was difficult to assess why Zn atoms prefer to segregate to the β/matrix interface and whether this segregation reduces the interfacial energy to such a level that ultimately leads to a significantly enhanced nucleation rate of the β phase. Again, the use of atomic-resolution EDS-STEM may provide vital clues to the accurate thickness and structure of the segregated Zn atoms in the β/matrix interface.

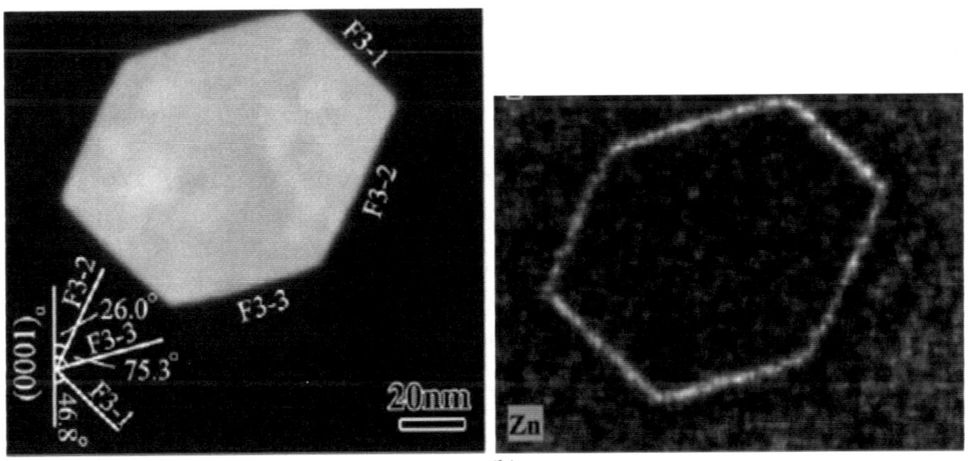

(a) (b)

Figure 5. (a) HAADF-STEM image, and (b) EDS-STEM showing Zn segregation at the β-Mg_2Sn/matrix in a Mg–Sn–Zn alloy [30].

4. Summary

The advent of aberration-corrected STEM techniques has greatly facilitated the atomic-scale characterization of precipitates in a group of scientifically and commercially important aluminium and magnesium alloys, and has advanced our understanding of the roles of micro-alloying elements in the precipitation process. However, it is still necessary to combine the HAADF-STEM imaging technique with spectroscopy techniques such as EDS-STEM or EELS-STEM in order to reveal the mechanism via which the segregation of solutes or micro-alloying elements stimulates the nucleation and formation of the key strengthening precipitate phases in these alloys.

Acknowledgements

The author is grateful for discussions with his students, postdocs, colleagues and collaborators including Barry Muddle, Ian Polmear, Laure Bourgeois, Sam Gao, Yuman Zhu, Kazuhiro Hono, Chaoqiang Liu and Houwen Chen in the past 20 years, and the support from the Australian Research Council.

References

[1] Hardy H K 1950-51 *J. Inst. Met.* **78** 169
[2] Hardy H K 1951-52 *J. Inst. Met.* **80** 483
[3] Silcock J M, Heal T J and Hardy H K 1955 *J. Inst. Met.* **84** 23
[4] Sankaran R and Laird C 1974 *Mater. Sci. Eng.* **14** 271
[5] Kanno M, Suzuki H and Kanoh O 1980 *J. Jpn. Inst. Met.* **44** 1139
[6] Kanno M, Suzuki H and Kanoh O 1979 *J. Jpn. Inst. Light Met.* **29** 223
[7] Suzuki H, Kanno M and Araki I 1975 *J. Jpn. Inst. Light Met.* **25** 413
[8] Ringer S P, Hono K and Sakurai T 1995 *Metall. Mater. Trans. A* **26A** 2207
[9] Bourgeois L, Nie J F and Muddle B C 2005 *Phil. Mag.* **85** 3487
[10] Bourgeois L, Dwyer C, Weyland M, Nie J F and Muddle B C 2012 *Acta Mater.* **60** 633
[11] Nie J F, Aaronson H I and Muddle B C 1999. *Proc Int. Conf. on Solid-solid Phase Transformations* ed M Koiwa, K Otsuka and T Miyazaki (Tokyo: The Japan Institute of Metals) p 157
[12] Nie J F and Muddle B C 1999 *Mater. Forum* **23** 23
[13] Muddle B C and Polmear I J 1989 *Acta Mater.* **37** 777
[14] Hono K, Sano N, Babu S S, Okano R and Sakurai T 1993 *Acta Metall. Mater.* **41** 829
[15] Reich L, Murayama M and Hono K 1998 *Acta Mater.* **46** 6053
[16] Hutchinson C R, Fan X, Pennycook S J and Shiflet G J 2001 *Acta Mater.* **49** 2827
[17] Kang S J, Kim Y W, Kim M and Zuo J M 2014 *Acta Mater.* **81** 501
[18] Pickens J R, Heubaum F H, Langan T J, Kramer L S 1989. *Proc. 5th Int. Conf. on Aluminum–Lithium Alloys* ed E A Starke and T H Sanders (Birmingham: Mater. Comp. Eng. Publications), p 1397
[19] Murayama M and Hono K 2001 *Scripta Mater.* **44** 701
[20] Araullo-Peters V, Gault B, de Geuser F, Deschamps A and Cairney J M 2014 *Acta Mater.* **66** 199
[21] Gault B, de Geuser F, Bourgeois L, Gabble B M, Ringer S P and Muddle B C 2011 *Ultramicroscopy* **111** 683
[22] Kang S J, Kim T H, Yang C W, Lee J I, Park E S, Noh T W and Kim M 2015 *Scripta Mater.* **109** 68
[23] Gumbmann E, Lefebvre W, de Geuser F, Sigli C and Deschamps A 2016 *Acta Mater.* **115** 104
[24] Gao X and Nie J F 2008 *Scripta Mater.* **58** 619
[25] Zhu Y M, Morton A J and Nie J F 2008 *Scripta Mater.* **58** 525
[26] Nie J F, Gao X and Zhu S M 2005 *Scripta Mater.* **53** 1049
[27] Nie J F, Oh-ishi K, Gao X and Hono K 2008 *Acta Mater.* **56** 6061
[28] Sasaki T T, Oh-ishi K, Ohkubo T and Hono K 2006 *Scripta Mater.* **55** 251
[29] Mendis C L, Bettles C J, Gibson M A and Hutchinson 2006 *Mater. Sci. Eng. A* **435-436** 163
[30] Liu C Q, Chen H W and Nie J F 2016 *Scripta Mater.* **123** 5

1 billion tons of nanostructure – segregation engineering enables confined transformation effects at lattice defects in steels

D Raabe, D Ponge, M-M Wang, M Herbig, M Belde, H Springer

Max-Planck-Institut für Eisenforschung, Max-Planck-Str. 1, 40237 Düsseldorf, Germany

Abstract. The microstructure of complex steels can be manipulated by utilising the interaction between the local mechanical distortions associated with lattice defects, such as dislocations and grain boundaries, and solute components that segregate to them. Phenomenologically these phenomena can be interpreted in terms of the classical Gibbs adsorption isotherm, which states that the total system energy can be reduced by removing solute atoms from the bulk and segregating them at lattice defects. Here we show how this principle can be utilised through appropriate heat treatments not only to enrich lattice defects by solute atoms, but also to further change these decorated regions into confined ordered structural states or even to trigger localized decomposition and phase transformations. This principle, which is based on the interplay between the structure and mechanics of lattice defects on the one hand and the chemistry of the alloy's solute components on the other hand, is referred to as segregation engineering. In this concept solute decoration to specific microstructural traps, viz. lattice defects, is not taken as an undesired effect, but instead seen as a tool for manipulating specific lattice defect structures, compositions and properties that lead to beneficial material behavior. Owing to the fairly well established underlying thermodynamic and kinetic principles, such local decoration and transformation effects can be tuned to proceed in a self-organised manner by adjusting (i) the heat treatment temperatures for matching the desired trapping, transformation or reversion regimes, and (ii) the corresponding timescales for sufficient solute diffusion to the targeted defects. Here we show how this segregation engineering principle can be applied to design self-organized nano- and microstructures in complex steels for improving their mechanical properties.

1. Introduction to the Segregation Engineering principle

The term segregation engineering (SE) describes a self-organized microstructure manipulation method that is based on the combination of (i) confined thermodynamically driven chemical Gibbs segregation via solute decoration to specific lattice defect traps, and (ii) the associated structural, order or phase changes triggered at these decorated sites. The concept of SE works particularly well for steels and Ti alloys owing to their rich spectrum of available allotropic phase transformations [1,2]. Here, we present a number of cases where local solute segregation, and the associated ordering phenomena at lattice defects, lead to confined martensite-to-austenite reversion in Fe-Mn and Fe-C steels.

In these examples the targeted SE effects occur in a spatially confined manner, namely, at lattice defects, which act as efficient and highly thermodynamically attractive solute traps, thus, entailing strong solute segregation when heat treated into regimes that allow for sufficient solute diffusion. The resulting combinations of local strain fields at such lattice defects on the one hand, and the locally altered chemical composition at these defects on the other hand, can initiate structural transitions, solute ordering, and phase transformation states in such cases where the sample is heated into a phase regime which is intercritical at the decorated and distorted defect region [3,4]. When aiming at the local modification of the microstructure, this SE approach is applied in such a way that these critical conditions are only fulfilled at the lattice defect regions, leaving the adjacent matrix essentially unaffected. These effects, which are essentially built on the decoration driving forces phenomenologically described by the Gibbs adsorption isotherm, are referred to as equilibrium segregation engineering, figure 1 [1].

Alternatively, it is also conceivable to exploit such effects for the case of non-equilibrium compositional gradients that are inherited either from a Scheil-type solidification inhomogeneity or from incomplete intercritical phase dissolution [5-7]. Such scenarios, which are based on kinetically acquired compositional inhomogeneities, are referred to as non-equilibrium segregation engineering.

When applying equilibrium SE specifically to grain boundaries we refer to this type of interface manipulation via solute decoration in conjunction with confined ordering or even transformation as 'grain boundary segregation engineering', figure 1 [1].

For creating self-organized microstructures by the SE approach one has to take certain thermodynamic and kinetic rules into account. The thermodynamic rules include the requirement to consider also the competition with the formation of alternative phases in the bulk matrix (rather than being confined to specific defects), the consideration of the equilibrium or non- equilibrium segregation coefficients, co-segregation effects, phase stability (relevant when designing confined states that can undergo spatially confined athermal phase transformations), as well as the interplay of chemical and mechanical driving forces [1-4]. The kinetic rules include consideration of the diffusion coefficients as well as of thermal versus diffusional phase transformation pathways [5-7].

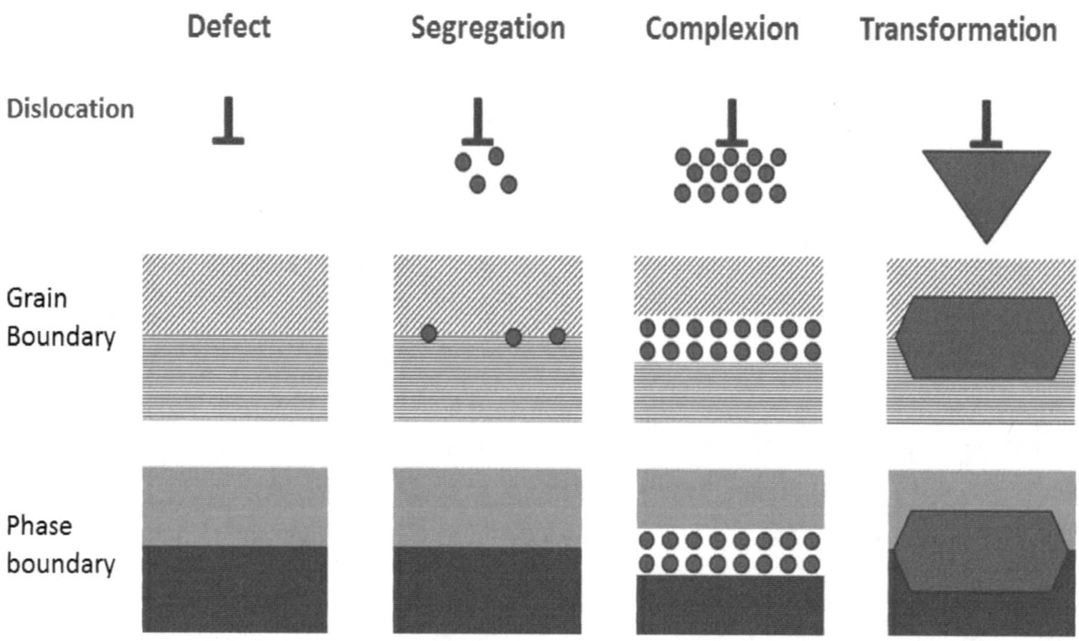

Figure 1. Some of the decoration and ordering states utilized at lattice defects in the field of SE. The chemo-mechanical interactions promoting solute decoration of lattice defects allow, through adequate heat treatment, to manipulate the structure and composition of the defect sites. This can lead to either conventional Gibbs equilibrium solute segregation, lowering the defect's self-energy, complexion or similar defect stabilized confined structure formation or even confined phase formation and transformation

The main idea behind the SE concept lies in the approach to consider solute decoration to defects not just as an undesired side effect associated with processing and heat treatment but to exploit it as a tool for the site-specific manipulation of interfaces with the aim to stabilize small grain sizes or to create intrinsically organized nano- or microstructures consisting of the initial matrix phase and segregation-induced ordered or transformation phase regions that can be rendered to follow the template of the landscape of the lattice defect structure, figure 2 [1].

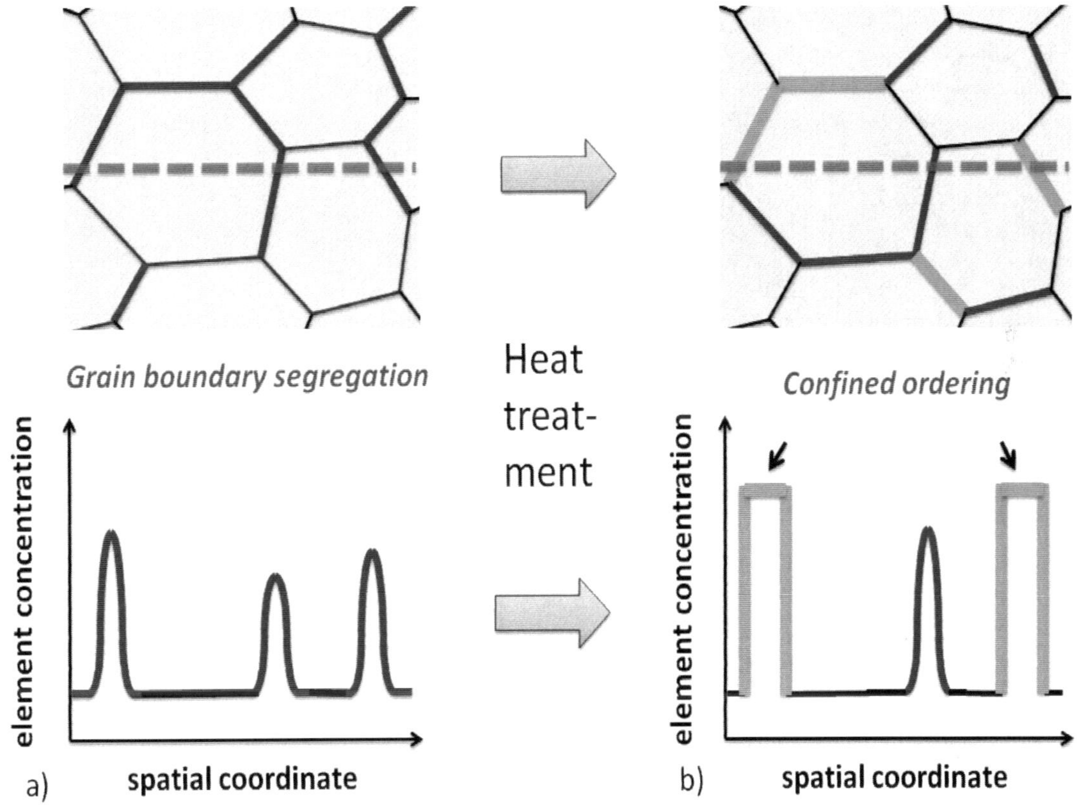

Figure 2. Example of grain boundary SE. Solute segregation to grain boundaries can, upon heat treatment, lead to phase transformation. The change in chemical composition, together with elastic distortions at interfaces, can be used for instance to create reversed austenite in a martensitic matrix.

2. Probing grain boundary segregation by correlative atom probe tomography and electron microscopy

The systematic manipulation of lattice defects through their trapping effect on solutes and associated means of modifying them in terms of confined ordering or transformation effects requires characterization methods that are suited to probe such phenomena at sufficiently high chemical, spatial and crystallographic resolution [3,8-13]. Such correlative characterization of lattice defects can be achieved by the use of atom probe tomography in concert with scanning electron microscopy, scanning transmission electron microscopy and/or transmission electron microscopy (TEM) [1,8,11].

Atom probe tomography enables revealing local solute decoration at defects with near ppm precision. Electron microscopy methods, with their different resolution features regarding imaging and diffraction, provide the required additional corresponding structure resolution to analyse these sites, figure 3. Practically all electron microscopy probing methods can nowadays be applied directly to atom probe tips, including high resolution atomic column imaging in corrected scanning transmission electron microscopy.

Hence, by (i) preparing atom probe tips via site specific focussed ion beam cutting, (ii) imaging and diffraction of these tips in electron microscope studies, and (iii) the final evaporation and compositional probing of the so characterized tips in the atom probe, a joint and fully correlative analysis of solute decorated or ordered states at lattice defects down to the atomic scale is enabled, figure 4 [3,8,9].

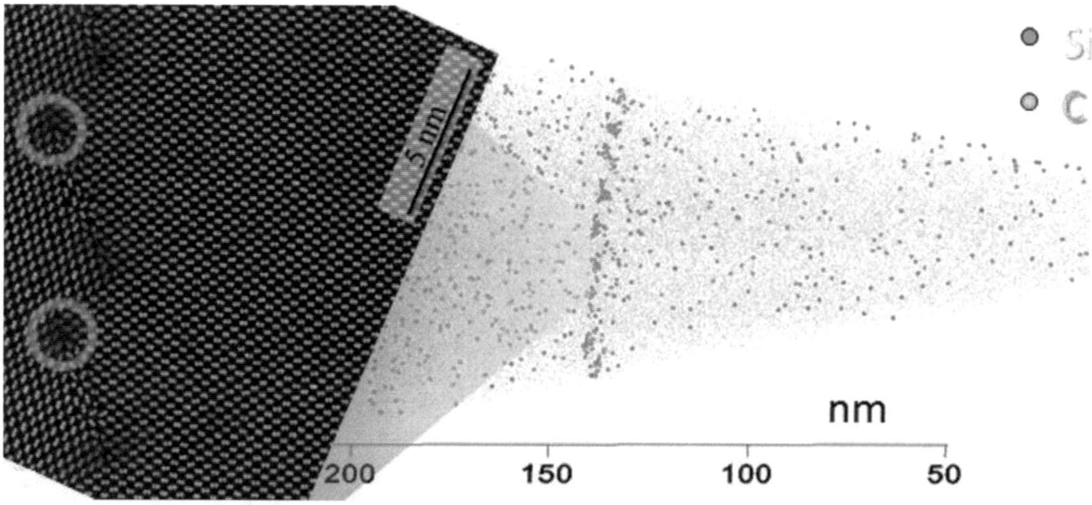

Figure 3. Correlative electron microscopy and atom probe tomography of grain boundary segregation in Si. Taken from the work of Stoffers et al. [13,14].

Figure 4 shows an example where Mn is strongly segregated to dislocations and grain boundaries. The Mn concentration of the Fe-Mn alloy has been mapped using correlated atom probe tomography in conjunction with TEM imaging conducted directly on the atom probe tip. Such measurements are capable of quantitatively revealing the extent of solid segregation at defects such as dislocations and grain boundaries [3].

Interestingly, one can observe in figure 4 that some of the dislocations that are clearly visible in the TEM image do not appear in the atom probe tomographic analysis in terms of any solute concentration envelope, indicating that these dislocations are obviously not sufficiently solute decorated.
In contrast, some of the dislocations that are highlighted in the atom probe tomographic analysis as a result of their local concentration enrichment are not observed in the TEM images, since the diffraction condition is not fulfilled locally, i.e. more beam conditions would need to be used for the full characterization of these dislocations.

The specimen from which these atom probe tomography tips were cut by using site specific extraction by focused ion beam instrument was a 50% cold rolled and subsequently 450°C/6h heat-treated Fe-9wt.%Mn model alloy. The data shown in figure 4 are taken from the work of Kuzmina et al. [3].

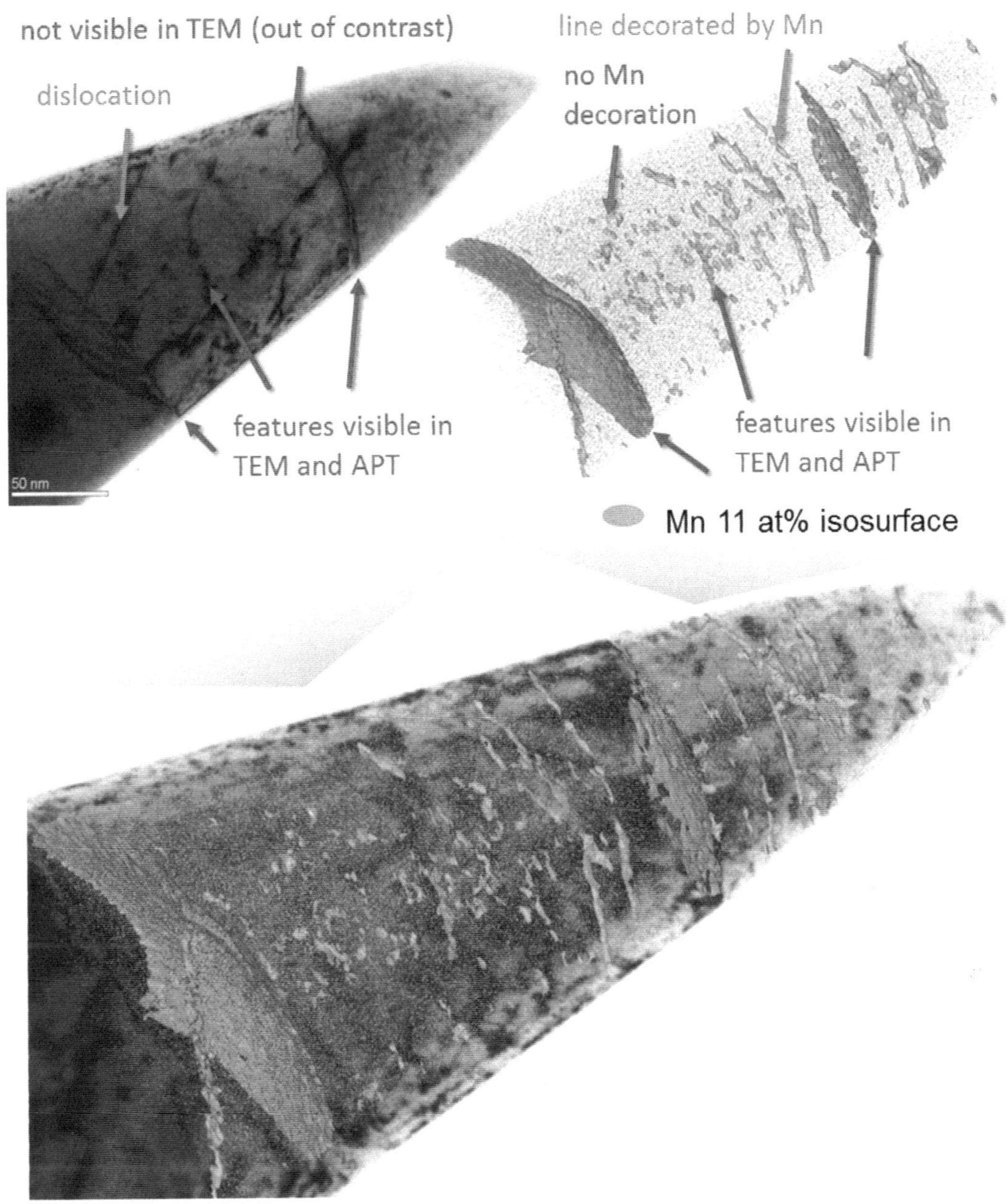

Figure 4. Mn concentration in a Fe-Mn alloy mapped using correlated atom probe tomography in conjunction with TEM imaging conducted directly on the atom probe tip. The data and some of the images are taken from [3].

3. Confined martensite-to-austenite reversion at lath interfaces creates self-organized two-phase nano-laminates

SE can be used for the self-organized design of two-phase or even multi-phase laminate microstructures. For this purpose we selected a steel with 9 wt.% Mn, owing to the high segregation tendency of Mn to lattice defects in Fe. After first heating the alloy into the solid solution austenitic regime, the material was subsequently quenched into the cubic martensite state. After this the steel was finally intercritically heat-treated at 600° C for 8 hours, figure 5.

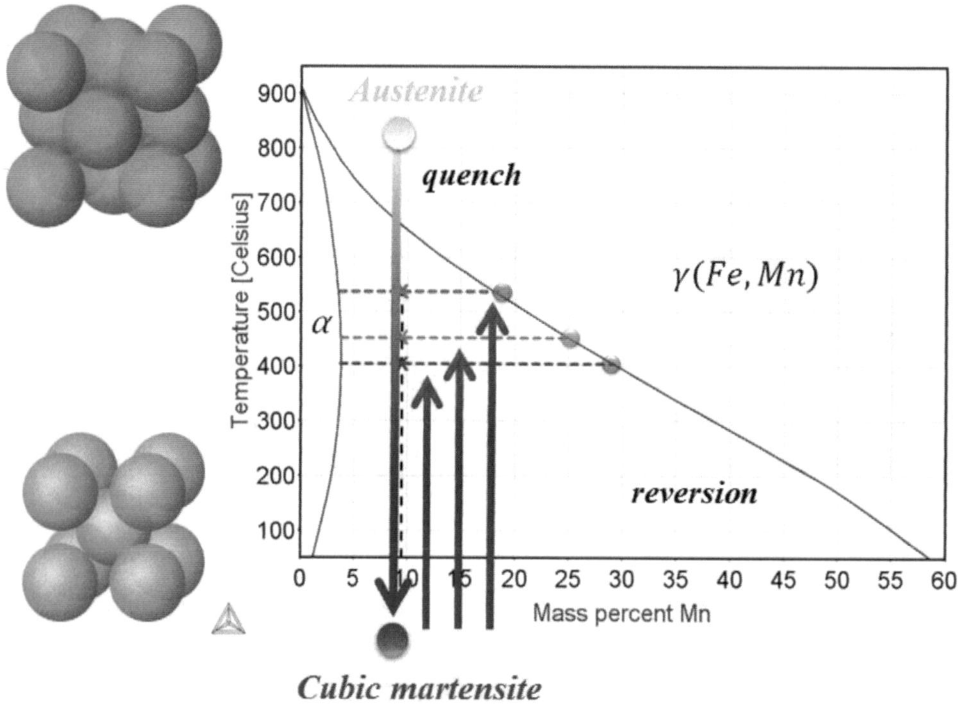

Figure 5. Martensite-to-austenite reversion heat treatment [1,3,4].

During this re-heating treatment the martensite lath grain boundaries, which formed during quenching, were gradually decorated by Mn following the Gibbs adsorption isotherm. At these decorated interfaces martensite-to-austenite reversion was observed owing to the chemical decoration in conjunction with the local elastic distortion. This process occurred in a very confined manner, leading to transformation of at first only the interface regions, figure 6 [15-17]. With further heat treatment these originally confined and only nm-thin martensite-to-austenite transformation zones at the former lath interfaces could be grown further, expanding and thickening sideways into the adjacent martensite matrix [15-19].

Next it was studied how such martensite – austenite nano-laminate steels can lead to improved mechanical properties such as high damage resistance, figure 7,8. These laminate microstructures were found to efficiently share the load by dynamic strain partitioning among the lamellar arranged phases, specifically when one of the phases undergoes load-driven phase transformation, such as the metastable austenite in the case shown in figure 6. The specific thermodynamic stability of the austenite laminates can be designed through the appropriate partitioning during heat treatment.

Recent experiments have also shown that such nano-laminate steels can exhibit increased crack resistance, particularly in such cases where one of the laminate phases, such as the austenite, is rendered metastable so that it can deform via a transformation induced plasticity (TRIP) effect when mechanically loaded [20]. Mechanical fatigue of complex engineering alloys is an ubiquitous phenomenon and it is often dealt with by simply including high safety-factors in structural design solutions. In a nano-laminate Fe9Mn3Ni1.5Al steel (in wt.%) we found the simultaneous activation of roughness-induced and of transformation-induced crack termination mechanisms [21].

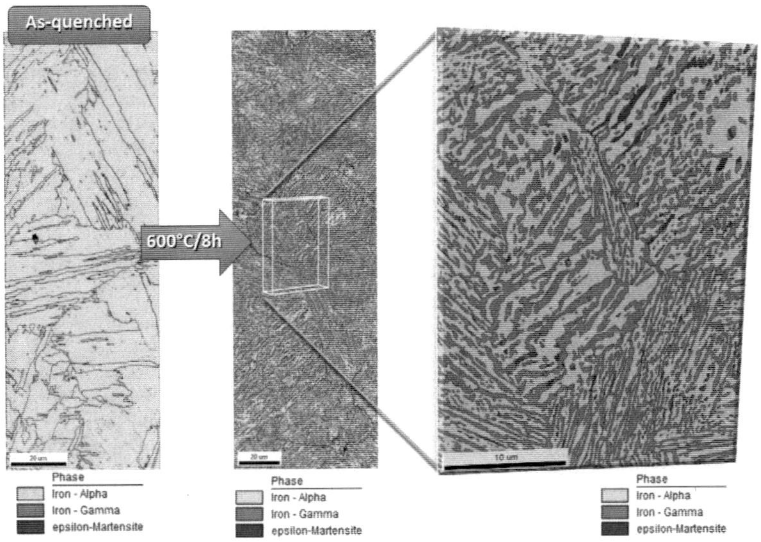

Figure 6. Example of a martensite-to-austenite reversion heat treatment. The EBSD map shows how the lath interface segregation in concert with a spatially confined martensite-to-austenite phase transformation has been used for the design of a laminate steel consisting of three phases. The process proceeds through a segregation plus reversion heat treatment after the initial quench from the austenite solid solution regime. The lamellar arrangement of the newly formed transformation microstructure is due to the fact that the Mn segregated to the martensite lath interfaces, entailing site-specific austenite reversion at these interfaces. The data and EBSD maps are taken from the work of Wang et al. [15-18].

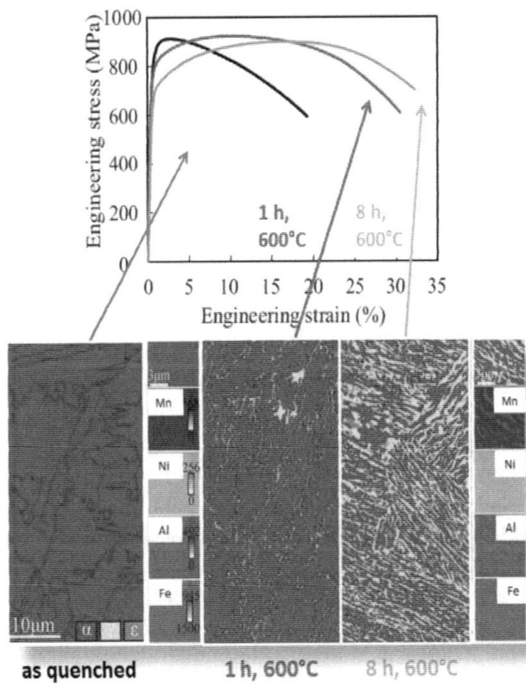

Figure 7. Example of the improvement of the stress-strain response of the laminate steel. The data are taken from the work of Wang et al. [15-18].

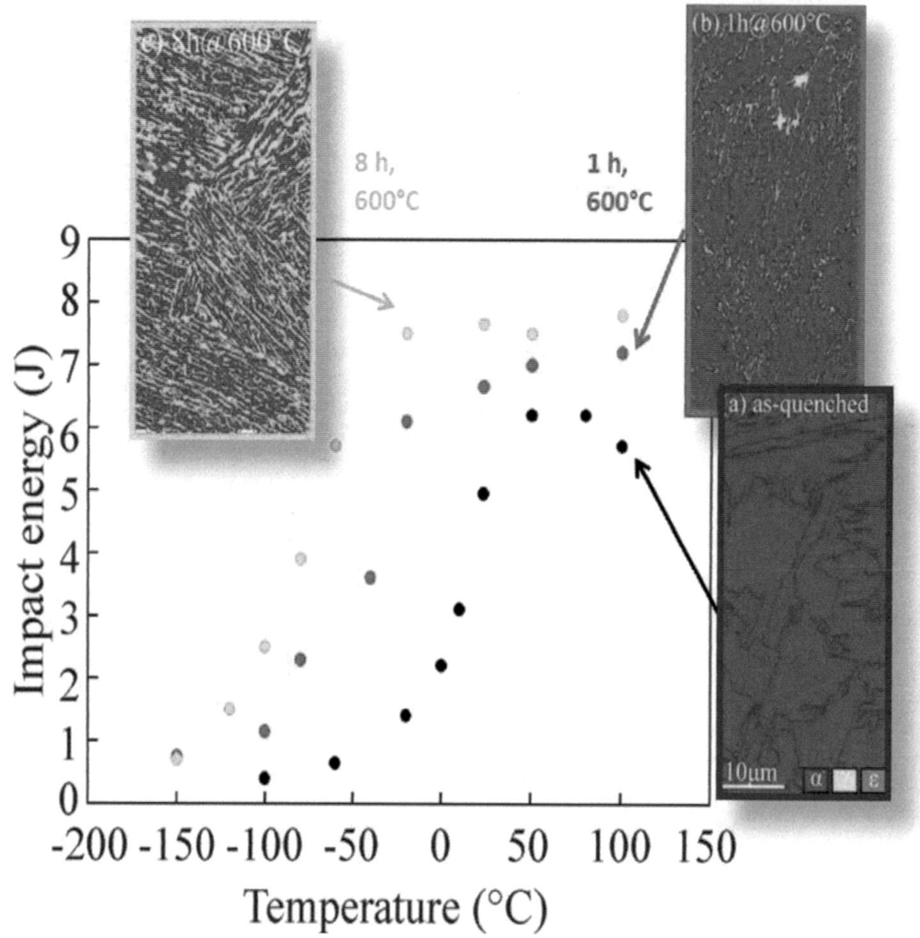

Figure 8. Example of the improvement in the impact toughness response of the laminate steel. The data are taken from the work of Wang et al. [15-18].

The alloy consists of a hierarchical structure comparable to bone, composed of laminate-arranged martensite and metastable austenite phases, and showed improved fatigue resistance compared to conventional high strength alloys as for example DP steels [21-23]. The formation of localized compressive stress fields, which were created from the volume-expanding deformation-induced phase transformation from the metastable face centered cubic γ phase into the near body centered cubic α'-martensite at crack tips, was observed to suppress crack nucleation and crack propagation, figure 9 [24].

4. Non-equilibrium segregation engineering through vessel phases: Tailored austenite retention around partially dissolved precipitates

Another approach to the utilization of confined transformation zones consists in the deliberate introduction of site specific and geometrically confined non-equilibrium segregation zones. This microstructure design method is based on the utilization of so-called vessel phases, which were recently introduced and applied to the design of austenite in multiphase steel microstructures [5-7,25],

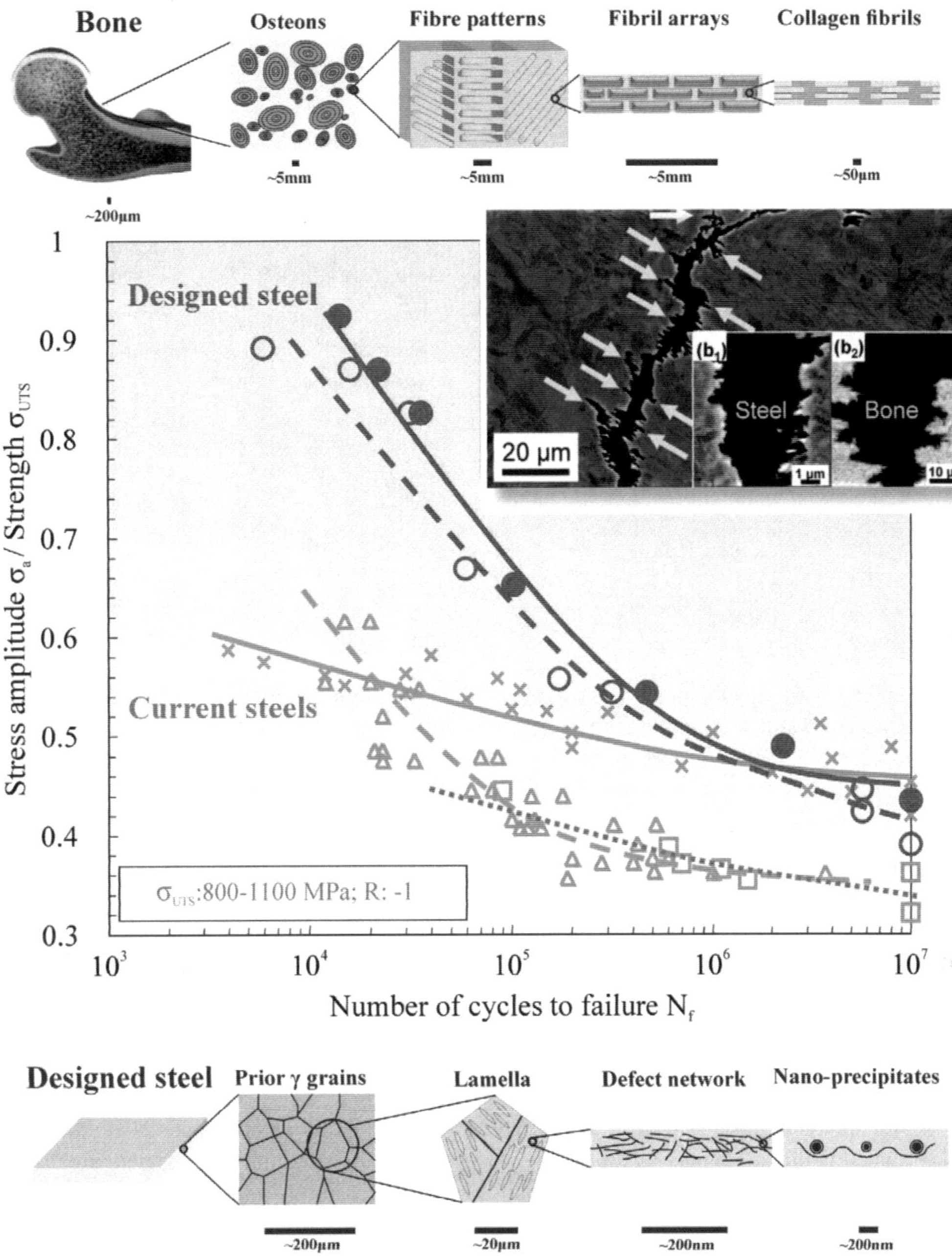

Figure 9. Bone like fatigue resistance observed in a nano-laminate martensite-austenite steel. The figures and data have been taken from the work of Koyama et al. [24].

figure 10, 11. Vessel phases are precipitations carrying certain elements that can – upon partial or complete dissolution – be released to create well defined local phase states, such as metastable austenite at those positions where these precipitates were originally located.

In complete equilibrium the confinement of these localized diffusion regions around the dissolving particles would be lost owing to far reaching diffusion. In the non-equilibrium case, however, this diffusion process toward equilibration is interrupted, so that confined regions with a composition different from the composition of the matrix can be formed.

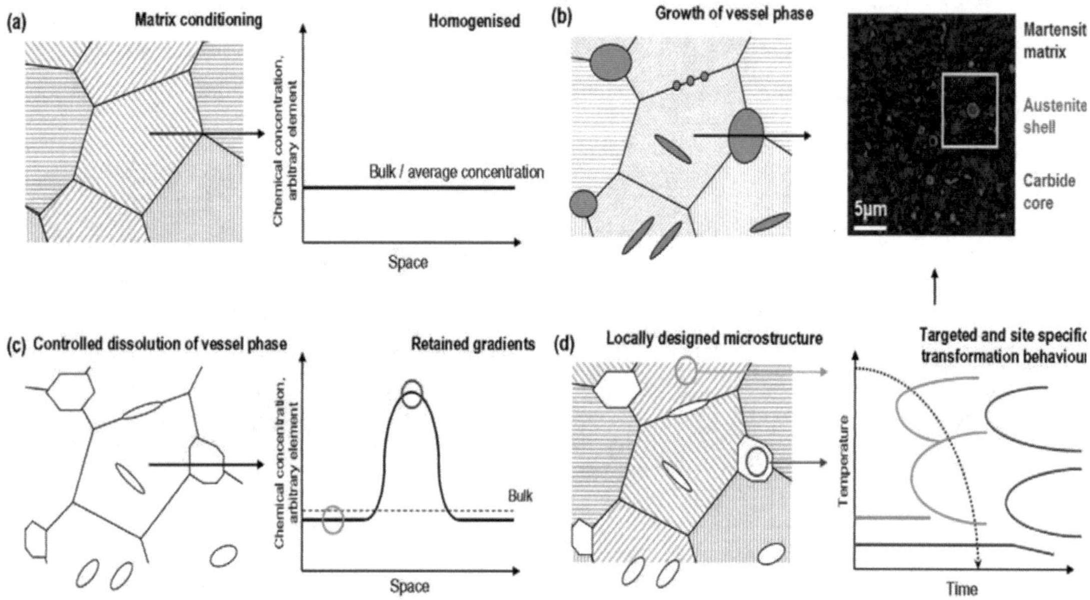

Figure 10. Schematic of site-specific phase transformations induced by use of vessel phases. These phases are precipitates that exhibit a high equilibrium enrichment factor before being made to dissolve upon heat treatment, creating local retained austenitic zones within an otherwise martensitic matrix upon quenching. The size and stability of these austenitic zones can be adjusted by tuning the diffusional fluxes during the dissolution treatment. The figure has been adapted from Belde et al. [5-7,25].

The approach consists of three main processing steps. The first step in this sequence is conducting an adequate tempering cycle to achieve sufficiently high compositional homogeneity of the matrix. This processing step is also referred to as the matrix conditioning step, as it is used to control the precipitation conditions of the vessel phase. The second processing step is a heat treatment for conducting the vessel phase precipitation. This step is termed the accumulation step, as it serves to build up high enrichment factors with respect to the bulk material. It is essential in this approach as the imposed temperature and time define the composition and stability of the precipitates that act as elemental containers to later provide the elements that help creating the confined metastable phase region upon dissolution.

The third step is the complete or incomplete vessel phase dissolution, referred to as the dissolution step. Controlling the vessel phase dispersion and morphology allows one to design retained austenite regions by using the site-specific chemical gradients that form around the dissolving vessel phases to induce a local reduction in the martensite start temperature.

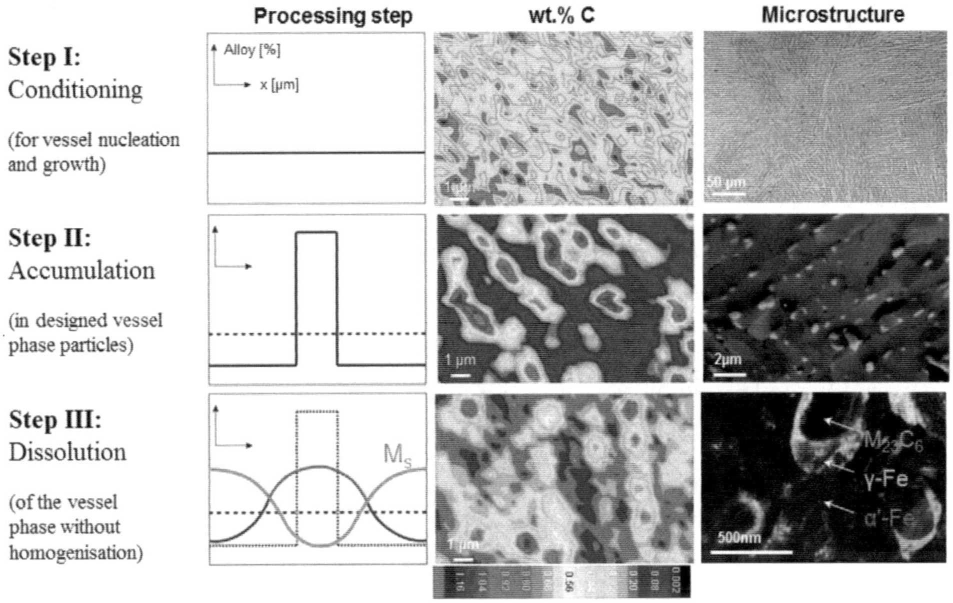

Figure 11. The three phases of vessel segregation engineering in schematic, chemical, and microstructural representations; adapted from the work of Belde et al [5-7,25].

5. Conclusions

In this contribution we have shown how the site specific equilibrium or non-equilibrium solute decoration of lattice defect traps, such as dislocations and grain boundaries, can be used to tune the microstructure by the development of confined regions with different composition, structure, phase state and / or stability within otherwise homogeneous microstructural landscapes.

Because the spatial arrangement of the used trapping pattern agents, viz., the distribution of the dislocations and grain boundaries, is often characterized by small average defect spacing (such the dislocation spacing, and the spacing between lath grain boundaries), SE allows self-organized structuring to be achieved at very small length scales.

Acknowledgments

The authors gratefully acknowledge funding of part of this work by the European Research Council under the umbrella of the EU 7th Framework Programme (FP7/2007-2013) through ERC Advanced Grant agreement 290998, SmartMet.

References

[1] Raabe D, Herbig M, Sandlöbes S, Li Y, Tytko D, Kuzmina M, Ponge D and Choi P-P 2014 Grain boundary segregation engineering in metallic alloys: A pathway to the design of interfaces *Current Opinion in Solid State and Materials Science* **18** 253

[2] Lejcek P 2010 Grain Boundary Segregation in Metals Springer Berlin Heidelberg, Berlin, Heidelberg

[3] Kuzmina M, Herbig M, Ponge D, Sandlöbes S and Raabe D 2015 Linear complexions: Confined chemical and structural states at dislocations *Science* **349** 1080

[4] Dmitrieva O, Ponge D, Inden G, Millán J, Choi P, Sietsma J and Raabe D 2011 Chemical gradients across phase boundaries between martensite and austenite in steel studied by atom probe tomography and simulation *Acta Materialia* **59** 364

[5] Springer H, Belde M and Raabe D 2013 Bulk combinatorial design of ductile martensitic stainless steels through confined martensite-to-austenite reversion *Materials Science and Engineering A* **582** 235

[6] Springer H, Belde M and Raabe D 2016 Combinatorial design of transitory constitution steels: Coupling high strength with inherent formability and weldability through sequenced austenite stability *Materials & Design* **90** 1100

[7] Belde M, Springer H, Inden G and Raabe D 2015 Multiphase microstructures via confined precipitation and dissolution of vessel phases: Example of austenite in martensitic steel *Acta Materialia* **86** 1

[8] Herbig M, Raabe D, Li Y J, Choi P P , Zaefferer S and Goto S 2014 Atomic-Scale Quantification of Grain Boundary Segregation in Nanocrystalline Material *Physical Review Letters* **112** 126103

[9] Herbig M 2017 Spatially correlated electron microscopy and atom probe tomography: Current possibilities and future perspectives. *Scripta Mater* [in print]

[10] Guo W, Jägle E A, Choi P P, Yao J, Kostka A, Schneider J M and Raabe D 2014 Shear-Induced Mixing Governs Codeformation of Crystalline-Amorphous Nanolaminates *Physical Review Letters* **113** 035501

[11] Li Y, Raabe D, Herbig M, Choi P P, Goto S, Kostka A, Yarita H, Borchers C and Kirchheim R 2014 Segregation stabilizes nanocrystalline bulk steel with near theoretical strength *Physical Review Letters* **113** 106104

[12] Herbig M, Choi P P and Raabe D 2013 Combining Structural and Chemical Information on the Nanometer Scale by Correlative TEM and APT *Microscopy and Microanalysis* **19** 948

[13] Stoffers A, Cojocaru-Mirédin O, Seifert W, Zaefferer S, Riepe S and Raabe D 2015 Grain boundary segregation in multicrystalline silicon: correlative characterization by EBSD, EBIC, and atom probe tomography, *Progress in Photovoltaics: Research and Applications* **23** 1742

[14] Stoffers A, Ziebarth B, Barthel J, Cojocaru-Mirédin O, Elsässer C and Raabe D 2015 Complex Nanotwin Substructure of an Asymmetric Σ9 Tilt Grain Boundary in a Silicon Polycrystal *Physical Review Letters* **115** 235502

[15] Wang M M, Tasan C C , Ponge D, Kostka A and Raabe D 2014 Smaller is less stable: Size effects on twinning vs. transformation of reverted austenite in TRIP-maraging steels *Acta Materialia* **79** 268

[16] Wang M M, Tasan C C , Ponge D, Dippel A Ch and Raabe D 2015 Nanolaminate transformation-induced plasticity–twinning-induced plasticity steel with dynamic strain partitioning and enhanced damage resistance *Acta Materialia* **85** 216

[17] Wang M M, Tasan C C , Koyama M, Ponge D and Raabe D 2015 Enhancing hydrogen embrittlement resistance of lath martensite by introducing nano-films of interlath austenite *Metallurgical and Materials Transactions A* **46** 3797

[18] Wang M M, Tasan C C , Ponge D and Raabe D 2016 Spectral TRIP enables ductile 1.1 GPa martensite *Acta Materialia* **111** 262

[19] Raabe D, Ponge D, Dmitrieva O and Sander B 2009 Nanoprecipitate-hardened 1.5 GPa steels with unexpected high ductility *Scripta Materialia* **60** 1141

[20] Li Z, Pradeep K G, Deng Y, Raabe D and Tasan C C 2016 Metastable high-entropy dual-phase alloys overcome the strength-ductility trade-off, *Nature* **534** 227

[21] Kemsley D S 1956 Interior deformation markings in copper fatigue specimens. *Nature* **178** 653

[22] Peterlik H, Roschger P, Klaushofer K and Fratzl O 2006 From brittle to ductile fracture of bone *Nature Materials* **5** 52

[23] Nalla R K, Kinney J H and Ritchie R O 2003 Mechanistic fracture criteria for the failure of human cortical bone *Nature Materials* **5** 164

[24] Koyama M, Zhang Z, Wang M M, Ponge D, Raabe D, Tsuzaki K, Noguchi H and Tasan C C 2017 Bone-like crack resistance in hierarchical metastable nanolaminate steels *Science* **355** 1055

[25] Belde M, Springer H and Raabe D 2016 Vessel microstructure design: A new approach for site-specific core-shell micromechanical tailoring of TRIP-assisted ultra-high strength steels *Acta Materialia* **113** 19

Atom probe study on microstructure change in severely deformed pearlitic steels: application to rail surfaces and drawn wires

Jun Takahashi

Advanced Technology Research Labs., Nippon Steel & Sumitomo Metal Corporation, 20-1 Shintomi, Futtsu-City, Chiba, 293-8511 Japan

E-mail: takahashi.3ct.jun@jp.nssmc.com

Abstract. Pearlitic steel is used as the material for high tensile steel wires, rails and wheels due to its high work hardening and wear resistance. These properties arise from a layered structure comprising deformable lamellar ferrite and hard lamellar cementite. This paper reviews the microstructural change in heavily drawn pearlitic steels wires and worn surfaces of pearlitic rails using atom probe tomography analysis. The cementite decomposition mechanism was elucidated for heavily drawn pearlitic steel wires. For pearlitic rail steels, atomic scale characterization of worn surfaces and of the white etching layer (WEL) were performed, and a mechanism for the formation of the WEL was proposed. The differences and similarities in microstructure and in the state of the cementite in these severely deformed pearlitic steels are discussed.

1. Introduction

Pearlitic steel exhibits high work hardening ability due to its layered structure comprising lamellar ferrite and lamellar cementite, wherein the former is a deformable soft phase and the latter is an undeformable hard phase. The microstructure of pearlitic steel is composed of pearlitic blocks (nodules) with the same crystallographic direction. Each block is composed of pearlitic colonies with the same orientation of pearlitic lamellae [1]. When pearlitic steel is drawn, these colonies rotate to be nearly parallel to the drawing direction and then the interlamellar spacing of pearlitic lamellae decreases. The increase in hardness and strength of pearlitic steels with reduction of the interlamellar spacing of pearlitic lamellae, and the yield strength of pearlitic steels has been explained by Hall–Petch relationship [2,3] or by an Orowan model [4], wherein the grain size corresponds to the interlamellar spacing.

High strength pearlitic steel wires are produced by heavily drawing high carbon pearlitic steels. Nowadays, steel wires with tensile strengths of greater than 4000 MPa have been commercialized by drawing hypereutectoid steel wire rods, and wires of tensile strength more than 6000 MPa can be produced experimentally [5]. High tensile strength wires are widely used for cable wires of bridges, wire ropes, saw wires, and steel-cords for tires. Developments of still higher tensile strength wires are required for weight reduction of automotive tires, which leads to a reduction of carbon dioxide emissions. However, the lamellar cementite forming the pearlitic structure is decomposed partially or fully by heavy drawing, and the mechanical properties of the wires are affected by the cementite decomposition [6,7]. The mechanism of the cementite decomposition has been discussed without a common consensus being reached [8].

IOP Conf. Series: Materials Science and Engineering **219** (2017) 012007 doi:10.1088/1757-899X/219/1/012007

On the other hand, pearlitic steels are also used for rails and wheels of trains, where high wear resistance is required [3,9]. It is reported that pearlitic steel has better wear resistance than steels with other microstructures, such as martensite, bainite, and ferrite containing spherical cementite [9]. However, the exact reason for the high wear resistance has not been sufficiently understood. The change in the microstructure and cementite state by wear needs therefore to be more thoroughly investigated. Furthermore, after use for a long time pearlitic rails often show a "white etching layer" (WEL) at the contact surface [10]. The WEL causes defects in the rail surface but the formation mechanism of the WEL has not been clarified.

We have investigated these traditional issues of pearlitic steel products by atomic scale characterization techniques using atom probe tomography (APT). In this paper we describe three research topics regarding the change in microstructure and cementite state of severely deformed pearlitic steel products [11–13].

2. Cementite decomposition of heavily drawn pearlitic steel wires [11]

2.1. Cementite decomposition

Cementite decomposition in cold-worked steels was first detected by Mössbauer spectroscopy in the 1970s [14]. Cementite decomposition in heavily drawn wires has been directly observed using transmission electron microscopy (TEM) and atom probe field ion microscopy (AP-FIM) since the second half of the 1990s [15–19]. It was reported based on APT analysis that most cementite was decomposed in drawn wires with true strains of 3.5–4.5 [16,17] and that the lamellar cementite was fully decomposed in wires heavily drawn to a true strain of 5.1 [18]. In contrast, Nam *et al.* reported using Mössbauer measurements that at most 50% cementite was decomposed in drawn wires with true strains of more than 3, and the ratio tended to saturate even as the true strain was increased, because of carbon saturation at dislocations located near the cementite/ferrite interfaces [20].

The proposed mechanisms for cementite decomposition can be classified into three hypothesis models. The first is the strong attractive interaction between carbon atoms and the high density of dislocations introduced by heavy deformation [6,19,20]. This interaction energy is larger than the binding energy between carbon and iron atoms in cementite (~0.8 eV/atom compared to ~0.5 eV/atom), and thus carbon atoms can move to the dislocations at temperatures where diffusion of carbon atoms is enabled. The second is the Gibbs–Thomson effect [15,17]. Due to the fragmentation of cementite by heavy deformation, the free energy of the interface of cementite increases and the local solubility of carbon increases, which leads to the decomposition of cementite. The third is the so-called "carbon drag effect." In this model, moving dislocations during deformation carry the trapped carbon atoms because of the large attractive interaction between carbon and dislocations [8,21]. The dislocations generated from the interface of lamellar cementite move to the lamellar ferrite with trapped carbon atoms, and carbon atoms remain in the ferrite matrix by pairwise annihilation of the dislocations, and consequently highly supersaturated solute carbon is formed.

One of the reasons for the difficulty in reaching a consensus on which of these mechanism are most dominant is that the characterized wires in previous studies in general do not have the same drawing and aging conditions. In particular, heavily drawn wires were significantly influenced by aging at room temperature (RT) after drawing, and in most of the previous reports, not much attention has been paid to this unintentional aging. We have investigated the mechanism of cementite decomposition and the sites of carbon atoms after decomposition using sample wires where such unintentional aging can be avoided.

2.2. Preparation of sample wires

The material used in this study is composed of hypereutectoid steel with composition Fe-0.92C-0.48Mn-0.22Si mass% (Fe-4.13C-0.47Mn-0.42Si at.%). A wire rod with a diameter of 5.5 mm was cold-drawn with a powder lubricant to a diameter of 2.0 mm and heated to 950°C, followed by patenting in a lead bath at 580°C. The patented wire was further cold-drawn to 0.2 mm in a liquid lubricant to a true strain of $\varepsilon = 4.61$, wherein true strain ε is defined as ln (A_i/A_f), where A_i and A_f are the initial and final cross-sectional areas, respectively. The speed of the final drawing was fixed at 30 m/min, which is significantly lower than that used in the commercial production process. The as-drawn wires were immediately cooled with dry-ice or immersed in liquid nitrogen, and afterwards were kept in a freezer below −30°C to prevent the wire from aging at RT. The aim of the lower-speed drawing and sample freezing is to distinguish between the heavy deformation effect during drawing and thermal effects after drawing. The as-drawn wires ($\varepsilon = 4.61$) were also intentionally aged for 30 min at each temperature from 100°C to 400°C in 50°C intervals.

Figure 1 shows the change in yield and tensile strengths of sample wires as a function of aging temperature. The tensile strength of the as-drawn wire was about 4480 MPa. The tensile strength increased with low temperature aging and showed a maximum strength after aging at 100–150°C. The peak aging region is represented by the shaded area in Figure 1. For aging above the temperature showing the maximum strength, the tensile strength decreased strongly with increasing aging temperature. The maximum increment in tensile strength due to aging of the as-drawn wire was estimated to be approximately 150 MPa. In carbon steels, such a strengthening through low temperature aging is caused by dislocation locking by carbon atoms (so-called Cottrell atmosphere formation) [22], and the lowering of the strength after the peak aging is caused by recovery and/or recrystallization [23]. We discuss whether a similar interpretation is suitable for heavily drawn steel wires in terms of the change in the microstructure and carbon state. For this detailed characterization using TEM, FIM, and APT was performed in the sample wires indicated by arrows in figure 1.

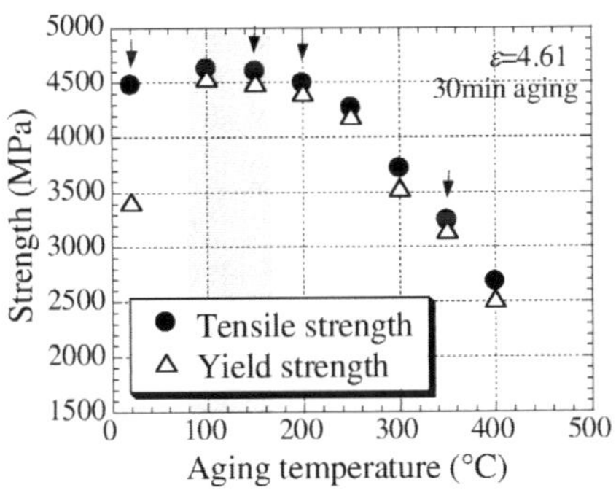

Figure 1. Change in the yield and tensile strengths of the sample wire ($\varepsilon = 4.61$) as a function of aging temperature [11].

2.3. TEM observations

Figure 2 shows example microstructures of sample wires ($\varepsilon = 4.61$) with and without aging in the longitudinal section by TEM bright-field observation. Cementite lamellae and ferrite lamellae in heavily drawn pearlitic wires are preferentially aligned along the drawing direction, namely, the wire axis. The drawn pearlitic lamellar structure was not uniform in interlamellar spacing, and the thickness of the ferrite lamellae was in the range of 5–20 nm. Large internal strains and dislocations were observed in the lamellar ferrite. Distinct lamellar cementite was not observed using TEM. The lamellar ferrite exhibited a strong <110> texture along the wire axis, but each lamella showed different contrast, which indicates that each ferrite lamella has a different crystallographic direction around the axis. Differences in microstructure were unobservable in wires aged at temperatures below 200°C. In contrast, a recovered microstructure was observed in the wire aged at 350°C, in which the morphology of the ferrite lamellae changed to a slightly rounded shape.

2.4. APT and FIM analyses

Needle-shaped specimens for APT analysis were fabricated directly from the sample wires using a standard two-stage electro-polishing technique [24]. In the preparation method, the tip position of the needle specimen always corresponds to the center region of the wire. Consequently, APT analysis is conducted in the center region of the wire along the probing direction parallel to the drawing direction (wire axis). Figure 3 shows APT analysis results of the sample wires (a) before aging, and after aging at (b) 150°C, (c) 200°C, and (d) 350°C, wherein the representative 3D carbon maps and concentration profiles are shown. In the as-drawn wire (before aging), a high carbon concentration (>20 at.%) and a low silicon concentration in the lamellar cementite remain, and thus the lamellar ferrite maintains a low carbon concentration. The results indicate that cementite decomposition hardly took place regardless of the heavy drawing ($\varepsilon = 4.61$).

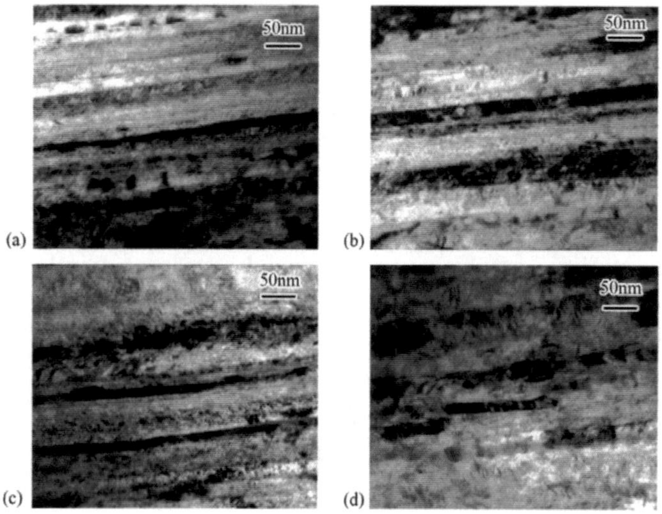

Figure 2. TEM bright-field images of longitudinal sections of the sample wires with (a) unaged, (b) 150°C aging, (c) 200°C aging and (d) 350°C aging [11].

In the wire aged at 150°C, which had the maximum tensile strength (peak aging), the carbon state significantly differs from that of the as-drawn wire. Carbon atoms are distributed almost homogeneously, with about 4–5 at.% in the analyzed volume, which concentration is almost coincident with the average carbon content in the wire. A similar special carbon distribution of was reported using APT in a heavily drawn pearlitic wire of $\varepsilon = 5.1$ by Hono *et al.*, where they suggested complete decomposition of cementite was due to the heavy deformation [18]. Such a uniform distribution of carbon atoms was observed in most of the analyzed volumes of the aged wire. The interlamellar spacing of the wire was estimated to be 5–20 nm, and thus it is considered that the analyzed volumes mostly include the region of the prior lamellar cementite and its interface. Therefore, the local carbon concentration was almost the same throughout the lamellar ferrite, the lamellar cementite, and their interface.

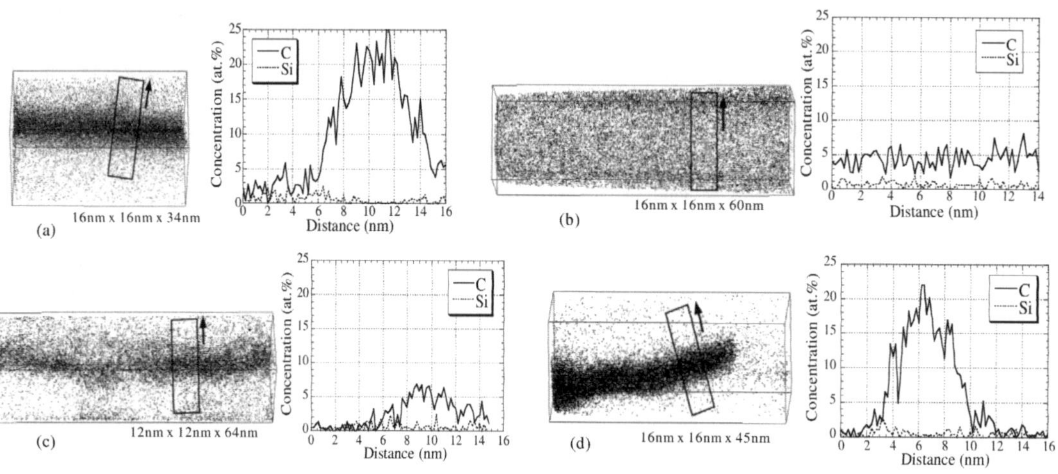

Figure 3. 3D carbon maps and concentration profiles of the sample wires (a) unaged, and after aging at (b) 150°C, (c) 200°C, and (d) 350°C [11].

In the wire aged at 200°C, carbon atoms are enriched in a wide band region with a much lower carbon concentration than the 25 at.% stoichiometric concentration of cementite. Thus, the enriched region is considered to be a prior lamellar cementite or lamellar ferrite boundaries with dislocations. The carbon concentration in the ferrite matrix except the enriched region is much lower than that of the 150°C aged wire. In the wire aged at 350°C, which had tensile strength much lower than that of the as-drawn wire, small features with a high carbon concentration appeared along ferrite boundaries (i.e. prior lamellar cementite) and the carbon concentration at these features is more than 20 at.%, whereas the silicon concentration is low within these features and is slightly higher near their edges. These results indicate that small spherical cementite or carbide was generated along such boundaries, wherein silicon atoms were discharged by the partitioning. In contrast, the carbon concentration in ferrite was much lower than that of the wires aged at lower temperatures.

Figure 4 shows FIM images of the sample wires without and with aging (compare with figure 3). The FIM images cover a much larger region than the APT analysis region, wherein bright and dark areas correspond to ferrite and cementite lamellae, respectively [24]. The FIM images significantly changed with aging temperature, which supports the APT results. In particular, it was surprising that both the APT carbon map and the FIM image changed by low temperature aging at 150°C for 30 min.

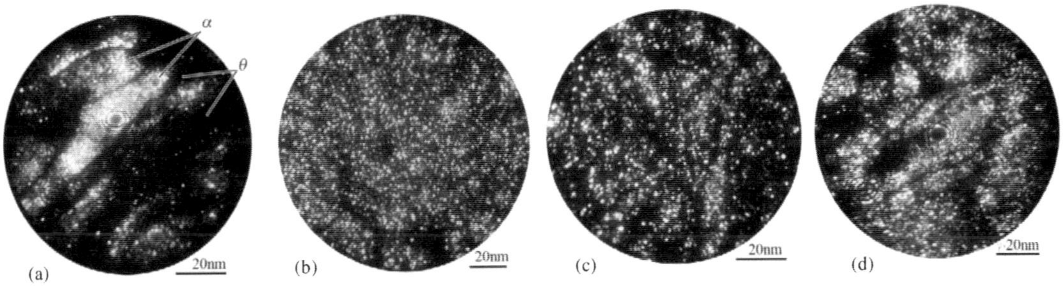

Figure 4. FIM images showing the change in the carbon state of the sample wires (a) unaged, and after aging at (b) 150°C, (c) 200°C, and (d) 350°C [11].

2.5. Mechanism of cementite decomposition

The carbon distribution in drawn pearlitic lamellae in the wire changed significantly, depending on the aging temperature, namely, the degree of aging. In our experiment, under the low-speed drawing and sample freezing after drawing conditions, the cementite decomposition hardly took place just after drawing. However, through low temperature aging at 150°C, the lamellar cementite was significantly decomposed and an almost homogeneous carbon distribution with high concentration corresponding to the average carbon content was formed. The results definitely indicate that the cementite decomposition took place during aging after drawing and not during drawing. Thus, the carbon drag effect wherein the moving dislocations carry carbon atoms from cementite to ferrite lamellae during drawing is not the main mechanism of the cementite decomposition. The fact that carbon atoms moved during aging after drawing suggests the possibility that it is either the large dislocation-carbon interaction energy effect or the Gibbs–Thomson effect, as both these effects can be limited by the diffusion of carbon atoms.

Furthermore, by increasing the aging temperature, the carbon concentration in ferrite decreased and carbon enrichment at ferrite boundaries (prior cementite lamellae) took place. This phenomenon is unexpected because the movement of carbon atoms changed to the opposite direction between 150°C and 200°C. In order to understand the phenomenon, temperature dependent factors associated with the cementite decomposition are taken into consideration. Based on the Gibbs–Thomson effect [25], it is calculated that fragmented hemi-cylindrical cementite with a radius of 0.5 nm showed a carbon content of about 0.41 at.% at 150°C. This value is one order of magnitude lower than the observed carbon concentration of 4–5 at.% in lamellar ferrite (figure 4b). If a high concentration of defects is introduced in the cementite phase, the solubility of carbon is predicted to increase more because the free energy of cementite is effectively increased to a higher level. Such a solid solution state is unstable, and thus the precipitation of carbon clusters and/or carbides should be expected. However, such clusters were not observed in either the APT and TEM investigations. Therefore, the cementite decomposition cannot be explained by the Gibbs–Thomson effect alone. The main mechanism of cementite decomposition is therefore presumed to be the interaction between carbon atoms and lattice defects introduced during heavy drawing.

2.6. Origin of the uniform distribution of carbon atoms

The experimental observations reveal that not only heavy drawing, but also aging, plays an important role in the cementite decomposition. Two factors are required for the decomposition of cementite: the first is diffusion of carbon atoms over a distance substantially greater than the interlamellar spacing, and the second is a large number density of more stable sites rather than the binding energy between carbon and iron in cementite [19,20]. The dislocations introduced by heavy drawing, as well as nterfaces between lamellar cementite and ferrite, have been proposed as such stable sites for carbon atoms.

The 3D carbon maps of the wire aged at 150°C showed almost an homogeneous carbon distribution of 4–5 at.% throughout the lamellar ferrite, lamellar cementite, and their interfaces. The high spatial resolution of the atom probe technique can sufficiently discriminate carbon atomic positions when they are located at dislocations and/or at dislocation walls. The diffraction analysis with TEM showed that there were dislocation walls at the interfaces of the lamellar ferrite and at the small angle boundaries across the lamellar ferrite [19]. The homogeneous carbon distribution observed in the 150°C aging wire cannot be explained therefore by carbon enrichment at dislocation walls in ferrite and/or at interfaces within the lamellar ferrite.

It has been reported that plastic deformation introduces excess vacancies in steel [26,27]. Vacancies of the order of 10^{-4} per bcc lattice site at a strain of about 20% were estimated from thermal desorption spectroscopy of tritium, and the concentration of vacancies was found to increase with decreasing temperature and by increasing the given amount of strain [27]. In addition, it has been reported that the vacancies in steel have a large attractive interaction with carbon atoms and that the interaction energy (~0.85eV/atom) is sufficiently larger than the binding energy between carbon and iron in cementite [28]. The change in the carbon state in the experiment is therefore explained using the presence of a high number density of excess vacancies. Carbon migration over a sufficient distance is difficult at RT although a high number density of vacancies was introduced into the drawing wire. In the aging at 150°C, the carbon atoms are able to migrate and become trapped at the vacancies distributed homogeneously in the matrix ferrite. Therefore, the cementite decomposition proceeded significantly at 150°C aging. Carbon enrichment at the ferrite boundaries (locations of prior lamellar cementite) was observed in the wire aged at 200°C, and this temperature is almost coincident with the decomposition region of the carbon-vacancy complex [28,29]. The complexes were decomposed and the dissolved carbon atoms segregated at the boundaries, as shown in figure 3. X-ray diffraction measurements have already confirmed that the lattice parameter of the ferrite phase was not affected by the cementite decomposition [18], which supports our proposed model that carbon atoms segregate at the lattice defects such as vacancies, dislocations, and boundaries in ferrite. In the aging at higher temperatures, the carbon atoms segregate at the boundaries and precipitate as small spherical cementite along the boundaries (figure. 3).

3. Rolling contact wear surface of pearlitic rail steel [12]

3.1. Wear resistance

The rails in modern railway systems are subjected to serious wear and rail corrugation from fast trains with massive axle loads that cause structural defects and modulation of the rail surfaces. In order to prolong rail life, rails that have high wear resistance must be developed. Increasing the hardness of rail materials has been studied as a means of improving wear resistance because the mechanism of rail wear is considered to be adhesive wear [30]. In addition, increasing the hardness of the contact surface through work hardening has improved wear resistance [3,9]. Pearlitic steel has high potential for the hardening of the contact surfaces due to its high work hardenability. To understand the mechanism, nanoscale characterizations of the worn surfaces have been performed.

3.2. Specimens produced by rolling-sliding wear tests
A hyper-eutectoid pearlitic steel with a chemical composition of Fe-1.0C-0.70Mn-0.40Si-0.25Cr mass% was used in this study. The steel was austenitized by annealing at 1100 °C for 5 min and was cooled to 600 °C at a cooling rate of 30 °C/s. The pearlite transformation was completed during isothermal holding at 600 °C for 5 min, and then the steel was quenched with nitrogen gas. The steel was fully pearlitic with interlamellar spacing of about 100 nm and a Vickers hardness of about 400 HV. Using cylindrical test specimens extracted from the sample steel, a two-cylinder rolling-sliding wear test was conducted on a Nisihara wear test machine capable of simulating the wear of the a rail and wheel system [9,31]. The test conditions to simulate the extreme wearing conditions of the railhead side on curved tracks were as follows: maximum Hertzian contact pressure of 640 MPa (load of 666 N), pearlitic wheel specimen hardness of 380 HV, rotational speed of 800 rpm, and slip ratio of 20% (wheel rotation speed of 640 rpm). The rolling contact surface was cooled with compressed air (0.1 Nm^3/min) to prevent temperature increases and to remove debris. The weight loss was measured to evaluate the wear property of the test specimens. The rate of weight loss was almost constant after more than 1 × 10^5 rolling cycles, i.e., a steady state was nearly achieved. Thus, test specimens with more than 1 × 10^5 cycles of wear were used in this study.

Figure 5 shows the hardness depth profiles estimated using either on a polished surface or on the contact surface [12]. The hardness significantly increases as the contact surface is approached from the depth range of 20–30 μm and reaches a value of more than 800 HV just beneath the surface. It should be noted however that the experimental data vary with the inhomogeneity of the microstructure.

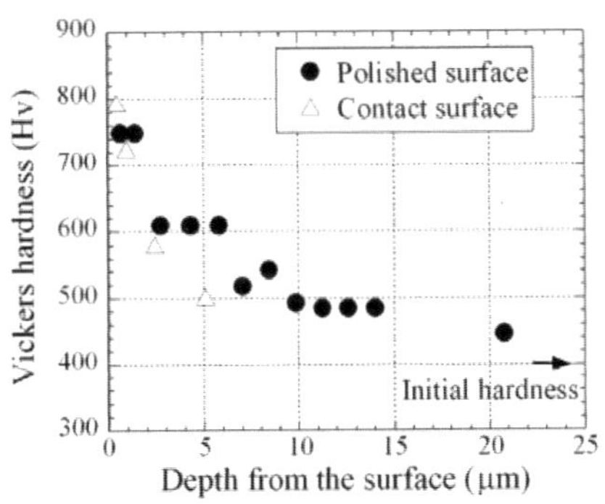

Figure 5. Micro-Vickers hardness profiles as a function of depth from the contact surface [12].

3.3. Microstructure of contact surface layer
Figure 6 shows TEM bright-field images of a longitudinal section from the surface layer of a specimen after 1 × 10^5 wear cycles. The arrows indicate the contact surface. The low-magnification image (figure 6a) of the entire area shows that pearlitic lamellae were plastically deformed and are increasingly refined with decreasing distance from the contact surface. The thinned lamellae were elongated in the direction opposite to the rolling direction, which indicates that the microstructure was formed by metal flow of the surface material, based on the tangential force caused by sliding friction between the rail and wheel specimens [3,9]. The high-magnification image just beneath the contact surface (figure 6b) shows that the surface region contains very fine lamellae parallel to the surface. The interlamellar spacing on average reached nearly 10 nm at the surface. Different domains having slightly coarse lamellae were also observed in the image. These domains most likely arose from pearlite colonies with an initial lamellar structure of a different orientation. Moreover, the high-magnification image at 4 μm-depth below the surface (figure 6c) shows that deformed pearlitic lamellae of about 50 nm interlamellar spacing extend obliquely upward and that the lamellar ferrite contains dislocations and cell boundaries.

Figures 7a and 7b show TEM bright-field images and diffraction patterns of a longitudinal section and a cross section, respectively, of the surface layer in a specimen after 7 × 10^5 cycles of wear. The diffraction patterns were taken from a selected area of about 500 nm diameter. The longitudinal section image shows clear fine lamellar structures parallel to the contact surface, as seen also in figure 6. In contrast the cross section image reveals shows a more broken up morphology rather than lamellar structures. The discontinuous contrast in the images of the elongated ferrite lamellae suggest that subgrain boundaries and/or large strains are present within each lamellae. The two diffraction patterns suggest that the deformed ferrite lamellae have strong [110] crystallographic texture along the elongating direction, with random rotation around the [110] axis. Such a texture was also observed in heavily drawn pearlitic wire [11,15,16]. Figure 7c shows a schematic of the grain structure just beneath the contact surface.

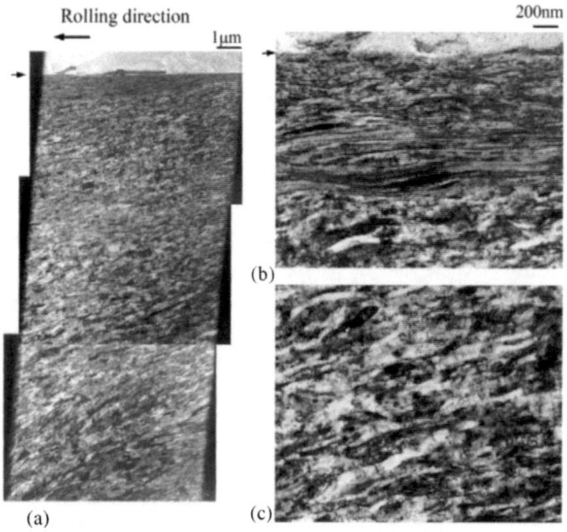

Figure 6. TEM bright-field images of a longitudinal section of the rolling-sliding wear surface layer; (a) entire area (0–15 μm depth), (b) contact surface region, and (c) 4 μm-depth region [12].

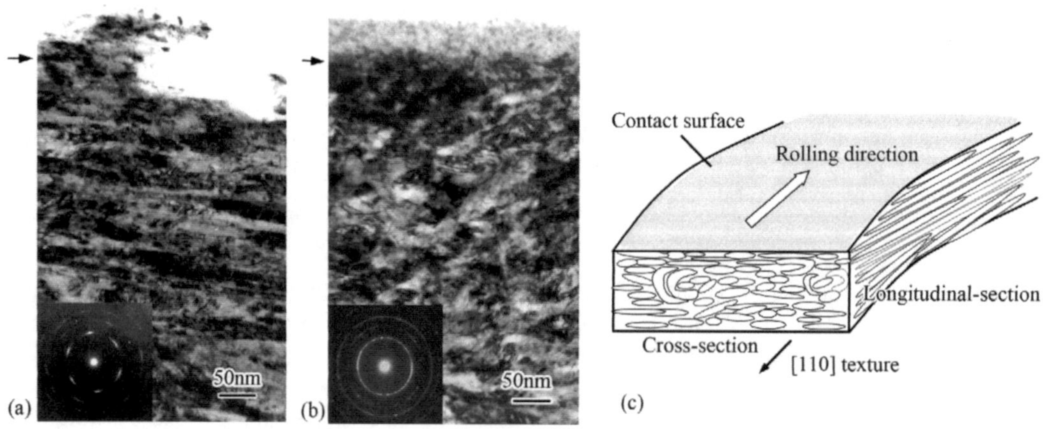

Figure 7. TEM bright-field images and diffraction patterns from (a) cross section and (b) longitudinal section of the rolling-sliding wear surface layer. (c) Schematic of the rolling contact surface [12].

The bright-field and dark-field TEM observations were not able to reveal the presence of cementite lamellae just beneath the contact surface. Therefore, the carbon distribution of the surface layer was investigated by APT. For this a needle tip was produced from the topmost surface in the direction normal to the contact surface by a FIB lift-out method. Figure 8a shows a FIM image of the region just beneath the contact surface in the specimen with 1×10^5 cycles of wear. The image mostly consists of regions of alternating bright and dark concentric rings, which suggests that ferritic lamellae pile up in the needle direction. However, the bright rings are divided by extra dark bands corresponding to boundaries, which indicate that the thinned ferritic lamellae are partially refined, as suggested by the TEM observations. Figure 8b shows the APT analysis results at the 100–200 nm depth region. The arrow indicates the direction of the contact surface. The carbon map shows that carbon-enriched regions forming plate-like structures appear at intervals of 5–20 nm, which correspond to thinned cementite lamellae deformed by the rolling-sliding wear test. The concentration profile of carbon atoms in a box selected nearly perpendicular to these regions shows peak concentrations of carbon are in the range of 15–25 at.%. The high carbon concentrations near the stoichiometry of 25 at.% for the carbon-enriched regions indicate that the lamellar cementite is slightly decomposed but most lamellar cementite is retained as thinned lamellae. In addition, there is an enrichment of chromium and manganese atoms in these regions, whereas silicon atoms are denuded in the same regions. Such a distribution of alloying elements is caused by partitioning during the pearlite transformation. Based on these results it can be concluded therefore, that a mixing of alloying elements with the iron matrix and/or a sufficient diffusion of alloying elements did not occur during the rolling-sliding wear testing.

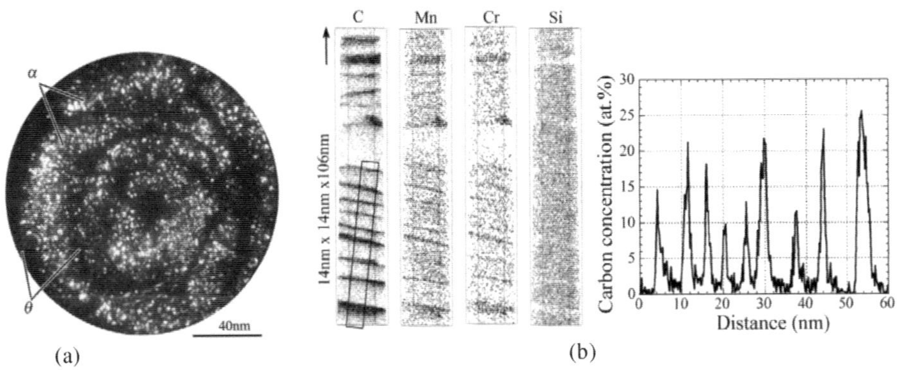

Figure 8. (a) FIM image of the region just beneath the contact surface. The tip direction is perpendicular to the contact surface. (b) 3D elemental maps and carbon concentration depth profile of the region just beneath the contact surface [12].

3.4. Hardening mechanism of the contact surface

In rolling-sliding wear, the microstructure of the contact surface is formed by the rate balance between the removal of surface material and the deformation of core material. Thus, a worn contact surface in a steady state was characterized in this study. The contact surface after the wear test was twice as hard as initially, which leads to better wear resistance in pearlitic steels. TEM observation showed that interlamellar spacing was reduced as the contact surface was approached, and it reached about 10 nm just beneath the surface. APT showed that the lamellar cementite in the deformed pearlitic structure mostly existed as thinned lamellae. Therefore, the increase in hardness in the surface layer can be explained basically by refinement of pearlitic lamellae. It has been reported that the hardness of pearlitic steels obeys the Hall–Petch rule written by

$$\mathrm{Hv} = 150 + 2.15 S_t^{-1/2}, \qquad (1)$$

where S_t (mm) is the interlamellar spacing [3]. For spacing of about 10 nm, the hardness of the contact surface is estimated to be 830 HV, which is nearly equal to the result of the Vickers hardness tests (figure 5). Furthermore, the TEM and FIM results suggested that the thinned ferrite lamellae just beneath the contact surface were refined by the boundaries. The refinement of the thinned lamellae and the presence of high-density dislocations should further increase surface hardness [32]. However, the exact hardness of the surface layer within 400 nm, corresponding to the topmost surface, is not measured in this study because the depth of indenters normally exceeds 1 μm. The surface hardness within 400 nm is predicted to be much higher than 800 HV [33].

In conclusion, a plastically deformed fine pearlitic structure with interlamellar spacing of about 10 nm was formed just beneath the contact surface by the rolling-sliding wear test. The reduction of the spacing considerably increased the hardness of the contact surface. However, lamellar cementite was sufficiently retained as thinned lamellae because of low temperature deformation.

4. Origin and formation mechanism of WEL in rail track surface [13]

4.1. WEL in rail track surface

The WEL is one of the defects caused in the surface of rail tracks [10,34–39]. It is a very hard thin layer and is considered to be a source of microcrack formation. Much researches has been conducted on the origin and the formation mechanism of WEL in the past decades. Newcomb *et al.* [34] first pointed out that although the WEL has a martensitic microstructure, it is formed by cementite decomposition without austenization due to repeated shear fatigue. Österle *et al.* [37] and Stadlbauer *et al.* [39] reported that martensite is formed after austenization due to a temperature rise caused by frictional heating at the rail-to-wheel contact areas. In contrast, Baumann *et al.* [35,36] reported that the WEL is composed of nano-ferrite crystals or is in the process of nano-ferrite crystallization caused by repeated severe plastic deformation. There still remain various questions and uncertainties therefore regarding the origin and formation mechanism of the WEL. In this study, we investigated the origin and formation mechanism of the WEL in terms of the distribution of alloying elements in the WEL using APT analysis. We also discuss the difference in the microstructure and the cementite state between worn surfaces and in the WEL.

4.2. Sample rail

For this study a sample rail having a WEL on its surface was extracted from a high speed rail track. The material was JIS60k pearlitic steel rail, made in 1984, and the accumulated passing tonnage was estimated to be about 150 million tons. The chemical composition of the rail was Fe-0.70C-0.93Mn-0.23Si-0.03Cr mass% (Fe-3.37C-0.46Si-0.95Mn-0.03Cr at.%). Figure 9a shows an image of the sample rail. The WEL can be visually recognized as the band of a slightly different contrast elongated in the track direction. This band, of about 5 mm in width, was located near the center of the rail head, at which location the rail and the wheel make contact. Figure 9b shows an optical micrograph of the surface of a longitudinal section perpendicular to the rail track direction after nital etching. The white featureless contrast region in the pearlite matrix corresponds to the WEL stripe, with thickness and width in the range of 20–30 μm and 300–500 μm, respectively.

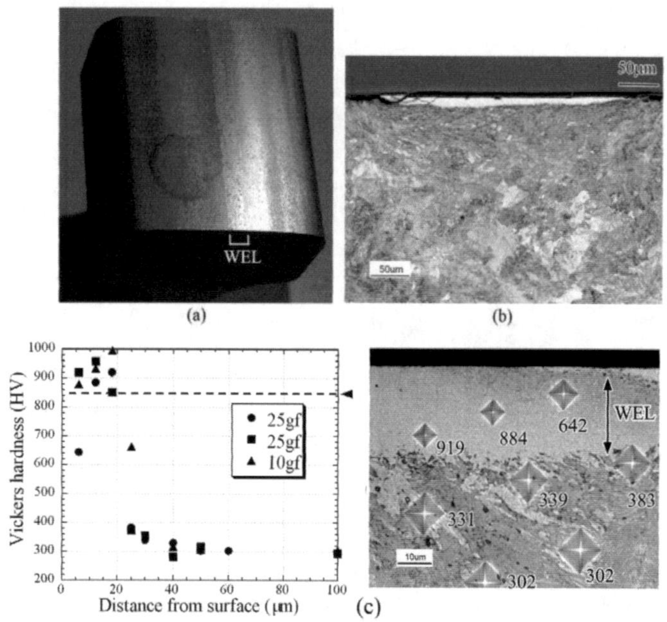

Figure 9. (a) Image of a sample rail with a WEL band. (b) Optical micrograph of a longitudinal section surface etched after nital etching. (c) Hardness profiles as a function of depth position and optical micrograph after the hardness test [13].

Figure 9c shows micro-Vickers hardness test results for the WEL and pearlite matrix regions of the rail sample, wherein the dashed line represents the hardness of the quenched martensite of the same material. The transverse section parallel to the track direction after nital etching was used in the test. In the figure the numerical number near each mark represents the hardness (HV) in that region. The hardness of the pearlite matrix is approximately 300 HV and it slightly increases close to the interface between the WEL and the matrix. The hardness discontinuously increases at the interface and reaches around 1000 HV. The value is higher than the hardness (840 HV) of the quenched martensite. It should be noted that the hardness within the WEL is dependent on the depth position [40].

4.3. TEM observation

Figure 10 shows cross-sectional TEM bright-field images of the surface region containing both the WEL and the matrix. A low-magnification image (figure 10a) shows the entire area containing the WEL, the matrix, and their interface region. The arrowheads indicate the approximate interface position. The microstructure of the WEL region is completely different from that of the matrix pearlite. Figures 10b and 10c show high-magnification images and selected area diffraction patterns of the matrix and WEL regions, respectively, from the locations indicated in figure 10a. The microstructure of the matrix shows a slightly deformed pearlitic structure with an interlamellar spacing of about 200 nm, and with high dislocation density in the ferrite lamellae. In contrast, the WEL region consisted of fine grains of a few hundred nanometers with very high dislocation density. The diffraction pattern, using an aperture diameter of 500 nm, shows discrete diffraction spots, not indicative of nano-ferrite crystals, and no lamellar cementite was observed in the region. Such a microstructure is very similar to lath martensite although the grain size is smaller than in a normal martensitic structure.

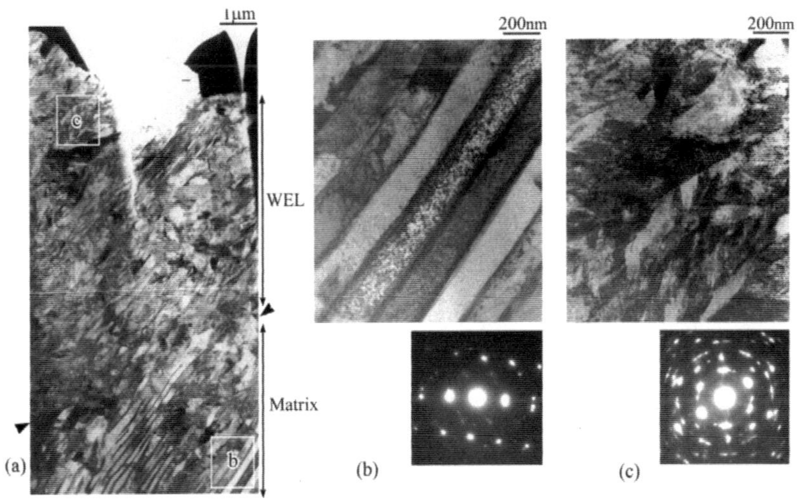

Figure 10. TEM bright-field images and selected area diffraction patterns from a longitudinal section of the rail surface; (a) entire area, (b) matrix region, and (c) WEL region [13].

4.4. APT analysis

Figure 11a shows the results of APT analysis of a volume at the topmost surface of the rail sample but outside the WEL region. The analysis position was about 1 mm away from the WEL band and 5 μm beneath the surface. The arrow represents the surface direction. The 3D elemental maps indicate definite pearlite features with lamellar cementite of 25 at.% carbon, suggesting that decomposition of lamellar cementite does not occur. In the cementite lamella, manganese is enriched and silicon is depleted due to the partitioning of these alloying elements [41]. The cementite lamella seen here is about 12 nm in thickness, which is comparable to those in the initial pearlite. This suggested that lamella thinning due to heavy deformation did not proceed to any considerable extent regardless of the topmost surface.

Figure 11b shows the results of APT analysis of a volume at the topmost surface (depth 2 μm) inside the WEL. From the carbon map it is seems that cementite lamellae were thinned by heavy deformation. However, the peaks of carbon concentration are in the range of 5–10 at.%, which is much lower than the cementite composition 25 at.%. A manganese-enriched and silicon-depleted zone is observed in the lower part of the box, which corresponds to a prior cementite lamella, although no cementite exists in this region. Therefore, the carbon-enriched layers in the carbon map are not deformed cementite lamellae but carbon segregation to boundaries and/or defects. In the WEL, the lamellar cementite is completely decomposed and carbon atoms are redistributed throughout the ferrite lamellae in solid solution and by segregation to defects. The manganese-enriched and silicon-depleted zone is slightly diffuse but its thickness indicates that lamella thinning does not significantly proceed even in the WEL. Fine carbide precipitates were also observed at a different location of the same depth region inside the WEL.

Figure 11c shows the APT analysis result for the lower region inside a WEL (depth 25 μm), just above the interface between the the WEL and the matrix, at which location a maximum hardness was found. The solute carbon concentration in the bottom region is higher than that in the topmost surface regions. The manganese-enriched and silicon-depleted zone is also much more distinct than seen in the topmost surface regions, which indicates that the diffusion length of manganese and silicon atoms (substitutional atoms) in the bottom region was smaller than in the topmost surface region after the decomposition of the lamellar cementite. The APT analysis result of the matrix region just beneath the interface (depth of 30 μm) showed the presence of deformed pearlite with high dislocation density as observed with TEM. The elemental maps indicated typical pearlitic features with the cementite lamellae having the carbon composition of around 25 at.%.

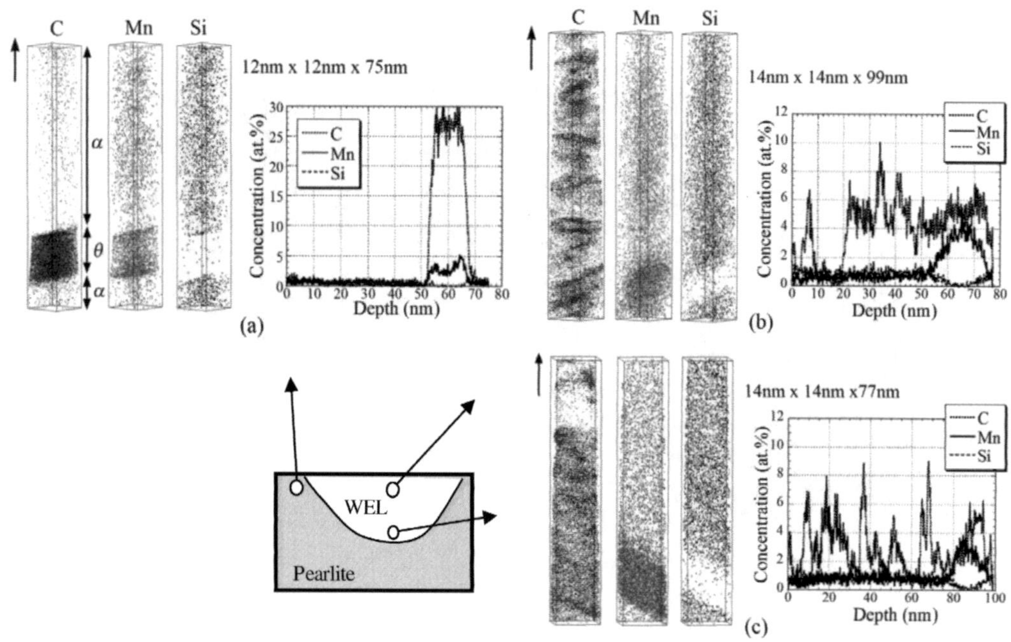

Figure 11. 3D elemental maps and concentration profiles of (a) the topmost surface of the matrix region outside the WEL (depth 5 μm), (b) the topmost surface inside the WEL (depth 2 μm) and (c) the lower region inside the WEL near the interface (depth 20 μm) [13].

4.5. Origin and formation mechanism of WEL

It was found that the cementite lamellae in the WEL decomposed completely and carbon atoms distributed throughout the lamellae in the forms of solid solution, segregation and precipitation. However, manganese-enriched and silicon-depleted zones corresponding to prior cementite lamellae were also observed, and the manganese-enriched zone at the bottom of the WEL was much more distinct than those at the topmost surface regions. It was found that lamella thinning did not considerably proceed at the surface around the WEL, which implies that the surface region did not undergo heavy deformation in the sample rail. These results suggest that the change of microstructure is caused not by severe plastic deformation but probably by phase transformation due to frictional heating. Therefore, the temperature change of the rail surface regions was investigated by thermal distribution calculations of frictional heating at the rail-to-wheel contact areas using the finite element method [42].

For the model the main calculation parameters an axle load of 10000 kgf (9.8 × 10^4 N) per wheel, a contact length and width both of 10 mm, and a train speed of 30 m/sec (108 km/hour). The dynamic friction coefficient was taken to be 0.4, a value one order higher than the rolling contact friction coefficient [43]. Figure 12 shows the temperature profiles calculated at different depth positions below the contact surface as a function of the time from the contact start. The temperature at each depth position increases rapidly at the contact start point and reaches a peak value at the contact end point. The temperature falls exponentially with the time after contact ceases, mainly due to thermal conduction as the rail functions as an efficient heat sink. The highest achieved temperature increased with decreasing distance from the contact surface. The rate of temperature rise was estimated to be 4 × 10^6 °C/s, which is significantly higher compared with normal heating methods. The time at which high temperatures were maintained was estimated to be approximately 0.5 ms, more or less corresponding to the duration of wheel-rail contact.

Temperatures at the depth positions of 2 μm and 20 μm increased considerably above the A_1 point, i.e., the thermal equilibrium austenization temperature (727°C in Fe-C alloy), and the temperature at the depth position of 60 μm almost reached the A_1 point. The actual austenization temperature is considered to be higher than the A_1 point. The result suggests that the austenization border region, namely, the interface between the WEL and the matrix, appears clearly at depth positions of less than 60 μm. In the calculations, however, the achieved temperature strongly depends on the parameters used. Therefore, it is difficult to quantitatively determine the temperature rise only from calculations because the calculation parameters are not well known. Evidence of the achieved temperatures based on the experiment results was therefore sought to enable a meaningful and quantitative discussion of the temperature rise.

For this we focused on the concentration profile of the manganese-enriched zone. The higher the temperature achieved and the longer the time after the decomposition of lamellar cementite, the more diffusive the concentration profile appears to be, due to the long diffusion length of manganese atoms. Diffusive concentration profiles of the manganese-enriched zone were not observed in the deformed pearlitic structure in which cementite lamellae remained (figure 11a). Thus, the actual achieved temperature can be estimated by analyzing the degree of change in the concentration profile of the manganese-enriched zone based on the diffusion length in austenite.

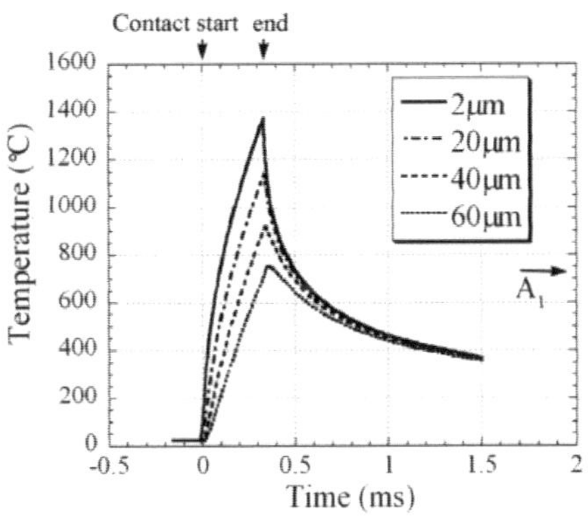

Figure 12. Temperature profiles of various depth positions obtained by frictional heating calculation at rail-to-wheel contact areas [13].

Figure 13a shows a schematic explaining the change in atomic concentration profile caused by diffusion, wherein the initial concentration profile is assumed to be a rectangle with the height C_0 and width $2h$ of the enriched zone [44]. The number of each line represents the ratio of the diffusion length to the zone width $(Dt)^{0.5}/h$, where D and t are the diffusion coefficient and time, respectively. As time passes, the rectangle changes to a diffusive shape. Figure 13b shows the actual concentration profile of the manganese-enriched zone, perpendicular to the prior cementite lamella in the topmost surface region within the WEL (figure 11b). With the assumption that C_0 is 3.6 at.% and $2h$ is 12 nm from the analysis result of the pearlite matrix, the value of $(Dt)^{0.5}/h$ is estimated to be nearly 3/4. The value indicates that the diffusion length $2(Dt)^{0.5}$ is approximately 9 nm at the topmost surface. From the calculations presented in figure 12, the maximum achieved temperature is estimated to be about 1300°C. Table 1 summarizes the achieved temperatures estimated by the shape analysis of the manganese-enriched zones observed at different depth positions. The achieved temperatures are in the range of 1300–1400°C at the topmost surface (depth of 2 μm) and lower than 1150°C at the bottom position (depth of 25 μm) inside the WEL. The results indicate that the temperature of the WEL region can increased significantly above the austenization temperature and that the achieved temperature increases more closer to the contact surface. The achieved temperatures estimated from our experiments support therefore the temperature rise profiles predicted by the frictional heating calculation.

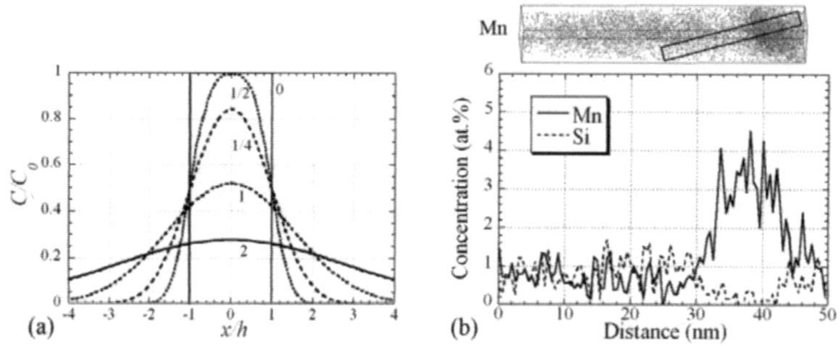

Figure 13. (a) Change in concentration profile from an initial rectangular shape based on diffusion and (b) actual concentration profile of the manganese-enriched zone of the topmost surface of the WEL (figure 11b) [13].

Table 1. Shape analysis results of the atomic concentration profiles of manganese-enriched zones at various depth positions of the rail sample [13].

Region	Depth position (μm)	Mn peak concentration (at.%)	Shape	$(Dt)^{0.5}/h$	$2(Dt)^{0.5}$ (nm)	Estimated achieved temperature (°C)
WEL	2	2.7±0.16	Diffused	~3/4	~9	~1300
WEL	2	2.2±0.14	Diffused	~1	~12	~1400
WEL	25	3.6±0.18	Slightly diffused	<1/4	<3	<1150
Pearlite	30	3.6±0.18	Not diffused			

Speich *et al.* [45] reported from experiments on rapid heating using laser radiation that the growth rate of austenite into pearlite was nearly 479 nm/ms at 829 °C in pearlitic steel of 0.96 mass% carbon and that the rate increased with increase in temperature. Furthermore, Ganeev [46] reported that the irradiation of a pulse laser with a pulse width of 6 ns and frequency of 20 Hz hardened a surface region of 3–8 μm in depth in pearlitic steel of 0.6 mass% carbon, which suggested that the nucleation of austenite occurred within the pulse width of 6 ns. These reports indicate that the nucleation and growth of austenite in pearlitic steel are sufficiently fast to complete the austenite transformation in the period during which a high temperature is maintained by frictional heating.

Based on the above we conclude that the WEL was generated by a martensitic transformation after rapid austenization, resulting from a temperature rise due to the contact frictional heating. It is known that the rail contact surfaces undergo heavy deformation and are subjected to hydrostatic pressure due to the passage of trains in addition to heating [47]. In our proposed model of the WEL formation, heavy deformation due to the passage of trains was not taken into consideration. Such a heavily deformed pearlite seems to transform more easily to austenite kinetically than undeformed pearlite because (1) the density of cementite interfaces that can act as nucleation sites of austenite grains increases and (2) the austenite temperature decreases further with an increase in the carbon concentration in ferrite. However, significant thinning of the pearlitic lamellae was not observed in the WEL region in our sample. This fact indicates that the transformation to austenite could take place in the surface region without requiring heavy deformation of pearlite.

Figure 14 shows a schematic explaining the change in hardness on the cross-sectional surface of the rail according to our proposed model. First, the passage of trains produces plastic deformation in the rail surface region due to the load from the trains, and consequently, the hardness of the surface region slightly increases due to work hardening. In rare case that the special condition of rail-to-wheel contact is satisfied, large frictional heating locally increases the surface temperature above the austenization temperature. The WEL is then generated in the contact surface region wherein martensite transformation occurs due to rapid cooling. The hardness of the WEL region increases uniformly just after the WEL formation. Then, the further passage of trains causes the surface region to warm up due to rolling contact friction and plastic deformation, which leads to a tempering of the WEL martensite in the surface region. The hardness of the surface region is thus lowered, depending on the degree of tempering. The final hardness profile observed in the sample rail is explained well by the model.

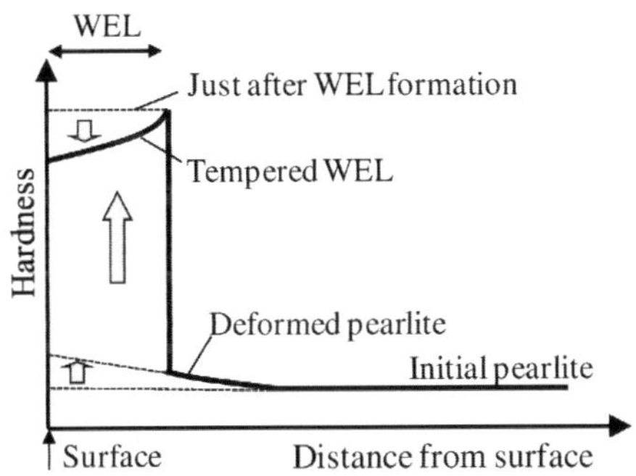

Figure 14. Schematic explaining the hardness change of the rail surface region due to WEL formation and tempering [13].

5. Microstructure change in pearlitic steels.

Very thin pearlitic lamellae of 5–20 nm in interlamellar spacing were observed both in the heavily drawn pearlitic steel wires and in the topmost wear surface of pearlitic rails. The reduction of the interlamellar spacing results in the high work hardening of pearlitic steels, which leads to high strength and high hardness, resulting in high wear resistance. The microstructures of the drawn wires and the topmost wear surface of the pearlitic steels were very similar, wherein the deformed ferritic lamellae have a strong [110] crystallographic texture along the elongation direction, with a random rotation around the [110] axis, and where most of the lamellar cementite retained in the form of thin layers when aging is prevented by keeping the material at a low temperature. Cementite decomposition proceeded during aging after drawing in the heavily drawn pearlitic steel wires. These results suggested that cementite decomposition results in a strong attractive interaction between carbon atoms and defects in steels. Therefore, the following two factors are required to enable cementite decomposition: (i) diffusion of carbon atoms over a distance substantially greater than the interlamellar spacing, and (ii) a large number density of more stable sites with lower energy compared to the binding energy between carbon and iron in cementite.

On the other hand, the WEL seen in rail-track surfaces was completely different from the microstructure of the deformed pearlitic steels. The experimental observations and analysis suggest that the WEL is generated by a martensitic transformation after rapid austenization, resulting from a temperature rise due to the contact frictional heating. The final microstructure and hardness characteristics of the WEL can be explained by additional tempering effects due to the further passage of trains after formation of the WEL. The likely temperature rise, predicted by finite element calculations, was actually confirmed by analysis of the distribution of manganese in prior lamella cementite, based on a diffusion model. It is considered that WELs are generated under special contact conditions when a sufficiently high load transfer takes place locally and/or the contact area becomes extremely small during the passage of a high speed trains. Such a condition may also be satisfied when the brakes of a train traveling at a high speed are applied.

6. Summary

In this paper we have reviewed a number of APT investigations of industrially important pearlitic steel products. The change in microstructure and state of cementite in the pearlitic steels significantly influences the properties in each case. An understanding of the mechanisms responsible for the differences in microstructure, and their change during operation, as well as their properties, is expected to bring improvements not only of pearlitic steel products in current use, but also in materials design for achieving high strengthening.

References

[1] Takahashi T, Nagumo M and Asano Y 1978 *J. Japan Institute of Metals and Materials* **42** 708
[2] Hyzak J M and Bernstein I M 1976 *Met. Trans.* **7A** 1217
[3] Perez-Unzueta A J and Beynon J H 1993*Wear* **162-164** 173
[4] Marder A R and Bramfitt B L 1976 *Met. Trans.* **7A** 365
[5] Li Y J, Choi P, Goto S, Borchers C, Raabe D and Kirchheim R 2012 *Acta Mater.* **60** 4005
[6] Tarui T, Takahashi T, Tashiro H and Nishida S 1996 *Metallurgy, processing and applications of metal wires* (TMS Warrendale PA) eds Paris HG and Kim DK p 87
[7] Maruyama N, Tarui T and Tashiro H 2002 *Scripta Mater.* **46** 599
[8] Umemoto M 2000 *Materia Japan* **39** 219
[9] Ueda M, Uchino K and Senuma T 2004 *Tetsu-to-Hagané* **90** 2023
[10] Eyr T S and Baxter A 1972 *Tribol. Int.* **5** 256
[11] Takahashi J, Kosaka M, Kawakami K and Tarui T 2012 *Acta Mater.* **60** 387
[12] Takahashi J, Kobayashi Y, Ueda M, Miyazaki T and Kawakami K 2103 *Mater. Sci. Technol.* **29** 1212
[13] Takahashi J, Kawakami K and Ueda M 2010 *Acta Mater.* **58** 3602
[14] Gridnev V N, Gavrilyuk V G, Dekhtyar I Y, Meshkov Y Y, Nizin P S and Prokopenko V G 1972 *Phys Stat Sol (a)* **14** 689
[15] Languillaume J, Kapelski G and Baudelet B 1997 *Acta Mater* **45** 1201
[16] Read H G, Reynolds Jr W T, Hono K and Tarui T 1997 *Scripta Mater.* **37** 1221
[17] Sauvage X, Correaux J, Danoix F and Blavette D 2000 *Philo. Mag. A* **80** 781
[18] Hono K, Ohnuma M, Murayama M, Nishida S, Yoshie A and Takahashi T 2001 *Scripta Mater.* **44** 977
[19] Tarui T, Maruyama N and Tashiro H 2005 *Tetsu-to-Hagané* **91** 265
[20] Nam W J, Bae C M, Oh S J and Kwon S-J 2000 *Scripta Mater.* **42** 457
[21] Korznikov A V, Ivanisenko Yu V, Laptionok D V, Safarov I M, Pilyugin V P and Valiev R Z, 1994 *Nanostruct. Mater.* **4** 159
[22] Cottrell A H and Bilby B A 1949 *Proc. Phys. Soc. A* **62** 49
[23] Yamada Y 1976 *Tran. ISIJ* **16** 417
[24] Miller M K, Cerezo A, Hetherington M G and Smith G D W 1996 *Atom probe field ion microscopy* (New York: Oxford University Press)
[25] Hinchliffe C E and Smith G D W 2001 *Mater. Sci. Technol.* **17** 148
[26] Mecking H and Estrin Y 1980 *Scripta Metall.* **14** 815
[27] Nagumo M, Ohta K and Saitoh H 1999 *Scripta Mater.* **40** 313
[28] Vehanen A, Hautojärvi P, Johansson J and Yli-Kauppila J 1982 *Phys. Rev. B* **25** 762
[29] Tapasa K, Barashev AV, Bacon D J and Osetsky Yu N 2007*Acta Mater.* **55** 1
[30] Tomlinson G A 1929 *Philos. Mag.* **7** 905
[31] Ueda M, Uchino K and Kobayashi A 2002 *Wear* **253** 107
[32] Dollar M, Bernstein I M and Thompson A W 1988 *Acta Metall.* **36** 311
[33] Katsuki F and Yonemura M 2007 *Wear* **263** 1575
[34] Newcob S B and Stobbs W M 1984 *Mater Sci. Eng.* **66** 195
[35] Baumann G, Zhong Y and Fecht H-J 1996 *Nanostruct. Mater.* **7** 237

[36] Baumann G, Fecht H-J and Liebelt S 1996 *Wear* **191** 133
[37] Österle W, Rooch H, Pyzalla A and Wang L 2001 *Mater. Sci. Eng. A* **303** 150
[38] Lojkowski W, Djahanbakhash M, Bürkle G, Gierlotka S, Zielinski W and Fecht H-J 2001 *Mater. Sci. Eng. A* **303** 197
[39] Stadlbeauer W, Loos W and Werner E A 2001 *Proc. 2nd World Tribology Congress* (Vienna: Austria)
[40] Zhang H W, Ohsaki S, Mitao S, Ohnuma M and Hono K 2006 *Mater. Sci. Eng. A* **421** 191
[41] Hultgren A and Kuo K 1953 *Mem. Sci. Rev. Met.* **50** 847
[42] Fischer F D, Werner E A, Knothe K and Angew Z 2001*Math. Mech.* **81** 75
[43] Zhang W, Chen J, Wu X and Jin X 2002 *Wear* **253** 82
[44] Crank J 1975 *The mathematics of diffusion,* 2th Ed (Oxford: Oxford university press) p15
[45] Speich GR, Demarest VA and Miller R L 1981 *Metall. Trans. A* **12A** 1419
[46] Ganeev R A 2002 *J. Mater. Proces. Technol.* **121** 414
[47] Bernasconi A, Filippini M, Foletti S and Vaudo D 2006 *Int. J. Fatigue* **28** 663

Reason for high strength and good ductility in dual phase steels composed of soft ferrite and hard martensite

Daisuke Terada[1,2], Gosuke Ikeda[2], Myeong-heom Park[2,3], Akinobu Shibata[2,3], Nobuhiro Tsuji *[2,3]

[1] Department of Advanced Materials Science and Engineering, Chiba Institute of Technology, Narashino, 275-0016, Japan
[2] Department of Materials Science and Engineering, Kyoto University, Yoshida Honmachi, Sakyo-ku, Kyoto 606-8501, Japan
[3] Elements Strategy Initiative for Structural Materials (ESISM), Kyoto University

E-mail: nobuhiro-tsuji@mtl.kyoto-u.ac.jp

Abstract. Dual phase (DP) steels in which the microstructures are composed of a soft ferrite phase and a hard martensite phase are known to show good strain-hardening, high strength and large elongation, but reasons for their superior mechanical properties are still unclear. In the present study, two types of DP structures, having either networked martensite or isolated martensite were fabricated in a low-carbon steel by different heat treatment routes, and their tensile deformation behavior was analyzed using the digital image correlation (DIC) technique. It was revealed that the DP specimens having networked martensite microstructures showed a better strength-ductility balance than the DP specimens with isolated martensite structures. The microscopic DIC analysis of identical areas showed that the strain distribution within the DP microstructures was not uniform and the plastic strain was localized in soft ferrite grains. The strain localized regions tended to detour around hard martensite but eventually propagated across the martensite. It was found also from the DIC analysis that the degree of strain partitioning between ferrite and martensite in the networked DP structure was lower than that in the isolated DP structure. The deformation became more homogeneous when the hard phase (martensite) was connected to form a network structure, which could be one of the reasons for the better strength-ductility balance in the networked DP structure compared to that in the isolated DP structure.

1. Introduction

Demands on high strength steels have been increasing more and more in recent years, based on two different requirements from our society. One requirement is to reduce the weight of transportation equipment thereby saving fossil fuels and reducing greenhouse gases. The other is to increase the safety of equipment and constructions in case of accidents such as collisions, and earthquakes. These two demands are conflicting, but strengthening of materials used for equipment and constructions is one of the promising solutions. In the field of automobile steels, for example, so-called advanced high strength steels (AHSS) have been energetically studied and developed [1]. One of the important AHSS is a dual phase (DP) steel in which the microstructure is composed of the soft phase (ferrite) and the hard phase (martensite or/and bainite). DP steels show good strain hardening ability, resulting in high tensile strength, good yield ratio, and fairly large tensile elongation. On the other hand, the exact reasons for the superior mechanical properties of DP steels are still unclear. One may say that the hard phase brings about high strength and soft phase gives large ductility, but this is not a scientific explanation. Since the microstructures of DP steels are micro-scale mixtures of ferrite and martensite, mechanical interactions between the soft and hard phases must be clarified for fundamentally understanding their good mechanical properties.

There are various microstructural parameters that can influence the deformation behavior of DP steels, namely, the volume fraction, size, hardness ratio, and distribution of the two phases [2-7]. In the present study, two different types of DP microstructures composed of soft ferrite and hard martensite are fabricated by different heat treatment routs in a low carbon steel. One type of microstructure has a networked martensite, while the other type has isolated martensite in the microstructure. For understanding the mechanical interaction between ferrite and martensite, local deformation in the microstructures is quantitatively analyzed by means of the digital image correlation (DIC) technique [8].

2. Experimental Procedures

A low carbon steel having a chemical composition of Fe-0.087C-0.79Si-1.77Mn-0.009P-0.002S-0.02Cr-0.01Mo-0.003N (mass%) was used in the present study. DP microstructures in proeutectoid steels can generally be obtained through interctitcal annealing at the ferrite + austenite two-phase temperature, followed by quenching, during which austenite transformes to martensite. When the phase structure prior to intecritical annealing changes, as is illustrated in figure 1, different types of DP microstructures having different distributions of martenste can be fabricated. In route I of figure 1 where a ferrite-based microstructure is intercritically annealed in low carbon steels, austenite preferentially forms along the ferrite grain boundaries (and on pearlite or cementite with small fractions), so that martensite obtained by quenching has a network structure. In route II, on the other hand, where the steel is firstly austenitized and then intercritically annealed, ferrite forms preferentially along grain boundaries of austenite, so that the martensite in the final DP microstructure is isolated. In the present study, some sheets 1 mm thick of the 0.087C-0.79Si-1.77Mn steel were firstly austenitized at 950°C for 1.8 ks and then air-cooled to room temperature to obtain a microstructure composed of ferrite and a small amount of pearlite (route I). The steel sheets with the ferrite-based microstructre were intercritically annealed in the ferrite + austenite two phase regions (760°C ~ 840°C) for 1.8 ks, followed by water-quenching to fabricate DP structures with a networked martensite. The other steel sheets were austenitized under the same condition as in route I (950°C, 1.8ks), then cooled down to temperatures from 720°C to 820°C, and intercritically annealed at the hold temperature for 1.8 ks, followed by water-quenching to obtain DP structures with isolated martensite (route II). All the heat treatments were carried out in a salt bath.

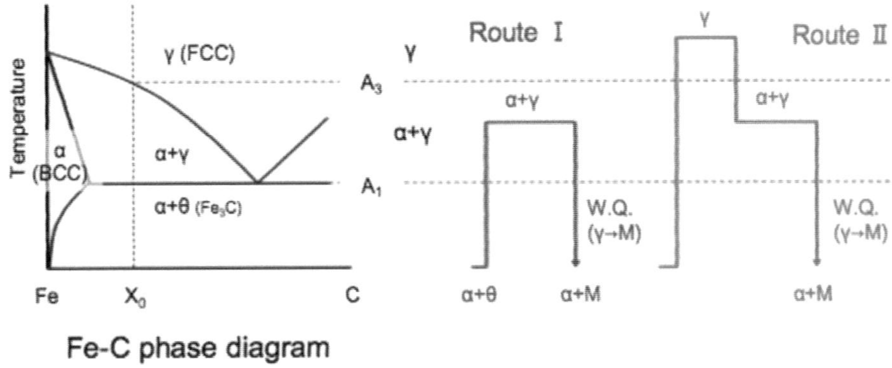

Figure 1. Schematic illustration of the Fe-C phase diagram and two kinds of heat-treatment route to obtain DP structures having different distribution morphologies of martensite.

Tensile specimens 5 mm in gage width, 10 mm in gage length and 1mm in thickness (1/5 of JIS-5 specimen) were cut from the heat-treated sheets. Tensile tests were conducted at room temperature at an initial strain rate of 8.3×10^{-4} s^{-1}. Microstructures in identical areas in the specimens were observed by scanning electron microscopy (SEM) before and after tensile testing to various strains. Prior to the tensile test and SEM observation, the broad surfaces of the specimens were mechanically polished using emery papers and then electro-polished in a 10% perchloric acid (100ml $HClO_4$ + 900ml CH_3COOH) solution. The electro-polished specimens were etched using a 3% nital (30ml HNO_3 +970ml C_2H_5OH) solution to reveal ferrite and martensite microstructures. The obtained SEM images of the identical areas were used as markers to track displacements of the same points in the microstructures. From the displacements of the markers (microstructural features), the two-dimensional strain tensor at each point in the microstructures was calculated, from which maps showing strain distributions in the microstructures were constructed.

3. Results and Discussion

Figure 2 shows SEM micrographs of the DP microstructures fabricated by heat treatment route I in figure 1. In the SEM micrographs, the ferrite is black, while martensite is seen as a white phase including substructures. In all micrographs of figure 2, networked martensite structures where martensite phase is connected to each other can be recognized. With increasing the intercritical annealing temperature, the volume fraction of austenite at the temperature increases, as is expected from the equilibrium phase diagram, so that the volume fraction of martensite (Vm) in the specimen after water-quenching increases. Mean grain sizes of ferrite (D_F) and martensite (D_M) are also indicated in figure 2, which are fairly fine and in the range from 2 μm to 5 μm.

Figure 3 shows SEM micrographs of the DP microstructures fabricated by heat treatment route II in figure 1. Different from figure 2, DP structures with isolated martensite are observed. The intercritical annealing temperatures shown in figure 3 were selected so as to have the volume fractions of martensite (Vm) similar to those of the DP structures with networked martensite shown in figure 2 (a-c). The volume fraction of martensite (Vm) increases with increasing the intercritical annealing temperature. The mean grain sizes of ferrite (D_F) are several micro-meters, although the mean grain sizes of martensite (i.e., the grain size of prior austenite) are larger than those in the network DP structures.

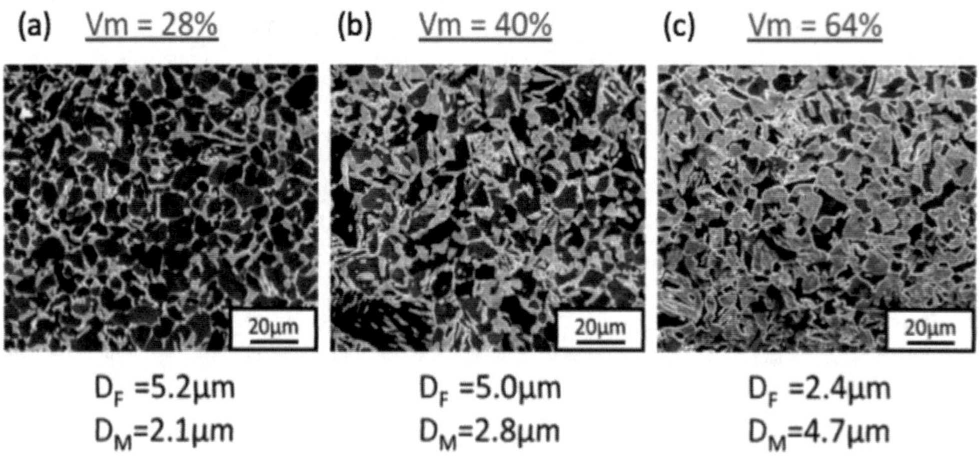

Figure 2. SEM microstructures of the DP specimens having networked martensite morphologies heat-treated using route I. Intercritically annealed at different temperatures: (a)

760°C; (b) 790°C; (c) 820°C. The volume fraction of martensite (Vm), mean grain size of ferrite (D_F) and mean grain size of martensite (D_M) are indicated.

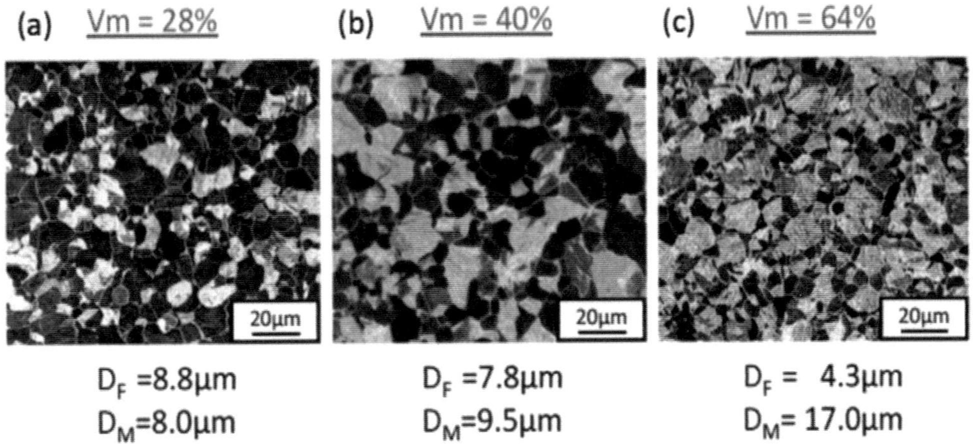

Figure 3. SEM microstructures of the DP specimens having isolated martensite morphologies heat-treated using route II. Intercritically annealed at different temperatures: (a) 740°C; (b) 760°C; (c) 780°C. The volume fraction of martensite (Vm), mean grain size of ferrite (D_F) and mean grain size of martensite (D_M) are indicated.

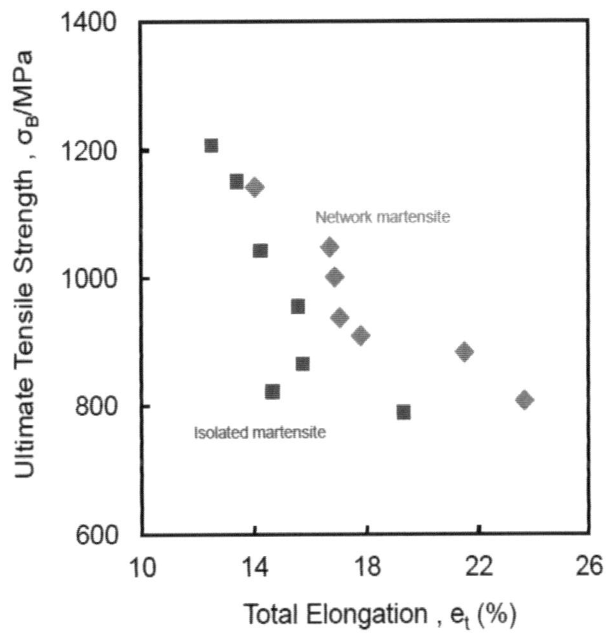

Figure 4. Ultimate tensile strength versus total elongation obtained from tensile tests showing the strength-ductility balance of the DP specimens with different volume fractions of martensite. Values of the DP specimens having networked martensite morphologies are plotted by red diamond symbols, while those of the isolated martensite specimens are plotted by blue square symbols.

Tensile tests of specimens having the two types of DP microstructures shown in figures 2 and 3 were conducted at room temperature. The obtained tensile strength and total elongation are plotted in figure 4. In both types of DP structures (with networked martensite or isolated martensite), the tensile strength decreases with increasing total elongation, which reflects a general trade-off balance between strength and ductility in most materials. However, it is noteworthy that the DP specimens having networked martensite show better strength-ductility balance than the DP structures having isolated martensite. To understand the reasons for the difference between the networked DP and the isolated DP, microscale DIC analysis was carried out for both types of DP structure.

Figure 5 shows SEM micrographs of an identical area in the specimen having the networked DP structure before the tensile test (a) and after deformation to various tensile strains (b-d). The volume fraction of martensite (Vm) in this specimen was 28%. With increasing the tensile strain, the microstructure gradually elongated along the horizontal direction, i.e., the tensile direction. Using those SEM microstructures, DIC analysis was conducted.

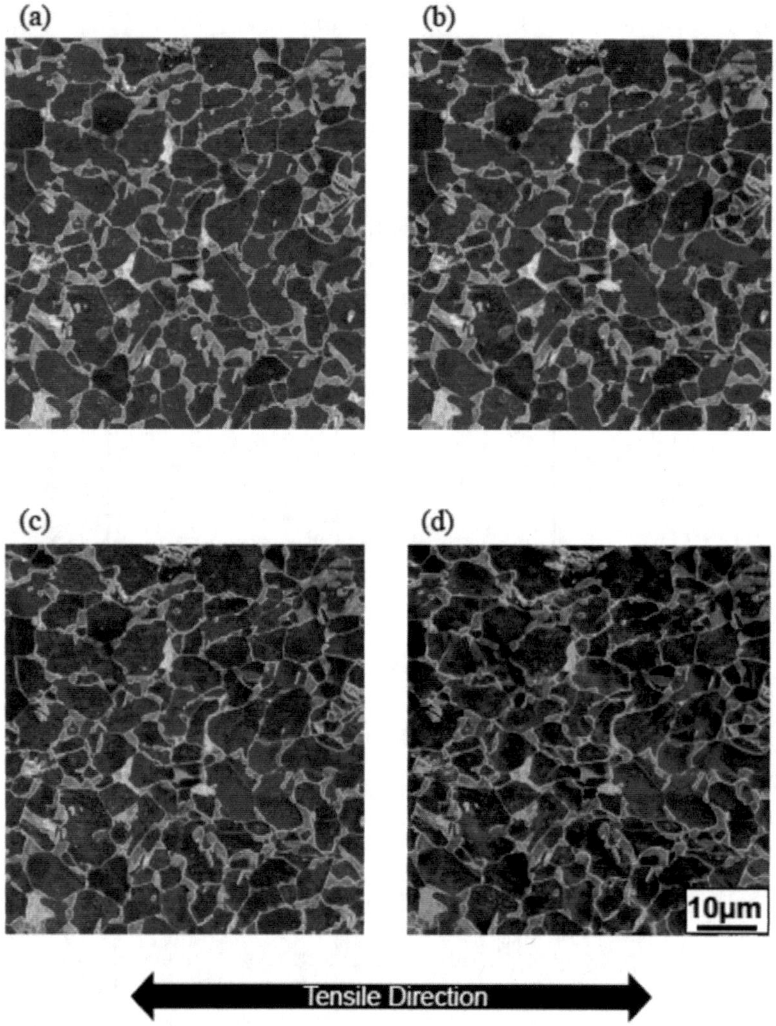

Figure 5. SEM microstructures of an identical area in the DP specimen having a networked martensite morphology. The volume fraction of martensite (Vm) was 28%. The specimen was deformed to various tensile strains: (a) before the tensile test; tensile tested to (b) 2.8%; (c) 6.5%; (d) 10.0% strain.

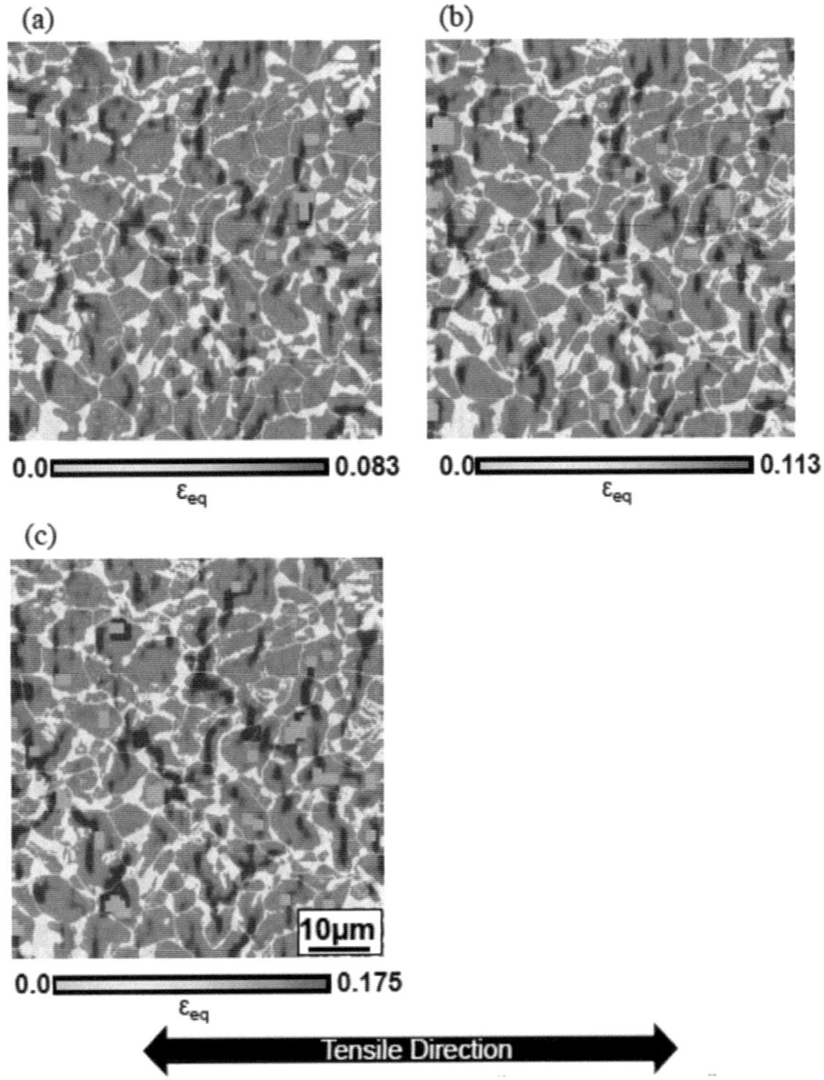

Figure 6. Local strain distribution maps obtained from the DIC analysis of the DP specimen having the networked martensite morphology. The volume fraction of martensite (Vm) was 28%. The specimen was deformed to various tensile strains: (a) tensile tested to 2.8% strain; (c) 6.5%; (d) 10.0%.

Local strain distributions obtained by the DIC analysis in the same area in the networked DP specimen deformed to three different tensile strains shown in figure 5 are plotted in figure 6. Color maps indicating the von Mises equivalent strain (ε_{eq}) at each point in the microstructure are superimposed on the DP micrographs where martensite and ferrite are drawn in white and black, respectively. The amount of the equivalent strain is quantitatively expressed by the colors according to the bars shown below each strain distribution map. As ferrite shows a fairly homogeneous contrast in the SEM images, it was sometimes difficult to track the positions, especially at center regions, of relatively coarse ferrite grains in the DIC analysis. Such uncertain regions are shown in gray in the strain distribution maps. The strain distribution within the DP microstructure was not uniform but

IOP Conf. Series: Materials Science and Engineering **219** (2017) 012008 doi:10.1088/1757-899X/219/1/012008

rather was localized even at small tensile strains. The strain-localized regions, indicated by yellow to red colors are mainly located within soft ferrite grains and tend to propagate across the hard martensite at higher strains. It should be noted that the DIC analysis was reset after each strain, so that the strain distributions in figure 6 (b) and (c) are those within the tensile deformation ranges from 2.8% to 6.5%, and from 6.5% to 10.0%, respectively. Nevertheless, it is interesting that the strain localized regions are observed at the same positions in figure 6 (a), (b) and (c).

Figure 7 shows strain distribution maps in the same area of a specimen having the isolated DP structure. The volume ftaction of martensite (Vm) of this specimen was same as that of figures 5 and 6, 28%. The distributions of von Mises equivalent strain are again not uniform but are localized mainly in the soft ferrite regions. Although the strain localized bands tended to be connected at higher strains, it is clear in figure 7 that the strain localized bands detoured around hard martensite.

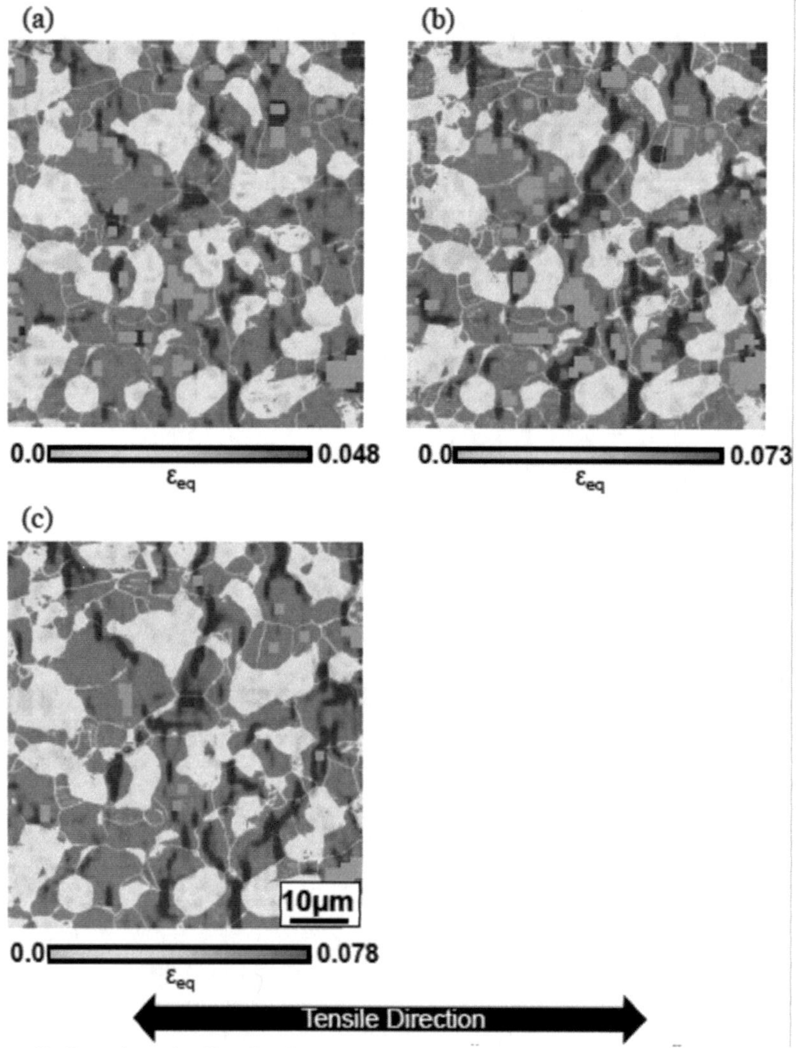

Figure 7. Local strain distribution maps obtained from the DIC analysis of the DP specimen having the isolated martensite morphology. The volume fraction of martensite (Vm) was 28%. The specimen was deformed to various tensile strains: (a) tensile tested to 1.1% strain; (c) 3.7%; (d) 6.4%.

Figure 8 shows frequency distributions of local equivalent strain within each phase in (a) the networked DP structure and (b) the isolated DP structure. The volume fraction of martensite (Vm) was 28% for both types of microstructures. As was shown in figures 6 and 7, local strains in the ferrite are larger than those in martensite in both types of DP structures, because of the strain localization in the softer ferrite. It is noteworthy, however, that the martensite in the networked DP structure has highr strain vales (i.e., is more deformed) than the martensite in the isolated DP structure. Also, the distributions of local strain in both the ferrite and martensite phases in the networked DP structure are broader than those in the isolated DP structure. The results indicate that the degree of strain partitioning between soft ferrite and hard martensite decreases when the hard phase (martensite) has a connected network structures. That is, the deformation becomes more homogeneous in the networked DP structure than in the isolated DP structure, which must contributes to the better strength-ductility balance in the networked DP structure than in the isolated DP structure shown in figure 4.

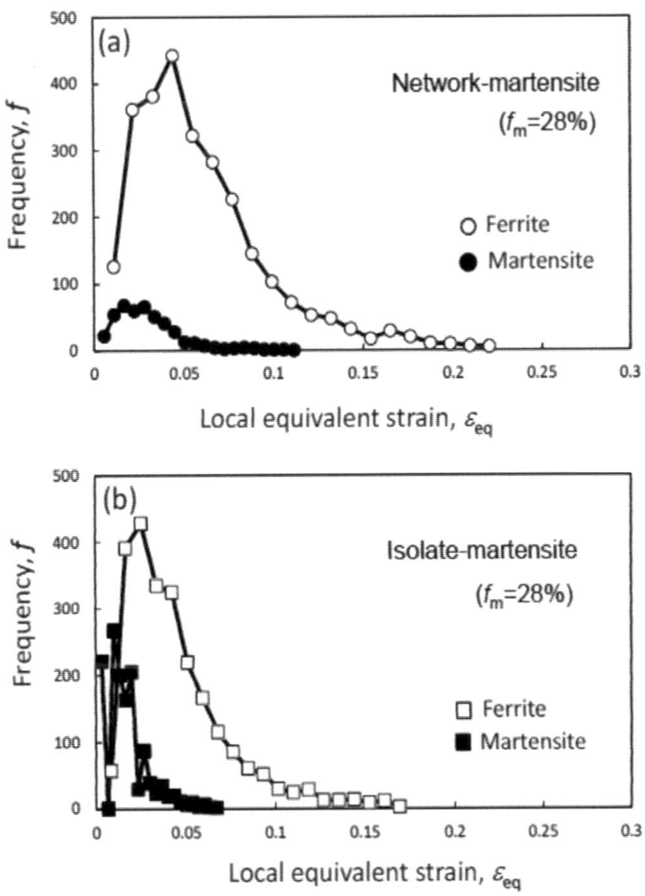

Figure 8. Frequency distribution of local strains in the DP specimens having (a) networked martensite morphology and (b) isolated martensite morphology. The volume fraction of martensite (Fm) was 28% for the both specimens.

4. Conclusion

Two types of dual phase (DP) microstructures, composed of a soft ferrite phase and a hard martensite phase, were fabricated in a low-carbon steel (Fe-0.087C-0.79Si-1.77Mn-0.009P-0.002S-0.02Cr-0.01Mo-0.003N in mass%) using two different types of heat treatment. When the starting microstructure prior to intercritical annealing in the ferrite + austenite two-phase region was ferrite, DP structures having a networked martensite were produced. In contrast, when the starting microstructure prior to intercritical annealing was austenite, DP structures with an isolated martensite morphology were obtained. Samples fabricated with these two types of DP structures, having various volume fractions of martensite, were tensile tested at room temperature. It was noteworthy that the DP structures with networked martensite showed a better strength-ductility balance than the DP structures with an isolated martensite morphology. The local strain distribution within the DP microstructures was quantitatively analyzed by digital image correlation (DIC) analysis during the tensile tests. The strain distributions within the DP microstructures were not homogeneous, but rather localized within soft ferrite grains. The strain-localized regions tended to detour around the hard martensite but eventually propagated across the martensite. It was found that the degree of strain partitioning between ferrite and martensite in the networked DP structure was lower than that in the isolated DP structure. That is, the deformation became more homogeneous when the hard phase (martensite) was connected to make a network structure. It is proposed that this is one of the reasons for the better strength-ductility balance in the networked DP structure compared to that in the isolated DP structure.

Acknowledgments

This work was financially supported by the Elements Strategy Initiative for Structural Materials (ESISM) and the Grant-in-Aid for Scientific Research (S) (No. 15H05767), both through the Ministry of Education, Culture, Sports, Science and Technology (MEXT), Japan. The authors gratefully appreciate the supports.

References

[1] Demeri M Y: *Advanced High-Strength Steels: Science, Technology and Applications*, 2016 ASM International, Ohio.
[2] Tomota Y and Tamura I 1973 *Tetsu-to-Hagané* **59** 96.
[3] Davies R G 1973 *Metal. Trans A* **9A** 671.
[4] Mondal D K and Dey R M 1992 *Mater. Sci. Eng.* **A149** 173.
[5] Sarwar M and Priestner R 1996 *J. Mater. Sci.* **31** 2091.
[6] Das D and Chattopadhyay P P 2009 *J. Mater. Sci.* **44** 2957.
[7] Hasegawa K, Toji Y and Minami H 2012 *Tetsu-to-Hagané* **98** 320.
[8] Chu T C, Ranson R F, Sutton M A and Peters W H 1985 *Experimental Mechanics* **25** 232.

Opportunities for aluminum-based nanocomposites

H Weiland
Arconic Inc., Arconic Technology Center, New Kensington PA15069, USA

hasso.weiland@arconic.com

Abstract. High performance aluminum alloys are conventionally made by heat treating alloys containing a variety of alloying elements in solid solution. Key performance attributes are controlled at the microstructural level by tailoring sizes and morphology of nano-sized second phases. This enabled the successful development of aluminum alloys having properties optimized in strength, damage tolerance and corrosion resistance. However, this process is naturally limited by the solubility of alloying elements in the aluminum matrix. In real world products, significant effort is deployed to achieve a homogeneous distribution of the alloying elements both at the macro and micro scales. Despite these efforts, heat treatable alloys can exhibit chemical gradients at grain boundaries, resulting in sub-optimized properties. Additionally, due to the very nature of the strengthening mechanisms, the properties of heat-treatable alloys are decreasing when exposed to elevated temperatures. To step outside the boundaries given by the solubility of alloying elements in the aluminum matrix, the extrinsic addition of nano-sized particles to the aluminum matrix is being evaluated.

1. Introduction

Improving the properties of aluminum alloys has a long history. A combination of the established metallurgical mechanisms, cold work, solid solution strengthening and precipitation strengthening, has enabled highly engineered materials to be developed for demanding customer requirements. These requirements typically focus on strength and damage tolerance, while at the same time being of low cost enabling wide spread applications. Many text books cover the underlying mechanisms and how they can be utilized to tailor a materials performance to an application.

For a material to be successful commercially, it needs to fulfill a spectrum of property requirements as well as meet a target cost point. Results of new materials or processes are frequently being reported as achievements in tensile strength; however, a successful material for most technical applications must meet not only strength but other properties such as damage tolerance [1] and/or formability. To date, aluminum alloys used for aerospace structures have driven most of the major advancements in material properties. With improvements to manufacturing process and cost structure, these alloys are now finding additional uses in other applications such as transportation. It is not the objective of this paper to review the state of advanced aluminum alloys.

Over the years, attempts have been made to step outside of the spectrum enabled by precipitation strengthening and other basic metallurgical methods by using a composite approach. This means adding an extrinsic material in small or large volume fractions to the aluminum matrix to not only improve strength but to also modify the modulus. This material class is commonly referred to as Metal Matrix Composites (MMCs) [2]. When MMCs were developed, most second phase materials that were combined with the aluminum matrix where particles with sizes of tens of microns in diameter. With today's availability of low cost inorganic particles with dimensions of less than one hundred nanometers in diameter, the development of new aluminum composites with enhanced properties can be envisioned. There is a need for composite materials which can be manufactured at low cost, are mass producible and achieve property combinations outside of the product portfolio of current aluminum alloys.

2. Motivation

A new class of materials are composites made of aluminum with added nanoparticles (Aluminum Nano Composites, AlNCs). Here non-metallic, thermodynamically stable oxide, carbide or nitride particles of submicron sizes are used to alter the properties of aluminum alloys. Related to this class are AlNCs based on carbon based nanoparticles, e.g. buckeyballs, carbon nanotubes (CNTs) or graphene platelets. The anticipated advantages over conventional aluminum alloys are:

1. The volume fraction of second phase particles is not controlled by the solubility of elements in the aluminum matrix. Thus, materials with very high room temperature strength could be developed.
2. Such particles do not change their geometric dimensions when exposed to temperatures below the melting point of aluminum. Therefore, a material strengthened by nanoparticles would keep a significant portion of its mechanical properties at operating temperatures. Current aluminum alloys lose their properties either due to recovery of the cold deformed structure or due to coarsening and dissolution of strengthening phases.
3. In precipitation strengthened alloys, solute from the matrix is used to form the strengthening phases in the matrix. However, at defects such as grain boundaries or in some systems around constituent phases, a depletion of solute occurs due to the formation of (near-)equilibrium phases, also known as Precipitation Free Zones (PFZs). The solute gradient and the (near-) equilibrium phases increase the susceptibility to corrosion as well as affecting crack nucleation and crack growth under monotonic or cyclic loading, with or without environmental influence. In a system where the strengthening phases are extrinsically added to the aluminum matrix (AlNC), it is expected that corrosion resistance and damage tolerance are improved.
4. The dissolution of precipitates in a heat affected zone from a welding operation will be eliminated, promising weldable, high strength aluminum materials.
5. Improving strength without solute additions to the aluminum matrix would allow for materials of both high strength and high electrical conductivity. Extrinsically added strengthening phases would not inherently reduce the conductivity of the aluminum matrix, making high strength, high conductive electrical connectors possible. The electrical conductivity would be reduced by the volume fraction of aluminum replaced with the second phases.
6. By eliminating the need to homogenize solute elements in the aluminum matrix with high temperature heat treatments or re-solutionizing solute after hot working operations, a lower cost production flow path is anticipated for aluminum nano-composites.

3. Target strength properties

Two mechanisms can affect the strength of a material when nanoparticles are added to the matrix. As such particles are incoherent and cannot be cut by dislocations, Orowan strengthening will occur if the particles are nearly homogenously distributed in the metal matrix. If the nanoparticles accumulate at dendrite or grain boundaries strengthening from a refined grain size according per the Hall-Petch relation will occur.

An approach for estimating yield strength for homogeneously dispersed second phases has been described in [3]. However, it is expected that the distribution in a commercial product is not homogeneous; therefore, a modification for random distribution of obstacles [4] was considered, resulting in an about 15% reduction in predicted yield strength.

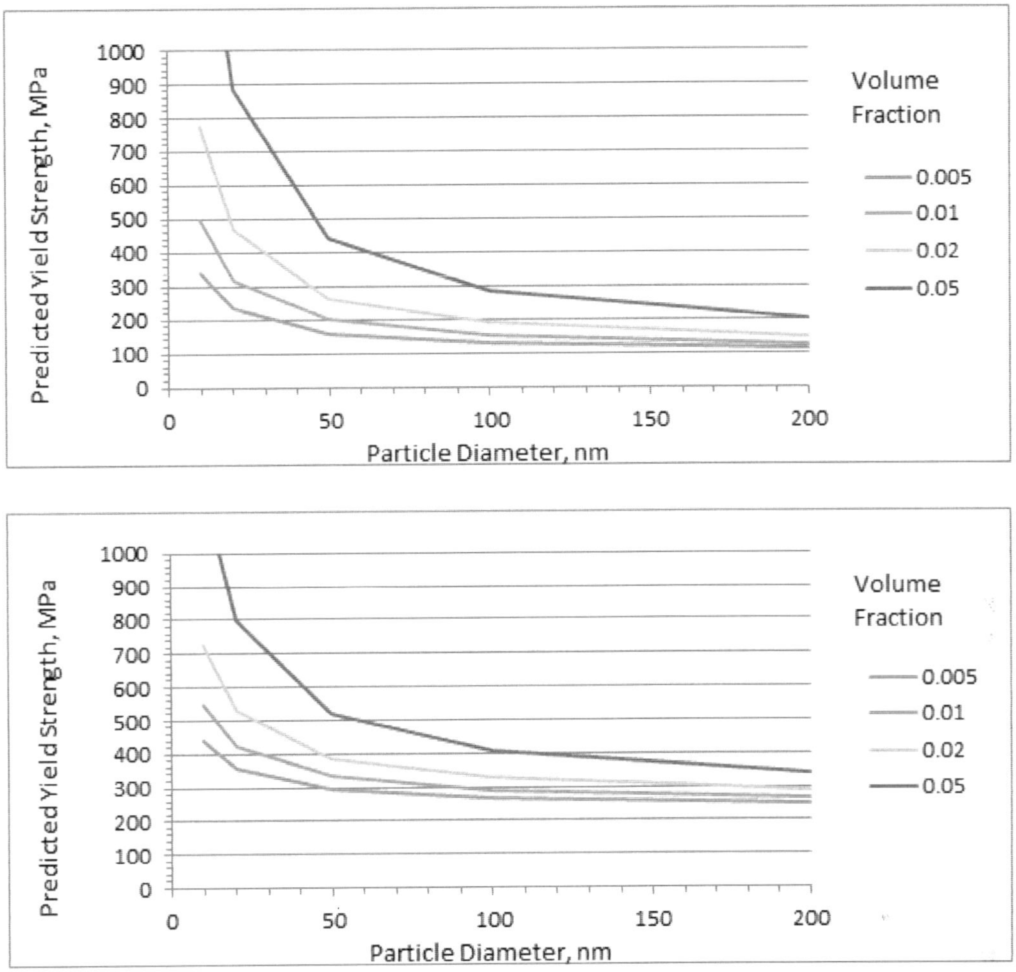

Figure 1. Predicted yield strength for particles 10 nm and larger in diameter and varies volume fractions of particles a) 1100-O b) 5083-H34

Assuming an aluminum product of commercial purity, also referred to as AA1100, and γ-Al2O3 as the nanoparticle phase, the resulting yield strength for a fully annealed material was estimated as a function of volume fraction and particle size (figure 1). Yield strength predictions for a strain hardening alloy, AA5083, with the same class of nano particles is also shown. For very large particles and at low volume fractions, such as constituent phases, the effect on the yield strength is negligible. However, for particles in the less-than 50 nm range, a significant effect on the yield strength can be expected, even at volume fractions in the 0.5 vol% range. If one assumes a strain-hardening material such as AA5083 for matrix, the simulations of the strain hardened material suggest that with 2 to 5 vol% of a second phase in the less than 50 nm range, a yield strength comparable to or even higher than current advanced aluminum alloys could be achieved.

The simulation indicates very large enhancements are possible for room temperature yield strength. With such potential for improvement, one needs to evaluate what potential fabrication paths could enable aluminum nanocomposites.

4. Processing options for making AlNC

Many processing routes can be considered to embed nanoparticles into an aluminum matrix, with several having been reported on [5-7]. For varies practical reasons, the use of CNTs and engineered nanoparticles (NPs) is separated. For one, CNTs with their high aspect ratio will have a different strength response than the nearly spherical particles, which are the bases of the strength simulations above. Additionally, CNTs will provide a different challenge than engineered nanoparticles in establishing a production method. This is due to the implications from the graphite character of its surface on interface bonding, to be discussed later.

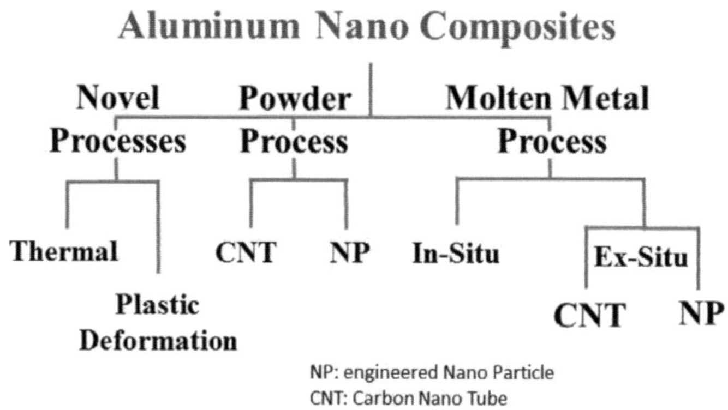

Figure 2. Processing Path for obtaining Aluminum Nanocomposites (AlNCs)

4.1. Powder Metallurgy

The most common method for making aluminum based nanocomposites is based on well-established powder metallurgical processes used for making MMCs. Many different approaches have been reported for mixing and milling, even sometimes under cryogenic temperatures [8], of an aluminum powder with nanoparticle or CNTs. Due to the high reactivity of aluminum powder, the metal particle size is usually in the tens of micrometers in diameter range. Adding nanoparticles in the sub-50 nm range provides a significant processing challenge to obtain a homogeneous powder distribution due to the fact that the second phase particles are several orders of magnitude smaller than the matrix powder itself. Nanoparticles will accumulate during mixing of both powders in voids between aluminum particles, causing an inhomogeneous distribution of the second phase in the matrix.

4.2. Ingot Metallurgy

While the emphasis of this discussion is on the addition of extrinsic nanoparticles, it is necessary to mention also an intrinsic process path, which makes use of the in-situ formation of small particles in the melt [9]. Utilizing the fact that the solubility of an alloying element is higher in a melt than in a solid, it has been demonstrated that a salt-based liquid reaction of a salt containing carbon particles with alloying elements in the molten metal can achieve large volume fractions of a second phase in the melt, resulting in a dispersion strengthened product. However, as the size of these second phase particles is controlled by the high reaction kinetics, it is difficult to consistently control the size of the particles formed in the melt to below one micrometer.

Adding particles directly to molten metal is an established technology [5, 6, 10]. Various technical approaches enable fabrication of an aluminum matrix containing extrinsically added second phase particles. The issue of segregation due to gravity as encountered in MMC fabrication is not anticipated in the case of nano-sized particles. However, other known challenges, both metallurgical and economical, are likely to occur. It is thus necessary to develop effective strategies to investigate effects and roles of NP's during both liquid-solid stage and post-solidification processing.

Consideration for the aluminum matrix is as follows: in an age-hardening aluminum system, it is very likely that the precipitation kinetics will be affected by the presence of nanoparticles. Nanoparticles will both act as a sink for vacancies as well as nucleation sites for strengthening phases, likely altering the precipitation kinetics and morphology. Thus, using a strain-hardening aluminum matrix promises better results.

One needs to anticipate many hurdles when considering a molten-metal route:

Most engineered particles are known to have a high wetting angle with regards to molten aluminum. However, wetting at the nanoscale is not fully understood. Engineered nanoparticles have a different surface layer in chemistry and crystallography than at the particle core. These surface layers could be oxides or carbon-contaminations, depending on the process used to fabricate nanoparticles. Additionally, processes for determining and controlling the largest particles in the distribution, those exceeding a few micrometers and thus could be crack nucleation sites, need to be developed.

After having overcome the issues of wetting, the next hurdle will be obtaining a homogeneous dispersion of nano-particles in the aluminum matrix. The properties of the final product will depend on those material volumes that shows either agglomerations or absence of nanoparticles. The dispersion is correlated to the wetting. Feed processes need to break-up any agglomerates of the feedstock and need to avoid agglomeration in the melt. While the latter are mechanical hurdles, the dispersion of nanoparticles in the solidified aluminum can also be affected by the solidification process: it is very much possible that the solidification front pushes the nanoparticles ahead of itself, resulting in an agglomeration at interdendritic spaces in the final material.

One needs also to consider the effects on processing such materials on existing equipment. As the target properties are primarily significant increases in strength, at room and elevated temperatures, processing equipment such as rolling mills and extrusion presses would experience higher loads. Consequently, continuous casting processes such as roll casting or similar, would be the preferred manufacturing methods.

Finally, one needs to ensure that all processing steps are compliant with environmental and health regulations. When full wetting can be achieved, the final product does not differ from a precipitation strengthened aluminum material and thus no engineered nanoparticle will be released during forming and/or machining operations. However, handling engineered nanoparticles from receiving, quality control to process additions requires new methods to ensure that such particles are not released into the environment.

5. Processing

Melt stirring is a common process. The use of ultrasonic agitation in assisting wetting and dispersion during melt stirring has been widely reported [e.g. 10]. However, in view of the need to adapt any process at the commercial scale, emphasis was taken on melt stirring without ultrasonic agitation. The material classes from which the nano particles were selected, were based on reported wetting characteristics (e.g. [11], that is boron carbide (B4C) and silicon carbide (SiC)). However, wetting at the nano scale could potentially behave differently than presently reported for micron-sized ceramic particles. Commonly available alumina and carbon nanotubes were evaluated, specifically the following particle classes of the respective, nominal particle diameter were used in this evaluation: B4C (50 nm), γ-Al2O3 (10 nm), SiC (60 nm) and nitrogen-doped SWCNTs (200 nm length).

The aluminum matrix was smelter grade aluminum of 99.5% purity, sometimes alloyed with magnesium. Magnesium extends the freezing range compared to unalloyed aluminum. The solidification front in an alloy with a longer freezing range provides more time for a reaction between the particle and the melt. Magnesium also is known to affect wetting of ceramic materials, but this alloying element was selected as it enhances work hardening. Additionally, halide salts, known for the capability to remove surface oxides and to increase wetting [12], were evaluated as wetting aids for the nanoparticles.

Varies methods were developed for safely adding nano-particles to the molten metal. Nano particles need to be delivered under the surface of the melt. Thus, nanoparticles and, when used, halide salts were encapsulated in aluminum containers which then were submerged in the melt. Upon melting of the container, the nanoparticles were released in to the stirring melt. Stirring needed to be slow enough not to create a vertex as the latter would pull in oxides from the surface, even when working under a protecting inert atmosphere.

The melt was cast into rectangular copper molds, allowing for solidification rates corresponding to those typically found for Direct Chill (DC) ingot casting, that is in the 30 K/s range. All ingots were extruded and heat treated before samples for metallurgical and mechanical evaluation were cut.

6. Results and Discussion

6.1. AlNCs based on Engineered Nanoparticles

Overall it was found that all the nanoparticles in evaluation were embedded in the solidified aluminum, especially when a halide salt were used as a wetting aid [13]. As expected the success rate was higher in the Mg-containing alloys. Figure 3 shows γ-alumina particles with a nominal particle diameter of 10 nm embedded in an Al-1.0 wt% Mg sample.

Figures 3a and 3b are the same area containing a grain boundary, taken under two different imaging conditions in the TEM, in Fig 3a both grains are in non-diffracting condition, while in Fig 3b, the grain on the right side is in strongly diffracting condition, indicated by the trace of the grain boundary. It is clear from these two images, that the nanoparticles are not pushed ahead of the moving solidification front and thus do not agglomerate at the interphase. Also, stirring is sufficient to have in this field of view a sufficiently good dispersion of the nanoparticles in the aluminum matrix. However, a similar sample with γ-alumina of nominal 60 nm in diameter observed with a larger field of view by SEM (figure 4), shows a similar dispersion in the aluminum matrix as observed in the TEM. Near a constituent phase particle (an iron-bearing phase), a higher density of nanoparticles can be detected. This indicates that during solidification not all nanoparticles are embedded in the matrix but that some accumulate in the remaining liquid at the end of the freezing range, resulting in an inhomogeneous distribution of the nanoparticles.

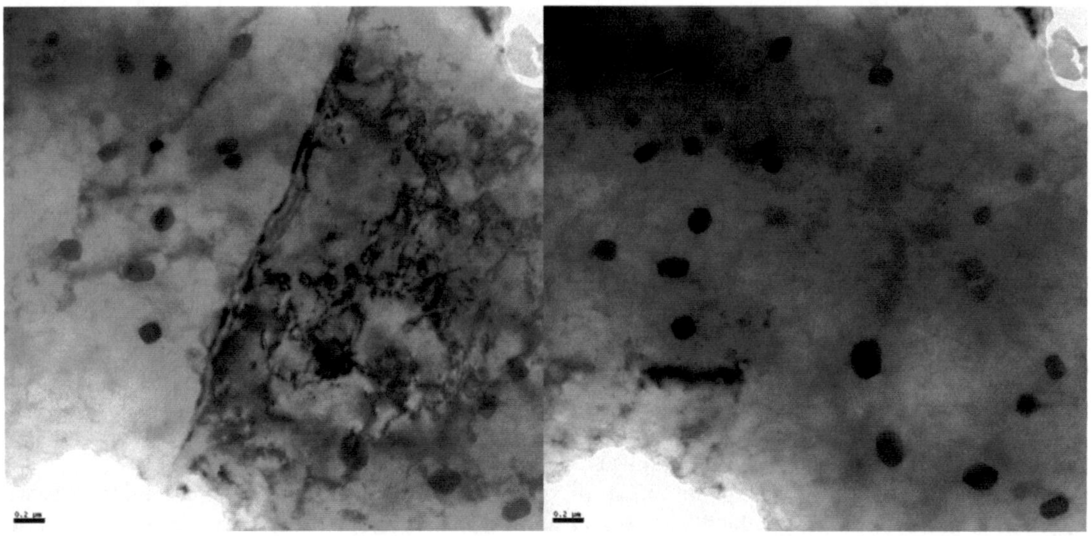

Figure 3. TEM micrograph of Al-1%Mg +1Vol% γ-alumina (10 nm) same area, two different orientations of TEM foil normal to electron beam to show position of grain boundary.

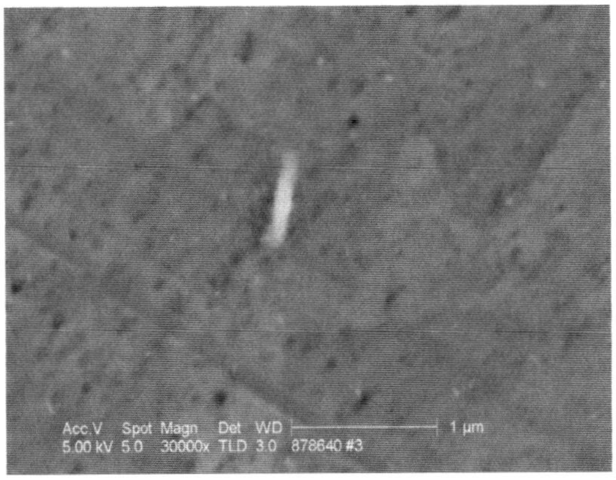

Figure 4. SEM micrograph of Al-1%Mg +1Vol% γ-alumina (60 nm) alumina in matrix and surrounding constituent particle.

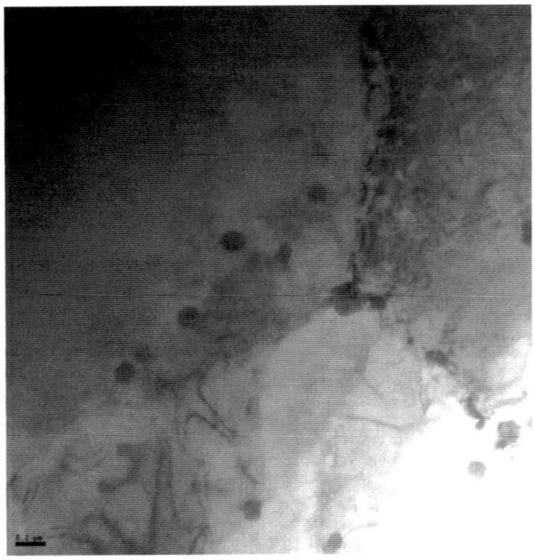

Figure 5. TEM micrograph of Al-1%Mg +1Vol% SiC (60 nm)

Similar results as with the γ-alumina were obtained for SiC particle addition having a nominal particle diameter of 60 nm (Fig 5). Also in this case the nanoparticles are not preferentially located at a grain boundary. A few particles can be observed at the triple junction and a grain boundary.

Figure 6. Aluminum matrix containing a carbon nanotube embedded by molten metal processing

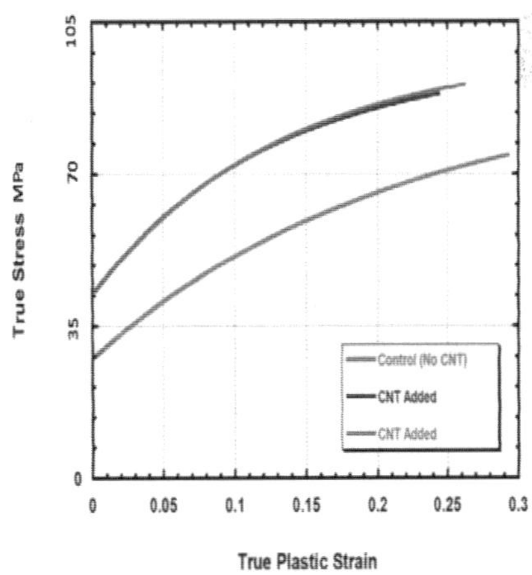

Figure 7. True stain – true stress curves for an AlNC with 0.1 wt% embedded CNTs

6.2. AlNCs based on CNTs

The graphite character of the surfaces of CNTs does not allow a CNT to physically bond with aluminum. On the other hand, if the whole surface would react with aluminum, aluminum carbide would form, detrimentally affecting corrosion and other properties. Thus, defect containing (nitrogen-doped) CNTs were selected for molten metal additions, with the particular CNTs having a bamboo structure. The concept was that aluminum can bond to the CNT at the defect site. Also here, salts were used as a fluxing material to enable embedding of CNTs to an aluminum matrix of smelter grade pure (99.5%). Assuming that no CNTs were lost during processing, the sample contains 0.1 wt% CNTs.

TEM analysis (figure 6) verified that CNTs were embedded in the aluminum matrix. As the sample had been plastically deformed, dislocations interacting with the CNT can be observed. Specifically note the dislocation loops in the center of the image around the CNT. It is important to note that at the given resolution, aluminum carbides at the interface between the aluminum matrix and the CNT were not detected.

Tensile testing (figure 7) shows an improvement of about 30% when compared to a control. The control has been processed the same way as the AlNC, incl. applying salt additions, however without the addition of CNTs.

7. Summary

With all the success in embedding nanoparticles in the solidified aluminum matrix without agglomeration, where are the issues? While large material volumes showed a distribution of nanoparticles similar to the TEM images shown above, some material volumes were found to not contain any nanoparticles. When nanoparticles were observed in the matrix, they were reasonably well dispersed. However, all the target properties discussed above require a near-homogeneous dispersion throughout the whole material volume. The objective of this approach was to achieve high strength aluminum materials similar to precipitation hardening materials, without the high solute content. Any material volume not achieving high strength due to the absence of strengthening nanoparticles will determine the macroscopic plastic behavior negatively. In the current work, the material behaved like a two-phase material, with the soft phase defining the plastic behavior. While initial results are promising, developing AlNCs with commercially relevant properties via a pure molten metal processing path remains a challenge.

References

[1] Denzer D K et al. *Proc 13th Intl Conf on Aluminum Alloys*, The Minerals, Metals, & Materials Society. Editors: Weiland H, Rollett AD, Cassada WA pp 586-592
[2] Rawal S 2001 *JOM,* **53**, 14-7
[3] Zhang Z and Chen D L 2006 *Scripta Materialia* **54** 1321–6
[4] Kocks U F 1966 *Phil. Mag.* **13** 541
[5] Borgonovo C and Apelian D 2011 *Materials Science Foru* **678** 1-22
[6] *Wikipedia*: https://en.m.wikipedia.org/wiki/Carbon_nanotube_metal_matrix_composites
[7] Bakshi, S R, Lahiri S and Agarwal A 2010 *International Materials Review* **55** 41-64
[8] Adams H and Dvorak M *US patent application* 20100189995 A1, July 29, 2010
[9] Chu M G *US patent* 6,398,882 B1, June 4, 2002
[10] Li X, Yang Y and Cheng X 2004 *Journal of Materials Science* **39** 3211-2
[11] Hashim J et al. 2001 *J of Mat Processing Technology* **119** 324-8
[12] Hanink D K, et al. US *Patent 2809423 A*, Oct 15, 1957
[13] Weiland H and Chu M G *US Patent* 9415440 B2, Aug 16, 2016

New mesoscopic constitutive model for deformation of pearlitic steels up to moderate strains

J. Alkorta[1,*], J.M. Martínez-Esnaola[1], P. de Jaeger[2] and J. Gil Sevillano[1]

[1]CEIT and TECNUN (University of Navarra), Pº Manuel de Lardizabal 15, 20018, Donostia-San Sebastian (Spain)
[2]NV Bekaert SA, Bekaertstraat 1, 8500 Zwevegem, Belgium

*jalkorta@ceit.es

Abstract. A new constitutive model for deformation of pearlitic steels has been developed that describes the mechanical behaviour and microstructural evolution of lamellar multi-colony pearlite. The model, a two-phase continuum model, considers the plastic anisotropy of ferrite derived from its lamellar structure but ignores any anisotropy associated with cementite and does not consider the crystal structure of either constituent. The resulting plastic constitutive equation takes into account a dependence on both the pearlitic spacing (arising from the confined slip of dislocations in the lamellae) and on strengthening from the evolving intra-lamellar dislocation density. A Kocks-Mecking strain hardening/recovery model is used for the lamellar ferrite, whereas perfect-plastic behaviour is assumed for cementite. The model naturally captures the microstructural evolution and the internal micro-stresses developed due to the different mechanical behaviour of both phases. The model is also able to describe the lamellar evolution (orientation and interlamellar spacing) with good accuracy. The role of plastic anisotropy in the ferritic phase has also been studied, and the results show that anisotropy has an important impact on both microstructural evolution and strengthening of heavily drawn wires.

1. Introduction

The eutectoid Fe-C pearlite is an aggregate of composite colonies constituted of bcc ferrite lamellae alternating with lamellae of orthorhombic cementite (Fe_3C iron carbide). Cementite occupies approx. 12% of the volume. The lamellae are approximately flat and continuous, although the real geometry always contains growth defects (lamellar discontinuities, and deviations from planarity). The as-transformed pearlitic spacing (λ) depends on the transformation temperature (on the undercooling with respect to the equilibrium eutectoid temperature), and in general falls in the range 50 nm $< \lambda <$ 600 nm. The colony size is often 10 or 20 times the pearlitic spacing or even more.

Each as-transformed colony is an intertwined two-phase bicrystal. It is well known that the strength of pearlitic steels is related to the interlamellar spacing through a Hall-Petch type empirical relationship. This is explained by the critical bowing of dislocation lines necessary to achieve long-range slip on its slip plane confined between two adjacent cementite lamellae [1-6].

The anisotropic behaviour of pearlite stems from both crystallographic and geometric origins. The two-phase continuum model proposed in this work deals strictly with the latter, assuming that plastic anisotropy of the ferrite arises only from its lamellar structure. The model is intended for practical use in plastic forming simulations of pearlitic steel. Its good predictive capabilities, even when strain-path

IOP Conf. Series: Materials Science and Engineering **219** (2017) 012010 doi:10.1088/1757-899X/219/1/012010

changes are assumed, indicate that in pearlite anisotropy from a geometrical origin dominates over that related to crystallographic texture.

2. Model

The model combines a Hill-based material description accounting for the plastic anisotropy of ferrite due to the confined slip of dislocations in the lamellae and an isotropic elastic- perfectly plastic von Mises plasticity model for cementite. The model assumes perfect adhesion between cementite and ferrite lamellae. The model also assumes identical isotropic elastic properties for both phases.

2.1. *Model for ferritic lamellae*

In pearlitic ferrite, the plastic deformation is controlled by dislocation glide confined between the two neighbouring cementite lamellae. The stress τ_{rss} allowing long-range slip is assumed to be composed of additive contributions from any kind of friction-like resistance (i.e. the true lattice friction stress and a contribution from solution strengthening), τ_0, as well from as the bowing of dislocation lines between the cementite walls or ferritic grain boundaries, $\Delta\tau_{bow}(s_{eff})$, where s_{eff} is the effective interlamellar spacing which, in principle, depends on the orientation of the applied stress, and from the resistance offered by the presence of a strain-dependent dislocation density, $\Delta\tau_\rho$, i.e.,

$$\tau_{rss} = \tau_0 + \Delta\tau_{bow}(s_{eff}) + \Delta\tau_\rho \tag{1}$$

The model accounts for this yield anisotropy with a simple Hill-model. The anisotropic yield locus proposed by Hill in 1948 [7] is the simplest extension of the von Mises yield criterion, i.e., a quadratic expression that can be written as

$$f = \sigma_e^2 - \sigma_0^2 = \tfrac{1}{2}\sigma^T P_H \sigma - \sigma_0^2 \tag{2}$$

where,

$$P_H = \begin{pmatrix} \alpha_{12}+\alpha_{31} & -\alpha_{12} & -\alpha_{31} & 0 & 0 & 0 \\ -\alpha_{12} & \alpha_{12}+\alpha_{23} & -\alpha_{23} & 0 & 0 & 0 \\ -\alpha_{31} & -\alpha_{23} & \alpha_{23}+\alpha_{31} & 0 & 0 & 0 \\ 0 & 0 & 0 & 6\alpha_{44} & 0 & 0 \\ 0 & 0 & 0 & 0 & 6\alpha_{55} & 0 \\ 0 & 0 & 0 & 0 & 0 & 6\alpha_{66} \end{pmatrix} \tag{3}$$

and

$$\begin{aligned} \alpha_{12} &= \left(\frac{\sigma_0}{\bar{\sigma}_{11}}\right)^2 + \left(\frac{\sigma_0}{\bar{\sigma}_{22}}\right)^2 - \left(\frac{\sigma_0}{\bar{\sigma}_{33}}\right)^2; \quad 3\alpha_{44} = \left(\frac{\sigma_0}{\bar{\sigma}_{12}}\right)^2 \\ \alpha_{23} &= \left(\frac{\sigma_0}{\bar{\sigma}_{22}}\right)^2 + \left(\frac{\sigma_0}{\bar{\sigma}_{33}}\right)^2 - \left(\frac{\sigma_0}{\bar{\sigma}_{11}}\right)^2; \quad 3\alpha_{55} = \left(\frac{\sigma_0}{\bar{\sigma}_{23}}\right)^2 \\ \alpha_{31} &= \left(\frac{\sigma_0}{\bar{\sigma}_{33}}\right)^2 + \left(\frac{\sigma_0}{\bar{\sigma}_{11}}\right)^2 - \left(\frac{\sigma_0}{\bar{\sigma}_{22}}\right)^2; \quad 3\alpha_{66} = \left(\frac{\sigma_0}{\bar{\sigma}_{31}}\right)^2 \end{aligned} \tag{4}$$

where $\bar{\sigma}_{ij}$ are the uniaxial/shear yield stresses. Particularized for the so-called "normal anisotropy" (transversely isotropic materials), i.e., planar isotropy in lamellar plane, e.g., the (1, 2) plane, $\bar{\sigma}_{11} = \bar{\sigma}_{22} = \sqrt{3}\bar{\sigma}_{12}$ and $\bar{\sigma}_{31} = \bar{\sigma}_{23}$. Thus, equation (2) reduces to

$$f = \sigma_e^2 - \overline{\sigma}_{11}^2 = \tfrac{1}{2}\sigma^T P_H \sigma - \overline{\sigma}_{11}^2 = 0 \tag{5}$$

with

$$P_H = \begin{pmatrix} 2 & N-2 & -N & 0 & 0 & 0 \\ N-2 & 2 & -N & 0 & 0 & 0 \\ -N & -N & 2N & 0 & 0 & 0 \\ 0 & 0 & 0 & 6 & 0 & 0 \\ 0 & 0 & 0 & 0 & 6M & 0 \\ 0 & 0 & 0 & 0 & 0 & 6M \end{pmatrix} \tag{6}$$

where $M = \overline{\sigma}_{12}^2 / \overline{\sigma}_{23}^2$ and $N = \overline{\sigma}_{11}^2 / \overline{\sigma}_{33}^2$. It is easy to see that as the active gliding dislocations under $\overline{\sigma}_{12}$ shear stresses are confined between two cementite lamellae, then a minimum $s_{eff} = s$ (s being the interlamellar spacing) should be considered in equation (1). Thus, we can anticipate that $M = \overline{\sigma}_{12}^2 / \overline{\sigma}_{23}^2 \geq 1$; a similar argument leads to $N = \overline{\sigma}_{11}^2 / \overline{\sigma}_{33}^2 \geq 1$. Implicitly, and compared with a crystal plasticity approach, the model assumes that slip may occur along any crystal direction and plane, and that the critical shear stress is minimum along lamellar planes ($\overline{\sigma}_{31} = \overline{\sigma}_{23} \leq \overline{\sigma}_{12}$). Particularly, when $M = N = 1$, pearlitic ferrite is fully isotropic (von Mises behaviour).

2.2. *Model for cementite lamella*

Cementite is considered to be mechanically isotropic with elastic- perfectly plastic behaviour (Y = 4500 MPa).

2.3. *Interphase boundaries*

Perfect adhesion is considered between cementite and ferrite. Thus, equivalent in-plane displacements and equivalent out-of-plane stresses are assumed on both sides of the boundary.

2.4. *Microstructural evolution*

The model assumes that plastic flow directly determines the orientation and spacing of the lamellae. Thus, the normal orientation of each lamella is given by

$$\hat{n} = \frac{F^{-T}\hat{n}_0}{\left\| F^{-T}\hat{n}_0 \right\|} \tag{7}$$

where $\hat{n}_0$ is the initial orientation of the lamella normal and F is the deformation gradient. Likewise, the interlamellar spacing is calculated as follows:

$$s = \frac{s_0}{\left\| F^{-T}\hat{n} \right\|} \tag{8}$$

where s_0 is the initial lamellar spacing. The model was implemented in a VUMAT subroutine for ABAQUS®/Explicit.

3. Uniform elongation with periodic boundary conditions

In order to check the capabilities of the proposed model and to verify the relevance of the anisotropic model for pearlitic ferrite, the same uniform elongation with periodic boundary conditions was simulated for a fully isotropic ferritic model ($M = N = 1$) and for an anisotropic ferritic model ($M, N > 1$) considering that gliding dislocations for $\overline{\sigma}_{23}$-type shear stresses are only confined by the colony size, i.e., $s_{eff} = c = 10$ μm in equation (1). Assuming a homogeneous initial inter-lamellar spacing of $s = 60$ nm, an initially cubic mesh with 25x25x25 elements was constructed with a Voronoi tessellation to mimic an equiaxed cluster of approximately 100 randomly oriented pearlite colonies (see figure 1).

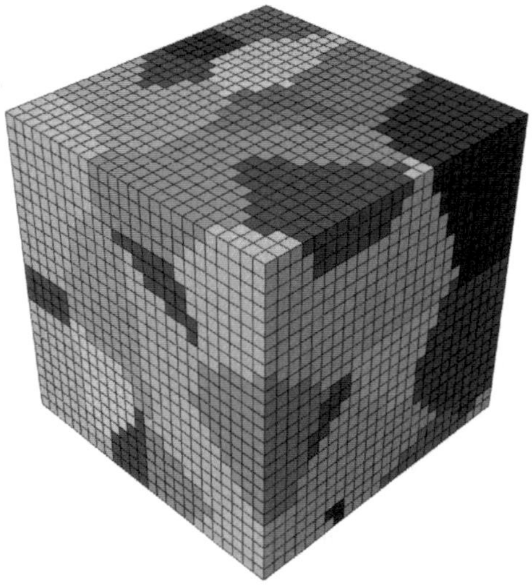

Figure 1. Coloured representation of the initial colony distribution in the designed mesh.

The virtual microstructure was uniaxially elongated with periodic boundary conditions up to an equivalent strain of about 4. Figure 2 shows a top view perpendicular to the loading direction (direction 3) after uniaxial elongation for the isotropic ferritic model (centre) and the anisotropic ferritic model (right). The results show that the deformation is quite homogeneous in the isotropic model and that colonies maintain a near self-similar shape even after severe elongation. In contrast, the anisotropic model leads to a heavily distorted mesh with colonies stretched in a way that reproduces the typical "van Gogh skies structure" observed in heavily drawn bcc wires [8, 9].

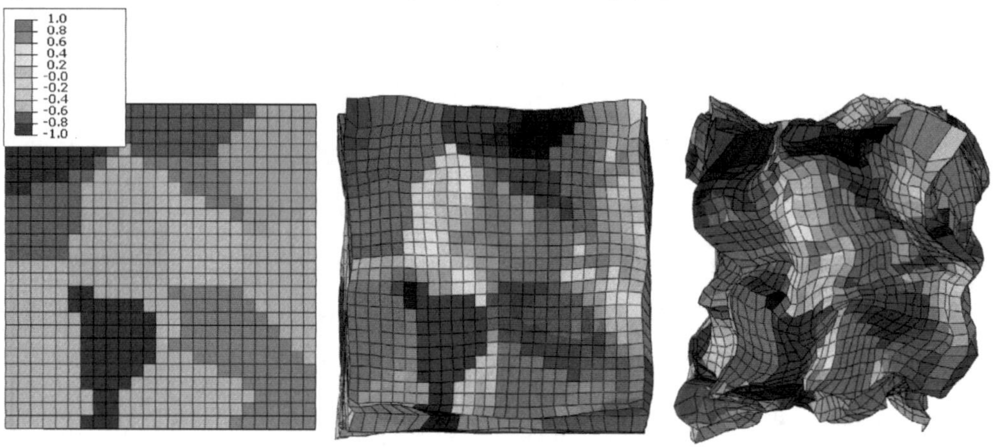

Figure 2. Top view perpendicular to the loading direction of the obtained mesh after a true strain about 4 (macroscopically axisymmetric elongation). Left: Initial mesh re-scaled; Centre: Mesh after maximum elongation for the isotropic ferritic model; Right: Idem for anisotropic ferritic model. Colour code represents the Y direction cosine of the normal to the lamellae.

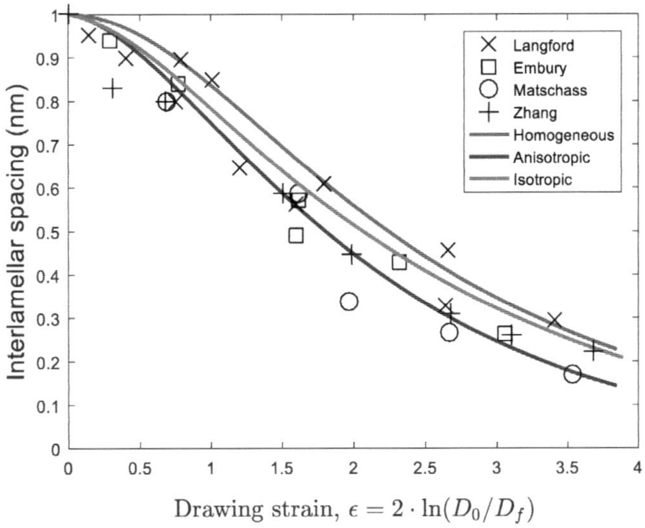

Figure 3. Evolution of normalized interlamellar spacing with applied equivalent strain for the simulated uniform elongation. Blue line: anisotropic ferritic model; red line: isotropic ferritic model; green line: evolution of mean interlamellar spacing of a randomly oriented distribution of pearlitic colonies assuming a homogeneous distribution of strain with equations (7) and (8). Symbols: experimental measurements of normalized interlamellar spacing for pearlitic wires drawn to different equivalent strains [1, 3, 10, 11].

With regard to the colony refinement, figure 3 shows the evolution of the normalized mean interlamellar spacing with equivalent strain for both models (red for the isotropic model, and blue for the anisotropic model). The results are compared with normalized experimental interlamellar spacings (symbols) at equivalent levels of drawing strain. The results show that the proposed anisotropic model (blue line) reproduces the evolution of spacing very accurately and better than the isotropic model. Moreover, the evolution for the isotropic model is very similar to that expected from homogeneous deformation (i.e., the evolution of mean interlamellar spacing considering a homogeneous deformation gradient and equations (7) and (8); see green line in figure 3).

The results in figures 2 and 3 implicitly show that the re-orientation of the colonies, and the evolution of the microstructure, is not as trivial as that described by the isotropic ferritic model. The proposed anisotropic model, in contrast, captures well the microstructural evolution.

Figure 4 shows a stereographic projection of lamellar normals at different equivalent applied strains. Again, the model reproduces the evolution of lamellar orientations as expected under uniaxial loading, namely that all the normals should re-orient perpendicular to the loading direction. The qualitative evolution of the orientation of the lamellae is quite similar for both models, but the anisotropic model promotes internal orientation spreading within colonies, and a random dispersion of orientations around the equator, which can be understood as an effect of colony curling.

With regard to the mechanical response, figure 5 shows the stress-strain curve and the work hardening slope (i.e., the derivative of the true stress with respect to the equivalent strain) for both models. The curves show three different regions. The very first part of the curves (see magnified region in figure 5) shows an elastic loading followed by a linear hardening until real yielding occurs. This "double yielding" is due to the different yield limits of ferrite and cementite. In a real experiment, this double yielding is difficult to see due to the heterogeneity of interlamellar spacings in pearlite that leads to different local critical shear stresses. The second part of the curve ($0.1 < \varepsilon_{eq} < 2$) shows a decreasing work hardening until it stabilizes around 400-500 MPa for the anisotropic model. At this stage, the ferrite work hardens following a Voce-type hardening, and the colonies re-orient as shown in figure 4. Finally, for $\varepsilon_{eq} > 2$, the colonies stretch as shown in figure 2 while the interlamellar spacing drops below 20 nm. The curve at this stage shows an increasing work hardening behaviour (>1 GPa for the anisotropic model).

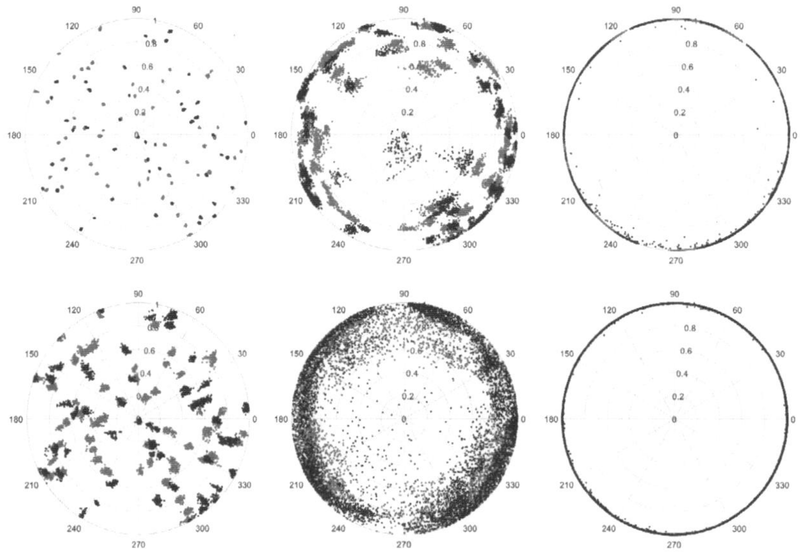

Figure 4. Stereographic projection of the lamellar normals for, ε_{eq}= 0.1, 1 and 4 (left to right). Upper row: Isotropic ferritic model; Bottom row: Anisotropic ferritic model. Blue and red points correspond to normals belonging to the upper and lower hemispheres, respectively.

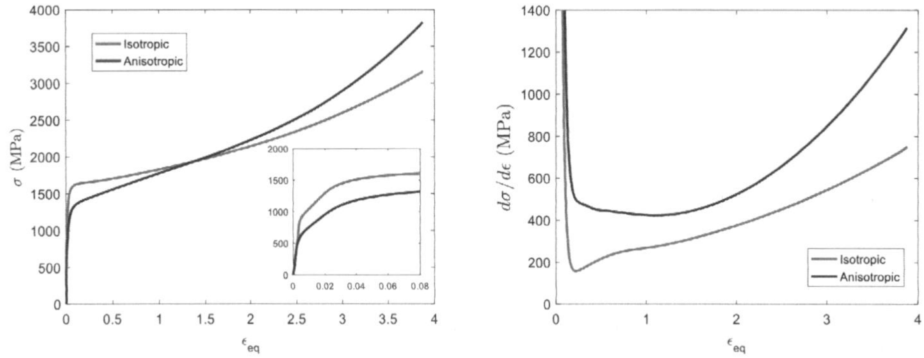

Figure 5. Left: True stress vs true strain curve for the simulated uniform elongation with periodic boundary conditions up to an equivalent strain of ~ 4. The box shows a magnified detail of the elastic region showing a "double yielding" behaviour. Right: Work hardening rate as a function of the applied equivalent strain.

Even though these three stages are seen in both the isotropic and anisotropic models, there are some clear differences in the stress-strain curves. Firstly, the curves intersect at $\varepsilon_{eq} \sim 1.3$. At low strains the isotropic model shows higher flow stresses than the anisotropic model as expected (note that for the anisotropic model, $N, M > 1$, i.e., the flow stress is lower for the out-of-plane components). In contrast, as deformation progresses, the anisotropic model shows a much higher work hardening rate than the isotropic model. This is due to the curling of the pearlitic colonies (figure 2), which reduces the mean interlamellar spacing (figure 3) and, thus, increases the strength of the pearlitic steel according to equation (1).

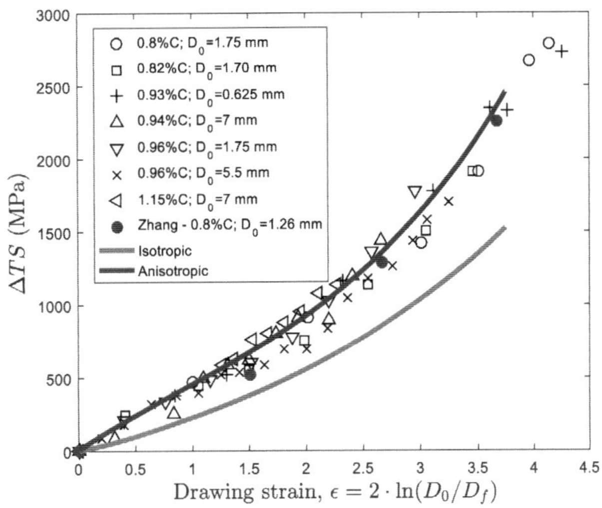

Figure 6. Difference between the ultimate tensile strengths of drawn wires over patented wires (ΔTS) vs the corresponding drawing strain. Open black symbols correspond to a compilation of experimental tests published by Lesuer et al. [12] and blue symbols correspond to experimental data obtained by Zhang et al. [13] on steel wires supplied by Bekaert). Superposed in blue (anisotropic) and red (isotropic), are the simulated stress curves of figure 5, from which the initial part of the curve has been subtracted.

The anisotropic model is in good correspondence with the experimentally observed evolution of strength of heavily drawn wires. Figure 6 shows the difference between the ultimate tensile strengths of drawn wires over patented wires (ΔTS) vs the corresponding drawing strain. Superposed in blue and red are the simulated stress curves from which the first part of the curve has been subtracted. The anisotropic model reproduces perfectly the strengthening of drawn wires even beyond the expected limits of the model (above equivalent strains of 2.5) where dissolution of cementite is expected.

4. Conclusions

- A new relatively simple mesoscopic numerical model has been proposed that describes accurately both the mechanical behaviour and the microstructural evolution of lamellar multi-colony pearlite. The model has been validated with experimental results obtained from literature in heavily drawn wires.
- The model confirms the key role of the anisotropic properties of the ferritic phase on the development of the microstructure, and on the strengthening ability, of heavily drawn wires.

Acknowledgements
This work has been carried out in the framework of a project funded by NV Bekaert SA.

References

[1] Embury A D and Fisher R M 1966 *Acta Metall.* **14** 147-159
[2] Langford G 1970 *Metall. Trans.* **1** 465-477
[3] Langford G 1977 *Metall. Trans A* **8A** 862-875

[4] Gil Sevillano J 1979 *ICSMA 5 Proc.* vol 2, ed P Haasen, V Gerold et al. (Oxford UK: Pergamon Press) p 719
[5] Gil Sevillano J 2011 *Wire Journal International* **44** 58-70
[6] Borchers C and Kirchheim R 2016 *Progr. Mater. Sci* **82** 405-444
[7] Hill R 1948 *Proc. Roy. Soc. London,* **193** 281–297
[8] Gil Sevillano J, Matey Muñoz L and Flaquer Fuster J 1998 *J. Physique IV* **8** 155
[9] Gil Sevillano J, García-Rosales C and Flaquer Fuster J 1999 *Phil. Trans. R. Soc. London A* **357** 1603-1619
[10] Matschass F 1988 *Wiss. Z. Techn. Univ. Magdeburg* **32** 21-28
[11] Zhang X, Hansen N, Godfrey A and Huang X 2016 *Acta Mater.* **114** 176-183
[12] Lesuer D R, Syn C K, Sherby O D and Kim D K 1966 *Lawrence Livermore Nat. Lab.* UCRL-JC-124374
[13] Zhang X, Godfrey A, Huang X, Hansen N and Liu Q 2011 *Acta Mater.* **59** 3422-3430

Characterisation of an Al-BN nanocomposite prepared by ball milling and hot extrusion

U Arlic[1]*, Z Drozd[2], Z Trojanová[2], O Molnárová[2] and A Kasakewitsch[3]

[1]Institute of Materials Science and Engineering, Clausthal University of Technology, 38678 Clausthal-Zellerfeld, Germany
[2]Faculty of Mathematics and Physics, Charles University, Ke Karlovu 3, 121 16 Prague 2, Czech Republic
[3]Weber Maschinenbau GmbH, 26506 Norden, Germany

*E-mail: uwe.arlic@tu-clausthal.de

Abstract. Aluminium-matrix-nanocomposites were manufactured by ball milling of microscale aluminium powder with BN nanoparticles in air, followed by subsequent consolidation by hot extrusion. The microstructure of the samples was studied using scanning electron microscopy. Vickers microhardness measurements were used to probe the mechanical properties of the samples. The amplitude dependent damping of the nanocomposites was measured at room temperature after thermal treatment of samples, and the linear thermal expansion was measured over a wide temperature range from room temperature up to 670K in the as-extruded state. The experimental results give a comprehensive picture of the behaviour of this nanocomposite system over the range of thermomechanical treatment conditions examined in this study. Based on these experimental data some possible influences of BN nanoparticles on the anelastic, plastic and thermal properties of microcrystalline aluminium are discussed.

1. Introduction

Aluminium based nanocomposites are potential materials for structural applications in the aerospace and automotive industries [1]. A ductile aluminium matrix may be reinforced with nanoparticles of oxides, carbides, nanotubes, nitrides or borides. Nanoparticles can be used to significantly improve mechanical properties, including specific stiffness (ratio of stiffness to density), specific yield strength, and the ultimate tensile/compression strength [2-4]. High temperature properties and creep resistance may also be substantially improved [5]. The extent to which a certain property is present in a nanocomposite (NC) can be controlled by selecting the appropriate amounts of matrix and reinforcement materials. Hence, NCs provide flexibility in terms of the properties required.

A main problem for fabrication of nanocomposites is achieving a uniform distribution of nanoparticles, which in general exhibit a tendency to form clusters. Ball milling is a very effective method to prepare not only aluminium nanocomposites but also many others [5, 6]. The resulting properties of NCs depend on the type, size and distribution of the nanoparticles in the matrix. Bonding between nanoparticles and the matrix metal (alloy) is a very important factor influencing mechanical and physical properties of NCs [7]. The standard precipitation hardening of aluminium alloys is limited by thermodynamic restrictions. In contrast strengthening by addition of nanoscale ceramic particles should show more stable behaviour especially at elevated temperatures. Hexagonal BN is a well-known ceramic material with a very high melting point of 3246K and a low density of $2.1x10^3kgm^{-3}$. It has a layered structure similar to graphite. Within each layer, boron and nitrogen atoms are bound by strong covalent bonds, whereas the individual layers are held together by weak van der Waals forces. Hexagonal BN nanoparticles should act as sliding layers as well as diffusion barriers in metal matrix composites [8].

In this study nanoscale hexagonal boron nitride (hBN) was used as the reinforcement phase in a microscale aluminium matrix (Al-nc). The microstructure, and microhardness, as well as the damping and thermal properties were investigated with the aim better to understand behaviour of Al-BN nanocomposites.

2. Experimental procedure

The matrix, based on a mixtures of nanoscale aluminium powder and small amounts of aluminium oxide (Al_2O_3), was combined with hBN dispersoids by a powder metallurgical route. Additionally a reference sample of pure aluminium matrix powder (µAl) was prepared using the same experimental parameters. The basic matrix powder consists of ball milled water atomized Al (99,7wt.%) powder (45µm mean diameter). This matrix system was supplemented by an addition of BN powder (500nm mean diameter), with the combined powders mixed in by a planetary ball mill (Retsch, PM400) under an air atmosphere. The powders were both mixed in corundum-coated steel milling cups using 300 g of corundum balls (11mm mean diameter) and a 2:3 powder to ball mass ratio. The resulting mixtures exhibited non-agglomerated flake-like particles with an average diameter of 20µm and 5µm thickness. Due to the production process the powder mixture also contains a small amount of Al_2O_3 dispersoids. The powders were enclosed in commercial grade Al capsules and hot extruded (Collin 630t, non-ferrous metal extruder) to rods of 20mm diameter at 620K after preheating for 3 h at the same temperature [9].

Samples for microstructural observations were prepared by the grinding, polishing and electrolytic etching in a Lectropol apparatus (using 33% HNO_3 + methanol electrolyte) at 248K and 15V for 5s (pure Al) or 2s (Al+BN). The grain size distribution and texture of samples were studied using a scanning electron microscope (FEI Quanta 200 FX) equipped with an EDAX EBSD (electron backscatter diffraction) camera. The EDAX OIM (orientation imaging microscopy) software was utilized for analysis of the EBSD observations. The microhardness of the samples was measured using an automatic microhardness tester (Qness Q10a) by Vickers indentation, using a load of 100gf (gramm force) and a dwell time of 10s.

The linear thermal expansion was measured over a wide temperature range from RT up to 670K for samples in the as-extruded state. The heating rate in the dilatometer (Netzsch 410) was 0.9K/min. The measurements were carried out in a protective atmosphere of purified Ar. Amplitude-dependent internal friction represented by the logarithmic decrement δ was measured as damping of the free vibrations in a bending mode (for details see [10]). The amplitude-dependent damping of the NCs was measured at RT after step by step thermal cycling between RT and an increasing upper temperature of the thermal cycle up to a maximum temperature of 773K. In the following the two sets of samples are referred to as µAl (Al 99.7wt.%) and Al+0.3BN (Al+0.3wt.% hBN).

3. Results and discussion

3.1. Microstructure of the Al-BN nanocomposite

Examples of the microstructure of both types of samples are shown in figures 1a (µAl) and 1b (Al+0.3BN). The micrographs, taken from the section perpendicular to the extrusion direction, show a very fine grain structure. The grain size distributions for these samples are shown in figure 2a,b, and confirm the presences of a very fine grain structure, as seen in figure 1. The mean grain size is about 2µm and it is approximately the same for both materials. For the case when materials are prepared by milling, refinement of the originally large grains into the microcrystalline regime is connected with a large plastic deformation. This large plastic deformation is partially recovered during consolidation of the powder.

From the EBSD data the grain boundary misorientation angles were also measured and the resulting distributions are shown in figure 3a,b. Both of the distributions (for μAl and Al+0.3BN) have a similar character, each with two local maxima – one at a very small misorientation angle and one at a high misorientation angle. Note the relatively high fraction of low angle boundaries. Both samples exhibit a texture typically observed in extruded fcc (face-centered cubic) materials [11, 12] as it is obvious from the inverse pole figure insets in figures 2a and 2b.

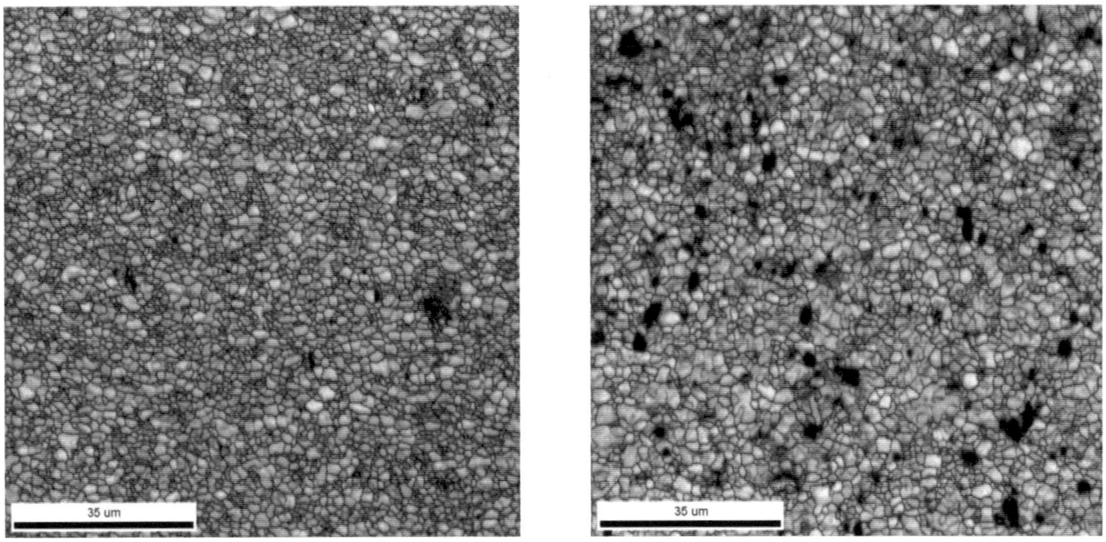

(a) (b)

Figure 1. Microstructure of μAl (a) and Al+0.3BN (b) taken from the section perpendicular to the extrusion direction.

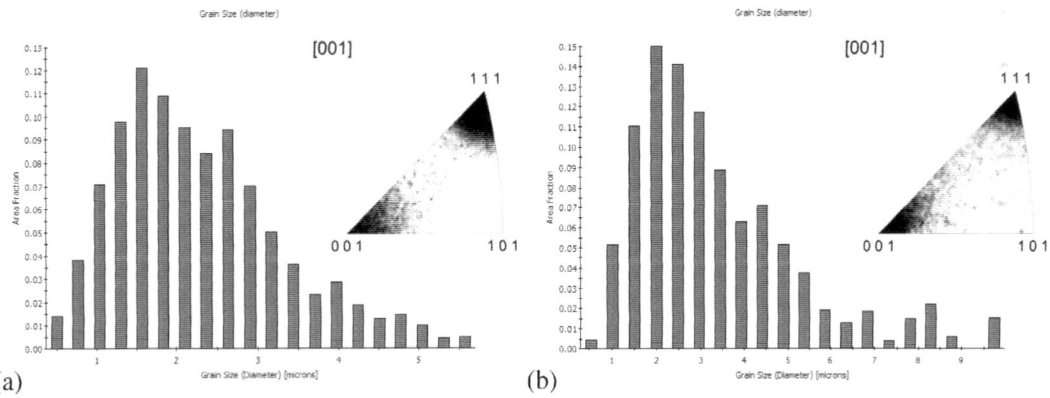

(a) (b)

Figure 2. Grain size distribution as determined from EBSD measurements for μAl (a) and Al+0.3BN (b). The inverse pole figure in the inserts document the existence of a texture in both materials.

3.2. Microhardness

The microhardness (HV0.1) was measured over the entire sample cross-section. The resulting microhardness maps are shown in figures 4a and 4b. A distinct difference between both materials is obvious from inspection of the microhardness maps. Note also the different scale in both pictures. The microhardness of the μAl sample exhibits relatively low values in the middle of cross-section while

IOP Conf. Series: Materials Science and Engineering **219** (2017) 012011 doi:10.1088/1757-899X/219/1/012011

higher values are observed at the edge. This variation most likely arises as a result of the additional friction deformation of rods in the extrusion tool. In contrast, the microhardness of the Al+0.3BN NC sample exhibits the highest values in the middle of the cross-section, and the lowest values in the edge area. These low microhardness values correspond to the mantle material used for the compacted powder. The real composite is inside of this mantle. The different values of microhardness determined for the μAl and Al+0.3BN samples indicate significant strengthening of the due to the hBN in the Al+0.3BN sample. The hBN particles are not visible in the electron microscope images collected in this study. According to our previous experience, nanoparticles in nanocomposites prepared by ball milling are not uniformly distributed in the matrix, but remain mainly in the vicinity of grain boundaries [13, 14].

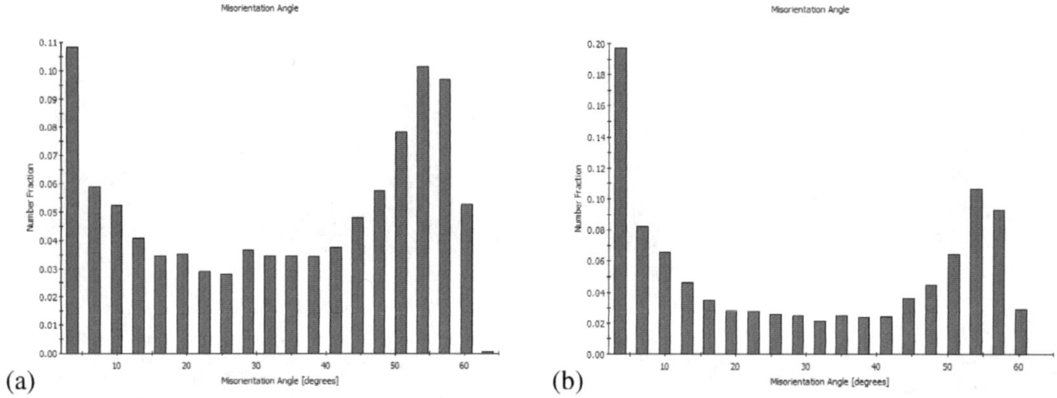

Figure 3. Distribution of grain boundary misorientation angles for (a) μAl, and (b) Al+0.3BN.

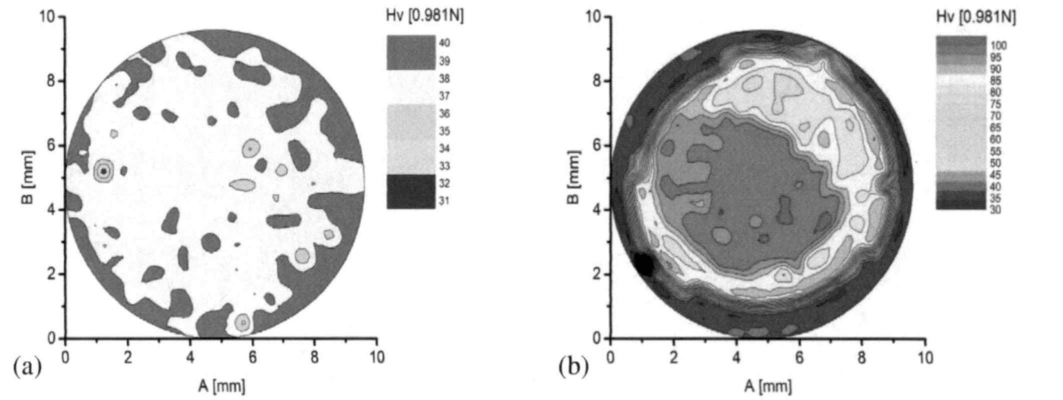

Figure 4. Maps showing the microhardness (HV0.1) measured in the cross-section of samples: (a) μAl, and (b) Al+0.3BN.

Several strengthening mechanisms in NCs can be considered, including Orowan strengthening [15] grain boundary strengthening [16], and increased dislocation density due to large differences in the thermal expansion coefficient between the matrix and ceramic dispersoids (e.g., BN and Al_2O_3) [17]. As mentioned above, bonding between both phases is very important factor. In the case of Al-BN NCs, high strength can be also achieved due to formation of a strong chemical bonding at the Al/BN interface. This chemical bonding provides effective load transfer from the matrix to particles [18].

3.3. Internal friction

The amplitude dependency of the logarithmic decrement are shown in figures 5a,b for lower and higher maximum strain amplitudes respectively. Note that the temperatures depicted in the figures are not measurement temperatures, but rather indicate the upper temperatures of the thermal cycles. The strain dependence of the decrement can be divided into a strain independent and a strain dependent component. For such experimental measurements the decrement can be expressed as

$$\delta = \delta_0 + \delta_\mathrm{H}\,(\varepsilon) \qquad (1)$$

where δ_0 is the amplitude independent component, found at low amplitudes. While several mechanisms may contribute to the amplitude independent component (dislocations, grain boundaries, thermoelastic effect, etc.) the strain amplitude dependent component δ_H is usually caused by the presence of dislocations in the material. The strain amplitude dependence of the logarithmic decrement suggests dislocation unpinning processes from pinning points (solute atoms, vacancies, small clusters of point defects). Interactions between dislocations and pinning points may be explained using the Granato-Lücke theory of dislocation damping [19, 20]. In this theory, vibration of shorter and longer dislocation segments is the main source of internal friction. The critical strain ε_{cr} at which the internal friction becomes amplitude dependent may be used to calculate the effective critical stress amplitude corresponding to the micro yield stress according to the following equation

$$\sigma_c = \mathrm{E}\,\varepsilon_{cr} \qquad (2)$$

where E is the Young's modulus. The stress necessary for a thermal break-away of dislocation loops σ_T at a finite temperature is given by [21]

$$\sigma_T = \sigma_M \left[1 - \left(\frac{kT}{U_1} \ln A \right)^{2/3} \right] \qquad (3)$$

where U_1 is an effective activation energy, A is model parameter and kT has its usual meaning. The stress necessary for break-away of shorter dislocation segments from their pinning points in the pure mechanical process is described by

$$\sigma_M = \frac{2F_m}{b(\ell_1 + \ell_2)} \qquad (4)$$

Here F_m is the maximum force between the dislocation and the pinning point, and ℓ_1 and ℓ_2 are the lengths of neighbouring dislocation segments.

The amplitude dependency of the decrement seen in figure 5 shows that the amplitude-dependent component of decrement δ_H increases with increasing upper temperature of the thermal cycle (see figure 5a). This tendency was reversed after cycling with the upper temperature higher than 633K (figure 5b). A similar tendency was observed also for the critical strain ε_{cr}, which decreases with increasing upper temperature of the thermal cycle up to 633K. Thermal cycling with an upper temperature higher than 633K results in an increase in the critical strain.

As a rule the coefficient of thermal expansion (CTE) of the ceramic reinforcement phase is lower than that of most the metallic matrix in most composites. This means that when the composite is subjected to a temperature change, thermal stresses will be generated. Even small temperature changes may generate thermal stresses. These internal stresses generated by thermal cycling of composites can be reduced by various relaxation mechanisms including dislocations generation and glide, decohesion or sliding of the matrix-reinforcement interface, and diffusion of solute atoms in the matrix.

IOP Conf. Series: Materials Science and Engineering **219** (2017) 012011 doi:10.1088/1757-899X/219/1/012011

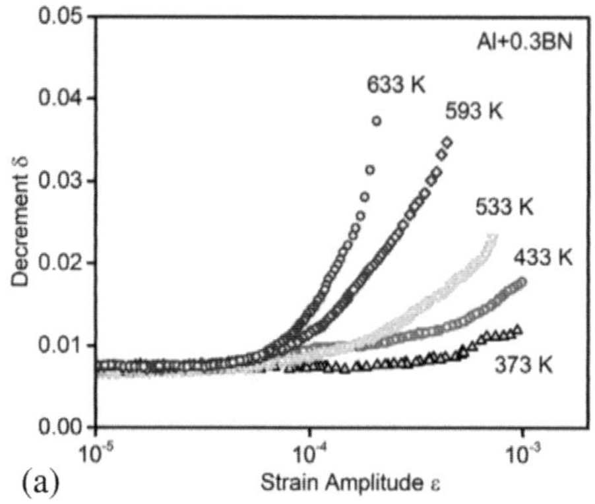

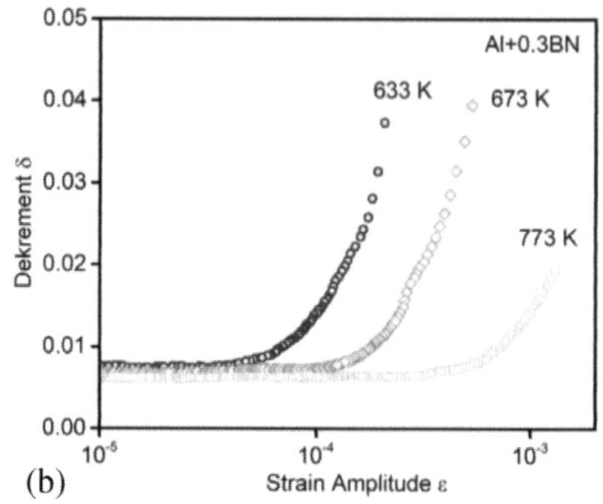

Figure 5. Amplitude dependency of decrement estimated for (a) lower, and (b) upper temperatures of the applied thermal cycles.

In the present samples thermal cycling of the Al-BN nanocomposite generated thermal stresses in the Al matrix in the vicinity of the BN and Al_2O_3 dispersoids. When the difference between room temperature and the upper temperature reached 200K new dislocations were induced into the Al matrix. Because the number of pinning points remained constant and dislocation density increased the effective length of dislocation segments ℓ and thus the critical strain decreased. The Al matrix was step by step plastified. When the radius plastic zones increases above a certain value, the plastic zones may overlap. In such overlapped zones dislocation loops formed near interfaces have the opposite sign on the both sides of the particle. Under internal stresses dislocations move and annihilate in the matrix. This means that the dislocation density after cycling at temperatures higher than 633K again decreases. Accordingly both logarithmic decrement and the critical strain strongly depend on the dislocation density. Chmelík and co-workers [22], measuring the acoustic emission from a thermally treated composite (without any applied stress), have shown that new dislocations are formed in the cooling part of the thermal cycle.

3.4. Thermal expansion

Temperature dependences of the thermal expansion coefficient for μAl and nc Al+0.3 BN are shown in figure 6. The measured CTE values for the NC sample are significantly lower than those for the μAl sample. The extrapolated CTE at 300K for Al+0.3BN is $23 \times 10^{-6} K^{-1}$ and for μAl is $23.8 \times 10^{-6} K^{-1}$ for. Using the rule of mixtures the CTE of a composite can be calculated as:

$$a_c = (1 - V_p)a_m + V_p\, a_p \tag{5}$$

where α_C, α_m and α_p represent the CTEs of the composite, matrix, and particle, respectively, and V_p is the volume fraction of the reinforcing phase particles. Taking a value for polycrystalline BN CTE(hBN) of $4.563 \times 10^{-6} K^{-1}$ [23] the value calculated using the rule of mixture CTE is practically the same as for μAl owing to the very small nanoparticle content. In the quasi-harmonic Debye model, the thermal expansion coefficient, α, is described by

$$\alpha = \frac{\gamma C_p}{K_s V} = \frac{\gamma C_v}{K_T V} \tag{6}$$

where γ is the Grüneisen parameter, V the volume, K_T the isothermal bulk modulus, K_S the adiabatic bulk modulus, and C_V and C_P are the heat capacities at constant volume and pressure, respectively [24]. The lower values of CTE than expected are very probably a consequence of the lower material density.

Material prepared by ball milling and hot extrusion may contain a certain fraction of pores [25]. Consolidation defects like pores and cracks could not fully be avoided by hot extrusion at different temperatures. Annealing experiments show that additional consolidation of the hot extruded materials occurs at higher temperatures above 673K but micro cracks could not completely be removed. Additionally, thermal cracks could be nucleated in powder metallurgical materials by quenching from high temperatures.

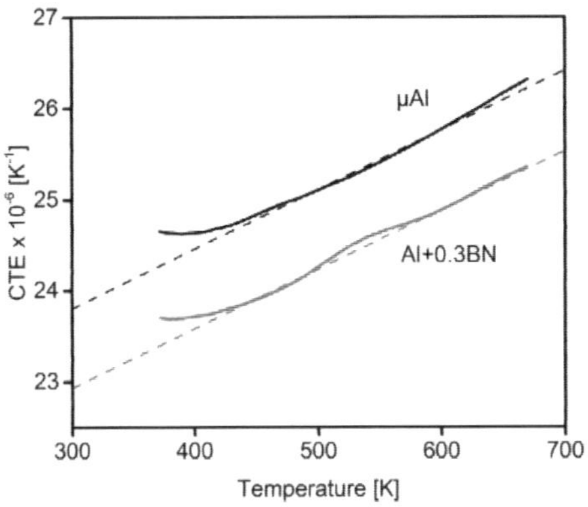

Figure 6. Temperature dependence of the thermal expansion coefficient for the μAl and Al+0.3BN samples prepared in this study.

4. Conclusions

An aluminium-based nanocomposite containing 0.3wt.% of BN nanoparticles was successfully prepared by ball milling of Al and BN powders followed by hot extrusion. The nanocomposite also contained a small fraction of Al_2O_3 dispersoids introduced by surface oxidation of the Al powder. For comparison pure Al samples were also prepared from milled and consolidated powder. The mechanical and physical properties of the as-extruded samples were investigated leading to the following main conclusions:

- Both NC materials exhibit a very fine grain structure, with mean grain size of approximately 2 μm;
- Microhardness measurements reveal significant strengthening of the nanocomposite as compared with the sample prepared from only Al powder;
- Internal friction measurements suggest that new dislocations are produced after thermal cycling of the nanocomposite;
- The nanocomposite exhibits a distinct reduction in thermal expansion coefficient compared with the value expected based on a rule of mixtures analysis. This lower value is attributed to porosity of the material prepared using a powder metallurgical route.

Acknowledgements

We are sorry that our acknowledgments came too late to Professor Werner Riehemann from the Clausthal University of Technology, who gave the impetus to this work.

References

[1] Koli D K, Agnihotri G and Purohit R 2015 Advanced Aluminium Matrix Composites: The Critical Need of Automotive and Aerospace Engineering Fields *Materials Today: Proceedings* **2** 3032–41

[2] Yamaguchi M, Bernhardt J, Faerstein K, Tsuchia K, Shtansky D, Bando Y, Golovin I, Sinning H and Goldberg D 2013 Fabrication and characteristics of melt-spun Al ribbons reinforced with nano/micro-BN phases *Acta Mater.* **61** 7604–15

[3] Saravanan L and Senthilvelan T 2015 Investigations on the hot workability characteristics and deformation mechanisms of aluminium alloy-Al_2O_3 nanocomposite, *Materials and Design* **79**

6–14
[4] Mokdad F, Chen D L, Liu Z Y, Ni D R, Xiao B L and Ma Z Y 2017 Hot deformation and activation energy of a CNT-reinforced aluminum matrix nanocomposite, *Mater. Sci. Eng.* **695** 322–31
[5] Hosseini Monazzah A, Simchi A and Seyed Reihani S M 2010 Creep behavior of hot extruded Al-Al2O3 nanocomposite powder *Mater. Sci. Eng.* **527** 2567–71
[6] Casati R, Fiocchi J, Fabrizi A, Lecis N, Bonollo F and Vedani M 2017 Effect of ball milling on the ageing response of Al2618 composites reinforced with SiC and oxide nanoparticles J. *Alloys Compd* **693** 909–12
[7] Trojanová Z, Ferkel H, Lukáč P and Riehemann W 2002 Two new high-damping magnesium composites *Phys. Stat. Sol. (a)* **193** 205–10
[8] Reischer F, Pippel E, Woltersdorf J, Stockel S and Marx G 2007 Carbon fibre-reinforced magnesium: Improvement of bending strength by nanodesign of boron nitride interlayers. *Materials Chemistry and Physics* **104** 83–7
[9] Kazakewitsch A and Riehemann W 2012 Amplitude Dependent Damping of Aluminum-Matrix-Nanoparticle-Composites *Solid State Phenomena* **184** 307–12
[10] Göken J and Riehemann W 2002 Thermoelastic damping of the low density metals AZ91 and DISPAL *Mater. Sci. Eng.* **324** 134–40
[11] Böhm H 1992 *Einführung in die Metallkunde* Bibliographisches Institut GmbH
[12] Gottstein G 2014 *Materialwissenschaft und Werkstofftechnik, Physikalische Grundlagen* Springer-Verlag Berlin Heidelberg
[13] Ferkel H and Mordike B L 2001 Magnesium strengthened by SiC nanoparticles *Mater. Sci. Eng.* A **298** 193–9
[14] Drozd Z, Trojanová Z, Arlic U, Kasakewitsch A and Dash K Effect of BN and graphite nanoparticles on mechanical and physical properties of magnesium *In this volume*
[15] Yeh Y H, Nakashima H, Kurishita H, Goto S, Yoshinaga H 1990 Threshold Stress for High-Temperature Creep in Particle Strengthened Al-1.5 vol%Be Alloys *Mater. Trans. JIM* **31** 284-292
[16] Hansen N 1977 The effect of grain size and strain on the tensile flow stress of aluminium at room temperature *Acta Metall.* **25** 863–79
[17] Arsenault J R and Shi N 1986 Dislocation generation due to differences between the coefficients of thermal expansion *Mater. Sci. Eng.* **81** 175–187
[18] Krasheninnikov A V, Berseneva N, Kvashnin D G, Enkovaara J, Björkman T, Sorokin P and Goldberg D 2014 *J. Phys. Chem.* **C118** 26894–901
[19] Granato A V and Lűcke K 1956 Theory of mechanical damping due to dislocations *J. Appl. Phys.* **27** 583–93
[20] Granato A V and Lűcke K 1956 J. Application of dislocation theory to internal friction phenomena at high frequencies *Appl. Phys.* **27** 789–805
[21] Granato A V and Lűcke K 1981 Temperature dependence of amplitude - dependent dislocation damping *J. Appl. Phys.* **52** 7136–42
[22] Chmelík F, Trojanová Z, Kiehn J, Lukáč P and Kainer K U 1997 Non-destructive characterisation of microstructure evolution in Mg based metal matrix composites submitted to thermal cycling *Mater. Sci. Eng.* **234-236** 774–7
[23] Brožek V, Dufek V, Harok V and Rohan P {http://konsys-t.tanger.cz/files/proceedings/metal_00/papers/135.pdf}
[24] Grüneisen E 1912 Theorie des festen Zustandes einatomiger Elemente *Ann. Physik* **39** 257–306
[25] Vijay A 2011 Analysis of Grüneisen parameter and its volume derivatives for NaCl and hcp iron *Indian J Pure & Appl. Phys.* **49** 825–8

Dislocation based controlling of kinematic hardening contribution to simulate primary and secondary stages of uniaxial ratcheting

S Bhattacharjee[1], S Dhar[1,*] and S K Acharyya[1]

[1]Jadavpur University, Kolkata, West Bengal, India, 700032

* Corresponding Author e-mail: sankar_dhar1@rediffmail.com, Telefax: +91 33 2414 6532

Abstract: The primary and secondary stages of the uniaxial ratcheting curve for the C-Mn steel SA333 have been investigated. Stress controlled uniaxial ratcheting experiments were conducted with different mean stresses and stress amplitudes to obtain curves showing the evolution of ratcheting strain with number of cycles. In stage-I of the ratcheting curve, a large accumulation of ratcheting strain occurs, but at a decreasing rate. In contrast, in stage-II a smaller accumulation of ratcheting strain is found and the ratcheting rate becomes almost constant. Transmission electron microscope observations reveal that no specific dislocation structures are developed during the early stages of ratcheting. Rather, compared with the case of low cycle fatigue, it is observed that sub-cell formation is delayed in the case of ratcheting. The increase in dislocation density as a result of the ratcheting strain is obtained using the Orowan equation. The ratcheting strain is obtained from the shift of the plastic strain memory surface. The dislocation rearrangement is incorporated in a functional form of dislocation density, which is used to calibrate the parameters of a kinematic hardening law. The observations are formulated in a material model, plugged into the ABAQUS finite element (FE) platform as a user material subroutine. Finally the FE-simulated ratcheting curves are compared with the experimental curves.

1. Introduction

Cyclic plastic deformation occurs when a material is subjected to an load with an amplitude beyond its yield limit. The experimental results form the framework of the material response under different cyclic loading conditions. When it comes to modeling of the cyclic plastic behavior, a key role is played by the kinematic hardening (KH) rule, to generate the cyclic hysteresis loops and to describe fundamental cyclic plastic phenomena such as the Bauschinger effect. There are other different material responses observed for balanced cyclic loading, in which cyclic hardening or softening behavior is observed experimentally.

During low cycle fatigue, dislocation substructures are formed surrounded by low angle sub-grain boundaries containing arrays of trapped edge dislocations [1]. Dislocation annihilations are likely to occur when moving dislocations interact with low angle subgrain boundaries. The annihilation of boundary dislocations can be a reasonable phenomenon to justify isotropic cyclic softening in low cycle fatigue (LCF) loadings [2].

On the other hand, when a material is subjected to unbalanced cyclic loading, a ratcheting phenomenon is observed, as the accumulation of plastic strain in loading direction. Transmission electron microscope (TEM) observations show that the formation of a substructure during asymmetric loading is difficult and generally delayed, due to the continuous generation of new dislocations triggered by the asymmetry in loading. Potentially, these new dislocations can pile-up and increase the hardening [3,4].

In this work, the isotropic softening phenomenon based on dislocation annihilation at low angle tilt boundaries is employed for the simulation of cyclic softening behavior. The Ohno-Wang kinematic hardening rule [5] is employed to simulate basic stress-strain hysteresis loops including a Bauschinger effect. For asymmetric loading, the extra hardening evolves due to the generation of an excess dislocation density corresponding to the ratcheting strain. The KH parameters of the Ohno-Wang law are calibrated with an appropriate dislocation function to incorporate this extra hardening. The developed material model is plugged into the ABAQUS CAE (v6.8) commercial finite element software in the form of a user defined material subroutine (UMAT) and the simulated results are compared with the experimental data.

2. Material

SA333 Grade-6 C-Mn steel is commonly used in primary heat transport pipes in Indian PHWR nuclear power plants. The chemical composition of SA333 Grade-6 C-Mn steel is shown in table 1.

Table 1. Chemical composition of SA333 C-Mn steel.

Element	C	Mn	Si	P	Fe
wt. (%)	0.18	0.90	0.02	0.02	Rest

The typical microstructure of the material is given in figure 1, showing pearlite bands (volume fraction 30%) in a ferrite matrix with an average ferrite grain size of 22μm. It is assumed in the following that all the deformation occurs mainly in the ferrite grains, because ferrite is softer and thus more susceptible to deformation than pearlite.

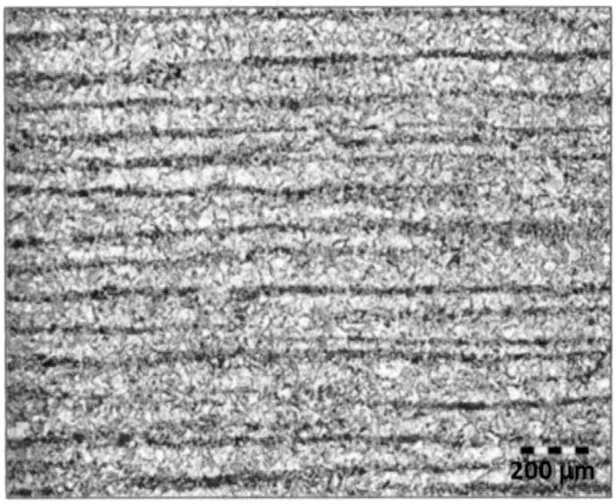

Figure 1. Optical micrograph of SA333.

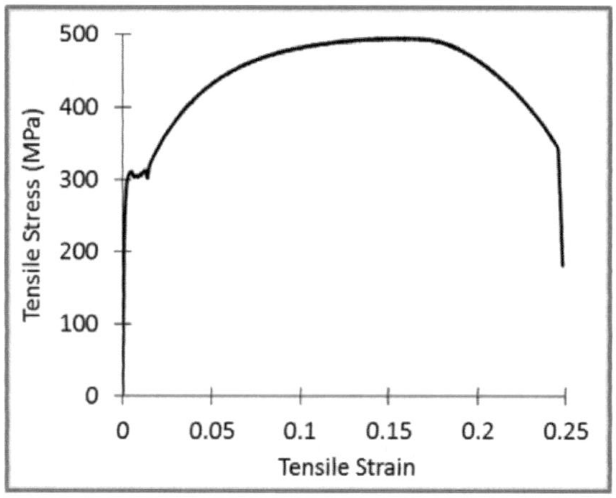

Figure 2. Tensile plot for SA333.

A typical tensile plot of the SA333 is presented in figure 2. The steel has moderate strength (YS = 304MPa, UTS = 495MPa) with elastic modulus of 207GPa, and a ductility (elongation) of 24.5%. It is important to note that during tensile testing the material developed large Lüders bands upto about 1.4% strain.

TEM images obtained using a Philips CM200 operated at 200kV are displayed in figure 3 for the material in the as received condition, as well as after a few cycles (10 cycles) and after failure of the samples.

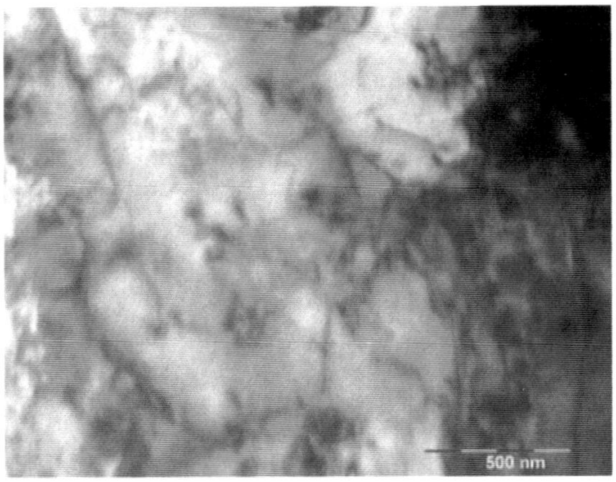

Figure 3a. TEM micrograph of SA333 in the 'as received' condition.

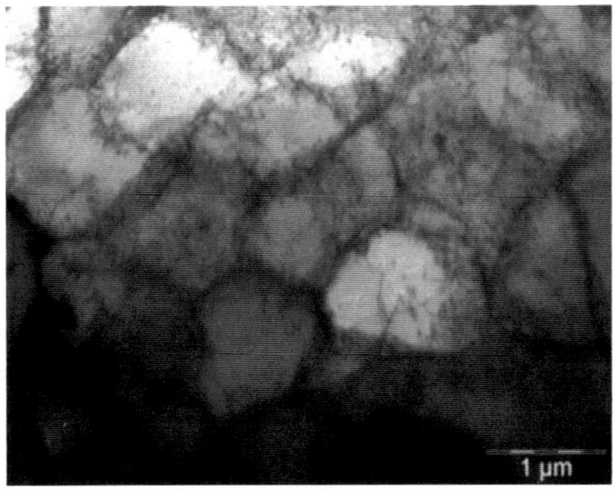

Figure 3b. TEM micrograph of SA333 after 10 cycles (1% strain amplitude).

Although the TEM micrographs of the material in the as received condition (figure 3a) show no definitive dislocation structure, there are nevertheless weak traces of dislocation structures formed during previous plastic deformation.

In contrast, the TEM micrographs of the sample loaded to 10 cycles (figure 3b), show direct evidence for the formation of dislocation substructures. Annihilations of the entrapped dislocations in low angle sub-grain boundaries will result in an isotropic softening stress (negative) and therefore, reduce the cyclic yield stress. Similarly, TEM micrographs of the sample after failure due to balanced cyclic loading (figure 3c) shows the presence of well-defined dislocation substructures, with an average cell size of 1.2 µm.

As a comparison, material subjected to asymmetric loading (mean stress 80MPa; stress amplitude 310MPa) shows delayed formation of a dislocation cell substructure as seen from the TEM micrographs in figure 4a-c. In this unbalanced loading condition, even at half of the lifetime to failure (500 cycles) no distinct sub structure is formed (figure 4a). Continuous generation of new dislocations leads to a bias in the dislocation interactions and entangling behavior is observed. Eventually however the generation rate of new dislocations gradually decreases and a sub-cell dislocation structure is formed (figure 4b).

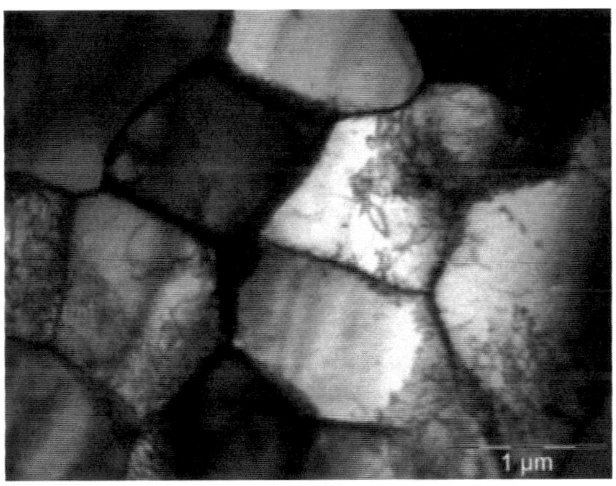

Figure 3c. TEM micrograph of SA333after failure (1% strain amplitude, 500 cycles).

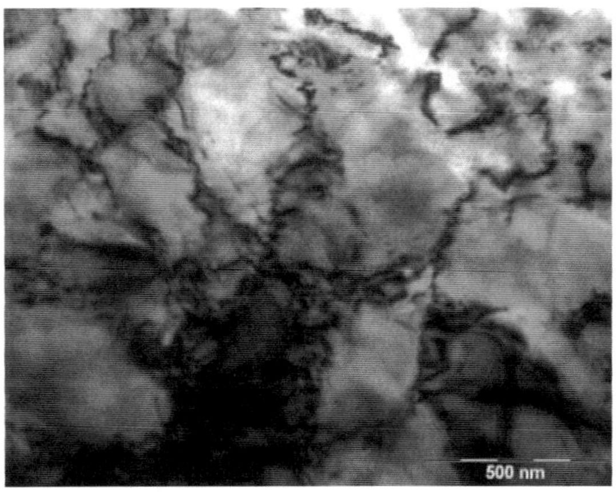

Figure 4a. TEM micrograph of SA333 after half lifetime (≈500 cycles).

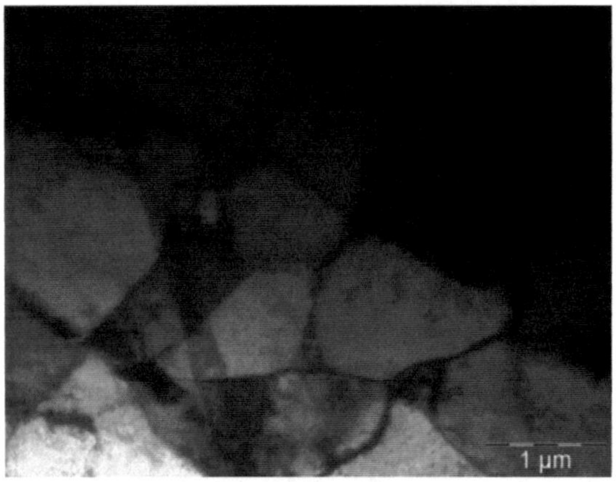

Figure 4b. TEM micrograph of SA333 after failure (longitudinal section of the specimen).

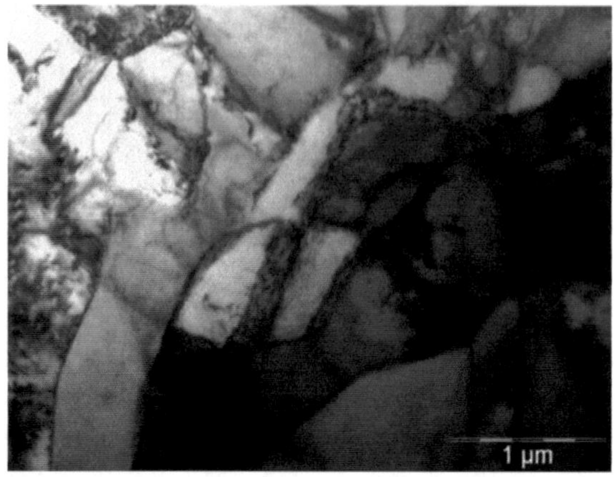

Figure 4c. TEM micrograph of SA333 after failure (cross section of the specimen).

3. Experiments

3.1. Uniaxial LCF experiment

Uniaxial low cycle fatigue tests were conducted by Bhabha Atomic Research Centre, India, at various strain amplitudes ranging from ±0.35% to ±0.75%. Typical cyclic plastic phenomena are observed in the stress-strain loops including a Bauschinger effect (figure 5a) and cyclic softening (figure 5b). The cyclic softening rate typically decreases with the number of cycles and saturates within the first 200 cycles. Properties of cyclic hardening/softening depend not only on material microstructure, but also on loading amplitude and more generally on the previous strain history. Such transient behavior of material requires accurate calibration of material parameters.

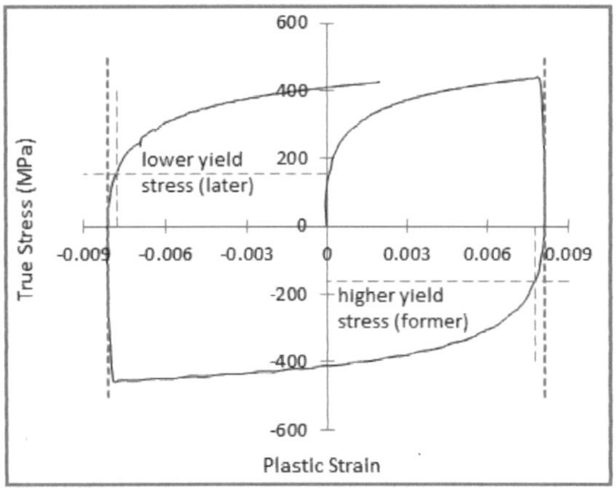

Figure 5a. Experimental results showing Bauschinger effect (0.75% strain amplitude).

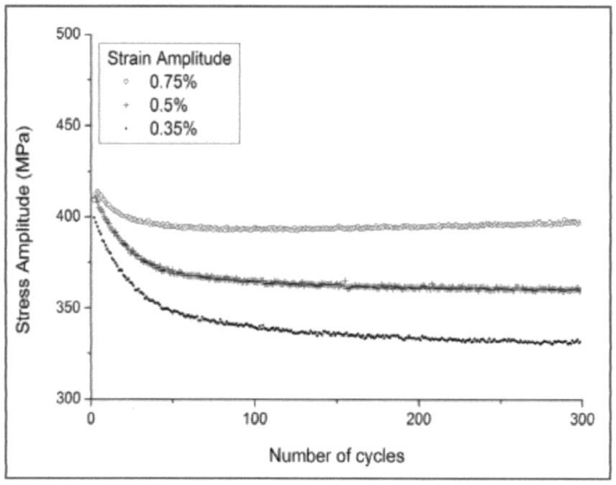

Figure 5b. Experimental results showing cyclic softening (strain amplitude 0.35% to 0.75%).

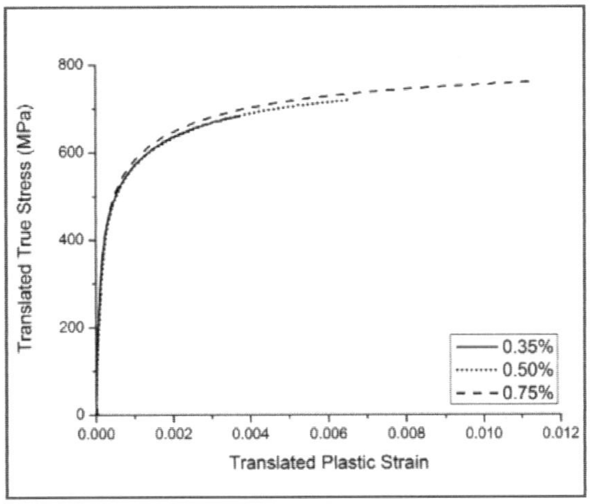

Figure 5c. Experimental results of SA333 showing Masing behavior.

3.2. Uniaxial Ratcheting tests

The evolution of the engineering ratcheting strain [given as $(\varepsilon_{max}+\varepsilon_{min})/2$] as a function of number of cycles during engineering stress-controlled experiments are shown in figure 6a-b for stress amplitudes of 310MPa, 350MPa and 390MPa and mean stresses of 40MPa, 80MPa, and 120MPa. It is observed that significant amounts of ratcheting strain are accumulated before failure.

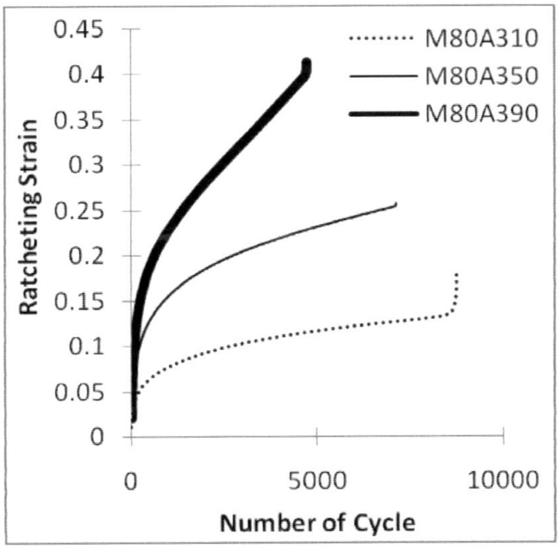

Figure 6a. Stress controlled ratcheting of SA333 showing a comparison between stress amplitudes of 310MPa, 350MPa and 390MPa
(all with mean stress of 80MPa).

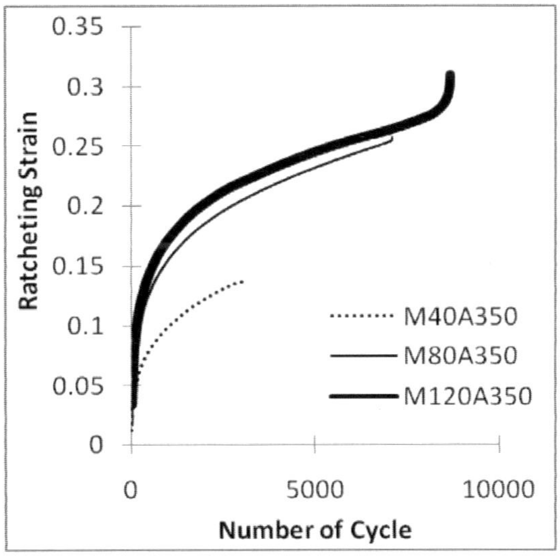

Figure 6b. Stress controlled ratcheting for SA333 Showing a comparison between mean stresses of 40MPa, 80MPa and 120MPa
(for a stress amplitude of 350MPa).

4. Material model

In order to model material plasticity a suitable yield criterion is required. In this present study, the von-Mises yield criterion is applied:

$$\Phi = \left[\frac{3}{2}\left(S_{ij}-\alpha_{ij}\right):\left(S_{ij}-\alpha_{ij}\right)\right]^{\frac{1}{2}} - \sigma_c = 0\,, \qquad (1)$$

Where, S_{ij} and α_{ij} are reduced stresses and deviatoric back stress components respectively. The current yield stress, $\sigma_{c,}$ can be additively decomposed into an initial yield stress and an isotropic softening stress.

The plastic modulus is formulated from the consistency condition conforming associative flow rule, incorporating the hardening laws as:

$$h = n_{ij} : \frac{dS_{ij}}{d\varepsilon_{eq}^{p}} = n_{ij} : \frac{d\alpha_{ij}}{d\varepsilon_{eq}^{p}} + \frac{d\sigma_c}{d\varepsilon_{eq}^{p}}, \tag{2}$$

where n_{ij} is the flow vector, and $d\varepsilon^{p}_{eq}$ is the equivalent plastic strain rate. Here n_{ij} is defined as:

$$n_{ij} = \frac{3}{2}\frac{S_{ij} - \alpha_{ij}}{\sigma_{eq}}, \tag{3}$$

Where σ_{eq} is the von-Mises equivalent stress and where the magnitude of the equivalent plastic strain rate is estimated as,

$$d\varepsilon_{eq}^{p} = \left[\frac{2}{3} d\varepsilon_{ij}^{p} : d\varepsilon_{ij}^{p}\right]^{\frac{1}{2}}. \tag{4}$$

Calculation of plastic modulus, h requires incorporation of hardening rules, the selection of which is very important for the model. The KH law selected for the developing the model is a six-segmented ($m = 6$) Ohno-Wang law [5]. The Masing behaviour of the material for the considered strain range is modelled by a single cyclic yield stress and one set of KH parameters. The total contribution of the six deviatoric back stress segments is:

$$\alpha_{ij} = \sum_{k=1}^{m} \alpha_{ij}^{k}. \tag{5}$$

Here, α_{ij}^{k} is the kth segment of the (deviatoric) back stress tensor, and for each segment, the back stress evolution law is as follows:

$$d\alpha_{ij}^{k} = \sqrt{\frac{3}{2}} C^{k} r^{k} \left[n_{ij} - \left(\frac{\left\| \alpha_{ij}^{k} \right\|}{r^{k}} \right)^{\chi^{k}+1} l_{ij}^{k} \right] d\varepsilon_{eq}^{p}. \tag{6}$$

In the above expression, l^{k}_{ij} is the direction of the corresponding back stress:

$$l_{ij}^{k} = \frac{\alpha_{ij}^{k}}{\left\| \alpha_{ij}^{k} \right\|} \qquad \textit{and,} \qquad \left\| \alpha_{ij}^{k} \right\| = \left[\frac{3}{2} \alpha_{ij}^{k} : \alpha_{ij}^{k} \right]^{\frac{1}{2}}. \tag{7}$$

C^{k}, and r^{k} are the material parameters for the KH law, where C^{k} are the hardening coefficients and r^{k} are the saturation values of α^{k}_{ij} for the kth segment.

4.1. Softening Model (Isotropic):

Recovery or softening by annihilation is always isotropic. The chosen scheme is to calculate the softening stress (negative) σ_2 and algebraically add this to the initial stress to determine the current yield stress, i.e.,

$$\sigma_c = \sigma_0 + \sigma_2. \tag{8}$$

The cyclic softening stress $\sigma_2$is calculated [6] from the dislocation annihilation model [2] at the low angle grain boundaries. This is not discussed in this article. The softening stress is calculated in an incremental form as:

$$\Delta\sigma_2 = -\frac{\alpha M^2 Gb}{S_n}\frac{\Lambda}{D}\lambda\frac{y_e}{b}\Delta\varepsilon^{p}, \tag{9}$$

where α is the interaction strength, M is the Taylor factor, G is the shear modulus, b is the magnitude of the Burger's vector, Λ is the dislocation density of edge dislocations at sub-grain boundaries per unit length, D is the average sub-grain size, λ is the slip band width, S_n is the number of independent slip systems (taken to be five to accommodate von-Mises plasticity), and y_e is the critical annihilation distance. If n is the number of annihilation sites per slip band interacting with low angle sub-grain boundaries then, $y_e = \lambda/2n$. If p is the probability of annihilation of dislocations of opposite sign, finally, the plastic modulus is formulated as:

$$h = \sum_{k=1}^{m} C^k r^k \left[1 - \left(\frac{\left|\alpha_{ij}^k\right|}{r^k} \right)^{\chi^k+1} l_{ij}^k : n_{ij} \right] - \frac{\alpha M^2 G b p}{S_n} \frac{\Lambda}{D} \frac{\lambda^2}{2nb}. \tag{10}$$

4.2.Ratcheting Model

It is observed after examination of the TEM micrographs of both ratcheted and LCF samples that dislocation sub-cell formation is usually delayed in case of ratcheting. The delay occurs due to a continuous generation dislocations due to asymmetry in loading. Experimental evidence shows that the dislocation density in this material can be fitted using an exponential growth model, hence from the Orowan equation, the increase in dislocation density is derived by Mecking and Kocks [7] as:

$$d\rho = \frac{M\,dq}{b\,L} - \frac{2y_e\,\rho\,M\,dq}{b\,S_n}, \tag{11}$$

where dq is the increase of the size of the plastic strain memory surface. Chaboche, Dang Van and Cordier [8] have introduced the plastic strain memory surface in strain space where it is given as:

$$F = \frac{2}{3}\left(\varepsilon_{ij}^p - \beta_{ij}\right):\left(\varepsilon_{ij}^p - \beta_{ij}\right) - q^2 \le 0. \tag{12}$$

Here, β_{ij} is the centre of the memory surface and q is the radius. For strain controlled LCF, $\beta_{ij}=0$ and q is a constant equal to the plastic strain amplitude. Therefore, dq is zero. For ratcheting, the plastic strain centre shifts due to the ratcheting strain and thereby the radius of the plastic strain memory surface increases. Chobache et. al. [8] introduced an expression to account for the increase in the size of the plastic strain memory surface, and later McDowell [9] modified the expression adding a threshold term, defined as:

$$dq = \left[\eta H(F) - \xi q\left(1 - H(F)\right)\right]\left\langle n_{ij} : n_{ij}^* \right\rangle d\varepsilon_{eq}^p. \tag{13}$$

Here, $H(F)$ stands for a heavy side function, n_{ij} is the flow direction, and n_{ij}* is the normal to the plastic strain memory surface is given as:

$$n_{ij}^* = \frac{2}{3}\frac{\varepsilon_{ij}^p - \beta_{ij}}{q}, \tag{14}$$

where η and ξ are tuning parameters. In this analysis, dq is considered as the measure of the ratcheting strain.

A function, defined by the dislocation density, is employed to control the intensity of hardening by modifying the r^k values of the Ohno-Wang KH law, motivated by the fact that with time, randomly distributed newly generated dislocations are susceptible to pile-up. The function is as follows:

$$f = 1 - \exp\left(-\frac{\rho - \rho_c}{\rho_c} \right). \tag{15}$$

There is a critical density of dislocations (ρ_c) corresponding to the first quarter cycle of the LCF loading, beyond which the unbalanced loading effect come into play. The critical dislocation density acts like a threshold for entry into the ratcheting domain. The function f acts like a scaling function, which operates between 0 to 1 denoting initial and saturated values of r^k respectively. According the r^k values are modified as:

$$r^k = r^k\left\{1+\left(\bar{\chi}-1\right)f\right\}. \tag{16}$$

The evolution of r^k values, shown above, is actually controlled by the function f and the parameter $\bar{\chi}$. The rate of hardening is modulated by the function f and the amount of hardening is governed by the parameter $\bar{\chi}$. For each segment of the KH rule the function and the parameter are kept unchanged for simplicity. The evolution of $\bar{\chi}$ with plastic strain amplitude is assumed to be exponential:

$$\bar{\chi} = \chi_0 + \chi_{sat}\left\{1-\exp\left(-B\,q\right)\right\}. \tag{17}$$

From the experimental observations, χ_0 and χ_{sat} are determined. χ_0 is taken as 1 to ensure the LCF phenomena in balanced loading where no change in r^k occurs .Within the critical dislocation density no evolution of r^k is allowed. Beyond the critical point, r^k evolve up to a saturation value of χ_{sat}. The values of χ_{sat} and the rate parameter B is found from the secondary ratcheting loops, and calibrated by trial and error.

5. Determination of Material Parameters

The required elastic parameters for the model (shown in table 2) are extracted from the uniaxial tensile tests following standard procedures. The cyclic yield stresses are extracted from the stress-plastic strain data of the uniaxial LCF tests for all strain amplitudes. The variation of yield stresses with strain amplitude is negligible due to the Masing behaviour of the material (figure 5c). The initial cyclic yield stress is found to be 195MPa.

As input to the Ohno-Wang KH Law, the parameters C^k, r^k are determined from the experimental results of uniaxial tension-compression and symmetric cyclic loading at (the highest) strain amplitude of 0.75%.The loading branch of the saturated stress-plastic strain hysteresis loop is divided in six segments (k = 1, 2 … 6) to obtain one set of the KH parameters. It is evident from experimental results that the material shows the Masing behavior in the plastic strain range of the present work. Therefore, one set of value for C^k and r^k (k = 1, 2 … 6) can accurately evaluate the kinematic hardening property of the material. The procedure, outlined in the work of Jiang and Kurath [10], has been followed to determine the KH Parameters. The values of C^k and r^k determined for the present material are shown in table3. Table 4 below provides the estimated values for the cyclic softening parameters. Some of the material parameters are determined from TEM observations and a few are obtained by calibration.

Table 2. parameters of SA333 C-Mn steel, used in simulations.

Young's modulus (GPa)	Poisson's Ratio
207	0.3

Table 3. C^k and r^k values.

C^1 = 16329.93	C^2 = 4082.48	C^3 = 1632.99	C^4 = 816.49	C^5 = 326.59	C^6 = 116.64
r^1 = 11.77MPa	r^2 = 17.20MPa	r^3 = 21.01MPa	r^4 = 26.14MPa	r^5 = 34.57MPa	r^6 = 36.68MPa

Table 4. Material parameter for dislocation annihilation at tilt boundary.

Parameter		α	M	G	b	D	Λ_0	λ	n (per λ)	S_n
Unit				MPa	nm	μm	m^{-1}	nm		
Strain Amplitudes	0.35%	0.30	2.99	80769	0.25	1.2	$1.0\text{x}10^9$	30	13.5	5
	0.50%								33	
	0.75%								90	

The evolution of the dislocation density is governed by equation (11). All necessary additionally required material parameters are listed in table 5. The value of the initial dislocation density is taken from literature, and the critical dislocation density (ρ_c) is back-calculated from the evolution law at the quarter cycle, subject to the corresponding symmetric loading amplitude. The values of χ_0, χ_{sat} and B are fitting parameters, chosen by trial and error.

Table 5. Material parameter for ratcheting model.

Plastic strain amplitude at quarter cycle	L µm	ρ_0 m^{-2}	ρ_c m^{-2}	χ_0	χ_{sat}	B
0.304%			1.5E12			
0.640%	22	1E8	3.2E12	1	0.7	133
1.059%			5.3E12			

6. Finite element simulation

The LCF tests were simulated using the ABAQUS software. Only the working (gauge length) portion of the actual specimens were modelled. Figure 7 shows schematic representation of the model mesh.

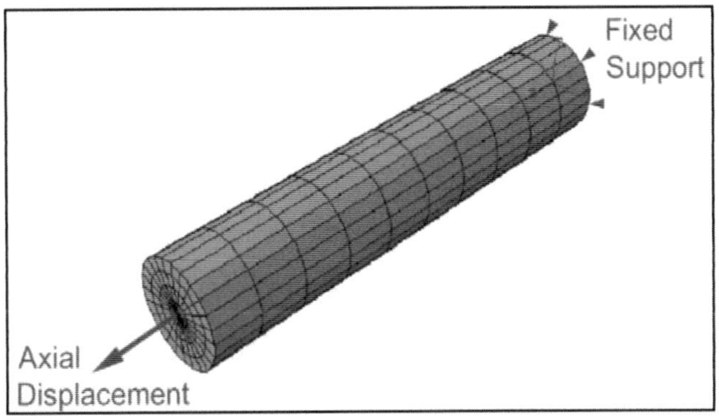

Figure 7. Mesh and loading arrangement for the finite element simulations.

The model for the specimen gauge is discretized using structural meshing. A total of 576 8-noded linear hexahedral elements along with at central zone of 144 6-noded linear wedge elements are used to maintain the radial orientation. For LCF loading, a displacement boundary condition is imposed according to the strain amplitudes using a balanced triangular amplitude function. For ratcheting, a point load corresponding to the peak stress is applied to a reference point connected to the body through kinematic coupling, and an asymmetric triangular amplitude function is employed. The material model is incorporated in the finite element analysis through the user material subroutine (UMAT) with a proper numerical integration scheme.

7. Results and discussion

7.1. Uniaxial low cycle fatigue – validation of the KH rule and softening model

Figure 8 shows a comparison of the experimental and simulated stress-strain hysteresis loops at the 30th cycle, as well as the form of the cycle-dependent softening of the peak stress for tension-compression cyclic loading subject to a strain amplitude of 0.35%.

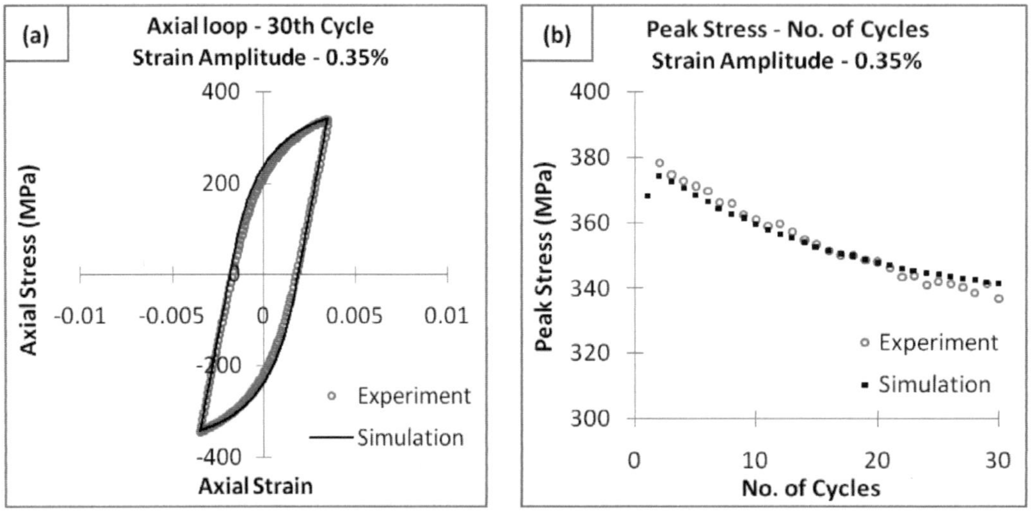

Figure 8. Comparison of uniaxial experimental and simulated results for 0.35% strain amplitude, (a) stress-strain hysteresis loops at 30th cycle, (b) peak stress vs. number of cycles.

7.2. Uniaxial ratcheting loading – validation of the ratcheting model

Qualitative and quantitative matching of both the hysteresis loops and the cyclic softening curvature assures the transferability of the model and necessary material parameters to complex loading conditions such as asymmetric ratcheting loading. Figure 9a-b shows a graphical representation of ratcheting experiments and simulated curves. It is observed that, the ratcheting strain increases with increasing mean stress and increasing stress amplitude. However, for higher stress values (both mean stress and stress amplitude) the model over-estimates the experimental results. Note that the influences of mean stress and stress amplitude are captured separately in the simulation.

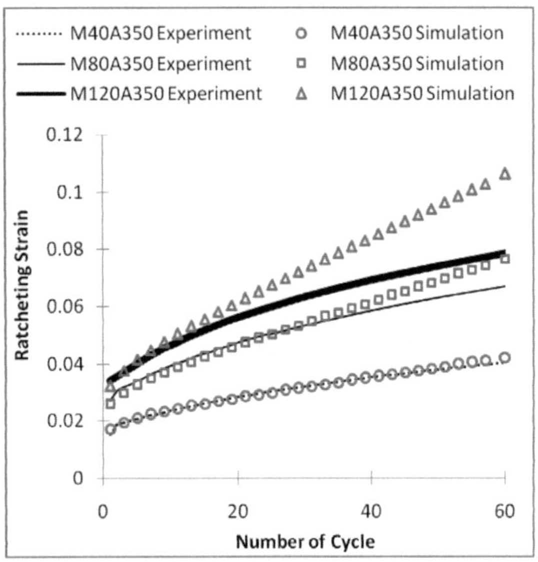

Figure 9a. Constant Stress Amplitude = 350MPa

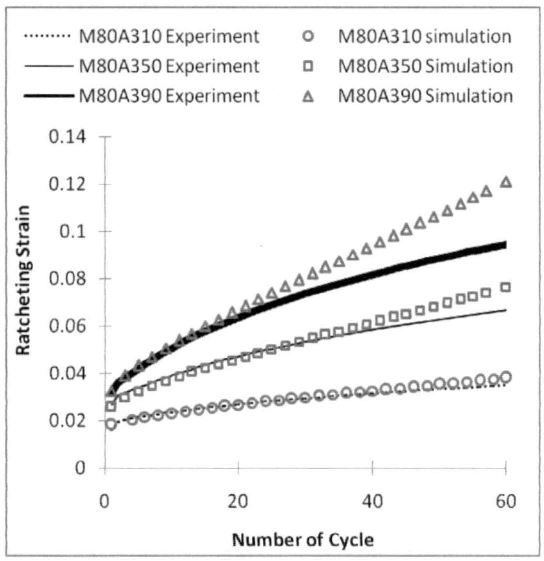

Figure 9b. Constant Mean Stress = 80MPa

8. Conclusion

The present work has examined four cyclic plastic phenomena, namely: i) steady-state hysteresis loops and the Bauschinger effect, ii) the Masing phenomena, iii) cyclic softening in initial transient cycles and, iv) ratcheting. The hysteresis loop shape and Bauschinger effect is well modelled by the Ohno-Wang kinematic hardening law. The parameters for the Ohno-Wang law have been calculated from experimental data on uniaxial LCF loading. It is found that for the material examined (SA333 Grade-6 C-Mn steel) one set of KH parameters is sufficient as the material shows Masing behaviour. Cyclic softening is effectively simulated using model assuming dislocation annihilation at low angle sub-grain boundaries. The ratcheting is simulated by modifying the kinematic hardening contribution by an additional increase in dislocation density due to the asymmetry in loading. Phenomenological modelling results using this model for cyclic softening and ratcheting show a reasonable agreement with the experimental results.

Acknowledgements

The authors acknowledge Bhabha Atomic Research Centre, Mumbai for financial assistance through collaborative project and National Metallurgical Laboratory, Jamshedpur for experimental support. The authors also acknowledge Dr. Surajit Kumar Paul, National Metallurgical Laboratory, Jamshedpur for TEM micrographs.

References

[1] Read W T and Shockley W 1950 *Physical Review* **78** 3 275
[2] Sauzay M, Brillet H, Monnet I, Mottot M, Barcelo F, Fournier B and Pineau A 2005 *Materials Science and Engineering A* **400-401** 241
[3] Cailletaud G and Sai K 2008 *Materials Science and Engineering A* **480** 1–2 24
[4] Taleb Lakhdar and Hauet Annie 2009 *International Journal of Plasticity* **25** 71359
[5] Ohno N and Wang J D 1993 *International Journal of Plasticity* **9** 3375
[6] SinclairCW, Poole W J and Bréchet Y 2006 *Scripta Materialia* **55** 8739
[7] Meckings H and Kocks U F 1981 *Acta Metallurgica* **29** 1865
[8] Chaboche J L, Dang Van K and Cordier G 1979 *SMiRT-5*, Berlin
[9] McDowell D L 1985 *Journal of Applied Mechanics* **52** 298
[10] Jiang Y and Kurath P 1995 *International Journal of Plasticity* **12** 3387

Investigation of the grain size effect on mechanical properties of Ti-6Al-4V alloy with equiaxed and bimodal microstructures

Yan CHONG[1*], Tilak BHATTACHARJEE[1,2], Akinobu SHIBATA[1,2], Nobuhiro TSUJI[1,2]

[1]Department of Materials Science and Engineering, Kyoto University, Kyoto, Japan
[2]Element Strategy Initiative for Structural Materials (ESISM), Kyoto University, Kyoto, Japan

*E-mail: chong.yan.6w@kyoto-u.ac.jp

Abstract. In this study, equiaxed microstructures with various grain sizes ranging from 6.0μm to 0.3μm and bimodal microstructures with grain sizes of primary α ranging from 5.0μm to 0.6μm were successfully fabricated by hot deformation and subsequent annealing in α+β two phase region in Ti-6Al-4V alloy having a martensite initial microstructure. The mechanical properties of both microstructures with different grain sizes were tested at room temperature. It was found that the strength (yield and tensile strength) of both microstructures increased with the decrease of the grain size. In the equiaxed microstructure, the uniform elongation gradually decreased with the decrease of the grain size, which was in consistent with the behavior of other metallic materials with ultrafine grains. However, in the bimodal microstructure, relatively large uniform elongation (~8%) was stably obtained regardless of the grain size. The unique and superior balance between strength and uniform elongation in bimodal microstructures was discussed, considering the contribution of interfaces between primary α grains (α_p) and transformed β areas (β_{trans}).

1. Introduction

Ti-6Al-4V is one of the most popular α (HCP) + β (BCC) titanium alloys, which has been widely used for structural application due to its high strength-to-weight ratio and exceptional corrosion resistance [1]. A variety of microstructures including lamellar/martensite microstructure, bimodal microstructure and equiaxed microstructure can be obtained in Ti-6Al-4V by designing different heat treatments or thermomechanical processing routes [1]. Compared with the lamellar/martensite microstructure, the equiaxed and bimodal microstructures had a better balance between strength and ductility, making them more widely used in the industrial applications.

Grain refinement is the only strengthening mechanism that can increase the strength and toughness of metallic materials at the same time. It has been, on the other hand, found that yield strength and tensile strength significantly increase due to grain refinement, whereas uniform and total elongation decrease when the grain size becomes ultrafine (smaller than a few micro-meters). This is attributed to the loss of work-hardening ability and resultant early plastic instability in ultrafine-grained (UFG) materials compared to its coarse-grained counterparts. One way to restore the work hardening ability of UFG materials is to make the UFG structures dual phase comprised of soft matrix phase and hard second phase. One such example is a dual phase steel composed of soft ferrite phase and hard martensite or bainite phase. Calcagnotto et al. [2] have shown that by reducing the ferrite grain size from 12.4μm to 1.2μm in a ferrite-martensite dual phase steel, both yield strength and tensile strength significantly increased while uniform elongation and total elongation were hardly affected. Similar results have also been reported in other dual phase steels [3-5].

Recently, Matsumoto and other researchers [6-11] proposed a method so-called warm deformation (T≤800°C) of martensite microstructure for obtaining ultrafine-grained (UFG) equiaxed microstructure in Ti-6Al-4V alloy, following the process using martensite initial microstructure in carbon steels [12]. Although the obtained UFG equiaxed microstructure showed a low-temperature or/and high strain rate superplasticity [7, 8], the corresponding room temperature mechanical properties suffered from poor uniform elongation (~1%) [11]. On the other hand, no efforts have been made on getting UFG bimodal microstructure as well as its mechanical properties. Therefore, the main scope of the present study is to fabricate equiaxed and bimodal microstructures with different fine grain sizes in Ti-6Al-4V alloy and then systematically investigate the influence of grain size on the mechanical properties of both microstructures.

2. Experimental methods

The chemical composition of the Ti-6Al-4V alloy used in this study was Ti-6.29Al-4.35V-0.155O-0.225Fe (in wt.%). Cylinder samples (8mm in diameter, 12mm in height) cut from the as-received billet were solution-treated at 1050°C for 30min and followed by water quenching to get a fully martensite microstructure shown in Fig. 1(a). There are two advantages in using martensite microstructure rather than lamellar microstructure as the initial microstructure of subsequent thermomechanical treatment. The first advantage is the much finer α-lamellae thickness in martensite microstructure than that in lamellar microstructure, which would enhance grain refinement of α phase in hot deformation. The average α-lamellae thickness in the martensite microstructure was measured to be ~1μm. The second advantage is the much larger area having uniform microstructures in hot-deformed specimens, which made mechanical testing of obtained equiaxed/bimodal microstructures possible and reliable. Hot compression experiments were conducted on a thermomechanical processing simulator (Thermecmastor-Z) at various deformation temperatures and strain rates. The processing route is illustrated in Fig. 1(b). The samples with the martensite microstructure were heated to the deformation temperatures at a heating rate of 10°C/s and held for 20min for homogenizing temperature of the specimen. After that, the samples were uniaxially hot compressed to a true strain of 0.8 at a certain temperature and strain rate. After the hot deformation, in order to get equiaxed microstructures, the samples were subsequently cooled at a slow cooling rate (10°C/min) to 600°C and followed by water quenching. On the other hand, in order to get bimodal microstructures, the samples were re-heated to 910°C and held for different periods of time (2~1200s) followed by water quenching.

After the hot compression and heat treatment, the cylindrical specimens were cut in half along a diameter. The sections parallel to the compression axis and including the center of the specimens were then ground and polished in standard metallographical methods. Electrical polishing in a solution of 10% perchloric acid and 90% methanol at the temperature of -30°C was conducted to reveal dual-phase microstructures. Microstructural observations of the hot-compressed and annealed specimens were carried out at the center of the specimens on the sections described above. Backscattered electron (BSE) observations were conducted using a JEOL 7800F scanning electron microscope (SEM). The accelerating voltage for the BSE experiments was 15kV. Electron backscattered diffraction (EBSD) investigations were conducted using TSL system attached to a field emission gun SEM (JEOL 7100F). The accelerating voltage was 15kV. The collected data were analyzed using OIM software. Micro-tensile specimens were cut from the center of the specimens, with the gauge length of 2mm, width of 1mm and thickness of 0.5mm. The broad face of the sheet-type tensile specimen was parallel to the compression axis, and the tensile direction of the specimen was perpendicular to the compression axis. SEM characterization for all the tensile samples were conducted before tensile experiments to confirm the microstructures. Tensile strains of the micro-specimens were precisely measured by digital image correlation (DIC) method. For each thermomechanical condition, at least two specimens were tested.

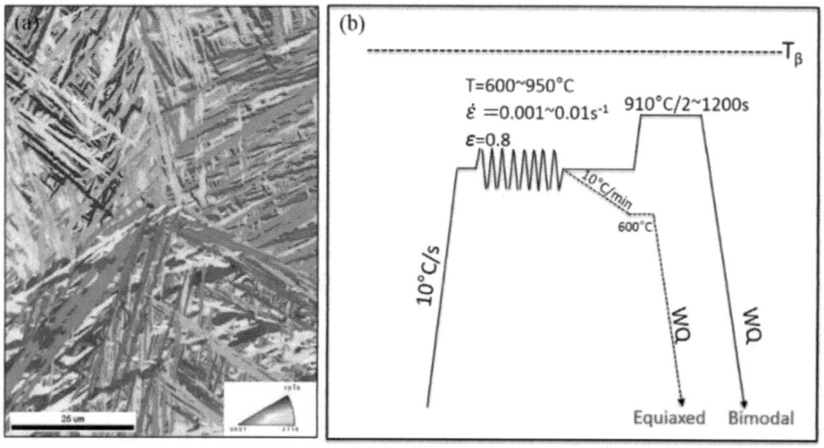

Figure 1. (a) EBSD inverse pole figure (IPF) map of the martensite initial microstructure used in this study; (b) Illustration of thermomechanical processing for obtaining equiaxed and bimodal microstructures with different grain sizes.

3. Results and discussion

Through the thermomechanical processes shown in Fig. 1(b), eight kinds of equiaxed microstructures with different average grain sizes ranging from 6.0μm to 0.3μm and eight kinds of bimodal microstructures with average grain sizes of equiaxed primary α ranging from 5.0μm to 0.6μm were successfully fabricated. Examples of some typical microstructures are shown in Fig. 2. In Fig. 2 (a)-(c), image quality + grain boundary maps are shown for the equiaxed microstructures with average grain sizes of 6.0μm, 2.1μm and 0.5μm, respectively. In these figures, blue and red lines represent high angle and low angle grain boundaries, respectively. Fairly uniform microstructures were observed in Fig.2 (a)-(c), and the fraction of retained β phase was about 5%. In Fig. 2 (d)-(f), BSE images are shown for the bimodal microstructures with average grain sizes of 5.0μm, 2.4μm and 0.8μm, respectively. The bimodal microstructures were comprised of primary α grains with equiaxed morphologies (highlighted by red dotted lines in Fig. 2(d)) and transformed β areas, i.e., the β areas till final cooling (highlighted by yellow dotted lines in Fig. 2(d)). The transformed β area was made of the fine secondary α-lamellae, which formed during the water quenching process after annealing at 910°C in Fig. 1(b), and a small amount of retained β between the secondary α-lamellae. It should be noted that the volume fractions of primary α grains were around 47% for all the bimodal microstructures with different grain sizes.

The mechanical properties of both kinds of microstructures with different grain sizes were tested at room temperature. The engineering stress-strain curves obtained are shown in Fig. 3. In the equiaxed microstructures (Fig. 3(a)), the yield/tensile strength increased with the decrease of the grain size, while the uniform elongation gradually decreased. When the grain size was smaller than 1μm, early necking of tensile specimens happened shortly after yielding, which indicated the early plastic instability due to the limited work-hardening ability of the UFG equiaxed microstructure. This result agreed with the previous studies on single-phased UFG materials [13-15]. There was also a tendency of slight decrease in total elongation with the grain refinement for the equiaxed microstructures. In the bimodal microstructures (Fig. 3(b)), the yield/tensile strength also increased with the decrease of the grain size. However, unlike the gradual decrease of uniform elongation in the equiaxed microstructures, the uniform elongation in the bimodal microstructure was maintained after the grain refinement. Even in the UFG bimodal microstructure with a grain size of 0.6μm, 7% uniform elongation could be achieved, which was much larger than that of the UFG equiaxed microstructure with a grain size of 0.5μm. There was no clear change and tendency of the total elongation with the grain refinement in the bimodal microstructures. It can be concluded, therefore, that the grain refinement of the bimodal microstructure of Ti-6Al-4V alloy provides a possibility to increase the strength without sacrificing the ductility.

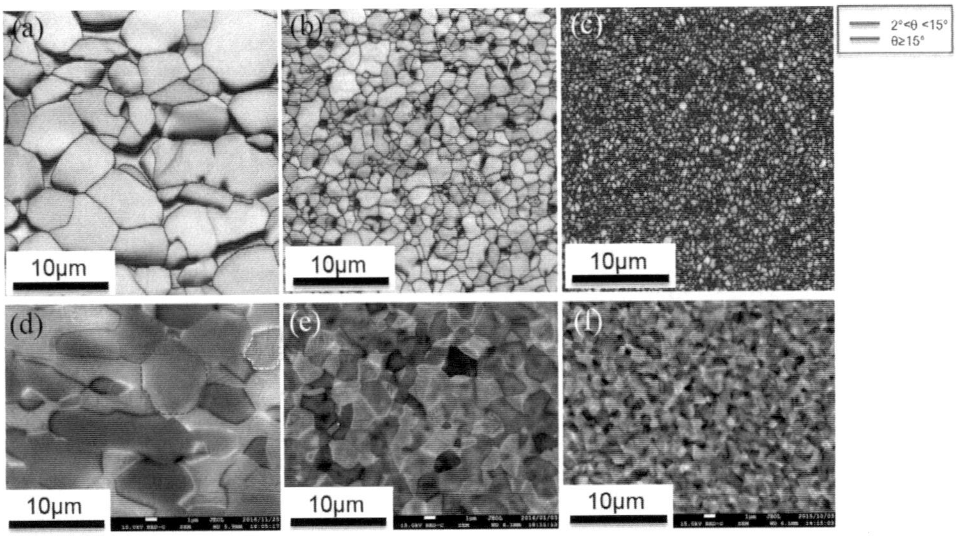

Figure 2. Examples of equiaxed and bimodal microstructures with different grain sizes obtained after thermomechanical processing shown in Fig.1(b): (a) equiaxed, average grain size D=6.0μm, (b) equiaxed, D=2.1μm, (c) equiaxed, D=0.5μm, (d) bimodal, average grain size of equiaxed primary α D=5.0μm, (e) bimodal, D=2.4μm, (f) bimodal, D=0.8μm.

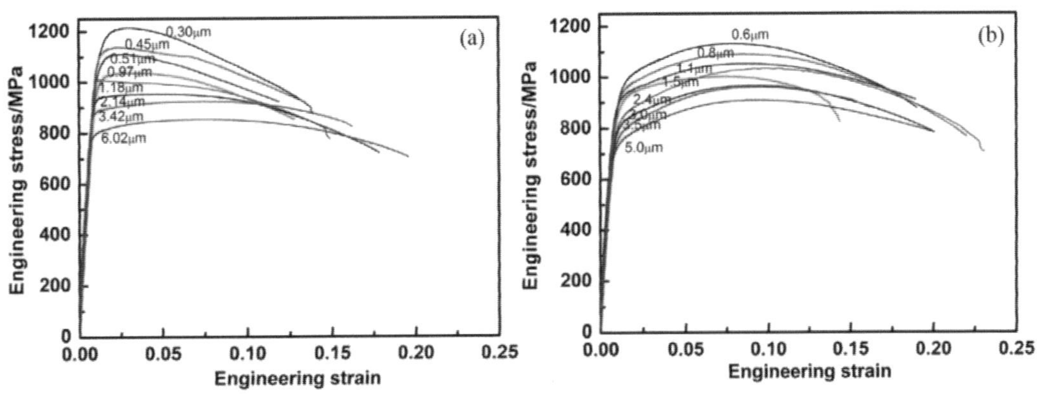

Figure 3. Engineering stress-strain curves of equiaxed (a) and bimodal microstructures (b) with different grain sizes of equiaxed α.

The bimodal microstructure can be technically considered as a composite of equiaxed α grains and martensite (transformed β areas), as shown in Fig. 4(a). Therefore, in order to evaluate the separate contribution of these two constituents to the excellent work-hardening ability of bimodal microstructure, the engineering stress-strain curves of fully equiaxed, martensite and bimodal microstructures are compared in Fig. 4(b). In this figure, the full martensite microstructure was obtained by solution treatment in β single phase region and followed by water quenching. The grain sizes of the equiaxed microstructure and bimodal microstructure were similar to each other (0.98μm and 0.80μm). It is interesting that neither the martensite nor the equiaxed microstructures showed good work-hardening ability and large uniform elongation, although their combination (bimodal microstructures) showed nice work-hardening and uniform elongation. In the equiaxed microstructure with a grain size of about 1μm, necking occurred shortly after yielding. In the martensite microstructure, the peak stress was reached shortly after yielding and the flow stress remained nearly the same till fracture. However, when the two constituents were combined together, the bimodal microstructure exhibited slightly lower yield strength than other two microstructures but much larger work-hardening ability with ~8% of uniform elongation. This might suggest that the strain partitioning behavior between primary α grains and transformed β areas (martensite) as well as the interfaces between these two constituents played important roles in obtaining the superior work-hardening ability in the bimodal microstructure.

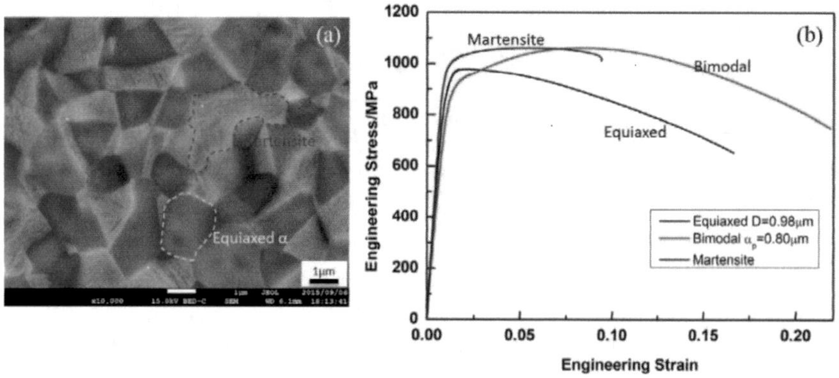

Figure 4. (a) BSE image of a bimodal microstructure, which can be considered as a composite of equiaxed α grains and martensite (transformed β areas); (b) Comparison of engineering stress-strain curves between fully equiaxed, martensite and bimodal microstructures.

To further investigate the deformation behavior of bimodal microstructures, interrupted tensile deformation was carried out on specimens with pre-polished surfaces. A bimodal microstructure with a mean grain size (of equiaxed primary α) of 2.4μm (Fig.5(a)) was used. Engineering stress-strain curve of the bimodal microstructure shown in Fig. 5 was divided into two regions separated by the peak stress. The tensile deformation was stopped at different tensile strains, i.e., in region (1) (before the peak stress), at the peak stress, and in region (2) (after the peak stress). BSE images of the specimens at different stages are shown in Fig.5 (b), (c), and (d), respectively. In region (1) (Fig. 5(b)), which corresponded to the uniform deformation region, a number of straight slip lines were observed within the primary α grains, as were pointed out by white arrows in the figure. No clear slip lines were observed within the transformed β areas, probably due to the much finer lamellar structure inside. At the peak stress (Fig. 5(c)), more slip lines with larger contrast were observed within the primary α grains, as indicated by a white arrow. Micro-shear bands formed within the transformed β areas, as pointed out by red arrows. After necking (Fig. 5(d) and (e)), both slip lines within the primary α grains and micro-shear bands within the transformed β areas became more severe, especially within necked part of the tensile specimen (Fig. 5(e)). Some micro-cracks were found either within the micro-shear bands or at the interface between primary α grains and transformed β areas.

The above results suggested that the deformation behaviors of primary α grains and transformed β areas differed significantly to each other. It is believed in this study that such difference would lead to a strain partitioning between primary α grains and transformed β areas, similar to the case of ferrite/martensite dual-phase steels [3-5]. The strain partitioning would also lead to a strain gradient near the interface, which would possibly activate unusual slip systems (like pyramidal) in addition to the basal and prismatic slip systems which are commonly observed in HCP-α of titanium alloys during deformation. With the decrease in the grain size in the bimodal microstructure, more interfaces between primary α grains and transformed β areas are introduced into the microstructure and therefore more fractions of pyramidal slip systems could be activated. This can provide additional work-hardening ability compensating the loss of work-hardening ability in the fine-grained structure which would lead to a stable uniform elongation in the fine-grained bimodal microstructures. Nevertheless, the detailed mechanism needs further investigation to show clearer evidences.

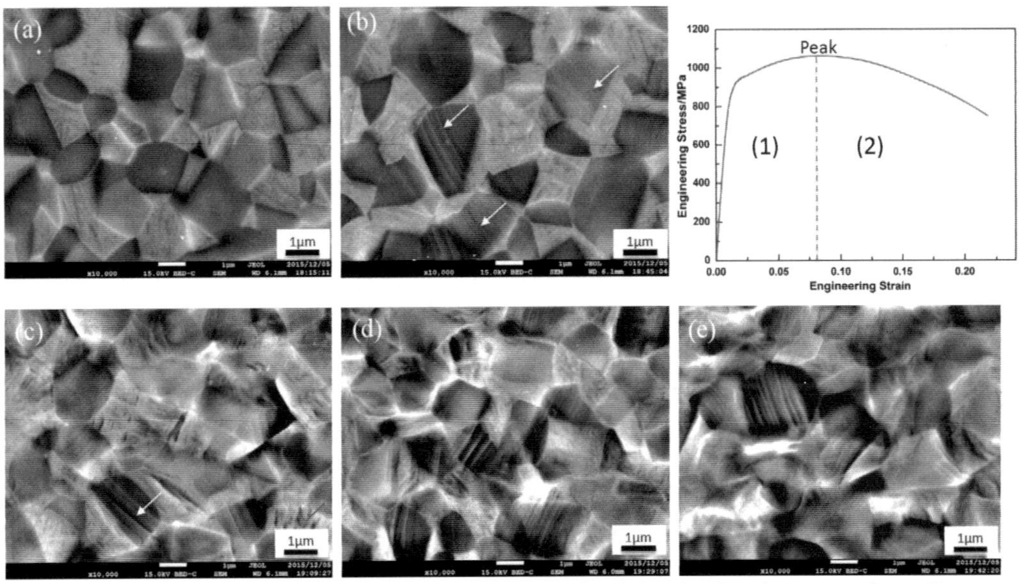

Figure 5. BSE images of the bimodal microstructure (D=2.4μm) after interrupted tensile tests: (a) Initial microstructure (undeformed); (b) interrupted before the peak stress, region (1) in the stress-strain curve inserted; only slip lines marked by white arrows were observed within the primary α grains, (c) interrupted at the peak stress; slip lines marked by white arrows were observed within the primary α grains and micro-shear bands marked by red arrows were observed within the transformed β areas; (d) interrupted after the peak stress, region (2), out of the necked region; (e) after the peak stress, region (2), out of the necked region. Some micro-cracks were observed at the interfaces between primary α grains and transformed β areas.

4. Conclusions

In this study, equiaxed and bimodal microstructures with various different grain sizes were successfully fabricated by thermomechanical processing starting from martensite initial microstructure in Ti-6Al-4V alloy. By tensile tests at room temperature, it was found that the bimodal microstructure had a superior balance between strength and ductility, in which the strength increased with the decrease of the grain size while the uniform elongation remained unchanged. The reason for the stable uniform elongation in the bimodal microstructure was briefly discussed. It was considered that the increased volume fraction of interfaces between primary α grains and transformed β areas with decreasing the grain size played an important role in enhancing work-hardening ability to maintain the uniform elongation in the UFG bimodal microstructures.

Acknowledgement

The authors were financially supported by the Cross-Ministerial Strategic Innovation Promotion Program on Structural Materials for Innovation (SIP-SM^{4}I) in Japan, and the support is gratefully appreciated.

References

[1] Lutjering G 1998 *Mater. Sci. Eng* **A243** 32-45
[2] Calcagnotto M, Adachi Y, Ponge D and Dierk Raabe 2011 *Acta Mater.* **59** 658-670
[3] Calcagnotto M, Ponge D and Dierk Raabe 2010 *Mater. Sci. Eng.* A **527** 7832-7840
[4] Son Y I, Lee Y K, Park K T, Lee C S and Shin D H 2005 *Acta Mater.* **53** 3125-3134
[5] Calcagnotto M, Ponge D and Dierk Raabe 2012 *ISIJ Int.* **52** 874-883
[6] Matsumoto H, Liu B, Lee S K, Li Y P, Ono Y and Chiba A 2013 *Metall. Mater. Trans.* **44A** 3245-3260
[7] Matsumoto H, Yoshida K, Lee S H, Ono Y and Chiba A 2013 *Mater. Let.* **98** 209-212
[8] Matsumoto H, Velay V and Chiba A 2014 Mater. Des. **66** 611-617
[9] Mironov S, Murzinova M, Zherebtsov S, Salishchev G A and Semiatin S L 2009 *Acta Mater.* **57** 2470-2481
[10] Zherebtsov S, Murzinova M, Salishchev G and Semiatin S L 2011 *Acta Mater.* **59** 4138-4150
[11] Zherebtsov S, Kudryavtsev E, Kostjuchenko S, Malysheva S and Salishchev G 2012 *Mater. Sci. Eng.* A **536** 190-196
[12] Tsuji N,Ueji R, Minamino Y and Saito Y 2002 *Scripta Mater.* **46** 305-310
[13] Tsuji N, Kamikawa N, Ueji R, Takata N, Koyama H and Terada D 2008 *ISIJ Int.* **48** 1114-1121.
[14] Tsuji N 2009 *J. Phys.* **165** 012010
[15] Tsuji N, Ito Y, Saito Y and Minamino Y 2002 *Scripta Mater.* **47** 893-899

Deformation processed Al/Ca nano-filamentary composite conductors for HVDC applications

C F Czahor[1,2,3], I E Anderson[1,2], T M Riedemann[1] and A M Russell[1,2]

[1] Ames Laboratory, U.S. Department of Energy, Ames, Iowa
[2] Materials Science and Engineering, Iowa State University, Ames, Iowa
[3] Wind Energy Science, Engineering, and Policy, Iowa State University, Ames, Iowa

E-mail: cczahor@iastate.edu

Abstract. Efficient long-distance power transmission is necessary as the world continues to implement renewable energy sources, often sited in remote areas. Light, strong, high-conductivity materials are desirable for this application to reduce both construction and operational costs. In this study an Al/Ca (11.5% vol.) composite with nano-filamentary reinforcement was produced by powder metallurgy then extruded, swaged, and wire drawn to a maximum true strain of 12.7. The tensile strength increased exponentially as the filament size was reduced to the sub-micron level. In an effort to improve the conductor's ability to operate at elevated temperatures, the deformation-processed wires were heat-treated at 260°C to transform the Ca-reinforcing filaments to Al_2Ca. Such a transformation raised the tensile strength by as much as 28%, and caused little change in ductility, while the electrical conductivity was reduced by only 1% to 3%. Al/Al_2Ca composites are compared to existing conductor materials to show how implementation could affect installation and performance.

1. Introduction

A 48% increase in worldwide energy demand is expected by 2040, which will require significant expansion of the electrical power transmission infrastructure [1]. Expanded long-distance transmission grids in China, the United States, and elsewhere are expected to make greater use of high-voltage direct current (HVDC) transmission, the preferred technology for long distances [2]. Conventional aluminum-conductor steel-reinforced (ACSR) cables are not well suited for HVDC transmission due to the presence of the heavy and poorly conducting steel core needed for strength and sag-resistance. Al/Ca composite conductors with monolithic construction have shown promise as a possible next-generation conductor for this application.

Deformation-processed metal-metal composites (DMMCs) provide an appealing approach for producing lightweight materials with both high strength and high conductivity. DMMCs can be produced by powder metallurgy and severe plastic deformation resulting in nano-filamentary reinforcement. Extensive study of this class of materials has shown that they exhibit strength that increases exponentially with true strain while maintaining conductivity close to the rule-of-mixtures (ROM) prediction [3]. Both materials in the composite must be highly ductile in order to withstand the extensive deformation processing without fracturing. The proposed conductor material utilizes Al as the primary phase, and Ca as the secondary reinforcement phase that deforms into sub-micron-thickness filaments during wire drawing, resulting in interface strengthening. Both Al and Ca are ductile fcc metals with high conductivities, similar mechanical properties, and low densities.

One key finding from previous study of Al/Ca composites was the identification of a transformation of the reinforcing Ca phase to the Al_2Ca intermetallic phase at temperatures as low as 175°C, a temperature sometimes reached by transmission conductors during periods of heavy electrical demand [4, 5]. This study examines the effect of intentionally converting the reinforcement phase to an intermetallic species to produce a material with high-temperature stability while retaining the superior performance properties that have been shown in the prior Al/Ca composite materials. A reduced initial volume loading of Ca was necessary to compensate for volume expansion during transformation and an expected increase in resistivity. A formulation with 11.5 vol. % Ca was chosen with the aim of producing a composite with 18 vol. % Al_2Ca, found in previous research to be a desirable DMMC reinforcement fraction [6]. The microstructure, temperature stability, conductivity, and mechanical properties were investigated for both Al/Ca and Al/Al_2Ca composite samples.

2. Experimental Procedure

An Al/Ca conductor sample was prepared starting from metal powders produced at Ames Laboratory. Al powder with 99.99% purity of particle sizes 45-75 μm and 99.5% pure Ca with particle size 75-125 μm were used in this study. Powders were weighed (155.45 g Al, 11.45 g Ca, which is 11.5vol.% Ca), blended with a multi-axis mixer (Turbula), and die compacted at a pressure of 61.2 MPa into cylindrical compacts (diameter = 72.1 mm). Each compact had a relative density of 83.8% of the theoretical value (from ROM calculation), in close agreement with the expected room temperature compressibility of Al powder [7]. The compacted powder "pucks" were stacked inside a pure Al (1100-H14 alloy) can with outer diameter 90.7 mm, inner diameter of 78 mm, and interior height of 196.8 mm. Indirect extrusion was performed at the TU Berlin Extrusion Research and Development Center at 285 °C, with an exit die diameter of 21 mm, giving an effective extrusion ratio of 14.9 when accounting for porosity and void space present in the billet. A rod of length 275 cm was sectioned into quarters, and the can material was machined from the rod's exterior with a lathe. The resulting diameter was reduced at room temperature by swaging to 0.69 mm, at which point wire drawing was used to further reduce the diameter in a series of steps to 0.11 mm. The excellent ductility of the components in the composite allowed all deformation processing to be performed without any stress-relief annealing. The properties investigated in this study were compared on the basis of deformation true strain, as shown in table 1. This is calculated according to equation (1), where d_i and d_f are the initial and final wire diameters, respectively, and η is true strain.

$$\eta = 2 \ln \frac{d_i}{d_o} \tag{1}$$

Table 1: Wire sizes and strain levels for study of 11.5 vol. % Al/Ca.

Diameter (mm)	2.97	2.16	0.94	0.50	0.40	0.31	0.20	0.11
η	6.11	6.74	8.41	9.67	10.1	10.6	11.5	12.7

The microstructure of the wire samples was studied using a FEI Teneo LoVac field–emission scanning electron microscope (FE-SEM). The electrical conductivity was determined by four-point probe measurements at room temperature following ASTM B193-16 [8] by applying a series of currents with a Keithley 6220 Precision Current Source and measuring the voltage difference with a Keithley 2182A nano-voltmeter. Tensile testing was done with multiple instruments to accommodate the various wire sizes that were examined. Larger samples ($\eta<7$) were tested using an Instron 3367 instrument at an elongation rate of 2 mm/min. For samples with $8< \eta<12$, the tensile strength was measured using a Zwick/Roell Z 2.5 instrument at the same elongation rate. For the 0.11 mm diameter (η =12.7) wire, tests were completed at Psylotech's facility using their μTS (under-microscope) universal test system, allowing for precision measurements at low forces. The ultimate tensile strength (UTS) was used to characterize the samples as it is more reproducible and reliable than yield strength for small specimens and is the strength value most often cited for transmission conductor materials [9].

3. Results and Discussion

3.1 Microstructural analysis of Al/Ca (11.5 vol.%) and Al/Al_2Ca (18 vol.%) composites

The microstructures of DMMCs have been studied in great detail in the past several decades [3, 6, 10] and in recent years for the Al/Ca composite system [4, 5, 11]. Figure 1 shows that the ribbon-shaped morphology seen in the prior generation of Al/Ca composites exists at lower strain levels, but is altered by extensive mechanical working. The resultant morphology is in contrast to the expectations of classic composite morphology theory, which predicts cylindrical filaments due to axi-symmetrical deformation of a fcc secondary phase in a fcc matrix. A possible explanation is a temporary crystal structure transformation in the Ca from fcc to bcc during extrusion leading to plane-strain deformation giving rise to a ribbon-like shape [4, 5, 6], but this has not been further explored in this study.

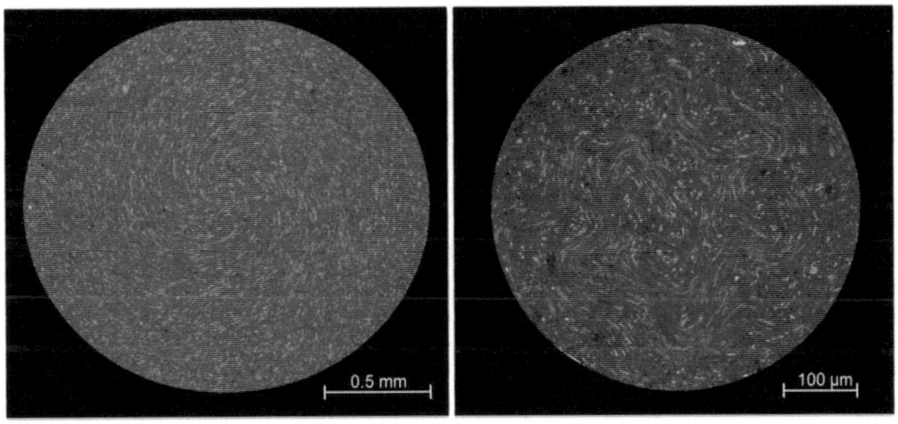

(a) (b)

Figure 1. Backscattered electron micrographs of Al/Ca (11.5 vol.%) composite transverse cross section with pure Ca filaments: (a) η=6.74, (b) η=9.67.

Following the conversion of the Ca reinforcement to Al_2Ca, the filament size increases, as shown in figure 2. One critical concern with the heat treatment process is spheroidization, which can lead to the breaking of filaments and consequently degradation in mechanical and electrical properties. Based on micrographs at various strain levels, it appears that this did not occur, and the filaments were intact.

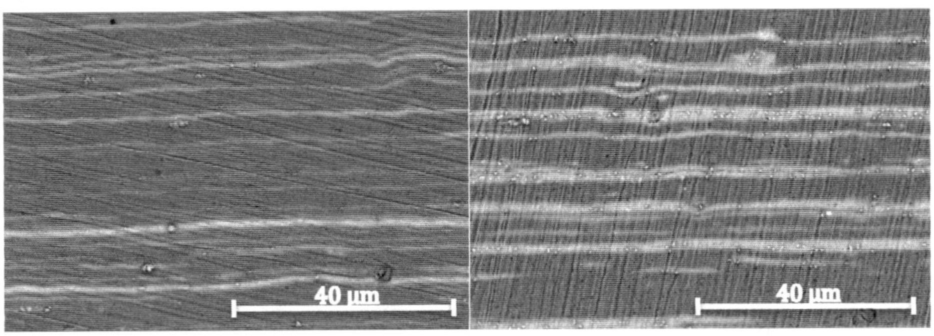

(a) (b)

Figure 2. Backscattered electron micrographs of Al/Ca (11.5 vol.%) composite (longitudinal cross section) at η=9.67: (a) pure Ca filaments, (b) Al_2Ca reinforcement after heat treatment.

The micrographs in figure 2 clearly show a change in the filament size, with a measured increase in thickness from 0.9µm to 1.5µm for the selected strain level. This difference is very close to that expected, indicating that the transformation to Al_2Ca is complete, as was also confirmed by separate differential scanning calorimetry (DSC) and X-ray ray diffraction studies [4, 5]. The filaments are able to deform with a consistent thickness as Al and Ca have nearly equal flow stresses, avoiding problems with a softer phase flowing around a harder second phase [6, 12]. The filament size and spacing are critical parameters that affect the performance of DMMCs and allow comparison of small lab scale specimens to potential full-size extrusions on the basis of true strain. DSC experiments and SEM observations confirmed that the transformation was complete but not excessive, enabling development of a heat treatment process and the study of modified performance properties.

3.2 Electrical Conductivity of Al/Ca (11.5 vol.%) and Al/Al_2Ca (18 vol.%) composites

Given the intended application of Al/Ca composites, conductivity is a critical property in evaluating the commercial viability of this material. The primary advantage of Al/Ca composites is the absence of a highly resistive steel core, needed for strength in conventional conductors (ACSR cables). The monolithic construction allows for high conductivity across the entire cross section with the direction of current flow being parallel to the filaments. Figure 3 shows results of conductivity measurements on both unconverted wires and those that have been transformed at various strain levels.

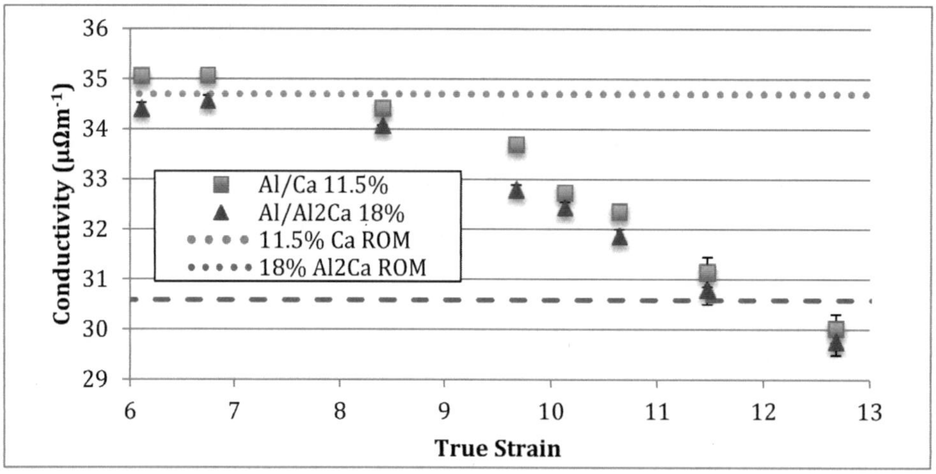

Figure 3. Electrical conductivity of Al/Ca (11.5% vol.) composite before and after transformation of filaments to Al_2Ca intermetallic at various strain levels.

At low strain levels the conductivity is very close to the rule of mixtures prediction for the given formulation, as current flows across the entire cross section and the filament size is significantly larger than the mean free path of the electrons. Tian et al. have studied the numerous scattering mechanisms in DMMCs and shown that by reducing the filament thickness and spacing, there is an increase in interface and grain boundary scattering that reduces the conductivity of the material [13].

The effect on conductivity of the transformation of the filaments to Al_2Ca is evident but was found to be small. For the wires subjected to the greatest level of deformation in this study ($\eta = 12.7$), the measured drop in conductivity is less than 1%. With a low loading of Ca in Al/Ca (11.5% vol.), the transformation has a minimal impact (maybe a slight increase) on interfacial area, and, therefore, the influence of interface scattering is not significantly affected. Thus, the very slight reduction observed does not agree with (and is well above) the ROM expectation for Al/Al_2Ca (18 vol.%), and is a pleasant surprise. Another factor that could cause a decrease in conductivity is a reduced volume fraction of the (highly conductive) Al matrix phase after Ca transformation to Al_2Ca, which has been found to leave the wire diameter essentially unchanged. However, the measured decrease also seems to be less than this effect would indicate, still leaving this beneficial observation unexplained. Therefore, the minimal decrease in conductivity seems to be a worthwhile sacrifice in exchange for the gain in strength and high-temperature stability that the intermetallic phase should provide. With complete transformation to Al_2Ca, the composite material is expected to have a significantly increased upper operating temperature limit, making it resilient during emergency overloading situations [9].

3.3 Tensile strength of Al/Ca (11.5 vol.%) and Al/Al_2Ca (18 vol.%) composites

While decreasing the filament size by mechanical working lowers the conductivity of the composite, the primary motivation for doing so is to benefit from the unique strengthening mechanism of DMMCs. Typically metal matrix composites exhibit strengths that are linearly dependent on the volume fraction of the reinforcement phase. The earliest Cu-Nb DMMCs produced by Bevk et al. showed that extensive deformation of reinforcements could lead to an exponential increase in strength greatly exceeding rule of mixture predictions [3]. The stress-strain relationship and fracture surfaces from tensile testing indicated that the composites of the present study still demonstrate ductile behavior upon conversion to intermetallic reinforcement while the ultimate tensile strength increases. Figure 4 shows the exponential relationship between deformation true strain and UTS for as-drawn and converted wire samples at various strain levels, with strength values for commercial cables included for comparison.

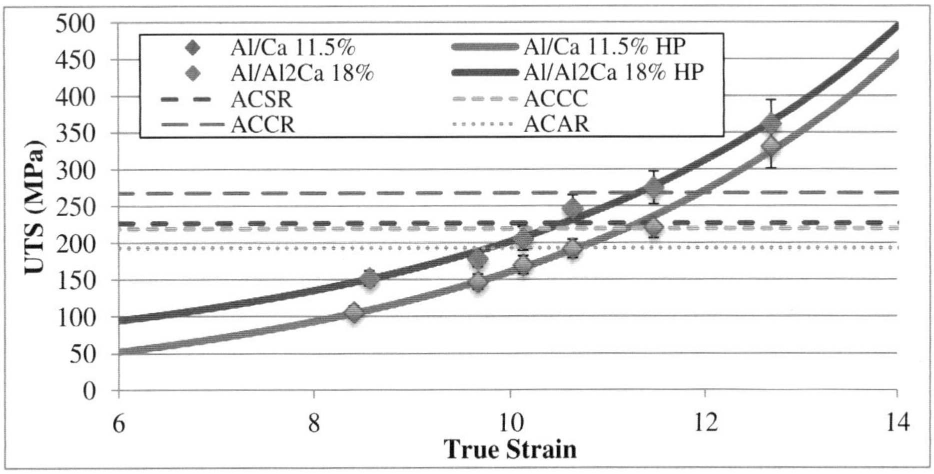

Figure 4. Tensile strength of Al/Ca (11.5% vol.) and Al/Al_2Ca (18% vol.) composites.

The influence of filament size and fractional loading suggests that interfacial area plays a major role in strengthening the composite. Study of early DMMCs has shown that interphase boundaries act as barriers to dislocation glide leading to interface strengthening [3]. This effect causes the strength to increase sharply at high strains, which has been successfully modeled by a modified Hall-Petch barrier model for the previous generations of Al/Ca composites [4, 5, 11]. Based on measurements of filament thickness (t), the dependence on true strain (η) was assumed to be ideal following the relationship of equation (2) where d_o is the starting Ca filament size.

$$t = d_o e^{-\frac{1}{2}\eta} \tag{2}$$

Using the calculated thickness from equation (2) and the UTS measurements, Hall-Petch relationships were determined for both as-drawn and transformed wires as shown by equations (3) and (4), respectively.

$$UTS = -10.7 + 148\, t^{-0.5} \tag{3}$$

$$UTS = 32.3 + 147\, t^{-0.5} \tag{4}$$

The evident strengthening effect from conversion of the reinforcement phase to intermetallic was expected, due to the increase in interfacial area upon volume expansion of the filaments. It has been shown that both the material and composition of the reinforcement phase in aluminum matrix DMMCs influence the ultimate tensile strength of the composite [4, 14, 15]. When compared to Al/Ca (11.5% vol.) it appears that the filament composition (Al_2Ca rather than Ca) has little effect, and that the volume of reinforcement is the dominant factor. The curves for these wires have a very similar shape and differ by an offset for all strain levels. This direct comparison of as-drawn and transformed wires however, fails to account for the impact of volume expansion on the effective true strain. To account for this the filament thickness calculated by equation (2) needs to be adjusted by a factor related to the volume expansion of the filaments, therefore reducing the effective true strain, which shifts the measured strength values to the left. The major takeaway from analyzing the tensile strength with this adjustment is that Al_2Ca is a more effective reinforcement material than pure Ca and would result in stronger wires at any given volume fraction of reinforcement.

Further study of the material properties of the Al_2Ca compound is necessary in order to utilize more advanced modeling techniques that incorporate the individual contributions from each phase to predict the strength [16]. From a practical standpoint, comparing unconverted and transformed wires on the basis of effective true strain is not ideal, as this would necessitate a larger starting extrusion billet size for converted wires that are of the same size. For composites including two ductile phases such as the Al/Ca system, strains as high as 16 can be reached without fracturing, but in so doing one must be mindful of the need to balance the trade-off between strength and conductivity. In addition to these considerations, the highest attainable strain level seems to be dictated mainly by the starting billet size and the ram force of the commercial extrusion capabilities that are available.

3.4 Comparison of Al/Al_2Ca (18 vol.%) composites to commercial cable technologies

Al/Al_2Ca composites have potential to be used as electrical conductors in applications with various demands due to the ability to tailor the performance properties of wires used for the construction of a monolithic cable. It has been shown that strength superior to other cable technologies is attainable while not sacrificing a large amount of conductivity. In addition to these properties, the proposed conductor material has a low density allowing for the use of larger conductors. Figure 5 shows a summary of the electrical conductivity and the specific strength (UTS/density) of the wires studied and of commercial wires that would be suitable for HVDC transmission.

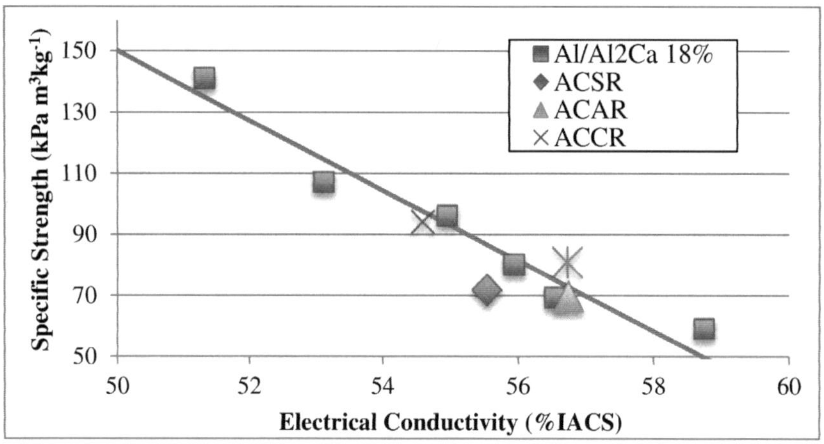

Figure 5. Summary of specific strength and electrical conductivity of Al/Al_2Ca composites and commercial cable technologies.

Selection of the best balance of strength and conductivity is dependent on whether an existing line is replaced or a new line is being constructed. Figure 5 shows that the conductivity is very similar to commercial wires for a given strength level at lower strain values. Being able to achieve high strength is appealing due to the impact that this can have on increasing the spacing of support towers. To further analyze the trade-offs in selecting a conductor, a 500kV, 3500 MW, 1360 km HVDC line, analogous to the Pacific DC Intertie, was considered [17]. Figure 6 shows the number of towers required and the annual cost of electrical losses ($0.05/kWh) for each conductor to deliver the same amount of power, given the same conductor sag as for the existing design that is based on ACSR cables.

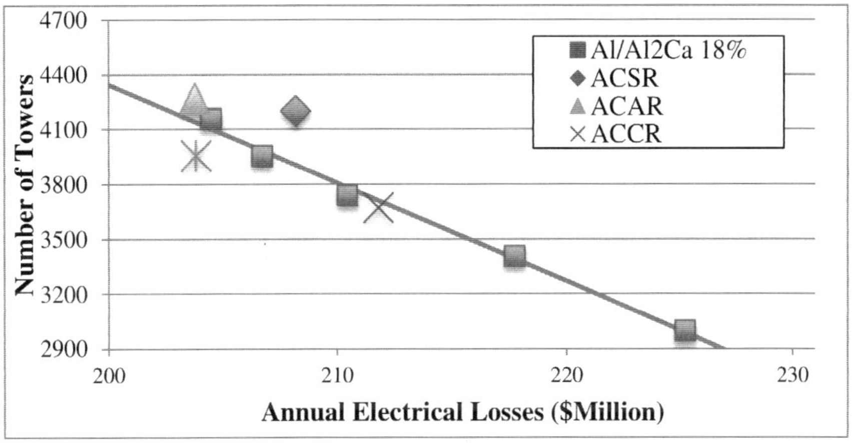

Figure 6. Comparison of initial investment in towers and electrical losses for Al/A12Ca composites and commercial cable technologies for 1360 km HVDC line.

Comparing the conductors in this fashion enables a decision on material selection to be made based on economics. Having a high strength can reduce the number of towers required, which has the potential to substantially lower installation costs. For the wires with the greatest strain (η = 12.7), 29% fewer towers would be needed, and the electrical losses would be 8.2% greater compared to using ACSR cables. Al/Al_2Ca composites could be especially useful as long-span conductors for river or canyon crossings where high strength is required. Moreover, the ability to alter the volume fraction and true strain allows for precise control of material properties, while cable construction is simplified by using a single material.

4. Conclusions

An Al/Ca (11.5% vol.) composite with nanofilamentary reinforcement was produced by powder metallurgy and deformation processing to a maximum true strain of 12.7 and the mechanical strength and electrical conductivity evaluated. Based on these investigations the following conclusions can be made:

- After controlled transformation of the reinforcement phase, volume expansion occurred producing a Al/Al_2Ca (18vol.%) composite. Longitudinal micrographs revealed no apparent filament instabilities or spheroidization.
- The electrical conductivity of the Al/Ca (11.5% vol.) composite decreased with reduced filament spacing at high true strains, consistent with previous results. The transformation of Ca filaments to Al_2Ca caused the conductivity to drop by less than 1% for wires subjected to the greatest level of deformation in this study (η = 12.7).
- The ultimate tensile strength of as-drawn Al/Ca (11.5vol.%) composite increased with deformation true strain due to the strong influence of interface strengthening, which is well

described by a modified Hall-Petch barrier model. The UTS increased upon transformation of the Ca reinforcement phase to Al_2Ca due to the increase in volume fraction and strength of the reinforcement phase, but retention of ductility was also observed.

- Al/Al_2Ca composite wires show promise to exhibit strength superior to current cable technologies combined with high electrical conductivity, low density, and likely high-temperature performance and stability, making them excellent candidates for use as HVDC transmission conductors.

Acknowledgements

The authors acknowledge the financial support of the U.S. Department of Energy-Office of Electricity, and Summit Technology Group. The contributions of Soeren Mueller at TU Berlin in supplying data and the contract access to their extrusion facility are appreciated. The Ames Laboratory is operated for the U.S. Department of Energy by Iowa State University under contract no. DE-AC02-07CH11358.

References

[1] International Energy Outlook 2016 U.S. Energy Information Administration
[2] Morhardt E 2017 Power transmission: the rise of the supergrid *The Economist* 71-72
[3] Bevk J, Harbison JP and Bell JL 1978 *J. Appl. Phys.* **49** 6031
[4] Tian L, Kim H, Anderson I and Russell A 2013 *Mat. Sci. Eng. A-Struct.* **570** 106-13
[5] Tian L, Russell A, Riedemann T, Mueller S and Anderson I 2017 *Mat. Sci. Eng. A* **690** 348-54
[6] Russell AM, Chumbley LS, and Tian Y 2000 *Adv. Eng. Mater.* **2** 11-22
[7] Neikov OD, Murashova IB, Yefimov NA and Naboychenko S 2009 Handbook of Non-Ferrous Metal Powders: Technologies and Applications (Oxford: Elsevier)
[8] ASTM Standard B193, 2016, "Standard Test Method for Resistivity of Electrical Conductor Materials", ASTM International, West Conshohocken, PA
[9] Tian L 2015 *Ph.D. Thesis* Iowa State University, Ames, IA
[10] Russell AM, Chumbley LS, Ellis TW, Laabs FC, Norris B and Donizetti GE 1995 *J. Mater. Sci.* **30** 4249-62
[11] Tian L, Anderson I, Riedemann T, Russell A and Kim H 2013 *Elec. Pow. Syst. Res.* **105** 105-14
[12] Frommeyer G and Wassermann G 1975 *Acta Metallurgica* **23** 1353-60
[13] Tian L, Anderson I, Riedemann T, and Russell A 2014 *Acta Mater.* **77** 151-61
[14] Xu K, Russell AM, Chumbley LS, Laabs FC, Gantovnik VB and Tian Y 1999 *J. Mater. Sci.* **34** 5955-59
[15] Russell AM, Lund T, Chumbley LS, Laabs FA, Keehner LL and Harringa JL 1999 *Composites Part A: Applied Science and Manufacturing* **30** 239-47
[16] Tian L, Russell A and Anderson I 2014 *J. Mater. Sci.* **49** 2787-94
[17] Eriksson C 2005 Critical Intertie *Power Engineering International*

Role of copper on Laves phase morphology in 9-12%Cr steels

Hilmar K Danielsen*[1], Fang Liu[2]

[1] Department of Wind Energy, Technical University of Denmark, DK4000 Roskilde, Denmark
[2] Department of Industrial and Materials Science, Chalmers University of Technology, SE-412 96 Gothenburg, Sweden

*E-mail: hkda@dtu.dk

Abstract. In this work the Laves phase was found to appear in two different morphologies, namely granular shapes and in an elongated shape. No difference in crystallography could be detected between these morphologies. The Laves phase was only observed in its elongated form in Cu-containing steels, where it was the primary morphology present after short term ageing. After long term ageing, the elongated Laves phase was replaced by the granular morphology. It is speculated that Cu precipitates act as nucleation sites for the elongated Laves phase, resulting in an unstable orientation relationship with the matrix, an in the meta-stable elongated morphology of Laves phase precipitates.

1. Introduction

9-12%Cr martensitic steels are extensively used in fossil fired steam power plants because of their high creep and fatigue strength, oxidation resistance and moderate cost. The long term creep strength of these steels is dependent upon precipitation strengthening, and utilizes carbide and nitride precipitates that have to be stable at operating temperatures up to 600°C for 20 to 30 years. 9-12%Cr steels are often alloyed with W and/or Mo additions for solid solution strengthening. After certain thermal exposure W/Mo can form Laves-phase $Fe_2(W,Mo)$ precipitates. Depending on whether the precipitation strengthening effect is higher than the solid solution strengthening, Laves phase formation can be either beneficial or detrimental for the creep properties, which are dependent on the size distribution and coarsening rate of the Laves phase [1-4]. The precipitation process of Laves phase in these steels is normally very long, and can continue thousands of hours during service at 600°C [2,3]. This intermetallic phase has previously been observed to appear with two different morphologies, most frequently with almost equiaxed granular shapes but sometimes with elongated needle shape [4].

In the present paper investigations were carried out on different Nb- and Ta-containing test steels with similar compositions, the main difference being the addition of Cu in some of the steels. An emphasis was placed on understanding the effects of Cu on Laves-phase formation and on the underlying mechanisms.

2. Experimental

The alloys investigated in this work were based on a very low carbon 12%Cr composition with Nb or Ta instead of vanadium, Z-phase strengthened steels [5,6], which are slightly different from most other 9-12%Cr steels. A high Co and Ni content ensured a martensitic structure with little or no delta ferrite. The steels were aged at 650°C in order to precipitate the Laves phase for the study.

X-ray powder diffraction was carried out using a Bruker D8 Advance powder diffractometer equipped with Cu-radiation. In order to avoid interference from matrix reflections, the precipitates were extracted from the bulk by electrolysis in an acidic solution of 5% HCl in 95% ethanol, dissolving the matrix while gathering the precipitates by sedimentation. Some of the very small precipitates may not have been recovered by this sedimentation process.

Transmission electron microscopy was carried out on carbon extraction replicas and thin foils in bright field and in high angle annular dark field (HAADF) mode. A FEI Titan instrument equipped with a spherical aberration corrector on the probe- forming lenses was used, in addition to a JEOL 3000F field emission gun transmisson electron microscope (FEGTEM). A Leo Ultra 55 FEG scanning electron microscope (FEGSEM) equipped with an Oxford Inca energy dispersive X-ray spectroscopy (EDX) system was used, operating at 20 kV.

Atom probe tomography (APT) analysis was carried out in an Imago LEAP 3000X HR local electrode atom probe instrument. Optimized APT acquisition parameters for the steels have been reported previously [7]. The specimen temperature during analysis was in the range of 60-70 K. To avoid premature specimen failure a pulsed laser mode was used, with a pulsing frequency of 200 kHz and a laser energy of 0.3 nJ. The acquired data were further analyzed using the IVASTM 3.4.1 software developed by Imago. In order to make site-specific APT analysis of selected features, such as elongated Laves phase precipitates, an in-situ lift-out technique was used to prepare APT samples. For this purpose an FEI Versa 3D dual beam workstation, which is a combined focused ion beam instrument (FIB) and a SEM, was used.

Table 1. Composition of the alloys used in this work given in wt%; balance Fe.

	Cr	C	V	Nb	Ta	N	W	Mo	Co	Ni	Mn	Si	Cu	B
12CrTa-Z4	11.79	0.005	-	-	0.39	0.033	2.90	-	7.30	0.50	0.48	0.30	-	0.004
12CrTaCu-L1	11.96	0.005	-	-	0.36	0.058	2.46	0.21	6.49	1.03	0.47	0.29	1.00	0.007
12CrTaCuCL3	12.1	0.06	-	-	0.36	0.049	2.47	-	3.50	0.19	0.21	0.31	1.95	0.006
12CrNbCu-Z2	11.6	0.003	-	0.26	-	0.04	2.42	0.20	4.82	0.46	0.47	0.29	0.92	0.003
12CrNb-Z3	11.64	0.005	-	0.26	-	0.036	2.82	-	5.4	1.47	0.50	0.30	-	0.004

3. Results

3.1. SEM invsestigations

In the steels investigated, full precipitation of Laves phase can take up to thousands of hours at 650°C. In figure 1 the Laves phase is clearly shown by using the contrast in backscattered SEM imaging, which arises mainly from the difference in atomic number. Bright contrast indicates regions with heavier tungsten atoms from the Laves phase, supported by EDX analysis revealing W enrichment in these regions. The 12CrTa and 12CrTaCu specimens in figure 1 have been aged at 650°C for 1000h each, and show a very different morphology of the Laves phase, with a granular Laves phase in the steel without copper, and an elongated Laves phase in the steel containing copper. The elongated Laves phase seems to have a preferential precipitation direction. Along with its shape this indicates a certain orientation relationship between this type of Laves phase and the matrix, which does not occur for the granular Laves phase.

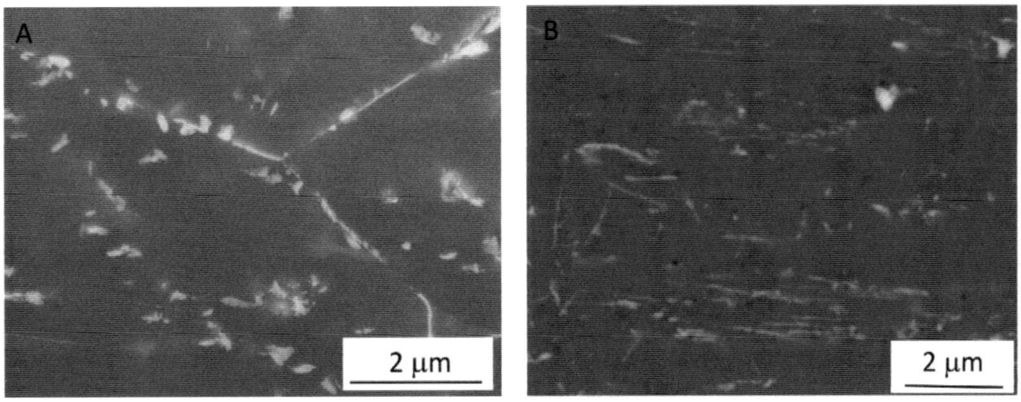

Figure 1. a) steel 12CrTa aged for 1000h/650°C contains granular type Laves morphology. b) 12CrTaCu aged for 1000h/650°C contains a large number of elongated type of Laves phase.

Specimens were sent to Siemens Industrial Turbomachinery for mechanical testing. The results showed a very marked difference in impact toughness: 3J for 12CrTa and 46J for 12CrTaCu. Investigations showed a high density of Laves phase at prior austenite grain boundaries after tempering in 12CrTa, since there are not any carbides in this steel. The distribution of Laves phase in 12CrTaCu is much more homogenous, with a larger portion of particles precipitating inside the grain. The difference in the distributions of Laves phase at the grain boundary is believed to be the cause of the low impact toughness in 12CrTa [8-10].

In the Cu-containing steels the Laves phase precipitated primarily in its elongated morphology, although some granular particles could also be observed. Since the 12CrTaCu and 12CrNbCu alloys contain Mo, a 12CrTaCuC alloy without Mo was used as a reference to show that Mo was not responsible for the elongated morphology.

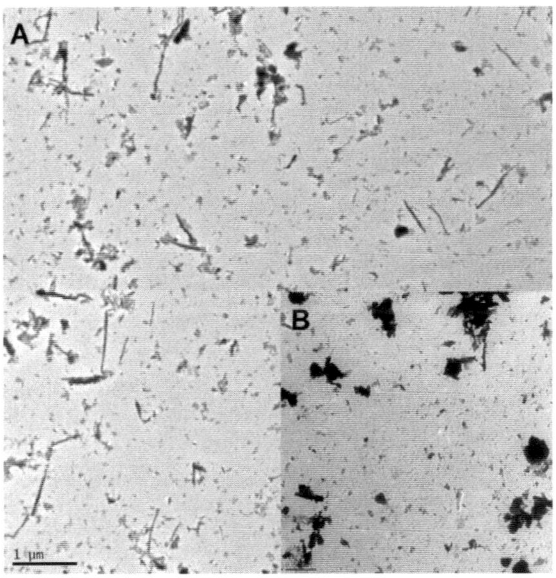

Figure 2. Carbon extraction replica of a) 12CrNbCu after 1000h/650°C ageing and b) 12CrNb after 300h/650°C ageing. Images scaled to same magnification.

3.2. TEM investigations

During exposure at 650°C of the alloys, the granular Laves phase was found to grow faster, the average particle volume becoming larger compared to the elongated Laves phase. Figure 2 shows a TEM image of carbon extraction replicas from 12CrNbCu and 12CrNb. Nitrides are seen as very small particles, while the larger elongated and granular particles are the Laves phase, confirmed by EDX. Comparing alloys with and without Cu, the Laves phase size distribution was much smaller in the Cu-containing alloys, and the elongated morphology should theoretically be the optimal shape for increasing creep strength. However, after about 10,000h of exposure at 650°C the elongated Laves phase had mostly been replaced by the granular morphology [11].

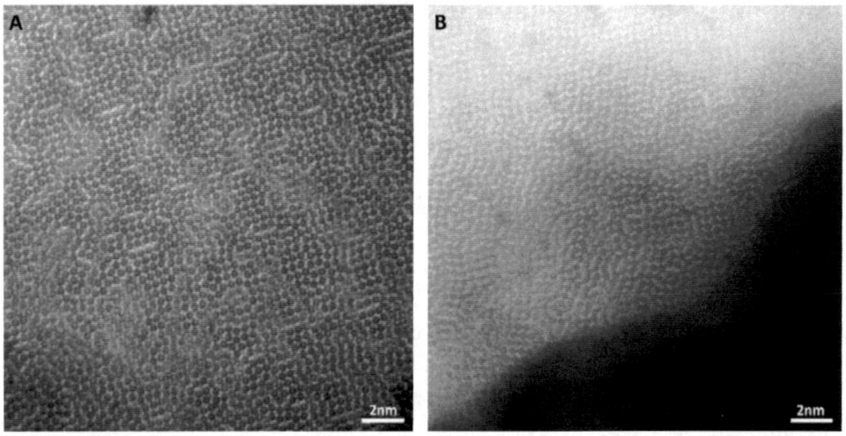

Figure 3. Atomic resolution TEM of the disordered crystal structure of Laves phase after 24h/650°C in a) 12CrNbCu and b) 12CrNb, showing primarily the heavy W atoms.

Atomic resolution TEM of both Laves phase morphologies was carried out on 12CrNbCu and 12CrNb after 24h ageing at 650°C, showing both morphologies to be very disordered, see figure 3. XRD measurements of both 12CrNbCu and 12CrNb showed a high degree of disorder during the first stages of precipitation, with very broad peaks, see figure 4. In spite of the different morphology and size distribution no difference was seen in the XRD data. After further ageing for 10,000h at 650°C, both morphology types developed significantly more well-defined peaks, indicating a more ordered crystal structure.

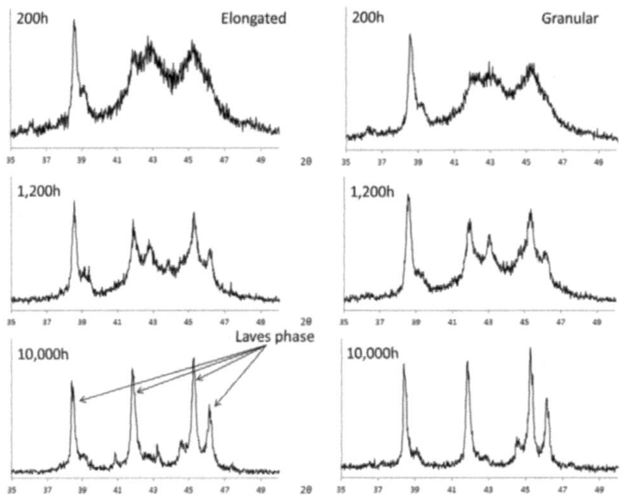

Figure 4. XRD of elongated Laves phase in 12CrNbCu (left) and granular Laves phase in 12CrNb (right) after ageing at 650°C.

3.3. APT investigations

Atom probe tomography was used to investigate the composition of the elongated Laves phase and Cu precipitates in the 12CrTaCu alloys after 300h/650°C, including the interface composition, to see if any Cu segregates to the interface between Laves phase and the steel matrix. The matrix concentration of several elements was found to decrease: Cu (0.24 at% vs. 0.89at%), W (0.23 at% vs. 0.76 at%), Mo (0.23 at% vs. 0.76 at%), and Cr (12.2 at% vs. 13.0 at%), and Ta and N concentration dropped to zero, indicating that precipitation of Laves, TaN and copper precipitates had occurred. Using an in-situ lift-out technique, specific precipitates were identified and investigated. The investigations revealed that the Laves phase itself contained no Cu, see table 2, nor had any Cu segregated at the Laves/matrix interface, see figure 5. A small amount of Cu was still present in the matrix although most had precipitated as pure copper particles, and Cu was not seen in in any other precipitates. The copper particles were almost pure Cu, see table 2, although they contained small amounts of Mn and Ni, the Mn content being especially pronounced at the interface with the matrix.

Table 2. Composition of matrix, elongated Laves Phase and copper precipitate after 300h/650°C in 12CrTaCu in at%.

	Fe	W	Cr	Co	Mo	Si	Mn	Ta	Ni	B	N	C	Cu
Laves	43.0	26.2	18.6	4.0	2.8	2.68	0.75	0.55	0.52	0.34	0.34	0.27	-
Copper	0.08	-	-	0.03	-	0.007	1.19	-	0.62	-	-	0.03	98.07
Matrix	78.8	0.23	12.2	6.3	0.05	0.59	0.56	-	1.0	-	-	-	0.24

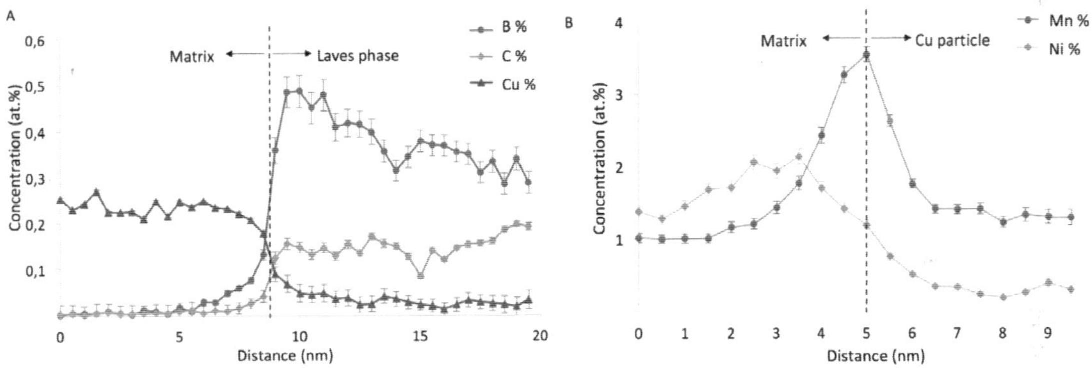

Figure 5. Atom probe tomography results of 12CrTaCu after 300h/650°C in at%: concentration profiles across a) an elongated Laves phase/matrix interface and b) a Cu/matrix interface.

4. Discussion

As in all alloys containing Cu, the Laves phase precipitates with an elongated morphology, and in the alloys without Cu the Laves phase precipitates with a granular morphology, and as there are no other appreciable differences in composition, it can be concluded that Cu must be responsible for the difference in morphology. However, Cu does not enter the Laves phase and is not part of its composition. Even though the sampling area is small, the measurements are very accurate. Furthermore there is no difference in crystallography between the two Laves phase morphologies. It can thus be speculated that Cu particles act as nucleation sites for Laves phase particles, which causes a different orientation relationship with the matrix, and this causes the difference in morphology. These many nucleation sites also cause a denser distribution of Laves phase inside the grains, unlike the granular Laves phase which nucleates primarily on prior austenite grain boundaries, which leads to a lower impact toughness.

The orientation relationship between the matrix and the elongated morphology does not appear to be stable, as granular Laves phases appear after prolonged ageing at 650°C and the elongated Laves phases either re-orient or are dissolved during coarsening. After 10,000h at 650°C there was nearly no elongated morphology left. The alloys were not investigated at 600°C, where the stability of the elongated Laves phase morphology could be different.

5. Conclusions

In the investigated steels Laves phase was found to appear in two different morphologies, namely as elongated form in steels containing Cu and a granular form in steels without Cu. There was no difference in crystallography or composition between these morphologies, and no Cu could be detected at the interface with the matrix in either case. In Cu-containing steels, the elongated shape was the primary morphology after short term ageing. After long term ageing, the elongated Laves phase was gradually replaced by the granular morphology. It can thus be speculated that Cu precipitates act as nucleation sites for the elongated Laves phase, resulting in an unstable orientation relationship with the matrix, and in the meta-stable elongated morphology of the Laves phase.

Acknowledgements

Fang Liu is partly funded via LIGHTer, a program financed within Strategic Innovation Areas, a mutual venture between VINNOVA, the Swedish Energy Agency, and Formas.

References

[1] Prat O, Garcia J, Rojas D, Sauthoff G and Inden G 2013 *Intermetallics* **32** 362-372
[2] Hald J, Korcakova L, Danielsen HK and Dahl KV 2008 *Mater. Sci. Tech.* **24** 149-158
[3] Hattestrand M and Andren 2001 *Mater. Sci. Eng. A* **318** 94-101
[4] Ku BS and Yu J 2001 *Scripta Materialia* **45** 205-211
[5] Danielsen HK and Hald J 2009 *VGB PowerTech* **89** 68-73
[6] Danielsen HK 2016 *Mater. Sci. Tech.* **32** 126-137
[7] Liu F and Andren HO 2011 *Ultramicroscopy* **111** 633-641
[8] Liu F and Andren HO 2014 *Advances in Mater. Tech. for Fossil Power Plants* 1116-1126
[9] Rashidi M, Johansson L, Andren HO and Liu F 2017 *Mater. Sci. Eng. A* **694** 57-65
[10] Morris JW Jr, Guo Z, Krenn CR and Kim YH 2001 *ISIJ International* **41** 599-611
[11] Danielsen HK 2014 *Mater. for adv. power engineering* 306-315

Synchrotron X-ray measurement of residual strain within the nose of a worn manganese steel railway crossing

S Dhar[1], Y Zhang[1], R Xu[2], HK Danielsen[1] and D Juul Jensen[1]

[1] Section for Materials Science and Advanced Characterization, Department of Wind Energy, Technical University of Denmark, Risø Campus, Roskilde, DK- 4000, Denmark

[2] Advanced Photon Source, Argonne National Laboratory, Argonne, IL, 60439-4800, USA

E-mail:sodh@dtu.dk, yubz@dtu.dk

Abstract. Switches and crossings are an integral part of any railway network. Plastic deformation associated with wear and rolling contact fatigue due to repeated passage of trains cause severe damage leading to the formation of surface and sub-surface cracks which ultimately may result in rail failure. Knowledge of the internal stress distribution adds to the understanding of crack propagation and may thus help to prevent catastrophic rail failures. In this work, the residual strains inside the bulk of a damaged nose of a manganese railway crossing that was in service for five years has been investigated by using differential aperture synchrotron X-ray diffraction. The main purpose of this paper is to describe how this method allows non-destructive measurement of residual strains in selected local volumes in the bulk of the rail. Measurements were conducted on the transverse surface at a position about 6.5 mm from the rail running surface of a crossing nose. The results revealed the presence of significant compressive residual strains along the running direction of the rail.

1. Introduction

Switches and crossings (S&Cs) are of great importance for any railway network as they allow trains to be directed from one track to another. The geometry, as well as the loading situations, in the nose of a crossing (see figure 1) differs from that of normal rails. It has been observed that the majority of track problems are associated with switches and crossings, which leads to higher maintenance costs than for any other part of the track. Previous studies performed on the deformation behavior of the crossing nose [1-5] clearly indicate that a very high contact stress occurs at the rail/wheel contact surface due to repeated passage of wheels. The stress levels are higher than the yield strength of the rail material, causing plastic deformation of the rail. The rail in the nose region is also subject to contact friction, and impact stresses, which lead to wear and rolling contact fatigue. Many defects can appear, including surface and subsurface cracks. The stress distribution (of both contact and residual stresses) plays an important role in generation of cracks in the material that may lead to failure. Residual stresses are generated in a rail during manufacturing as well as while in service. There are quite a few studies on the measurement of residual stresses in normal rail heads in the literature, including both modelling and experimental work [6-15]. Mostly tensile residual stresses are present in the rail head after the straightening process during manufacturing, however in service the stress state changes due to the variable loading situations resulting in the generation of a complex stress field.

There are several different experimental techniques to measure the residual stress in any material. Non-destructive methods include the use of X-rays or neutrons. However, ordinary X-rays are limited in terms of intensity and penetration depth. This means they typically give a mean stress averaged over a penetration depth only within a small area. In recent years, neutron diffraction has been used to determine residual stresses in the rail head [8-11]. Neutrons have a significantly larger penetration depth compared to ordinary X-rays, but due to their relative low intensity only stresses averaged over fairly large volumes in the mm^3-cm^3 range can be determined. Alternatively, with the advent of high energy synchrotron radiation techniques, local residual stresses can be determined in the bulk of the samples. Measurements of residual stresses on worn rails [12-14] as well as roller-straightened new rail heads [15] using synchrotron X-ray diffraction method have been published.

The novelty of the present work is that we focus on strains in the heavily loaded nose of an S&C, and that the selected S&C is made of manganese steel, optimized to improve wear and fatigue resistance compared to conventional pearlitic steel. Moreover the manufacturing process of manganese steel railway crossings is different from normal rails. Specifically, they are cast as a single piece and then explosion hardened.

The purpose of this work is therefore to apply synchrotron X-ray measurements to the nose of a rail crossing, and to determine the depth profiles of the residual strains. The crossing rail sample used in the experiment was taken from the rail network after being in service for five years on a major railway line in Denmark, carrying both freight and passenger trains.

2. Experimental Procedure

The residual strains in the nose of manganese steel crossing were investigated in this study. Visual inspection of the nose revealed severe damage with cracks and spallation. The chemical composition and mechanical properties of the steel are given in table 1 and table 2, respectively. The sampling area of the nose for synchrotron measurements is described in figure 1. The slice for the measurement was 5 mm thick and strain measurements were made on the transverse surface at various depths along the longitudinal direction. To avoid mechanical strains induced during grinding and cutting of the slice, the free surface was electropolished before measurements.

Table 1. Chemical composition of the manganese steel (wt%).

C	Mn	Si	Cr	P	S	Fe
1.14	12.81	0.26	0.12	0.05	0.05	balance

Table 2. Mechanical properties of the manganese steel.

Tensile strength	Yield Strength	Elongation	Hardness
~750 MPa	~365 MPa	~20%	~220 HV

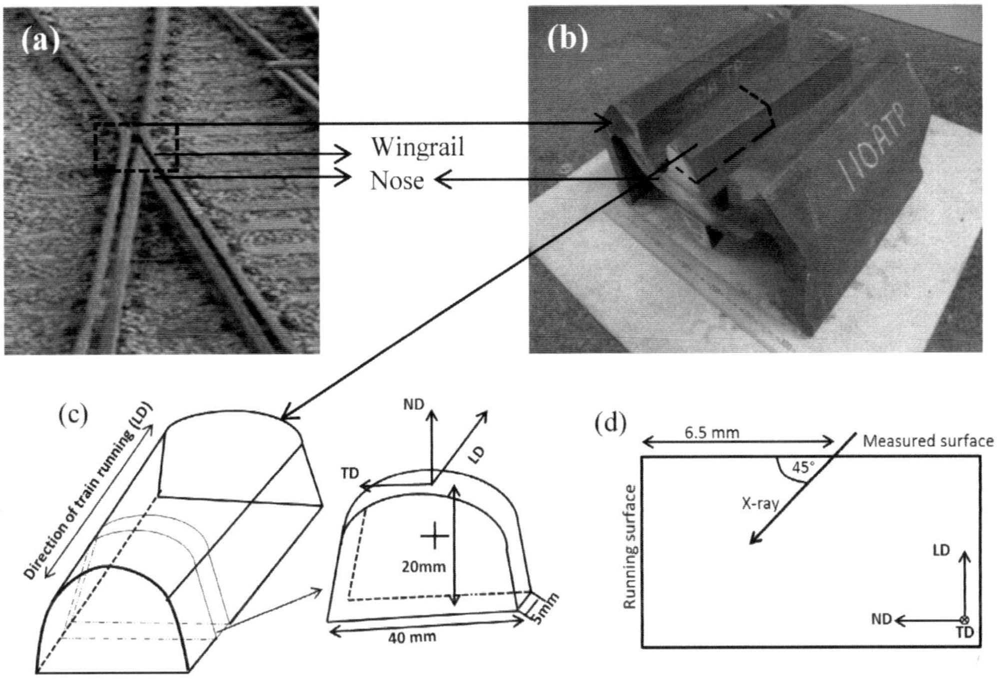

Figure 1. Nose specimen used for the synchrotron strain measurement: (a) photograph of typical crossing (with nose and wingrail); (b) part of a manganese steel crossing cut out from the track for investigation; (c) schematic showing the final sampling slice for strain measurements indicating also the normal direction (ND), transverse direction (TD) and longitudinal direction (LD). The black cross marks the position for synchrotron measurements. (d) Sketch showing the experimental set-up for the synchrotron measurements.

Synchrotron measurements were conducted on the transverse section at a position about 6.5 mm from the rail running surface of the crossing nose (see figure 1d) at beam line 34-ID-E at the Advanced Photon Source (APS), Argonne National Laboratory [16].

A focused polychromatic beam was first used to determine the orientations of the matrix grains. The use of non-dispersive Kirkpatrick-Baez (K-B) focusing mirrors helps in focusing of the beam on to the specimen. The resulting microbeam had a full-width half maximum of ~0.5 μm. The X-ray microbeam was scanned over the specimen, which was mounted on a holder at an inclination of 45° to the incoming beam (see figure 1d). The Laue diffraction patterns from the polychromatic scans were recorded on an area detector mounted in 90° reflection geometry 510.3 mm above the specimen. Diffraction patterns from different depths were obtained by the use of a Pt-wire of 100 μm diameter as a differential aperture. The Laue patterns at each depth were reconstructed by the use of the LaueGo software [17] available at APS beamline 34-ID-E. The patterns were indexed, from which the (hkl) indices of individual spots as well as their corresponding X-ray energies were determined. From these data, a spot with high intensity and its corresponding plane parallel to the specimen surface was selected for monochromatic energy scanning to determine the absolute lattice spacing. Thereby the strain (in one direction) was measured as:

$$\varepsilon = \frac{d - d_0}{d_0}, \tag{1}$$

where d is the lattice spacing of the sample and d_0 is the lattice spacing of the sample in the stress-free state. For the present study, d_0 was determined based on the lattice parameter measured at the base of the nose (at a depth of 20 mm from the running surface) using laboratory X-ray measurements assuming that, apart from manufacturing stresses, this location is free from stresses induced due to wheel-rail interactions. These measurements were done at the Department of Materials and Manufacturing Technology at Chalmers University of Technology, Sweden. A chromium source (X-ray wavelength λ = 2.2897 Å) was used. The lattice parameter determined from the (220) peak at a diffraction angle of 126.85° is a = 3.621 Å, which is used for to obtain the reference d_0 value.

3. Results and Discussion

The crystallographic orientations measured using polychromatic beam differential aperture X-ray diffraction (DAXM) at four locations, 0.5 mm apart, are shown in figure 2. The grain orientations within a depth of 0-50 μm from the measurement surface are shown. The result shows that the grain at a position of 6.5 mm from the running surface has the smallest orientation spread (see figure 2b) and the highest indexed percentage. This grain is therefore chosen for the monochromatic energy scan.

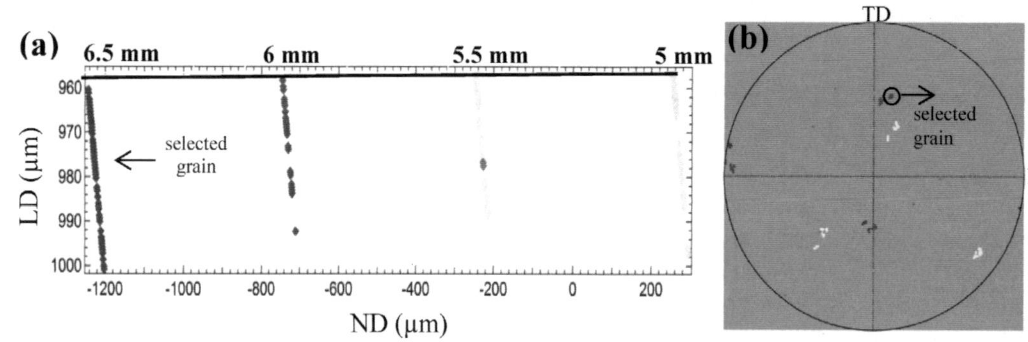

Figure 2. Grain orientations at different depths and different locations on the transverse surface (a) together with the corresponding {001} pole figure (b). The grain selected for monochromatic energy scan for determination of lattice absolute spacing is marked. The black line in (a) marks the approximate position of the measurement surface. The colors in (a) and (b) correspond to the crystallographic orientations.

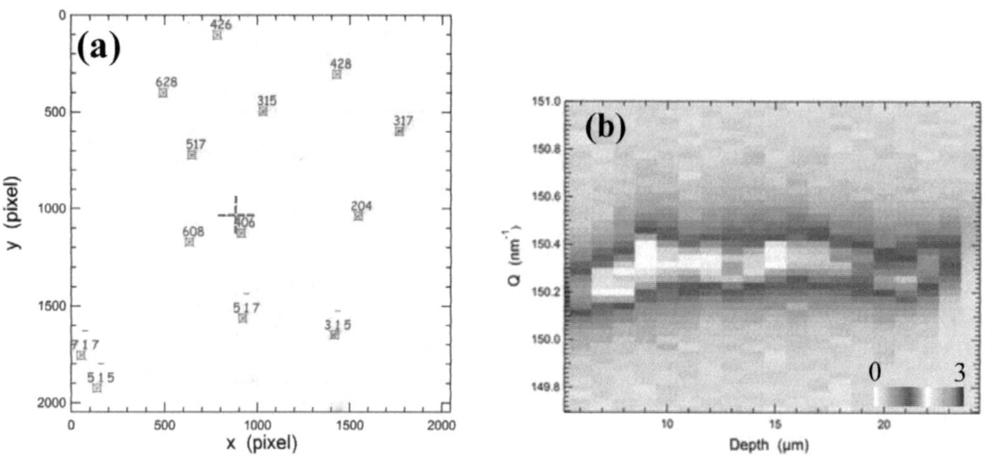

Figure 3. (a) Indexing of a polychromatic beam depth-resolved Laue diffraction pattern; (b) intensity distributions as a function of diffraction vector Q at each depth.

Figure 3a shows an indexed Laue pattern obtained at one of the depths from the polychromatic beam diffraction data for the selected grain, where the numbers indicate the assigned (hkl) indices of each spot. Based on this diffraction pattern the (517) spot was selected for monochromatic energy scanning, as the intensity of this spot is relatively high and the corresponding X-ray energy is also high, at about 21 keV. The normal of the chosen (517) plane is about 2° away from sample normal direction, which is marked by the dashed cross in figure 3a.

From the monochromatic energy scan, the diffraction images around the (517) spot at each energy step for each depth were reconstructed. As each energy corresponds to a specific diffraction vector Q ($Q = 2\pi/d$), the integrated intensity of each diffraction image as a function of the diffraction vector can be determined at all X-ray penetration depths (see figure 3b). A few examples of the Q distribution at different depths are given in figure 4a, which clearly shows that the strain distribution varies within the measured volume from tensile to compressive strains. The dashed line marks the Q value corresponding to d_0 of the (517) crystallographic plane. A shift to the right relative to the dashed line indicates compressive strains and vice versa for tensile strains. At each depth, the Q distribution was fitted using a Gaussian function and the center of the distribution, Q_c was used to determine the crystallographic plane spacing, *d*.

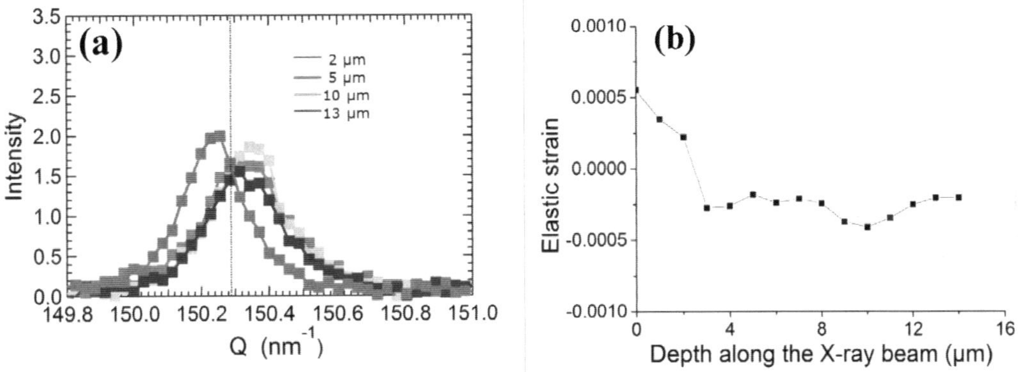

Figure 4. (a) Intensity variation as a function of Q at different depths within the sample (The dashed line marks the Q value for d_0 of the (517) crystallographic plane). (b) The residual strain along LD as a function of depth along the incoming X-ray.

Figure 4b shows the elastic strain component normal to the transverse surface of the sample developed in the nose crossing at different depths from the measuring surface. It can be seen that in the bulk of the material there are residual compressive strains in the range of 2×10^{-4} up to 4.11×10^{-4}.

Whereas theoretically the out-of-plane normal strain at a free surface is zero, tensile strains are seen within the first 3 μm from the free surface in figure 4b. Two explanations for this tensile strain are possible. Firstly, the reference d_0 may not be strain free. It is very likely that the grains even at depth of 20 mm below the running surface are also under compression, thus the dashed line in figure 4a should be shifted to the left, and the curve in figure 4b should move down towards more compressive strain. Secondly, due to small experimental errors, the free surface may be wrongly identified by about 2-3 μm from the currently assumed position. If the measurement surface in figure 4b (i.e. the 0 point) changes about 2-3 μm to the right, the tensile strain at the measurement surface is reduced. Irrespective of this, we can conclude the bulk of the measured volume is under compressive residual strains.

The problem of railway damage and failure can be caused by multiple mechanisms but it is evident that interaction between residual stresses and the defects (cracks etc.) can lead to severe problems. Generally compressive stresses are beneficial because they are supposed to inhibit crack propagation whereas tensile stresses are conducive for crack propagation. Although we find the longitudinal stresses are compressive, the normal or the transverse component of the stress may have a tensile component which could lead to crack evolution and failure. Also, generally the presence of a significant compressive stress is counterbalanced by a tension zone underneath. A complete map of all the components of the stresses in the three directions at different depths from the running surface would be helpful to understand the interaction of residual stress with damage mechanisms. This will be investigated in future work. Also the synchrotron data will be compared with those measured using standard laboratory X-rays.

4. Conclusion

The residual strain distribution within a 5 mm thick slice of the nose of a manganese railway crossing was analyzed in the longitudinal direction at a distance of 6.5 mm from the rail surface. To our knowledge this is the first time synchrotron radiation has been employed to measure the strains in a nose of a manganese railway crossing. It was found that compressive strains exist along the longitudinal direction (with some small oscillations in the values) even at a depth of 6.5 mm from the wheel contact surface. Rail-wheel interactions at the nose of the crossing induces severe plastic deformation of the nose, the effect of which is thus evident even at a depth of 6.5 mm from the contact surface.

Acknowledgement

The authors gratefully acknowledge support from the Innovation Fund Denmark through the project "INTELLISWITCH – Intelligent Quality Assessment of Railway Switches and Crossings" (Grant no 4109-00003B) and grant from Interreg ESS & MAX IV: Cross Border Science and Society (Project number DTU-009). Use of the Advanced Photon Source was supported by the U.S Department of Energy, Office of Basic Energy Sciences, under Contract No.DE-AC02-06CH11357. The authors would also like to thank the Department of Materials and Manufacturing Technology at Chalmers University of Technology, Sweden for helping with the laboratory X-ray measurements.

References

[1] Johansson A, Pålsson B, Ekh M, Nielsena JCO, Ander MKA, Brouzoulis J and Kassa E 2011 *Wear* **271** 472–481
[2] Wiest M, Daves W, Fischer FD and Ossberger H 2008 *Wear* **265** 1431–38
[3] Guo SL, Sun DY, Zhang FC, Feng XY and Qian LH 2013 *Wear* **305** 267-273
[4] Lv B, Zhang M, Zhang FC, Zheng CL, Feng XY, Qian LH and Qin XB 2012 *Int J Fatigue* **44** 273-278
[5] Harzallah R, Mouftiez A, Felder E, Hariri S and Maujean JP 2010 *Wear* **269** 647–654
[6] Xiao JH, Zhang FC and Qian LH 2014 *Fatigue Fract Engng Mater Struct* **37** 219–226
[7] Kaiser R ,Stefenelli M , Hatzenbichler T, Antretter T, Hofmann M, Keckes J and Buchmayr B 2015 *J Strain Anal Eng* **50(3)** 190–198
[8] Sasaki T, Takahashi S, Kanematsu Y, Satoh Y, Iwafuchi K, Ishida M and Morii Y 2008 *Wear* **265** 1402–07
[9] Jun TS, Hofmann F, Belnoue J, Song X, Hofmann M and Korsunsky AM 2009 *J Strain Anal Eng* **44(7)** 563-568
[10] Luzin V, Prask HJ, Herold TG, Gordon J , Wexler D , Rathod C, Pal S , Daniel W and Atrens A 2013 *Neutron News* **24(3)** 9-13
[11] Rathod C, Wexler D, Luzin V, Boyd P and Dhanasekar M 2014 *Mater Sci Forum* **77** 213-218
[12] Webster PJ, Hughes DJ, Mills G and Vaughan GBM 2002 *Mater Sci Forum* **404-407** 767-772
[13] Pyzalla A, Wang L, Wild E and Wroblewski T 2001 *Wear* **251** 901–907
[14] Kelleher JF, Buttle DJ, Mummery PM and Withers PJ 2005 *Mater Sci Forum* **490-491** 165-170
[15] Kelleher J, Prime MB, Buttle D, Mummery PM, Webster PJ, Shackleton J and Withers PJ 2003 *J Neutron Res* **11(4)** 187-193
[16] Liu W, Zschack P, Tischler JZ, Ice GE and Larson BC 2010 *AIP Conf. Proc.* **1365** 108-111
[17] Tishchler JZ 2014 Reconstructing 2D and 3D X-ray orientation map from white beam Laue *Strain and Dislocation Gradients from Diffraction* eds R Barabash and GE Ice (London: Imperial College Press) chapter 10 pp 358-375.

Effect of hexagonal boron nitride and graphite nanoparticles on the mechanical and physical properties of magnesium

Z Drozd[1], Z Trojanová[1], U Arlic[2], A Kasakewitsch[3] and K Dash[4]

[1]Faculty of Mathematics and Physics, Charles University, Ke Karlovu 3, 121 16 Prague 2, Czech Republic
[2]Institute of Materials Science and Engineering, Clausthal University of Technology, 38678 Clausthal-Zellerfeld, Germany
[3]Weber Maschinenbau GmbH, 26506 Norden, Germany
[4]Institute of Materials Engineering, Indian Institute of Science, Bengaluru, Karnataka 560012, India

E-mail: zdenek.drozd@mff.cuni.cz

Abstract. Magnesium nanocomposites reinforced with 3 vol.% of hexagonal boron nitride (hBN) and ultrafine-grained magnesium with 3 vol.% of graphite (Gr) nanoparticles were prepared by milling followed by hot extrusion. The influence of the hBN and Gr nanoparticles on the mechanical and physical properties of magnesium were subsequently investigated. The microstructure of the nanocomposites was analysed by electron microscopy. The nanocomposites were deformed in compression over a wide temperature range from room temperature to 300 °C and true stress-true strain curves were determined. The linear thermal expansion coefficient was measured over a wide temperature range from room temperature up to 400 °C. Pre-deformation in compression was used to estimate the influence of twins on the thermal expansion coefficient. The results are analysed and possible physical mechanisms for the observed mechanical and physical properties are discussed.

1. Introduction

Wrought magnesium and magnesium alloys have recently been the focus of researchers working with extruded and rolled materials. Such materials exhibit a finer microstructure and better mechanical properties in comparison with cast materials. A further possibility for improvement of mechanical properties is the preparation of composite materials with a reinforcing phase having different physical and mechanical properties [1-8]. The resulting properties of such composites can be tailored according to the requirements of the technical praxis. If the size of the reinforcing phase particles or fibres extends down to the nanometer region (1-100 nm) it is possible to speak about nanocomposites. Nanocomposites based on magnesium or magnesium alloys have significant advantages including very good mechanical properties. The reinforcing phase content is usually lower than 5 vol.%. The disadvantages of magnesium nanocomposites are mainly the complicated and expensive preparation techniques, and the anisotropy of these materials. This anisotropy arises from the anisotropy of the hexagonal close packed magnesium crystal structure (hcp) with consequences for both mechanical and physical properties [9]. Magnesium-based composites prepared with various amounts of nanoparticles have been studied by several authors [6]. The resulting properties of such nanocomposites depend on volume fraction of the reinforcing phase particles, their mechanical, physical and chemical properties and the distribution of particles in the matrix. An important role is also played by the bonding between the nanoparticles and magnesium matrix [10]. Hexagonal boron nitride (hBN) ceramics and graphite exhibit a similar layered structure in which atoms within each layer (C, and B with N) are bound by strong covalent bonds, whereas the layers are held together only by weak van der Waals forces. Both (Gr and hBN) have high melting point (T_M(Gr)=3642 °C, T_M(hBN)=2973 °C) and low density (ρ(Gr)=2.2x10^3 kgm^{-3}, ρ(hBN)=2.1x10^3 kgm^{-3}). Both nanoparticles should act as sliding layers as well as diffusion barriers in metal matrix composites [15]. In spite of several attempts there is no satisfying model describing the mechanical and physical properties of nanocomposites. This is due in part to the complex nature of this problem, including many factors which influence the nanocomposites behaviour. In the paper we contribute to the solving of this problem by comparing the mechanical and physical properties of two magnesium nanocomposites reinforced with particles of similar structure.

2. Experimental procedure

Nanocomposites were prepared by the powder metallurgical route. Microcrystalline magnesium was used as a matrix for the nanocomposites. Hexagonal boron-nitride or graphite nanoparticles were used as a reinforcement phase.

The first nanocomposite – ultrafine-grained (UFG) magnesium containing 3 vol.% of graphite nanoparticles (UFG-Mg+3nGr) – was obtained using Mg powder prepared by gas atomization of a high purity magnesium melt, in an argon atmosphere containing 1% of oxygen for powder passivation. The mean particle diameter was approximately 40 μm. The graphite powder used had a median particle size of 1-2 μm. The Mg powder was mixed with 3 vol.% of graphite powder in an asymmetrically moving mixer for 8 h. Afterwards the powder mixtures were milled for 8 h at 200 rpm in a planetary ball mill (Retsch, PM 400) in a sealed argon atmosphere. The milling vessel of 500 ml volume was made of corundum (Al_2O_3) and the milling balls (11 mm diameter) were made of hardened steel (100Cr6). The ball-to-powder weight ratio was 10:1. The composite was encapsulated in an evacuated Mg container (70 mm in diameter), degassed at 350 °C, and extruded at the same temperature with a 400 ton horizontal extrusion press (Collin 630t, extrusion die, 14 mm outlet diameter). Analyses of the extruded material in a light spark analyser (Spectrolab, Spectro Analytical Instruments) revealed no contamination of the composite by elements such as Fe from the extruder tools or milling balls, or Al from the milling vessel [8]. The mean grain size of specimens used was estimated, using transmission electron microscopy and X-ray profile analysis, to be about 200 nm.

The second nanocomposite – UFG magnesium with 3 vol.% of BN nanoparticles (UFG-Mg+3nBN) – was prepared by ball milling of commercially available magnesium powder with 45 μm mean diameter particle size, mixed and milled together with hexagonal boron nitride nanoparticles of 500 nm mean particle size in a planetary ball mill (Retsch, PM 400) under air atmosphere. A corundum-coated steel milling cup with a volume of 500 ml was filled with the powder mixture and 7 mm diameter and 13 mm diameter light corundum balls. The ratio of the mass of powder, the 7 mm balls and the 13 mm balls was 1:1.5:1.5. The powder volume in the milling cup, being about 20% of the whole volume, was milled for 6 h at 150-200 rpm. Oxidation, plastic deformation, and fracture of the magnesium powder particles took place during the milling. The graphite-like structure of the fine hBN nanoparticles was deformed during the milling and created inter-grain material layers. The milled powder was filled into aluminium capsules of 70 mm diameter, 2 mm wall thickness, and about 220 mm height, closed at one end with a circular plate of 5 mm thickness. The powder was multistage cold pre-pressed at a pressure of 65 MPa. The capsules were repeatedly filled to about one half of the remaining height after successively pressings up to 210 mm height of the pressed powder. Afterwards the capsule was closed by another circular plate of 5 mm thickness with a small hole of 2 mm diameter in the middle using fitting threads in the capsules and plates. The small hole acted as gas outlet during hot extrusion. The pre-compacted samples were hot extruded to rods of 12 mm diameter (1:5.8 extrusion ratio) using a horizontal 630 t hot extrusion press. The extrusion temperature and speed were 350 °C and 1.5 mm/s, respectively.

The texture of the samples was studied using a FEI Quanta 200 FX scanning electron microscope equipped with an EDAX electron backscatter diffraction (EBSD) camera. Orientation imaging microscopy (OIM) software was utilized to analyze the EBSD data sets. The grain size was estimated in the planes perpendicular to the extrusion direction to be about 600 nm.

Specimens were deformed in compression in an Instron machine in the temperature range from room temperature to 300 °C at a constant crosshead speed, giving an initial strain rate in the order of 10^{-3} s^{-1}. True stress-true strain curves were obtained from the measurements. Characteristic stresses – compression yield stress (CYS) and ultimate compression strength (UCS) were estimated from the stress-strain curves.

Specimens for transmission electron microscopy (TEM) analysis were prepared by cutting slices from the center of compression test specimens with diameter of 3 mm and thickness of 0.8 mm. These slices were ground and then dimpled to about 15 μm thickness in the center. The final step of the specimen preparation was ion milling using a precision ion polishing system (Gatan 691) under conditions of 4.5 kV and an incident angle of 4 - 3.5°. The specimens were examined using a Philips CM12 transmission electron microscope operated at 120 kV.

The linear thermal expansion of specimens was measured in an argon protection atmosphere using a Netzsch 410 dilatometer, over a temperature range from RT up to 400 °C. Four temperature cycles (heating-cooling) were performed using heating and cooling rates of 0.9 K/min.

The Young's modulus was measured using a resonant frequency and damping analyser (RFDA). Samples were excited to vibrations in the resonant frequency using a small striker. Free vibrations of the sample were registered with a microphone and a fast Fourier transform was used for estimation of the resonant frequency.

3. Results and discussion

True stress-true strain curves obtained for UFG-Mg+3nGr and UFG-Mg+3nBN are given in figures 1a and 1b. Characteristic features of the curves are the small local maxima observed at the beginning of the curves at RT and at 100 °C. Such a small maxima is usually observed in deformation curves of magnesium samples exhibiting a basal texture prepared by hot extrusion [11].

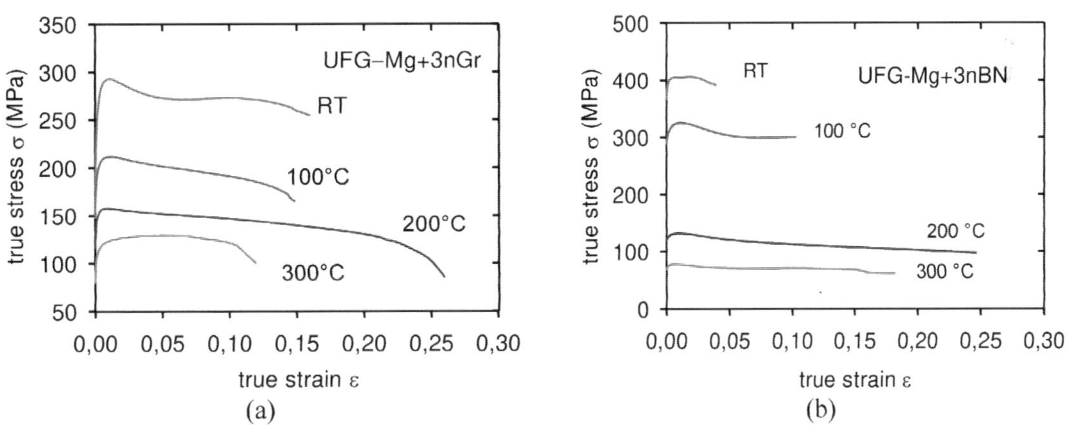

Figure 1. True stress-true strain curves determined at various temperatures in compression for UFG-Mg+3nGr (a) and UFG-Mg+3nBN (b).

The temperature dependency of the compression yield stress (CYS) and the ultimate compression stress (UCS) are shown in figures 2a and 2b. The stress (CYS and UCS) required for deformation at RT and 100 °C are much higher for the nanocomposite containing BN particles than for the composite reinforced by Gr nanoparticles. Both characteristic stresses decrease with increasing deformation temperature. This increase is stronger for the nanocomposite containing BN nanoparticles. At 200 °C and 300 °C both the CYS and UCS are higher for UFG-Mg+3nGr, providing evidence for the better thermal stability of the composite with Gr. The plasticity of UFG-Mg+3nBN sample is very low at RT, but increases to comparable amounts in both materials at elevated temperatures. The highest plasticity was estimated for both materials at 200 °C. Such behaviours were observed in other magnesium composites [11]. The main glide system in polycrystalline hexagonal magnesium alloys and composites is basal slip of <*a*> dislocations with Burgers vector of $1/3[11\bar{2}0]$ on the (0001)) basal planes. According to the von Mises criterion, the activity of five independent slip systems is necessary for a general compatible deformation [12]. However for basal slip in magnesium only two systems are independent. The Von Mises condition may be fulfilled by the additional slip of <*a*> dislocations in prismatic systems or by activity <*c*+*a*> dislocations with a Burgers vector of 1/3 $[11\bar{2}3]$ in non-basal slip systems (prismatic or pyramidal). In magnesium several potential twinning systems may also be found, among which the $\{10\bar{1}2\}\langle 10\bar{1}1\rangle$ twinning system is most often observed.

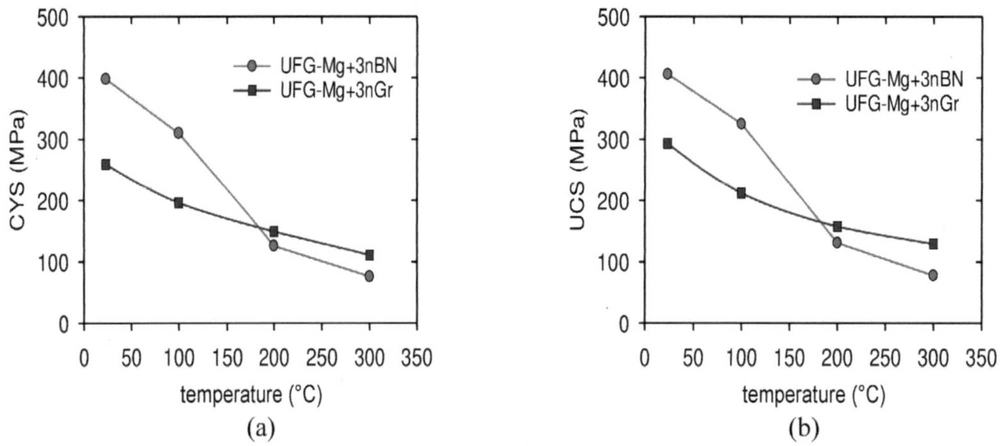

Figure 2. Dependence of mechanical properties on temperature for both nanocomposites: (a) compression yield stress, CYS, and (b) ultimate compression strength, UCS.

Hexagonal magnesium materials prepared with the hot extrusion technique exhibit a pronounced texture with the basal planes parallel to the extrusion direction. Such a fibre texture was found also in UFG-Mg+3nBN, as shown in figure 3. Texture of magnesium reinforced with BN particulates synthesised using microwave-assisted sintering technique, were studied by Seetharaman and co-workers [13]. They estimated weakening of the formed basal texture with an increase in nano BN particles.

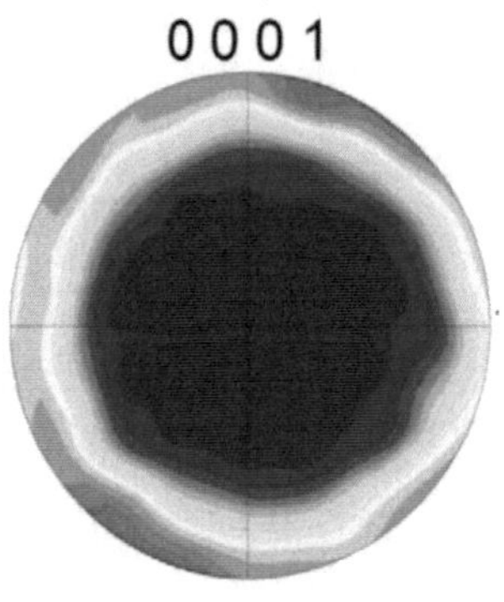

Figure 3. Basal texture of the as-extruded UFG-Mg+nBN nanocomposite.

Preferential orientation of the basal planes in the extrusion direction does not allow deformation by twinning during tensile deformation. However, tensile twins are activated during compression experiment along <0001> where the Schmid factor for twinning is high. Such twinning causes a misorientation of 86.3° between the twinned and the untwinned lattice [14], as shown schematically in figure 4. Twinning deformation is exhausted after a plastic strain of several percent, and further deformation continues with dislocation activity in the reoriented grains [15]. New twin boundaries can act as obstacles for dislocation motion and therefore rapid hardening may be observed. The twinning mechanism is not a thermally activated process whereas dislocation motion depends strongly on the temperature. At higher temperatures the critical resolved shear stress for <***c*** + ***a***> dislocations in the non-basal planes decreases up to the value necessary for twinning and therefore deformation at temperatures above 200 °C continues via this non-basal dislocation mode.

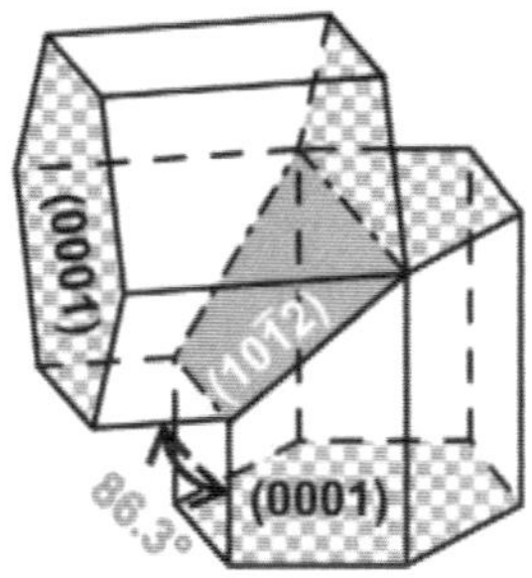

Figure 4. Schematic illustration of misorientation associated with an extension twin.

TEM micrographs given in figure 5 show the substructure of the composite with Gr nanoparticles after deformation at (a) room temperature, and (b) 300 °C. In both cases nanoparticles situated in the grain boundaries are clearly visible. Only a few particles were found inside of the grains. This situation is likely to result from the layered structure of the reinforcement. The sliding layers of the nanoparticles were deformed with the grain boundaries and surrounded by the magnesium grains. Owing to similar structure of hBN particles we may consider a similar behaviour also in the BN nanocomposite. After deformation at room temperature the dislocation density is low. In comparison the dislocation density is higher after deformation at 300 °C (see figure 5b).

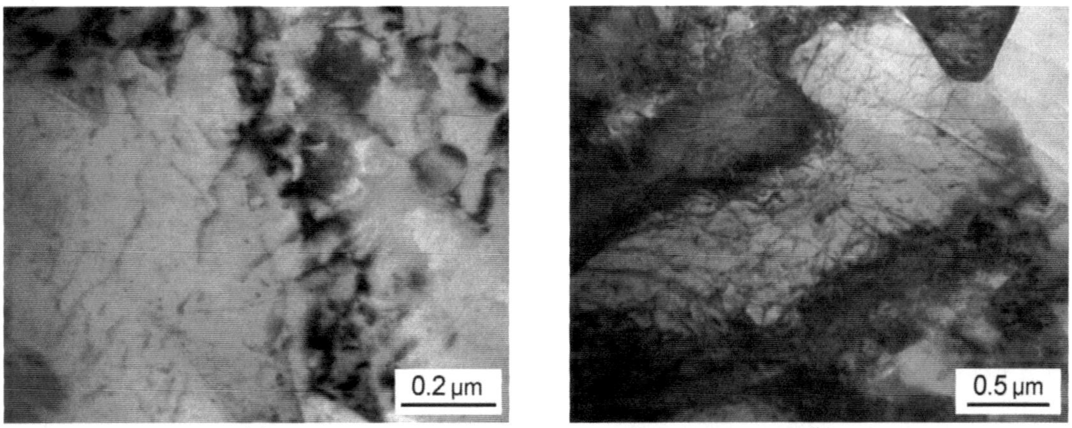

(a) (b)

Figure 5. TEM micrographs of UFG-Mg+3nGr taken after deformation at RT (a) and 300 °C (b).

From figure 1 and 2 it follows that the deformation stresses at 100 and 200 °C are lower for the composite containing Gr nanoparticles although the particle content and preparation technique are the same. Furthermore, the Gr nanocomposite has a smaller grain size in comparison with the BN nanocomposite. Two factors may account for such behaviour:

- The mean grain size of the UFG+3nGr specimens used was estimated to be about 200 nm, compared to the UFGMg+3nBN samples where a value of 600 nm was found. It is possible to consider that for such small grain sizes a negative deviation from the Hall-Petch relationship may be found.
- The bonding between the Gr particles and the Mg matrix is weaker than that for BN nanoparticles. In particular, it is known that Mg does not form carbides and therefore the binding between the matrix and Gr particles is expected to be weak.

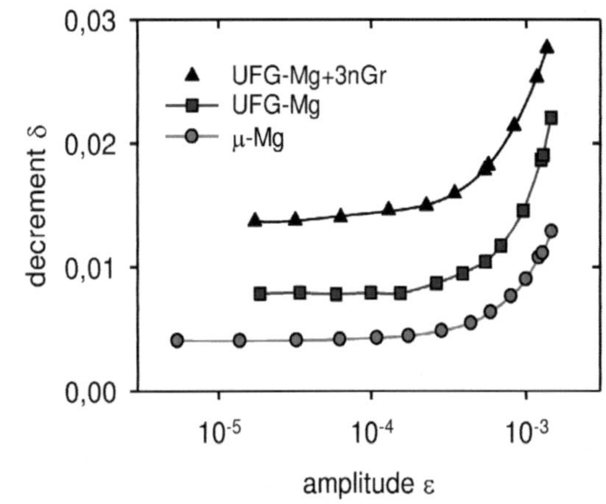

Figure 6. Amplitude dependence of decrement estimated for various materials.

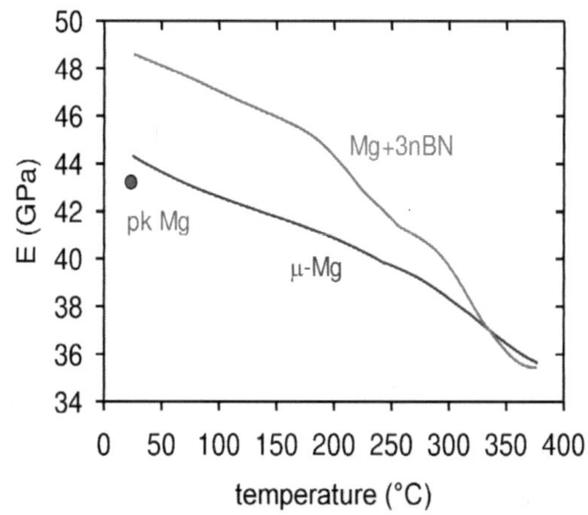

Figure 7. Temperature dependence of Young's modulus.

These observations are further illuminated by the data in figure 6 showing the amplitude dependence of the internal friction characterised with the logarithmical decrement. The amplitude dependence of the decrement can be divided into two regions. For small amplitudes the decrement does not depend on the amplitude, whereas in the higher amplitudes region the decrement rapidly increases with increasing strain amplitude. While the decrement in the amplitude-independent region is influenced by various mechanisms (including dislocations, grain boundaries, and thermoelastic effects) the amplitude dependent component of the decrement is only due to the presence of dislocations in a material. The amplitude dependency of the decrement, seen in figure 6, shows the influence of grain boundary sliding on the internal friction. The lower decrement values were observed for microcrystalline Mg (μMg) comparing with values measured for ultra fine-grained Mg (UFG-Mg) with the grain size ~ 150 nm. A higher density of grain boundaries increases the decrement due to grain boundary sliding. In the case for the Gr nanocomposite the decrement is very high – in the order 10^{-2}. These high values of the decrement are due to a high contribution of grain boundary sliding as a consequence of the weak bonding between the Mg and Gr nanoparticles. Such behaviour also influences the mechanical properties. Note that all materials were prepared with the same technique – hot extrusion.

The temperature dependence of the Young's (E) modulus is given in figure 7 for both the BN nanocomposite and pure Mg (prepared also by hot extrusion). It can be seen that the relatively small nanoparticle content significantly increased the modulus. A more quantitative comparison, for example by application of the rule of mixtures, is somewhat problematic in this case because we do not know the influence of texture on the modulus values. The high anisotropy of Mg and especially hBN, together with the sample texture, complicate, therefore, any calculations. We can only conclude that the nanocomposite modulus rapidly decreases with increasing temperature. At temperatures higher than 325 °C values of the E-modulus in Mg and UFG-Mh+3nBN are practically the same.

The temperature dependence of the thermal expansion coefficient (CTE) estimated for UFG-Mg+3nBN is shown in figure 8a, together with the variation of the CTE for microcrystalline Mg (μMg) also prepared by hot extrusion. The presence of nanoparticles results in a slight decrease in the CTE. For comparison the dependence for polycrystalline Mg (pk Mg) is also introduced. The different course of the temperature dependence estimated for pk Mg and μMg may be explained by the existence of texture in the extruded μMg. The CTE of Mg single crystals exhibit significant anisotropy, with a value in the *a* direction for CTE(*a*) = 7.045 x 10^{-6} K^{-1}, and a much higher value in the *c* direction of CTE(*c*) = 11.758 x 10^{-6} K^{-1} [16]. Both μMg and UFG-Mg+3nBN exhibit a basal texture with preferential orientation in the long sample axis in which the CTE in the one dimensional approximation is measured.

Pre-deformation in compression results in substantial increase in the CTE at lower temperatures, with less of an effect at higher temperatures, as demonstrated in figure 8b. As it was mentioned above, the compressive plastic deformation of textured magnesium materials is accompanied by the mechanical twinning. As a result parts of grains are reoriented so that the *c*-axis of the hexagonal magnesium lattice is reoriented nearly parallel with the sample axis. The situation is schematically depicted in figure 8b. Because the CTE is higher in the *c*-axis direction, reorientation of grains due to twinning increased the sample thermal expansion. Calculated thermal expansion coefficients using rule of mixture (RoM) are slightly higher than the measured values (figure 8a).

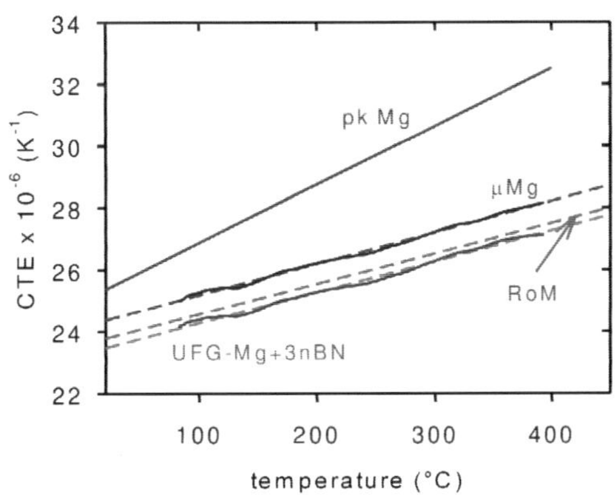

(a)

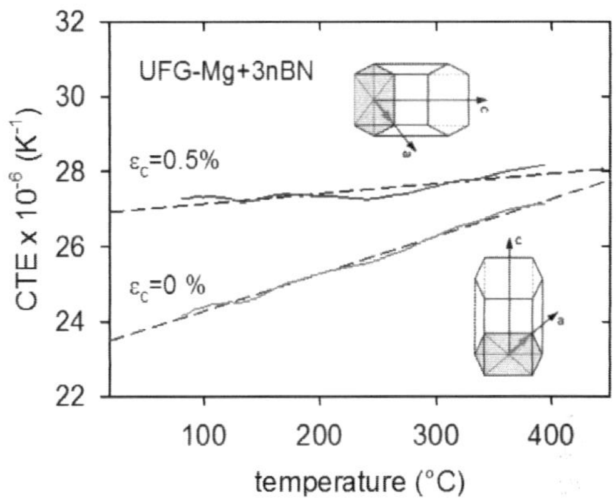

(b)

Figure 8. (a) Temperature dependence of CTE for the as-extruded UFG-Mg+3nBN sample, as well as for microcrystalline magnesium (μMg) and coarse grain polycrystalline magnesium (pk Mg); (b) effect of pre-straining in compression (0.5%) on the CTE.

4. Conclusions

Magnesium based nanocomposites were successfully prepared by a powder metallurgical route using mixing, followed by milling and then hot extrusion. The reinforcing nanoparticles of either graphite or hexagonal boron nitride exhibit a similar layered structure. Compression tests were performed at elevated temperatures for both nanocomposites, and the temperature dependency of the Young's modulus and of the thermal expansion coefficient were measured in the composite containing BN nanoparticles. Based on the experimental observations the following conclusions can be made:

- The compressive stress-strain curves of both materials exhibit a similar shape.
- In the low temperature regime the reinforcing effect of BN nanoparticles is much higher in comparison with that for Gr nanoparticles.
- Nanocomposites containing Gr particles are more stable that those containing BN at temperatures higher than 200 °C.
- Weak bonding between Gr nanoparticles and Mg matrix contribute to high damping due to grain

boundary sliding.
- Addition of BN nanoparticles to Mg results in an increased Young`s modulus.
- The thermal expansion coefficient of UFG-Mg+3nBN is strongly texture dependent.

Acknowledgements
Authors gratefully acknowledge the financial support from the Czech Science Foundation under Grant 107-15/11879S.

References

[1] Goh C S, Wei J, Lee L C and Gupta M 2007 Properties and deformation behaviour of Mg–Y_2O_3 nanocomposites *Acta Mater.* **55** 5115–5121
[2] Habibi M K, Joshi S P and Gupta M 2010 Hierarchical magnesium nano-composites for enhanced mechanical response *Acta Mater.* **58** 6104-6114
[3] Tun K S, Wong W L E, Nguyen Q B and Gupta M 2013 *Materials* **6** 1826-1839
[4] Drozd Z and Trojanová Z 2008 Plastic Properties of Microcrystalline Mg with Ceramic Nanoparticles *Materials Science Forum* **567-568** 189-192
[5] Gupta M and Wong W L E 2015 Magnesium-based nanocomposites: Lightweight materials of the future *Mater. Characterization* **105** 30–46
[6] Dieringa H 2011 Properties of magnesium alloys reinforced with nanoparticles and carbon nanotubes: A review *J. Mater. Sci.* **46** 289–306
[7] Trojanová Z and Lukáč P 2006 Anelastic and plastic properties of microcrystalline magnesium with ceramic nanoparticles *Proc. of International Symposium on Magnesium Technology in the Global Age (Montréal)* Canad. Inst. of Mining, Metallurgy and Petroleum, Canada pp 131-145
[8] Trojanová Z, Makowska-Mielczarek A, Riehemann W and Lukáč P 2012 Influence of thermal and mechanical cycles on the damping behaviour of Mg based nanocomposite *(Fatigue behaviour of fiber reinforced composites Experiments and simulations)* eds W Yao *et al* (Lancaster USA: Destech Publications, Inc.) pp 239-248
[9] Máthis K, Čapek J, Zdražilová Z and Trojanová Z 2011 Investigation of tension–compression asymmetry of magnesium by use of the acoustic emission technique *Mater. Sci. Eng.* A **528** 5904–5907
[10] Trojanová Z, Ferkel H, Lukáč P and Riehemann W 2002 Two new high-damping magnesium composites *Phys. Stat. Sol.* (a) **193** 205-210
[11] Trojanová Z, Lukáč P, Minárik P, Németh G, Gupta M, Seetharaman S and Di Fabio G 2016 Influence of inconel particles om mechanical properties of an extruded magnesium *Proc. Int. Conf.* Metal 2016, 25th Anniversary International Conference on Metallurgy and Materials, (Ostrava: Tanger) pp 1248-1253
[12] von Mises R 1928 Mechanik der plastischen Formänderung von Kristallen *Zeitschrift für Angewandte Mathematik und Mechanik* **8** 161-185
[13] Seetharaman S, Subramanian J, Tun KS, Hamouda A S and Gupta M 2013 Synthetis and characterization of nano boron nitride reinforced magnesium composites produced by the microwave sintering method *Materials* **6** 1940−1955
[14] Klimanek P and Pötzsch A 2002 Microstructure evolution under compressive plastic deformation of magnesium at different temperatures and strain rates *Mater. Sci. Eng.* A, **324** vol 1-2 145-150
[15] Wang Y N, Huang J C 2007 The role of twinning and untwining in yielding behavior in hot-extruded Mg-Al-Zn alloy *Acta Mat.* **55** 897−905
[16] Lukáč P 2006 *Metallurgy, Design Data, Applications (Magnesium Technology)* eds H Friedrich and B L Mordike (Berlin: Springer–Verlag) pp 63–107

Electron tomography of dislocations in an Al-Cu-Mg alloy

Z.Q. Feng [1,2], **C.W. Lin** [1], **T.T. Li** [1], **X. Luo** [1], **G.L. Wu** [1], **X.X. Huang** [1]
[1] College of Materials Science and Engineering, Chongqing University, Chongqing 400044, China
[2] Electron Microscopy Center of Chongqing University, Chongqing University, Chongqing 400044, China

E-mail: zqfeng@cqu.edu.cn

Abstract. Dislocation loops and helices are typical dislocation structures in quenched Al-Cu-Mg alloys. In this study, we combine electron tomography and Burgers vector analysis to characterize the spatial arrangement and crystallographic features of such dislocation loops and helices. It is found that the dislocation loops are fully edge in character and on {011} planes. The axis of each dislocation helix formed during quenching is found to be parallel to the Burgers vectors of the helix, which is always of <011> type.

1. Introduction

Dislocation loops and helices are two typical types of intrinsic dislocations generated during water quenching from the solid solution treatment temperature in precipitation-hardenable Al alloys [1]. The presence of these dislocations and their morphological and crystallographic characteristics will affect the aging behaviour of the alloys. Previous investigations have indicated that dislocation loops formed during quenching are usually full dislocations of mixed edge and screw character on {111} planes, and that they can form through the reaction of a moving Shockley partial dislocation (a/6<112>) with Frank partial dislocation loops (a/3<111>), formed by the condensation and collapse of quenched-in vacancies on the close-packed {111} planes [2]. However, in a study of the dislocation sources in a quenched Al-7%Mg alloy, Embury et al. [3] observed that the condensation and collapse of quenched-in vacancies actually occurred on $\{011\}_{Al}$ planes, and that the resultant dislocation loops possessed an a/2<110> type Burgers vector. Moreover, recent transmission electron microscopy (TEM) investigations of a quenched Al-Cu-Mg alloy revealed that all the dislocation loops are of pure edge character and lie on {011} planes [4]. As to the helical dislocations, they can form via the reaction of prismatic dislocation loops with straight screw dislocation lines, or the through climb of a mixed dislocation segment with its ends pinned by solute atoms or some other obstacles [2,5]. The Burgers vector of a helical dislocation is always parallel to its helical axis [2,6]. The helical axis was often observed to be approximately parallel to the <011> direction, but a <001> helix axis direction has also been reported [5,7-9]. These different observations of the crystallographic characteristics of dislocation loops and helices could be caused by the two-dimensional (2D) analysis of the crystallographic and morphological features of the dislocation loops and helices. A more precise characterization is thus required to fully understand the nature of such dislocation loops and helices.

Dislocation electron tomography has been developed and applied to study the three-dimensional (3D) morphology of dislocation structures [10,11] and to visualize the details of the interaction of moving dislocations with grain boundaries and precipitate particles [12]. In this study, we combine electron tomography and Burgers vector analysis to characterize the 3D configuration and crystallography of dislocation helices and loops in a quenched Al-Cu-Mg alloy.

2. Experimental

A commercial AA2024 aluminum alloy sheet with the nominal composition of Al-4.2Cu-1.5Mg-0.6Mn-0.5Fe-0.5Si (wt%) was chosen for this study. Several small specimens were cut and solid solution-treated at 495 °C for 1 h, followed by water quenching. To prepare thin foil samples for TEM observations, the treated specimens were mechanically ground and punched into disks of 3 mm in diameter. The small disks were finally thinned by twin-jet electropolishing with a solution of 30% nitric acid and 70% methanol at a temperature below -25 °C at 15V.

TEM observations were performed using a 200kV Tecnai 20 electron microscope equipped with a fully automated TEM tomography system. A single tilt tomography holder (Fischione model 2020) was used to acquire several tilt series of dark field (DF) images over a tilt angle range from -70° to +70° with an increment of 1°. A precision sample mounting system developed in house was used to control the diffraction contrast conditions during the entire tilting process for data acquisition. The tomography data were aligned and reconstructed using the Inspect 3D software, and finally visualized in 3D using the Avizo 9.0 software. All the dislocations, irrespective of their 3D morphology, were recognized as full dislocations, and their Burgers vectors were determined according to the invisibility criterion ($\boldsymbol{g}\cdot\boldsymbol{b} = 0$) for full dislocations in typical face centered cubic (FCC) crystals.

3. Results and discussion

Figure 1 shows two examples of 2D projections of the dislocation structure observed in the quenched Al-Cu-Mg alloy. A high density of dislocations, especially dislocation helices and loops (figure 1b), are observed within the Al grain matrix. Note that some of the dislocations overlap with each other in these 2D images, leaving an open question as to whether or not they are actually in contact and interacting, or if they are spatially separated. To characterize the crystallographic features and spatial arrangements of these dislocations, detailed TEM crystallographic analysis, as well as electron tomography, was carried out. For the dislocations in the area of interest as shown in figure 1b, five reflections, namely $\boldsymbol{g}$=[020], $[00\bar{2}]$, $[0\overline{22}]$, $[11\bar{1}]$and $[\overline{111}]$ that are related to the two crystal zone axes of $[1\bar{1}0]$ and [100] were chosen to obtain DF images of the dislocations (figure 2). Based on the dislocation invisibility criterion, the Burgers vectors for all dislocations seen in this area (figure 1b) were fully determined (see table 1).

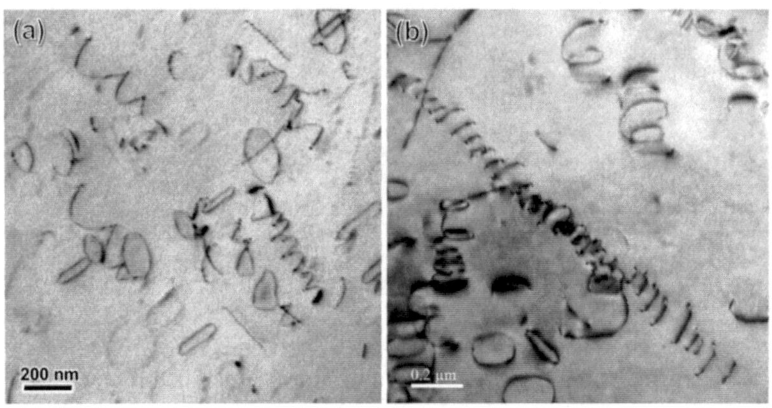

Figure 1. TEM images showing typical observations of dislocation helices and loops in an Al-Cu-Mg alloy after solid solution treatment and water quenching. A detailed analysis of dislocation Burgers vectors and tomography was carried out for the dislocations shown in (b) (for details see figure 2 and figure 3).

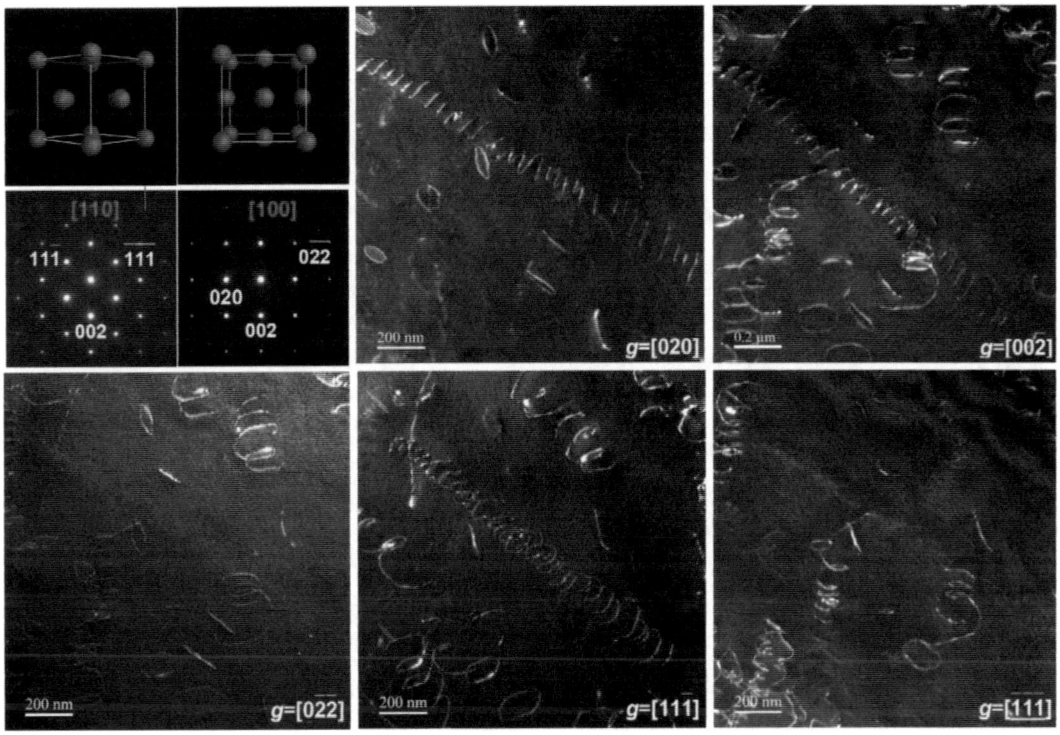

Figure 2. Top-left: $[1\bar{1}0]$ and [100] oriented FCC crystal units and corresponding diffraction patterns with self-consistent indexing; Rest: DF images using different reflections $\boldsymbol{g}$=[020], $[00\bar{2}]$, $[0\bar{2}\bar{2}]$, $[11\bar{1}]$ and $[\bar{1}\bar{1}\bar{1}]$ as labelled in the figures.

Table 1. Visibility analysis of possible full dislocations in FCC crystals for five selected reflection vectors $\boldsymbol{g}$ (★ indicates that the dislocation is visible).

g \ *b*	***b* (×a/2)**					
	[110]	$[1\bar{1}0]$	[101]	$[10\bar{1}]$	[011]	$[01\bar{1}]$
[020]	★	★			★	★
$[00\bar{2}]$			★	★	★	★
$[0\bar{2}\bar{2}]$	★	★	★	★	★	
$[11\bar{1}]$	★			★		★
$[\bar{1}\bar{1}\bar{1}]$	★		★		★	

To understand the dislocation morphology and configuration in 3D, DF-TEM electron tomography was employed. With the aid of a precision sample mounting system, a constant two-beam condition ($\boldsymbol{g}$ = [002]) was well controlled during a single tilt data acquisition process. A tilt series of DF dislocation images as well as corresponding diffraction patterns from -70° to +70° were then obtained and are shown in Figure 3, together with the corresponding unit cell geometry. It can be seen that the diffraction contrast conditions for dislocations during the image acquisition process were rather stable, and most of the images including the ones taken at high tilt angles are in good contrast. A few images taken when the beam directions were close to zone axes are of poor contrast; such images were not

IOP Conf. Series: Materials Science and Engineering **219** (2017) 012018 doi:10.1088/1757-899X/219/1/012018

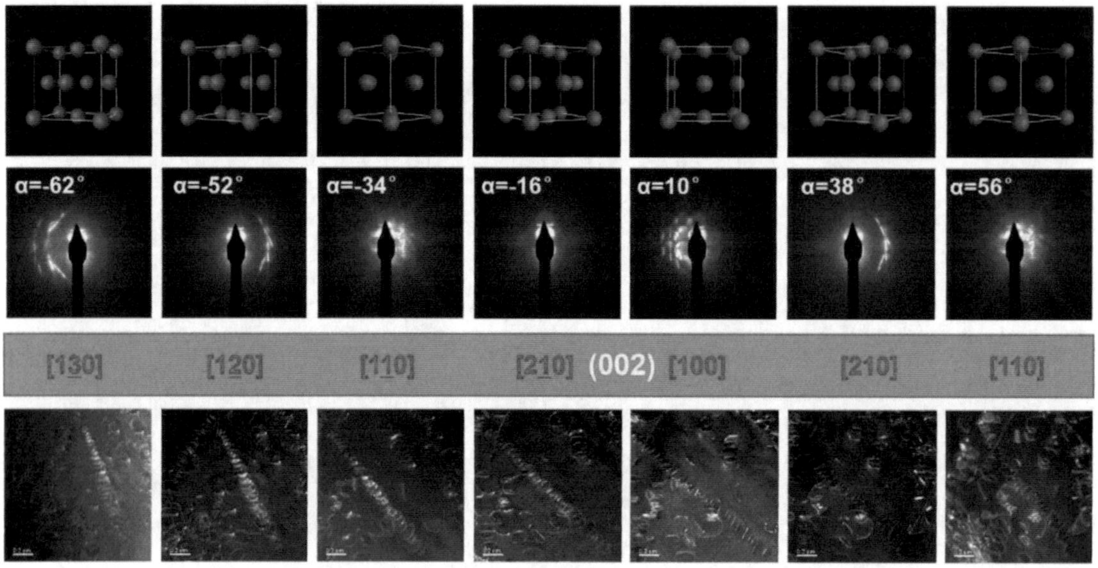

Figure 3. The typical crystal orientations and corresponding diffraction patterns and DF images in the tilt series from -70° to 70° by using reflection $\boldsymbol{g}$=[002].

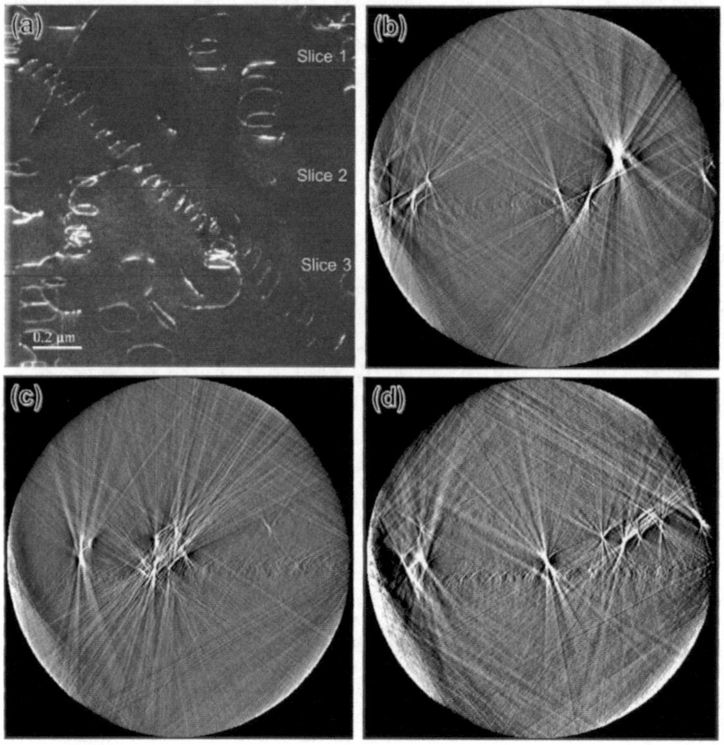

Figure 4. (a) DF image at zero tilt; (b-d) the back-projected cross-sections corresponding to slices 1-3 in (a) showing the quality of tilt axis alignment.

used for the 3D reconstruction. Image alignment using cross correlation algorithm [13-15] exhibited good quality of position and tilt axis alignment, which can be seen more clearly in three back-projected cross-sections as shown in figure 4. In particularly, it can be noted that the projections of some dislocations show circular-arc cross-sections (see arrows), and some of them are too close to be clearly distinguished, which may affect the resolution in the projection direction.

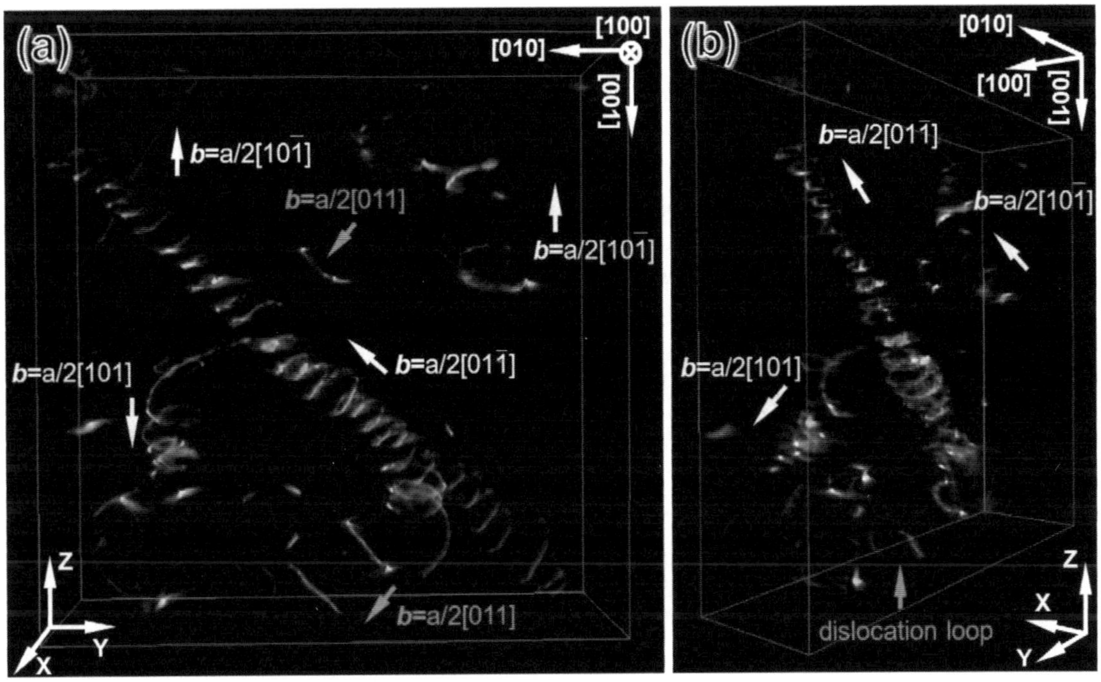

Figure 5. Two projections of 3D reconstruction of the dislocation structure in Figure 2b. Note that (a) is tilted 10° to transform from original sample coordinate system to the crystal coordinate system.

The weighted back projection algorithm [14,15] was used for data reconstruction. In the voxel-projection of the 3D reconstruction, shown in figure 5a, most of the dislocation helices and loops can be clearly observed, and the image fit well with the original 2D dislocation structure observations (figure 1b). By referring to the lattice orientation used during acquisition process, a 10° tilt about z axis was performed to adjust the reconstructed voxel close to the $[100]_{Al}$ direction, which allows the crystallographic orientations of the axes of helical dislocations to be readily determined. Combined with the Burgers vector analysis results (figure 2), the crystallography and morphology of the dislocation loops and helices could thus be fully characterized. All the visible dislocation loops were identified to be pure edge type full dislocations on {011} planes (marked by green arrows). For the three helical dislocations seen in this area, the 3D analysis revealed that their helical axes are parallel to [101], [01$\bar{1}$], and [10$\bar{1}$], respectively. No <001>-type axes of helical dislocations were observed. To be noted is that caution is needed in determining the axes of dislocation helices. For example, when viewing the dislocation helices from the [100] direction (figure 5a) both the helices of Burgers vectors [101] and [10$\bar{1}$] (marked yellow) appear to have their axes parallel to [001] direction. It is concluded that all the dislocation helices have their axes parallel to their Burgers vectors. These results are in good accordance with previous results [2,4,6,7].

The above results also indicate a possibility for a full 3D characterization of crystallography and morphology of dislocation structures by transforming the dislocation electron tomography in the sample coordinate system to the crystal coordinate system. More detailed analysis of the crystallographic features of dislocation structures are to be presented in a forthcoming paper.

4. Conclusions

The morphological and crystallographic features of dislocation loops and helices in a water-quenched Al-Cu-Mg alloy were characterized by means of dislocation tomography and Burgers vector analysis. The dislocation loops were identified to be pure edge type dislocations on {011} planes, while the axes of dislocation helices were determined to be parallel to their Burgers vectors, which were all of <110> type.

Acknowledgements

Financial supports from the National Natural Science Foundation of China (51501022, 51327805, 51471039), State Key Research and Development Program of MOST of China (2016YFB0700401), Natural Science Foundation of CQ (cstc2016jcyjA0266) and Fundamental Research Funds for the Central Universities (CDJZR14130002) are gratefully acknowledged.

References

[1] Totten G E and MacKenzie D S 2003 *Handbook of Aluminum: Physical Metallurgy and processes* vol I, ed Marcel Dekker (New York)
[2] Hull D and Bacon D J 2001 *Introduction to Dislocations* 4th ed Butterworth-Heinemann (Oxford)
[3] Embury J D and Nicholson R B 1963 *Acta Metall.* **11** 347
[4] Feng Z Q, Yang Y Q, Huang B, Luo X, Li M H, Han M and Fu M S 2011 *Acta Mater.* **59** 2412
[5] Wilson R N 1969 *J. Inst. Met.* **97** 80
[6] Thomas G and Whelan M J 1959 *Phil. Mag.* **40** 511
[7] Hutchinson C R and Ringer S P 2000 *Metall. Mater. Trans. A* **31A** 2721
[8] Raviprasad K, Hutchinson C R, Sakurai T and Ringer S P 2003 *Acta Mater.* **51** 5037
[9] Wilson R N, Moore D M and Forsyth P J E 1967 *J. Inst. Met.* **95** 177
[10] Barnard J S, Sharp J, Tong J R and Midgley P A 2006 *Science* **313** 319
[11] Tanaka M, Higashida K, Kaneko K, Hata S and Mitsuhara M 2008 *Scripta Mater.* **59** 901
[12] Liu G S, House S D, Kacher J, Tanaka M, Higashida K and Robertson I M 2014 *Mater. Charact.* **87** 1
[13] Amat F, Castano-Diez D, Lawrence A, Moussavi F, Winkler H and Horowitz M 2010 *Methods in Enzymology* **482** 343
[14] Jose-Jesus Fernandez 2012 *Micron* **43** 1010
[15] Jose-Jesus Fernandez 2012 *Current Opinion in Solid State and Materials Science* **17** 93

Quantum-mechanical study of tensorial elastic and high-temperature thermodynamic properties of grain boundary states in superalloy-phase Ni_3Al

Martin Friák[1], Monika Všianská[2,1], David Holec[3] and Mojmír Šob[2,1,4]

[1] Institute of Physics of Materials, Academy of Sciences of the Czech Republic, v.v.i., Žižkova 22, CZ-616 62 Brno, Czech Republic

[2] Central European Institute of Technology, CEITEC MU, Masaryk University, Kamenice 5, CZ-625 00 Brno, Czech Republic

[3] Department of Physical Metallurgy and Materials Testing, Montanuniversität Leoben, Franz-Josef-Strasse 18, A-8700 Leoben, Austria;

[4] Department of Chemistry, Faculty of Science, Masaryk University, Kotlářská 2, CZ-611 37 Brno, Czech Republic

E-mail: `friak@ipm.cz`

Abstract. Grain boundaries (GBs), the most important defects in solids and their properties are crucial for many materials properties including (in-)stability. Quantum-mechanical methods can reliably compute properties of GBs and we use them to analyze (tensorial) anisotropic elastic properties of interface states associated with GBs in one of the most important intermetallic compounds for industrial applications, Ni_3Al. Selecting the $\Sigma5(210)$ GBs as a case study because of its significant extra volume, we address the mechanical stability of the GB interface states by checking elasticity-based Born stability criteria. One critically important elastic constant, C_{55}, is found nearly three times smaller at the GB compared with the bulk, contributing thus to the reduction of the mechanical stability of Ni_3Al polycrystals. Next, comparing properties of $\Sigma5(210)$ GB state which is fully relaxed with those of a $\Sigma5(210)$ GB state when the supercell dimensions are kept equal to those in the bulk we conclude that lateral relaxations have only marginal impact on the studied properties. Having the complete elastic tensor of $\Sigma5(210)$ GB states we combine Green's-function based homogenization techniques and an approximative approach to the Debye model to compare thermodynamic properties of a perfect Ni_3Al bulk and the $\Sigma5(210)$ GB states. In particular, significant reduction of the melting temperature (to 79-81% of the bulk value) is predicted for nanometer-size grains.

1. Introduction

Grain boundaries (GBs) are extended defects which are critically important for a number of properties of solid-state materials, including, e.g., their macroscopic strength [1-6]. GBs often possess highly distorted atomic configurations and, therefore, it is advantageous to use reliable theoretical tools, such as quantum-mechanical (so-called first-principles or *ab initio*) calculations, when studying them [7-22]. Their properties are very sensitive to compositional changes [23-36]. GB properties are very sensitive to various (e.g., thermodynamically-driven) compositional changes Also the impurities, even in very low concentrations (ppm), can dramatically change their characteristics (see, for example, Refs. [37, 38]). Intergranular embrittlement, which is usually associated with segregation of impurities on the GB, is accompanied by a significant reduction of the ductility and strength. Impurities can also drastically reduce local magnetic properties at grain boundaries in magnetic materials, as documented e.g. in elemental Ni (Refs. [39, 40]). In particular, in ultra-fine grained (UFG) or nano-granular materials, GBs can become even more important than the grain interior itself and it is possible to achieve unprecedented properties not available in bulk samples. When reducing the grain sizes to truly nano-scale levels, even elastic properties may become significantly altered by GBs as compared with the bulk.

The principal goal of our research is assess high-temperature thermodynamic properties, in particular the melting temperature, of those material regions which are affected by GBs (the concept of complexions is worth mentioning in this context [41-62]) employing the knowledge of their tensorial elastic properties. Our recent paper [63] as well as the present one thus complement many previous GB-related studies that were focused predominantly on scalar characteristics (energy, strength, changes in inter-atomic bonds) as well as electronic structure and atomic configuration, see e.g. Refs. [24, 64-78]. Tensorial properties require rather computationally demanding approaches (see e.g. [79]) but this disadvantage is compensated by the wealth of insight which is gained. The knowledge of a full elastic tensor allows for assessing the mechanical stability of the studied phases via generic Born stability criteria [80, 81]. Further, elastic properties also play a crucial role in thermodynamic stability as they determine long-wave phonon frequencies and can be employed within classical approaximative model of Slater [82] to predict (high-temperature) thermodynamic properties of materials (for example the melting temperature as discussed below).

Here we analyze thermodynamic and tensorial elastic properties of the $\Sigma5(210)$ grain boundaries in one of the most important intermetallics, Ni_3Al, which appears e.g. as the γ' phase in Ni-based superalloys. In particular, we study the $\Sigma5(210)$ GB (see figure 1) which is one of the smallest GBs exhibiting an extra volume (when compared with the bulk), similarly as in many non-special GBs.

2. Methodology

Our *ab initio* calculations were performed within the framework of density functional theory [83, 84] using the Vienna Ab initio Simulation Package (VASP) [85, 86]. The exchange and correlation energy was treated within the generalized gradient approximation as parametrized by Perdew, Burke, and Ernzerhof [87] and implemented in projector augmented wave pseudopotentials [88]. We used a plane-wave energy cut-off of 500 eV with a $5 \times 17 \times 3$ Monkhorst-Pack k-point mesh for the 64-atom supercells, reducing the forces on atoms under 1 meV/Å.

The two different GB chemical compositions corresponding to either both Ni and Al or solely Ni atoms at the GB plane are shown in figures 1a,b. As periodic boundary conditions apply, the supercells are periodic approximants of the real GB-associated interface states. Elastic constants were computed by the stress-strain method outlined in Refs. [89, 90]. For the lower-energy GB variant we compared the results for two different types of simulations (and strain state). First, we determined thermodynamic, structural and tensorial elastic properties when all atomic positions as well as the shape and the volume of the supercell were fully relaxed. Second, we relaxed all the above mentioned degrees of freedom except for the supercell deimensions within the GB plane which we keep equal to the equilibrium values predicted for the bulk Ni_3Al.

Finally, when simulating the application of external strains to determine the elastic constants (see Refs. [89, 90]), we also relaxed the atomic positions. Our tests show that the most critical shear elastic constant, C_{55}, that could be as low as 15 GPa in case when only Ni atoms are located at the interface (see more details in our recent paper [63]), is much higher (64 GPa) if these atomic relaxations are omitted.

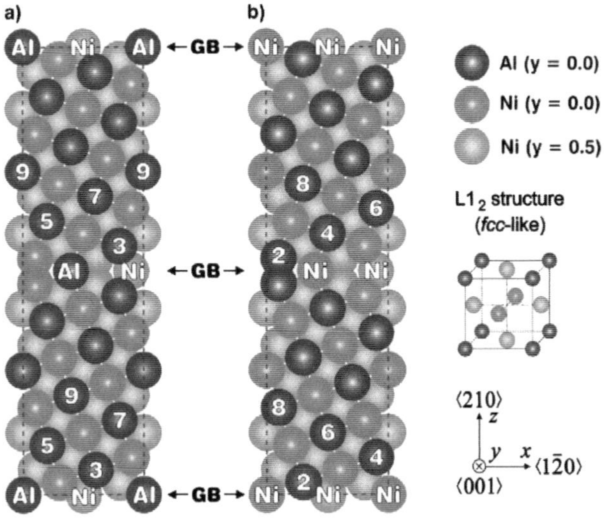

Figure 1. Schematic visualization of two supercells used in our *ab initio* calculations of interface states associated with the $\Sigma5(210)$ grain boundaries in Ni_3Al with different interface stoichiometries, $\Sigma5(210)^{Al,Ni}$ with both Al and Ni atoms at the GB plane (a) and $\Sigma5(210)^{Ni,Ni}$ with only Ni atoms at the GB plane (b). The blue spheres represent Ni atoms and the orange ones Al (or Si) atoms. The numbers shown at selected atomic positions label the atomic layers with respect to the grain boundary (which would be marked as the number 1).

3. Results

Our recent publication [63] summarizes most of our results related to the elasticity of GB states. The current paper uses these findings to assess high-temperature thermodynamic properties of GB state, in particular differences in the melting temperature when compared to the bulk Ni_3Al. A few selected facts from Ref. [63], which are critically important for the thermodynamic modeling discussed below, will be nevertheless briefly discussed here. In particular, our calculations show that the $\Sigma5(210)^{Al,Ni}$ with both types of atoms (Al and Ni) at the interface (figure 1a) has a lower GB energy of 1.38 J/m^2 (it is 1.66 J/m^2 in the case of the $\Sigma5(210)^{Ni,Ni}$ with only Ni atoms at the interface - see figure 1b). The lower energy variant $\Sigma5(210)^{Al,Ni}$ has the additional volume per unit GB area 0.29 Å (Å^3/Å^2), i.e. lower than 0.32 Å computed for the higher energy variant $\Sigma5(210)^{Ni,Ni}$. Our values related to $\Sigma5(210)^{Ni,Ni}$ are in excellent agreement with the previous theoretical results of 1.7 J/m^2 and 0.36 Å [8]. Due to the lower energy predicted for the $\Sigma5(210)^{Al,Ni}$ variant, we will focus only on this GB state from now on. Calculated elastic constants of $\Sigma5(210)^{Al,Ni}$ interface states are given in table 1 together with bulk Ni_3Al elastic constants in the same coordination system. The table contains three sets of elastic constants for $\Sigma5(210)^{Al,Ni}$ GB states. Two of them are computed for both a fully relaxed state and an interface-constrained state. In the latter case the supercell dimensions within the GB interface are kept constant and equal to the equilibrium values computed for the bulk Ni_3Al but the atomic positions, the overall volume, and the supercell size in the direction perpendicular to the GB interface are relaxed so as to minimize the total energy of the studied system. The last set of elastic constants is obtained with the help of a linear-elasticity approach. In order to visualize how the interface states associated with GBs respond to uniaxial loading along different crystallographic directions, we exhibit directional dependencies of the Young's modulus and linear compressibility in figure 2.

IOP Conf. Series: Materials Science and Engineering **219** (2017) 012019 doi:10.1088/1757-899X/219/1/012019

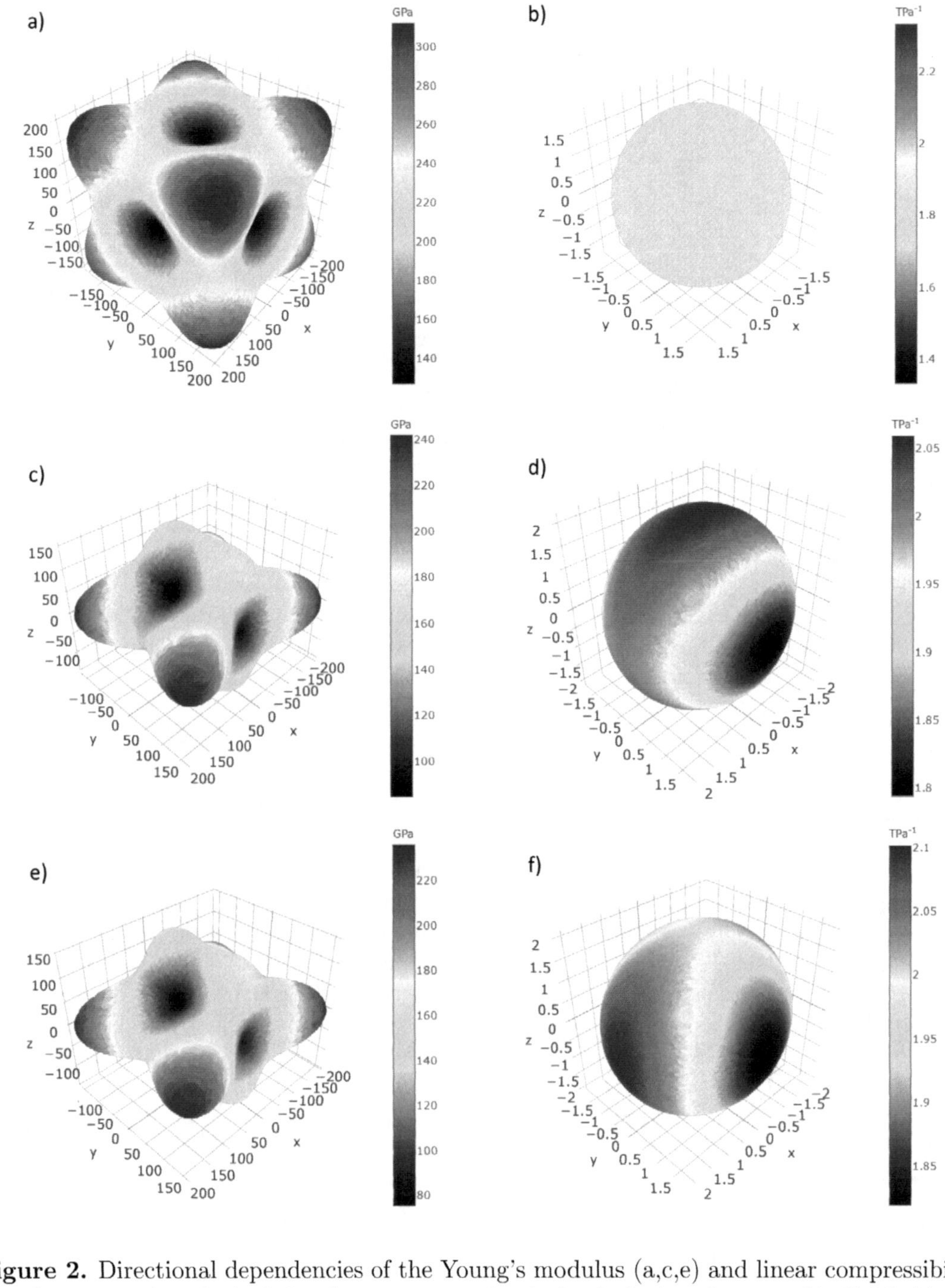

Figure 2. Directional dependencies of the Young's modulus (a,c,e) and linear compressibilities (b,d,f) computed with the help of quantum-mechanical calculations in case of bulk Ni_3Al (a,b), fully relaxed $\Sigma5(210)^{Al,Ni}$ (c,d) and the $\Sigma5(210)^{Al,Ni}$ with the supercell dimensions kept equal to those in the bulk (e,f). Please note that the sub-figures (c-f) have their coordination systems defined in figure 1 while those visualizing properties of the bulk Ni_3Al (a,b) have the orthonormal coordination system matching the unit vectors of the elementary cell of Ni_3Al.

Table 1. The *ab initio* computed elastic constants (all in GPa) of (i) the Ni_3Al bulk in the coordination system of the studied GBs (shown in figure 1), (ii) Ni_3Al $\Sigma5(210)$ GBs states (see figure 1) both fully relaxed and interface-constrained, and (iii) prediction based on the linear-elasticity approach from Ref. [91].

Ni_3Al states:	C_{11}	C_{12}	C_{13}	C_{22}	C_{23}	C_{33}	C_{44}	C_{55}	C_{66}
bulk $\langle 1\bar{2}0\rangle$ $\langle 001\rangle$ $\langle 210\rangle$	298	152	100	246	152	298	128	76	128
$\Sigma5(210)^{Al,Ni}$ fully relaxed	246	152	114	218	148	246	61	29	116
$\Sigma5(210)^{Al,Ni}$ interface-constrained	243	146	112	214	152	244	60	25	115
Lin. elast. [91]	278	152	119	246	152	278	127	71	127

The studied $\Sigma5(210)^{Al,Ni}$ GB interface states have an orthorhombic symmetry and comparison of sub-figures figure 2a,b based on *ab initio* results obtained for the bulk Ni_3Al with those obtained for the $\Sigma5(210)^{Al,Ni}$ GB states (figure 2c-f) clearly shows how very different the elasticity of GB states is compared with the bulk (see the impact of other grain boundaries and twins on the Young's modulus reported in Ref. [92]). The stiffest directions with the Young's modulus of about 250 GPa are within the same crystallographic plane as the GB interface while the softest directions with the Young's modulus lower than 90 GPa are inclined to the GB interface plane. When checking the computed elastic constants listed in table 1, it is important to note that the obtained shear elastic constants C_{44} and C_{55} of interface states associated with the $\Sigma5(210)$ GBs are considerably lower than those in the bulk. In particular, the elastic constant C_{55} is nearly three times lower (29 and 25 GPa) when compared with the bulk value (76 GPa in the same coordination system). To assess these changes and their effect on the mechanical stability, we apply Born and Huang's stability criteria [80] that connect the mechanical stability with the positiveness of leading principal minors of the matrix of elastic constants. As the diagonal elements C_{44}, C_{55} and C_{66} must be positive, the drop predicted for C_{55} clearly identifies the weakest link. Methodologically, it is also important to note that the two different simulation approaches, i.e. full relaxation and the restricted relaxation when the GB interface dimensions in the GB plane are kept equal to the bulk values (as a model for the constraining impact of the underlying grain interior), gave us very similar results (compare the sub-figures 2c,e and 2d,f) as well as the actual values of the elastic constants in table 1.

Our *ab initio* study is devoted to tensorial properties of interface GB states, i.e. to material formed by only a few atomic layers of atoms close to a particular extended defect. In order to study them, the size of the supercells is carefully chosen to contain only a minimum amount of the bulk material (the grain interior; more details may be found in Ref. [63]). Complementarily, we also address another limiting case when we have such a high amount of bulk grain material that the impact of the interface states related to grain boundaries becomes negligible. To do so we employ a linear-elasticity approach derived for coherent multilayers (so-called superlattices) by Grimsditch and Nizzoli [91]. This method processes as the input the tensor of elastic constants for bulk Ni_3Al (see the values in table 1) that is rotated [93] by two different angles so as to mimic the rotations of the two interfacing grains which form the studied $\Sigma5(210)$ GB. The elastic constants obtained in this way essentially correspond to a coherent two-phase multilayer system (stacked along the z-axis in our coordinate system). The resulting elastic constants are listed in table 1 and visualized in figure 3 in the form of a directional dependence of the Young's modulus. We can conclude that the pronounced softening of the Young's modulus under 100 GPa, which we discussed above in case of the GB states, is not present.

IOP Conf. Series: Materials Science and Engineering **219** (2017) 012019 doi:10.1088/1757-899X/219/1/012019

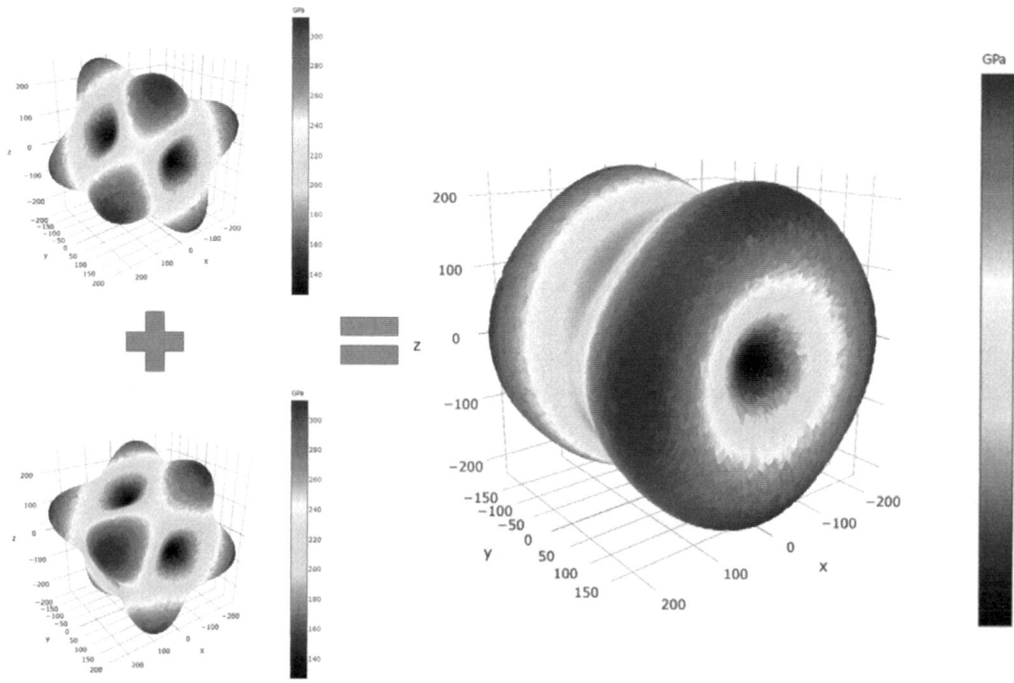

Figure 3. Schematics of the application of the linear elasticity theory by Grimsditch and Nizzoli [91] illustrated by the directional dependencies of the Young modulus (in GPa) of both grains (the directional dependences are rotated in the same way as are rotated the two grains forming the grain boundary) and the combined grain boundary $\Sigma5(210)$ state (the GBs interface is the x-y plane).

As we have now the complete elastic tensor of $\Sigma5(210)^{\mathrm{Al,Ni}}$ GB state (both fully relaxed and interface-constrained for comparison) we can use the Debye model [94] to access the impact of different elasticities of GB states on some of high-temperature thermodynamic properties of the studied states. In particular, we employ approximative Slater's approach [82] which was successfully used, e.g., in the case of Ti-Ta alloys as reported in the recently published paper by Chakraborty *et al.* [95]. In this modeling scheme the Debye temperature Θ_{D} is estimated using the following equation:

$$\Theta_{\mathrm{D}} = \frac{\hbar}{k_{\mathrm{B}}}\left(6\pi^2 V^{1/2} n\right)^{1/3} f(\nu)\sqrt{\frac{B_S}{M}}, \qquad (1)$$

which is valid for an isotropic solid and, therefore, we will below used integral homogenized elastic characteristics corresponding to a polycrystalline aggregate (which are isotropic). The function $f(\nu)$ is defined as

$$f(\nu) = \left\{3\left[2\left(\frac{2}{3}\frac{1+\nu}{1-2\nu}\right)^{3/2} + \left(\frac{1}{3}\frac{1+\nu}{1-\nu}\right)^{3/2}\right]^{-1}\right\}^{1/3} \qquad (2)$$

IOP Conf. Series: Materials Science and Engineering **219** (2017) 012019 doi:10.1088/1757-899X/219/1/012019

where $\hbar$ is the reduced Planck constant, k_B is the Boltzmann constant, V is molecular volume and mass per formula unit and M is a logarithmically-averaged molecular mass per formula unit containing n atoms. The function $f(\nu)$ depends on the value of the homogenized Poisson's ratio. The adiabatic bulk modulus B_S we approximate here by the homogenized value of the static bulk modulus $B(V) = V(d^2E(V)/dV^2)$ (as the measure of triaxial compressibility).

Both integral elastic characteristics, the homogenized bulk modulus B and the Poisson ratio, are determined for each set of elastic constants (for the bulk Ni_3Al and both fully relaxed and interface-constrained $\Sigma5(210)^{Al,Ni}$ GB states, see table 1) by the self-consistent T-matrix solution for the effective medium. The T-matrix approach is based on the multiple scattering theory and was originally applied to determine elastic properties of single-phase polycrystals with cubic symmetry by Zeller and Dederichs [96]. This concept was generalized by Middya and Basu [97] to the case of single-phase crystals of non-cubic symmetry and further extended by Middaya *et al.* [98] to multi-phase composites. The method is implemented in the SC-EMA [99-101]. The resulting values of the integral (polycrystalline) isotropic Young's modulus Y, bulk modulus B and the Poisson's ratio as well as the volume $V/V_{\rm bulk}$ (relative to the bulk) are summarized in table 2.

As the next step, according to the approximative Lindemann's relation there is a linear connection between the value of the Debye temperature and the melting temperature (see, e.g., Ref. [102]) and we suppose a linear coefficient of this relation to be constant for all studied states. Analyzing equation (1) we can see that in case of the bulk Ni_3Al and the prediction of interface states by the linear-elasticity method (a limit of infinitely big grains) the Debye temperature (and thanks to the linear relations also the melting temperature) is predicted to be nearly identical because their molar masses and volumes are identical and their Poisson's ratios are nearly the same (the differences being caused mostly by numerical rounding errors).

The situation is very different in the case of both quantum-mechanically studied $\Sigma5(210)^{Al,Ni}$ GB states. Their molar mass is identical too (the computational supercells are stoichiometric containing an integer number of Ni_3Al formula units) but they exhibit very significant changes in the Poisson number ν and the bulk modulus B (to a lesser extent also those of the volume) which determine the values of the Debye temperature and, subsequently, the melting temperature. The numerical analysis of equation (1) then results in a drastic reduction of the melting temperature to 80.7 % in case of fully-relaxed $\Sigma5(210)^{Al,Ni}$ GB state and 78.7 % in case of the interface-constrained one.

Table 2. The calculated integral (polycrystalline) isotropic Young's modulus Y (in GPa), bulk modulus B (in GPa) and the Poisson's ratio as well as the volume $V/V_{\rm bulk}$ (relative to the bulk) based on the parameters obtained for (i) the Ni_3Al bulk, (ii) Ni_3Al $\Sigma5(210)^{Al,Ni}$ GBs states (see figure 1a) both fully relaxed and interface-constrained, and (iii) a coherent stack of infinitely big grain assessed by the linear-elasticity approach from Ref. [91].

Ni_3Al states:	Y (in GPa)	B (in GPa)	ν	$V/V_{\rm bulk}$
bulk	223	183	0.2966	1.0000
$\Sigma5(210)^{Al,Ni}$ fully relaxed	147	172	0.3569	1.0111
$\Sigma5(210)^{Al,Ni}$ interface-constrained	141	169	0.3608	1.0128
Lin. elast. [91]	220	183	0.2996	1.0000

The above estimated reductions of the melting temperature seem to be too large but it is worth reminding that the studied $\Sigma5(210)^{Al,Ni}$ GB state (as computed by the supercell in figure 1a) would model ultra-fined grains with the size of about 1 nm. For these grain sizes, and in the case of nano-particle systems in general, it is known that the melting temperature may be significantly reduced (see, e.g., Ref. [103]). Further, the predicted drop of the melting temperature as a function of the particles size is especially steep for nano-particles smaller than about 10 nm (see the reported reduction to 88% of the meting temperature in case of 5 nm nano-particles in Ref. [104]). Therefore, our predictions have the correct order of magnitude.

4. Conclusions

We have applied quantum-mechanical calculations to estimate high-temperature thermodynamic properties (changes in the melting temperature) of the interface states in Ni_3Al associated with the $\Sigma5(210)$ grain boundary and to determine their tensorial elastic properties. We find the elastic constants of GB states to be very different from those of the bulk Ni_3Al and nearly three-fold reduction of the elastic constant C_{55} is identified as the crucial weakest link for the mechanical stability of the studied GB states. Having the complete elastic tensor of $\Sigma5(210)$ GB states we combine a Green's-function-based homogenization technique and Slater's approximative approach to the Debye model to compare thermodynamic properties of a perfect Ni_3Al bulk and the $\Sigma5(210)$ GB states. As a result, significant reduction of the melting temperature (80.7 % and 78.7 % compared with the bulk value) is predicted. Our study thus clearly shows the importance of elastic-constant analysis for, studies of interface states close to GBs when determining their mechanical (in-)stability as well as thermodynamic properties based on *ab initio* calculations (see, e.g., Refs. [105, 106]). These studies should be ideally extended by phonon spectrum calculations, in-plane translation analysis (see, e.g., Ref. [107]) and simulations of larger deformations (see, e.g., Refs. [108-117]) in the future.

Acknowledgments

This research was supported by the Academy of Sciences of the Czech Republic through the Fellowship of J. E. Purkyně (M.F.), by the Ministry of Education, Youth and Sports of the Czech Republic under the Projects CEITEC 2020 (LQ1601) [M.F., M.Š. and M.V.] and the research infrastructure IPMINFRA (LM2015069) as well as by the Czech Science Foundation, Project GA 16-24711S (M.F., M.Š. and M.V.). Computational resources were supplied by the Ministry of Education, Youth and Sports of the Czech Republic under the Projects CESNET (Project No. LM2015042), CERIT-Scientific Cloud (Project No. LM2015085), and IT4Innovations National Supercomputer Center (Project No. LM2015070) provided under the program Projects of Large Research, Development and Innovations Infrastructures. We gratefully acknowledge software development contributions of Dr. T. Rebok, Bc. D. Lago and Bc. Š. Brynych from the Faculty of Informatics of the Masaryk University in Brno, Czech Republic, that allowed us to visualize the directional dependencies of Young's moduli and linear compressibilities in figures 2-6. We also highly appreciate discussions with Prof. Ing. Pavel Lejček, DrSc., from the Institute of Physics of the Academy of Sciences of the Czech Republic in Prague, Czech Republic.

References

[1] Duscher G, Chisholm M F, Alber U and Ruhle M 2004 *Nature Mater.* **3** 621–626
[2] Lu G H, Deng S H, Wang T M, Kohyama M and Yamamoto R 2004 *Phys. Rev. B* **69** 134106–1–134106–9
[3] Pokluda J, Černý M, Šandera P and Šob M 2004 *J. Comp.-Aided Mat. Design* **11** 1–28
[4] Kohyama M 2002 *Phys. Rev. B* **65** 184107–1–184107–11
[5] Ogata S, Umeno Y and Kohyama M 2009 *Model. Sim. Mat. Sci. Eng.* **17** 013001–1–013001–33
[6] Pokluda J, Černý M, Šob M and Umono Y 2015 *Progress in Materials Science* **73** 127–158
[7] Mishin Y, Mehl M J, Papaconstantopoulos D A, Voter A F and Kress J D 2001 *Phys. Rev. B* **63** 224106–1–224106–16

[8] Lu G, Kioussis N, Wu R and Ciftan M 1999 *Phys. Rev. B* **59** 891–898
[9] Yazyev O V and Louie S G 2010 *Phys. Rev. B* **81** 195420–1–195420–7
[10] Muller D A, Singh D J and Silcox J 1998 *Phys. Rev. B* **57** 8181–8202
[11] Wang Y and Li J 2010 *Acta Materialia* **58** 1212–1235
[12] Mrovec M, Groeger R, Bailey A G, Nguyen-Manh D, Elsässer C and Vitek V 2007 *Phys. Rev. B* **75** 104119–1–104119–16
[13] Yan Y F, Noufi R and Al-Jassim M M 2006 *Phys. Rev. Lett.* **96** 205501–1–205501–4
[14] Lojkowski W and Fecht H J 2000 *Prog. Mat. Sci.* **45** 339–568
[15] Dawson I, Bristowe P D, Lee M H, Payne M C, Segall M D and White J A 1996 *Phys. Rev. B* **54** 13727–13733
[16] Imaeda M, Mizoguchi T, Sato Y, Lee H S, Findlay S D, Shibata N, Yamamoto T and Ikuhara Y 2008 *Phys. Rev. B* **78** 245320–1–245320–12
[17] Tanaka S and Kohyama M 2001 *Phys. Rev. B* **64** 235308–1–235308–12
[18] McKenna K P and Shluger A L 2009 *Phys. Rev. B* **79** 224116–1–224116–11
[19] Kohyama M 2002 *Model. Sim. Mater. Sci. Eng.* **10** R31–R59
[20] Oba F, Nishitani S R, Adachi H, Tanaka I, Kohyama M and Tanaka S 2001 *Phys. Rev. B* **63** 045410–1–045410–10
[21] Ochs T, Beck O, Elsässer C and Meyer B 2000 *Phil. Mag. A* **80** 351–372
[22] Lejček P, Šob M and Paidar V 2017 *Progress in Materials Science* **87** 83–139
[23] Ackland G J, Mendelev M I, Srolovitz D J, Han S and Barashev A V 2004 *J. Phys. Cond. Mat.* **16** S2629–S2642
[24] Lu G H, Zhang Y, Deng S, Wang T, Kohyama M, Yamamoto R, Liu F, Horikawa K and Kanno M 2006 *Phys. Rev. B* **73** 224115–1–224115–5
[25] Yan M, Sob M, Luzzi D E, Vitek V, Ackland G, Methfessel M and Rodriguez C 1993 *Phys. Rev. B* **47** 5571–5582
[26] Zhang L, Da Silva J L F, Li J, Yan Y, Gessert T A and Wei S H 2008 *Phys. Rev. Lett.* **101** 155501–1–155501–4
[27] Braithwaite J S and Rez P 2005 *Acta Materialia* **53** 2715–2726
[28] Christensen M and Wahnstrom G 2003 *Phys. Rev. B* **67** 115415–1–115415–11
[29] Domain C and Becquart C S 2005 *Phys. Rev. B* **71** 214109–1–214109–12
[30] Du Y A, Ismer L, Rogal J, Hickel T, Neugebauer J and Drautz R 2011 *Phys. Rev. B* **84** 144121–1–144121–13
[31] Zhang L, Shu X, Jin S, Zhang Y and Lu G H 2010 *J. Phys. Cond. Mater.* **22** 375401–1–375401–5
[32] Asta M and Hoyt J J 2000 *Acta Materialia* **48** 1089–1096
[33] Thomson D I, Heine V, Payne M C, Marzari N and Finnis M W 2000 *Acta Materialia* **48** 3623–3632
[34] Wachowicz E, Ossowski T and Kiejna A 2010 *Phys. Rev. B* **81** 094104–1–094104–9
[35] Šesták P, Černý M and Pokluda J 2011 *Intermetallics* **19** 1567–1572
[36] Wachowicz E and Kiejna A 2008 *Comp. Mater. Sci.* **43** 736–743
[37] Sutton A P and Balluffi R W 1995 *Interfaces in crystalline materials* (Oxford: Oxford University Press)
[38] Lejček P 2010 *Grain boundary segregation in metals* (Heidelberg: Springer)
[39] Všianská M and Šob M 2011 *Prog. Mat. Sci.* **56** 817–840
[40] Všianská M and Šob M 2011 *Phys. Rev. B* **84** 014418–1–014418–5
[41] Tang M, Carter W C and Cannon R M 2006 *Phys. Rev. B* **73** 024102–1–024102–14
[42] Rohrer G S 2011 *J. Mater. Sci.* **46** 5881–5895
[43] Cantwell P R, Tang M, Dillon S J, Luo J, Rohrer G S and Harmer M P 2014 *Acta Materialia* **62** 1–48
[44] Rohrerw G S 2011 *J. Amer. Ceram. Soc.* **94** 633–646
[45] Dillon S J and Harmer M P 2008 *J. Amer. Ceram. Soc.* **91** 2304–2313
[46] Raabe D, Herbig M, Sandlöbes S, Li Y, Tytko D, Kuzmina M, Ponge D and Choi P P 2014 *Curr. Opin. Sol. Sta. Mater. Sci.* **18** 253–261
[47] Dillon S J and Harmer M P 2008 *J. Amer. Ceram. Soc.* **91** 2314–2320
[48] Protasova S G, Kogtenkova O A, Straumal B B, Zieba P and Baretzky B 2011 *J. Mater. Sci.* **46** 4349–4353
[49] Dillon S J, Harmer M P and Luo J 2009 *JOM* **61** 38–44
[50] Luo J 2009 *Appl. Phys. Lett.* **95** 071911–1–071911–4
[51] Shi X and Luo J 2011 *Phys. Rev. B* **84** 014105–1–014105–5
[52] Kundu A, Asl K M, Luo J and Harmer M P 2013 *Scr. Mater.* **68** 146–149
[53] Kuzmina M, Herbig M, Ponge D, Sandlöbes S and Raabe D 2015 *Science* **349** 1080–1083
[54] Rheinheimer W, Baeurer M, Chien H, Rohrer G S, Handwerker C A, Blendell J E and Hoffmann M J 2015 *Acta Materialia* **82** 32–40
[55] Bojarski S A, Ma S, Lenthe W, Harmer M P and Rohrer G S 2012 *Metall. Mater. Trans. A* **43A** 3532–3538
[56] Rickman J M, Chan H M, Harmer M P and Luo J 2013 *Surf. Sci.* **618** 88–93
[57] Ma S, Cantwell P R, Pennycook T J, Zhou N, Oxley M P, Leonard D N, Pennycook S J, Luo J and Harmer

M P 2013 *Acta Materialia* **61** 1691–1704
[58] Bojarski S A, Harmer M P and Rohrer G S 2014 *Scrip. Mater.* **88** 1–4
[59] Kogtenkova O A, Straumal B B, Protasova S G, Gornakova A S, Zieba P and Czeppe T 2012 *JETP Lett.* **96** 380–384
[60] Frazier W E, Rohrer G S and Rollett A D 2015 *Acta Materialia* **96** 390–398
[61] Zhou N and Luo J 2015 *Acta Materialia* **91** 202–216
[62] Moghadam M M, Rickman J M, Harmer M P and Chan H M 2015 *J. App. Phys.* **117** 045311–1–045311–9
[63] Friák M, Všianská M, Holec D, Zelený M and Šob M 2017 *Science and Technology of Advanced Materials* **18** 273–282
[64] Ogata S, Umeno Y and Kohyama M 2009 *Model. Sim. Mater. Sci. Eng.* **17** 013001–1–013001–33
[65] Pang X Y, Janisch R and Hartmaier A 2016 *Model. Sim. in Mater. Sci. Eng.* **24** 015007–1–015007–16
[66] Razumovskiy V I, Ruban A V, Razumovskii I M, Lozovoi A Y, Butrim V N and Vekilov Y K 2011 *Scripta Materialia* **65** 926–929
[67] Tahir A M, Janisch R and Hartmaier A 2014 *Mater. Sci. Eng. A* **612** 462–467
[68] Tahir A M, Janisch R and Hartmaier A 2013 *Model. Sim. in Mater. Sci. Eng.* **21** 075005–1–075005–16
[69] Razumovskiy V I, Vekilov Y K, Razumovskii I M, Ruban A V, Butrim V N and Mironenko V N 2011 *Phys. Sol. State* **53** 2189–2193
[70] Hristova E, Janisch R, Drautz R and Hartmaier A 2011 *Comp. Mat. Sci.* **50** 1088–1096
[71] Janisch R, Ahmed N and Hartmaier A 2010 *Phys. Rev. B* **81** 184108–1–184108–7
[72] Razumovskiy V I, Lozovoi A Y and Razumovskii I M 2015 *Acta Materialia* **82** 369–377
[73] Scheiber D, Razumovskiy V I, Puschnig P, Pippan R and Romaner L 2015 *Acta Materialia* **88** 180–189
[74] Janisch R and Elsässer C 2009 *Int. J. Mater. Res.* **100** 1488–1493
[75] Janisch R and Elsässer C 2008 *Phys. Rev. B* **77** 094118–1–094118–9
[76] Gemming S, Janisch R, Schreiber M and Spaldin N A 2007 *Phys. Rev. B* **76** 045204–1–045204–10
[77] Janisch R and Elsässer C 2003 *Phys. Rev. B* **67** 224101–1–2241011–11
[78] Razumovskiy V I, Lozovoi A Y and Razumovskii I M 2016 *Acta Materialia* **106** 401–402
[79] van de Walle A 2008 *Nature Materials* **7** 455–458
[80] Mouhat F and Coudert F X 2014 *Phys. Rev. B* **90** 224104–1–224104–4
[81] Černý M 2007 *Mat. Sci. Eng. A* **462** 432–435
[82] Slater J C 1939 *Introduction to Chemical Physics* (McGraw-Hill, New York)
[83] Hohenberg P and Kohn W 1964 *Phys. Rev.* **136** B864–B871
[84] Kohn W and Sham L J 1965 *Phys. Rev.* **140** A1133–A1138
[85] Kresse G and Hafner J 1993 *Phys. Rev. B* **47** 558–561
[86] Kresse G and Furthmüller J 1996 *Phys. Rev. B* **54** 11169–11186
[87] Perdew J P, Burke K and Ernzerhof M 1996 *Phys. Rev. Lett.* **77** 3865–3868
[88] Blöchl P E 1994 *Phys. Rev. B* **50** 17953–17979
[89] Yu R, Zhu J and Ye H Q 2010 *Comp. Phys. Comm.* **181** 671–675
[90] Zhou L, Holec D and Mayrhofer P H 2013 *J. Appl. Phys.* **113** 043511–1–043511–8
[91] Grimsditch M and Nizzoli F 1986 *Phys. Rev. B* **33** 5891–5892
[92] Šesták P, Černý M and Pokluda J 2009 *ESOMAT 2009 - 8th European Symposium on Martensitic Transformations* ed Sittner, P and Paidar, V and Heller, L and Seiner, H
[93] Ting T C T 1996 *Anisotropic elasticity* (New York, Oxford University Press)
[94] Debye P 1912 *Ann. Phys.* **39** 789–839
[95] Chakraborty T, Rogal J and R D 2016 *Phys. Rev. B* **94** 224104–1–224104–9
[96] Zeller R and Dederichs P H 1973 *Phys. Stat. Solid. B* **55** 831–842
[97] Middya T R and Basu A N 1986 *J. Appl. Phys.* **59** 2368–2375
[98] Middya T R, Paul M and Basu A N 1986 *J. Appl. Phys.* **59** 2376–2381
[99] Titrian H, Aydin U, Friák M, Ma D, Raabe D and Neugebauer J 2013 *MRS Proc. Vol. 1524*
[100] Friák M, Counts W A, Ma D, Sander B, Holec D, Raabe D and Neugebauer J 2012 *Materials* **5** 1853
[101] Zhu L F, Friák M, Lymperakis L, Titrian H, Aydin U, Janus A M, Fabritius H O, Ziegler A, Nikolov S, Hemzalová P, Raabe D and Neugebauer J 2013 *J. Mech. Behav. Biomed. Mat.* **20** 296–304
[102] Poirier J P 2000 *Introduction to the Physics of the Earth's Interior* (Cambridge University Press)
[103] Zhu Y F, Lian J S and Jiang Q 2009 *The Journal of Physical Chemistry C* **113** 16896–16900
[104] Kroupa A, Káňa T, Buršík J, Zemanová A and Šob M 2015 *Phys. Chem. Chem. Phys.* **17**(42) 28200–28210
[105] Houserová J, Friák M, Šob M and Vrest'al J 2002 *Comp. Mat. Sci.* **25** 562–569
[106] Houserová J, Vrest'ál J, Friák M and Šob M 2002 *CALPHAD* **26** 513–522
[107] Salehinia I, Shao S, Wang J and Zbib H M 2015 *Acta Materialia* **86** 331–340
[108] Šob M, Wang L G and Vitek V 1998 *Kovove materialy - Metallic Materials* **36** 145–152
[109] Friák M, Šob M and Vitek V 2001 *Phys. Rev. B* **63** 052405–1–052405–4

[110] Wang L G, Šob M and Zhang Z Y 2003 *Journal of Physics and Chemistry of Solids* **64** 863–872
[111] Legut D, Friák M and Šob M 2010 *Phys. Rev. B* **81** 214118–1–214118–19
[112] Černý M, Pokluda J, Šob M, Friák M and Šandera P 2003 *Phys. Rev. B* **67** 035116–1–035116–8
[113] Šesták P, Černý M and Pokluda J 2008 *Mat. Sci. Eng. A* **481-482** 247–249
[114] Šesták P, Černý M and Pokluda J 2008 *Strength of Materials* **70** 12–15
[115] Černý M, Šesták P, Pokluda J and Šob M 2013 *Phys. Rev. B* **87** 014117–1–014117–4
[116] Friák M, Šob M and Vitek V 2003 *Phys. Rev. B* **68** 184101–1–184101–10
[117] Černý M, Šesták P, Řehák P, Všianská M and Šob M 2016 *Mat. Sci. Eng. A* **669** 218–225

Microstructures and mechanical property of a Fe-Ni-Al-C alloy containing B2 intermetallic compounds

Si Gao[*, 1], **Yu Bai**[1,2], **Akinobu Shibata**[1,2], **Nobuhiro Tsuji**[1,2]

[1]Department of Materials Science and Engineering, Kyoto University, Sakyo-ku, Kyoto, 606-8501, Japan
[2]Elements Strategy Initiative for Structural Materials (ESISM), Kyoto University, Yoshida Honmachi, Sakyo-ku, Kyoto 606-8501, Japan

[1]E-mail address: gao.si.8x@kyoto-u.ac.jp

Keywords: high strength steel, intermetallic compounds, mechanical property

Abstract. Specimens of a Fe-24Ni-6Al-0.4C alloy with ultra-fine grained dual phase microstructures composed of FCC austenite and B2 phases were fabricated by cold rolling followed by annealing treatments in the temperature range from 850°C to 1100°C. Tensile tests of these specimens revealed a similar yield stress while the tensile strength was found to vary significantly. The different tensile strength of the samples was considered to be caused by the different volume fractions of the hard B2 phase.

1. Introduction

It is well-known that intermetallic compounds are usually avoided in designing metallic alloys because their brittleness leads to poor ductility of the whole material [1]. However, recently some research has shown that some alloys containing intermetallic compounds as relatively coarse second-phase particles can exhibit a good combination of strength and tensile ductility [2-4]. It is believed that the intermetallic compound plays an important role in such superior mechanical properties of these materials, but the detailed mechanisms responsible for this have not yet been studied and clarified. In the present study, the microstructure and tensile behavior of a Fe-24Ni-6Al-0.4C alloy, containing a NiAl-type B2 intermetallic compound, were investigated in order to understand the mechanism responsible for good mechanical properties in this new type of alloy.

2. Experimental procedures

An alloy having a chemical composition of Fe-24Ni-6Al-0.4C (mass%) was used in the present study. An as-received thick plate of 10 mm in thickness was homogenized at 1280°C for 5 hours in argon atmosphere, quenched into water, and then cold rolled to 1.15 mm in thickness, corresponding to an 88.5% rolling reduction. In order to obtain different microstructures, the cold-rolled sheets were annealed at different temperatures ranging from 850°C to 1100°C for 10 minutes using either a salt bath or a vacuum furnace and then quenched into water. Tensile test specimens of 10 mm in gauge length, 5 mm in gauge width and 1 mm in thickness were cut from the as-homogenized plate or from the cold rolled and subsequently annealed sheets using electron discharge machining. The tensile tests were carried out at room temperature at an initial strain rate of 8.3×10^{-4} s^{-1}. Prior to tensile testing the surface of each specimens was covered by white paint and then by a dispersion of black spots, in order to conduct digital image correlation (DIC) analysis. Images were taken during tensile testing by a CCD camera and then analyzed using the VIC2D software allowing a precisely measurement of both the tensile elongation and the strain distribution in the specimen during tensile deformation.

Microstructures of the specimens were observed in a scanning electron microscope (SEM) using back-scattered electron (BSE) contrast, and by electron back-scatter diffraction (EBSD) on longitudinal sections taken perpendicular to the transverse direction (TD) of the sheets. Average grain sizes of the specimens were determined by the linear intercept method from EBSD grain boundary maps.

3. Results and discussion

An EBSD grain boundary map and phase map of the as-homogenized specimen are shown in figure 1, which reveals an almost single austenitic (FCC) phase, fully recrystallized microstructure having a very coarse grain size around 150 μm. Some small B2 particles can be observed on grain boundaries of the austenite matrix, although the fraction of this phase is very small.

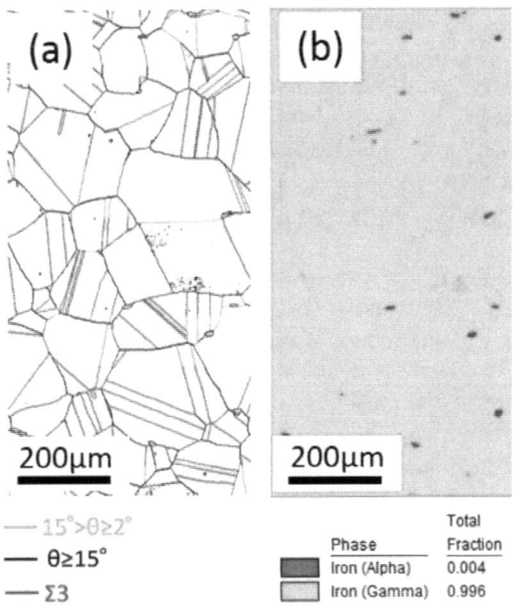

Figure 1. EBSD boundary map (a) and phase map (b) of the as-homogenized specimen. Low angle grain boundaries, high angle grain boundaries and twin boundaries are denoted by green, black and red colors, respectively in (a). The austenite phase (gamma) and B2 phase (recognized as alpha phase) are denoted by green and red colors in (b).

Example microstructures of the cold rolled and subsequently annealed specimens are shown in figure 2. Austenite and B2 phases can be easily distinguished from their dark and white contrast from the SEM-BSE images. It is seen in figure 2 (a) that the specimen annealed at 850°C has a fully recrystallized dual phase microstructure, characterized by equiaxed grains of the austenite phase with very fine grain sizes, and B2 particles that are mostly dispersed on grain boundaries and triple junctions of the austenite phase. From the EBSD observations, the measured average grain size of the austenite matrix was determined as 0.45 μm. The area fraction and average grain size of the B2 phase in this specimen were 61.8% and 0.17 μm, respectively. With increasing annealing temperature up to 1000°C, the area fraction of the B2 phase greatly decreased to 35%, accompanied by an overall microstructural coarsening, as seen figure 2(b). At 1000 °C the average grain sizes of the austenite matrix and B2 phase were increased to 0.32 μm and 0.7 μm, respectively. With further increase of annealing temperature to 1100°C, the area fraction of the B2 phase decreased to 16.6%, accompanied by a slight increase in the average grain size of austenite matrix and B2 phase to 0.37 μm and 1.1 μm, respectively. It should be noted that the material maintained an ultrafine-grained microstructure at elevated temperatures as high as 1100°C, which has been scarcely observed in conventional single or dual phase alloys [5-7]. The stability of the ultrafine microstructure is probably because the B2 phase does not dissolve even at high temperature, owing to the excellent thermal stability of such intermetallic compounds. As such The B2 particles dispersed on grain boundaries and triple junctions greatly inhibited the grain growth of the FCC (austenite) matrix by exerting a dragging force on migrating grain boundaries, thus allowing ultrafine grained microstructures to be maintained even after annealing at high temperatures [8].

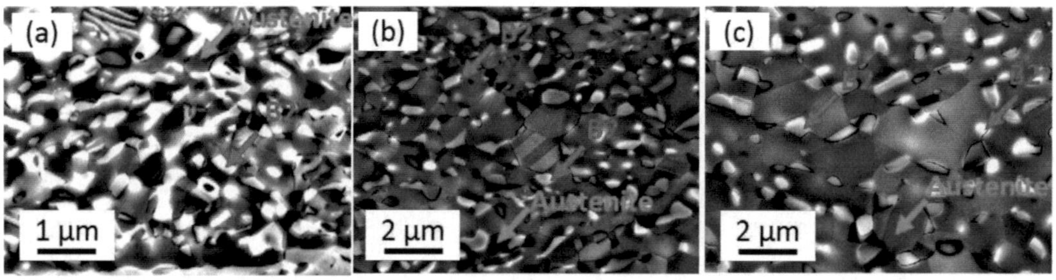

Figure 2. Microstructures of the cold-rolled and subsequently annealed specimens observed by SEM-BSE; (a) 850°C for 10 minutes, (b)1000°C for 10 minutes and (c) 1100°C for 10 minutes. Austenite (FCC) matrix in dark contrast and B2 phase in bright contrast are pointed out by green and red arrows, respectively. A decrease in the amount of B2 phase with increasing the annealing temperature can be readily observed.

Example tensile stress-strain curves of the as-homogenized, and the cold rolled and subsequent annealed specimens are shown in figure 3. The yield stress, tensile strength and total elongation of the as-homogenized specimen were 226 MPa, 600 MPa and 23%, respectively. After 88.5% cold rolling, the specimen showed a dramatically increased yield strength and tensile strength of 1543 MPa and 1770 MPa, respectively, but a quite limited total elongation of only 3%. Such tensile properties have often observed in heavily deformed austenitic steel, due to the highly deformed microstructures and associated low strain hardening capability [9,10]. On the other hand, it is noteworthy that the cold rolled and subsequently annealed specimens exhibited quite different tensile properties depending on the annealing temperature. The 850°C annealed specimen showed a significantly reduced yield strength of 558 MPa but an even higher tensile strength of 1923 MPa and a restored tensile elongation of 10%, compared with those of the cold rolled specimen. With increasing the annealing temperature to 1000°C and 1100°C, the tensile strength decreased to 1618 MPa and 1124 MPa, with the total elongation gradually increasing to 18% and 21%, respectively. It should be emphasized that although the annealed specimens having ultrafine grain microstructures and B2 particles exhibited moderate yield strength and total elongation, they showed significant larger strain hardening and ultimate tensile strengths, compared with other ultrafine grain materials having single austenite phase [6,7].

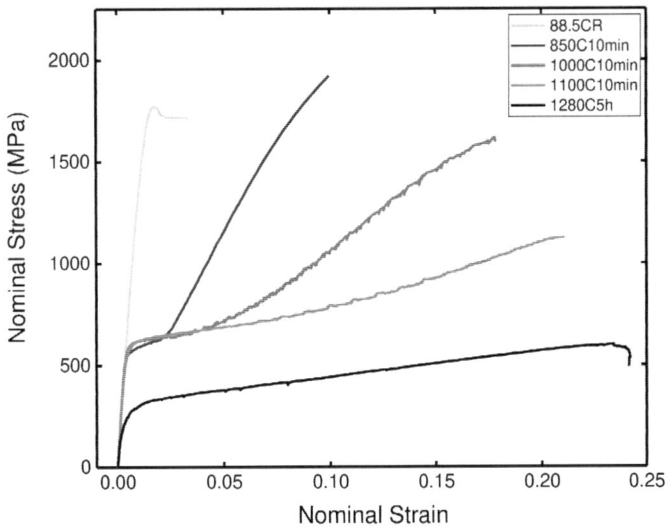

Figure 3. Nominal stress-strain curves of the as-homogenized(1280°C for 5 hours), 88.5% cold rolled, cold rolled and subsequently annealed Fe-24Ni-6Al-0.4C specimens.

Serrated flow can be clearly observed on stress-strain curves of the 1000°C and 1100°C annealed specimens, which indicates the occurrence of the deformation induced martensitic transformation that is often observed in Fe-Ni-C alloy systems [11]. It is very interesting that the annealed specimens have an approximately identical value of the yield strength (~600 MPa) and exhibit similar deformation behavior at early stage of tensile deformation. More specifically, the stress-strain curves of the annealed specimens exhibited a smooth transition from elastic deformation to plastic deformation (continuous yielding), followed by a nearly plateau-like part with a low strain hardening rate, which continued to about 2.5% plastic strain. After this plateau, the specimens showed pronounced differences in strain hardening and eventually fractured without necking, indicating an early fracture before plastic instability. Using the digital image correlation analysis, it was confirmed that the plateau-like flow curve corresponded to the formation and propagation of localized deformation bands, similar to Lüders-band deformation characteristic after discontinuous yielding in carbon steels (results to be published in another paper). Discontinuous yielding has been well-known to occur in low carbon ferritic steels due to a strong interaction between interstitial atoms (carbon and nitrogen) and dislocations. Moreover, it has been found recent years that discontinuous yielding is an universal mechanical behavior in most metallic materials with ultrafine grain sizes [12], although the mechanism of the interesting phenomena in UFG metals is still not fully understood. Considering the fine grain sizes and high carbon content of the specimens used in the present study, it is not surprising to observe discontinuous yielding and a low strain hardening rate at early stages of the tensile tests. However, discontinuous yielding phenomena are generally characterized by a distinct yield drop at the elasto-plastic transition, which is not observed in the present material. It is considered that the absence of the distinct yield drop and the plateau-like portion at the early stage of the tensile test in the present material is due to the exceptionally high strain hardening rate, which can partially compensate the stress drop during yielding, resulting in only the plateau flow in the beginning of tensile deformation.

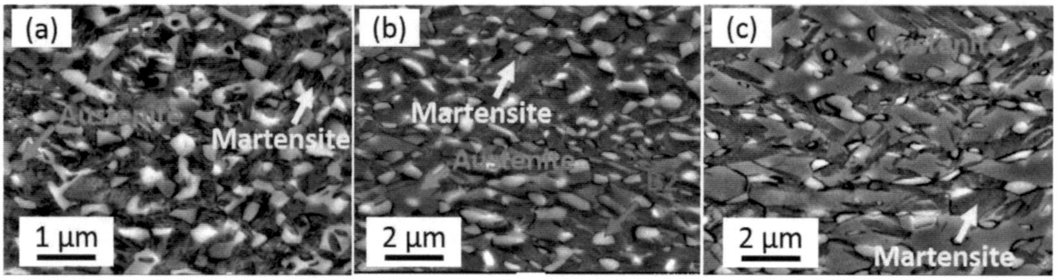

Figure 4. SEM micrographs of (a) 850°C annealed, (b) 1000°C annealed and (c) 1100°C annealed specimens after tensile fracture. The austenite phase, B2 phase and deformation induced martensite phase are denoted by green, red and yellow arrows, respectively.

After the early plastic strain stage with low strain hardening rate, the stress strain curves start to exhibit higher strain hardening rates. To explain the large difference in the strain hardening behavior of the present material containing B2 phase, microstructures of the annealed specimens after tensile deformation were observed and are shown in figure 4. Martensite phase can be clearly distinguished by the dark contrast on the micrographs of all three specimens, indicating that deformation induced martensitic transformation occurred during the tensile tests. It is seen that after the tensile deformation, a large fraction of austenite still remained in the 1100°C annealed specimen while only smaller amounts of austenite remain in the 1000°C and 850°C annealed specimens. The area fractions of martensite after the tensile deformation were also quantitatively measured by EBSD. Since EBSD mapping analysis cannot distinguish BCC martensite and B2 phases, the area fraction of martensite was calculated as the difference of the BCC phases before and after tensile deformation.

Table 1. Phase fractions of austenite, B2, and martensite measured by SEM-EBSD before and after tensile deformation.

Specimen	Phase fraction before tensile test (%)		Phase fraction after tensile test (%)		
	Austenite	B2	Austenite	B2	Martensite
850°C10min	38.1	61.8	7.6	61.8	30.6
1000°C10min	64	35.5	34.7	35.5	29.3
1100°C10min	83.4	16.6	51.3	16.6	32.1

The area fraction of each phase before and after tensile testing is summarized in table 1. Here it was assumed that the fraction of B2 phase did not change by the deformation. It should be noted that the fraction of the deformation induced martensite after tensile deformation was almost the same, approximately 30% in all the three specimens, although the initial area fraction of austenite was quite different depending on the annealing temperature. Thus the differences in the phase fraction among these three specimens after the tensile deformation mainly came from the amount of B2 phase and remaining austenite. It is well known that austenite is much softer than martensite or B2 phases that have a BCC or ordered BCC crystal structure, thus the contribution of the remaining deformed austenite to the overall tensile strength is considered to be small. Considering that the fraction of martensite is nearly identical in all of the three specimens, while B2 phase fraction is quite different, the large difference in the ultimate tensile strength is likely to arise from the different fraction of the B2 phase in each specimen. In other words, the large area fraction of B2 in the 850°C annealed specimen contributes predominantly to its high strain hardening and high tensile strength. As the area fraction of the B2 phase decreases with increasing the annealing temperature, the strain hardening and tensile strength of the specimens also decrease with increasing annealing temperature.

4. Conclusion

1. Fe-24Ni-6Al-0.4C alloy specimens having ultrafine grained and dual-phase microstructures composed of a FCC matrix and B2 phase were fabricated by cold rolling followed by subsequent annealing. The amount of B2 phase decreased from 61.8% to 16.6% by increasing the annealing temperature from 850°C to 1100°C. The dispersed B2 particles greatly impeded the microstructure coarsening process during annealing, allowing an ultrafine grained FCC matrix to be maintained even after annealing at 1100°C.
2. Tensile tests revealed a moderate strength and total elongation in the as-homogenized specimen having a coarse-grained microstructure, but very high strain hardening and ultimate tensile strengths in the ultrafine grained specimens containing B2 phase.
3. EBSD investigations revealed that the amount of deformation-induced martensite was almost same in all the annealed specimens. It was considered therefore that the volume fraction of B2 phase plays an important role in the strain hardening behavior and high tensile strength of the UFG specimens annealed at different temperatures.

Acknowledgments
This work was financially supported by the Elements Strategy Initiative for Structural Materials (ESISM), and the Grant-in-Aid for Scientific Research (S) (No. 15H05767), all through the Ministry of Education, Culture, Sports, Science and Technology(MEXT), Japan.

References

[1] Deevi S C and Sikka V K 1996 *Intermetallics* **4** 357–375
[2] Furuta T, Kuramoto S, Ohsuna T, Oh-Ishi K and Horibuchi K 2015 *Scr. Mater.* **101** 87–90
[3] Kim S H, Kim H and Kim N J 2015 *Nature* **518** 77–79
[4] Yang M X et al. 2016 *Acta Mater.* **109** 213–222
[5] Sun G S et al. 2015 *Mater. Charact.* **110** 228–235
[6] Dini G, Najafizadeh A, Ueji R and Monir-Vaghefi S M 2010 *Mater. Des.* **31** 3395–3402
[7] Tian Y Z et al. 2017 *Mater. Charact.* **126** 74–80
[8] Munoz-Morris M A, Calderon N and Morris D G 2008 *J. Mater. Sci.* **43** 3674–3682
[9] Miura H et al. 2017 *Scr. Mater.* **133** 33–36
[10] Ondobokova M, Belyakov A, Enikeed N, Molodov D A and Kaibyshev R 2017 *Mater. Sci. Eng.* A **689** 370–383
[11] Suh D W, Park S J, Lee C H and Kim S 2009 *J. Metall. Mater. Trans. A Phys. Metall. Mater. Sci.* **40** 264–268
[12] Tsuji N, Ito Y, Saito Y and Minamino 2002 *Scr. Mater.* **47** 893–899

IOP Conf. Series: Materials Science and Engineering **219** (2017) 012021 doi:10.1088/1757-899X/219/1/012021

Role of interfaces on microstructure refinement and mechanical properties of severe plastically deformed copper and copper-silver eutectic

P Ghosh, K S Kormout and R Pippan

Erich Schmid Institute of Materials Science, Austrain Academy of Sciences, Leoben, 8700 Austria

Abstract. The process of microstructural refinement during deformation by high pressure torsion (HPT) of commercially pure copper (Cu-B), copper powders (Cu-P) and cast copper silver eutectic alloy (CuAg-E) were investigated. The presence of a surface oxide layer or of immiscible silver lamellae leads to much finer grain size. A saturation in microstructural refinement was attained in both Cu-B and Cu-P samples, while the CuAg-E sample showed a linear hardening propensity down to a 30 nm interlamellae spacing. The possible mechanisms for strengthening in the CuAg-E alloy are further discussed in view of the evolution of interlamellae spacing with applied shear strain.

1. Introduction

Design of suitable microstructure is key to the development of high strength advanced metals. In this regard, severe plastic deformation (SPD) techniques have played an important role for synthesis of bulk ultrafine grained, (100 nm < d < 1 μm) and nanocrystalline (NC; d < 100nm) metals [1]. During SPD of coarse grained metals different types of dislocation cell walls are generated. These cell walls can initially be characterized as incidental dislocation boundaries with lower misorientation angle (1^0-5^0) and geometrically necessary boundaries with higher misorientation angle (>5^0) [2]. However, with increasing strain the misorientation between neighboring cells gradually increases until most of the boundaries (≥ 70%) have a misorientation angle >15^0 [2, 3]. In conjunction with the evolution of cell wall angle, the average dislocation cell size also reduces with increasing strain [4 ,5] until a saturated dynamic equilibrium size is attained. Generally, during high pressure torsion (HPT) of pure metals the saturated grain size is attained within a von Mises strain (ε_{vM}) of 10-20 [3]. Several studies have shown that addition of secondary phases can significantly reduce the saturation grain size [3]. In addition to the grain refinement, the deformation processes and the mechanisms for hardening in NC metals have been the focus of many investigations [6-11].

It is the objective of present study to investigate the role of surface oxide layers and a secondary silver phase on grain refinement and their subsequent influence on the hardening of copper composites. For this study, samples were SPD-deformed by HPT. The shear strain (γ) and the von Mises strain (ε_{vM}) in HPT samples are expressed as

$$\gamma = \sqrt{3}\,\varepsilon_{vM} = \frac{2\pi rN}{t} \qquad (1)$$

where r is the radius, N represents the number of revolutions and t is the thickness of the disc. As the shear strain is linearly related with the radius (see equation (1)), a wide range of deformation states can be investigated from one sample.

2. Experimental Procedure

Commercially pure bulk copper (Cu-B), copper powder (Cu-P) with particle size ~30 - 50 μm from Alfa Aesar, and arc-melted $Cu_{30}Ag_{70}$ (in wt %) eutectic alloys (CuAg-E) were processed by constrained HPT at room temperature (300 K) [3]. The bulk Cu-B and CuAg-E samples were first compressed and then deformed by torsion, while the Cu-P powders were first filled in a hollow copper tube, pressed to a solid lump and then deformed by torsion. A compressive stress of 4 GPa was applied during torsion of all the samples. The Cu-B and Cu-P samples were deformed by 10 revolutions and the CuAg-E sample was deformed by only 1 revolution. The final deformed samples were 30 mm in diameter and 2.5 mm in height.

Vickers hardness measurements of the HPT-deformed samples were carried out using a Buehler Micromet 5100 microindenter under a static load of 300 g applied for 10 seconds. The indents were made on the center half cross section of the deformed discs at four different positions for each radius (figure 1). A Carl Zeiss Leo 1525 FE scanning electron microscope (SEM) and a Philips CM-12 transmission electron microscope (TEM) operating at 120kV were used for microstructral analysis. Back-scatter SEM images were taken in both the radial and tangential direction, while selected TEM images were taken in the radial direction.

3. Results

3.1. Hardness

Figure 1 shows the evolution of hardness as a function of applied shear strain for all the materials. The hardness of the Cu-B and Cu-P samples was saturated after 10 revolutions. It is interesting to note that the Cu-P sample requires a higher strain ($\gamma \approx 200$) in order to attain a saturation in hardness. In contrast, the CuAg-E alloy showed pronounced, nearly linear strain hardening, at significantly lower applied strains.

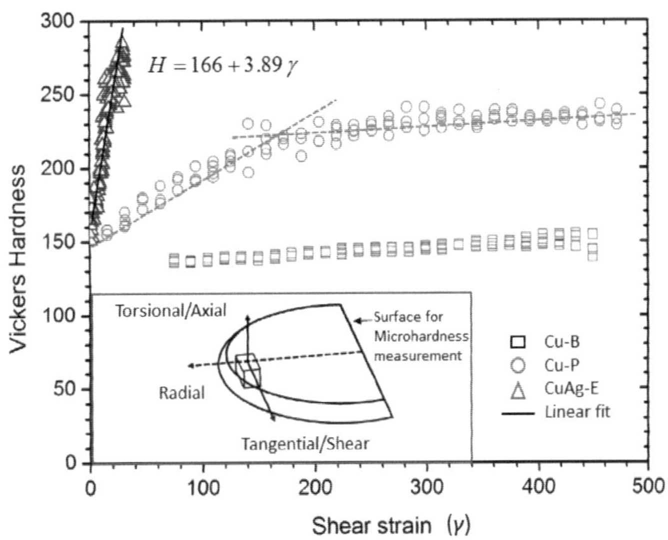

Figure 1. Evolution of Vickers hardness of Cu-B, Cu-P and CuAg-E with applied shear strain. A schematic showing the different sample coordinates is given in the inset.

IOP Conf. Series: Materials Science and Engineering **219** (2017) 012021 doi:10.1088/1757-899X/219/1/012021

3.2 Microstructure

Figure 2 shows example SEM and TEM micrographs in the radial direction for the deformed metals. Both the Cu-B and Cu-P samples show uniform microstructure with a saturation grain size of about 250 nm and 50 nm, respectively. Moreover, for the Cu-P sample several twins are visible inside the grains (marked with red circles). In contrast to the copper samples, the SEM image of the CuAg-E sample shows a non-uniform structure. Further high resolution SEM and TEM revealed a nano-lamellae structure with alternating Cu and Ag lamellae of 10 nm and 20 nm, respectively. Thus, the presence of surface oxides in the Cu-P sample and immiscible Ag in the CuAg-E alloy caused significant microstructural refinement. It is important to point out that unlike Cu-B and Cu-P (initial grain size, d_0 ~ 5-20 μm) the initial thickness of the Cu lamellae in the CuAg-E alloy was only ~100 nm.

In order to further understand the higher strain hardening characteristic of the CuAg-E alloy, figure 3 shows the evolution of microstructure with applied strain in tangential direction. In the as-cast state, the colonies of alternating Cu and Ag lamellae were oriented randomly. However, with increasing shear strain the Cu-Ag lamellae became aligned along the shear plane (perpendicular to the torsion axis). It is interesting to note that similar to the observations of rolling and wire drawing of eutectic alloys a gradient microstructure is developed during HPT of the Cu-Ag eutectic [12, 13]. Colonies of lamellae that were nearly parallel to the shear plane refined at a faster rate, while the lamellae that were oriented at larger angles were first bent and then aligned along the shear plane. This suggests that the Cu-Ag interphase boundary provides strong barrier for transfer of dislocation slip between neighboring phases. A linear intercept method was used to estimate the interlamellae spacing (λ) with increasing shear strain. Figure 3 (e) shows the evolution of the interlamellae spacing normalized by the initial interlamellae spacing (λ_0) in the as-cast state. The normalized lamellae spacing followed a nearly inverse relation with an exponent of (-0.9).

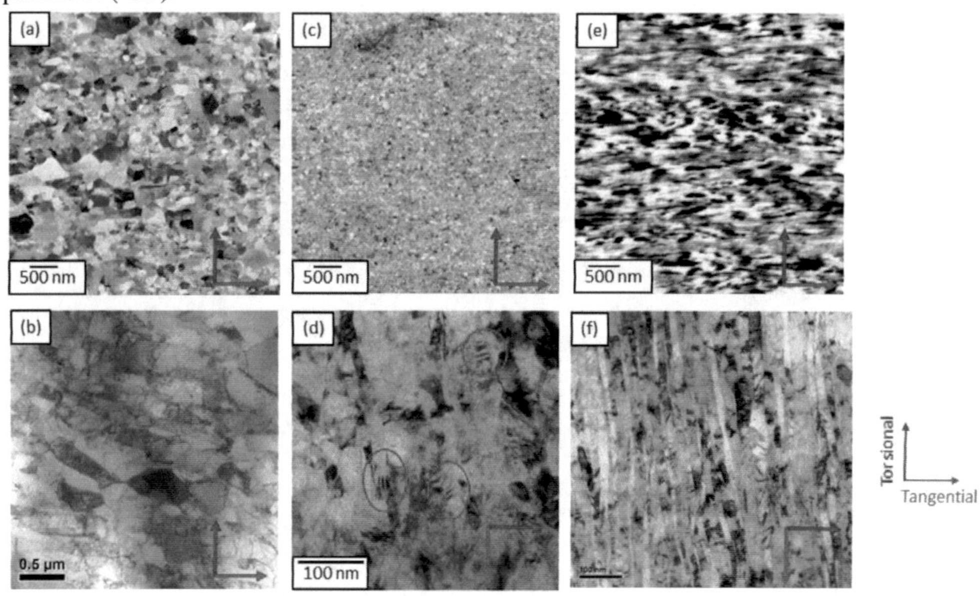

Figure 2. SEM (a, c, e) and TEM (b, d, f) micrographs of Cu-B (a, b), Cu-P (c, d) and CuAg-E (e, f) in the radial direction. A few grains containing twins are marked in (d) with red circles.

4. Discussion

4.1. Microstructure refinement in Cu-B and Cu-P

Investigations of the correlation between structure and properties of HPT deformed metals have shown that the saturation grain size is strongly dependent on the deformation temperature and presence of alloying elements and secondary phases [3]. For pure metals, the saturation grain size reduces with decreasing deformation temperature [3, 14]. However, below a critical temperature the saturation grain size does not show any significant reduction. These observations clearly show that the structural recovery mechanisms at saturation have both thermal and athermal components for grain boundary (GB) and triple junction (TJ) migration. Evidence for deformation-assisted grain coarsening has also been observed during HPT deformation and indentation of NC nickel [15, 16].

A recent study by Renk *et al.* [17] has demonstrated that the balance between thermal driven and deformation-assisted GB and TJ motion can lead to a variety of microstructures ranging from elongated to equiaxed grains. For Cu-B deformed at 300 K, the dynamic equilibrium among various recovery processes led to a saturation grain size of 250 nm and an aspect ratio of 2-3. Most of the grain boundaries were high angle with dislocation debris inside the grains. However, for Cu-P the presence of surface oxides caused further grain refinement. A larger amount of shear strain was needed to attain the saturation grain size. The presence of oxide particles was confirmed by selected area diffraction and X-ray photoelectron spectroscopy. These results will be presented in a subsequent publication. It is well established that oxide particles can significantly reduce the thermal mobilities of GBs and TJs [18]. Thus, it appears that the presence of oxide particles can shift the dynamic equilibrium between the thermal- and deformation-induced GB and TJ motion to a more refined grain size.

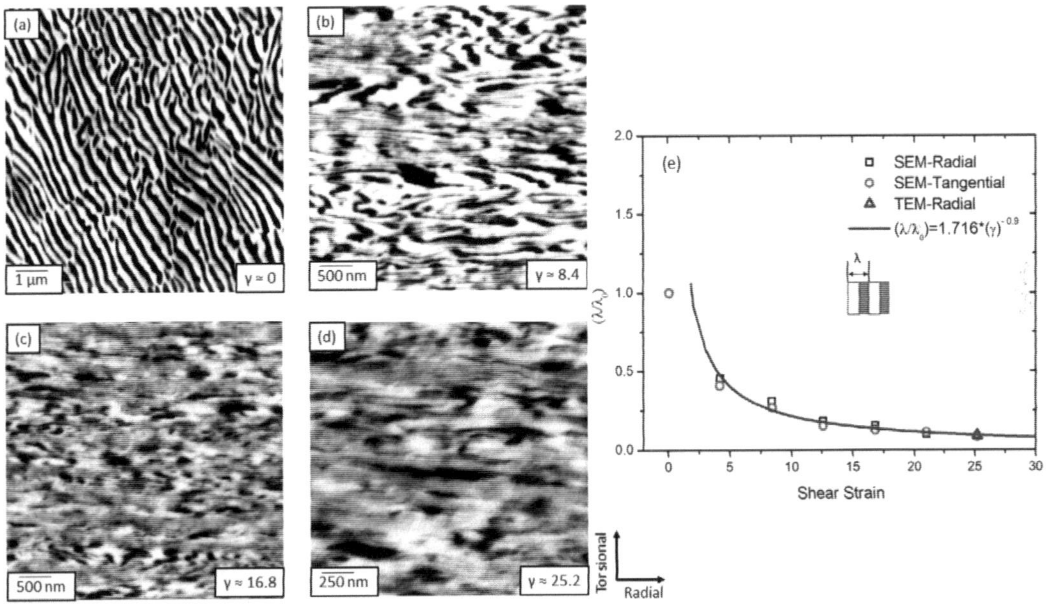

Figure 3. SEM micrographs of CuAg-E in tangential direction at (a) $\gamma = 0$, (b) $\gamma = 8.4$, (c) $\gamma = 16.8$, (d) $\gamma = 25.2$. (e) The evolution of normalized interlamellae spacing with applied shear strain. The blue curve shows the power law fit.

Another interesting feature of the Cu-P microstructure is the presence of several deformation twins inside the grains. In a detailed analytical calculation, Zhu *et al.* [11] have suggested that the propensity for twinning in nanocrystalline metals follows an umbrella curve, with the number of twins first increasing and then decreasing with reduction of grain size. The optimum grain size for twinning in copper was estimated to be 46 nm, which is close to the grain size of Cu-P. Although these calculations were made for pure metals and the influence of GB oxygen atoms is not clear, the presence of a significant number of grains with twins in Cu-P is noteworthy. It is also interesting to note that Cu-P develops a bulk crystallographic texture similar to low stacking fault Cu-15 wt% Zn brass. Thus, the twins in Cu-P grains appear to influence the deformation pattern significantly. A detailed comparison of the microstructural evolution of Cu-P will be presented in a subsequent paper.

4.2 Microstructure refinement and hardness correlation in CuAg-E
During HPT the grains are deformed in simple shear. Thus, at large strains (γ>5) the thickness of the grains can be correlated with the shear strain (γ) according to,

$$d \approx \frac{d_0}{\gamma} \tag{2}$$

Comparing equation (2) with figure 3 suggests that the individual lamellae of the Cu and Ag followed nearly perfect strain transfer between the individual phases without any significant slip (like GB sliding) at the interphase boundary. A similar correlation between the externally applied strain and interlamellae spacing was observed during drawing of eutectoid pearlitic steel [12, 19]. Under equilibrium conditions, the Cu and Ag phases show negligible solid solubility. However, as both Cu and Ag have face centered cubic structure and maintain a crystallographic correlation in the as-cast state, with their (111) planes parallel to the interphase boundary, dislocation pile-up in one phase can cause nucleation of dislocations in the neighboring phase [20-22]. In fact Wang *et al.* [20] have discussed the possibility for interface-facilitated deformation twinning in Cu lamellae and correlated this with the development of a Ag-type texture in Cu lamellae.

Further details about the deformation mechanisms can be obtained by comparing the hardness with the thickness of the Cu and Ag lamellae. In an early study, Frommeyer and Wassermann [23] have shown that the ratio of the thickness of Cu and Ag lamellae remains nearly constant, thus in this study we compare the hardness with the interlamellae spacing. Figure 4 shows the evolution of the hardness values with interlamellae spacing from both the previous [23, 24] and present study. It is interesting to note that although Frommeyer and Wassermann [23] used wire drawing of directionally cast CuAg eutectic alloys, their hardness and interlamellae spacing measurements correlate well with the present study. In contrast, the hardness values for vapor-deposited CuAg multilayers are higher [24]. One possible reason for this could be the difference in overall composition of vapor-deposited multilayers, where Cu and Ag layers of equal thickness were used, while for the CuAg-E alloy, the Cu to Ag layer thickness ratio is ~0.5.

In previous studies on multilayer composites, dislocation pile-up based Hall-Petch strengthening [25] and dislocation loop based Orowan strengthening [24, 26] were considered to be the primary possible deformation mechanisms. The Vickers hardness and the corresponding interlamellae spacing of CuAg-E (except the initial lamellae spacing) from the present study (figure 4) were plotted in different functional forms in order to check the possibilities for these mechanisms. Table 1 shows the various fitting parameters for these strengthening relations. In fact both Hall-Petch and Orowan strengthening mechanisms give reasonable fit. However, a logarithmic fit suggests a best fit exponent of -0.28. Thus, it appears that further quantitative structural analysis of Cu-Ag interfaces is required before ascertaining the primary deformation mechanism in CuAg-E alloys.

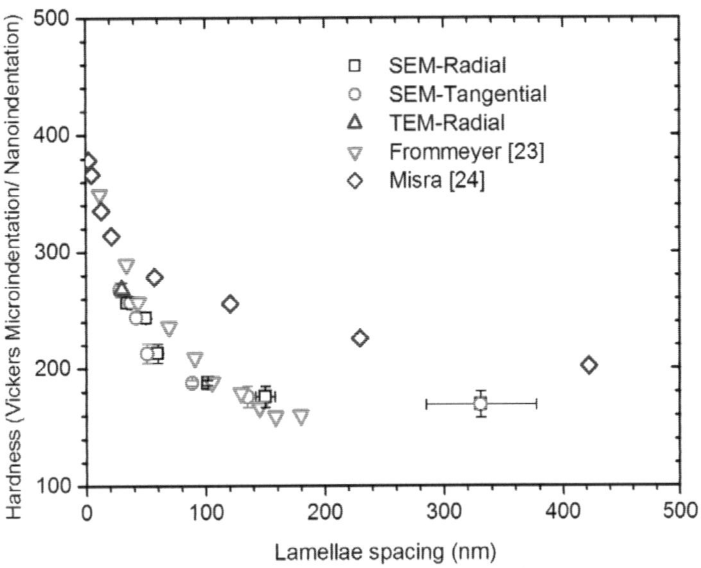

Figure 4. Evolution of hardness of CuAg multilayers with decreasing interlamellae spacing. For the vapor-deposited CuAg multilayers the nanoindentation hardness values were divided by 1.25 to get an equivalent Vickers microhardness [27].

Table 1. Fitting parameters for different hardening laws fitted with hardness in HV and λ in nm for CuAg-E. The initial interlamellae size is excluded in the fit. The error in the fitted parameter is given in parentheses (Err) and R represents the linear correlation coefficient.

Fitted variables (Y-X)	Linear fit Y = A + BX		
	A (Err)	B (Err)	R
Ln(H) – ln(λ)	6.52 (0.08)	-0.28 (0.02)	0.98
H – (1/λ)	155 (6)	3444 (268)	0.97
H – $(1/\lambda)^{0.5}$	95 (9)	940 (63)	0.98

5. Conclusion

The microstructural refinement of commercially pure copper (Cu-B), copper powders (Cu-P) and cast CuAg eutectic alloy (CuAg-E) were investigated after high pressure torsion. The results show that a saturation microstructure with constant hardness can be achieved in the Cu-B and Cu-P samples. However, the presence of surface oxides in the Cu-P samples allows a reduction in the saturation grain size to 50 nm. At such finer grain size significant twinning activity commences, changing the deformation pattern towards that of low stacking fault energy metals such as CuZn alloys. The CuAg-E alloy, on the other hand, undergoes severe strain hardening with the CuAg interlamellae spacing reducing almost inversely with applied shear strain. The correlation of microhardness and interlamellae spacing suggest that both Hall-Petch and Orowan strengthening are equally possible and further in-depth microstructural analysis is necessary in order to understand the primary deformation mechanisms in such CuAg lamellar composites.

Acknowledgment
The authors are thankful to EU for funding the work under ERC-USMS grant 340185.

IOP Conf. Series: Materials Science and Engineering **219** (2017) 012021 doi:10.1088/1757-899X/219/1/012021

6. References

[1] Estrin Y and Vinogradov A 2013 *Acta Mater.* **61** 782-817
[2] Hughes DA, Liu Q, Chrzan DC and Hansen N 1997 *Acta Mater.* **45** 105-112
[3] Pippan R, Scheriau S, Taylor A, Hafok M, Hohenwarter A and Bachmaier A 2010 *Ann. Rev. Mater. Res.* **40** 319-43
[4] Pippan R, Wetscher F, Hafok M, Vorhauer A and Sabirov I 2006 *Adv. Eng. Mater.* **8** 1046-56
[5] Godfrey A and Hughes DA 2000 *Acta Mater.* **48** 1897-1905
[6] Kumar KS, Swygenhoven HV and Suresh S 2003 *Acta Mater.* **51** 2003 5743-74
[7] Hansen N 2004 *Scripta Mater.* **51** 801-6
[8] Morris DG 2010 *Rev. Metal. Madrid* **46** 173-86
[9] Hafok M and Pippan R 2010 *Int. J. Mater. Res.* **101** 1097-104
[10] Rupert TJ, Trenkle JC and Schuh CA 2011 *Acta Mater.* **59** 1619-31
[11] Zhu YT, Liao XZ and Wu XL 2012 *Prog. Mater. Sci.* **57** 1-62
[12] Langford G 1977 *Metall. Trans.* A **8A** 861-75
[13] Zhang X, Godfrey A, Hansen N, Huang X, Liu W and Liu Q 2010 *Mater. Char.* **61** 65-72
[14] Ghosh P, Renk O and Pippan R 2017 *Mater. Sci. Engg.* A **684** 101-9
[15] Zhang K, Weertman JR and Eastman JA 2005 *Appl. Phys. Lett.* **87** 061921
[16] Yang B, Vehoff H, Hohenwarter A, Hafok M and Pippan R 2008 *Scripta Mater.* **58** 790-3
[17] Renk O, Ghosh P and Pippan R 2017 *Scripta Mater.* **137** 60-63
[18] Gottstein G and Shvindlerman LS 2010 *Grain boundary migration in metals*, CRC Press
[19] Zhang X, Godfrey A, Hansen N and Huang X 2013 *Acta Mater.* **61** 4898-4909
[20] Wang J, Beyerlein IJ, Mara NA and Bhattacharya D 2011 *Scripta Mater.* **64** 1083-6
[21] Tian YZ and Zhang ZF 2013 *Scripta Mater.* **68** 542-5
[22] Zheng SJ, Wang J, Carpenter JS, Mook WM, Dickerson PO, Mara NA and Beyerlein IJ 2014 *Acta Mater.* **79** 282-91
[23] Frommeyer G and Wassermann G 1975 *Acta Metall.* **23** 1353-60
[24] Misra A, Hirth JP and Kung H 2002 *Phil Mag* A **82** 2935-51
[25] Spitzig WA, Pelton AR and Laabs FC 1987 *Acta Metall.* **35** 2427-42
[26] Verhoeven JD, Chumbley LS and Laabs FC, Spitzig WA 1991 *Acta Metall. Mat.* **39** 2825-34
[27] Sawa T 2010 IMEKO 2010 TC3, TC5 and TC22 Conferences 171-4

IOP Conf. Series: Materials Science and Engineering **219** (2017) 012022 doi:10.1088/1757-899X/219/1/012022

Selective Laser Melting of Hot Gas Turbine Components: Materials, Design and Manufacturing Aspects

Stergios Goutianos

Department of Wind Energy, Section of Composites and Materials Mechanics, Technical University of Denmark, Risø Campus, DK-4000 Roskilde, Denmark

E-mail: gout@dtu.dk

Abstract. Selective Laser Melting (SLM) allows the design and manufacturing of novel parts and structures with improved performance e.g. by incorporating complex and more efficient cooling schemes in hot gas turbine parts. In contrast to conventional manufacturing of removing material, with SLM parts are built additively to nearly net shape. This allows the fabrication of arbitrary complex geometries that cannot be made by conventional manufacturing techniques. However, despite the powerful capabilities of SLM, a number of issues (e.g. part orientation, support structures, internal stresses), have to be considered in order to manufacture cost-effective and high quality parts at an industrial scale. These issues are discussed in the present work from an engineering point of view with the aim to provide simple quidelines to produce high quality SLM parts.

1. Introduction

In the field of heavy duty gas turbines there is a push toward reducing the cost of electricity, and due to legislation to lowering NOx emission levels. To lower the cost of electricity, the efficiency of gas turbines has to be increased [1]. This translates to higher turbine inlet temperature and reduced cooling consumption. To achieve higher temperatures, nickel-based superalloys can be replaced by new ceramic matrix composites with higher temperature capability [2, 3]. An alternative solution is to develop new designs incorporating complex and more efficient cooling schemes. However, such complex designs can be realised only with new manufacturing technologies such as selective laser melting (SLM) [1].

Selective laser melting is an additive manufacturing (AM) technology allowing direct production of metallic parts from powder materials. Thin layers of powder, typically between 20 and 60 μm, are deposited in a bed and locally melted by a laser beam according to a computer aided design model. The powder bed is then lowered and a new layer of powder is deposited. These steps are continuously repeated until the part is built to nearly net shape [4]. This is why AM is also called solid freeform fabrication or digital manufacturing. With SLM parts of arbitrary complexity can be manufactured, such as internal complex cavities which cannot be made by traditional manufacturing of removing material due to constraints such as tooling and physical access to surfaces for machining.

While SLM is one of the most rapidly developing manufacturing techniques [5], it is also faced with a number of issues such as cost, non-optimal processing parameters, and high internal stresses. These topics will be analysed in the following sections.

2. Cost of Selective Laser Melting

Figure 1 shows the cost of SLM relative to conventional (casting) manufacturing. SLM is competitive over conventional manufacturing when the production volume is low, as can be seen in Figure 1a. The cost per part is independent of the volume for SLM, while for casting it rapidly decreases as the volume increases. However, due to continuous technological developments (*e.g.* the cost of SLM machines decreases and the build speed increases using machines with more laser beams), the cost per part decreases over time, as indicated by the dashed line.

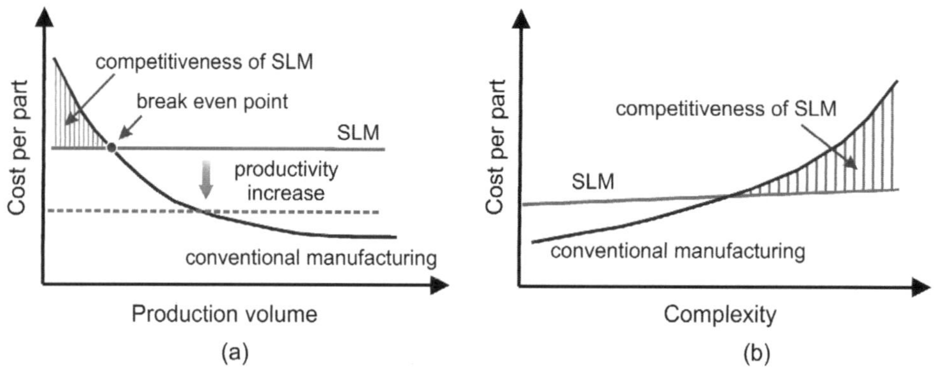

Figure 1. Cost of selective laser melting *vs* Conventional (casting) manufacturing.

Aerospace or gas turbine components have more than one function. A structural component such as a turbine blade also has an internal structure for passing coolant through it. As mentioned above, to improve the efficiency, advanced cooling schemes with complex internal features are required. Fabrication of such parts require expensive molds and more complex process planning if conventional casting is used. This results in a significant increase of the cost per part as shown in Figure 1b. On the other hand, the cost per part for SLM is almost independent of the part complexity (complexity for free). Furthermore, with SLM many parts can be consolidated into one integrated design [6, 7] avoiding assembly operations and increasing reliability, and thus reducing the cost per part over its lifetime.

3. Design for Selective Laser Melting

The unique capabilities of SLM offer new opportunities for the design of gas turbine components with significant improvements. However, manufacturing difficulties related to the SLM manufacturing process should be considered in the design process in order to produce high quality parts.

3.1. Overhaning structures

One of the most common issue is overhanging structures [8, 9] shown in Figure 2. When the layer build has the same dimensions as the previous layers, as shown in Figure 2a (no overhanging structures - $\alpha = 90^o$), the heat is dissipated to the previous built layers and the geometrical distortion of the part is minimal. In practice however, when building parts of complex geometries the layers have different dimensions and overhanging structures, described by the inclination angle α, cannot be avoided.

When $45^o < \alpha < 90^o$ (Figure 2b), the overhanging structure is relatively small. The previous built layers provide a conductive heat support and the distortion of the part is usually within

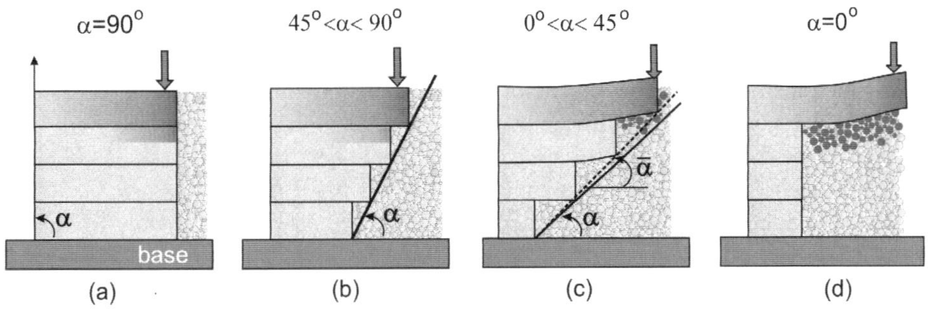

Figure 2. Overhanging structures and dross formation for large inclination angles

accepted tolerances. When α is less than 45°, then the overhang-length of the layer build is supported by a powder zone which has significantly smaller conductivity than the bulk conductivity. As a result, such overhanging structures are distorted and the final component does not meet the design tolerances. The extreme case occurs when $\alpha = 0^o$ where a large amount of the powder below the layer is also melted. Apart from the part distortion, the surface quality of the overhanging structure is low, which is a significant drawback. The surface roughness can significantly influence for example the flow of the coolant in the internal cavities of a turbine component.

To minimise the presence of overhanging structures, the building direction has to be carefully chosen. Figure 3a, shows a simple part with a overhang structure ($\alpha = 0^o$). By rotating the part as shown in Figure 3b, the inclination angle is increased ($\alpha > 45^o$) and the quality of the built part can be significantly improved. Alternatively, a different build direction can be chosen as shown in Figure 3c with $\alpha = 90^o$ giving the optimal quality. In practice when building complex gas turbine components, it is typically not possible to find a build direction that contains no overhanging structures. In this case, the build direction has to be optimised to result in bigger inclination angles for the most critical surfaces of the component.

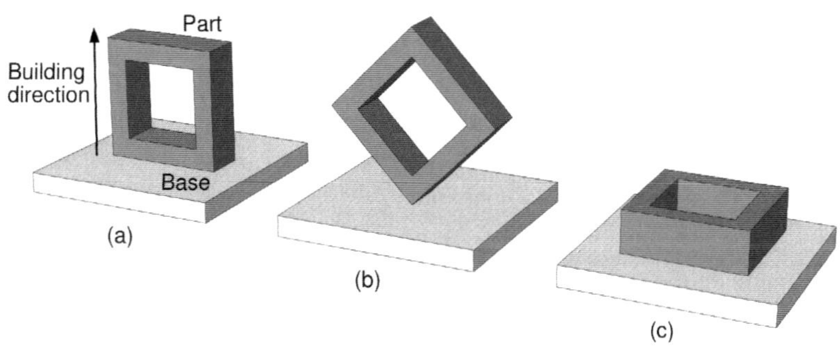

Figure 3. Dependence of the overhanging structures with the building direction [9].

3.2. Support structures

Alternatively, instead of producing one integrated component, several parts can be printed separately and then joined together. In this case it may be easier to define an optimal building direction and thus obtain better quality components. In this solution there are issues with the joint strength, which is critical for structural gas turbine components. A different approach is to

IOP Conf. Series: Materials Science and Engineering **219** (2017) 012022 doi:10.1088/1757-899X/219/1/012022

use support structures as shown in Figure 4. To avoid the critical ($\alpha = 0^o$) overhang structure for the building direction (Figure 4a), the empty volume can be filled with the same material (Figure 4b - block support) during the SLM process, preventing the collapse of melted metal to the powder bed below. Such a support structure is inefficient since it slows significantly the build time due to its large volume and results in an increased cost per part. It is used rarely, for example, to support massive overhanging structures. This is one of the reasons that support structures are typically lattice structures, as shown in Figure 4c. Additionally, the support structures are removed after the build through post-machining operations. It is easier and less costly to remove lattice support structures than block support structures. To ease the removal of support structures, the support structure is decomposed in two parts: teeth that connect the main support with the part, and the main support [9].

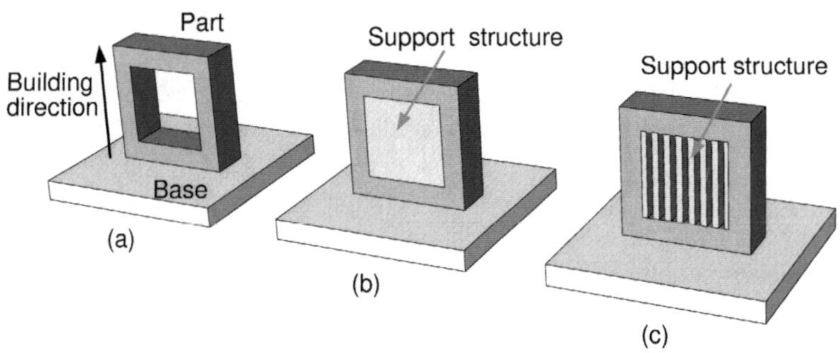

Figure 4. a) Unsupported overhang structure, b) solid support structure and c) lattice structure as support structure

There are many factors that have to be considered in designing support structures, such as the additional build time, post-machining operations and the mechanical strength of the supports to constrain the deformation of the part. In practice, the build direction is optimised in such a way as to minimise any overhanging structures and to minimise the use of support structures [10, 11].

3.3. Self-supporting structures

From the above, it is clear that overhanging structures are in general present when building complex parts, and that the use of support structures should be minimised in order not to increase the cost of the part.

After the build direction is defined, there are several options in the design process to minimise the use of support structures by using SLM-friendly design features. In Figure 5, a circular hole (*e.g.* a cooling hole in a turbine component) with a critical overhanging structure is replaced with an elliptical hole that is a self-supporting structure. Sharp edges (see Figure 6) should be avoided by using fillets or chamfers that have a self-supporting shape.

3.4. Topology optimisation

As mentioned above the build time has a significant impact on the cost per part. Topology optimisation [12, 13] can be applied to reduce the weight/volume of the part and thus reduce the cost (the build time is proportional to the weight of the part). First a finite element mode of the part is created and then a material density value equal to 1 ($1 \rightarrow$ fully dense) is assigned to each element. The optimisation algorithm determines the optimal material density distribution to achieve the optimisation objective (*e.g.* strain energy) and constraints (*e.g.* final volume).

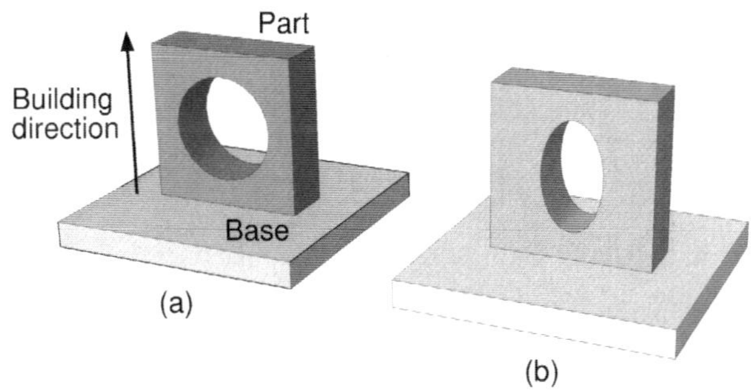

Figure 5. Design features with self-supporting structures.

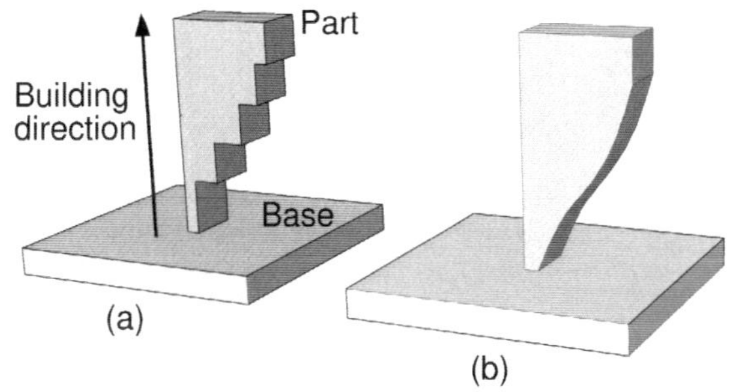

Figure 6. Design features with self supporting shapes.

A simple example is given in Figures 7 and 8. A cantilever plate is loaded with a force on its right side. Figure 8 shows the stresses in the optimised (reduced weight) part. The dark colours indicate regions of high stresses.

Figure 7. Simple topology optimisation example.

Figure 8. Contour of Von Mises Stresses after optimisation with dark colours indicating highly stressed regions.

The corresponding material density contour is given in Figure 9. In this optimisation problem, the material density is allowed to vary between 1 (initial density) and 0 (no material is present / material is removed). As can be seen, there are regions with intermediate densities between 0 and 1. These regions cannot be manufactured in practice as they indicate that a material with lower density should be used, which is unphysical. Therefore, a density of 1 is assigned to these regions before manufacturing the part. If the volume of the intermediate density regions is large then the manufactured part is not fully optimised in terms of weight reduction. It should be mentioned that modern optimisation algorithms minimise these regions by penalising the occurrence of regions with intermediate densities. A different approach is shown in Figure 10 where the minimum material density is set to 0.2. As a result almost the entire part has an intermediate density. With SLM it is feasible to replace these artificial volumes with lattice structures having the same density. Care should be taken that the mechanical response of the lattice structures is very close to the mechanical response of the material that they replace.

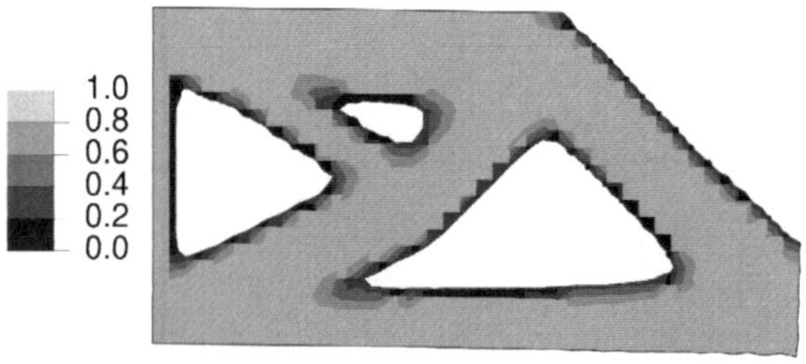

Figure 9. Contour of material density, ρ_m, when ρ_m is allowed to vary between 1 and 0.

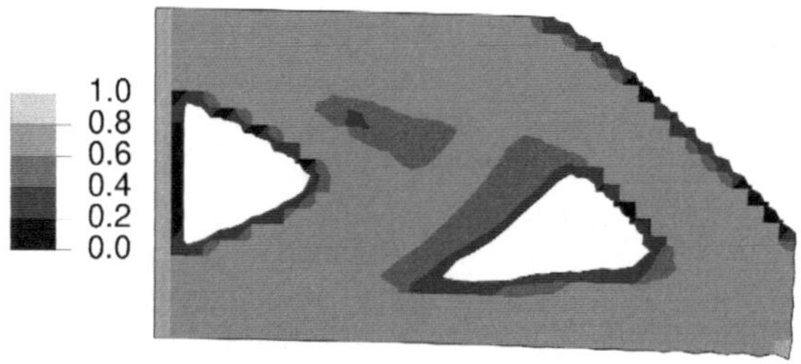

Figure 10. Contour of material density, ρ_m, when ρ_m is allowed to vary between 1 and 0.2.

Topology optimisation has few drawbacks when applied in practice. It is computationally demanding, especially when multiple load cases have to be analysed. Also, additional surfaces are created (see Figure 9 and 10), leading to an increased number of overhanging structures. Optimised geometries of complex parts are not manufacturable directly as they contain many irregular surfaces that need to be smoothed, as well as regions of high local stresses (stress concentrations). These high stresses can be reduced by shape optimisation but usually topology optimisation results are used as a guide to create a novel design.

A secondary but important benefit of topology optimisation is that it leads to designs with thin walls and less residual stresses. This reduces the occurrence of cracks in parts after the SLM process or after subsequent heat treatments.

4. Materials

High temperature alloys such as Ni-based alloys are used for the hot gas path components of gas turbines or jet engines. Many Ni-based alloys, such as MarM247 [14] and CM247LC [15, 16], crack under high solidification rates and thus there is a continuous effort to develop improved powders for these alloys. It was shown for example that the powder chemical composition has a strong influence on the hot cracking of IN738LC [17]. So far, several nickel-based alloys such as Hastelloy X [18] or IN718 [19] have been successfully processed by SLM. It should be added that the chemical composition and the powder manufacturing process affects the porosity of the SLM manufactured parts.

The grain structures of the SLM processed alloys are different from those processed by other manufacturing processes (*e.g.* casting) due to the high solidification rate. The high thermal gradients cause crystals to grow preferentially in well defined directions. As a result, the mechanical properties are different between the build direction and the direction perpendicular to the build direction (mechanical anisotropy). In many cases when a cast material is replaced by the corresponding SLM material, this anisotropy is not desired and it is usually reduced by high temperature heat treatments [20]. Optimising the laser scanning strategy during the SLM process can also reduce the anisotropy [21].

Removal or reduction of the anisotropy stems mainly from the replacement by SLM of existing components manufactured by investment casting, whose lifetime calculations are based on isotropic material properties. In the future, and as the SLM process matures, it may be feasible to apply complex scanning strategies to tailor the texture [22] and thereby the anisotropic mechanical properties spatially during the SLM process. The local optimisation of material properties could result in an increase of the lifetime of SLM-manufactured components.

5. Concluding remarks and Outlook

Selective laser melting has the potential to change the design of the hot gas path components of gas turbines. Novel designs and concepts can be realised, leading to more efficient gas turbines. The differences in the design process and in the production of hot gas path components by SLM compared to conventional manufacturing processes have been highlighted. It was shown that:

- Several parameters, such as building direction, overhanging structures, support structures, self-supporting structures and topology optimisation, have to be considered or used to manufacture high quality components with lower cost. This high degree of flexibility enables the designer to explore the trade-offs in order to find the parameters that best meet the final goal/design.
- The material properties used of SLM-manufactured parts differ from the material properties of the cast counterparts. The main difference is the anisotropic nature of the SLM processed materials. Methods to remove or reduce the anisotropy are available and may also lead to new design optimisation.

Further progress and standardization in materials, process developments and design practices will establish selective laser melting as a manufacturing process for serial production of structural components in gas turbines. Non-destructive methods to efficiently assess the quality and dimensional accuracy of the produced parts are needed to increase the confidence in serial production with SLM.

References

[1] Schurb J, Hoebel M, Haehnle H, Kissel H, Bogdanic L and Etter T 2016 *Proceedings of ASME Turbo Expo 2016: Turbomachinery Technical Conference and Exposition* (Seoul, South Korea) p 57262
[2] Luthra K L 1997 *Journal of the American Ceramic Society* **80** 3253–3257
[3] Zok F W 2016 *American Ceramic Society Bulletin* **95** 22–28
[4] Gu D D, Meiners W, Wissenbach K and Poprawe R 2012 *International Materials Reviews* **57** 133–164
[5] Goossens M, Mittelbach F and Samarin A 2015 *Additive manufacturing technologies: 3D printing, rapid prototyping and direct digital manufacturing* (2nd Edition, New York, Heidelberg, Dordrecht, London: Springer)
[6] Schiller G J 2015 *IEEE Aerospace Conference* (Big Sky, MT)
[7] Herderick E D 2017 *The Journal of The Minerals, Metals & Materials Society* **69** 437–438
[8] Vayre B, Vignat F and Villeneuve F 2012 *Mechanics & Industry* **13** 89–96
[9] Calignano F 2014 *Materials and Design* **64** 203–213
[10] Alexander P, Allen S and Dutta D 1998 *Computer-Aided Design* **30** 343–356
[11] Das P, Chandran R, Samant R and Anand S 2015 *Procedia Manufacturing* **1** 343–354
[12] Bendsøe M P 1989 *Structural Optimization* **1** 193–202
[13] Bendsøe M P and Sigmund O 1999 *Archive of Applied Mechanics* **69** 635–654
[14] Gasser A, Backes G, Kelbassa I, Weisheit A and Wissenbach K 2010 *Laser Technik Journal* **7** 58–63
[15] Carter L N, Attallah M M and Reed R C 2012 *Superalloys 2012* ed Huron E S, Reed R C, Hardy M C, Mills M J, Montero R E, Portella P D and Telesman J (John Wiley & Sons, Inc., Hoboken, NJ, USA) pp 577–586
[16] Carter L N, Martin C, Withers P J and Attallah M M 2014 *Journal of Alloys and Compounds* **615** 338–347
[17] Engeli E, Etter T, Hövel S and Wegener K 2016 *Journal of Materials Processing Technology* **229** 484–491
[18] Wang F 2012 *International Journal of Advanced Manufacturing Technology* **58** 545–551
[19] Wang Z M, Guan K, Gao M, Li X Y, Chen X F and Zeng X Y 2012 *Journal of Alloys and Compunds* **513** 518–523

[20] Etter T, Kunze K, Geiger F and Meidani H 2015 *IOP Conf. Series: Materials Science and Engineering* **82** 012097
[21] Thijs L, Kempen K, Kruth J P and Van Humbeeck J 2013 *Acta Materialia* **61** 1809–1819
[22] Geiger F, Kunze K and Etter T 2016 *Materials Science & Engineering A* **661** 240–246

Anisotropy of mechanical and thermal properties of AZ31 sheets prepared using the ARB technique

K Halmešová[1*], Z Trojanová[2], J Džugan[1], Z Drozd[2], P Minárik[2] and M Knapek[2]

[1]Comtes FHT a.s., Průmyslová 995, 334 41 Dobřany, Czech Republic

[2]Charles University, Faculty Mathematics and Physics, Ke Karlovu 3, 121 16 Praha 2, Czech Republic

[*]Corresponding e-mail: kristyna.halmesova@comtesfht.cz

Abstract. In the accumulative roll bonding (ARB) technique, repeated stacking of material followed by conventional roll-bonding is carried out. For this process the surfaces are cleaned with ethanol and then joined together by rolling. The rolled material is then cut into two halves, again surface treated and roll-bonded. This process may be repeated several times. For the magnesium alloy AZ31 (Mg-3Al-1Zn) rolling at an elevated temperature of 400 °C is necessary for ARB because of the low plasticity of hexagonal magnesium alloys at lower temperatures. Samples for this study were prepared using 1 to 3 ARB passes through the rolling mill. It was found that the ARB substantially refined the grain size of sheets to the micrometer scale. The microstructure and texture of the deformed samples were studied by light and electron microscopy. The mechanical properties of the ARB samples were explored using tensile test-pieces cut from the sheets with the tensile axis taken either parallel or perpendicular to the rolling direction, where a significant anisotropy in both mechanical properties and Young's modulus was found. Anisotropy is explained on the basis of the specific microstructure and texture formed during the ARB process.

1. Introduction

Accumulative roll bonding is very effective approach to refine the grain structure and improve mechanical properties of a metallic material. Relatively big sheets may be prepared using this technology. In comparison with ECAP this method is also cheaper [1]. Severe plastic deformation of magnesium alloys is real problem because of the low plasticity at lower temperatures of hexagonal magnesium materials. For compatible deformation of polycrystals the activity of five independent slip systems is necessary [2]. In the hexagonal close packed cell with the basal slip ($(0001)/\langle 11\bar{2}0\rangle$) only two independent crystallographic equivalent systems exist and therefore non-basal slip (prismatic, pyramidal) or mechanical twinning must be activated. Shortage of easy glide slip systems in magnesium is the reason of the limited plasticity of hexagonal magnesium alloys. Originally, the ARB process should be performed at temperatures below the recrystallization temperature. Because of low plasticity in magnesium materials this is however not possible and a higher process temperature must be used to achieve good formability. Despite this, substantial grain refinement and improvement of mechanical properties has been achieved via ARB in pure Mg [3], and in magnesium alloys of the AZ and AM series [4-7]. The ARB process influences not only the mechanical properties of magnesium materials but also other physical properties and a distinctive anisotropy is observed. The objective of this paper is to show this anisotropy and to attempt to explain observed phenomena.

2. Experimental results

2.1. Material and methods

Commercially available sheets of AZ31 magnesium alloy (nominal composition 3Al-1Zn-0.2Mn-balance Mg in wt%) were used in this study. Two sheets were rolled together in a four-high rolling mill. The rolling speed was 0.4 ms^{-1}. Prior to rolling, the sheets were ground and cleaned. The resulting surfaces were thus free from oxides and had sufficient roughness, which is essential to making a high-quality joint. In addition, the sheets were riveted together along one side. The initial sheet thickness was 2 mm and the rolling reduction was 50% in each pass. The rolling process took place at 400 °C. We performed one, two and three passes through the rolling mill. Samples were hereafter designed as ARB_0 (as received sheet) ARB_1 (after one pass through the rolling mill), ARB_2 (two passes) and ARB_3 (three passes). Samples from sheets were cut so that the longer sample axis was either parallel (L orientation) or perpendicular (T) to the rolling direction.

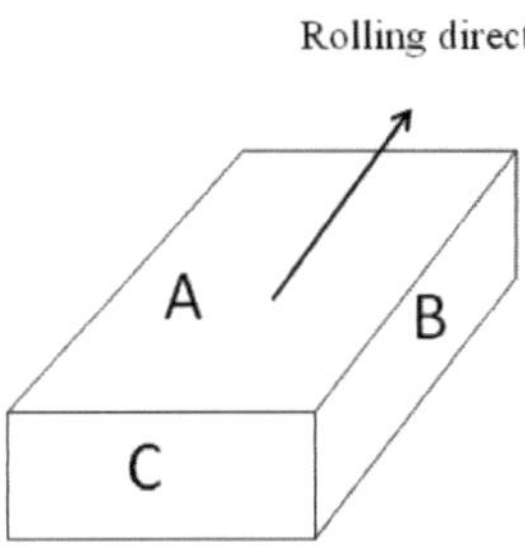

Figure 1. Three planes used for microstructure observation.

The microstructure of the samples was studied using light microscopy (OLYMPUS and NEIPHOT) and using a FEI Quanta 200 FX scanning electron microscope (SEM) equipped with EDAX EBSD system; OIM software was utilized for to analyse EBSD data. The step size used for EBSD measurements was 0.8 μm. Intermetallic phases were analysed in a TESCAN VEGA LMU II SEM.

Tensile tests were performed in an INSTRON 5486 machine with an initial strain rate in the order 10^{-3} s^{-1} at room temperature on miniature tensile test specimens using a recently developed test method [8-15].

The thermal diffusivity was measured with the laser-flash method (Linseis LFA 1000) in the temperature range between 20°C and 350°C. The measurements were done in vacuum with a Nd:YAG laser; the temperature rise on the sample surface was measured with a liquid-nitrogen cooled infrared detector. Thermal diffusivity coefficient a corresponds to the halftime of temperature increase according to the following formula: $a = 0.1388 \cdot \ell^2 / t_{0.5}$, where, ℓ, is the thickness of the specimen and $t_{0.5}$ is the time at 50% of temperature increase.

The Young's modulus was measured in resonant frequency and damping analyser (RFDA) system. Samples were excited to vibrations in the resonant frequency using a small striker. Free vibrations of the sample were registered with a microphone and a fast Fourier transform was used for estimation of the resonant frequency.

2.2. Microstructure of samples

The microstructure of samples was studied on the three principal planes indicated in figure 1. An EBSD map of the as prepared sheet is shown in figure 2(a). The microstructure (estimated from the A plane) of the as prepared sheet (figure 2(a)) contains a non-uniform grain structure - bigger grains are surrounded by small grains. A typical feature of this rolled sheet is an abundance of twins. Substantial refinement was achieved by the ARB procedure. Very fine grains are visible in figure 2(b, c, d). The similar colour of grains in the map indicates that during the rolling process a certain texture was developed. The grain size decreased to several μm. The microstructure of the as-prepared sheet taken from the B plane is shown in figure 3. Bigger particles elongated in the rolling direction were analysed using line EDX spectroscopy, as shown in figure 5. Bright particles represent an Al–Mn binary phase. In the literature, these particles were identified as Al_8Mn_5 and $Al_{11}Mn_4$ phases [16,17]. Also, particles containing Si were found. No particles of γ-phase $Mg_{17}Al_{12}$ were found in ARB sheets. Bonding of sheets was visible on both B and C plane. After three passes through the rolling mill 7 new interfaces were formed as seen in figure 4 (viewed on the C plane).

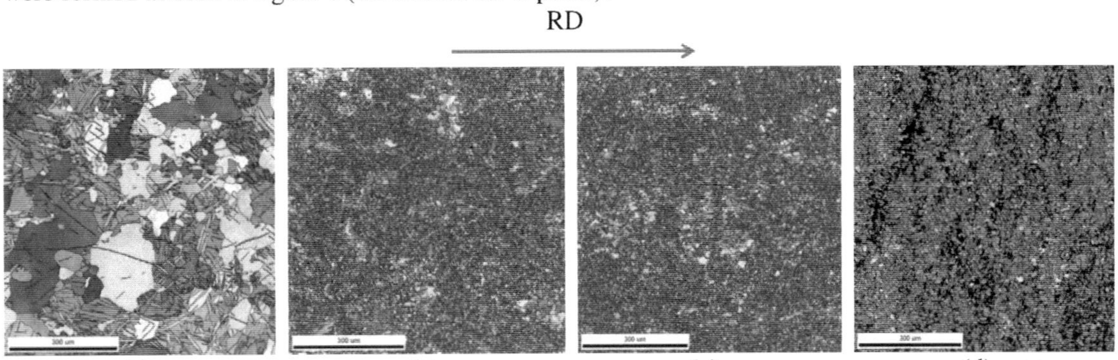

(a) (b) (c) (d)

Figure 2. Microstructure of sheets taken from plane A: ARB_0 (a), ARB_1 (b), ARB_2 (c), ARB_ 3 (d).

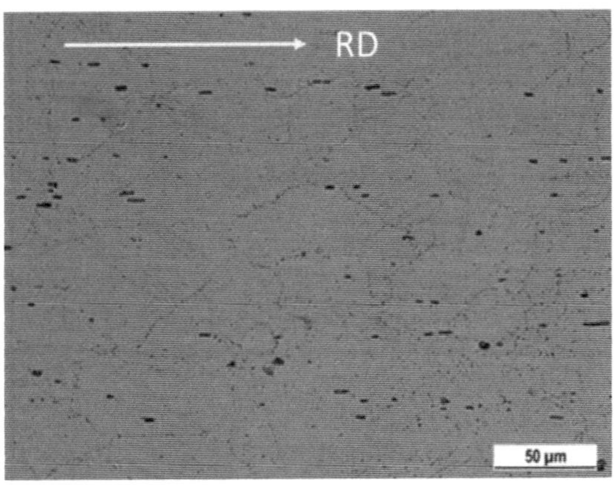

Figure 3. SEM micrograph taken from plane B of the ARB_0 sample.

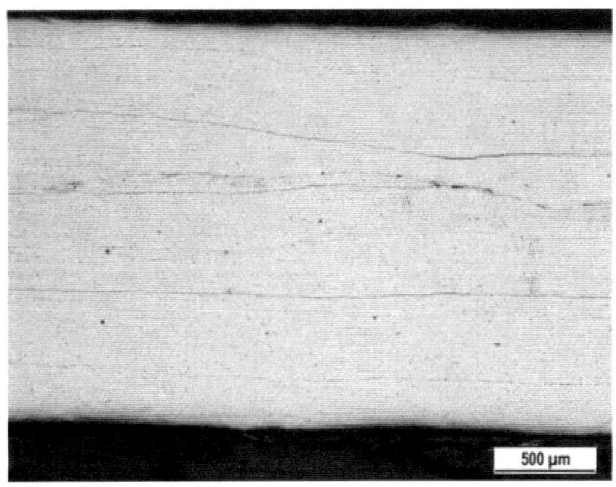

Figure 4. New interfaces formed in the ARB_3 sample visible on the C plane.

Pole figures documenting the texture developed during rolling and ARB procedure in samples are shown in figure 6. The calculated textures show that the basal planes (0001) of the grains are preferentially oriented parallel to the rolled sheet surface (A plane) with the $\langle 11\bar{2}0 \rangle$ direction oriented along the rolling direction.

2.3. Mechanical tests

True stress-true strain curves obtained at room temperature for ARB_0, ARB_1 and ARB_2 samples cut in the L orientation (tensile axis parallel to the rolling direction) are shown in figure 7(a). ARB results in an increased deformation stress. The difference between curves for ARB_1 and ARB_2 samples is smaller compared with the difference between ARB_0 and ARB_1. Similar curves were obtained for samples cut in the T orientation (see figure 7(b)). The strengthening effect of the ARB process is more significant for this orientation. The strengthening effect of the ARB process is also obvious from table 1. Both the yield and ultimate tensile strength increase with the number of passes for each sample orientation. Note, that the ductility of samples is relatively high, much higher compared with cast materials.

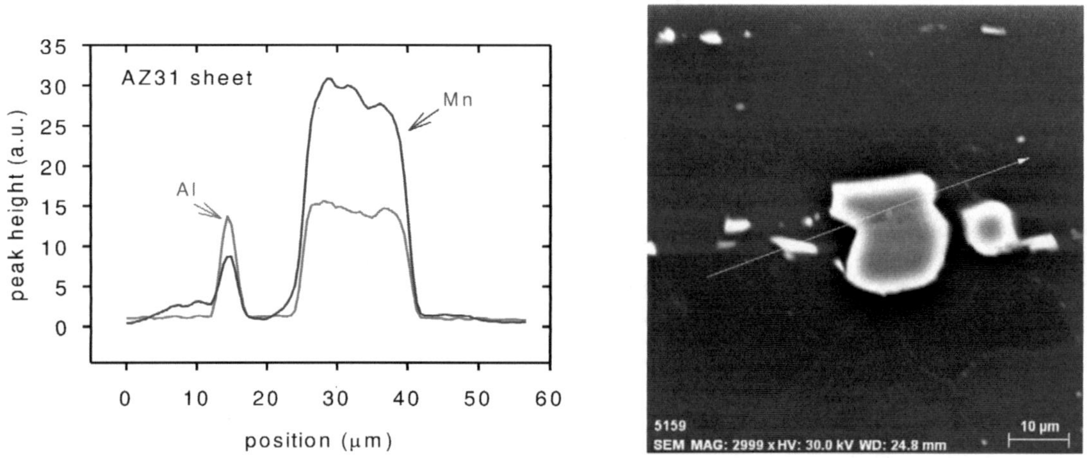

(a) (b)

Figure 5. Mapping of chemical elements in bright particles (a) visible in (b).

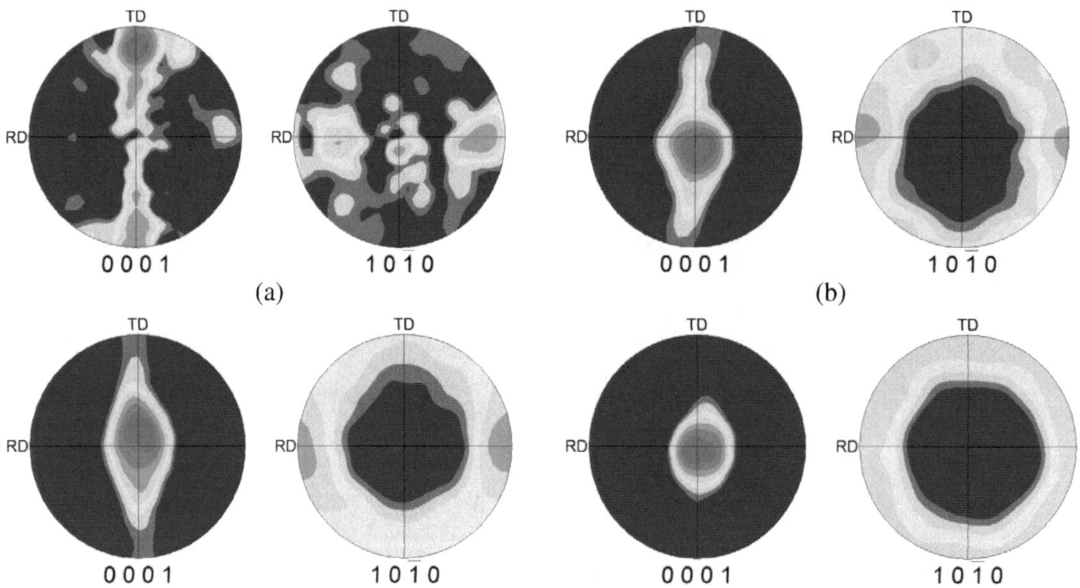

(a) (b) (c) (d)

Figure 6. Pole figures for ARB_0 (a), ARB_1 (b), ARB_2 (c) and ARB_3 (d); samples, taken from the A plane.

Table 1. Yield stress (YS) and ultimate tensile strength (UTS) estimated for T and L sample orientations.

Number of passes	YS (L)	YS(T)	UTS(L)	UTS(T)
0	162.5	85.0	307.0	297.0
1	223.8	172.4	344.9	365.6
2	227.0	212.0	356.6	370.0

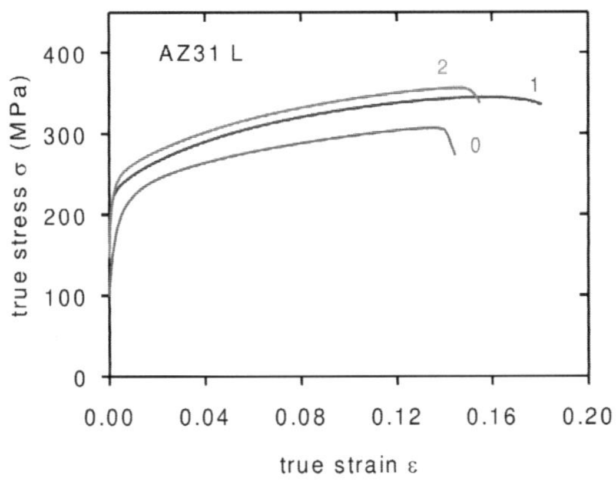

(a)

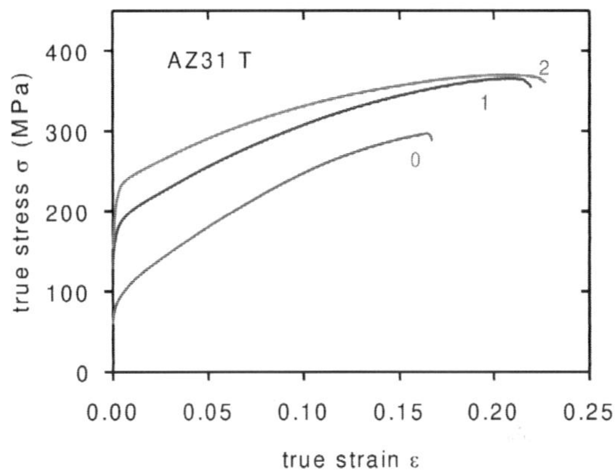

(b)

Figure 7. True stress-true strain curves obtained for ARB_0, ARB_1 and ARB_2 samples in the L orientation (a) and T orientation (b).

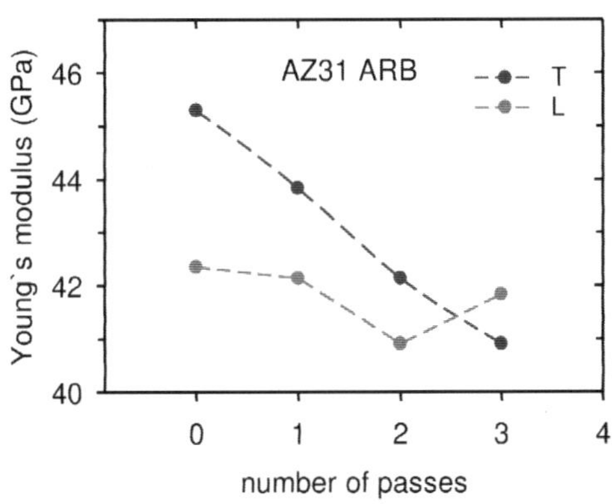

Figure 8. Dependence of Young`s modulus on the number of passes for samples in the L and T orientations.

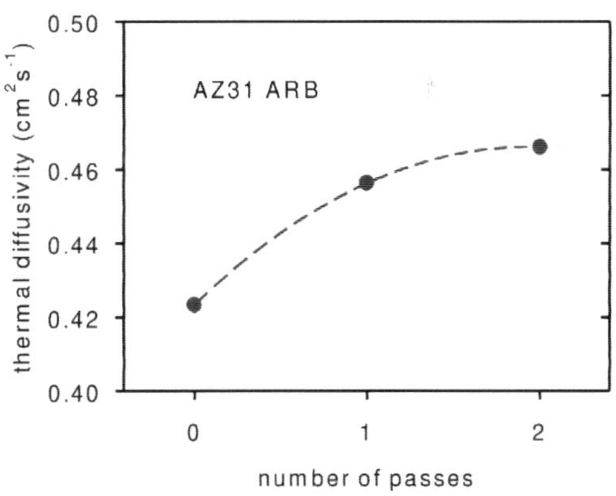

Figure 9. Dependence of thermal diffusivity on the number of passes.

2.4. Young's modulus

The dependence of the Young`s modulus on the number of passes is shown in figure 8 for samples in both the L and T orientations. A significant planar anisotropy is obvious. While the modulus decreases with in general increasing number of passes, the modulus for the L orientation decreases only for 1 and 2 passes, and after the third pass in the rolling mill it increases again.

2.5. *Thermal diffusivity*

Thermal diffusivity was measured perpendicular to the surface of sheets, i.e. perpendicular to the L-T plane. The capability of the measuring apparatus did not enable the measurements in the B and C planes to be made. Results of the diffusivity dependence on the number of passes are shown in figure 9. Surprisingly thermal diffusivity increases with increasing number of passes even though new interfaces were introduced into samples during ARB.

3. Discussion

The von Mises equivalent strain, ϵ, accumulated during the ARB process can be calculated [18] based on fact that the von Mises strain for 50% rolling is 0.8, for three passes the equivalent strain is relatively high at a value of 2.4. The grain size decreases with the number of passes. The majority of this decrease is observed after the first pass. The influence of the second and third pass is weaker, and the grain size decreased down to a stabilised value of 4.5±0.5 μm. The microstructure after ARB process was significantly more homogenous than in the as-prepared sheet. Grain refinement contributed to the observed increase of deformation stresses due to Hall-Petch strengthening:

$$\sigma = m\,(\tau_0 + k_s\, d^{-1/2}), \tag{1}$$

where m is the Taylor orientation factor and k_s gives a measure of the shear stress concentration required in the average instance to propagate the plastic flow across grain boundaries [19, 20]. The main grain refinement occurred between the ARB_0 and ARB_1 sheets and therefore the highest strength increase was observed after the first ARB pass. Subsequent passes in the roll mill refined the grains structure only slightly and the strength increase is also only moderate. The grain refinement influenced deformation stresses of samples cut in the rolling as well as perpendicular directions. The grain refinement may be described as continuous rotational dynamic recrystallisation (RRX). This mechanism involves dynamic polygonisation of rotated lattice regions adjacent to the grain boundaries [21, 5]. Deformation characteristics, introduced in table 1, show significant planar anisotropy. Values estimated for the T orientation are lower compared with the L orientation. This anisotropy decreases with the number of passes. While the yield stress in the T orientation increased after first pass two times, in the L orientation it increased only by 23 %.

The observed planar anisotropy is a consequence of the deformation texture [22, 23]. Mechanisms of plastic deformation strongly depend on the initial texture [24]. The series of pole figures for ARB samples given in figure 6(a-d) show the texture development during the ARB process. AZ31 magnesium alloys sheets usually exhibit a typical rolling texture in which most grains have their basal planes parallel to the sheet surface. Normally the basal plane texture increases with increasing rolling reduction [25]. From figure 6(b-d) it follows that the basal texture is not ideal; the distribution of basal poles shows a basal pole tilted away from the normal direction towards the transverse direction. The deformation process starts in each grain by the activation of basal slip. Balík et al., studying the in-plane deformation of AZ31 alloy sheets, estimated that rapid hardening took place in the basal plane at the beginning of plastic deformation [22]. The deformation continues in the secondary regime where the double prismatic slip is the dominant mechanism. On the other hand deformation twinning as a prevailing mechanism may be considered at the beginning of deformation process in the transverse direction. Acoustic emission measurements performed in situ confirmed the formation of twins during deformation of samples in the T orientation. The twinning activity depends on the grain size and decreases with decreasing grain size [26]. This is also reason why the planar anisotropy decreases after the first and second pass through the rolling mill.

The Young`s modulus measurements in figure 8 also show significant planar anisotropy. This anisotropy decreases with increasing number of passes. While the modulus measured for the L orientation was lower in comparison with the T orientation, after three passes in the rolling mill the tendency is opposite. This behaviour can be related to the texture development during ARB. Calculation of the elastic modulus for textured materials using Reuss or Voigt models requires a detailed analysis of texture [27, 28]. The elastic modulus of pure Mg and alloys with addition of Al and Zn atoms was studied by Sumimoto et al. in [29]. They estimated the Young`s modulus of pure

Mg to be 43.2 GPa, and that it did not depend on the grain size. Al and Zn solutes with concentration lower than 2 at.% increased the elastic modulus up to approximately 44 GPa. This value is between two limiting values estimated for the samples in L and T orientations. A more developed texture decreases the elastic moduli in both orientations for the first and third rolling pass. The third pass in rolling mill increased again the modulus in the L orientation.

Thermal conductivity may be calculated from thermal diffusivity using the simple relation:

$$\kappa = a_d \, \rho \, c_h, \tag{2}$$

where a_d is the thermal diffusivity, ρ is the density and c_h specific heat. The thermal conductivity can be attributed to heat transported along with the conduction of electrons and from phonons travelling through the lattice: $\kappa = \kappa_e + \kappa_{ph}$. The lattice conductivity κ_{ph} decreases with increasing disorder in the lattice, i.e., with increasing concentration of lattice defects (vacancies, dislocations, grain boundaries) [30]. In metals the electron thermal conductivity $\kappa_e > \kappa_{ph}$. Bass [31] estimated that the electrical conductivity of a magnesium single crystal along the *c* axis is higher than in the *a* direction, i.e., $\kappa_{el}(c)/\kappa_{el}(a)>1$. At given temperature the thermal conductivity is proportional to the electrical conductivity, i.e., the same proportionality is expected for the thermal conductivity. In the textured sheet the hexagonal cells are mostly oriented such, that the *c* axis is perpendicular to the sheet surface. Thermal conductivity was measured in the same orientation, i.e., mostly in the *c* direction. The texture strengthening after the first and second passes increased the number of grains oriented with the *c* axis perpendicular to the sheet surface, thus also resulting in an increase in the thermal conductivity.

4. Conclusions

The microstructure and mechanical properties of an AZ31 magnesium alloy deformed by ARB to up to 3 passes have been examined in detail. Typical features of the ARB sheets are a fine grain structure and a developed crystallographic texture. Together these features account for the main influence on the material properties. The following key points can be concluded from the study:

- The ARB process substantially refined the grain structure.
- A planar anisotropy of mechanical properties is observed with the tensile yield strength for the transverse direction lower than that for tension parallel to rolling direction. This anisotropy was less pronounced after the second pass through the rolling mill.
- The Young`s modulus measured in the rolling direction is lower in comparison with the modulus measured in the transversal direction. This tendency was reversed after the third pass through the rolling mill.
- The thermal diffusivity of ARB sheets increases with the number of passes. This is due to development of a pronounced rolling texture, where the materials *c* axis is mostly perpendicular to the sheet surface.

Acknowledgements. This work was realised with the support of the Grant Agency of the Czech Republic under the contract 107-15/11879S.

References

[1] Pan F, Wang Q, Jiang B, He J, Chai Y and Xu J 2016 *Mater. Sci. Eng.* A **655** 339-345
[2] von Mises R 1928 *Zeitschrift für Angewandte Mathematik und Mechanik* **8** 161-185
[3] Chang H, Zheng YH, Wu K, Gan WM, Tong LB and Brokmeier H-G 2010 *Mater. Sci. Eng.* A **527** 7176-7183
[4] Zhan MY, Li YY and Chen WD 2005 *J. Iron Steel Res. Inst.* **14** 167-172
[5] Del Valle JA, Pérez-Prado M-T, Ruano OA 2005 *Mater. Sci. Eng.* A **410** 353-357
[6] Pérez-Prado M-T, Del Valle JA, Contreras JM and Ruano OA 2004 *Scripta Mater.* **50** 661-665
[7] Schwarz F, Eilers C and Kruger L 2015 *Mater. Char.* **105** 144-153
[8] Džugan J, Procházka R and Konopík P 2014 *Small Specimen Test Techniques*, 6th Volume, STP 1576, Mikhail A. Sokolov and Enrico Lucon, Eds 12-29

[9] Konopik P, Džugan J and Rund M *Metal 2014* ISBN:978-80-87294-54-3
[10] Konopík P, Džugan J and Procházka R *Metal 2013* ISBN:978-80-87294-41-3
[11] Rund M, Procházka R, Konopík P, Džugan J and Folgar H 2015 *Procedia Engineering* Vol.**114** 410-415
[12] Procházka R, Džugan J, Kövér M 2015 *Archives of Mater. Sci. Eng.* Vol. **76** Iss. 2 134-139
[13] Konopik P and Džugan J 2012 *2nd International Conference SSTT* 319-328
[14] Konopik P, Rund M and Džugan J 2015 *Proceedings of the 6th international conference on mechanics and materials in design* 277-284
[15] Džugan J, Konopík P, Rund M and Procházka R 2015 Pressure Vessels and Piping Conference Vol.1A: *Codes and Standards*, Boston USA ISBN: 978-0-7918-5692-5, Paper No. PVP2015-45958, pp. V01AT01A066
[16] Ohno M, Mirkovic D and Schmid-Fetzer R 2006 *Acta Mater.* **54** 3883–3891
[17] Braszczyńska-Malik KN 2009 *J. Alloys Compd.* **477** 870-876
[18] von Mises R 1913 *Göttin. Nachr. Math. Phys.* 582–592
[19] Armstrong RW 1967 *Acta Metall.* **15** 667-668
[20] Armstrong RW and Walley SM 2008 *International Materials Reviews* **53** 105-128
[21] Ion SE, Homphreys FJ and White HS 1982 *Acta Metall.* **30** 1909-1919.
[22] Balík J, Lukáč P and Kužel R 2012 *Acta Physica Polonica A* **122** 435-438
[23] Bohlen J, Chmelík F, Dobroň P, Letzig D, Kaiser F, Lukáč P and Kainer KU 2004 *J. Alloys Compd.* **378** 207-213
[24] Balík J, Dobroň P, Chmelík F, Kužel R, Drozdenko D, Bohlen J, Letzig D and Lukáč P 2016 *Int. J. Plast.* **76** 166-185
[25] Pérez-Prado MT, Del Valle JA, and Ruano OA, 2004 *Scripta Mater.* **50** 667-671
[26] Máthis K, Čapek J , Zdražilová Z and Trojanová Z 2011 *Mater. Sci. Eng.* A **528** 5904–5907
[27] Reuss A 1929 *Z. Angew. Math. Mech* **9** 49-58
[28] Voigt W 1910 *Lehrbuch der Kristallphysik* Leipzig
[29] Sumimoto T, Cáceres CH and Veidt M 2002 *J. Light Metals* **2** 49-56
[30] Berman R 1976 *Thermal conduction in solids* Claredon Press
[31] Bass J, Fisher KH 1982 *Landolt-Börnstein New Series* **III/15a**

Analysis of bearing steel exposed to rolling contact fatigue

K T Hansen[1], S Fæster[2*], A Natarajan[2], O V Mishin[2], H K Danielsen[2], D Juul Jensen[2] and P Klit[1]

[1] Department of Mechanical Engineering, Technical University of Denmark 2800 Kgs. Lyngby, Denmark
[2] Department of Wind Energy, Technical University of Denmark, Risø Campus, 4000 Roskilde, Denmark

*E-mail: sfni@dtu.dk

Abstract. The objective of this work is to characterize fatigue damage in roller bearings under conditions of high load and slippage. A test rig constructed for rolling contact fatigue tests of rings is described, and test results are presented for rings taken from two spherical roller bearings. The preparation of the rings and the loading situation are explained. Test conditions are chosen with the aim of achieving pitting formation at the contacting surfaces. During testing the contact pressure, torque and the rotational speed are monitored and recorded. After testing the tested rings have been characterized using X-ray tomography and scanning electron microscopy. The observations confirm that rolling contact fatigue testing at high loads leads to pitting failure at the contacting surfaces. The pitting mostly appears on one side of the contact, attributed to a non-uniform contact pressure in the axial direction.

1. Introduction

Bearings in wind turbine drivetrains operate under highly dynamic conditions, even during normal operation. This is due to the interaction of the wind turbine rotor with wind turbulence, wind shear and yaw, which result in varying cyclic loads on the main shaft. The drivetrain configuration can be based on either a 3-stage high speed geared transmission, a medium speed transmission or a direct drive (gearless). In all cases, the main shaft is connected to the rotor hub through one or two main bearings. The conventional drivetrain configuration has the generator and converter behind the tower [1]. In some large direct drive gearless configurations, the main bearing, along with the generator, is bolted to the hub, thus front-mounting the entire drivetrain to reduce deflections [1]. Nevertheless, the main bearings and/or gearbox bearings are always loaded dynamically in the axial and radial directions.

Several tests have been carried out both in test facilities [2] as well through direct measurements on wind turbines [3] regarding drivetrain loading conditions for the validation of load simulations and the design process. The Gearbox Reliability Collaborative [2] instigated by the National Renewable Energy Laboratory in the USA, has been a multi-year effort, where loads at the bearings of a 3-stage planetary gearbox of a 750 kW wind turbine drivetrain were measured in a test facility under different transient conditions. While the tests validated multi-body model predictions of drivetrain loads, they also revealed the role played by bearing lubrication and slippage of rollers inside the bearings on accelerating bearing damage and roller fatigue.

Wind turbine gearbox and bearings are designed as per the IEC 61400-4 standard [4], following the load conditions prescribed in IEC 61400-1 [5]. Several design load cases, such as normal stops, emergency stops, grid outage, and start-ups, as realistically experienced by a turbine, result in highly transient loads in the gearbox bearings. The roller bearings in such cases may be momentarily underloaded, i.e., a situation where there is insufficient contact force between the rollers and the inner raceway, resulting in sliding of the rollers and subsequent impact upon re-contact [6]. Other extreme load situations, depending on the action of the turbine control system during extreme events, can result in extreme contact stresses inside bearings [7], which may exceed the permissible contact stress limits of the bearing.

This contact between the roller and the raceway is usually modelled as a line contact for roller bearings [7]. Accordingly the maximum contact stress depends on the radial load, the diameter of the raceway, the diameter of the roller and its width. The contact stiffness of the bearing increases with increasing bearing size, if the turbine is upscaled. Therefore, for a fixed turbine size and rating, increasing the diameter of the bearing can reduce the contact stress experienced by the bearing during extreme events. As per the IEC 61400-4 standard, the contact stresses should be below 2.5 GPa for bearings up to a 500 mm inner raceway diameter. While this is the ultimate contact stress limit, the rolling contact fatigue (RCF) load limit depends on the number of load cycles, the rotational speed variations of the bearing and the wear of the surface. The roller or raceway surface wear is difficult to quantify numerically, since it depends on lubrication quality, surface roughness, pre-existing defects etc., but the presence of wear may significantly reduce the rolling fatigue lifetime.

The aim of the present work is to set up a laboratory facility for the testing of RCF of gearbox bearings under controlled load, sliding and lubrication conditions. In this paper the test set-up is described, and preliminary results are presented. The surface of a bearing ring subjected to RCF in the new test facility also is characterized using X-ray tomography and scanning electron microscopy.

2. Experimental

2.1. *Description of the rolling contact fatigue testing machine*

A test rig for RCF of bearing steel under conditions similar to those in modern wind turbines has been constructed at the Department of Mechanical Engineering at the Technical University of Denmark. In the test rig, two drive shafts are forced together by a hydraulic cylinder (figure 1). By positioning a ring on each of the two individual shafts the outer surfaces of the rings can be brought together. The surface of both rings therefore corresponds to the inner ring in a bearing. The revolutions per minute of the two shafts are controlled individually to create rolling, as well as sliding, conditions. A lubrication inlet is positioned just above the position at which the two rings contact each other, and the amount of lubricant flowing into the contact is controlled. The interfacial torque generated between the two surfaces is measured by a torque transducer attached to drive shaft 2 (see figure 1).

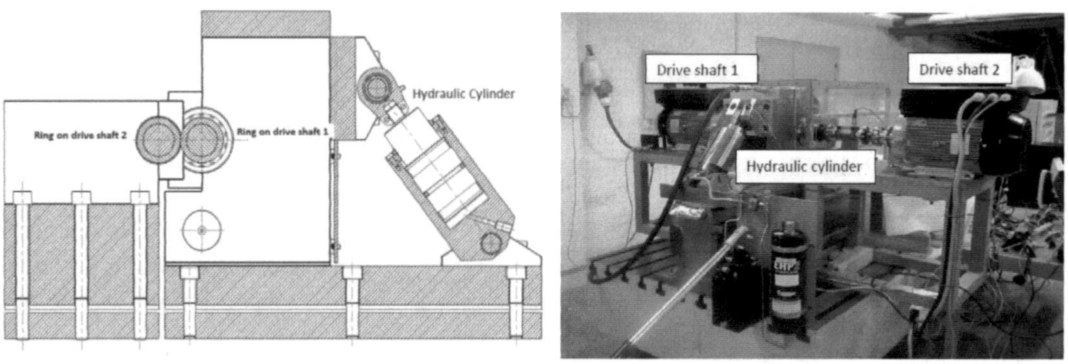

(a) (b)

Figure 1. The test rig used in the present experiment: (a) schematic illustration; (b) photographic image.

2.2. *Material and rolling contact fatigue testing conditions*
The rings used in the present experiment were taken from two spherical roller bearings received from SKF (figure 2). Both rings were removed by breaking the roller cage and removing enough rollers to loosen the inner ring. The desired contact area shown in figure 2(a) was ground on both rings. One ring was ground to a width of 5 mm, while the other was ground to a width of 3 mm. This was done to prevent two opposing edges rolling against each other. These rings are in the following termed the 5-mm ring and the 3-mm ring, respectively. The contact diameters of the rings were 75.84 mm for the 3-mm ring and 75.33 mm for the 5-mm ring. The cross-section of the intended line contact is shown in figure 2(b). This contact is comparable to the contact in the roller bearing.

Both rings were pressed onto their respective shafts until a diametrical expansion of 0.12 mm was achieved on each ring, as measured by the contact diameter. This expansion approximately corresponds to a fitting pressure of 132 MPa for the 5-mm ring and 135 MPa for the 3-mm ring at 2000 rpm and 20 °C [8]. The difference in fitting pressure is due to the difference in the unexpanded contact diameter, as the ring with the larger diameter (3-mm ring) requires a higher fitting pressure in order to expand the outer diameter by 0.12 mm. Both shafts were installed in the test rig, and their axial positions were adjusted using shim washers and wave springs to achieve about 1 mm of clearance on each side of the 3-mm ring. After the shafts were installed, the roundness, measured using a 0.01 mm dial indicator, was 0.02 mm for the 5-mm ring and 0.01 mm for the 3-mm ring.

The lubricant used in the present experiment was the fully synthetic Mobil Delvac gear oil 75W-140, which is a heavy duty drivetrain lubricant with a viscosity of 182 cSt at 40 °C and 25 cSt at 100 °C. Droplets of the lubrication fell freely into the contact zone between the two rings. The target contact pressure for the tests performed was chosen to be 3.5 GPa to ensure relatively fast pit formation.

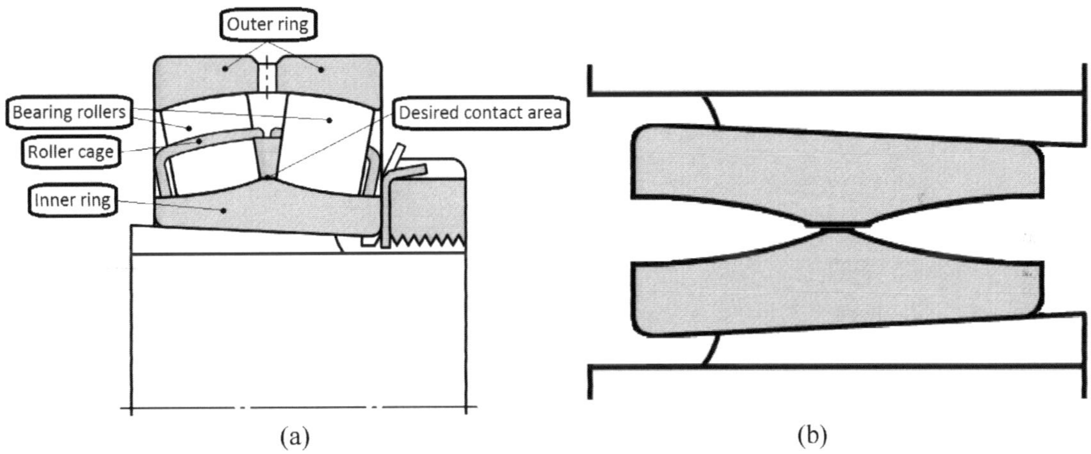

Figure 2. Schematic illustration of bearings applied in the present experiment: (a) cross-section of the bearing; (b) cross-section of the inner rings ground to the desired contact width.

2.3 Roughness and hardness measurements, tomography and microscopy
One 5-mm ring containing a coarse pit after RCF testing was characterized in the present work. The surface roughness Ra was first measured over a length of 5 mm on the contact surface, avoiding pitted areas. Vickers hardness measurements were then made both on the contact surface and along the mid-thickness in the cross-section of the ring. The load applied was 1 kg, with a dwell time of 15 s.

Two samples were then machined from the ring in the location of the pit and from the opposite side of the ring, and 3D X-ray tomography scanning was performed on both samples using a Zeiss Xradia 520 Versa. The scans were performed using a polychromatic beam with energies up to 160 kV from a tungsten target. 3D density maps were reconstructed by a standard filtered back projection based on a Feldkamp algorithm [9] from 3201 projections obtained during a 360° rotation of the sample. The reconstructions covered a 2k × 2k × 2k voxel volume, with a voxel size of 7.67 μm. 3D volumes were segmented and visualized using the Avizo software package. The outer surface of the samples was also inspected using a Zeiss Supra 35 scanning electron microscope (SEM).

3. Results and Discussion

3.1. Rolling contact fatigue test data

The force applied from the hydraulic loading arrangement during the RCF test is shown in figure 3. In this figure, the target force is represented by a black line, and a ±4.62 MPa range below and above the target is marked, which corresponds to ±50 N. The moving average (MA), calculated over a duration of 300 seconds, is mostly within the target range over the entire dataset. However as it is necessary to adjust the force manually during testing, the force occasionally drifts out of this range. The abrupt vertical increases and drops seen in figure 3 correspond to starting up and shutting down of the test rig. Based on the MA, it is concluded that the rings were mostly subjected to ~3.5 GPa contact pressure.

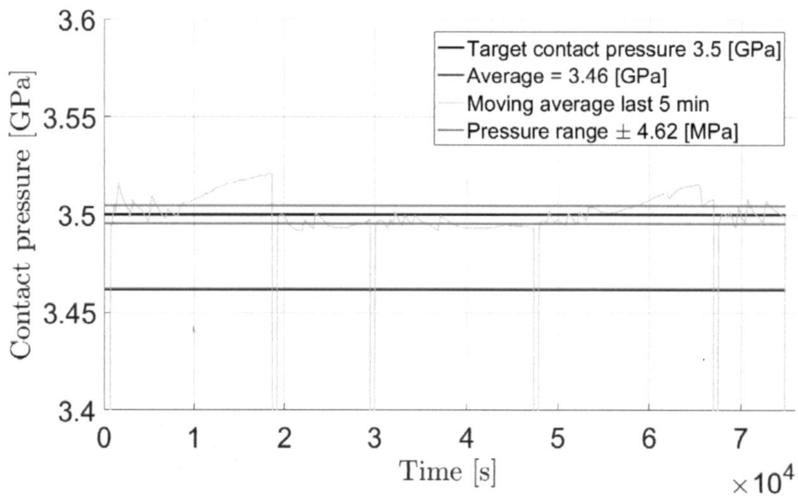

Figure 3. The applied contact pressure as a function of time.

Figure 4 shows the complete torque history for the entire RCF test. Although the torque should ideally be zero for pure rolling, there are small variations in the operating speed of each drive. These variations originate from the frequency converters controlling the drives, as they do not supply a steady frequency. Again, the steep changes seen periodically in the MA of the torque are due to start-up and shut-down of the test rig. It is clear from figure 4 that the test rings experienced some slip during the test. However, considering that the maximum deviation seen in figure 4 is only 0.6 Nm, the level of slip is small.

The rotational speed of each drive is shown in figure 5. It is evident there was a difference in the speed of the two drives, arising from the different contact diameters of the test rings. The rotational speed of each drive is, however, considered to be close enough to the goal of 2000 rpm. The periodic sudden increases in speed originate from start-up of the test rig, and are seen to decrease within a short period of time as the steady state operating speed is achieved. The graphs in figure 5 demonstrate similar patterns, which in combination with the applied torque data (figure 4), indicates that during the test the conditions of the rings were fairly close to pure rolling.

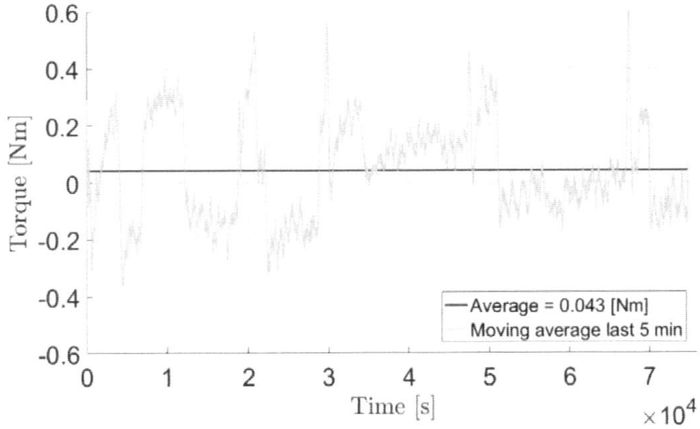

Figure 4. The torque transmitted as a function of time.

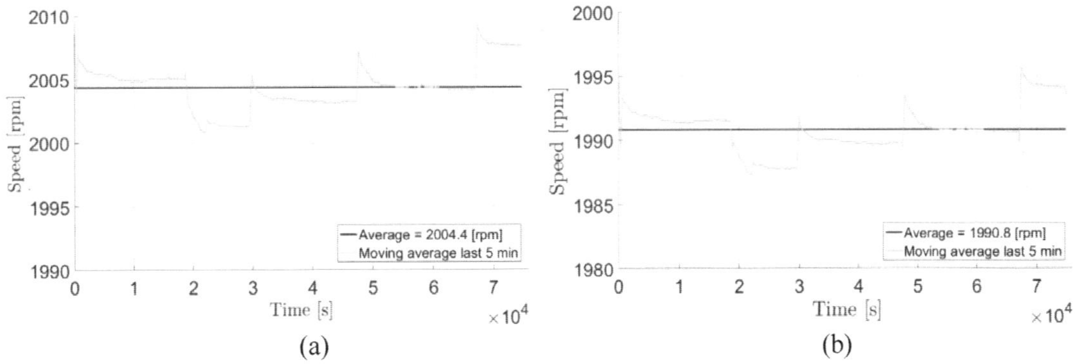

Figure 5. Rotational speed as a function of time: (a) drive 1; (b) drive 2.

Table 1 shows the total contact time of the ring pair during test operation before pitting failure occurred, as well as the total number of revolutions of the individual rings, the mean of each adjustable parameter, and the surface roughness Ra measured after the test at locations away from the pitted area. The Ra values indicate that the surfaces (except for the pitted areas) are similar to those after the initial grinding, for example, Ra values measured on the 5-mm ring before and after the test are 0.6 μm and 0.69 μm, respectively. As can be seen in table 1, the total number of revolutions exceeded 10^6, which indicates that the test pieces entered the high cycle fatigue range. Thus, the load conditions for the laboratory experiment are comparable to actual loading conditions for bearings in operation.

Table 1. Data for the entire test run to maintain the maximum contact pressure of 3.5 GPa.

Parameter	Number	Unit
Total time	20:47:53	hh:mm:ss
Total no. of revolutions (drive 1)	2,526,662	r.
Total no. of revolutions (drive 2)	2,509,550	r.
Mean force	18,667	N
Mean torque	0.04	Nm
Mean speed (drive 1)	2,004	rpm
Mean speed (drive 2)	1,991	rpm
Ra (5-mm ring)	0.69	μm
Ra (3-mm ring)	0.52	μm

3.2. Hardness data and defect observations

The average Vickers hardness measured on the contact surface is 800 HV1, which is greater than that measured along the ring mid-thickness of 740 HV1. Thus, the contact surface was hardened considerably during the test.

X-ray tomography images from the outer surface of the 5-mm ring are shown in figure 6. A circumferential groove with a depth of 35 μm is observed (marked by arrows) near one edge of the contact surface both in the pit-free sample (figure 6(a)) and in the sample with the coarse pit (figure 6(b)). Apparently, this groove extends over the entire ring circumference. The groove originates from the contact between the edge of the 3-mm ring and the 5-mm ring, and results from an angular misalignment between the two ring surfaces caused during the grinding process, which leads to conical contact between the two rings mainly along this groove. This was confirmed by inspecting another pair of rings, where it was found that the centre axis of the grinding stone was not properly aligned with the centre axis of the test piece. This resulted in a conical shape of the contacting surface with an inclination angle of 0.88°. Some degree of variation between the slopes is common since each ring is ground separately, but this variation is small compared to the magnitude of the misalignment. The conical shape greatly changes the contact situation, which ideally should have been a 3-mm wide line contact. The contact pressure was thus much higher at one end of the line contact than originally planned. As a result of the high contact pressure, it is likely that plastic yielding occurred in both rings, widening the contact area during the test.

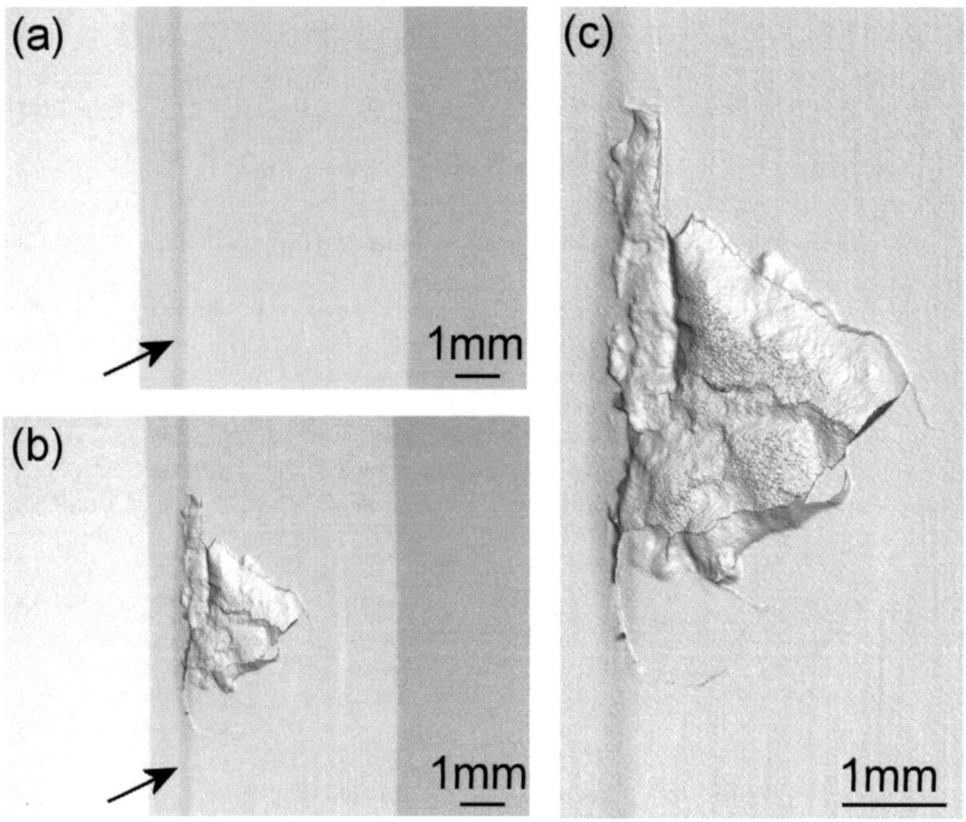

Figure 6. X-ray tomography images from two samples machined from opposite sides of the tested 5-mm ring: (a) pit-free sample; (b,c) sample with a coarse pit, which is shown at low (b) and higher (c) magnifications. The groove corresponding to the high pressure region extending over the entire ring circumference is marked by arrows in (a,b).

The size of the pit shown in figure 6(b) is ~3 mm, and the depth is 420 μm. Since the pit is widest in the vicinity of the circumferential groove, it is suggested that the pit starts propagating from this groove. A SEM image in figure 7(a) shows the central portion of the coarse pit, where multiple cracks are seen both near the pit and on the pit surface. Cracks are also observed at some distance from the pit, but still close to the edge of the contact surface (see figure 7(b)). The defects observed on the contact surface appear to be examples of damage associated with classical pitting failure.

The RCF test was stopped when the vibration level increased significantly, due to the development of the coarse pit shown in figure 6(b,c) and figure 7(a). The cracks along the contact zone show, however, that more pits had started to develop before the RCF test was stopped. It is suggested that the cracks started below the surface, where the shear stress had its maximum value. Furthermore, SEM inspection reveals cracks on the contact surface even in the pit-free sample. Figure 8 shows an X-shape crack with arms extending over at least 0.5 mm, observed near the circumferential groove marked in figure 6(a). This crack is apparently very shallow and thus not observable in the X-ray tomography image in figure 6(a). Such cracks are considered to be precursors of pits.

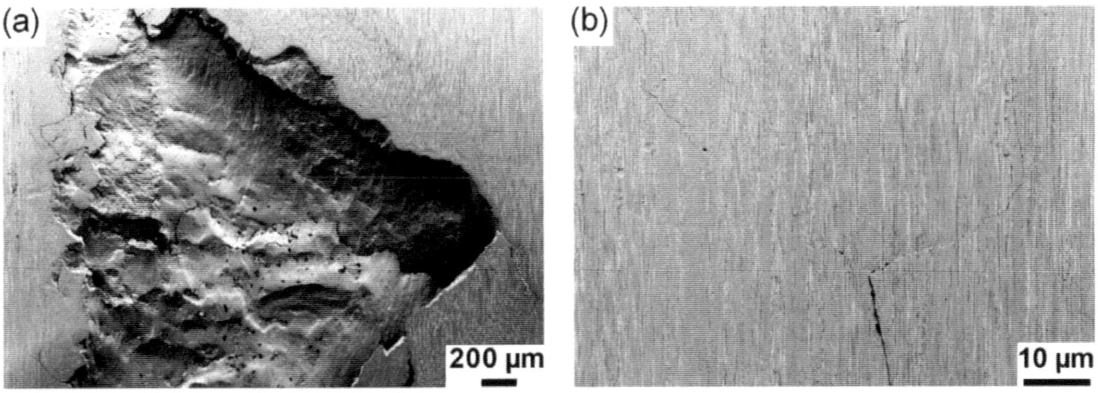

Figure 7. SEM images from the sample containing the coarse pit: (a) central part of the pit; (b) cracks far away from the pit, but still near the groove.

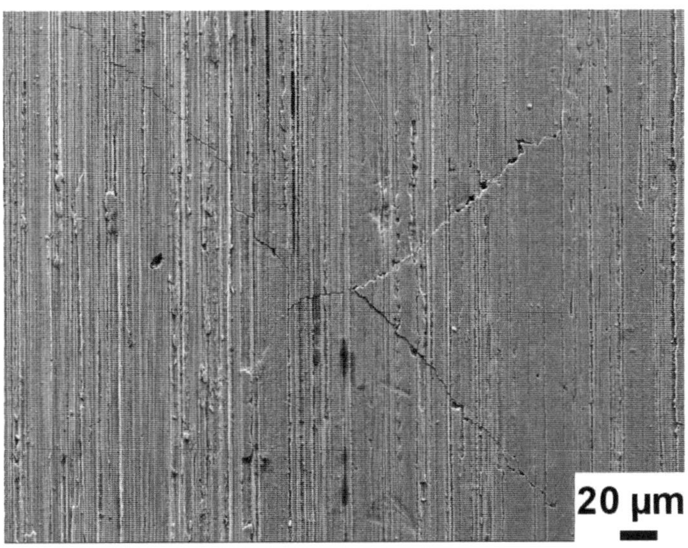

Figure 8. SEM image showing an X-shape crack near the groove in the pit-free sample.

4. Conclusions and outlook

A test rig has been constructed at the Department of Mechanical Engineering at the Technical University of Denmark and it has been shown that the rig can be effectively used for studies of rolling contact fatigue of bearing steels under well-controlled conditions. In this test rig, the loading on the rings can be adjusted to obtain a specified maximum contact pressure, and the rotational speed of the contacting rings can be adjusted to obtain specified slip conditions. The circumferential stress in the rings can be achieved by adjusting the axial ring position on the conical shaft.

Initial results using the test rig on samples taken from two spherical roller bearings using rings of 3 mm and 5 mm width lead to the following conclusions:

- Rolling contact fatigue testing under a high load results in pitting failure at the contacting surfaces. Pitting was preferably observed on one side of the contact, attributed to a non-uniform contact pressure along the axial direction. Due to the conical shape of the contacting ring surfaces the contact pressure along the line contact was sufficiently high so that plastic deformation occurred, indicated by the fact that the 3-mm wide ring produced a groove in the 5-mm wide ring.
- A progression in damage at the contact surface, stemming from the grinding process used in the present experiment, was observed over several thousand revolutions of the test piece, finally resulting in severe vibrations. This situation is also typical in wind turbine applications, where manufacturing defects and misalignments can lead to premature failures of the roller bearings in gear boxes.

Future work will include experiments with different loads and slippage conditions.

Acknowledgement

The work was supported by the Strategic Research Center "REWIND – Knowledge based engineering for improved reliability of critical wind turbine components", grant no. 10-093966 from the Danish Research Council for Strategic Research.

References

[1] http://evolution.skf.com/us/new-challenges-for-rotor-bearings-in-the-8-mw-offshore-category
[2] LaCava W, Keller J and McNiff B 2012 Gearbox reliability collaborative: test and model investigation of sun orbit and planet load share in a wind turbine gearbox. *53rd Structures, Structural Dynamics and Materials Conference* (Honolulu, Hawaii)
[3] Guo Y, Bergua R, van Dam J, Jove J and Campbell J. (2015) *Wind Energy* **18** 2199
[4] IEC 61400-4. Wind turbines - Part 4: Design requirements for wind turbine gearboxes 2012 *International Electrotechnical Commision*
[5] IEC 61400-1. Wind turbines - Part 1: Design of wind turbines, Edition 3 2005 *International Electrotechnical Commision*
[6] Dabrowski D and Natarajan A 2017 *Wind Energy* DOI: 10.1002/we.2098
[7] Gallego Calderon J F, Natarajan A and Cutululis N A. 2017 *Wind Energy* **20** 325
[8] Klit P, Casper K and Pedersen N L. *Machine Elements Analysis and Design.* Polyteknisk Forlag, Kgs. Lyngby, First edition, 2009
[9] Feldkamp L A, Davis L C and Kress J W 1984 *J Opt Soc Am A* **1** 612

IOP Conf. Series: Materials Science and Engineering **219** (2017) 012025 doi:10.1088/1757-899X/219/1/012025

Microstructural and hardness gradients in Cu processed by high pressure surface rolling

Q Y He[1, 2], M Zhu[1], Q S Mei[3], C S Hong[2], G L Wu[1], X Huang[2]

[1] College of Materials Science and Engineering, Chongqing University, Chongqing 400045, China

[2] Section for Materials Science and Advanced Characterization, Department of Wind Energy, Technical University of Denmark, DK-4000 Roskilde, Denmark

[3] School of Power and Mechanical Engineering, Wuhan University, Wuhan 430072, China

E-mail:qyhe@cqu.edu.cn

Abstract. The surface of an annealed Cu plate was processed by a high pressure surface rolling (HPSR) process. It is found that the deformed surface layer in the Cu plate after HPSR can be as thick as 2 mm and is characterized by a gradient microstructure, with grain sizes varying from the nanoscale in the topmost surface to the microscale in the bulk. The hardness varies from 1.37 GPa at the topmost surface to about 0.6 GPa in the coarse-grained matrix. The results of the investigation demonstrate that the HPSR process shows good potential for the generation of thick gradient microstructures on the surface of bulk metallic materials.

1. Introduction

In the past two decades, various surface deformation techniques, such as surface mechanical attrition treatment (SMAT) [1, 2], surface mechanical grinding treatment (SMGT) [3, 4] and high energy shot-peening [5], have been developed to generate surface gradient layers in metallic materials. Using these techniques, a gradient microstructure with a variation in grain size along the depth direction is typically formed in the surface region of initially coarse-grained material [6]. Such a surface gradient microstructure significantly increases the hardness [2], and improves both wear resistance [7] and other surface properties [3], leading to an extended service lifetime [8]. A superior combination of strength and ductility can also be achieved by introducing a surface gradient nanostructure [9]. However, the thickness of the deformed surface layer with a gradient nanostructure is often quite limited, being a few hundred micrometers in most cases [4, 9]. To further explore the potential of gradient microstructured materials, techniques that produce thicker layers of gradient microstructure are highly desirable. In the present study, a technique called high pressure surface rolling (HPSR) is presented. To demonstrate this technique HPSR has been applied to a Cu plate and the resulting microstructure and hardness are examined and discussed.

2. Experimental

The material used in the present study was a commercial purity copper (99.95%) disk plate with a diameter of 60 mm and a thickness of 6 mm. The disk was fully recrystallized with a grain size of few hundred micrometers after annealing at 500 °C for 2 h. A newly developed technique, namely the high pressure surface rolling (HPSR) process, was used to deform the copper disk plate. The HPSR procedure is schematically illustrated in figure 1a. Three groups of cylinder rollers (GCr15, Ø8 mm ×

10 mm), containing six rollers in each group, are pressed into the sample surface under a load. Then the sample platform is rotated, leading to rotation of the rollers and rolling deformation of the sample surface. The repeated rolling of the rollers on the surface of the sample leads to highly localized plastic strain in the surface layer of the sample. As a result the grains are gradually refined and a depth dependent microstructure is generated. Detailed information about the HPSR process can be found in Ref. [10]. In this study, a load force of 20 kN and a rotation speed of 2 rpm were used. Each session of the process was interrupted after a 5 min treatment time and the sample then cooled in liquid nitrogen to reduce the temperature increase during HPSR. In total six sessions were applied to the plate to effectively refine the grains in the surface and increase the thickness of the nanostructured surface layer.

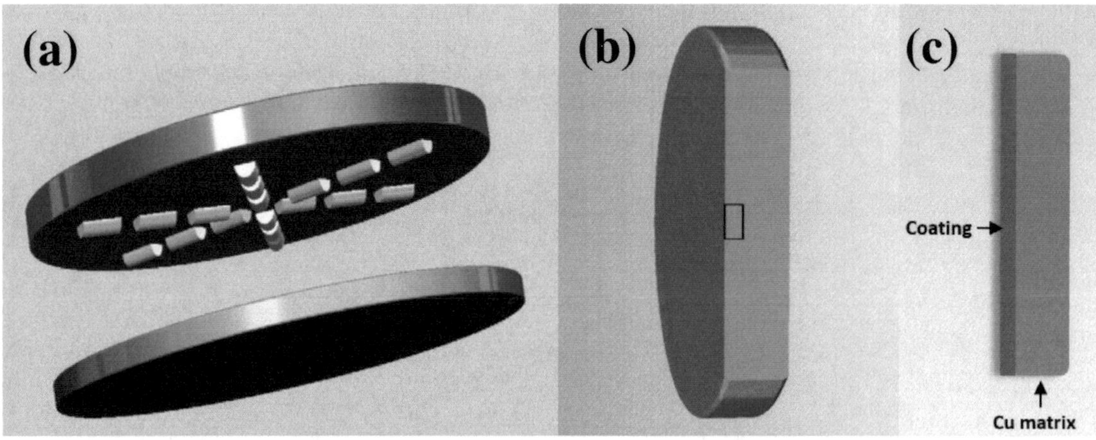

Figure 1. Schematic illustration of (a) the experimental set-up for high pressure surface rolling of a Cu plate; (b) the cross-section used for microstructural observations and hardness tests; (c) the coating of a protective Cu layer on the treated surface.

A specimen was cut from the rolled plate at a distance of 25 mm to the center of the plate (as indicated by the black box in figure 1b), and then a protective Cu layer was electrodeposited on the treated surface (as indicated by darker shading in figure 1c). The hardness variation along the normal direction from the surface was measured using a load of 20 g and a loading duration of 10 s. In the top 250 μm of the sample hardness indents were taken at intervals. The nearest indent to the top surface was taken at a depth of ~15 μm. At least 10 indents were measured at each depth. The cross-sectional microstructure of the HPSR sample was characterized by electron channeling contrast (ECC) and electron backscatter diffraction (EBSD) in a TESCAN MIRA 3 scanning electron microscope. Samples for EBSD were mechanically polished and then electropolished in a solution of phosphoric acid (25%), alcohol (25%) and deionized water (50%) at room temperature. To explore the deformed microstructure in different locations, a step size of 20 nm was used for depths less than 300 μm, and a large step size (100-200 nm) was used at larger depth. The variation in boundary spacing as a function of depth from the surface was also determined from the EBSD data. Values of boundary spacing were taken from line scans along the normal direction using a minimum misorientation of 1° for identification of boundaries.

3. Results and discussion

The deformed microstructure after HPSR is shown in figure 2. It is clearly seen in the low magnification ECC image (figure 2a) that a deformed layer of about 2 mm in thickness is formed on the coarse grained substrate without a clear interface between the deformed layer and the matrix. A typical gradient structure along the depth from the surface is clearly observed although the detailed microstructure in the deformed surface is too fine to resolve by ECC at this magnification. Figure 2b shows the hardness variation as a function of depth from the surface. The plate shows a steep hardness gradient to a depth of 2000 μm from the surface. The average hardness is as high as 1.37±0.03 GPa in the topmost surface, which is significantly higher than that of the initial coarse-grained (CG) condition (0.60±0.02 GPa), as indicated by the horizontal dashed line. The hardness decreases gradually with increasing distance from surface, and approaches that of the un-deformed coarse grains at a depth of 2000 μm. A plateau in hardness values is observed at a depth of 300-600 μm.

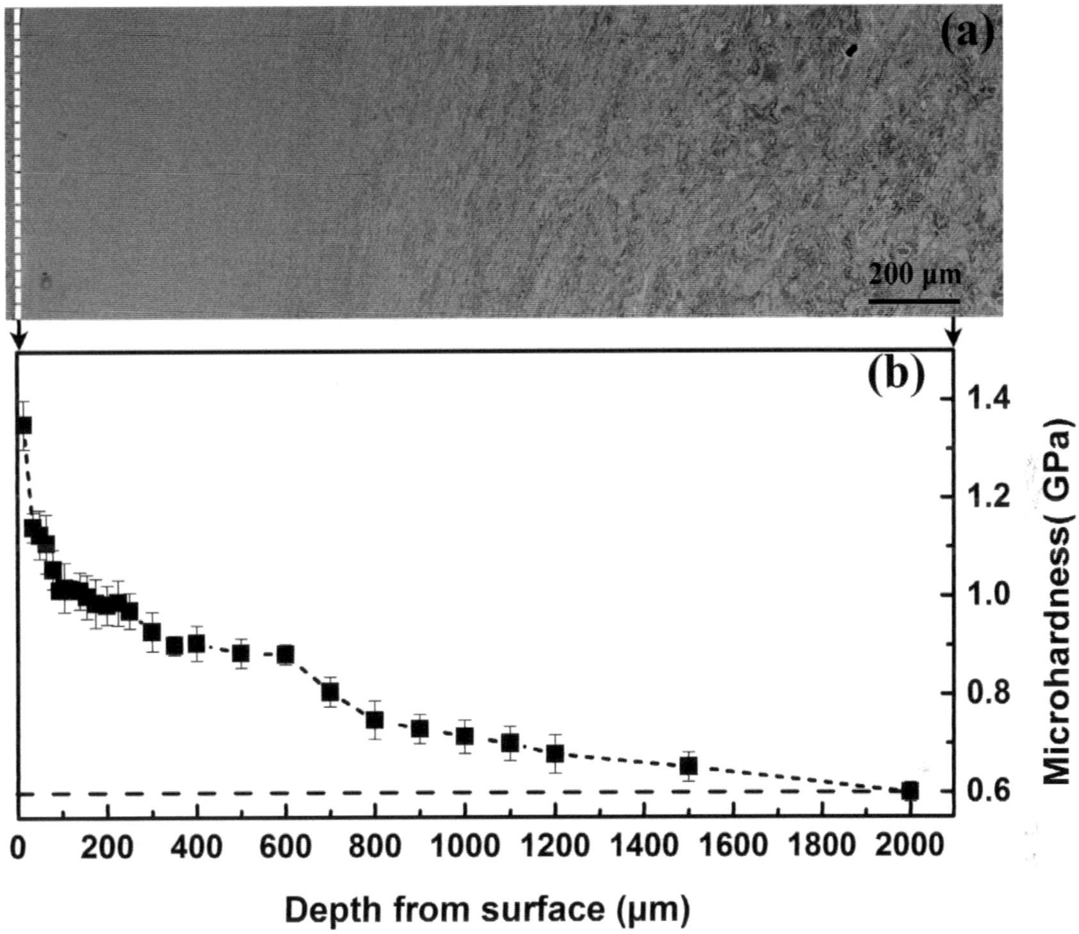

Figure 2. (a) ECC image in the longitudinal section of the HPSR Cu sample. The white dashed line indicates the position of the surface. (b) Vickers hardness as a function of depth from the surface. The horizontal dashed line indicates the initial hardness of 0.6 GPa. Note the correspondence between the image in (a) and (b) as marked by a pair of arrows.

To reveal the microstructure in the topmost surface layer of the HPSR processed Cu, high magnification ECC observation was also performed. In the topmost ~1 μm a nanolaminated structure is developed, as shown in figure 3. Most of the nanolaminated structure is elongated along the direction of the plastic deformation, which is parallel to the surface. The average boundary spacing is not fully resolved in the SEM ECC image, but the thickness scale is approximately 50 – 100 nm. In the depth range from 1-10 μm, alternating layers of ultrafine laminated (UFL) structure and nanolaminated (NL) structure are observed. The ultrafine laminated structures are slightly elongated and have an aspect in the range 2 – 6. The average boundary spacing of the UFL structure is about 130±20 nm. The morphology of each NL regions is similar to that in the topmost 1 μm, but the laminated structure is slightly thicker, with an average boundary spacing of about 105 nm. It is found that the nanolaminated structure in the topmost surface is different from that previously reported in Cu processed by SMAT [1] and SMGT [4, 9], where slightly elongated nanograins are produced in the topmost surface It should be noted that the nanostructures in the gradient layer are generally unstable [11, 12]. The grains may grow either under load or thermal treatments. A substantial temperature rise is unavoidable during the repeated HPSR processing, and this may be high enough to induce noticeable structural coarsening. The formation of the UFL-NL mixed layers in the HPSR processed Cu could be a result of such growth of the nanolaminated structure during HPSR processing.

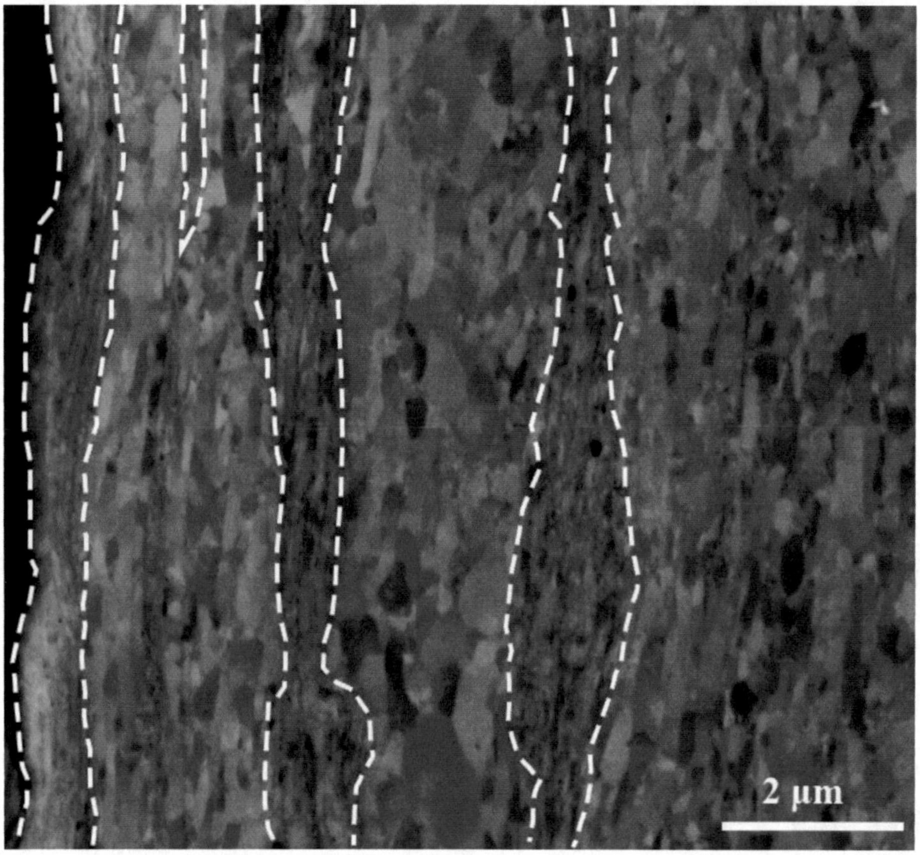

Figure 3. A cross-sectional ECC image of the deformation microstructure in the topmost ~10 μm. The surface is at the left edge. The white dashed line indicates the position of nanolaminated boundaries.

The detailed microstructure in the deformed surface layer was further investigated by EBSD. A gradient microstructure, varying from the topmost surface to the coarse grained bulk is clearly observed in figure 4a-d. The black layer in the left part of figure 4a corresponds to the microstructure in the top ~10 μm in figure 3, where high strain concentration and fine grains make it impossible for EBSD to resolve the details of the microstructure. In depths from 10-300 μm, well-defined nanometer-thick laminated structures parallel to the surface are developed, as shown in figure 4a. The laminated structures are up to several microns in length with thickness ranging from 110±20 nm at a depth of ~20 μm to 270±20 nm at a depth of ~300 μm. A regular lamellar deformation microstructure is observed at depths of 300-600 μm (figure 4b). The boundary spacings vary from hundreds of nanometers to a few micrometers with increasing depth from the surface. As shown in figure 4c, a transition region composed of a lamellar structure and a slightly refined grain layer can be observed at the depths of 600-1200 μm. The right part of figure 4d shows that at depths larger than 2000 μm the grains are slightly deformed but otherwise approximately similar to the annealed microstructure.

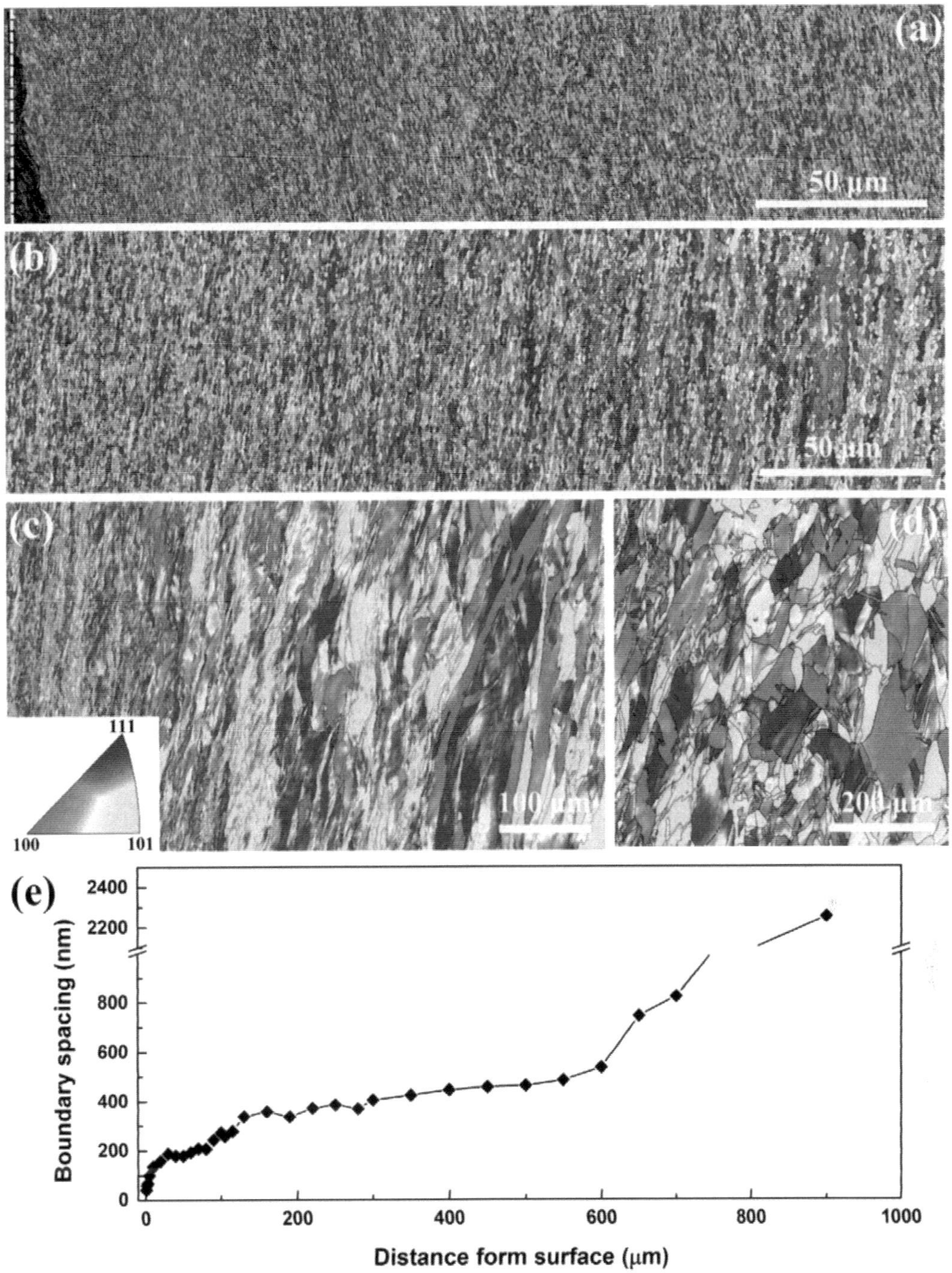

Figure 4. Typical EBSD maps at various depths from the surface (a) 0-300 μm, (b) 300-600 μm, (c) 600-1200 μm and (d) >1200 μm. The white dashed line in (a) indicates the position of the surface. (e) Variation of boundary spacings as a function of distance from the surface, as measured from EBSD maps.

Figure 4e shows the evolution of the boundary spacing as a function of depth. A good accordance is found between the boundary spacing and the hardness results. In the range of depths from 0-150 μm, the boundary spacing increases significantly with increasing depth, corresponding to a relatively steep decrease in hardness. In the depth range of 150-600 μm, a relatively homogeneous ultrafine structure with boundary spacing of about 350±50 nm is developed, which corresponds well to a relatively slow decrease in hardness in this range. At depths larger than 600 μm, the boundary spacing increases with increasing depth, corresponding to a further gradual decrease in hardness. Taken together, the microstructure introduced by HPSR is characterized by a gradient structure from nanostructured grains, to deformed coarse grains, to un-deformed grains, and correspondingly to a gradient of hardness.

Surface mechanical attrition treatment [2], surface mechanical grinding treatment [4, 9], as well as other mechanical treatments [13], have also been used to produce a surface gradient structure in Cu. The refinement of microstructure is attributed to the gradient distribution of strain and strain rate on the sample surface during surface plastic deformation [3]. Although the thickness of the gradient layer is strongly dependent on the materials and treatment parameters, such as rotation speed and tool geometry, the previously reported surface layer thickness values are mostly limited to about 1 mm. However, the present results (figure 2a and figure 4) show that the thickness of the overall deformed surface layer is ~2 mm, which is much thicker than previously reported. In the HPSR process, a load force as high as 20 kN is utilized, which produces a high pressure on the sample surface. Hence, large plastic strains over a broad area and reaching to large depth are achieved during the HPSR deformation, resulting in a thick deformed layer. The HPSR technique provides a new approach for the generation of gradient microstructures extending over a large depth on the surface of bulk metallic materials. This new technique requires only simple procedures and it is expected that the process can easily be scaled-up and adapted to for industrial production and applications.

4. Conclusions

In this study a new technique, high pressure surface rolling (HPSR), has been applied to refine the grain structure on the surface of coarse-grained pure Cu. A deformed surface layer about 2 mm thick is generated, characterized by gradient in grain size, and correspondingly in a micro-hardness gradient, from the surface to the sample interior. The grain size varies from the nanometer scale at the top surface to the micrometer scale in un-deformed interior volumes. The hardness varies from 1.37 GPa near the topmost surface to about 0.6 GPa in the un-deformed coarse-grained matrix. The study demonstrates that HPSR is a promising process for producing a thick layer of gradient microstructure on the surface of bulk metallic materials.

Acknowledgments

The authors thank the support from the Fundamental Research Fund of Central Universities of China (Grant No. 106112017CDJPT130005). QYH also thanks the support from Graduate Student Research Innovation Project of Chongqing University (No. CYB14007).

References

[1] Lu K and Lu J 1999 *J. Mater. Sci. Technol.* **15** 193
[2] Lu K and Lu J 2004 *Mater. Sci. Eng. A* **375-377** 38
[3] Liu X C, Zhang H W and Lu K 2013 *Science* **342** 337
[4] Li W L, Tao N R and Lu K 2008 *Scr. Mater.* **58** 546-9
[5] Zhang X, Hansen N, Gao Y and Huang X 2012 *Acta Mater.* **60** 5933-43
[6] Lu K 2014 *Science* **345** 6203-1455
[7] Wang Z B, Tao N R, Li S, Wang W, Liu G, Lu J and Lu K 2003 *Mater. Sci. Eng. A* **352** 144
[8] Huang H W, Wang Z B, Lu J and Lu K 2015 *Acta Mater* **87** 150
[9] Fang T H, Li W L, Tao N R and Lu K 2011 *Science* **331** 1587-90
[10] Mei Q S, Tsuchiya K and Gao H 2012 *Scr. Mater.* **67** 1003
[11] Ruper T J, Gianola D S, Gan Y and Hemker K J 2009 *Science* **326** 1686
[12] Zhang K, Weertman J R and Eastman J A 2004 *Appl. Phys. Lett.* **85** 5197
[13] Hong C, Huang X and Hansen N, 2015 *IOP Conf. Series: Materials Science and Engineering* **89** 012026

Change in the microstructure and mechanical properties of drawn pearlitic steel with low-temperature aging

D Hirakami[1], K Ushioda[2, 3], T Manabe[1], K Noguchi[4], K Takai[5], Y Hata[6], S Hata[7], H Nakashima[7]

[1] Bar & Wire Rod Research Lab., Steel Research Laboratories, Nippon Steel & Sumitomo Metal Corporation, JP
[2] Technical Research & Development Bureau, Nippon Steel & Sumitomo Metal Corporation, JP
[3] Graduate School of Natural Science and Technology, Kanazawa University
[4] Graduate Student, Sophia University, JP
[5] Department of Engineering and Applied Science, Faculty of Science and Technology, Sophia University, JP
[6] Graduate Student, Kyushu University, JP
[7] Department of Electrical and Materials Science, Faculty of Engineering Sciences Kyushu University, JP

E-mail: hirakami.9fc.daisuke@jp.nssmc.com

Abstract. Hydrogen embrittlement is a serious problem in high-strength steels. Drawn pearlitic steel shows excellent resistance to hydrogen embrittlement despite its high strength, and aging treatment at a low temperature can simultaneously improve its strength and hydrogen-embrittlement resistance. To clarify the mechanism for this we have used thermal desorption analysis (TDA) and the newly developed precession electron diffraction analysis method in the transmission electron microscope. After agingat 100 °C for 10 min, the amount of hydrogen seen amount on the TDA curve reduced at around 100 °C. In contrast, when aging was performed at 300 °C, the hydrogen amount further reduced at around 100 °C and the unevenly deformed lamellar ferrite zone was locally recovered. For the samples that were aged at the low temperature, we confirmed that their yield strength and relaxation stress ratios increased simultaneously with improvement in the hydrogen-embrittlement property. We infer that segregation of carbon or formation of very fine carbide in dislocations during aging is the cause of these behaviors.

1. Introduction

Drawn pearlitic steel shows excellent resistance to hydrogen embrittlement [1]. However, the mechanism through which its hydrogen-embrittlement resistance is improved remains unclear. Pearlitic steel drawn to large strains has a lamellae structure oriented in the drawing direction. Since a high-density of dislocations are introduced in the ferrite, it is important to clarify the relationship between these microstructural factors and the hydrogen-embrittlement resistance.

Thermal desorption analysis (TDA) is a well-known experimental technique for analyzing the state of hydrogen in steel [2]. TDA of the drawn pearlitic structure shows one hydrogen peak near 100 °C

IOP Conf. Series: Materials Science and Engineering **219** (2017) 012026 doi:10.1088/1757-899X/219/1/012026

and a second peak near 300 °C. The observation of two peaks means that there are multiple hydrogen trapping sites in the structure [3]. Many studies on the relationship between aging and the hydrogen-embrittlement susceptibility of drawn pearlitic steel have been conducted [3-5], and have focused on microstructural changes in severely worked pearlitic steel [7-9]. However, changes in the microstructure, mechanical properties, and hydrogen-embrittlement resistance of such steel with aging at a relatively low temperature of 100–300 °C for a relatively short time of 10 min after wire drawing with a strain of ~2, as done in the present study, have not yet been elucidated. Especially, a detailed microstructural analysis has not been conducted regarding the manner of recovery and carbide precipitation to dislocations, among other things.

Therefore, the present study endeavors to deepen the understanding of the effect of aging on the mechanical properties and hydrogen-embrittlement susceptibility of drawn pearlitic steel from the viewpoint of microstructural change. More specifically this study was designed to allow the observation of dislocation structures and carbides in as-drawn and subsequently aged pearlitic steel specimens using transmission electron microscopy (TEM) and to perform grain orientation analysis [10] of small regions using the nanobeam precession electron diffraction method.

2. Experiment

2.1. Material

SWRS82B, a high-carbon steel (0.8 mass% C steel), was used as the experimental material. First, a 122 × 122 mm^2 billet of SWRS82B was heated to 1100 °C and hot-rolled into a rod with diameter ϕ = 13.0 mm. The microstructure of the as-rolled material was pearlite. The chemical composition is given in table 1.

The pearlitic steel rod was then heated at 880 °C for 10 min and held at 530 °C for 120 s to induce the pearlitic transformation. The heat-treated rod was pickled, zinc-phosphated, and dry-drawn to ϕ = 5.0 mm at room temperature. Some of the drawn wires were examined in the as-drawn state. The rest were aged at 100 °C for 10 min (BL100) or at 300 °C for 10 min (BL300). Table 2 shows the tensile strength of the specimens. For tensile testing a ϕ = 5 mm round-bar specimen with a gauge length (GL) 90 mm was used at a tensile speed of 1 mm/min. It is clear that the 100 and 300 °C aging treatments significantly increased the yield strengths of the BL100 and BL300 specimens.

Table 1. Chemical composition of the steel used (mass%).

	C	Si	Mn	P	S	N	O
SWRS82B	0.82	0.19	0.78	0.017	0.015	0.0035	0.0015

Table 2. Tensile strength of the as-drawn and subsequently aged wire for 10 min at 100 °C (BL100) and 300 °C (BL300).

Sample	0.2% Yield strength (MPa)	Tensile strength (MPa)
as-drawn	1530	1943
BL100	1700	2029
BL300	1920	2038

2.2. Evaluation of dislocation stability

To investigate the influence of aging conditions on the stability of dislocations, a stress relaxation test was performed using a smooth, round bar specimen with ϕ = 5 mm. An initial stress of 0.7 σ_B (ultimate tensile strength) was applied at a test temperature of 30 °C, and the decrease in the stress was measured in a constant grasping interval. The ratios of stress relaxation (relaxation stress/initial stress) were compared for the as-drawn and subsequently aged specimens at different aging temperatures.

2.3. Methods for evaluating hydrogen embrittlement of drawn pearlitic steel

To investigate the effect of aging treatment on hydrogen embrittlement, the prepared $\phi = 5$ mm specimens, described in Section 2.1, were tested using a constant-load test. These specimens were prepared by cutting a round rod to a length of 300 mm and then notching the rod circumferentially at 140 mm from the end to a depth of 0.4 mm, an angle of 60°, and a radius of curvature of 0.12 mm. These were polished with emery paper to remove the oxide film and then pre-charged with hydrogen and immediately loaded the specimen, and the time to fracture was then measured. To keep the hydrogen concentration constant during testing, tests were conducted under continuous hydrogen charging conditions, using a solution of 0.1 N NaOH + 5 g/L NH_4SCN, a current density of 2 A/m^2, and a temperature of 30 °C.

2.4. Method for analyzing hydrogen in steel

Hydrogen analysis was conducted using gas chromatography. The specimens were ultrasonically washed in acetone before measurement. Each specimen was dried and placed in the quartz tube of a heating chamber. After the atmosphere in the quartz tube was replaced by argon carrier gas, the measurement was started. The heating rate was 100 ºC/h, and the measurement was performed at temperatures of up to 600 °C.

2.5. Microstructural observation using TEM and the grain-orientation mapping method

The drawn and subsequently aged specimens were observed using TEM to investigate the changes in their microstructures with aging. Both the as-drawn and subsequently aged specimens were cut to ϕ 5.0 mm × 1 mm and punched into $\phi =$ 3.0 mm disks, with the disk centers located at a depth of 1.5 mm from the outer circumference. Thin-film specimens were prepared via mechanical polishing and twin-jet electrolytic polishing. The nanobeam precession electron diffraction method was used to obtain grain orientation maps of the α-Fe phase. In this method, an electron beam is precessed on the specimen (beam-irradiation diameter of ~10 nm and incident angle of 0.6°). That is, an electron beam with a spot size of 10 nm was automatically scanned at a step size of 10 nm to obtain electron diffraction spots representing grain orientations. The TEM observation surface included the longitudinal drawing direction.

2.6. Measurement of dislocation density using X-ray diffraction

The dislocation density was measured at the quarter depth of the specimens cut longitudinally using the Williamson–Hall method.

3. Experimental Results

3.1. Effect of the aging temperature on the stress–strain curve and dislocation stabilization

Figure 1 shows the results of the tensile test. The 0.2% yield stress increased with the aging temperature up to 300 °C. Furthermore, the largest total elongation was recorded for the specimen with an aging temperature of 300 °C (BL300 °C).

Figure 2 shows the results of the stress-relaxation test. The ratios of stress relaxation were 3.4% for the as-drawn material, 3.2% for the specimen aged at 100 °C (BL100 °C), and 1.6% for the BL300 °C sample. These results suggest that the dislocations were stabilized by the aging treatment.

3.2. Change in the hydrogen-embrittlement characteristics of drawn pearlitic steel with aging treatment

The hydrogen-embrittlement characteristics were evaluated using a constant-load tensile test. We previously reported that aging treatment at temperatures as high as 300 °C improves hydrogen-embrittlement resistance [11].

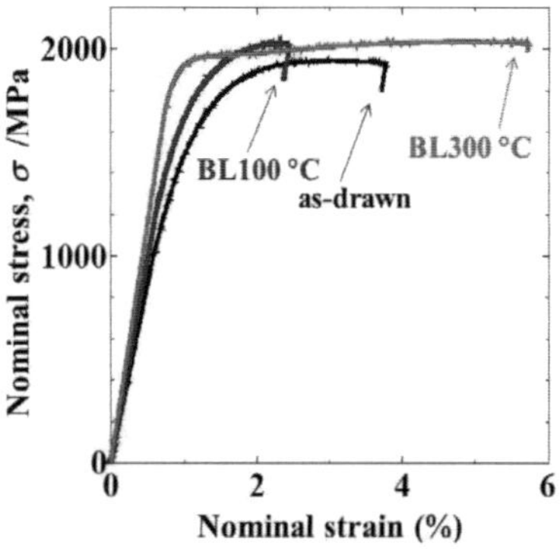

Figure 1. Stress–strain curves of the as-drawn, BL100°C, and BL300 °C specimens.

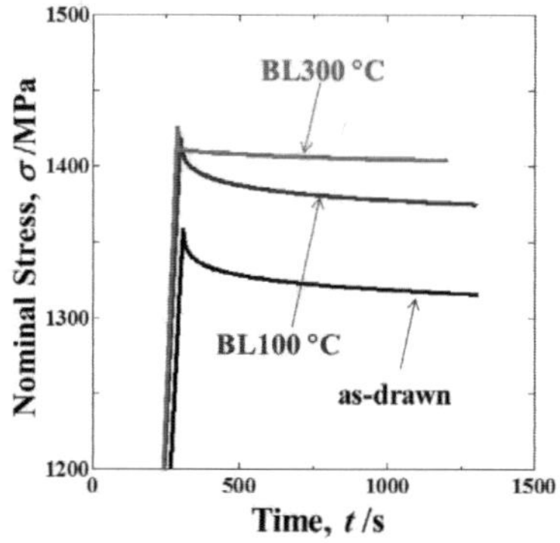

Figure 2. Stress-relaxation curves of the as-drawn, BL100 °C, and BL300 °C specimens.

3.3. Change in the hydrogen trapping state of drawn pearlitic steel with aging treatment

Figure 3 shows the TDA profiles of the specimens subjected to hydrogen charging for 144 h. The hydrogen concentration is expected to reach an equilibrium value up to the sample center. Two desorption peaks at 100 °C and 300 °C were observed, where the first peak, near 100 °C, corresponds to the hydrogen trapped in the stress field of dislocations, the second peak, near 300 °C, corresponds to hydrogen trapped in dislocation cores [12]. As the aging temperature increased, the height of both peaks decreased, indicating a decrease in trapper hydrogen content.

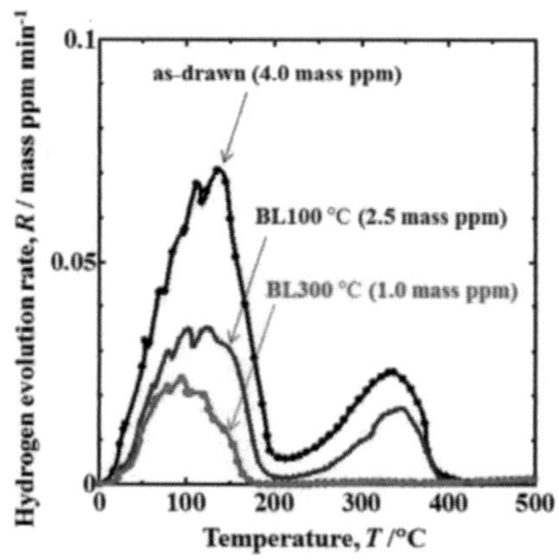

Figure 3. Thermal desorption profiles of the as-drawn, BL100°C, and BL300 °C specimens charged with hydrogen for 144 h.

3.4. Microstructural changes in drawn pearlitic steel with aging treatment

Figure 4 shows bright-field TEM images of the as-drawn specimen as well as of specimens subsequently aged at 100 °C and 300 °C. The drawing direction is indicated by white arrows. Dislocations were observed in the lamellar ferrite in each specimen. The width of the lamellar ferrite was 50–100 nm in each specimen. No clear changes were observed either in the ferrite-phase width or the cementite structure as a result of the aging treatments. The specimens aged at 300 °C, however, yielded images suggesting some dislocation rearrangement in lamellar ferrite, as indicated by the dashed-line circles in Fig. 4(c). A contrast suggesting the precipitation of carbides was also observed, as indicated by black arrows in Fig. 4(c). Strain contrast arising from the localized presence of high-density dislocations was also observed, as indicated by red circles in Figs. 4(a)–(c).

Figure 5 shows the detailed information of cementite in the as-drawn specimen and in the specimen subsequently aged at 300 °C. The dark-field image in Fig. 5(d) indicates the commencement of spheroidization of very fine carbides at an aging temperature of 300 °C.

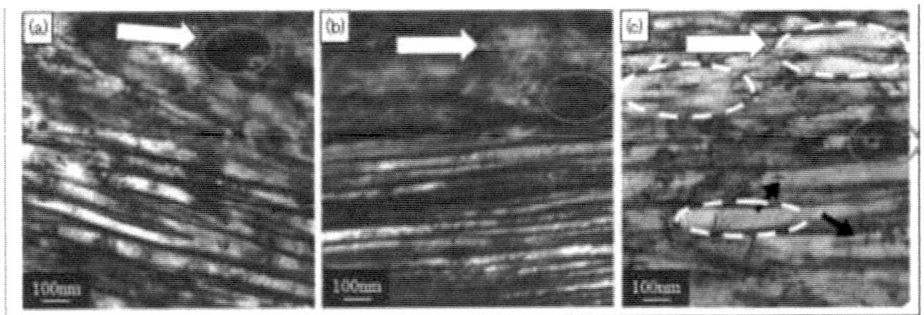

Figure 4. TEM images showing the dislocation structure in (a) the as-drawn specimen and in the specimens subsequently aged at (b) 100 °C and (c) 300 °C.

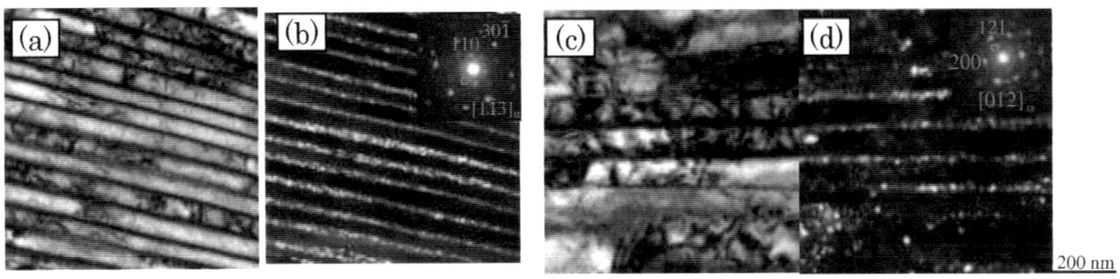

Figure 5. TEM images showing the lamellar cementite structure (a) the bright field image of the as-drawn specimen, (b) dark field image of the as-drawn specimen, (c) bright field image of the specimen aged at 300 °C, and (d) dark field image of specimen aged at 300 °C.

Figure 6 shows ferrite-phase orientation maps obtained by subjecting the as-drawn specimen and the specimens subsequently aged at 100 °C and 300 °C to precession electron diffraction. In the grain-orientation map of each specimen, grains of the same color are elongated in the drawing direction. A region with the same color is inferred to be an initial block with a width of approximately 1–2 μm. The gradation in the grain orientation in a block is inferred to reflect a packet. In this way, the inhomogeneity of the microstructure of the drawn pearlitic steel is related to the grain orientations of the blocks before the drawing operation. It is worthwhile to note that the specimen aged at 300 °C revealed a region with a fine microstructure having high-angle grain boundaries (Fig. 6(c)). This suggests that the aging treatment at 300 °C locally decreased the dislocation density through rearrangement of dislocation, resulting in subgrain formation. The magnified TEM orientation maps shown in Fig. 7 evidently imply that dislocation rearrangement proceeds during aging at 300 °C.

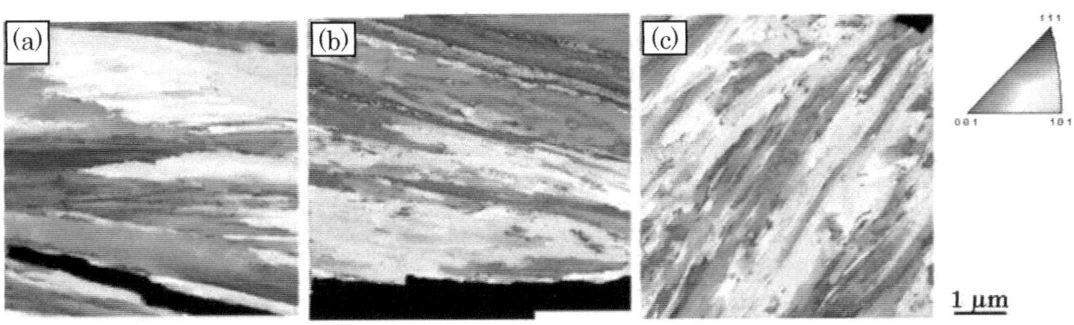

Figure 6. TEM grain-orientation maps of specimens: (a) as-drawn, (b) subsequently aged at 100 °C, and (c) subsequently aged at 300 °C.

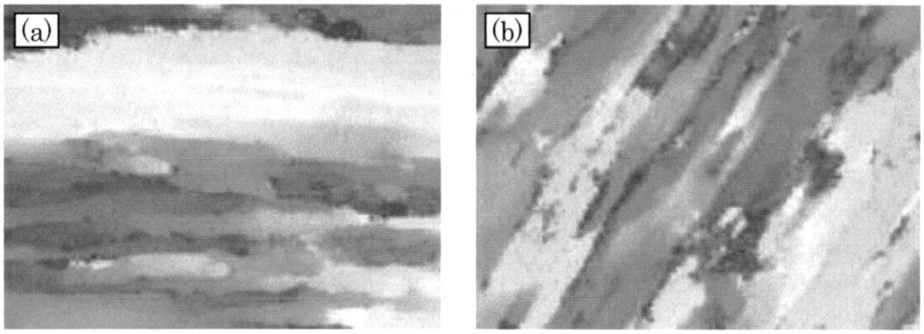

Figure 7. Magnified TEM grain-orientation maps of specimens: (a) as-drawn and (b) subsequently aged at 300 °C.

4. Discussion

4.1. Change in the yield strength and relaxation-stress ratio with aging treatment

The yield strength increased upon the application of the aging treatment, more significantly at 300 °C than at 100 °C. Dislocations during the relaxation test were also stabilized by the aging treatment in a similar manner to that observed in the tensile test, where again, the aging treatment at 300 °C had a stronger effect than that at 100 °C. During the drawing process for pearlitic steel at room temperature, lamellar cementite is expected to be partly dissolved as solute carbon in the ferrite matrix together with an introduction of dislocations. These dislocations are supposed to be locked or pinned by segregated carbon or by very fine carbides formed during the aging treatment. Taking into account the fact that recovery simultaneously occurs during aging at 300 °C, the increment in the pinning force for dislocation movement by segregated carbon or very fine carbides is inferred to be much larger than the softening effect of recovery.

4.2. Change in hydrogen embrittlement with aging treatment

Takai and Watanuki [3] confirmed the presence of two peaks for hydrogen near 100 °C and 300 °C, respectively, in the TDA curves of drawn pearlitic steel. Doshida and Takai [4] explained that plastic deformation of drawn pearlitic steel increases hydrogen-embrittlement susceptibility and that this condition is correlated to the increase in the magnitude of the lower temperature hydrogen peak. Meanwhile, Chida et al. [6] reported that aging the drawn pearlitic steel results in a decrease in hydrogen content associated with the lower temperature peak. They considered that during aging, carbon is preferentially trapped in dislocations, thereby decreasing the number of hydrogen trapping sites. Figure 3 in the present study shows results similar to those obtained by Chida et al. The hydrogen content of the lower temperature peak significantly decreased after aging at 100 °C for 10 min, accompanied by a further decrease in size of both peaks after aging at 300 °C. Aging for only a relatively short period of 10 min clearly increased the yield strength. The ultimate tensile strength of the drawn pearlitic steel was, however, barely affected (see table 2).

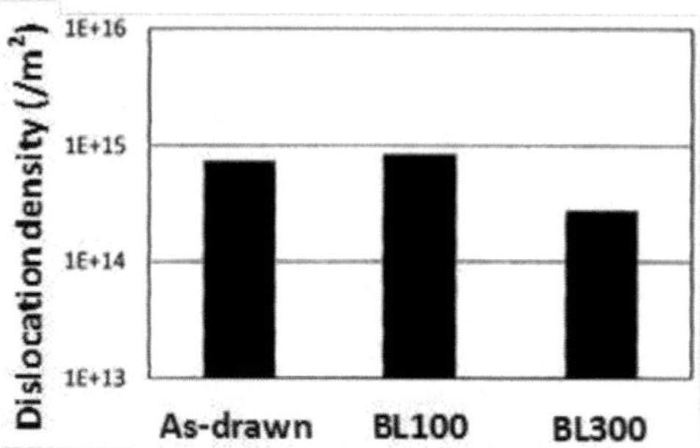

Figure 8. Dislocation density of the following specimens: as-drawn and subsequently aged at 100 °C (BL100) and 300 °C (BL300).

Simultaneously, aging for a short duration resulted considerably decreased hydrogen-embrittlement susceptibility [11].

Microstructural changes during the aging treatment are considered to have a positive effect against hydrogen embrittlement. The TEM grain-orientation maps of the minute regions of the as-drawn specimen and those of the specimen subsequently aged at 300 °C (figures. 7(a) and (c), respectively) not only revealed the grains elongated in the drawing direction but also indicated the occurrence of dislocation rearrangement more prominently during aging at 300 °C. Furthermore, the TEM grain orientation maps revealed for the first time the formation of a fine microstructure with high-angle grain boundaries wherein recovery progressed and subgrain formation apparently proceeded. This microstructural change is expected to soften the material. A slight hardening actually occurred, however, during aging at 300 °C. This is presumably caused by the further strengthening resulting from the segregation of carbon or the precipitation of very fine carbides, as shown in figures. 5(a) and (b).

Figure 8 shows the results of the dislocation-density measurements performed using X-ray diffraction to indirectly evaluate the aging-induced recovery behavior. The dislocation density hardly changed with aging at 100 °C for 10 min, suggesting the occurrence of only the decoration of dislocations with carbon. Meanwhile, aging at 300 °C for 10 min slightly decreased the dislocation density. The results obtained via X-ray diffraction are considered to be in good agreement with the TEM grain-orientation maps.

4.3. Proposed mechanism for the change in the microstructure and properties with aging treatment

The change in microstructure of drawn pearlitic steel during low-temperature aging is schematically illustrated in figure 9. Pearlitic steel in the as-drawn stage consists of an elongated lamellar structure consisting of ferrite and cementite. Since cementite is considered to be partly dissolved as solute carbon during the drawing process, solute carbon is expected to exist in dislocated lamellar ferrite. When drawn pearlitic steel is aged at 100 °C for 10 min, the solute carbon segregates to the dislocations (figure 9 (a)). Therefore, the yield stress is considered to increase as a result of the Cottrell atmosphere. Moreover, hydrogen-embrittlement susceptibility is expected to decreasem, because the amount of harmful hydrogen segregated to the dislocations decreases owing to the preferential segregation of carbon to the dislocations. Moreover, the mobility of dislocation, which was initially promoted by the HELP theory owing to the segregated hydrogen, is inferred to be reduced by the presence of segregated carbon in the dislocations.

When drawn pearlitic steel is aged at 300 °C for 10 min, carbon further segregates to the dislocations and very fine carbide precipitation may occur (figure 9 (b)). On the other hand, aging at 300 °C induces local recovery or subgrain formation (see the regions enclosed with white broken circles in figure

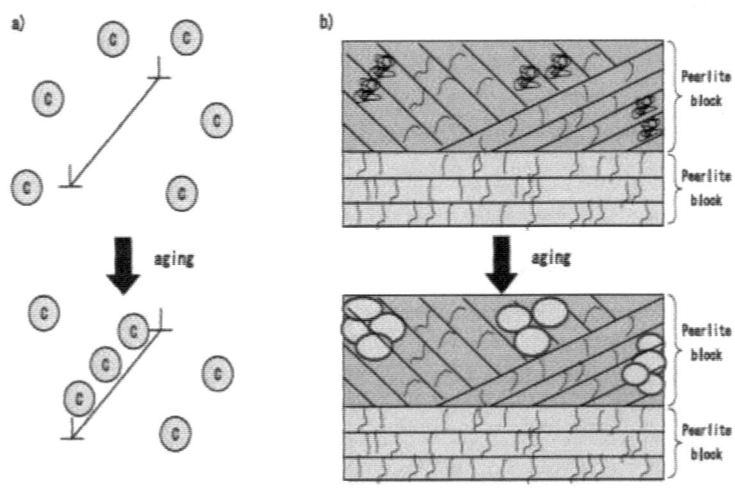

Figure 9. Schematic illustrations showing the change in microstructure and the mechanism of the improvement in mechanical properties upon aging the as-drawn specimen at a) 100 °C and b) 300° C .

4(c) and the circles in the lower part of figure 9(b)). Despite the softening effect induced by recovery, the yield strength increased. This significant increase in the yield strength with aging at 300 °C is expected to be caused by the very strong pinning force exerted by fine carbides. As for hydrogen-embrittlement susceptibility, it should be essentially high after wire drawing, as pointed out by Doshida and Takai [4], owing to the presence of hydrogen segregation at dislocations, and due to the deterioration of toughness associated with the work-hardened structure, particularly in high-strained regions where dislocations are tangled in a complex manner (see figure 4 and the upper part of figure 9(b)). However, aging at 300 °C improved hydrogen-embrittlement resistance. This presumably occurred because dislocations are strongly pinned by the very fine carbides and the toughness is improved by the recovery.

5. Conclusions

The changes in the microstructure and mechanical properties, such as strength, stress relaxation, and hydrogen-embrittlement susceptibility, resulting from low temperature aging (at either 100 °C or 300 °C) were investigated for a drawn 0.8 mass% C pearlitic steel. The microstructure was evaluated using the newly developed nanobeam precession electron diffraction technique in the tranasmission electron microscope to obtain grain-orientation maps. The following findings were obtained:

(1) The yield strength of the specimen increased upon aging, and the increase was more prominent after aging at 300 °C compared to 100 °C.

(2) The stress-relaxation ratio increases with the aging temperature in a manner similar to that observed for yield strength, which implies that the dislocations were stabilized by aging of the drawn wire.

(3) Aging treatment, especially at 300 °C, decreased the hydrogen-embrittlement susceptibility.

(4) The height of the lower temperature peak seen in TDA measurements (associated with hydrogen trapped in dislocation stress fields) decreased with aging. The peak height of this peak and that of the higher temperature peak (associated with hydrogen trapped at dislocation cores) further decreased with aging at 300 °C.

(5) The aging treatment at 300 °C decreased the dislocation density and accelerated localized recovery, leading to the rearrangement of dislocations.

(6) Solute carbon released by dissolution of cementite during wire drawing was assumed to segregate to dislocations and to precipitate as very fine carbides during aging. The microstructural changes in terms of the state of carbon, hydrogen, and the dislocation substructure are concluded to be responsible for improving the balance of strength and hydrogen-embrittlement resistance during low-temperature aging.

References

[1] Takai K, Seki J and Honma Y 1995 *Tetsu-to-Hagane* **81** 1025
[2] Suzuki H and Takai K 2012 *ISIJ Int.* **52** 174
[3] Takai K and Watanuki R 2003 *ISIJ Int.* 43 520
[4] Doshida T, NAKAMURA M, SAITO H, SAWADA T and Takai K 2014 *Acta Mater* **61** 7755
[5] Chida T, Kosaka M, Kubota M and Tarui T 2006 *CAMP-ISIJ* **19** 1158
[6] T.Chida,M.Kosaka,M.kubota and T.Tarui 2005 *CAMP-ISIJ* **18** 559
[7] Daitou Y and Hamada T 2000 *Tetsu-to-Hagane* **86** 105
[8] Makii K and Ibaraki N 1998 *Workshop of STX-21* (NIMS: Ibaraki) p 100
[9] Hono K, Ohmura M, Murayama M, Nishida S, Yoshie A and Takahashi T 2001 *Scripta Mater.* **44** 977
[10] Moeck P, Rouvimov S, Häusler I, Neumann W and Nicolopoulos S 2013 *Ultramicroscopy* **128** 68
[11] Hirakami D, Ushioda K, Manabe T, Noguchi K, Takai K, Hata Y, Hata S, Nakashima H 2016 *ISIJ International* **56** 893
[12] Enomoto E, Hirakami D and Tarui T 2012 *Metall. and Materi. Trans. A* **43A** 572

IOP Conf. Series: Materials Science and Engineering **219** (2017) 012027 doi:10.1088/1757-899X/219/1/012027

Characterization of voids in shock-loaded Al single crystal by combining X-ray tomography and electron microscopy

Chuanshi Hong[1,*], Søren Fæster[1], Niels Hansen[1], Xiaoxu Huang[1] and Rozaliya I. Barabash[2]

[1] Section for Materials Science and Advanced Characterization, Department of Wind Energy, Technical University of Denmark, Risø Campus, 4000 Roskilde, Denmark
[2] Materials Science & Technology Division, Oak Ridge National Laboratory, P.O. Box 2008, Oak Ridge, TN 37830, USA

*E-mail: chho@dtu.dk

Abstract. A combination of X-ray tomography and electron backscatter diffraction (EBSD) was applied to investigate both the shape of voids and the plastic deformation around voids in an Al single crystal shock-loaded in the <1 2 3> direction. The combination of these two techniques allows the addition of crystallographic information to X-ray tomography and allows the addition of three-dimensional information to EBSD data. It is found that the voids are octahedral with {1 1 1} faces and that regular patterns of lattice reorientation exist around individual voids. The results provide new insights to the process of void growth during shock loading, which is important for both civil and military applications.

1. Introduction

Ductile fracture of metals characterizes their mechanical behavior and is of importance for their application in society. Ductile fracture is known to occur via nucleation, growth and coalescence of voids [1–3], phenomena which have been investigated by analytical and numerical modelling. Modeling of void growth on the continuum level has been well documented [4–9], from which void shape and plastic anisotropy are two factors known to affect void growth to the first order. Atomic simulations have shown that void growth is governed by dislocation slip [10–14]. Experiments on void growth have traditionally focused on the measurement of statistical quantities, such as void sizes and void volume fraction [15–18], whereas experimental insights on the underlying mechanisms are scant.

In the present work we applied a unique combination of X-ray tomography and electron backscatter diffraction (EBSD) to investigate voids in an Al single crystal shock loaded in the <1 2 3> orientation. This crystal has been examined previously by polychromatic X-ray microdiffraction [19]. In the present study new insights were obtained by combining three-dimensional (3D) information (tomography) with a crystallographic analysis in two dimensions (EBSD).

2. Experimental

Shock loading of the Al single crystal was performed using a gas gun assembly [19,20]. The initial dislocation density of the crystal was less than 10^7 cm^{-2} as measured by X-ray diffraction, which is typical of nearly perfect metallic single crystals. The crystal, in the form of a disk of 25 mm in diameter and 5 mm in thickness, was shock loaded in the <1 2 3> direction, parallel to the axis of the disk. The impact velocity was 165 $m \cdot s^{-1}$ and corresponds to a strain rate of the magnitude of 10^5 s^{-1} and a pressure in the range of 10–20 GPa.

Characterization of the shock loaded sample was carried out on a core disk of 5 mm in diameter cut from the center of the initial disk. X-ray tomography was carried out using a Zeiss Xradia 520 Versa micro-CT (computerized tomography) system. A polychromatic conical beam with X-ray energies up to 140 keV was used for the CT scans. In total 1401 image projections were recorded over a rotation of 360° during the scan. EBSD samples were prepared in the section containing the axis of the core disk. The samples were mechanically polished and then electro-polished. EBSD was conducted using a Zeiss Supra 35 scanning electron microscope, operated at a high tension of 20 kV and with a mapping carried out using a step size of 0.5 μm.

3. Results and analysis

Due to the impact loading, micro voids form in a spall region at about one third the thickness of the sample from the rear surface of the sample. The voids are already visible in the metallographic section as shown in figure 1a. X-ray tomography (figure 1b) reveals that the voids distribute rather homogeneously in the spall region, and that the diameters of voids range from a few to a few hundred micrometers with the mode of the size distribution being about 65 μm. The formation of spallation voids under shock is well known to be a result of the interaction of shock waves.

Remarkably, the voids consistently show the shape of an octahedron with somewhat rounded edges and vertices (figure 1c). By combining X-ray tomography with EBSD, the geometry of the void shape was measured in the coordinate system of the single crystal, revealing that the faces of the octahedral voids are on {1 1 1} planes of the single crystal and that the axes connecting opposite vertices of the octahedrons are along the <1 0 0> directions.

Regular patterns of lattice reorientation were detected on the metallographic section examined by EBSD (figure 2). Morphologically, the reorientation patterns can be classified into four groups. In two groups (figure $2a_1$ and b_1), a void is visible in the center of the each reorientation pattern. The morphology of the patterns in these two groups is essentially the same but are inverted (i.e. rotated by 180° around the normal of the EBSD section) relative to each other. In the other two groups (figure $2c_1$ and d_1), the patterns are observed on a smooth surface without visible voids. The morphology of these two groups also exhibits 180° rotational symmetry.

Each reorientation pattern is composed of zones of different rotations. In figure $2a_1$ of the first group, sixteen zones are identified, as labelled 0–15. Significant lattice rotation relative to the mean matrix orientation is observed in all identified zones except for Zone 0, where the rotation is negligible. A similar magnitude of lattice rotation is observed for Zones 1–15. The direction of lattice rotation is illustrated by spin rotors in figure $1a_2$. It is mainly the difference in the rotation direction that distinguishes the zones, which is generally larger than 90° between neighboring zones. The zones are located with mirror symmetry about the bisector of the vertex angle of the isosceles shape of the sectioned void (i.e. the vertex angle near Zone 0). The rotation axes in mirrored zones are mirrored-and-inverted relative to each other.

The reorientation pattern in figure $2b_1$ and b_2 of the second group has all the characteristics described above for the pattern of the first group. All zones in figure $2a_1$ have counterparts in figure $2b_1$. The direction of lattice rotation in each zone in figure $2b_1$ is largely the same as that of the counterpart zone in figure $2a_1$.

The reorientation patterns in figure $2c_1$ and d_1 have an apparent three-fold symmetry. Three pairs of zones are identified in each pattern, labelled as 1–2, 3–4 and 5–6, which are bounded by three high angle boundaries. The three boundaries coincide with the {1 1 0} traces of the mean matrix orientation. As shown in figure $2c_2$ and d_2, the direction of lattice rotation changes substantially between neighboring zones of the different pairs bordered by high angle boundaries.

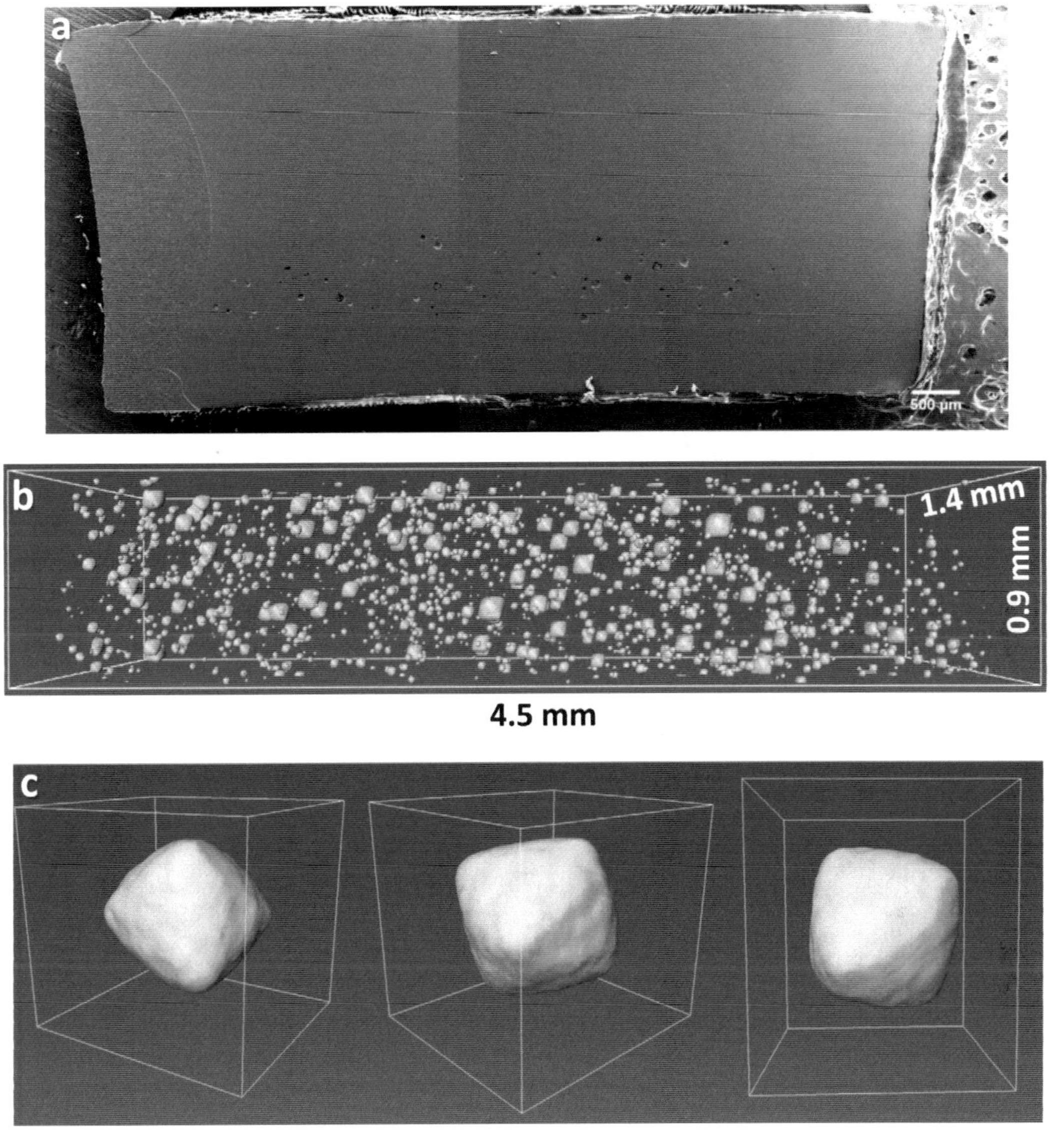

Figure 1. (a) A secondary electron image showing the cross-section of the shock-loaded sample. The shock direction is vertical. (b) Reconstruction of voids in the spall region. (c) Close-ups of a single void viewed along different directions.

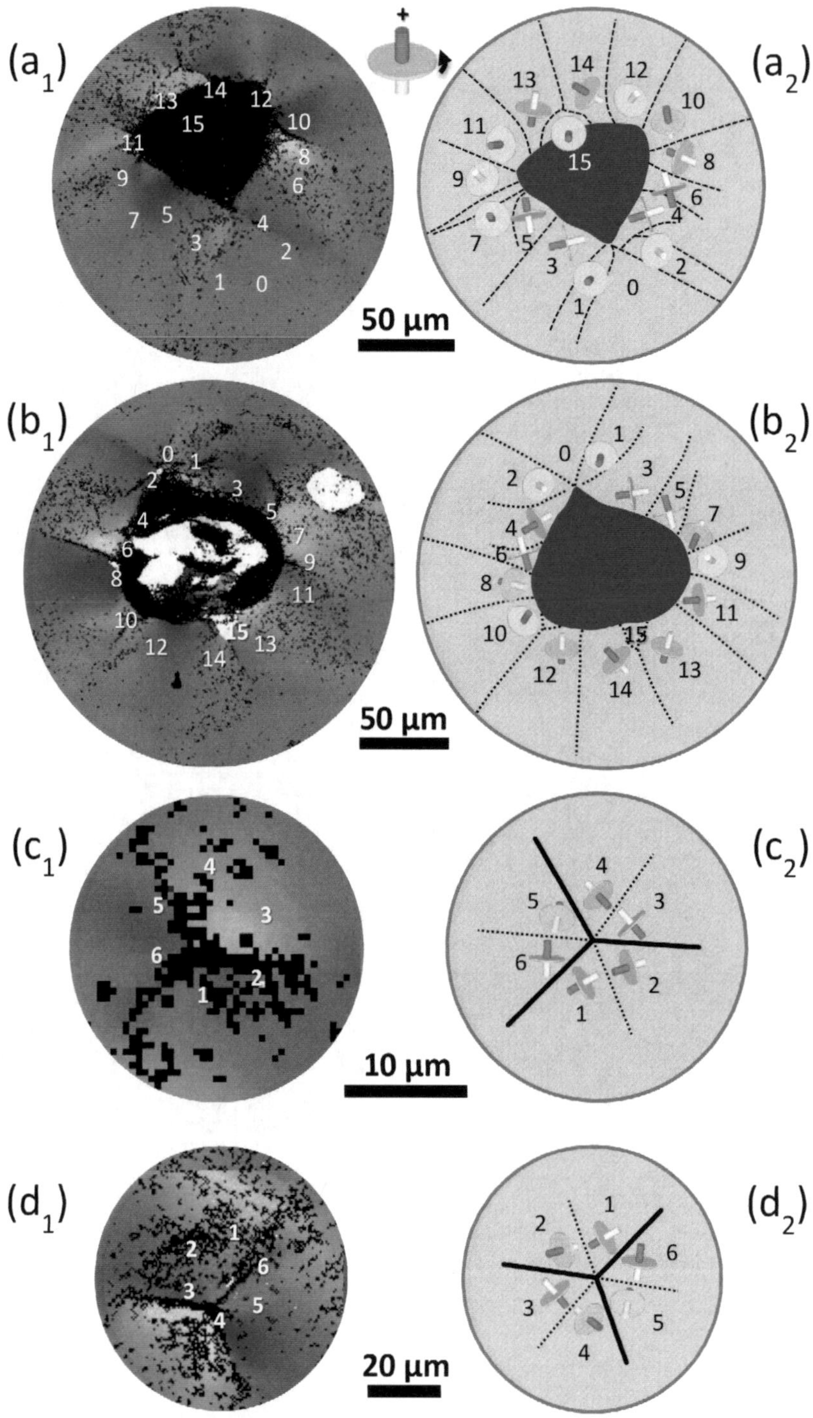

Figure 2. Examples of the four typical morphologies of lattice reorientation patterns observed on the cross-section in figure 1a. (a_1–d_1) EBSD maps (Euler angle coloring). The numbers indicate individual zones of lattice reorientation. (a_2–d_2) Axes of lattice rotation relative to the mean matrix orientation in each individual zone, illustrated by spin-rotors in the sketches of the reorientation patterns. The positive direction of the spin-rotors is defined by the inset between (a_1) and (a_2). The solid lines mark visible boundaries and the dashed lines mark approximately the borders of neighboring zones.

A void in the shape of an octahedron is expected to have different profiles on a metallographic section depending on the sectioning plane and position. Figure 3 shows the profile of a reconstructed void on a section parallel to the EBSD section as a function of the sectioning position. A profile of the void similar to figure $2b_1$ is obtained when the void is sectioned at the upper half, and one similar to figure $2a_1$ is obtained when the void is sectioned at the lower half.

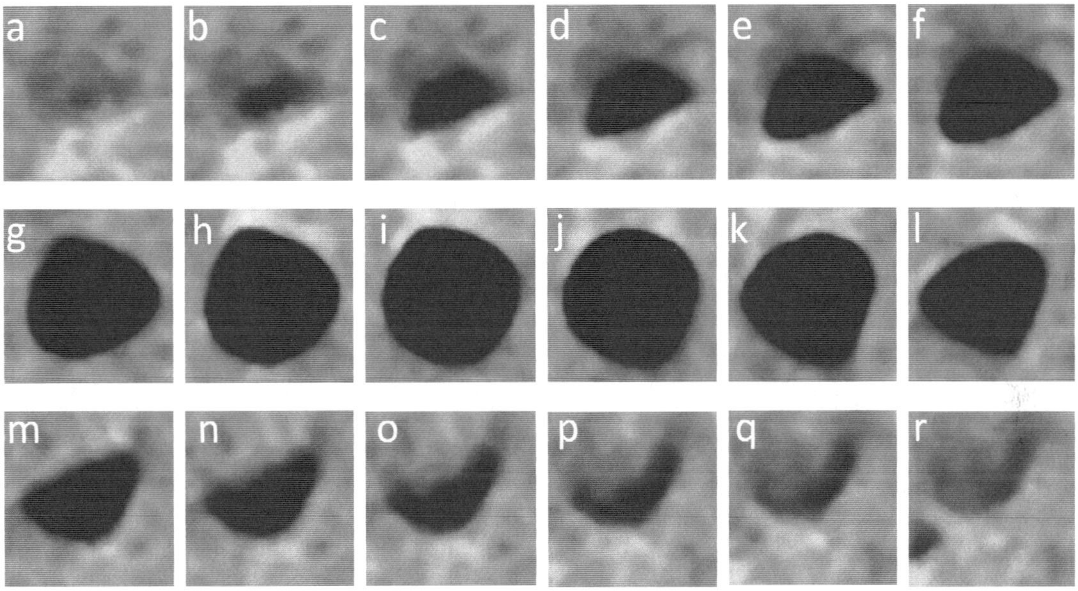

Figure 3. (a–r) Evolution of the sectioned morphology of a void as the virtual section (which is parallel to the EBSD section) moves through a void in the 3D reconstruction.

Moreover, the four voids associated with the four reorientation patterns in figure 2, and the exact position of the EBSD section, were identified in the three dimensional (3D) reconstruction (figure 4a). It is therefore concluded that the four reorientation patterns shown in figure 2 correspond to voids sectioned at typical positions, as summarized in figure 4b.

4. Discussion

The combination of X-ray tomography and EBSD provides an integrated approach for 3D characterization of voids with crystallographic information. On one hand, EBSD provides crystallographic information for volumes sampled by X-ray tomography. On the other hand, X-ray tomography provides the missing information in the third dimension conventional EBSD on 2D sections. Combined the two techniques allows the four groups of reorientation patterns to be unified. This combination of techniques provides an approach for 3D characterization that is less resource demanding compared 3D XRD techniques or serial-sectioning EBSD, and is in particular suitable for studies of single crystals.

The unique approach employed in the present work shows novel results on void growth. The 3D shape of the void is clearly shown to be crystallographic in the present work, which validates previous results on void shape in two dimensions [1,21–23]. Although X-ray tomography has been used to investigate voids [16–18,24,25], the focus was on statistical quantities, such as void sizes and volume fraction, and only limited attention was paid to the shape of individual voids. Recently, octahedral voids were observed by X-ray tomography in a polycrystalline material [26], but no effort was made to determine the crystallographic characteristics of their shape. Regular reorientation patterning around voids growing inside a metal matrix during ductile fracture has not been reported previously, although a high density of dislocations or a large magnitude of lattice rotations around voids have been shown by a few experiments [1,19,22,26].

The formation of a crystallographic octahedral shape of the voids is believed to be caused by the anisotropy of plastic flow related to active slip systems in the original single crystal. It is suggested that a highly symmetric configuration of active slip systems is needed for the evolution of the voids into such a shape, which is consistent with the EBSD results. Lattice rotation during plastic deformation is well known to be determined by dislocation slip. The systematic variation in the rotation direction in individual zones in the reorientation patterns indicates a systematic variation of active slip systems. The regular reorientation patterns with 3D symmetry indicate a highly symmetric configuration of these systems. An in-depth analysis of the reorientation patterning based on active slip systems can be found in [27], also including a TEM characterization and a preliminary analysis of the dislocation structures in the reoriented zones around voids.

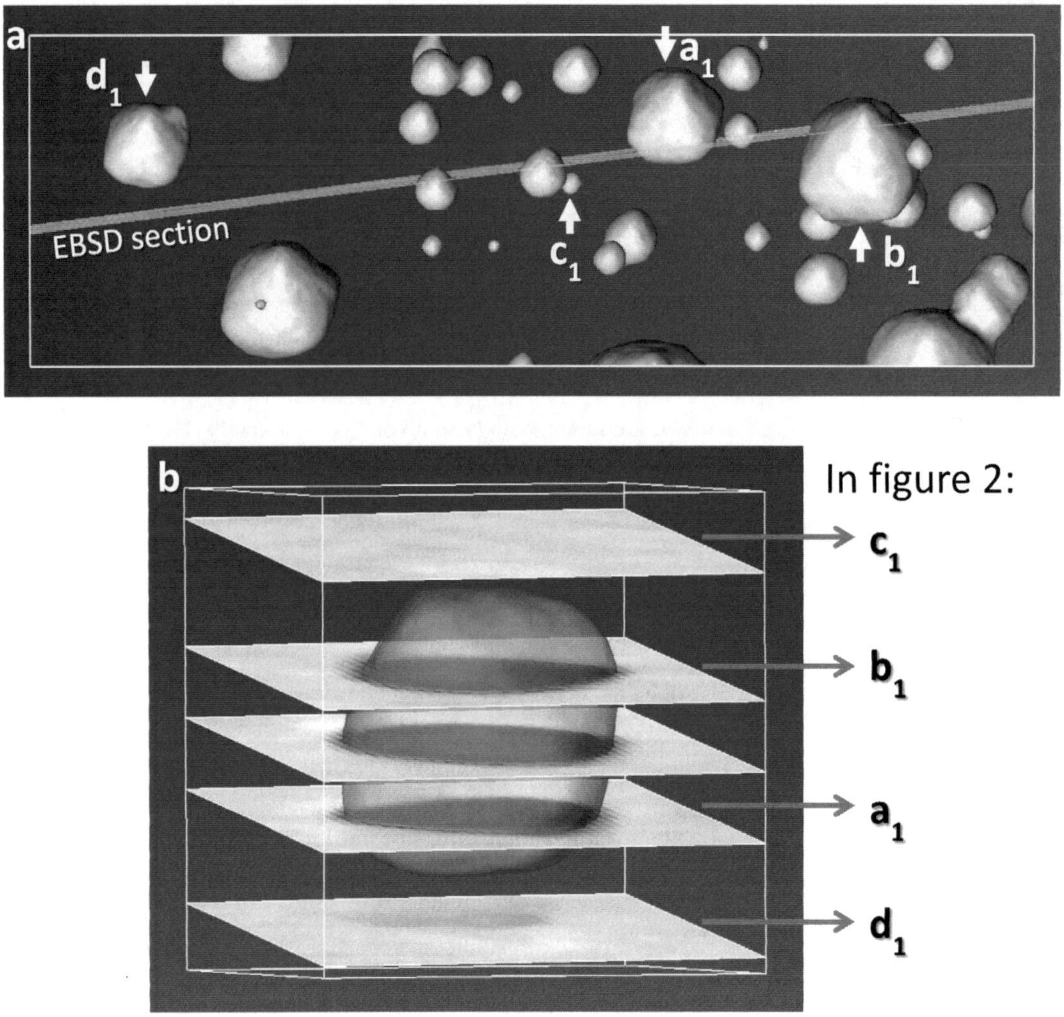

Figure 4. (a) Voids associated with the four reorientation patterns, a_1, b_1, c_1 and d_1 in figure 2, identified in the 3D reconstruction. The exact position of the EBSD section in the 3D reconstruction is also displayed. (b) Summary of the relation between the sectioning position relative to the void and the reorientation patterns observed in figure 2.

5. Conclusions

By combining X-ray tomography and EBSD, a unique approach has been applied in the present work to investigate both the void shape and the pattern of lattice reorientation around voids in a shock-loaded Al single crystal. The approach allows the void shape observed by X-ray tomography to be related to crystallography, and allows the positions of individual voids associated with the lattice reorientation patterns observed in EBSD to be determined, based on which the following can be concluded:

1) The voids have a consistent octahedral shape with faces parallel to the {1 1 1} planes of the single crystal.

2) Regular patterns of lattice reorientation exist around the voids. The patterns consist of zones of large lattice rotation with high angle boundaries between some neighboring zones.

3) The lattice reorientation patterns can be classified into four groups based on morphology, which is an artefact of sectioning of the same 3D reorientation pattern at different positions.

4) The analysis underpins basic and engineering studies of ductile fracture caused by shock loading (strain rate ~10^5 s^{-1}), important for both civil and military applications of metallic components.

6. References

[1] Meyers M A and Taylor Aimone C 1983 Dynamic fracture (spalling) of metals *Prog. Mater. Sci.* **28** 1–96

[2] Garrison W M and Moody N R 1987 Ductile fracture *J. Phys. Chem. Solids* **48** 1035–74

[3] Benzerga A, Leblond J-B, Needleman A and Tvergaard V 2016 Ductile Failure Modeling *Int. J. Fract.* **201** 29–80

[4] Tvergaard V 1981 Influence of voids on shear band instabilities under plane strain conditions *Int. J. Fract.* **17** 389–407

[5] Tvergaard V 1989 Material Failure by Void Growth to Coalescence *Adv. Appl. Mech.* **27** 83–151

[6] Lemaitre J 1985 A Continuous Damage Mechanics Model for Ductile Fracture *J. Eng. Mater. Technol.* **107** 83–9

[7] Castañeda P P and Zaidman M 1994 Constitutive models for porous materials with evolving microstructure *J. Mech. Phys. Solids* **42** 1459–97

[8] Benzerga A A and Leblond J-B 2010 Ductile Fracture by Void Growth to Coalescence *Advances in Applied Mechanics* vol 44pp 169–305

[9] Tvergaard V and Needleman A 1984 Analysis of the cup-cone fracture in a round tensile bar *Acta Metall.* **32** 157–69

[10] Lubarda V A, Schneider M S, Kalantar D H, Remington B A and Meyers M A 2004 Void growth by dislocation emission *Acta Mater.* **52** 1397–408

[11] Tang Y, Bringa E M and Meyers M A 2012 Ductile tensile failure in metals through initiation and growth of nanosized voids *Acta Mater.* **60** 4856–65

[12] Rudd R E 2009 Void growth in bcc metals simulated with molecular dynamics using the Finnis–Sinclair potential *Philos. Mag.* **89** 3133–3161

[13] Marian J, Knap J and Ortiz M 2005 Nanovoid deformation in aluminum under simple shear *Acta Mater.* **53** 2893–900

[14] Marian J, Knap J and Ortiz M 2004 Nanovoid Cavitation by Dislocation Emission in Aluminum *Phys. Rev. Lett.* **93** 165503-1–4

[15] Benzerga A A, Besson J and Pineau A 2004 Anisotropic ductile fracture: Part I: experiments *Acta Mater.* **52** 4623–38

[16] Maire E, Bordreuil C, Babout L and Boyer J-C 2005 Damage initiation and growth in metals. Comparison between modelling and tomography experiments *J. Mech. Phys. Solids* **53** 2411–34

[17] Ueda T, Helfen L and Morgeneyer T F 2014 In situ laminography study of three-dimensional individual void shape evolution at crack initiation and comparison with Gurson-Tvergaard-Needleman- type simulations *Acta Mater.* **78** 254–70

[18] Lieberman E J, Lebensohn R A, Menasche D B, Bronkhorst C A and Rollett A D 2016 Microstructural effects on damage evolution in shocked copper polycrystals *Acta Mater.* **116** 270–80

[19] Barabash R I, Ice G E, Kumar M, Ilavsky J and Belak J 2009 Polychromatic microdiffraction analysis of defect self-organization in shock deformed single crystals *Int. J. Plast.* **25** 2081–93

[20] Gray G T 1993 Influence of Shock-Wave Deformation on the Structure/Property Behavior of Materials *High-Pressure Shock Compression of Solids* (New York, NY: Springer New York) pp 187–215

[21] Stevens A L, Davison L and Warren W E 1972 Spall fracture in aluminum monocrystals: a dislocation-dynamics approach *J. Appl. Phys.* **43** 4922

[22] Perez-Bergquist A G, Cerreta E K, Trujillo C P, Cao F and Gray G T 2011 Orientation dependence of void formation and substructure deformation in a spalled copper bicrystal *Scr. Mater.* **65** 1069–72

[23] Crépin J, Bretheau T and Caldemaison D 1996 Cavity growth and rupture of β-treated zirconium: A crystallographic model *Acta Mater.* **44** 4927–35

[24] Maire E, Bouaziz O, Di Michiel M and Verdu C 2008 Initiation and growth of damage in a dual-phase steel observed by X-ray microtomography *Acta Mater.* **56** 4954–64

[25] Bingert J F, Henrie B L and Worthington D L 2007 Three-Dimensional Characterization of Incipiently Spalled Tantalum *Metall. Mater. Trans. A* **38** 1712–21

[26] Qi M L, Yao Y, Bie B X, Ran X X, Ye W, Fan D and Li P 2015 Nucleation and growth of damage in polycrystalline aluminum under dynamic tensile loading *AIP Adv.* **5** 37116

[27] Hong C, Fæster S, Hansen N, Huang X and Barabash R I 2017 Non-spherical voids and lattice reorientation patterning in a shock-loaded Al single crystal, *Acta Mater. in press*

Strengthening and toughening of layered Ti-Al metal composites by controlling local strain contribution

M Huang[1], J S Chen[1], H. Wu[1], G H Fan[1, 2] and L Geng[1]

1School of Materials Science and Engineering, Harbin Institute of Technology, Harbin, 150001, P. R. China
2Center for Analysis and Measurement, Harbin Institute of Technology, Harbin, 150001, P. R. China

E-mail: ghfan@hit.edu.cn (G.H. Fan)

Abstract. Layered Ti-Al metal composites (LMCs) with different thickness ratios of the Ti and Al layers were fabricated by hot-rolling and annealing. To study the effect of layer thickness on the mechanical properties of LMCs from the viewpoint of local strain distribution, the strain evolution of LMCs was investigated via in-situ tensile testing. It is found that the mechanical properties of LMCs are correlated with the degree of strain localization. Suppressing strain localization during plastic deformation is crucial to achieve the goal of both strengthening and toughening in LMCs. Additionally, the layered structure can facilitate the redistribution of strain localization, and the transfer of strain localization can be effectively controlled by changing the thickness ratio of the Ti and Al layers.

1. Introduction

Considering the increasing demands of industrial technology application, such as energy reduction and in-service security, a new challenge for the combined strengthening and toughening of metal matrix composites (MMCs) has been recognized. However, the traditional trade-off between strength and ductility is difficult to overcome. Nevertheless, in recent years, researchers have found that a combination of strength and ductility may be accomplished by modifying the distributions of microstructures and components, such as layered, bi-continuous and gradient structures [1-8]. This research has indicated that one feasible strategy is that a combination of strengthening and toughening achieved by configuration design [9-11].

It is well known that damage of a metal may be attributed to strain localization and the following crack nucleation [12]. Recently, an effect of localized strain on the strengthening and toughening of nano-metals and amorphous alloys has been investigated [13]. Control of the strain capacity by preventing strain localization is thus crucial to enhance plastic deformability. A direct visualization of local strain by digital image correlation (DIC) is useful to understand strengthening and toughening mechanisms in MMCs with different configuration designs. In this paper, we show by following the strain evolution it is possible to understand how the design of the microstructural configuration can be used to achieve a combination of strength and ductility.

In this study, layered Ti-Al metal composites (LMCs) were selected to characterize the correlation between configuration design and strengthening/toughening mechanism via localized strain distribution analysis. The deformation behavior was investigated by DIC during in-situ tensile testing to reveal the effect of local strain distribution on mechanical properties.

2. Experimental methods

2.1 Sample preparation
Commercially pure Ti sheets (TA1) with an original thickness of 100 μm and pure Al sheets (1060) with an original thickness of 100 μm and 50 μm, were selected as raw materials. They were cut into several squares with surface areas of 100 mm width by 100 mm length, and then chemically eroded using 10 vol.% HF solution and 10 wt.% NaOH solution, respectively, in order to remove the surface oxide layer. The as-prepared sheets were alternately stacked, hot-pressed in vacuum under 40 MPa at 500 °C for 1 h, then hot-rolled at 500 °C to a total thickness reduction of 30%, and then finally annealed at 500 °C for 1 h.

2.2 Microstructure characterization and tensile tests
The microstructure of the layered composites was examined by a FEI HELIOS Nanolab600i scanning electron microscope (SEM). The sample surface used for SEM observation was the transverse-normal direction plane (TD-ND), prepared by mechanical grinding and polishing.

The tensile tests were performed on an Instron-1186 Universal Testing Machine at room temperature (RT) at a strain rate of 5×10^{-4} s^{-1}. Flat dog-bone shape tensile specimens with a gauge dimension of 5 mm × 18 mm × 2 mm were prepared by electrical discharge machining and mechanical grinding. A total of three specimens for each condition were used to achieve representative statistics.

2.3 Strain evolution process
Dog-bone shape tensile specimens, 2mm in thickness, 1.5mm in width, and 10mm in gauge length, were prepared and loaded in-situ using Kammrath-Weiss micro-tensile stage mounted in a FEI HELIOS Nanolab600i SEM. The tensile displacement rate was 5 μm/s, corresponding to a strain rate of 5×10^{-4} s^{-1}. High resolution micrographs covering 900 × 800 pixels were captured at different tensile strains. The VIC-2D software was utilized for data post-processing and for the calculation of the local strain distribution.

3. Results

3.1 Microstructure and mechanical properties
Two kinds of layered Ti/Al composites were successfully prepared by hot pressing and hot rolling of alternately stacked Ti and Al foils. In each case, the thicknesses of the Ti foils after rolling was 70 μm, while the Al foils were 70 μm or 35 μm in thickness, respectively, referred to in the following as samples R70 and R35. The microstructures of these two composites are shown in figure 1, indicating also the uniform deformation behavior of the Ti and Al foils during the multi-pass hot rolling (without local necking). Additionally, the high-magnification backscatter electron (BSE) micrographs in figure 1b and d, show a well-bonded interface and that no cracking or debonding was found at the interface between the Ti and Al layers.

Figure 2 shows the engineering stress-strain curves of samples R35 and R70, both exhibiting a superior strength-ductility synergy imparted by layered structure. The improved yield strength and ultimate tensile strength of the R35 sample can be easily understood based on a rule of mixtures strengthening. However, its tensile elongation to fracture is 46%, compared to 38% for the R70 sample. This finding is counterintuitive, and also implies that the thickness of the Al layers can significantly change the mechanical behavior of LMCs. To further investigate this, we applied the DIC method for direct visualization of full-field strain distribution, to determine how the strain evolution process was affected by the Al layer thickness.

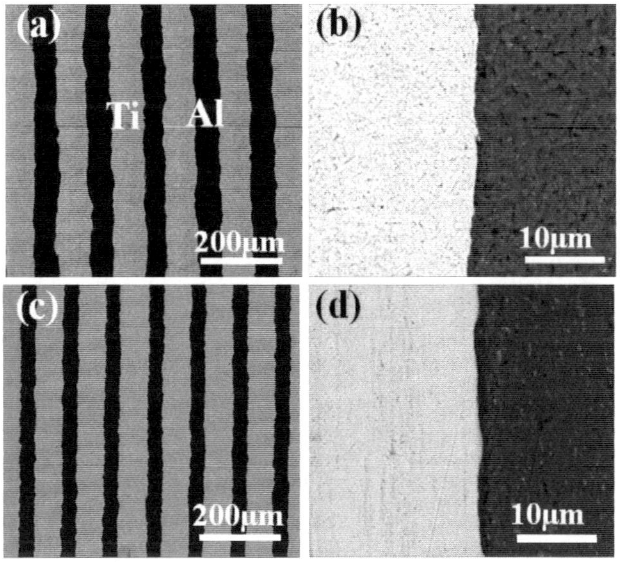

Figure 1. The microstructures of LMCs with the same Ti layer thicknesses of 70 μm and different Al layer thicknesses of 70 μm and 35 μm (R70 and R35). The low magnification image: (a) R70 and (c) R35; The high magnification image: (b) R70 and (d) R35.

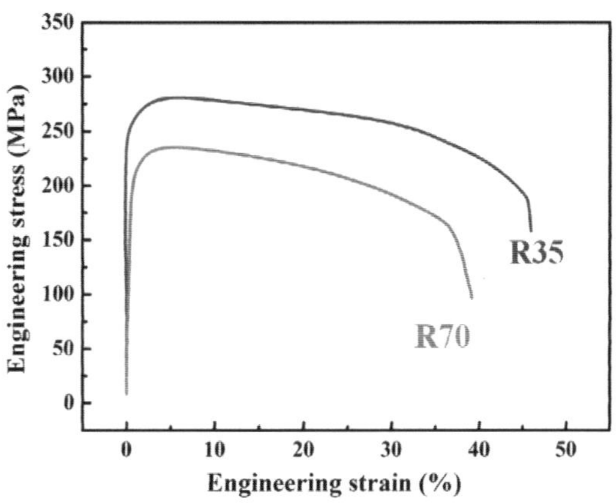

Figure 2. The engineering stress-strain curves of LMCs with different Al layer thicknesses. The loading direction is parallel to TD.

3.2 Strain evolution process

Combining in-situ loading and DIC analysis was used to reveal the dependence of the Al layer thickness on the strain evolution process during tensile deformation. The engineering stress-strain curves of the R35 and R70 samples are shown in figure 3a, where the up-and-down serrations are due to the interrupted loading mode necessary for capturing images for DIC analysis at different tensile macro-strains of 4.0%, 6.0%, 9.0%, and 15.0%.

For layered Ti/Al composites, the soft Al layer should yield first compared to the adjacent hard Ti layer. The non-synchronous deformation behaviors of Ti and Al layers produce the stress/strain partitioning, which may explain the generation of compressive strains within the Al layer or at the interface between the Ti and Al layers at a tensile macro-strain of 4.0% as shown in figure 3b. As the plastic deformation progresses, the initial compressive strains gradually disappear and are replaced by severe strain localization occurring at the Ti layer or at the interface (figure 3c and d). However, the early strain localization of Ti layer is effectively relieved by deformation in the adjacent Al layer via stress-transfer, presenting a strain non-localization trend even as the tensile macro-strain is increased up to 15% (figure 3e). In other words, in the R35 sample localized strains present during the plastic deformation can be efficiently transferred to other deformation regions, thus allowing the R35 sample to plastically deform over wider deformation region. For comparison, we examined the ε_{xx} local strain evolution of the R70 sample under otherwise identical conditions, and found that the premature strain localization occurred and quickly developed in the R70 sample even at a very low strain level of 6.0% (figure 3g). Careful examination of strain evolution process of the R35 and R70 samples indicates that the thin Al layer obviously alleviated the degree of strain localization of the Ti layer at the same strain levels.

Figure 4 shows the quantitative measurements of ε_{xx} local strain with increasing tensile macro-strain. Two scalar indicators were introduced to evaluate the degree of strain localization during the tensile deformation. The first is the average ε_{xx} local strain of the Ti layer (figure 4a), and the other is the average ε_{xx} local strain at the region of interest (ROI) normalized by the macroscopic engineering strain (figure 4b). It is found that the R70 sample sustains a higher average/macroscopic strain ratio compared to the R35 sample (figure 4b). Therefore, the plastic deformation of the R70 sample may take place at a very limited region, and the decrease in the Al layer thickness significantly alleviates the degree of strain localization. Additionally, an inflection point appeared at a macroscopic strain of ~15.0%, corresponding to the start of severe strain localization (figure 3i). It is thus concluded that strain localization leads to deformation instability of layered composites, which is experimentally supported by the sharply increased magnitude of the average/macroscopic strain ratio in figure 4b.

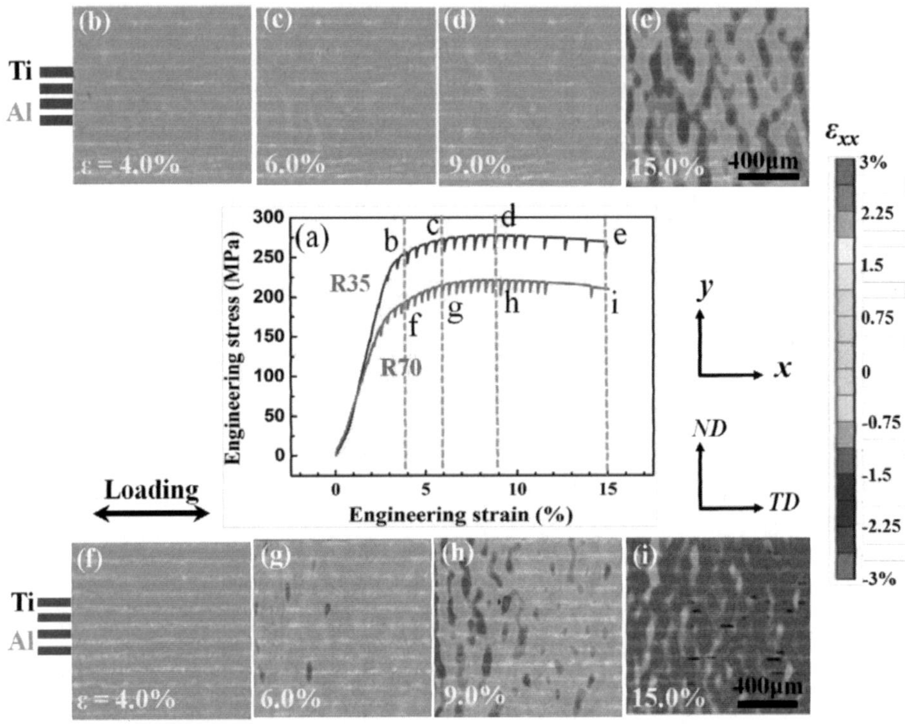

Figure 3. The evolution of ε_{xx} local strain for LMCs with different Al layer thicknesses along the loading direction at different tensile strains. (a) Engineering stress-strain curves; (b-i) Strain evolution mapping of (b-e) R35 and (f-i) R70 samples. The tensile macro-strains at which the SEM images were captured (used for strain analysis) are labeled by yellow font in (b-i).

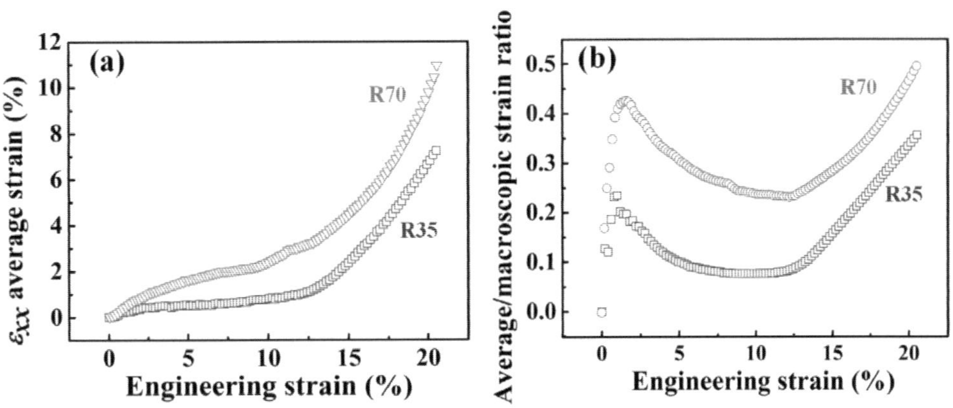

Figure 4. Quantitative statistics of ε_{xx} local strains with increasing tensile strains. Two scalar indicators were introduced to represent the degree of strain localization during the deformation: (a) the average ε_{xx} local strains of component Ti layer; (b) the ratio of average ε_{xx} local strains to macroscopically engineering strains.

4. Discussion

The confined strain evolution process of the layered Ti/Al composites, as shown in figure 3 and 4, is intimately correlated with their microstructure. After hot rolling, the hard Ti layer has a strong <0001>//ND basal texture compared to the typical recrystallized condition of the Al layer, as shown in figure 5. Our recent work has demonstrated that the plastic deformation of LMCs is actually dominated by the hard Ti layer [14], and easy prismatic slip is expected under such a crystallographic characteristic of a strong basal texture. During tensile deformation, strain localization of the Ti layer may occur when prismatic slip in the Ti can no longer accommodate the imposed strain. Generally, the onset of strain localization should induce the start of necking. However, the latter is delayed by the layered structure via the strain-transfer behavior, as shown in figure 3. Our work also indicates that the degree of strain-transfer can be controlled by adjusting the thickness of the soft Al layer.

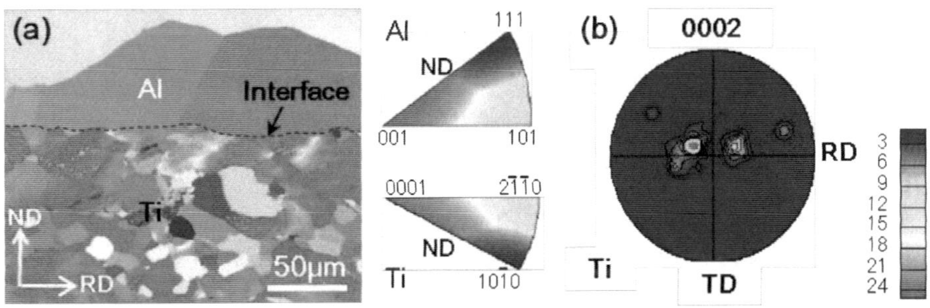

Figure 5. Crystal orientation maps of LMC obtained by electron backscatter diffraction (EBSD): (a) inverse pole figure for normal direction (ND); (b) (0001) pole figure for Ti in LMC. The Image was reproduced with permission from Ref. [14].

The two-dimensional DIC mapping provides insights in understanding the relationship between microstructure design and in-plane strain evolution during tensile deformation. The different layered structures (therefore thickness ratio) change the deformation behavior of the LMCs. For example, localized strains within the Ti layer can be effectively transferred to a wider deformation region as a consequence of the adjacent Al layers. In addition, the mechanical properties and textures of layered components should be taken into consideration for the strain-transfer behavior [14, 15]. Firstly, LMCs with different Al layer thicknesses have similar crystallographic textures due to the similar processing parameters during the sample preparation. Therefore the influence of crystal orientation is negligible. Secondly, the difference in the elastic modulus and yield strength between Ti and Al results in a deformation incompatibility of the layered composites and the formation of compressive stresses in the Al layer. Finally, the plastic deformation of LMCs is essentially a stress-mediated strain evolution process, and the layered structure has a significant influence on the redistribution of local strain [4]. The strain localization of the Ti layer will be aggravated if the accumulated stresses cannot be relaxed by local plastic deformation. To some extent, the thickness reduction of the Al layer facilitates an effective transfer of any accumulated stresses in the Ti layer to adjacent Ti layers across the thin Al layer [4, 16]. As a result, the strain localization of Ti can be effectively relieved by a thinner Al layer.

Similarly, the mechanical properties of MMCs are influenced by the spatial distribution of the component phases, the grain size, the chemical constituents and the morphology of the reinforcements [15]. In the past few years, integrating these factors has been constantly investigated for performance optimization. For example, the layered configuration design of component phase significantly enhances the capability of strain transfer [14]. Based on the layered configuration design of MMCs, we conclude that (i) suppressing strain localization during the plastic deformation is crucial to achieve the strategy of strengthening and toughening for MMCs, and (ii) the mechanical properties of MMCs depend on the degree of strain localization. Good deformation stability can be obtained if strain localization or stress concentration in one area can be effectively transferred to another area. In MMCs the transfer of local strain can be achieved by microstructural configuration design, especially layered configuration design.

5. Conclusions

To study the effect of configuration design on the mechanical properties of MMCs from the viewpoint of local strain distribution, LMCs have been designed and prepared using pure Ti and Al sheets. From in-situ tensile testing combined with DIC analysis, it is found that the mechanical properties of the LMCs are essentially governed by the degree of strain localization. Suppressing strain localization during the plastic deformation is crucial to achieve the goal of combined strengthening and toughening in LMCs. Moreover, a transfer of strain localization can be achieved by the use of a layered configuration design, where the strain transfer is found to be more effective with decreasing Al layer thickness. This finding may provide new insights in guiding the design of high-performance MMCs.

Acknowledgements

This work was financially supported by the National Natural Science Foundation of China (Grant No. 51571070, 51571071) and Key Laboratory of Micro-systems and Micro-structures Manufacturing of Ministry of Education, Harbin Institute of Technology (Grant No. 2015KM002).

References

[1] Wu X L, Yang M X, Yuan F P, Wu G L, Wei Y J, Huang X X, Zhu Y T 2015 *Proc. Natl. Acad. Sci. USA* **112** 14501

[2] Fang T H, Li W L, Tao N R, Lu K 2011 *Science* **331** 1587

[3] Deville S, Saiz E, Nalla R K, Tomsia A P 2006 *Science* **311** 515

[4] Wu H, Fan G H, Huang M, Geng L, Cui X P, Xie H L 2017 *Int. J. Plast.* **89** 96

[5] Huang L J, Wang S, Dong L J, Zhang Y Z, Pan F, Geng L, Peng H X 2012 *Mater. Sci. Eng.* A **545** 187

[6] Peng H X, Fan Z, Evans J R G 2001 *Mater. Sci. Eng.* A **303** 37

[7] Launey M E, Munch E, Alsem D H, Saiz E, Tomsia A P, Ritchie R O 2010 *J. R. Soc. Interface.* **7** 741

[8] Rodriguez-castro R, Wetherhold R C, Kelestemur M H 2002 *Mater. Sci. Eng.* A **323** 445

[9] Liu G, Zhang G J, Jiang F, Ding X D, Sun Y J, Sun J, Ma E 2013 *Nat. Mater.* **12** 250

[10] Meyers M A, Mishra A, Benson D J 2006 *Prog. Mater Sci.* **51** 427

[11] Huang L J, Geng L, Peng H X 2015 *Prog. Mater Sci.* **71** 93

[12] Antolovich S D, Armstrong R W 2014 *Prog. Mater Sci.* **59** 1

[13] Kou H N, Lu J, Li Y 2014 *Adv. Mater.* **26** 5518

[14] Huang M, Fan G H, Geng L, Cao G J, Du Y, Wu H, Kang H J, Wang T M, Du G H, Xie H L 2016 *Sci. Rep.* **6** 38461

[15] Mortensen A, Llorca. Annu J 2010 *Rev. Mater. Res.* **40** 243

[16] Li Y P, Zhu X F, Tan J, Wu B, Zhang G P 2009 *Philos. Mag. Lett.* **89** 66

Steady state and a general scale law of deformation

Yan Huang

BCAST, Brunel University London, Uxbridge, UB8 3PH UK

Email: yan.huang@brunel.ac.uk, Tel: +44 1895 266976

Abstract. Steady state deformation has been characterized based on the experimental results for dilute single-phase aluminium alloys. It was found that although characteristic properties such as flow stress and grain size remained constant with time, a continuous loss of grain boundaries occurred as an essential feature at steady state. A physical model, which takes into account the activity of grain boundary dislocations, was developed to describe the kinetics of steady state deformation. According to this model, the steady state as a function of strain rate and temperature defines the limit of the conventional grain size and strength relationship, i.e., the Hall-Petch relation holds when the grain size is larger than that at the steady state, and an inverse Hall-Petch relation takes over if grain size is smaller than the steady state value. The transition between the two relationships relating grain size and strength is a phenomenon that depends on deformation conditions, rather than an intrinsic property as generally perceived. A general scale law of deformation is established accordingly.

1. Introduction

From early studies on face-centred cubic (fcc) crystalline metals and alloys it is known that stress–strain curves exhibit well-defined stages and microstructure dependencies [1-3]. Extensive investigations have been conducted since that time to verify, characterize and model the mechanical response and microstructural evolution in different stages of deformation. One of the widely addressed topics is work hardening behaviour, from the initial hardening stage (stage I) in association with easy glide by single slip and/or inhomogeneous slip due to Piobert-Lüders band propagation, to athermal linear hardening (stage II) and dynamic recovery related hardening (stage III and IV) [4-6]. Plasticity of crystalline materials is traditionally separated into two regimes, namely cold deformation at one end and hot deformation, including creep, at the other. Although the hardening in stage II does not usually take place at elevated temperatures, for both conditions the resistance to plastic deformation, i.e., flow stress (σ) and strain rate ($\dot{\varepsilon}$) at a given temperature (T), reaches a steady state. In some metals and alloys, such as copper, nickel, and austenitic steels, stress oscillation may occur in the early stages of deformation due to recrystallization, but a steady state eventually follows and prevails. Irrespective of whether dynamic recrystallization occurs, the strain rate and temperature dependence of the steady-state stress is often described by an empirical equation [7, 8]:

$$\dot{\varepsilon} = A(\sinh \alpha\sigma)^n \exp\left(-\frac{Q}{RT}\right), \quad (1)$$

where A, α and n are temperature-independent constants, and Q is the activation energy. At low homologous T and high σ, equation (1) reduces to a power relation,

$$\dot{\varepsilon} = A'\sigma^n \exp(-\frac{Q}{RT}) \tag{2}$$

Similarly at high stresses it equation (1) reduces to an exponential relation

$$\dot{\varepsilon} = A'' exp(\beta\sigma) \exp\left(-\frac{Q}{RT}\right) \tag{3}$$

The constants α and n are related by β = αn. The transition between the high and low T regimes is called power-law breakdown, which is generally explained by the transition in the governing kinetics of dislocation storage and annihilation from climb at high T to glide-controlled spontaneous behaviour at low T. It is worth noting that the steady state is usually difficult to achieve under conventional tensile testing conditions since sample typically fail by cracking or necking before reaching this stage, but it can generally be achieved under hot rolling or torsion conditions. Creep is a case where a steady state can be obtained, particularly at elevated temperatures. In the last two decades, the exploitation of severe plastic deformation technologies has made it clear that steady state deformation occurs exclusively even at the liquid nitrogen temperature [9, 10].

It is often observed that the average dislocation density and subgrain size are constant with straining at the steady state [11]. There have been a number of models describing the evolution of microstructures at the steady state. For instance, the model proposed by Mecking and Kocks [12] and the one developed by Gottstein and Argon [13] predict the average dislocation density and the flow stress at the steady state of both constant strain rate test and creep test. A three parameter approach proposed by Nes [5] and Marthinsen and Nes [14] describes the steady state in terms of the subgrain size, the dislocation density inside the subgrains, and the subgrain boundary dislocation density or the subgrain boundary misorientation. There are also models that take into account dynamic recrystallization in the deformation processes, which involve stress oscillation due dynamic recrystallization [15-17]. An essential assumption in all these approaches is that the dislocation density at steady state is constant.

Although the steady state microstructure only depends on the instantaneous values of strain rate and temperature, not on their history, the dynamic response of microstructure to strain on the pathway to the steady state is strongly dependent on the characteristic grain size of the material. For coarse-grained materials, the yield stress is inversely proportional to the square root of grain size as described by the famous Hall-Petch (H-P) relationship [18,19]. The relationship holds for a wide range of polycrystalline materials with grain sizes ranging from the millimetre scale down to the sub-micrometer scale, and also holds for the dependence of flow stress on the deformed grain/subgrain structure during deformation [20]. However, an inverse H-P relationship has alsobeen reported for nano-grained materials [21, 22], where the strength of material is found to decrease with a decreasing grain size. Evidence for this switch from a normal H-P relationship to an inverse H-P relationship has been frequently observed in nanograined materials and it is widely accepted that the inverse H-P relationship is related to the intrinsic behaviour of crystalline materials as the grain size approaches a true nanoscale [23-26]. The explanation of the two regimes for the strength and grain size relationship has been mainly associated with the "high energy" character and high volume fraction of grain boundaries at nanoscale. The switchover from one regime to the other is often arbitrarily related to an absolute grain size rather than rationalized through mechanism analysis.

Steady state deformation presents a dynamic but definite correlation between flow stress and microstructure. This is probably the main reason for its attraction to materials scientists as it is amenable to detailed microstructural analysis and mathematical modelling. However, the nature for the establishment of a steady state is not properly understood, the characteristic microstructural features are not properly described and its fundamental importance is ignored. This paper aims to demonstrate that the essential and widely accepted assumption that the steady state is determined by a balance between dislocation accumulation due to work hardening and elimination due to dynamic recovery, based on which most analyses and modelling of steady state deformation are developed, is incorrect. A new model of steady state deformation will be put forward in consideration of the operation of grain boundary dislocations, which further leads to the establishment of a scale law of deformation.

2. Grain boundary loss at steady state

Steady state deformation occurs as a dynamical balance between the applied stress and microstructure is approached, where the microstructure is characterized by a constant grain/subgrain size and constant dislocation density. However, the total grain boundary area is not constant and, in fact, there is a constant loss of grain boundaries at steady state. This phenomenon has been long observed and recognized [27] but its paramount importance is overlooked and ignored in understanding the active deformation mechanisms. In plastic deformation, there is always a loss of dislocations from the interior to the surface, which makes the physical deformation possible. At steady state, the widely accepted concept that the generation of dislocation due to deformation is balanced by dynamic recovery leads to the miscalculation of a significant amount of dislocations. The amount of discrepancy is the total amount of dislocations moved out to the surface.

Geometrically, grains are required to change shape in proportion to the external shape change during plastic deformation. In the case of plane strain compression (PSC), for example, the geometrically required grain size in the compression direction (D_G) can be related to its initial value of D_0 and the accumulative true strain (ε_t) by [28]

$$D_G = D_0 exp(-\varepsilon_t) \tag{4}$$

The actual grain size tends to follow the above relation in the early stages of deformation, particularly at cryogenic temperatures, but rapidly deviates from it at high strains. Steady state deformation is an extreme condition where grain size remains constant and independent of strain. This means that there must be a loss of grain boundaries in order to compensate the grain size reduction due to geometrical compression described by Equation 4. The fraction of the lost HAB area (f_L) relative to the initial grain boundary area is determined by the total strain (ε_t),

$$f_l = 1 - \left(\frac{D_0}{D}\right) \exp(-\varepsilon_t), \tag{5}$$

where D is the instantaneous grain size. The grain boundary area loss at steady state is

$$f_L = 1 - \exp(-\varepsilon_s) \tag{6}$$

where ε_s is the accumulative strain at the steady state. It can be extrapolated from equation (6) that at very high strains, f_L is close to 1, i.e. most grain boundaries are eliminated from the material. This can be understood by envisaging a case where the workpiece thickness is equal to the grain size and there are no grain boundaries between the surfaces, except those aligned parallel to the compression direction.

The constant dislocation density assumption apparently can neither explain the constant grain structure nor the loss of grain boundaries. Thermally activated competitive coarsening between grains may do and needs to be examined. In order to remain at a constant grain size, grain boundaries must migrate during deformation at a rate equal to that of their compression, V_C, which can be related to strain rate, $\dot{\varepsilon}$, by

$$V_C = dD/dt = D\dot{\varepsilon} \tag{7}$$

The analysis of data obtained from an Al-0.1Mg alloy showed that the velocity was significantly too high to be explained by any possible mechanisms of grain boundary migration [29, 30]. In a previous study, an Al-0.1Mg alloy was firstly deformed to a true strain of 10 by equal channel angular extrusion at room temperature and then deformed by PSC to strains where a steady state was well established at temperatures from 77K to 473K and at a strain rate of 10^{-2} s^{-1}. The steady state grain size was measured by EBSD and the "boundary migration rate" was estimated according to equation (7), which is plotted against inverse temperature in figure 1. In the range of 77-273K, the apparent activation energy, determined from the slope of this plot, was as low as ~ 1 kJ mol^{-1}. Above 373K the slope gives an activation energy of ~ 60 kJ mol^{-1}. Velocities predicted by all possible boundary migration mechanisms, including curvature driven, stress and vacancy assisted migration and solute breakaway mechanism, are also plotted in the figure but none of these fit the measured values.

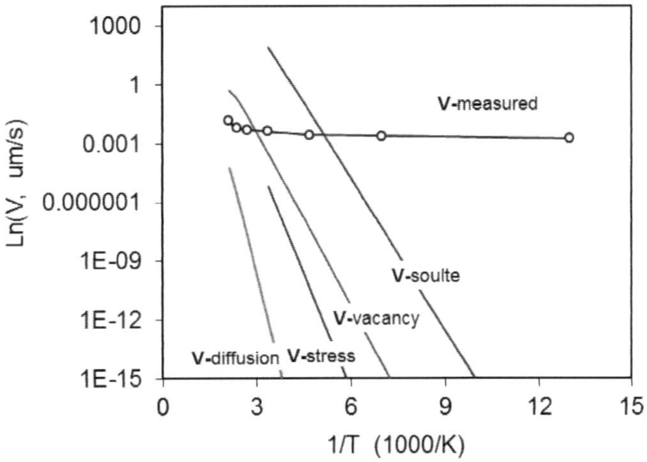

Figure 1. Estimated high angle boundary migration rate in the compression direction as a function of reciprocal temperature during PSC of an Al-0.1Mg alloy.

The boundary migration rate (V) is usually assumed to be proportional to the driving pressure (P), i.e., V = MP, where M is the boundary mobility. For conventional diffusion controlled boundary migration driven by boundary curvature [11],

$$V = \frac{\gamma}{4r} M_0 \exp\left(-\frac{Q}{RT}\right), \quad (8)$$

where r is the radius of curvature, γ the boundary energy, M_o the mobility coefficient and Q the activation energy. Using parameters determined from static grain growth experiments at 423 -773K for the same alloy (M_o = 0.1 $m^4J^{-1}s^{-1}$, Q = 130 $kJmol^{-1}$) [31, 32], taking γ = 0.32 J/m^2 and r = D/2, velocities calculated from equation (8) are lower than the experimentally determined rates (V_C) by a factor of 10^{10} at 298K, and by progressively larger amounts at lower temperatures, being ~ 10^{74} times slower at 77K. The effect of elastic energy differences between neighbour grains has been discounted in the literature due to the low anisotropy of Al crystals. Estimates using variations in the Taylor factor result in a driving pressure 10-30 times below that due to boundary tension. Equally, dislocation density is low in ultra-fine grained materials and does not provide a significant additional pressure for boundary migration. In any event, no realistic increase in the average driving pressure can account for the many orders of magnitude difference in boundary migration rates between observations and predictions at low temperatures. It has been claimed that high transient vacancy concentrations can be generated during severe deformation [33, 34] and enhance boundary migration. However, an abundant flux of vacancies would only reduce the vacancy formation part of the activation energy [35] and cannot explain the behaviour below room temperature. An additional possibility is that highly disordered grain boundary structures under dynamic conditions could have increased boundary mobility [36]. However, much higher apparent activation energies should be observed if such effects took place substantially.

Overall, none of the above reviewed mechanisms can explain the measured dynamic recovery rate, particularly at low temperatures. The conventional thermally activated boundary migration model therefore seems inappropriate under these conditions.

3. Grain boundary dislocation in operation

The operation of grain boundary dislocations is considered to be a feasible mechanism that can explain both the deformation and microstructural evolution at steady state. Dislocation emission from GBs is a long recognized and an experimentally verified phenomenon [37-41]. GB dislocation emission can lead to the decrease in boundary misorientation and even the elimination of the boundary. The merging of two adjacent grains due to orientation convergence can be an example of this, which gives rise to the development of textures, in addition to well addressed dislocation density dependency of the misorientation of low angle boundaries [11].

Figure 2 shows microstructural evolution during *in-situ* tensile deformation at 315°C and 2.5×10^{-4} s^{-1} in an Al-0.05Si alloy that was previously deformed in compression in a channel die to a reduction of 70% at 300°C. The electron backscatter images in figure 2 were taken from the same area during deformation between strains of 0.24 to 0.28, where steady state deformation was established. In this small strain interval, grain coalescence and subdivision is observed together with other features such as grain rotation. It can be seen from the figure that grain boundaries A and B disappeared *in-situ* without the involvement of migration and grain C was subdivided into several grains. The two mechanisms operated simultaneously and the steady state was maintained. The range of boundary misorientation involved was not clear but the observation confirms that strain-induced misorientation reduction and boundary elimination does occur.

Figure 2. Electron backscattered images of grain structures taken from the same area during steady state deformation of an Al-0.05Si alloy at 315°C and $2.5\times10^{-4}s^{-1}$ to strains of: a) 0.24, b) 0.26 and c) 0.28.

4. A new model of steady state deformation

It is the author's hypothesis that GB dislocation emission takes place at all stages of deformation including initial yielding. Before approaching steady state, dislocations emitted from grain boundary sources interact with dislocations inside grains and other obstacles. Their further glide and multiplication is controlled by the Frank-Read source mechanism [42]. The flow stress is primarily determined by the dislocation mean free path L, which is the distance travelled by a dislocation segment of a certain length before it is stored by interaction with the microstructure [43],

$$\tau = \tau_0 + \alpha Gb/L \tag{9}$$

where the constant $\alpha = 1$ for edge dislocations and 0.5 for screw dislocations, G is the shear modulus and b the Burger's vector. The mean free path, L, is related to average dislocation density (ρ) by $\rho = L^{-2}$. Apparently, if L and the square root grain size are linearly correlated, the observed H-P relationship can be explained more straightforwardly.

At the steady state, the dislocation mean free path and average grain size are convergent and the former reaches its minimum. For grains that are smaller than the average size, dislocation multiplication from a F-R source is not possible because the applied stress cannot overcome the tension of a dislocation segment that is smaller than the minimum dislocation mean free path. Dislocations emitted from grain boundary will be free to glide. The boundary dislocations are ultimately pushed into the surface of the material and consequently deformation takes place. As discussed above, a grain boundary reduces it misorientation and vanishes as a result of dislocation emission. The elimination of grain boundary due to dislocation emission will result in grain size increase, compensating the effect of external shape change. On the other hand, grains that are larger than the average steady state size can still accommodate F-R sources and deformation is dominated by dislocation glide and multiplication via the bowing out mechanism from the F-R sources. Their sizes decrease following equation (3) approximately if no new boundaries are generated within them. It is thus the balance between the operation of intragranular and GB dislocation sources that gives rise to a constant grain structure at the steady state. A new model of steady state is therefore proposed with the features described below.

- The steady state is the ultimate destination of flow stress and microstructure.
- The average grain size at the steady state (D_s) is determined by the minimum dislocation mean free path for given strain rate and temperature.
- There is always a distribution in grain size. For grains $D > D_s$, F-R sources operate while for grains $D < D_s$, GB dislocation sources dominate.
- The operation of F-R sources results in the decrease in grain size, while the operation of GB dislocations leads to the loss of GBs and an increase in grain size.
- F-R and GB sources are balanced, giving rise to a constant flow stress and grain size.

5. A new model of GB dislocation glide

It is proposed that a segment of dislocation emitted from grain boundary glides by the migration under the applied stress of its terminating points at the grain boundary or grain boundaries, from which it originates, i.e. its GB ends,. This is a new mechanism for dislocation movement, only suitable for GB dislocation sources as there is no involvement of dislocation multiplication in the process. Figure 3 shows the essential elements involved in this dislocation glide process, including a segment of dislocation line L ending at grain boundary points A and B, and intersecting with the grain boundary at an angle (β), the half opening angle relative to the dislocation glide direction (ϕ), and the curvature

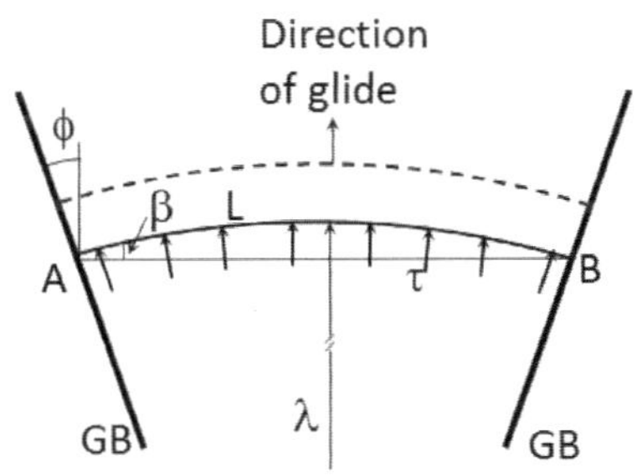

Figure 3. Geometrical model of dislocation glide by the migration of its GB ends.

of the dislocation line (λ). The glide of a dislocation in the configuration depicted in figure 3 will result in the extension of the dislocation line. Since under a given applied stress the curvature of the dislocation line will remain constant, the extension of the dislocation will be realized only by transforming the GB dislocation into the grain. If the angle ϕ is negative, alternatively, which is realistically possible, the dislocation line will retract as the dislocation sinks into grain boundary. . The dislocation, including its GB ends, should have the same Burgers vector with the same combination of edge and screw components along its line. With the help of the principle of virtual work, the stress required to operate the dislocation segment by the migration of its GB ends can be derived as a function of geometrical and material parameters and a simplified relation is given as

$$\tau = \tau_r + \frac{2w}{L}\tau_r^{gb} - \frac{\alpha Gb}{\lambda}\frac{2c-L/\lambda}{L}, \tag{10}$$

where τ_r is the stress required to overcome lattice, solute and other distribution resistances inside the grain, w is the width of grain boundary, τ_r^{gb} is the stress required to overcome resistance from the GB ends of the dislocation, and c = 1+sinβ-ϕ is a geometrical factor. The length of dislocation segment L is of the same scale as the grain size and its curvature λ is always larger than the steady state grain size. If 2c-L/λ is positive, which is safely met when ϕ is less than 0.5, the third term on the right hand side of equation (10) will be positive itself and it increases with decreasing L or grain size. Thus equation (10) predicts a proportional relationship between flow stress and grain size, i.e. that the stress increases with increasing grain size.

The model presented here demonstrates the basic concept of GB dislocation operation as a key mechanism of deformation at steady state, or in the pathway to the steady state, when the grain size is smaller than the steady state size. There are, however, several fundamental issues yet to be addressed to complete the full picture of this process. These include the activation and initiation of GB dislocations, and the relationship between GB dislocation density and misorientation. The annihilation of GB dislocations must occur, but dislocation emission is considered to contribute to deformation. There is also an issue whether all GBs are consumable by dislocation emission/annihilation as not all of them are "dislocation boundaries". Indeed, most GBs cannot be simply constructed by the configuration of dislocations as basic elements. However, the total loss of GB area in the Al-0.1Mg alloy deformed at room temperature by PSC to a true strain of 2.8, as part of the investigation on the same alloy as described above, was estimated by equation (5) to be about 90%. It is thus not unreasonable to assume that all GBs are consumable by dislocation operation.

6. A universal scale law of deformation

Equation (10) is considered to be a special form of an inverse Hall-Petch relationship, with respect to the dependence of material strength on grain size at its lower end. It has been widely claimed that an inverse H-P relationship only occurs when the grain size is refined to a real nanoscale, for example 10 nm. Logically, it is unconvincing to describe a behavioural change as a function of scale using the scale as a criterion. Actually, a strength-grain size dependence similar to the inverse H-P relation has been observed in creep and superplastic flow where the grain size covers a wide range of scales from nanometres to millimetres. Investigations on this topic have not yet produced convincing evidence to demonstrate substantial differences for the proportional strength and grain size relationships observed in monocrystalline materials and most other situations such as creep and superplastic deformation.

It is generally accepted and also the basis of the present discussion that dynamic responses of microstructure to plastic deformation are strongly dependent on grain size. Grain subdivision takes place during deformation of coarse grained materials, whereas equation (10) suggests that, starting from the other end with a grain structure smaller than that of the steady state, grain structure coarsening occurs during deformation. Equation (10) has a universal nature as it suggests that the switchover point, which is at the steady state, is determined by the deformation conditions, mainly strain rate and temperature but actual grain size, instead of concrete grain size. For a certain starting grain size, hardening occurs if the steady state size defined by the applied strain rate and temperature is smaller; otherwise, softening takes place.

Summary

A novel description of steady state deformation is provided in light of grain boundary loss with the support of experimental evidence obtained from the severe plastic deformation of an Al-0.1Mg alloy and SEM in-situ straining of an Al-0.05Si alloy. An analytical model of novel dislocation glide mechanism via the migration of grain boundary ends is proposed, which leads to the establishment of a universal scale law of deformation.

Acknowledgments

Financial support from EPSRC LiME Hub (UK) is gratefully acknowledged.

References

[1] Mitchell T E 1963 *Prog. Appl. Mater. Res.* **6** 117
[2] Nabarro F R N 1986 *Strength of metals and alloys* vol 3 eds H J McQueen et al (Pergamon Press) p 1667
[3] Gil S J 1993 *Plastic deformation and fracture of materials* ed H Mughrabi (VCH) p 19
[4] Zehetbauer M and Seumer V 1993 *Acta Metall. Mater.* **41** 577
[5] Nes E 1997 *Prog. Mater. Sci.* **41** 129
[6] Kocks U F and Mecking H 2003 *Prog. Mater. Sci.* **48** 171
[7] Kassner M and Perez-Prado M T 2000 *Progr. Mater. Sci.* **45** 1
[8] Frost H J and Ashby M F 1982 *Deformation-Mechanisms Maps,* (Oxford: Pergamon Press)

[9] Borner I and Eckert J 1997 *Mater. Sci. Eng.* **A226–228** 541
[10] Zebarjad S M and Sajjadi S A 2006 *Mater. Design* **27** 684
[11] Humphreys F J and Hatherly M 2004 *Recrystallization and related annealing phenomena* (Oxford: Elsevier)
[12] Mecking H and Kocks U F 1981 *Acta Metall.* **29** 1865
[13] Gottstein G and Argon A S 1987 *Acta Metall.* **35** 1261
[14] Marthinsen K and Nes E 2001 *Mater. Sci. Technol.* **17** 376
[15] Sandstro¨m R and Lagneborg R. 1975 *Acta Metall* **23** 387
[16] Roucoules C, Pietrzyk M and Hodgson P D 2003 *Mater. Sci. Eng.* **A339** 1
[17] Huang M, Rivera-D´az-del-Castillo P E J, Bouaziz O and Van der Zwaag S 2009 *Acta Mater.* **57** 3431
[18] Hall E O 1951 *Proc. Phys. Soc. Lond.* **B64** 747
[19] Petch N J 1953 *J. Iron Steel Inst.* **174** 25
[20] Hansen N 2005 *Mater. Sci. Eng.* **A409** 39
[21] Chokshi A H, Rosen A, Karch J and Gleiter H 1989 *Scr. Metall.* **23** 1679
[22] Nieh T G and Wadsworth J 1991 *Scrip. Metall. Mater.* **25** 955
[23] Gleiter H 2000 *Acta Mater.* **48** 1
[24] Meyers M A, Mishra A and Benson D J 2006 *Prog. Mater. Sci.* **51** 427
[25] Zhu Y and Langdon T G 2005 *Mater. Sci. Eng.* **A409** 234
[26] Hahn E N and Meyers M A 2015 *Mater. Sci. Eng.* **A646** 101
[27] Hughes D A and Hansen N 2000 *Acta Mater.* **48** 2985
[28] Jazaeri H and Humphreys F J 2004 *Acta Mater.* **52** 3239
[29] Huang Y and Prangnell P B 2008 *Acta Mater.* **56** 1619
[30] Huang Y 2013 *J. Mater. Sci.* **48** 4484
[31] Jazaeri H , Humphreys FJ and Bate SP 2006 *Mater Sci Forum* **519-521** 153
[32] Huang Y and Humphreys F J 2001 *Proc. 1st Int. Conf. on Recrystallization and Grain Growth* eds G Gottstein and D A Molodov (Springer) p 409
[33] Zhang K, Weertman J R and Eastman JA 2005 *App. Phys. Letts.* **87** 061921
[34] Zehetbauer M J, Steinter G, Schafler E, Korznikov A and Korznikova E 2006 *Mater. Sci. Forum* **503-504** 57
[35] Balarin M 1975 *Phys. State Solid* **A31** 111
[36] Hasnaoui A, Van Swygenhoven H and Derlet P M 2002 *Acta Mater.* **50** 3927
[37] Murr L L and Hecker S S 1979 *Scrip Metall.* **3** 167
[38] Shen Z, Wagoner R H and Clark W A T 1988 *Acta Metall.* **36** 3231
[39] Gemperle A, Gemperlová J and Zárubová N 2004 *Mater. Sci. Eng.* **A387-389** 46
[40] Sangid M D, Ezaz T, Sehitoglu H and Robertson I M 2011 *Acta Mater.* **59** 283
[41] Ovid'ko L A, Sheinerman A G and Valiev R Z 2014 *Scrip. Mater.* **76** 45
[42] Frank F C and Read Jr W T 1950 *Phys. Rev.* **79** 722
[43] Devincre B, Hoc T and Kubin L 2008 *Science* **320** 1745

IOP Conf. Series: Materials Science and Engineering **219** (2017) 012030 doi:10.1088/1757-899X/219/1/012030

Interfaces between a fibre and its matrix

H Lilholt and B F Sørensen

Section of Composites and Materials Mechanics, Department of Wind Energy, Technical University of Denmark, Risø Campus, Roskilde, Denmark

E-mail: hali@dtu.dk

Abstract. The interface between a fibre and its matrix represents an important element in the characterization and exploitation of composite materials. Both theoretical models and analyses of experimental data have been presented in the literature since modern composite were developed and many experiments have been performed. A large volume of results for a wide range of composite systems exists, but rather little comparison and potential consistency have been reached for fibres and/or for matrices. Recently a materials mechanics approach has been presented to describe the interface by three parameters, the interfacial energy [J/m2], the interfacial frictional shear stress [MPa] and the mismatch strain [-] between fibre and matrix. The model has been used for the different modes of fibre pull-out and fibre fragmentation. In this paper it is demonstrated that the governing equations for the experimental parameters (applied load, debond length and relative fibre/matrix displacement) are rather similar for these test modes. A simplified analysis allows the direct determination of the three interface parameters from two plots for the experimental data. The complete analysis is demonstrated for steel fibres in polyester matrix. The analysis of existing experimental literature data is demonstrated for steel fibres in epoxy matrix and for tungsten wires in copper matrix. These latter incomplete analyses show that some results can be obtained even if all three experimental parameters are not recorded.

1. Introduction

In fibrous composite materials the fibres and matrix interact to produce the resulting properties of the composites. This interaction takes several forms. The fibres and matrix separately contribute with their "own" properties (often in a fraction-weighted way), and these properties of the constituents are both the stiffness, the stress and the strain (elongation). The fibres and the matrix also interact with each other through the interface between them and contribute additional behavior and properties. Normally composite materials are composed of (one type of) fibres of cylindrical geometry with one long dimension and two small transverse dimensions, which normally form a roughly equiaxed cross section, that is often represented by a circular cross section. The representative model for the composite is thus a straight circular fibre in a block of matrix. The interface parameters are the *interfacial energy* G_{iic} [J/m^2], which is assumed to represent the chemical nature and bonding of the fibre surface to the matrix, the frictional *interfacial shear stress* τ_s [MPa], which is assumed to represent the topography and frictional sliding of the fibre relative to the matrix, and the *mismatch strain* Δe^T [-], which is caused by the thermo-elastic difference between fibre and matrix.

The first analysis of the interface was suggested by Kelly and Tyson [1], who for the system of metal wires (tungsten) in a metal matrix (copper) introduced and used the concept of a constant maximum interfacial shear stress, representing the sliding of the interface or the yielding of the near-by matrix. This concept was in the following years used in several (simple) models and related experiments, (see e.g. conference reports [2] [3]), and applied to many other combinations of fibres (e.g. glass, carbon, aramid) and matrices (e.g. polyester, epoxy, and other polymers). The one experimental method is typically the pull-out of a single fibre from a block of matrix, where the load on the fibre and the displacement of the fibre relative to the matrix are recorded. A plot of fibre load versus displacement typically shows a steeply rising initial part to a maximum load, followed by a load drop and a slowly decreasing load towards the final pulling out of the fibre from the matrix block. The maximum shear stress at the interface is typically calculated as the maximum fibre load divided by the interfacial area (equal to fibre circumference times embedded fibre length). The other experimental method is the single fibre fragmentation test, where a (mini)composite with one (brittle) fibre is loaded in tension along the fibre direction; the very low fibre volume fraction will cause multiple fracture of the fibre into a number of (short) segments, ideally of lengths between the critical fibre length and one half of the critical fibre length. This situation is the result of the concept of a constant interfacial shear stress [1], and can be used to derive the shear stress from the segment lengths and knowledge of the fibre fracture stress.

Recently, a micromechanical model including all three interface parameters has been proposed, and it has been applied to two cases of pull-out tests [4] [5] and to single fibre fragmentation test [6].

2. Micromechanical model for pull-out and single fibre fragmentation

The details of the model, its assumptions, its principles and the derivation of relevant equations have been given in the above references, and shall not be repeated here. The model includes the interface energy G_{iic}, the interface shear stress τ_s, and the mismatch strain for fibre / matrix Δe^T, as the characteristics of the interface in (fibrous) composites. The related experiments aim to record the load P_f on the fibre (pull-out cases) or the load (stress σ_c) of the composite (single fibre fragmentation case), the debond length l_d along the interface, and the relative displacement δ between fibre and matrix.

The concepts implied in the model are the *force balance* for the fibre and matrix, both at the debonding zone and far away from the zone, and including residual stresses, the *potential energy change* [7] during debonding, and the relative displacement between fibre and matrix obtained by *integration of the strain difference* along the debonded length at the interface.

The three test cases are the pull-out test with the matrix end clamped, called PO-1, the pull-out test with support of the matrix at the fibre end, called PO-2, and the single fibre fragmentation test, called SFFT. The comparative schematics for the three cases are shown in figure 1 of [5].

The model, as it has been developed [4] [5] [6], gives equations for the relations between the *experimental parameters*, the load (on fibre or on composite), the debond length, and the relative displacement. The equations include the three *interface parameters*, interface energy, interfacial shear stress and the mismatch strain. The equations also include *materials parameters* for the fibre and the matrix, as well as the geometry of the test specimens. The parameters are listed:

Interface parameters	$Giic$	interface energy, J/m^2
	τ_s	frictional shear stress between fibre and matrix, MPa
	Δe^T	thermo-elastic mismatch strain between fibre and matrix, [-]
Material parameters	E_f	fibre stiffness, GPa
	r	fibre radius, mm
	α_f	fibre thermal expansion coefficient, K^{-1}
	E_m	matrix stiffness, GPa
	α_m	matrix thermal expansion coefficient, K^{-1}
Experimental parameters	P_f	load on the fibre, N
	σ_c	stress on the composite, MPa
	l_d	debond length along the interface, mm
	δ	relative displacement between fibre and matrix, mm
	A_c	cross sectional area of matrix block / composite, mm^2
	ΔT	test temperature minus manufacturing temperature, ^{o}C

In the original development of the model the equations were typically derived and presented as relations for the debond length as a function of relevant parameters, and for the relative displacement as a function of relevant parameters. In order to demonstrate the similarity for the three test cases, pull-out case PO-1 and PO-2, and the single fibre fragmentation case SFFT, the equations will be rewritten, to present (i) load (fibre load or composite stress) as a function of debond length and (ii) displacement/debond length ratio as a function of debond length. The first relation will give the interfacial shear stress, and the second relation will give the interfacial shear stress and the interface energy, and in combination the two relations will give the mismatch strain.

This rewriting of the original equations, the related graphical plots and the analysis of these plots will be illustrated in the following for the three test cases.

3. Common concepts and relations

Some of the concepts and related equations are identical for all three test cases, and will presented here, even if they originally were derived in slightly different form in the individual papers. These concepts are the general *force balance* and *equality of strains* at any cross section of the composite (because the composite has unidirectional and long fibres), and the *thermal stresses and strains,* as well as the *fibre stresses* and the *matrix stresses* in the uncracked composite (ahead of the debond crack tip) and in the debond region, and the *fibre strain* and the *matrix strain* in the debond region.

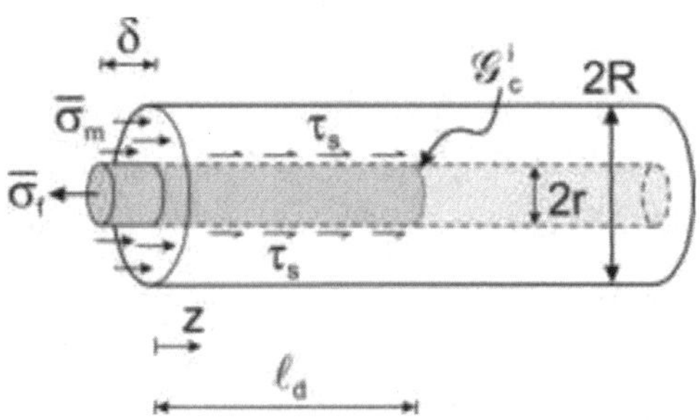

Figure 1. Geometry and loading with relevant parameters (case PO-2)

The *force balance* for the fibre and the matrix is, with sign convention as shown in figure 1:

$$P_c + (-P_f) + (-P_m) = 0$$

$$\sigma_c - V_f \cdot \sigma_f - V_m \cdot \sigma_m = 0$$

where σ is stress and V is volume fraction, with indices c for composite, f for fibre and m for matrix.

The three test cases are compared in figure 4 of [5], and the condition valid for each case is as follows:

PO-1: condition $\sigma_m = 0$, and thus $\sigma_c - V_f \cdot \sigma_f = 0$

PO-2: condition $\sigma_c = 0$, and thus $V_f \cdot \sigma_f + V_m \cdot \sigma_m = 0$

SFFT: condition $\sigma_f = 0$, and thus $\sigma_c - V_m \cdot \sigma_m = 0$

The *equality of strains* for fibre, matrix and composite implies that $e_f = e_m = e_c$.

The *thermo-elastic strains and stresses* are caused by the different stiffness (E) and thermal expansion coefficient (α) of the fibres and the matrix , respectively, and by the fact that the composite (test specimen) is manufactured at (typically) a higher temperature than the test temperature. The thermal strains (which are stress-free) of each component are

$$e_f^T = \alpha_f \cdot \Delta T \qquad \text{and} \qquad e_m^T = \alpha_m \cdot \Delta T$$

The fact that α_f and α_m (normally) are different, and assuming perfect contact at the fibre / matrix interface, means that the resulting strain for the composite is e_c^T, different from both e_f^T and e_m^T. This leads to residual stresses in fibre and matrix:

$$\sigma_f^{res} = E_f \cdot \left(e_c^T - e_f^T\right) \qquad \text{and} \qquad \sigma_m^{res} = E_m \cdot \left(e_c^T - e_m^T\right)$$

For the unloaded composite, these stresses are in balance by the volume fraction weighted relation:

$$V_f \cdot \sigma_f^{res} + V_m \cdot \sigma_m^{res} = 0$$

Inserting the stress expressions and solving for e_c^T, this expression for e_c^T is re-inserted into the stress relations:

$$\sigma_f^{res} = E_f \cdot \left(-\Delta e^T\right) \cdot \frac{V_m E_m}{E_c}$$

$$\sigma_m^{res} = E_m \cdot \left(\Delta e^T\right) \cdot \frac{V_f E_f}{E_c}$$

$$\Delta e^T = e_f^T - e_m^T = \left(\alpha_f - \alpha_m\right) \cdot \Delta T$$

This parameter Δe^T is called the mismatch strain. It is assumed to be unknown and can be derived from the model and its equations.

The *stresses in the composite* are analyzed for two regions, the region ahead of the interface crack tip (the uncracked composite) called the up-stream region, and in the region behind the crack tip (the debond region) called the down-stream region. For the up-stream region the total stress on fibre or matrix is the sum of residual stress and mechanical stress:

$$\sigma^+ = \sigma^{res} + \sigma^{mech}$$

where the test case condition and the related stress balance (see above) as well as the equality of strains is used to derive σ^{mech}. The details for each test case are presented below. For the down-stream region the total stress σ^- on fibre or matrix is a function of the position z (figure 1) and is found from the balance of forces, accounting for the stress transfer between fibre and matrix via the constant interfacial shear stress τ_s. For the fibre the force balance is applied at the fibre surface, for the matrix the force balance is applied at the inner surface of the matrix hole in which the fibre sits. The down-stream stresses depend on z-position and fulfil the force balance at any z-position. For the down-stream region we further need the strains for fibre and for matrix, these strains are found from the relation:

$$e^-(z) = e^{mech}(z) + e^T = \frac{\sigma^-(z)}{E} + e^T$$

where the (stress-free) thermal strain (for fibre or matrix) e^T must be added to the mechanical strain. The details for each test case are presented below.

A relation between the load (on fibre or on composite) and the debond length is established via the concept of *potential energy change* [7] during debonding. This requires the up-stream stresses for fibre and matrix and the down-stream stresses for fibre and matrix in the form:

$$\sigma_f^+ - \sigma_f^-(z) \qquad \text{and} \qquad \sigma_m^+ - \sigma_m^-(z)$$

The derivations are explained in detail in the original papers [4] [5] [6].

A relation between the load (on fibre or on composite), the debond length and the relative displacement between fibre and matrix in the debond region (down-stream) is established by *integration of the strain difference* along the debonded length at the interface from $z = 0$ to $z = l_d$. This requires the down-stream strains for fibre and matrix in the form:

$$e_f^-(z) - e_m^-(z)$$

The derivations are explained in detail in the original papers [4] [5] [6].

4. Pull-out case PO-1

This pull-out case represents the most often used test geometry. The model was presented [4] [8] with a detailed analysis and comments to the assumptions and concepts used to obtain the model. Here the results are presented and rewritten into a form, which is comparable to the other test cases. The load condition is $\sigma_m = 0$, and the force balance is thus $\sigma_c - V_f \cdot \sigma_f = 0$.

The up-stream stresses are

$$\sigma_f^+ = E_f \cdot \left(-\Delta e^T\right) \cdot \frac{V_m E_m}{E_c} + E_f \cdot \frac{\sigma_c}{E_c}$$

$$\sigma_m^+ = E_m \cdot \left(+\Delta e^T\right) \cdot \frac{V_f Ef}{E_c} + E_m \cdot \frac{\sigma_c}{E_c}$$

The down-stream stresses are

$$\sigma_f^-(z) = \sigma_f - 2 \cdot \frac{\tau_s}{r} \cdot z$$

$$\sigma_m^-(z) = \frac{V_f}{V_m} \cdot 2 \cdot \frac{\tau_s}{r} \cdot z$$

The down-stream strains are

$$e_f^-(z) = \frac{\sigma_f}{E_f} - 2\cdot\frac{\tau_s}{E_f\cdot r}\cdot z + e_f^T$$

$$e_m^-(z) = \frac{V_f E_f}{V_m E_m}\cdot 2\cdot\frac{\tau_s}{E_f\cdot r}\cdot z + e_m^T$$

The relation between stress (load) on the fibre σ_f and debond length l_d is [4] [8]:

$$\frac{l_d}{r} = \left(\frac{E_f}{2\tau_s}\right)\cdot\left(\frac{V_m E_m}{E_c}\right)\cdot\left(\frac{\sigma_f}{E_f} + \Delta e^T\right) - \frac{E_f}{\tau_s}\cdot\sqrt{\frac{V_m E_m}{E_c}\cdot\frac{G_{iic}}{E_f\cdot r}} \qquad (1)$$

The load on the fibre P_f ($= \sigma_f \cdot \pi\, r^2$) is written as a function of l_d:

$$P_f = \left[2\cdot\pi\, r^2\cdot E_f\cdot\sqrt{\frac{E_c}{V_m E_m}\cdot\frac{G_{iic}}{E_f\cdot r}} - \pi\, r^2\cdot E_f\cdot\Delta e^T\right] + \left[\frac{E_c}{V_m E_m}\cdot 2\cdot\pi\, r\cdot\tau_s\right]\cdot l_d \qquad (2)$$

This indicates a linear plot of P_f *vs* l_d, with slope and cut-off on y-axis given as

$$\text{slope} = \text{SL1} = \left[\frac{E_c}{V_m E_m}\cdot 2\cdot\pi\, r\cdot\tau_s\right]$$

$$\text{cut-off} = \text{CO1} = \left[2\cdot\pi\, r^2\cdot E_f\cdot\sqrt{\frac{E_c}{V_m E_m}\cdot\frac{G_{iic}}{E_f\cdot r}} - \pi\, r^2\cdot E_f\cdot\Delta e^T\right]$$

The relation between stress (load) on the fibre σ_f, the debond length l_d, and and the relative displacement between fibre and matrix δ is [4] [8]:

$$\frac{\delta}{r} = \left(\frac{\sigma_f}{E_f} + \Delta e^T\right)\cdot\frac{l_d}{r} - \frac{\tau_s}{E_f}\cdot\frac{E_c}{V_m E_m}\cdot\left(\frac{l_d}{r}\right)^2$$

The ratio δ/l_d is rewritten as a function of l_d, using the term $(\sigma_f/E_f + \Delta e^T)$ from eq (1):

$$\frac{\delta}{l_d} = \left[2\cdot\sqrt{\frac{E_c}{V_m E_m}\cdot\frac{G_{iic}}{E_{f\cdot r}}}\right] + \left[\frac{E_c}{V_m E_m}\cdot\frac{\tau_s}{E_f\cdot r}\right]\cdot l_d \qquad (3)$$

This indicates a linear plot of δ/l_d *vs* l_d, with slope and cut-off on y-axis given as

$$\text{slope} = \text{SL2} = \left[\frac{E_c}{V_m E_m}\cdot\frac{\tau_s}{E_f\cdot r}\right]$$

$$\text{cut-off} = \text{CO2} = \left[2 \cdot \sqrt{\frac{E_c}{V_m E_m} \cdot \frac{G_{iic}}{E_{f \cdot r}}} \right]$$

The two (linear) equations (2) and (3) allow the three interface parameters to be determined from two linear plots. Eq (2) gives the interface shear stress τ_s from the slope SL1, and yields a *combination* of G_{iic} and Δe^T from the cut-off CO1. Eq (3) gives the interface shear stress τ_s from the slope SL2, and the interface energy G_{iic} from the cut-off CO2. Inserting this value for G_{iic} into CO1 gives a value for Δe^T:

$$\Delta e^T = CO2 - \frac{CO1}{\pi \, r^2 \cdot E_f}$$

5. Pull-out case PO-2

This pull-out case has been used in some tests and implies a clamping device to support the matrix block at the fibre end of the specimen. The practical geometry has often been a drop of matrix (polymer) on the fibre, which is pulled out of this drop. The macroscopic specimen geometry of this droplet test is not well defined. The present model (which implies a circular cylindrical geometry) was presented [5] with a detailed analysis and comments to the assumptions and concepts used to obtain the model. Here the results are presented and rewritten into a form, which is comparable to the other test cases. The load condition is $\sigma_c = 0$, and the force balance is thus $V_f \cdot \sigma_f + V_m \cdot \sigma_m = 0$.
The up-stream stresses are

$$\sigma_f^+ = E_f \cdot \left(-\Delta e^T \right) \cdot \frac{V_m E_m}{E_c}$$

$$\sigma_m^+ = E_m \cdot \left(+\Delta e^T \right) \cdot \frac{V_f Ef}{E_c}$$

It should be noted that the second term with σ_c / E_c is not present because the condition is $\sigma_c = 0$.
The down-stream stresses are

$$\sigma_f^-(z) = \sigma_f - 2 \cdot \frac{\tau_s}{r} \cdot z$$

$$\sigma_m^-(z) = \frac{V_f}{V_m} \cdot \left(2 \cdot \frac{\tau_s}{r} \cdot z - \sigma_f \right)$$

The down-stream strains are

$$e_f^-(z) = \frac{\sigma_f}{E_f} - 2 \cdot \frac{\tau_s}{E_f \cdot r} \cdot z + e_f^T$$

$$e_m^-(z) = \frac{V_f E_f}{V_m E_m} \cdot \left(2 \cdot \frac{\tau_s}{E_f \cdot r} \cdot z - \frac{\sigma_f}{E_f} \right) + e_m^T$$

The relation between stress (load) on the fibre σ_f and debond length l_d is ([5] eq 18):

$$\frac{\sigma_f}{E_f} = 2 \cdot \sqrt{\frac{V_m E_m}{E_c} \cdot \frac{G_{iic}}{E_f \cdot r}} - \frac{V_m E_m}{E_c} \cdot \Delta e^T + 2 \cdot \frac{\tau_s}{E_f \cdot r} \cdot l_d \tag{4}$$

The load on the fibre P_f ($= \sigma_f \cdot \pi r^2$) is written as a function of l_d:

$$P_f = \left[2 \cdot \pi r^2 \cdot E_f \cdot \sqrt{\frac{V_m E_m}{E_c} \cdot \frac{G_{iic}}{E_f \cdot r}} - \frac{V_m E_m}{E_c} \cdot \pi r^2 \cdot E_f \cdot \Delta e^T \right] + \left[2 \cdot \pi r \cdot \tau_s \right] \cdot l_d \tag{5}$$

This indicates a linear plot of P_f *vs* l_d, with slope and cut-off on y-axis given as

$$\text{slope} = \text{SL1} = \left[2 \cdot \pi r \cdot \tau_s \right]$$

$$\text{cut-off} = \text{CO1} = \left[2 \cdot \pi r^2 \cdot E_f \cdot \sqrt{\frac{V_m E_m}{E_c} \cdot \frac{G_{iic}}{E_f \cdot r}} - \frac{V_m E_m}{E_c} \cdot \pi r^2 \cdot E_f \cdot \Delta e^T \right]$$

It should be noted that eq (5) is similar to eq (2) of case PO-1, with the scaling factor $V_m E_m / E_c$, such that loads (stresses) on the fibre relate *by* $P_f(PO\text{-}2) = V_m E_m / E_c \cdot P_f(PO\text{-}1)$.

The relation between stress (load) on the fibre σ_f, the debond length l_d, and the relative displacement between fibre and matrix δ is ([5] eq 11):

$$\frac{\delta}{r} = \left(\frac{E_c}{V_m E_m} \cdot \frac{\sigma_f}{E_f} + \Delta e^T \right) \cdot \frac{l_d}{r} - \frac{E_c}{V_m E_m} \cdot \frac{\tau_s}{E_f} \cdot \left(\frac{l_d}{r} \right)^2$$

The ratio δ / l_d is rewritten as a function of l_d, using the term $(E_c / V_m E_m \cdot \sigma_f / E_f + \Delta e^T)$ from eq (4):

$$\frac{\delta}{l_d} = \left[2 \cdot \sqrt{\frac{E_c}{V_m E_m} \cdot \frac{G_{iic}}{E_f \cdot r}} \right] + \left[\frac{E_c}{V_m E_m} \cdot \frac{\tau_s}{E_f \cdot r} \right] \cdot l_d \tag{6}$$

This indicates a linear plot of δ / l_d *vs* l_d, with slope and cut-off on y-axis given as

$$\text{slope} = \text{SL2} = \left[\frac{E_c}{V_m E_m} \cdot \frac{\tau_s}{E_f \cdot r} \right]$$

$$\text{cut-off} = \text{CO2} = \left[2 \cdot \sqrt{\frac{E_c}{V_m E_m} \cdot \frac{G_{iic}}{E_f \cdot r}} \right]$$

It should be noted that eq (6) is identical to eq (3) of case PO-1.

The two (linear) equations (5) and (6) allow the three interface parameters to be determined from two linear plots. Eq (5) gives the interface shear stress τ_s from the slope SL1, and yields a *combination* of G_{iic} and Δe^T from the cut-off CO1. Eq (6) gives the interface shear stress τ_s from the slope SL2, and the interface energy G_{iic} from the cut-off CO2. Inserting this value for G_{iic} into CO1 gives a value for Δe^T:

$$\Delta e^T = CO2 - \frac{CO1}{\pi r^2 \cdot E_f \cdot \frac{V_m E_m}{E_c}}$$

It should be noted that although eq (3) for case PO-1 and eq (6) for PO-2, respectively, are identical, the relations to find Δe^T are not identical, but (partly) scaled by the factor $V_m E_m / E_c$.

6. Single fibre fragmentation case SFFT

This test case represents a different test geometry and specimen geometry, as well as a different progress of the experiment. In the pull-out cases, one specimen gives one set of data. In the SFFT experiment, the (single) fibre will break several times (multiple fracture), and each fibre break gives two fibre ends which retract into the hole of the matrix, giving two sets of data for load, debond length and relative fibre/matrix displacement. This event is repeated at the next fibre break, and this continues until the composite fails. The (potentially) many fibre breaks cause concern over the possible interaction of the debond lengths from nearby fibre breaks. No experiments exist to illustrate the importance of this (possible) course of events.

The model was presented [6] with a detailed analysis and comments to the assumptions and concepts used to obtain the model. Here the results are presented and rewritten into a form, which is comparable to the other test cases. The load condition is $\sigma_f = 0$, and the force balance is thus $\sigma_c - V_m \cdot \sigma_m = 0$.

The up-stream stresses are

$$\sigma_f^+ = E_f \cdot \left(-\Delta e^T \right) \cdot \frac{V_m E_m}{E_c} + E_f \cdot \frac{\sigma_c}{E_c}$$

$$\sigma_m^+ = E_m \cdot \left(+\Delta e^T \right) \cdot \frac{V_f Ef}{E_c} + E_m \cdot \frac{\sigma_c}{E_c}$$

The down-stream stresses are

$$\sigma_f^-(z) = 2 \cdot \frac{\tau_s}{r} \cdot z$$

$$\sigma_m^-(z) = \frac{\sigma_c}{V_m} - \frac{V_f}{V_m} \cdot 2 \cdot \frac{\tau_s}{r} \cdot z$$

The down-stream strains are

$$e_f^-(z) = 2 \cdot \frac{\tau_s}{E_f \cdot r} \cdot z + e_f^T$$

$$e_m^-(z) = \frac{\sigma_c}{V_m E_m} - \frac{V_f E_f}{V_m E_m} \cdot 2 \cdot \frac{\tau_s}{E_f \cdot r} \cdot z + e_m^T$$

The relation between stress on the composite σ_c and debond length l_d is ([6] eq 22):

$$\frac{\sigma_c}{E_c} = 2 \cdot \sqrt{\frac{V_m E_m}{E_c} \cdot \frac{G_{iic}}{E_f \cdot r}} + \frac{V_m E_m}{E_c} \cdot \Delta e^T + 2 \cdot \frac{\tau_s}{E_f \cdot r} \cdot l_d \tag{7}$$

The stress on the composite σ_c is written as a function of l_d:

$$\sigma_c = \left[2 \cdot E_c \cdot \sqrt{\frac{V_m E_m}{E_c} \cdot \frac{G_{iic}}{E_f \cdot r}} + V_m E_m \cdot \Delta e^T \right] + \left[E_c \cdot 2 \cdot \frac{\tau_s}{E_f \cdot r} \right] \cdot l_d \tag{8}$$

It should be noted, in contrast to the pull-out cases, that for the SFFT case (only) the composite stress (load) can be recorded experimentally, and thus eq (8) must be written in terms of σ_c .This indicates a linear plot of σ_c *vs* l_d, with slope and cut-off on y-axis given as

$$\text{slope} = \text{SL1} = \left[E_c \cdot 2 \cdot \frac{\tau_s}{E_f \cdot r} \right]$$

$$\text{cut-off} = \text{CO1} = \left[2 \cdot E_c \cdot \sqrt{\frac{V_m E_m}{E_c} \cdot \frac{G_{iic}}{E_f \cdot r}} + V_m E_m \cdot \Delta e^T \right]$$

The relation between stress on the composite σ_c , the debond length l_d, and the relative displacement between fibre and matrix δ is ([6] eq 27):

$$\frac{\delta}{r} = \left(\frac{\sigma_c}{V_m E_m} - \Delta e^T \right) \cdot \frac{l_d}{r} - \frac{E_c}{V_m E_m} \cdot \frac{\tau_s}{E_f} \cdot \left(\frac{l_d}{r} \right)^2$$

The ratio δ/l_d is rewritten as a function of l_d, using the term $(\sigma_c / V_m E_m - \Delta e^T)$ from eq (7):

$$\frac{\delta}{l_d} = \left[2 \cdot \sqrt{\frac{E_c}{V_m E_m} \cdot \frac{G_{iic}}{E_f \cdot r}} \right] + \left[\frac{E_c}{V_m E_m} \cdot \frac{\tau_s}{E_f \cdot r} \right] \cdot l_d \tag{9}$$

This indicates a linear plot of δ/l_d *vs* l_d, with slope and cut-off on y-axis given as

$$\text{slope} = \text{SL2} = \left[\frac{E_c}{V_m E_m} \cdot \frac{\tau_s}{E_f \cdot r} \right]$$

$$\text{cut-off} = \text{CO2} = \left[2 \cdot \sqrt{\frac{E_c}{V_m E_m} \cdot \frac{G_{iic}}{E_f \cdot r}} \right]$$

It should be noted that eq (9) is identical to eq (3) of case PO-1, and to eq (6) of case PO-2. The two (linear) equations (8) and (9) allow the three interface parameters to be determined from two linear plots. Eq (8) gives the interface shear stress τ_s from the slope SL1, and yields a *combination* of G_{iic}

and Δe^T from the cut-off CO1. Eq (9) gives the interface shear stress τ_s from the slope SL2, and the interface energy G_{iic} from the cut-off CO2. Inserting this value for G_{iic} into CO1 gives a value for Δe^T:

$$\Delta e^T = -CO2 + \frac{CO1}{V_m E_m}$$

It should be noted that although eq (3) for case PO-1 and eq (6) for PO-2 and eq (9) for SFFT are identical, the relations to find Δe^T are not identical, but (partly) scaled by the factor $V_m E_m$.

7. Experimental test and analysis

The three test cases have similarities and some marked differences. The two pull-out cases give one set of experimental data (fibre load P_f, debond length l_d, and relative fibre/matrix displacement δ) for each specimen. In the SFFT case, one specimen (potentially) gives several (double) sets of data. The (potentially) many fibre breaks cause concern over the possible interaction of the debond lengths from nearby fibre breaks. No experiments exist to illustrate the importance of this (possible) course of events.

There are two equations for each test case, both giving a linear plot, where the slope and the cut-off values form the basis for calculating the (three) interface parameters. The experimental data and their analysis are summarized here.

7.1 Pull-out case PO-1

The plot of P_f *vs* l_d gives the interface shear stress from the slope SL1 as

$$\tau_s = \frac{V_m E_m}{E_c} \cdot \frac{1}{2\pi r} \cdot SL1$$

The plot of δ/l_d *vs* l_d gives the interface shear stress from the slope SL2 as

$$\tau_s = \frac{V_m E_m}{E_c} \cdot E_f \cdot r \cdot SL2$$

and the interface energy G_{iic} from the cut-off CO2 as

$$G_{iic} = \frac{V_m E_m}{E_c} \cdot E_f \cdot r \cdot \frac{(CO2)^2}{4}$$

and the mismatch strain Δe^T as

$$\Delta e^T = CO2 - \frac{CO1}{\pi r^2 \cdot E_f}$$

7.2 Pull-out case PO-2

The plot of P_f *vs* l_d gives the interface shear stress from the slope SL1 as

$$\tau_s = \frac{1}{2\pi r} \cdot SL1$$

The plot of δ/l_d *vs* l_d gives the interface shear stress from the slope SL2 as

$$\tau_s = \frac{V_m E_m}{E_c} \cdot E_f \cdot r \cdot SL2$$

and the interface energy G_{iic} from the cut-off CO2 as

$$G_{iic} = \frac{V_m E_m}{E_c} \cdot E_f \cdot r \cdot \frac{(CO2)^2}{4}$$

and the mismatch strain Δe^T as

$$\Delta e^T = CO2 - \frac{CO1}{\pi r^2 \cdot E_f \cdot \frac{V_m E_m}{E_c}}$$

7.3 Single fibre fragmentation case SFFT

The plot of σ_c *vs* l_d gives the interface shear stress from the slope SL1 as

$$\tau_s = \frac{1}{2 \cdot E_c} \cdot E_f \cdot r \cdot SL1$$

The plot of δ/l_d *vs* l_d gives the interface shear stress from the slope SL2 as

$$\tau_s = \frac{V_m E_m}{E_c} \cdot E_f \cdot r \cdot SL2$$

and the interface energy G_{iic} from the cut-off CO2 as

$$G_{iic} = \frac{V_m E_m}{E_c} \cdot E_f \cdot r \cdot \frac{(CO2)^2}{4}$$

and the mismatch strain Δe^T as

$$\Delta e^T = - CO2 + \frac{CO1}{V_m E_m}$$

It should be noted that the equations governing the calculations from the experimental analyses, are rather similar and in some situations identical.

8. Discussion

8.1 Model characteristics

The model is basically one and the same, and is applied to slightly different loading cases. This leads to the rather high degree of similarity for the various equations for load vs debond length and for displacement/debond length vs debond length, and to the scaling for the pull-out cases PO-1 and PO-2 as indicated above. It should be noted that the load vs debond length equation is individual for the three test cases, while the delta/debond length vs debond length equation is identical for all three test cases and is independent of load. This is expected because the delta and the debond length are both connected to the geometry of the debond process. This process is essentially a shear crack (mode II crack) of the interface, which is characterized by the interface energy (chemical nature) and the interfacial shear stress (interface topography).

This combined situation gives two linear equations, which are the basis for the simple and "smart" linear plots, leading to establishing the three characteristic interface parameters, as described above in the analyses.

8.2 Case $V_f = 0$

The equations have all been written with the term V_mE_m/E_c as a practical term, entering the equations where required. This term is normally very close 1, because normally one rather thin fibre is embedded in a fairly large block of matrix, which makes the fibre volume fraction very low, and thus the matrix volume fraction V_m very close to 1 and E_c very close to E_m. This fact can be used to simplify the equations by setting $V_mE_m/E_c = 1$. This situation is not always present as the experimental examples below illustrates.

8.3 Traditional experiments

Many interface characterizations and related *pull-out experiments* have been reported in the literature, e.g. [9] [10] [11] for the end-gripped test geometry PO-1, and [12] [13] for the droplet test geometry PO-2. The tests have normally been performed by using several specimens with different embedded lengths of fibre. The curve for pull-out force (stress) vs fibre displacement is described in detail in [5] and figure 2. In most cases an *apparent* interfacial shear stress has been calculated from the maximum pull-out force P_{max} and the embedded fibre length l_{emb}.

$$\tau_{app} = \frac{P_{max}}{2 \cdot \pi\, r \cdot l_{emb}}$$

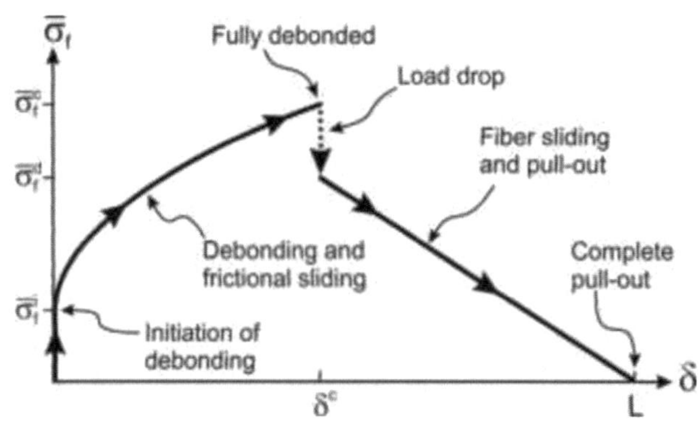

Figure 2. Sketch (not to scale) of fibre stress vs displacement between fibre and matrix for a traditional pull-out experiment with embedded fibre length L

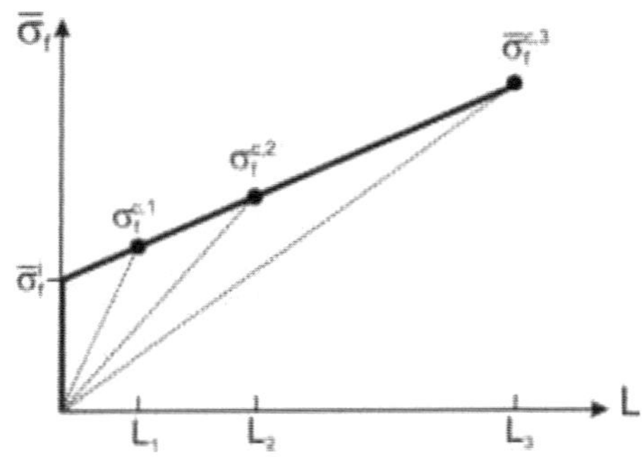

Figure 3. Sketch of the relation between fibre stress and debond length according to eq (2) and (5) for traditional pull-out experiments, the dotted slopes indicate the calculations of the *apparent* shear stress, which for this case is *decreasing* with increasing embedded fibre length (termed L)

It has often been noted in the individual publications in the literature, that this apparent shear stress was dependent on the embedded length, as discussed in [5] and illustrated in figure 3. This dependency is caused by the fact that the (maximum) fibre load is linear with but not directly proportional to embedded length. For the sake of clarity the eqs (2) and (5) for test cases PO-1 and PO-2, respectively, are written with the term $V_m E_m / E_c = 1$, making the two test cases and equations identical:

$$P_f = \left[2 \cdot \pi \, r^2 \cdot E_f \cdot \sqrt{\frac{G_{iic}}{E_f \cdot r}} - \pi \, r^2 \cdot E_f \cdot \Delta e^T \right] + \left[2 \cdot \pi \, r \cdot \tau_s \right] \cdot l_d \tag{10}$$

At the maximum fibre load P_{max} the embedded fibre has debonded completely, as sketched in figure 2, and starts to pull out of the matrix block (against frictional sliding). Therefore, at $l_d = l_{emb}$ the fibre load is $P_f = P_{max}$. From this equation (10) the *apparent* shear stress can be written

$$\tau_{app} = \frac{P_{max}}{2 \cdot \pi \, r \cdot l_{emb}} = \frac{[constant]}{l_{emb}} + \tau_s$$

The constant is controlled by G_{iic} and Δe^T (and material parameters), and only if this constant happens to be zero, will the *apparent* shear stress give the *true* interfacial shear stress τ_s. For all other situations the apparent shear stress will depend on the embedded length, with increasing or decreasing values dependent on the sign of the constant, as discussed in [5] and figure 3. The equation (10) suggests a direct way of making a correct analysis of simple pull-out experiments, based on various embedded fibre lengths and recorded maximum force from the pull-out curve, (figure 2): by plotting the P_{max} *vs* l_{emb} and using a linear fitting equation, the slope will give the true interfacial shear stress τ_s, and the cut-off will give a constant including a combination of G_{iic} and Δe^T, but it will not allow a determination of these two parameters individually. Examples will be given below to illustrate this incomplete analysis of pull-out experiments from the literature.

9. Practical examples

The model and its method of analysis will be illustrated by three examples. The one example is a complete experiment for steel fibres in polyester matrix, where all three experimental parameters load, debond length and relative displacement are recorded. The other two examples are incomplete and are from the literature, one is steel fibres in epoxy and one is tungsten wires in copper matrix. The relevant properties for the materials systems are collected in Table 1.

Table 1. Materials properties for the practical examples

Fibre-Matrix system	Steel-Polyester	Steel-Epoxy	Tungsten-Copper
Fibre stiffness E_f, GPa	210	200	400
Fibre radius r, mm	0.155	0.075	0.25
Fibre thermal expansion, α_f, 10^{-6} K^{-1}	11.7	11.7	4.3
Matrix stiffness E_m, GPa	3	3	100
Matrix thermal expansion, α_m, 10^{-6} K^{-1}	70	50	16.5
Specimen area A_c, mm^2	25	20	28
$V_m E_m / E_c$	0.825	0.944	0.973

9.1 Steel fibres in polyester matrix (complete analysis)

The pull-out process is case PO-1 with gripping of the matrix block. These experimental data from Prabhakaran et al [8] have been analyzed earlier, using a slightly more complex procedure [8], where the specimens and the experimental test fixture is described in detail. From the video recording of the debond process a total of 30 pictures were used to measure the debond length l_d at the corresponding fibre load P_f. From the (continuous) recordings of the relative fibre/matrix displacement, (30) values for δ were selected at these fibre loads. These (30) simultaneous sets of data are the basis for establishing the plot of P_f *vs* l_d in figure 4 and the plot of δ / l_d *vs* l_d in figure 5. The experimental and material parameters are listed in Table 1. It should be noted that with the fairly thick fibre of radius 0.155 mm the fibre volume fraction of 0.003 clearly is not close to zero, and thus the parameter $V_m E_m / E_c$ is about 0.825 and must be included in the calculations. The straight lines in the plots show rather good fit (R^2 is ca. 0.98). From the slopes and cut-offs of the straight lines the three interface parameters are calculated using the equations for case PO-1 from section 7.1. The results are shown here:

SL1	1.496	N/mm
CO1	12.259	N
τ_s	1.3	MPa
SL2	0.00006164	mm^{-1}
CO2	0.002153	
τ_s	1.6	MPa
G_{iic}	31	J/m^2
Δe^T	0.0014	

It is noted that the two plots give slightly different values for the interface shear stress τ_s , it is not clear whether this is caused by experimental scatter, or whether the model and/or the analysis has a systematic deviation. The experimental mismatch strain might be compared to an estimate from its definition, this requires knowledge of the thermal expansion coefficients for fibre and matrix and also the temperature difference between manufacturing temperature and test temperature. With reasonable assumptions as listed in Table 1, the estimated value is: $\Delta eT = (12 - 70) \cdot 10^{-6} \cdot (20 - 50) = 0.0017$.

A comparison with the results of the previous analysis [8] shows that the shear stress is very closely the same, while the interface energy and the mismatch strain deviate significantly.

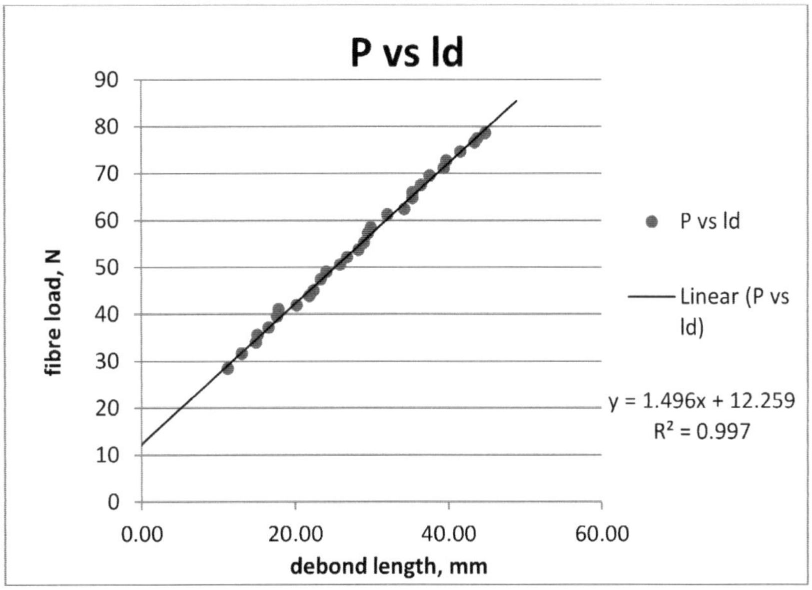

Figure 4. Experimental data for steel fibre in polyester

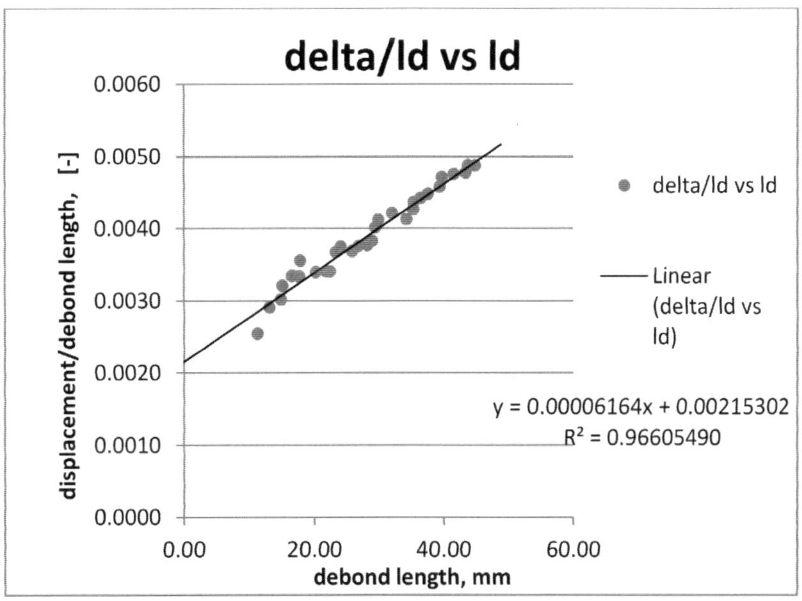

Figure 5. Experimental data for steel fibre in polyester

IOP Conf. Series: Materials Science and Engineering **219** (2017) 012030 doi:10.1088/1757-899X/219/1/012030

9.2 Steel fibres in epoxy matrix (incomplete analysis)

The pull-out process is case PO-2 with a clamp at fibre end of the specimen (droplet test). The experiments follow the traditional pull-out procedure with measurement of the maximum force for various embedded fibre lengths. The data originate from Gorbatkina [14] and were presented by Zhandarov and Mäder [15] in their figure 4b. The (7) sets of data for maximum force and embedded fibre length are analyzed according to the procedure described above for traditional, incomplete data and plotted as P_{max} *vs* l_{emb} in figure 6. The experimental and material parameters are listed in Table 1. It should be noted that with the moderately thick fibre of radius 0.075 mm the fibre volume fraction of 0.0009 clearly is not close to zero, and thus the parameter V_mE_m/E_c is about 0.944 and must be included in the calculations. The straight line in figure 6 shows rather good fit (R^2 is ca. 0.94). From the slope for the straight line, the interface shear stress is calculated using the equation for P_{max} *vs* l_{emb} in the full form eq (5) for case PO-2 from section 7.2. The results are shown here:

SL1	8.847	N/mm
CO1	15.533	N
τ_s	18.8	MPa

As emphasized above the cut-off CO1 only gives a combination of G_{iic} and Δe^T. If the mismatch strain is estimated from its definition and the data in Table 1, the value is $\Delta eT = (12 - 50) \cdot 10^{-6} \cdot (20 - 120) = 0.0038$, and with this value the interface energy is ca 250 J/m^2.

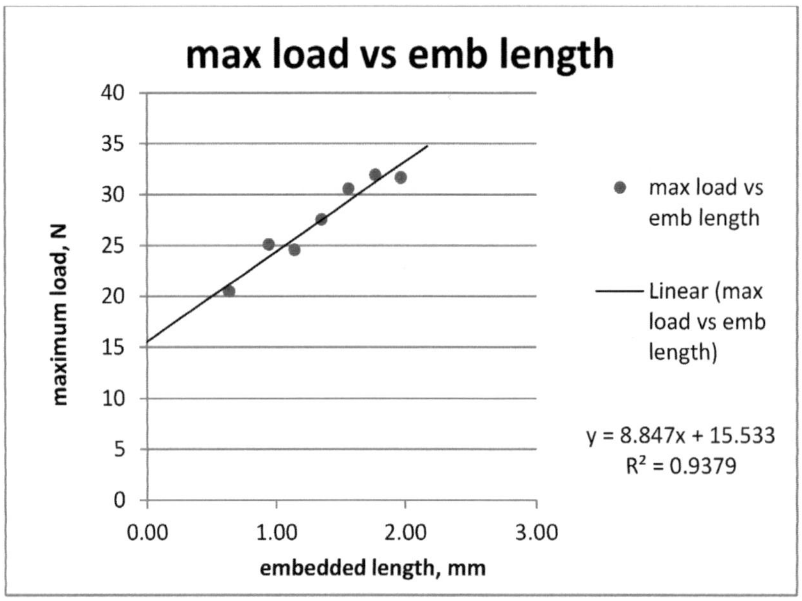

Figure 6. Experimental data for steel fibre in epoxy matrix

9.3 Tungsten wires in copper matrix (incomplete analysis)

The study of the interface in composites was (probably) initiated in the early 1960´s, and Tyson [16] performed some of the first pull-out tests for tungsten wires in blocks of copper matrix. One of these test series, which is presented by Kelly [17] figure 5.18, is performed at 600 °C, and includes 8 sets of data for embedded fibre length and maximum force for the pull-out experiment. The test case is PO-1. The 8 sets of data are analyzed according to the procedure described above for traditional, incomplete data and plotted as P_{max} *vs* l_{emb} in figure 7. The experimental and material parameters are listed in Table 1. It should be noted that with the moderately thick fibre of radius 0.25 mm the fibre volume fraction of 0.007 clearly is not close to zero, and thus the parameter $V_m E_m / E_c$ is about 0.973 and must be included in the calculations. The straight line in figure 7 shows rather good fit (R^2 is ca. 0.94). From the slope for the straight line the interface shear stress is calculated using the equation for P_{max} *vs* l_{emb} in the full form eq (2) for case PO-1 from section 7.1. The results are shown here:

SL1	47.863	N/mm
CO1	28.326	N
τ_s	29.6	MPa

As emphasized above the cut-off CO1 only gives a combination of G_{iic} and Δe^T. If the mismatch strain is estimated from its definition and the data in Table 1, the value is $\Delta eT = (4.3 - 16.5) \cdot 10^{-6} \cdot (600 - 800) = 0.0024$, and with this value the interface energy is ca 190 J/m^2. It must be emphasized that this rough calculation is based on an effective manufacturing and stress-free state for the copper matrix at about 800 °C.

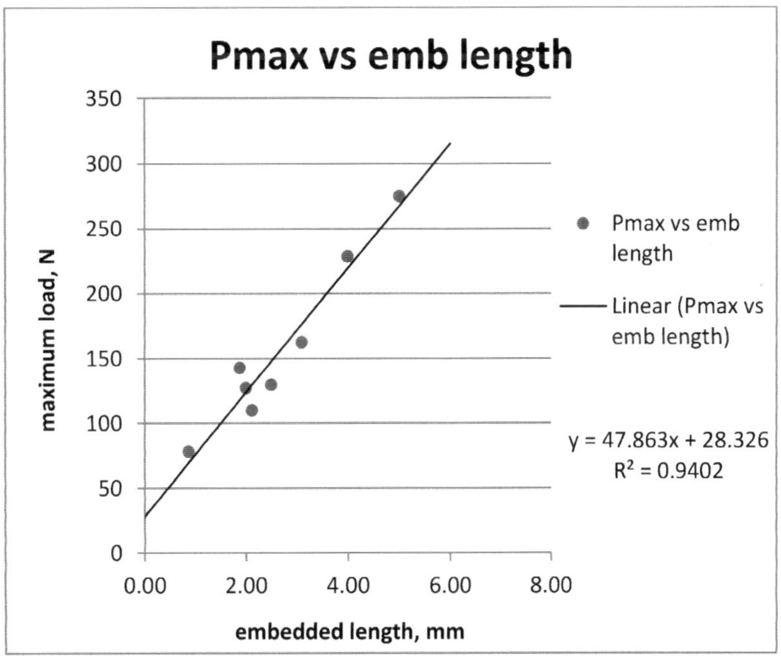

Figure 7. Experimental data for tungsten wire in copper matrix

10. Summary

The *characterization of the interface* between fibre and matrix in composites has been discussed and three interface parameters have been defined: the interface energy G_{iic} , the frictional interfacial shear stress τ_s and mismatch strain between fibre and matrix Δe^T.

The (new) *mechanical model* for the interface has been presented, and it has been based on the (three) test cases: pull-out of a fibre from a block of matrix, (i) supported by gripping at the matrix end of the specimen, called case PO-1, (ii) supported by clamping at the fibre end of the specimen, called PO-2, and (iii) fibre fragmentation in the single fibre composite, called SFFT. The model is shown to give rather similar but not identical results for the three cases. This facilitates and simplifies the equations for the test cases, which are presented in the same or comparable format.

Two *governing equations* are derived for each test case, one equation relating the fibre load or composite stress to debond length, and the other equation relating the relative displacement/debond length ratio to the debond length. The first equation is very similar for the three cases, and the second equation is identical for the three cases. This latter derivation is a new observation.

These two equations allow two *simple linear plots* and the direct determination of the three interface parameters from the slope and cut-off values of these plots. This is possible (only) when the three experimental parameters are recorded: fibre load, debond length and relative fibre/mtatrix displacement. This is illustrated for the material system of a steel fibre in polyester matrix.

Many *experiments exist* in the literature, which (only) measure the maximum force for pull-out of fibres with different embedded lengths. The mechanical model and the equation for fibre load vs debond length can be used to establish the *true* interfacial shear stress, and (only) a *combination* of interface energy and mismatch strain. This is illustrated for the material system of a steel fibre in epoxy matrix, in test case PO-2, and for the material system of a tungsten wire in copper matrix, in test case PO-1.

It is believed that the *results for the interface parameters* obtained here are correct, but few or no reference values exist to compare and validate the numerical results.

References

[1] Kelly A and Tyson W R 1965 *J. Mech. Phys. Solids* **13** 329

[2] Jones F R 1989 *Interfacial Phenomena in Composite Materials '89* (London: Butterworths)

[3] Verpoest I and Jones F R 1991 *Interfacial Phenomena in Composite Materials* (Oxford: Butterworth-Heinemann)

[4] Hutchinson J W and Jensen H M 1990 *Mechanics of Materials* **9** 139

[5] Sørensen B F and Lilholt H 2016 *Proc. 37th Risø Int. Symp. on Mat. Science,* eds B Madsen et al (Roskilde Denmark : Dept. of Wind Energy, Techn. Univ. of Denmark) p 115 *also* IOP Conf. Series: Materials Science and Engineering **139** 012009

[6] Sørensen B F 2017 *Mechanics of Materials* **104** 38

[7] Budiansky B, Hutchinson J W and Evans A G 1986 *J. Mech. Phys. Solids* **34** 167

[8] Prabhakaran R T D, Gupta M, Lilholt H, Sørensen B F and Mahajan P 2013 *Proc. 34th Risø Int. Symp. on Mat. Science,* eds B Madsen et al (Roskilde Denmark : Dept. of Wind Energy, Techn. Univ. of Denmark) p 341

[9] Piggott M R 1991 *Interfacial Phenomena in Composite Materials*, eds I Verpoest and F R Jones (Oxford: Butterworth-Heinemann) p 3

[10] Meretz S, Nowak H, Hampe A, Hinrichsen G, Schumacher K and Sernow R 1991 *Interfacial Phenomena in Composite Materials*, eds I Verpoest and F R Jones (Oxford: Butterworth-Heinemann) p 73

[11] Auvray M H, Chéneau-Henry P, Leroy F H and Favre J P 1994 *Composites* **25** 776

[12] Scheer R J and Nairn J A 1995 *J. Adhesion* **53** 45

[13] Gorbatkina Yu A and Ivanova-Mumjieva V G 2000 *Mechanics of Composite Materials* **36** 435
[14] Gorbatkina Yu A 1992 *Adhesive Strength of Fiber-Polymer Systems* (New York : Ellis Horwood)
[15] Zhandarov S and Mäder E 2003 *J. Adhesion Sci. and Techn.* **17** 967
[16] Tyson W R 1964 *Fibre Reinforcement of Metals* (Cambridge UK: Thesis)
[17] Kelly A 1973 *Strong Solids* (Oxford: Clarendon Press) second edition p 203

Local stress field induced by twinning in a metastable β titanium alloy

Fengxiang Lin[1], Matthieu Marteleur[1], Jon Alkorta[2], Pascal J. Jacques[1], Laurent Delannay[1]

[1]Université catholique de Louvain, Institute of Mechanics, Materials and Civil Engineering (iMMC), 1348 Louvain la Neuve, Belgium
[2]CEIT and Tecnun, University of Navarra, San Sebastian, Spain

E-mail: fengxiang.lin@uclouvain.be

Abstract. A crystal plasticity based finite element model incorporating deformation twinning is applied in 3D to simulate the stress field around a twin lamella in a metastable β titanium alloy. Compared with conventional 2D simulations, the stress field determined by the 3D simulation shows better agreement with that measured experimentally using high resolution electron backscattered diffraction. Reasons leading to the different results in 2D and 3D simulations are discussed. It is also discussed in which circumstances 2D simulation is sufficient, and in which 3D is necessary.

1. Introduction

Titanium (Ti) alloys have many unique properties, such as low density, and high corrosion resistance, which makes them suitable for many engineering applications. However, the traditional α Ti or dual-phase Ti alloys have either low strength or low ductility, which limits their usage. Recently, several metastable β Ti alloys have been designed to exhibit both high strength and ductility thanks to the enhanced work hardening ability arising from simultaneous TRIP (TRansformation Induced Plasticity) and TWIP (TWinning Induced Plasticity) effects [1, 2].

Twinning plays an important role during deformation of metastable β Ti alloys. Deformation twinning imposes local strains, which must be accommodated by neighboring grains. This will thus lead to local stress concentration, which significantly affects the microscopic plastic yielding under both monotonic and cyclic deformation. Experimentally, the local stresses can be determined using high resolution electron backscattered diffraction (HREBSD) [3], and the stress field around twins has been measured for a tensile deformed Zircaloy-2 sample [4]. Stresses around twins depend on many factors, including the twin shear strain, the morphology of the twin, the orientations of the parent and twin grains, and the orientations of the neighboring grains. To understand the measured stresses, and the influence of various parameters, it is necessary to build a crystal plasticity model that incorporates twinning. In a series of studies by Kumar et al. [5, 6, 7], a crystal plasticity model with a prescribed twin domain is used to simulate the stress field around a twin lamella for metals with hexagonal close-packed (hcp) structure. They found that the stresses depended on twin thickness, the orientations of the neighboring grains, and the elastic and plastic anisotropy. Twinning and stresses around twins have not been

Figure 1. Orientation map showing the microstructure of the longitudinal section after 1% elongation. The tensile direction is horizontal. The colors represent grain orientations. White points are un-indexed points, and the white lines in the figure are likely to be the martensitic laths. The two arrows mark $\{332\}\langle 113\rangle$ twins.

investigated for metastable β Ti, which has body centered cubic symmetry. Models for a hcp structure (e.g. [5, 8, 9, 10]) cannot be directly applied to metastable β Ti. Moreover, most of the models reported in the literature consider 2D twins or quasi-2D twins with twins extending across the entire thickness, and real 3D simulations incorporating twinning are still rare.

The objective of this work is to build a model that helps understanding of stresses around twins in metastable β Ti. We first construct a 3D crystal plasticity finite element model (CPFEM) that allows consideration of the twin shear strain. This model is then applied to simulate stress fields around a deformation twin in a metastable β Ti alloy. The simulation results are compared with experimental measurements using HREBSD. As a comparison, a 2D finite element (FE) simulation is also conducted, in order to understand in which circumstances 2D simulation is sufficient, and in which 3D is necessary. In the end, we discuss further improvements needed for the model.

2. Experimental results

The material considered in this work is a binary Ti-12 wt% Mo alloy. The ingot was first rolled at room temperature to a thickness reduction of 95%, and then annealed at 1173 K for 15 min and water quenched to obtain a single β phase fully recrystallized microstructure. This single β phase material was used as the starting material for further tensile deformation. The initial grain size was around 50 μm.

Uniaxial tensile deformation was applied along the previous rolling direction (RD). Figure 1 shows the microstructure obtained on the longitudinal section using EBSD after 1% elongation. After applying 1% elongation, twins were developed in a few grains, as marked by the two arrows in figure 1. The activated twinning system in this alloy is $\{332\}\langle 113\rangle$, according to the orientation relationship of the twins and their parent grains. Apart from twins, in figure 1, white lines/lamellae with similar thicknesses as twins present in many grains. They are not indexed

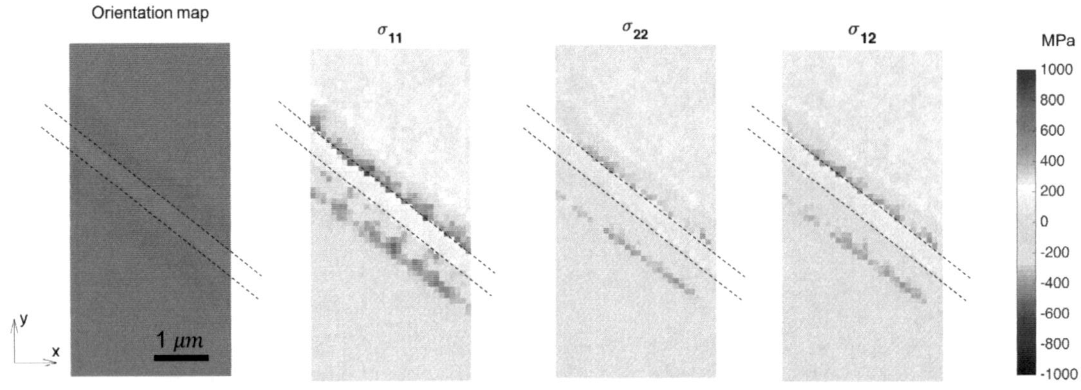

Figure 2. Stress field around a twin measured using HREBSD. The twin is marked by two parallel dashed lines. The stresses are transformed to a coordinate system where direction 1 is the twin shear direction $\vec{b}$ and direction 2 is the normal direction of the twin plane $\vec{n}$. The orientation map is colored according to the crystallographic orientation of the x axis.

in the EBSD measurements. These white lines are likely to be α'' martensite laths. Martensite laths have been observed using transmission electron microscopy in the same material after a similar tensile strain [2]. The α'' phase is not considered when indexing the EBSD patterns, and this is why they appear as un-indexed lines in the EBSD map. At this initial stage of deformation, both martensite transformation and $\{332\}\langle 113\rangle$ twinning contribute to plastic deformation. This is consistent with the observations in other metastable β Ti alloys [11, 12].

In the sample elongated by 1%, one grain with a twin was selected for HREBSD measurement (not shown in figure 1). The elastic strains and stresses around the twin lamella were determined using HREBSD, as shown in figure 2. The stresses are transformed to a coordinate system where direction 1 is the twin shear direction $\vec{b}$ and direction 2 is the normal direction of the twin plane $\vec{n}$. In the parent grain around the twin, compressive normal stresses along $\vec{b}$ (σ_{11}) are found, and normal stresses along $\vec{n}$ (σ_{22}) are also compressive, but weaker than σ_{11}. Negative shear stresses σ_{12} are found around the twin. Details of the HREBSD measurements can be found elsewhere [3, 13, 14, 15].

3. Model description

A 3D mesh containing 27 Voronoi grains is constructed with a volume of $1\times1\times1$. The geometry of the polycrystal structure is shown in figure 3a. Periodic boundary conditions are applied. A twin domain is prescribed inside one grain. The twin domain is defined as a plate with a thickness of 0.1, and the twin plate extends through the entire grain. The twin domain represents 3.35% of the grain volume. The twin is visible in a 2D section perpendicular to the macroscopic z direction (see figure 3b). The parent grain and the twin are assigned orientations the same as those measured by HREBSD. The twin interfaces correspond to the $\{332\}$ plane shared by the parent and the twin orientations. In the 2D section shown in figure 3b, the twin has the same inclination as the experimentally characterized section (figure 2), and this 2D section will be used to compare with the experimental results. The mesh is refined around the twin domain to better capture stresses around the twin. All the other grains are assigned random orientations. A 2D geometry is obtained using the 2D section shown in figure 3b. With this treatment, it is possible to compare 2D simulations with 3D ones viewed in the same section, so that grains have the same shape. However, it has to be noted that in 2D simulations, each grain is considered

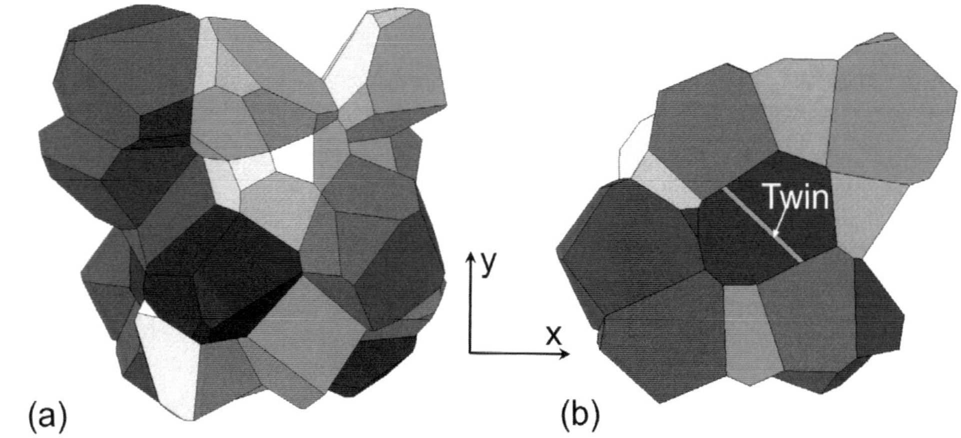

Figure 3. Representation of grains used in the study. (a) The initial 3D polycrystal structure with periodic boundary conditions. (b) A 2D section of the 3D geometry showing the prescribed twin in one grain. This section is also used for the 2D simulation for comparison.

to be a columnar grain along the z direction. The twin lamella is thus perpendicular to the simulated surface in the 2D simulation. The orientations of the parent grain and the twin for the 2D simulation are recalculated to make the {332} twin plane normal and the {113} twin shear direction on the simulated surface. We consider the 2D mesh elements as plane strain elements, i.e. ϵ_{zz} is the same for all elements in the 2D simulation.

A user material subroutine (UMAT) based on [16] is developed to incorporate twinning. At each time increment, we calculate the elastic and plastic strain increment as:

$$\Delta\epsilon = \Delta\epsilon^{el} + \Delta\epsilon^{pl} + \Delta\epsilon^{tw} \tag{1}$$

where $\Delta\epsilon$ is the total strain increment from the finite element (FE) solver, $\Delta\epsilon^{el}$ is the elastic strain increment, $\Delta\epsilon^{pl}$ is the plastic strain increment due to dislocation slip, and $\Delta\epsilon^{tw}$ is the twin shear strain. $\Delta\epsilon^{tw}$ is zero except for the twin domain during the process of twin formation. The rate of the plastic strain $\dot{\epsilon}^{pl}$ is calculated as:

$$\dot{\epsilon}^{pl} = \sum_{k} \mu^{(k)}\dot{\gamma}^{(k)}, \mu^{(k)} = sym(\vec{b}^{(k)} \otimes \vec{n}^{(k)}) \tag{2}$$

where $\mu^{(k)}$ is the symmetric part of the Schmid tensor for the slip system k, $\dot{\gamma}^{(k)}$ is the slip rate on the slip system k, $\vec{b}^{(k)}$ and $\vec{n}^{(k)}$ are the slip direction and the normal of the slip system k. A rate dependent equation is used to calculate the slip rates:

$$\dot{\gamma}^{(k)} = \dot{\gamma}_0 \left|\frac{\tau^{(k)}}{\tau_c^{(k)}}\right|^{1/m} sign\left(\tau^{(k)}\right) \tag{3}$$

where $\dot{\gamma}_0$ and m are two constants, representing the reference slip rate and the sensitivity exponent, respectively. $\tau^{(k)}$ and $\tau_c^{(k)}$ are the resolved shear stress on the slip system k and the critical resolved shear stress (CRSS) of this slip system, respectively. τ_c increases due to strain hardening. The Swift hardening law is used in the present study, and we assume an identical τ_c for all the slip systems:

$$\tau_c = \tau_{c0}\left(1+\frac{\Gamma_{tot}}{\Gamma_0}\right)^n, \Gamma_{tot} = \int_0^t \sum_k \left|\dot{\gamma}^{(k)}\right| dt \tag{4}$$

where τ_{c0}, Γ_0 and n are material parameters. In this work, the values for τ_{c0}, Γ_0 and n are assumed to be 50 MPa, 0.15 and 0.203. These values are not fine-tuned to match the experimental tensile curve. The present work shows therefore only qualitative results.

The twin shear strain is applied incrementally as a local transformation until the characteristic twin shear strain of the $\{332\}\langle 113\rangle$ twinning, 0.3536, is fully applied into the twin domain:

$$\Delta\gamma^{tw} = \frac{0.3536}{t_{tw}}\Delta t, \Delta\epsilon^{tw} = \mu^{tw}\Delta\gamma^{tw} \tag{5}$$

where t_{tw} is the total time used for applying the twin shear strain, assumed to be 1 s, Δt is the time increment defined by the FE solver, and μ^{tw} is the Schmid tensor for the prescribed twin variant.

The simulation is performed in two steps as follows:

- Uniaxial tensile deformation along the x direction to 1% elongation. Except for the twin domain, plastic deformation is achieved through dislocation slip in 12 $\{110\}\langle 111\rangle$ slip systems and 12 $\{112\}\langle 111\rangle$ slip systems. These 24 slip systems are assumed to have the same CRSS. For the twin domain, $\Delta\epsilon^{tw}$ is imposed in this step.
- Macroscopic elastic unloading, and partial elastic unloading to account for zero out-of-plane stress at the free surface of the sample.

4. Simulation results and discussion

Results of the 2D simulation are shown in figure 4a. The stresses have been transformed to a coordinate system defined by the twin shear direction $\vec{b}$ and the twin normal $\vec{n}$. Stresses from the 2D simulation are similar to those reported in literature [5, 10]. For σ_{11}, a typical butterfly pattern is observed, and σ_{22} is relatively small. For σ_{12}, negative shear stresses, i.e. along the inverse direction of the twin shear direction, are observed inside and around the twin. σ_{12} is one of the most important parameters determining the propagation and thickening of a twin. When a twin forms, σ_{12} is expected to be lower than before twinning both inside and in the vicinity of the twin, as the present simulation shows. The decrease of σ_{12} is termed backstress [5]. Results from the 2D simulation are, however, not in agreement with the HREBSD measurements shown in figure 2. Simulated stresses from the 3D simulation are shown in figure 4b in the same plane as the 2D simulation. Compared with the 2D results, the 3D results have better agreement with experiments, especially for σ_{11}.

Differences between the 2D and 3D simulation results are firstly due to the fact that the twin measured by HREBSD is highly inclined with respect to the observation plane. Figure 5 illustrates the inclination of the twin in 3D. The plane in blue corresponds to the observation plane for the HREBSD measurement, as well as the plane shown in figure 4. The twin plane $\vec{n}$ is inclined 36° with respect to the observation plane. The twin shear direction $\vec{b}$ is not on the observation plane either, but forms an angle of 54° to the observation plane. When a 2D simulation is used, $\vec{n}$ and $\vec{b}$ should be in the simulated plane, because components of ϵ^{tw} out of plane cannot be considered in 2D simulations. This explains why the 2D simulation and the experimental results are significantly different. For many hcp materials, a strong fiber texture is present in the initial material, for which it is possible to find a 2D section where most of the twins have $\vec{b}$ and $\vec{n}$ more or less in the plane. For these situations, a 2D simulation may give satisfactory agreement with experiments (e.g. [4]), whereas a 3D model is needed to simulate inclined twins and/or twin shear out of the plane in more general cases.

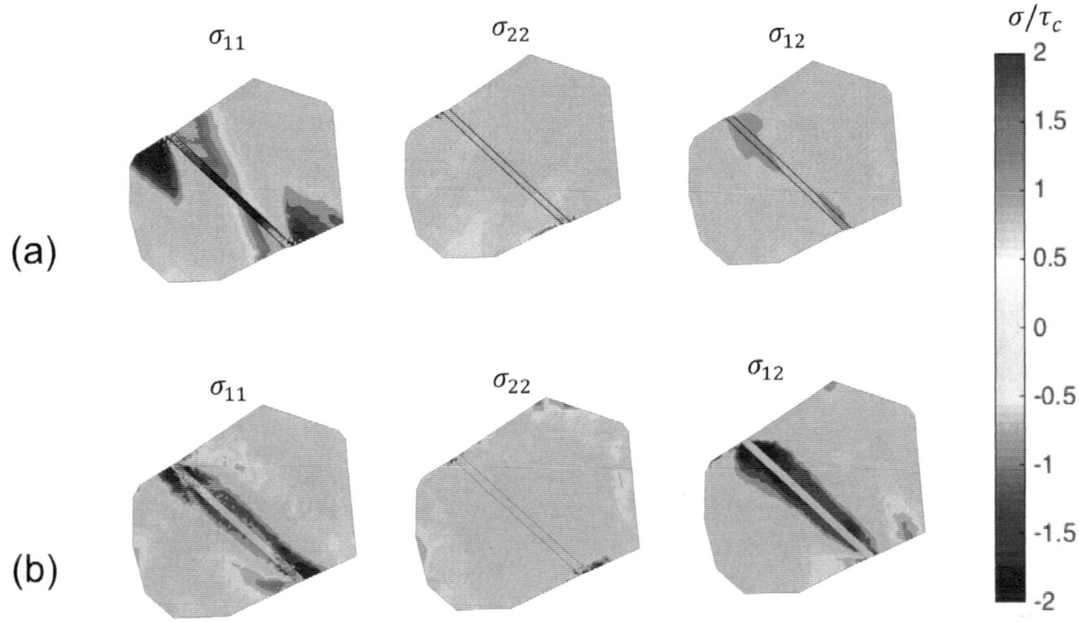

Figure 4. 2D (a) and 3D (b) simulations of the stress field around the twin. For the 3D simulation, the same 2D section as the 2D simulation is shown. The stresses are transformed to a coordinate system where direction 1 is the twin shear direction $\vec{b}$ and direction 2 is the normal direction of the twin plane $\vec{n}$ used for the simulations. Notice that $\vec{b}$ amd $\vec{n}$ are on the simulated plane for the 2D simulation, whereas that $\vec{b}$ amd $\vec{n}$ correspond to those observed experimentally for the 3D simulation.

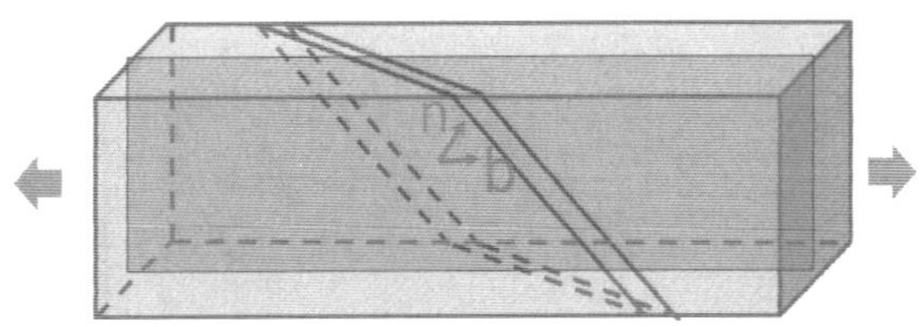

Figure 5. Sketch showing the inclination of the twin plane and twin shear direction for the grain shown in figure 2.

With the 3D model, it is also observed that stresses around the twin show through-thickness variations. Figure 6 shows the variation of σ_{11} in three sections parallel to each other. Figure 6b is approximately at the center of the grain, which is also shown in figure 4b. At the center of the grain, compressive σ_{11} values are found near both sides of the twin. Figures 6a and c are sections close to the edge of the grain, where the compressive σ_{11} at one side of the twin changes to tensile stresses. This change can be related to the influence of neighboring grains. In 2D simulations, only 2 neighboring grains more or less perpendicular to $\vec{b}$ can be considered, whereas in 3D, all neighbors are considered, including those at the lateral side of a twin plate.

The 3D simulation results shown in figure 4b agree qualitatively with the measurements. To get quantitative agreement, improvement of the model is needed. First, the hardening law needs to be fine-tuned. Second, except for the twin regions, it is considered that dislocation slip is the only mode for plastic deformation, however, according to the experimental results, martensitic

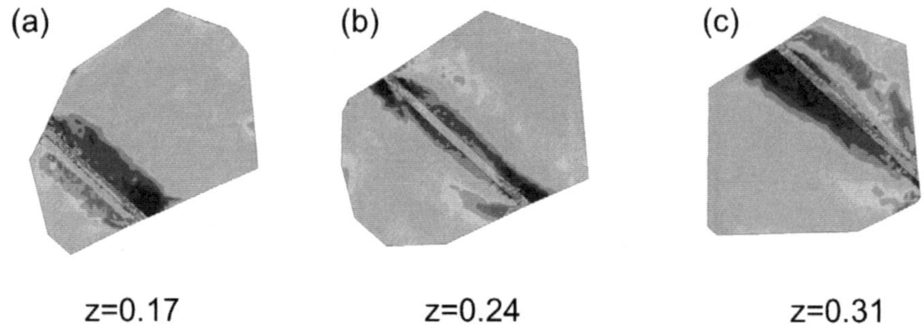

Figure 6. Variation of σ_{11} around the twin in three parallel sections according to the 3D simulation.

transformation and twinning occur simultaneously with dislocation slip. It is therefore necessary to consider the effects of martensitic transformation and twinning in neighboring grains, as well as in the parent grain. These effects will be taken into account in future work.

5. Conclusion

A 3D crystal plasticity finite element model is constructed and it is demonstrated that this model is able to capture the stress field around a mechanical twin. Conventional 2D simulations only give reliable results when the twin is perpendicular to the 2D plane and the twin shear direction is also in the 2D plane. For all the other cases, a 3D model is needed to achieve satisfactory simulations.

Acknowledgements

FXL and LD are mandated by the Belgian National Fund for Scientific Research (FSR-FNRS). Funding was obtained from the Communauté française de Belgique through the contract ARC C1.31105.001. This work was carried out within the framework of the IAP7/21 programme of the Belgian Science Policy.

References

[1] Marteleur M, Sun F, Gloriant T, Vermaut P, Jacques P J and Prima F 2012 *Scripta Materialia* **66** 749–752
[2] Sun F, Zhang J, Marteleur M, Gloriant T, Vermaut P, Laillé D, Castany P, Curfs C, Jacques P and Prima F 2013 *Acta Materialia* **61** 6406–6417
[3] Wilkinson A J, Meaden G and Dingley D J 2006 *Ultramicroscopy* **106** 307–313
[4] Abdolvand H and Wilkinson A J 2016 *International Journal of Plasticity* **84** 160–182
[5] Kumar M A, Kanjarla A, Niezgoda S, Lebensohn R and Tomé C 2015 *Acta Materialia* **84** 349–358
[6] Kumar M A, Beyerlein I J and Tomé C N 2016 *Acta Materialia* **116** 143–154
[7] Kumar M A, Beyerlein I J, McCabe R J and Tome C N 2016 *Nature Communications* **7** 13826
[8] Abdolvand H and Wilkinson A J 2016 *Acta Materialia* **105** 219–231
[9] Barnett M R, Stanford N, Ghaderi A and Siska F 2013 *Acta Materialia* **61** 7859–7867
[10] Zhang R Y, Daymond M R and Holt R A 2008 *Materials Science and Engineering: A* **473** 139–146
[11] Lai M, Tasan C C and Raabe D 2016 *Acta Materialia* **111** 173–186
[12] Ahmed M, Wexler D, Casillas G, Savvakin D G and Pereloma E V 2016 *Acta Materialia* **104** 190–200
[13] Alkorta J 2013 *Ultramicroscopy* **131** 33–38
[14] Marteleur M 2017 *On the plasticity mechanisms of β metastable Ti alloys, towards enhanced work hardening* Ph.D. thesis UCL-Université Catholique de Louvain
[15] Alkorta J, Marteleur M, and Jacques P J 2017 *Ultramicroscopy* **doi:10.1016/j.ultramic.2017.06.015**
[16] Dancette S, Delannay L, Renard K, Melchior M and Jacques P 2012 *Acta Materialia* **60** 2135–2145

IOP Conf. Series: Materials Science and Engineering **219** (2017) 012032 doi:10.1088/1757-899X/219/1/012032

Orientation and length scale effects on dislocation structure in highly oriented nanotwinned Cu

Qiuhong Lu[1], Xiaoxu Huang[2], Niels Hansen[2], Lei Lu[1*]

[1] Shenyang National Laboratory for Materials Science, Institute of Metal Research, Chinese Academy of Sciences, Shenyang, 110016, PR China
[2]Section for Materials Science and Advanced Characterization, Department of Wind Energy, Technical University of Denmark, Risø Campus, DK-4000, Roskilde, Denmark

*Email: llu@imr.ac.cn

Abstract. Highly oriented nanotwinned Cu has been compressed to 6% strain in directions 90°, 0° and 45° with respect to the twin boundaries of the almost parallel twins. In the 90° and 0° compressed samples Mode I and Mode II dislocations and their interactions with twin boundaries dominate the deformation of twin/matrix (T/M) lamellae with thickness less than 500 nm. In 45° compressed samples, Mode III dislocations, especially partial dislocations moving along the twin boundaries, dominate the deformation of fine T/M lamellae with thickness less than 100 nm, while dislocations from slip Modes I, II and III are identified in T/M lamellae more than 100 nm thick, where these dislocations extensively interact in the T/M lamellae with thicknesses more than 200 nm. Dislocation cells are observed in a twin lamella with a thickness of about 500 nm.

1. Introduction

Plastic deformation of metallic materials involves glide, interactions, and storage of dislocations. Important parameters are the structural characteristics and the loading conditions determining the number of potential slip systems, and which of these active [1]. For fcc and bcc single crystals and poly-crystals deforming by multi-slip, the structural evolution follows a universal pattern of structural subdivision by the formation of cell blocks and cells [2]. For small strains, the size of these features are in the micrometer/submicrometer range and their characteristics depend on the crystallographic orientation of the material in which they form and evolve [2, 3]. However, the structural scale is also important as there appears to be a limit of about 5 μm below which a well-defined cell block structure does not form. Examples are deformed and recrystallized Cu with a grain size of about 4 μm [4] and spark plasma sintered Al with a grain size of about 5 μm [5]. This limit may, however, be lower as cell blocks have been observed in twin/matrix (T/M) lamellae with a thickness of 0.5-1 μm in rolled nano-twinned (nt) Cu [6] processed by electrodeposition. The effects of orientation and structural scale on the evolution of deformation microstructure have therefore been chosen as the topic of the present research. The material is electrodeposited nt-Cu with T/M lamellar thickness ranging from tens to hundreds of nanometers [6] which has been compressed in three directions: 90°, 0° and 45° with respect to the twin plane. Samples have been compressed up to a strain of 6% where different slip modes and combinations of slip modes will be activated [7-13]. The deformed microstructure has been analyzed by TEM focusing on individual dislocations, dislocation-dislocation interactions and dislocation-twin boundary (TB) interactions. The T/M lamellar thickness covers different length scales, namely <100 nm, 100-200 nm and <500 nm.

2. Experimental

High-purity Cu sheets with nanoscale growth twins were synthesized by means of direct-current electro-deposition from an electrolyte of $CuSO_4$. More details about the deposition parameters are described in Ref. [7]. The nt-Cu sheet was deposited to a final thickness of more than 1.5 mm by carefully controlling the deposition parameters. The microstructure the as-deposited Cu specimens has been observed and is shown in figure 3 in Ref. [7]. The as-deposited Cu sample consists of micron/submicron-sized columnar grains with {111} out-of-plane texture. The grain size exhibits a distribution from 1 to 6 μm (with an average value of about 3 μm). These columnar grains are subdivided by nanoscale twin lamellae, most of which are parallel to the deposited sheet plane. Statistical analysis indicates that the twin thickness is about 30 nm on average, and varies from a few nanometers to about 500 nm.

Compression samples were cut from the as-deposited Cu sheets and then mechanically polished to a rectangular cuboid of $0.8 \times 0.8 \times 1.2$ mm^3. Uniaxial compression tests at room temperature were performed on an Instron 5848 microtester with a 2 kN load cell at a strain rate of 1×10^{-3} s^{-1}. The compression direction is parallel to the TBs (0° compression), perpendicular to the TBs, and at 45° with respect to the TBs (45° compression). The cross-head displacement was used to determine the imposed strain, with corrections for machine compliance. In this study, the total compression strain is ~6%.

Microstructures of specimens before and after compression were characterized using a FEI Tecnai F20 transmission electron microscope at an accelerating voltage of 200 kV. To examine the dislocation configurations and contributing Burgers vectors $\boldsymbol{b}$ of dislocation structures, two-beam diffraction contrast experiments were carried out in the TEM employing different diffraction vectors, $\boldsymbol{g}$, using the $\boldsymbol{g}\cdot\boldsymbol{b} = 0$ invisibility criterion [14].

Slip systems in nt-Cu can be divided into three categories according to their geometrical relations to the twin boundary plane: Mode I, where both slip plane and slip direction of the active slip systems are inclined to the TB, Mode II, where the slip plane is inclined to TB but the slip direction parallel to the TB, and Mode III, where both the slip plane and slip direction are parallel to the TB. According to the two beam TEM analysis, dislocations from different slip modes can be distinguished based on their visibility under different two beam conditions. With $\boldsymbol{g}_M$=$\boldsymbol{g}_T$=111, Mode I dislocations with Burgers vectors $\boldsymbol{b1}$=1/2[011], $\boldsymbol{b2}$=1/2[110], $\boldsymbol{b3}$=1/2[101] are all visible, while the Mode II or Mode III dislocations $\boldsymbol{b4}$=1/2[$1\bar{0}1$], $\boldsymbol{b5}$=1/2[$01\bar{1}$], $\boldsymbol{b6}$=1/2[$1\bar{1}0$] are all invisible, enabling separation of Mode I dislocations from Mode II and Mode III dislocations. Dislocations from Mode III and II slip systems have the same Burgers vectors ***b4, b5*** and ***b6***, but the dislocation lines of Mode III dislocations must be parallel to the TBs.

3. Results

Figure 1a shows typical observations of dislocation structures from the 90° compressed nt-Cu sample, which was taken with a $\boldsymbol{g}$ vector of $\boldsymbol{g}_M = \boldsymbol{g}_T$=[111]. These dislocations are all visible under $\boldsymbol{g}_M = \boldsymbol{g}_T$= [111], indicating that they have Burgers vectors of ***b1, b2*** or ***b3*** and belong to Mode I slip systems according to above analysis. Most dislocations are accumulated at the TBs and only a few dislocation segments (red arrows in figure 1a) are retained within the matrix and twin lamellae. The dislocation densities are not related very much to the thickness scale of the T/M lamellae. But when the T/M lamellae are very thin, the density of TBs is high, and presents a morphology with very high dislocation density in the fine twin lamellae ("F" in figure 1a). It should be noted that few dislocation interactions or tangling are seen inside the twin lamellae even in the twin lamellae with a thickness of about 200 nm ("C" in figure 1a).

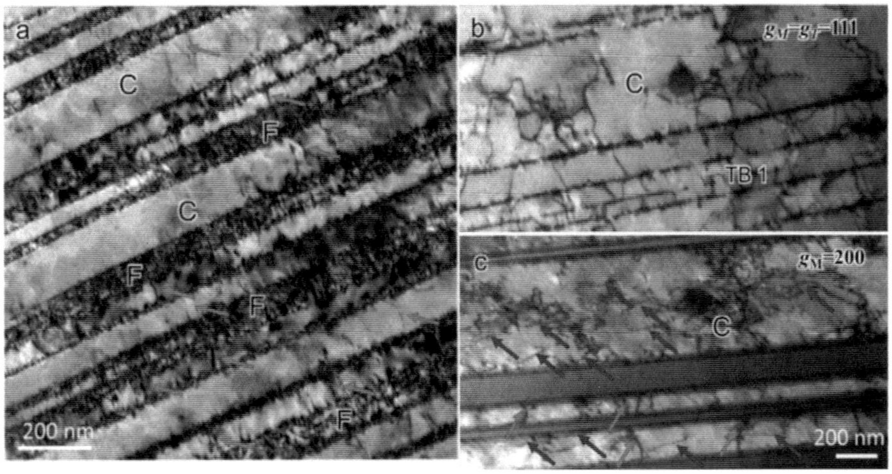

Figure 1. Two-beam diffraction images of dislocations under $\boldsymbol{g}_M = \boldsymbol{g}_T = [111]$ in 90° compression (a) and under $\boldsymbol{g}_M = \boldsymbol{g}_T = [111]$ (b) and $\boldsymbol{g}_M = [002]$ (c) in 0° compression of nt-Cu. The red and blue arrows point out the Mode I and II dislocations, respectively. The fine lamellae with a thickness less than 100 nm are marked "F"; the coarse lamellae with a thickness more than 100 nm are marked "C".

Figures 1b and 1c are typical two-beam diffraction TEM images of the nt-Cu under 0° compression analyzed under $\boldsymbol{g}_M = \boldsymbol{g}_T = [111]$ and $\boldsymbol{g}_M = [200]$, respectively. Observing using $\boldsymbol{g}_M = \boldsymbol{g}_T = [111]$ (figure 1b), Mode I dislocations are clearly seen both in the twin lamellae and on the TBs. The dislocations accumulated on the TBs between fine T/M lamella ("TB 1" in figure 1b) are much less numerous than those between the fine T/M lamellae in 90° compression ("F"s in figure 1a). Observing from $\boldsymbol{g}_M = [200]$ (figure 1c), many short dislocation lines slightly inclined to the TBs (blue arrows in figure 1c) are seen. These dislocations are all invisible under $\boldsymbol{g}_M = \boldsymbol{g}_T = [111]$, indicating that they have Burgers vectors of ***b4*** or ***b6*** and belong to Mode II slip systems. For the twin lamellae with a thickness of less than 200 nm, Mode II dislocations inside the twin lamellae appear to be as numerous as Mode I dislocations. Higher magnification observations in our former investigations [5] indicate, however, that the density of Mode II dislocations is much higher than that of Mode I dislocations but can only be identified by the tails on the TBs. In the twin lamellae with a thickness of more than 500 nm ("C" in figure 1b and 1c), a high density of Mode II dislocations and a few Mode I dislocations are sene and a small extent of dislocation interactions between Mode I and Mode II dislocations is observed. The TEM observations in figure 1 indicate that Mode I and Mode II dislocations dominate the deformation of T/M lamellae with thickness less than 500 nm in 90° and 0° compressions, respectively. Few dislocation interactions happen in twin lamellae less than 500 nm thick when compressed in either the 90° or 0° directions.

Figures 2a and 2b present the typical dislocation structures of T/M lamellae with a thickness less than 200 nm thick in the 45° compressed nt-Cu sample under $\boldsymbol{g}_M = [200]$. Few dislocations appear inside the fine T/M lamellae less than 100 nm thick ("F" in figure 2). As for the twin lamellae of 100-200 nm thick ("C1" in figure 2), different from the observations in 90° and 0° compressed samples, more dislocations inside the T/M lamellae and some dislocation interactions are observed. The dislocations types can be simply distinguished by their morphologies. Dislocation lines cutting through the TBs are Mode I dislocations (red arrows in figure 2a). Mode II dislocations (blue arrows in figure 2a) have dislocation lines slightly inclined to the TBs. However no Mode III dislocations are found inside the T/M lamellae. It is interesting to see that there are plenty of dislocations appearing on the TBs of all the T/M lamellae in shown in figure 2. Additional analysis has shown that these dislocations are all invisible under $\boldsymbol{g}_M = \boldsymbol{g}_T = [111]$, indicating that these are Mode III dislocations with Burgers vectors of ***b4*** or ***b6*** slipping on the TBs. Careful observations show that the dislocations on the

IOP Conf. Series: Materials Science and Engineering **219** (2017) 012032 doi:10.1088/1757-899X/219/1/012032

TBs present two different morphologies. A few dislocations have sharp and clear dislocation lines (arrows "1" in figure 2b), but most dislocations observed show a morphology of wide dark layers (arrows "2" in figure 2b), which should be the twinning partial dislocations. Twinning partials on the TBs have Burgers vectors of δA or δB or δC, can be classified as Mode III dislocations, and may also be visible under $\boldsymbol{g}_M$= [200] ($\boldsymbol{g}\cdot\boldsymbol{b}$=1/3, 2/3 and 1/3, respectively). It is obvious that dislocations from Mode III slip systems, especially partial dislocations slipping on the TBs, dominate the deformation of twin lamellae less than 200 nm thick in 45° compression. These observations are consistent with the previous investigations [7].

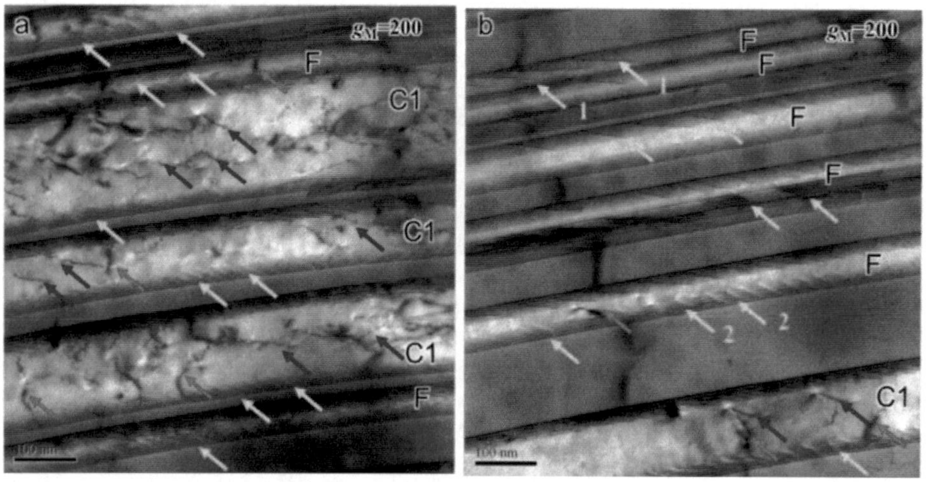

Figure 2. Enlarged two-beam diffraction images showing dislocations in coarse (a) and fine (b) twin lamellae of 45° compressed nt-Cu under under $\boldsymbol{g}_M$=200. The red, blue and green arrows point out the Mode I, II and III (Shockley partial) dislocations, respectively. Arrows "1" marked out two perfect Mode III dislocations on the TB. Arrows "2" marked out two partial Mode III dislocations on the TB. The fine lamellae with a thickness less than 100 nm are marked "F"; the coarse lamellae with a thickness of 100-200 nm are marked "C1".

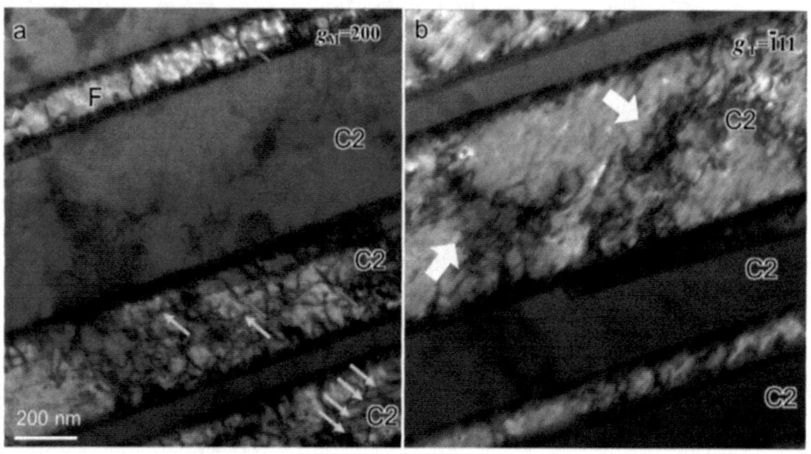

Figure 3. Two-beam diffraction images of dislocations in coarse twin lamellae of 45° compressed nt-Cu under $\boldsymbol{g}_M$=002 (a) and $\boldsymbol{g}_M = \bar{1}11$ (b). The red, blue and green arrows point out the Mode I, II and III dislocations, respectively. Dislocations from slip Modes I, II and III interacting in the twin lamella "C2"; and dislocation cells formed in twin lamella "C2" (about 500 nm) pointed out by white arrows.

Figures 3a and 3b show the typical dislocation structures of coarse T/M lamellae with a thickness of more than 200 nm ("C2" in figure 2) in the nt-Cu sample under g_M= [200] and g_T =[$\bar{1}$11] compressed in 45° direction with respected to the TBs. Compared with the coarse lamellae of 100- 200 nm thick ("C1" in figure 2), more Mode I (red arrows) and Mode II (blue arrows) dislocations appear inside these lamellae. Furthermore, Mode III dislocations (green arrows in figure 3a) with dislocation lines parallel to the TBs are observed inside the lamellae. Most dislocations are tangled inside the T/M lamellae. In a twin lamella with a thickness of about 500 nm dislocations are seen to tangle extensively and appear to be evolving into dislocation cell structures (white arrows in figure 3b).

4. Discussion

When analyzing the dislocation structure one must take into account that the samples have been cold-compressed to the strain of 6%. The microstructure therefore represents a work hardened state. In a previous study [7], samples have been analyzed after 2% and 6% compression, where a significant effect of strain on the microstructure has been found. An example is that slip transmission across twin boundaries is observed after 2% compression both in the 90° and in the 0° compressed samples, most pronounced in the former. By raising the strain to 6%, the frequency of slip transmission is reduced.

Based on such observations and taking the Schmid factor into account (Table 1) the microstructure will be analyzed in the following, separately for the three loading directions.

Table 1. Active slip systems and corresponding Schmid factors for the 90°, 0° and 45° compression.

Loading direction	90°	0°	45°
Active slip systems	Mode I dominant	Mode I and II; (Mode II dominant)	Mode I, II and III (Mode III dominant)
Maximum Schmid factors [7]	Mode I: 0.272; Mode II, III: 0	Mode I: 0.408; Mode II: 0.470; Mode III: 0	Mode I: 0.484; Mode II: 0.360; Mode III: 0.5

In 90° compressed samples, slip Mode I dominates with 6 active slip systems. In 0°compressed samples, slip Modes I and II are active, involving totally 9 systems. In 45°compressed samples, Modes I, II and III are active with 12 systems. As there are 12 systems in the twin and 12 in the matrix, the total number of systems is 21 as 3 systems in the slip plane are common to the twin and the matrix.

To analyze the microstructural characteristics in relation to the active slip systems, the slip distance, which is the distance from a dislocation source to a dislocation barrier (in the present case, a twin boundary), must also be taken into account. Dislocation sources can be grain boundaries, with a spacing of about 3 μm, and twin boundaries, with an average spacing of 30 nm. In general, it is expected that the flow stress decreases with increasing slip distance, which has been verified previously in [8] where the yield stress decreases with an expected increasing in slip distance when the loading direction is changed from 90° to 0° and from 0° to 45°. It follows that in the 90° compressed samples, the slip distance is small, of the order of the twin thickness. Dislocation-dislocation interaction is negligible and at 6% strain, the majority of dislocations are stored at or near the twin boundaries. Therefore a small effect of the structural scale is expected as is observed. In 0° compressed samples, the slip modes are I and II and the slip distance is only slightly larger than that in 90°, and therefore there is not a significant effect on the structural scale. Dislocations are stored at or near the twin boundaries and dislocation-dislocation interaction is only occasionally observed at the larger length scale of 200-300 nm. In 45° compression, with 3 active slip modes and a large slip distance, the effect of the structural scale is significant. With T/M lamellar thickness in the range 200-500 nm dislocation-dislocation interactions are frequent and dislocations can move in 3 dimensions as shown by the presence of dislocation cells (see Fig 3). However, when the structural scale is reduced to 50-200 nm, mode III will dominate with dislocations moving at, and parallel with, the twin boundaries, and the resulting structure in the T/M lamellae contains only isolated dislocations.

In the present analysis, the relationship between active slip systems and the deformation microstructure shows a strong effect of loading direction. A similar effect on the flow stress and work hardening behavior is also expected, which has been observed [8] and is to be further analyzed in a forthcoming paper.

5. Conclusions

Highly oriented nt-Cu, with the T/M lamellar thickness varying from a few nanometers to about 500 nm and being 30 nm on average, has been compressed to 6% strain in directions of 90°, 0° and 45° with respect to the TBs of the almost parallel twins. The dislocation structures in T/M lamellae with different thicknesses have been characterized. It is found that the active slip systems and dislocation configurations are strongly dependent on the loading direction.

In the 90° and 0° compressed samples, mode I and mode I + mode II, respectively, dominate the deformation of T/M lamellae with thicknesses in the range of 50-500 nm. Some dislocations are observed in the T/M lamellae but most dislocations are accumulated at the twin boundaries.

In the 45° compressed sample, a clear length scale effect is observed. In the range 200-500 nm, all three slip modes (I, II, and III) are active, and dislocation interactions can take place in the T/M lamellae resulting in the formation of dislocations cells. In the range of 50-200 nm, slip mode III dominates with dislocations moving at, and parallel to, the twin boundaries, with some isolated dislocations stored in the T/M lamellae.

Acknowledgements

This research was supported by National Nature Science Foundation of China (Nos. 51420105001, 51371171, 51101163, 51401211 and U1530146).

References

[1] Kocks UF, Canova GR 1981, in *Proc. of 2nd Risø Inter. Sym. on Mater. Sci. (Roskilde)*, 35–44
[2] Hansen N, Huang X, Pantleon W, Winther G 2006 *Philos. Mag.* **86** 3981–94
[3] Huang X, Winther G 2007 *Philos. Mag.* **87** 5189–214
[4] Huang X, Hansen N 2004 *Mater. Sci. Eng.* **387–389** 186–90
[5] Le GM, Godfrey A, Hansen N, Liu W, Winther G, Huang X 2013 *Acta Mater.* **61** 7072–86
[6] Lu QH, Sui ML, Huang X, Li DX Hansen N 2014 *Philos. Mag.* 94 2262–80
[7] Lu QH, You ZS, Huang X, Hansen N, Lu L 2017 *Acta Mater.* **127** 85–97
[8] You ZS, Li XY, Gui LJ, Lu QH, Zhu T, Gao HJ, Lu L 2013 *Acta Mater.* **61** 217–27
[9] Li XY, Wei YJ, Lu L, Lu K, Gao HJ 2010 *Nature* **464** 877–80
[10] Zhang X, Misra A 2004 *J. Appl. Phys.* **96** 7173–8
[11] Wang J, Li N, Misra A 2013 *Philos. Mag.* **93** 315–27
[12] Hodge AM, Wang YM, Barbee TW, Jr. 2008 *Scr. Mater.* **59** 163–6
[13] Deng C, Sansoz F 2009 *Nano Letters* **9** 1517–22
[14] Williams D, Carter CB, 2009 *Transmission Electron Microscopy*, Springer US, 3–22

EBSD characterization of deformed lath martensite in IF steel

Z Lv[1,2*], X Zhang[2], X Huang[2], N Hansen[2]

[1]Key Laboratory of Advanced Forging & Stamping Technology and Science (Ministry of Education of China), Yanshan University, Qinhuangdao 066004, China
[2]Section for Materials Science and Advanced Characterization, Department of Wind Energy, Technical University of Denmark, Risø Campus, DK-4000 Roskilde, Denmark

*E-mail: zqlv@ysu.edu.cn

Abstract. Rolling deformation results in the transformation of a lath martensite structure to a lamellar structure characteristic to that of IF steel cold-rolled to medium and high strains. The structural transition takes place from low to medium strain, and electron backscatter diffraction analysis shows that the frequency of medium angle boundaries with misorientation angles of 5-10° decreases with increasing strain, while the frequencies of boundaries with angles in the ranges of 1-5° and 10-25° increase, resulting in the evolution of a bimodal misorientation angle distribution. The microstructural evolution and the strength are characterized for lath martensite rolled to a thickness reduction of 30%, showing that large changes in the misorientation take place, while the strain hardening rate is low.

1. Introduction

Martensitic transformation (MT) is an efficient grain refinement process, which introduces high densities of high angle boundaries, low angle boundaries and dislocations. Ferrous martensite shows a variety of morphologies [1-3], namely lath, butterfly, lenticular and thin plate shapes. Among these, lath martensite (LM) has the most obviously industrial significance because it appears in quenched commercial steels, such as plain low-carbon steels, low-carbon and low-alloy steels, maraging steels, and interstitial free (IF) steels [3-5]. In low carbon (below 0.4 wt % C) steels, the martensitic structure is divided at decreasing scale into packets, blocks, sub-blocks and laths, as shown in figure 1. In the case of the K-S orientation relationship, there are 24 variants (from V1 to V24), with different direction and parallel plane relationships in the martensitic structure (see reference [2] for the definition of the 24 variants). A packet consists of parallel blocks [3], and the laths in the blocks of a given packet can have up to 6 variants (e.g. V1-V6), with different parallel direction relationships on the same conjugate parallel close packed plane [2, 3, 6]. Each block consists of laths of two specific K-S variant groups (sub-blocks), which are misoriented by small angles of about 10 degrees, such as V1 and V4, V2 and V5, and V3 and V6 [2, 3].

Plastic deformation (PD) is another efficient grain refinement process that can further improve the mechanical properties of metallic materials. A combination of MT and PD can further enhance the grain refinement, however, the hardening behavior of LM is different from that of the ferrite structure, especially at low strains [7]. Huang et al. found that the hardness of deformed LM increases by < 10% after cold rolling to 30%, whereas the hardness of deformed ferrite (with different original grain sizes) increases by 60 ~ 80% after a similar deformation strain [7]. There is a transition range at low strains (< 30%) where the work hardening is limited. It is therefore important to analyze the structural evolution of martensite during the early stages of cold rolling.

The present study follows previous research [7], with an aim of characterizing the deformed LM martensite at low strains (up to 30% rolling reduction) using electron backscatter diffraction (EBSD). The structure evolution is analyzed based on the changes of misorientation angles, as obtained from the EBSD data, and the evolution of mechanical properties with deformation is discussed.

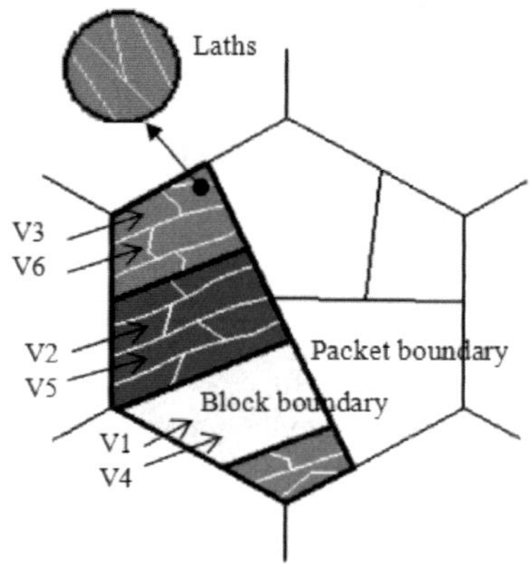

Figure 1. Schematic illustration showing a typical lath martensite structure in low carbon (0-0.4%) steels.

2. Material and experimental procedures

An IF steel (0.0026 C, 1.48Mn, 0.046Ti, 0.015Al, 0.0026 B, Fe balance, mass %) was used in this study. After austenitizing at 1473 K for 0.6 ks, specimens were quenched into iced brine to obtain a LM structure. The specimens with LM structure were cold rolled (CR) to 10% and 30% thickness reductions. Microstructures were characterized using a Zeiss Supra-35 scanning electron microscope (SEM) equipped with an Oxford Instruments EBSD detector. The SEM was operated at 20 kV. For the EBSD measurements, the samples were prepared by mechanical grinding and polishing with a final step using 1.0 μm diamond paste. The samples were then electrochemically polished at 30 kV for 20s in a methanol solution with 10 vol% perchloric acid. The EBSD maps were taken using a step size of 0.2 μm. Orientation data were collected using the Channel 5 acquisition software. Misorientation angles were calculated for statistical analysis where the misorientation angles of < 1° were excluded considering the angular resolution of the EBSD technique. To allow reliable statistics data for more than 1000 misorientation angles were collected in different areas of the EBSD maps for each sample.

3. Results and discussion

3.1 Un-deformed martensite

Here we use an inverse pole figure (IPF) coloring in the EBSD maps to show the microstructure of the initial sample without deformation, as shown in figure 2. The boundaries of the prior austenite grains (black lines), packets (blue lines) and blocks (red lines) can all be identified; an example is shown in figure 2b.

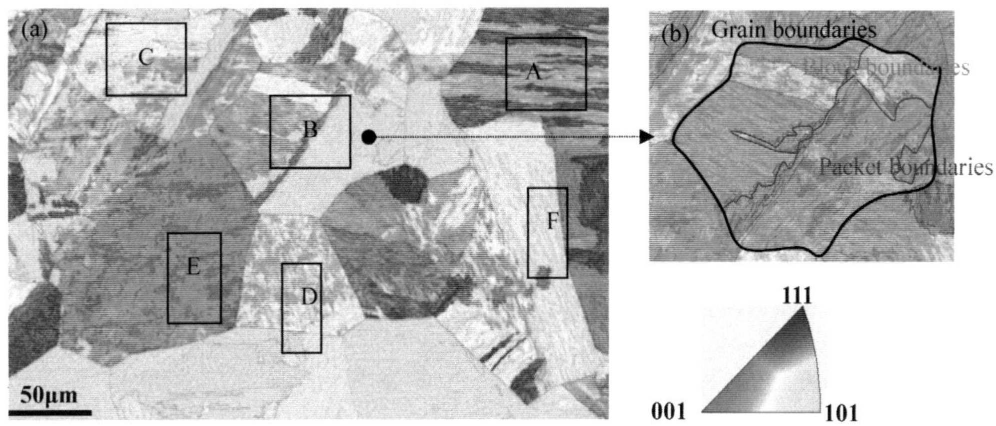

Figure 2. EBSD map of the un-deformed as-transformed martensite.

Figure 3a shows the distribution of misorientation angles between the 24 variants in the un-deformed LM. The misorientation angles between neighboring points were collected within individual prior austenite grains in the areas marked by A, B and C in figure 2a. Figure 3b lists the misorientation angles between V1 and other variants (V2-V24) forming the K-S orientation relationship [2]. If martensite variants are randomly distributed during the lath martensite formation process, boundaries with misorientation angles of > 15° should be much more frequent than boundaries with misorientation angles of <15°, as shown in figure 3b. However, the measured frequency of high angle boundaries is much smaller than the frequency of medium angle boundaries in the present case, as shown in figure 3a. The percentage of misorientations between 5-10° is about 24 %, which indicates the presence of a high number of sub-block boundaries (such as pairs with V1 and V4, V2 and V5, and V3and V6) in the LM, in agreement with previous observations [2,3,7].

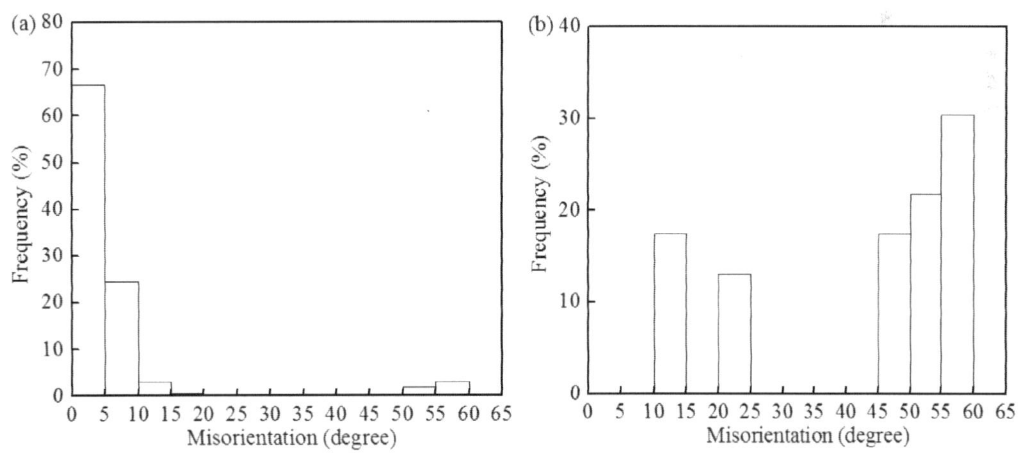

Figure 3. Misorientation angle distribution in un-deformed martensite: (a) from EBSD maps, excluding prior austenite grain boundaries; (b) between V1 and other variants (V2-V24) based on the K-S orientation relationship.

The misorientation angles between neighboring points were also measured in different blocks (such as areas D, E and F as marked in figure 2a). The resulting misorientation distribution based on all the measurements is shown in figure 4. The experimental results show the presence of sub-block boundaries represented by a weak peak at 7° to 8° in the un-deformed LM, which agrees with previous findings [7]. The percentage of misorientations below 5°, related to the lath boundaries, is 70%, and that from 5° to 15°, related to sub-block boundaries, is 30%.

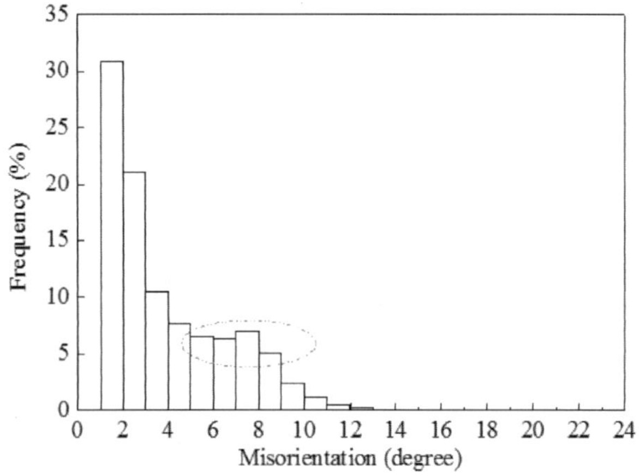

Figure 4. Misorientation distribution in blocks of un-deformed LM (excluding block boundaries).

3.2. Deformed microstructure

Figure 5 shows example EBSD maps of deformed martensite after 10% and 30% cold rolling. The rolling introduces some deformed structure, seen in area H of figure 5a and in area M of figure 5b, and is most pronounced in the 30% cold-rolled sample, where a certain structural alignment can be observed. The effect of cold rolling to low-strain on the shape of the prior austenite grains is observed to be small.

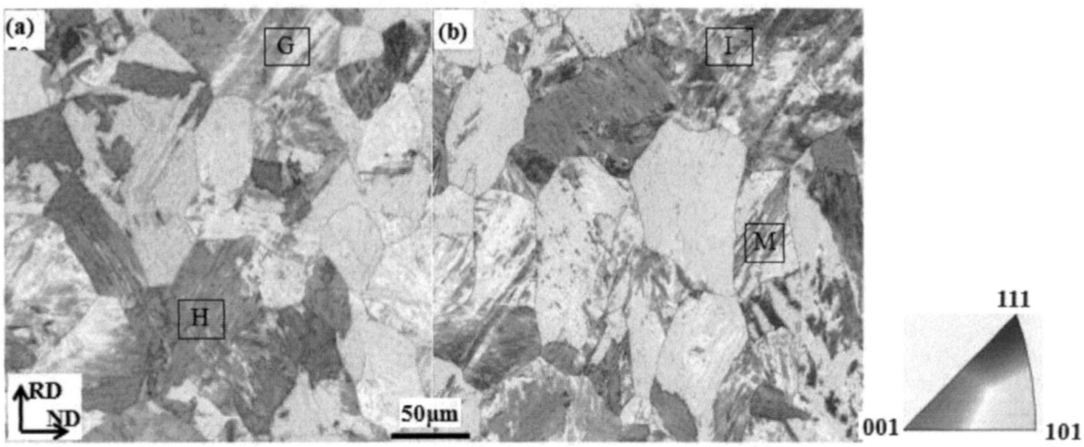

Figure 5. EBSD maps of cold rolled martensite: (a) 10% and (b) 30% thickness reduction.

Figure 6 shows the misorientation distribution from EBSD maps of the deformed LM after 10% and 30% cold rolling. The misorientation angles between neighboring points area again collected only from within the prior austenite grains.

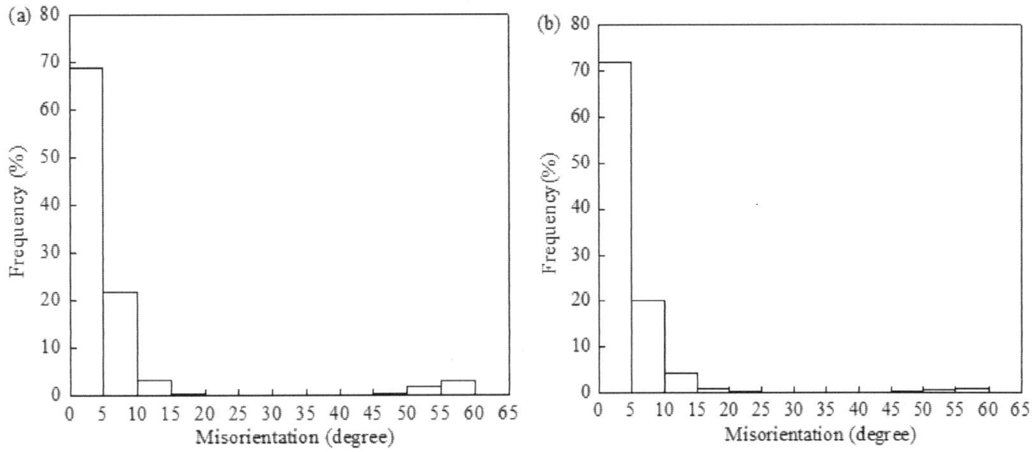

Figure 6. Misorientation distribution of deformed LM: (a) 10% CR. and (b) 30% CR.

The distributions of misorientation angles measured within different blocks at the two rolling reductions (such as areas G, H, I, and M in figure 5), considering all misorientations >1°, are illustrated in figure 7. The results show that high misorientation angles appear after 10% and 30% cold rolling. Misorientation angles up to 17° are observed in the 10% CR sample and to about 21° in the 30% CR sample. The percentage of misorientations below 5° is 68%, from 5° to 10° about 27%, and above 10° about 5% in the 10% cold-rolled sample. In the 30% CR sample, the percentage of misorientations below 5° is 74%, from 5° to 10° about 20%, and above 10° about 6%. The frequency of medium angles (5-10°) gradually decreases with increasing rolling reduction, and is accompanied by an increase in the frequency of lower (1-5°) and higher (> 10°) angles in the blocks.

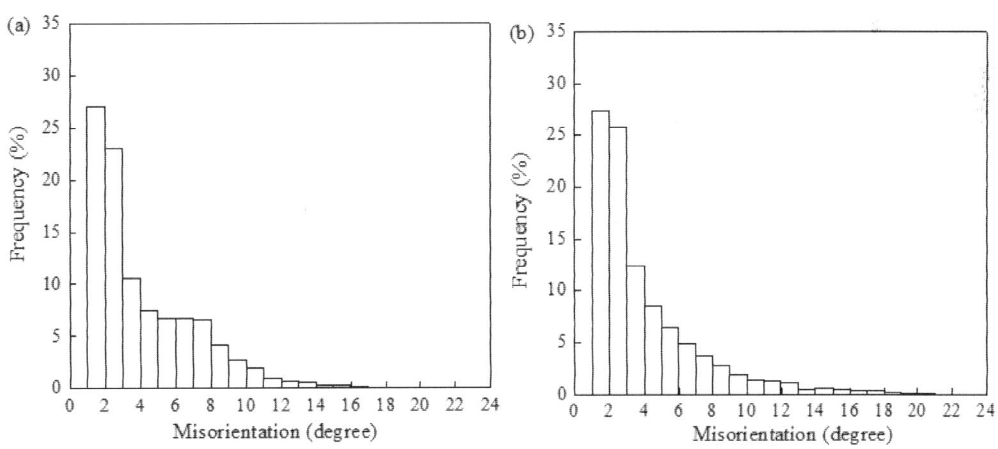

Figure 7. Misorientation distribution in blocks of deformed martensite (a) 10 % CR, (b) 30 % CR.

4. Discussion

The changes in structure and strength of lath martensite when cold rolled up to 80% (ε_{vM} = 1.86) have been analyzed in a recent paper [7]. It was found that the martensite structure broke down and transformed into a lamellar cell block structure characteristic of plastically deformed bcc and fcc metals [8-11]. It was also found that changes in strength are small for reductions up to 30%, followed by a stage of almost constant hardening rate typically characteristic of stage IV hardening [12]. The previous structural analysis based on transmission electron microscopy (TEM) is in the present study replaced by the EBSD analysis, covering only the transition region up to a rolling strain of 30%. In parallel to the structural analysis, the changes in flow stress and hardness have been determined. The structural analysis shows significant changes in the distribution of misorientation angles across dislocation boundaries that subdivide the martensite structure, with the fraction of boundaries with misorientation angles 5-10° decreasing and the fractions of boundaries of smaller angle (1-5°) and larger angles (10-25 °) increasing with increasing cold-rolling reduction, as shown in table 1.

Table 1. Percentage of misorientation angles in different ranges in the LM samples

Sample	1-5° (%)	5-10° (%)	10-25° (%)	25-45° (%)	45-65° (%)
Ideal K-S OR	0	0	30.4	0	69.6
Un-deformed LM	66.7	24.5	3.5	0	5.3
10% CR	68.9	21.8	3.8	0.1	5.4
30% CR	71.8	20.1	5.8	0.1	2.2

The misorientation angle distribution thereby evolves towards a bimodal distribution characteristic of bcc and fcc metals rolled to medium and high strains [13]. The strength analysis (not shown here) shows a relative small increase in hardness and flow stress in the transition region (0-30% thickness reduction). The hardening rate is low and of the same order as observed previously in martensite and IF steel rolled to reductions in the same range [7]. It can be assumed that the strengthening mechanisms in rolled martensite primarily are dislocation strengthening and boundary strengthening. It should be noted, however, that the contribution of these mechanisms to the strength cannot be quantified exactly based only on EBSD data, as they exclude dislocation boundaries with an angle < 1°. Qualitatively, however, the changes in the misorientation angle distribution during rolling suggest that contributions from both hardening mechanisms may increase, but not to a large extent. The evolution of structure and strength of martensite when rolled up to 30% shows a near balance of hardening and softening mechanisms. The hardening mechanisms relate to an increase in the misorientation angle across dislocation boundaries and to a decrease in boundary spacing. This will lead to strengthening, which is counteracted by dislocation annihilation and boundary removal, e.g. by triple junction motion [14]. The result is a remarkable structural change with only a small effect on the strength and the strain hardening behavior. The contributions of hardening and softening mechanisms in the transition region are under investigation by EBSD and TEM, but a major challenge is the structural heterogeneity caused by the operation and interaction of dislocation glide processes, which affects the microstructural evolution, where an example is the formation of micro-shear bands observed as S-bands [12,13].

5. Conclusions

An IF steel with a lath martensite structure has been cold rolled 10% and 30% thickness reductions and the structural evolution has been followed by EBSD analysis. The following conclusions can be drawn.

(1) The frequency of medium angle boundaries (5-10°) in martensite blocks decreases with increasing strain, while the frequencies of boundaries of smaller angles (1-5°) and of larger angles (> 10°) increase. The overall result is the evolution towards a bimodal misorientation angle distribution, characteristic for IF steel rolled to medium and large strains.

(2) The structural morphology changes from a lath martensite structure to a lamellar cell block structure characteristic of cold rolled IF steel. The strain hardening rate is low and the experiment demonstrates that significant structural changes can take place with only a limited effect on the flow stress and hardness. This can be understood in terms of a balance between strain hardening and dynamic

softening by recovery.

Acknowledgements
The authors would like to thank D. Juul Jensen for many helpful discussions. This work was supported by the Natural Science Foundation of Hebei Province for Distinguished Young Scholar (E2017203036).

References

[1] Maki T 1990 *Mater. Sci. Forum* **56-58** 157
[2] Morito S, Tanaka H, Konishi R, Furuhara T and Maki T 2003 *Acta Mater.* **51** 1789-1799
[3] Morito S, Huang X, Furuhara T, Maki T and Hansen N 2006 *Acta Mater.* **54** 5323-5331
[4] Krauss G and Marder A R 1971 *Metall. Trans. A* **2** 2343-2347
[5] Maki T, Tsuzaki K and Tamura I 1980 *Trans. ISIJ* **20** 207-211
[6] Stormvinter A, Miyamoto G, Furuhara T, Hedstrom P and Borgenstam A *2012 Acta Mater.* **60** 7265-7274
[7] Huang X, Morito S, Hansen N and Maki T 2012 *Metall. Mater. Trans. A* **43** 3517-3531
[8] Hughes D A and Hansen N 2000 Acta Mater. **48** 2985-3004
[9] Li B L, Godfrey A, Meng Q C, Liu Q and Hansen N 2004 Acta Mater. **52** 1069
[10] Liu Q, Huang X, Lloyd DJ, Hansen N 2002 Acta Mater. **50** 3789
[11] Zhang X, Hansen N, Gao Y and Huang X 2012 *Acta Mater.* **60** 5933
[12] Hansen N 2001 *Metall. Mater. Trans. A* **32** 2917-2935
[13] Hughes D A and Hansen N 1997 *Acta Mater.* **45** 3871
[14] Yu T, Hansen N, Huang X and Godfrey A 2014 *Mater. Res. Lett.* **2** 160

IOP Conf. Series: Materials Science and Engineering **219** (2017) 012034 doi:10.1088/1757-899X/219/1/012034

Recrystallization texture in nickel heavily deformed by accumulative roll bonding

O V Mishin[1], Y B Zhang[1], and A Godfrey[2]

[1] Section for Materials Science and Advanced Characterization, Department of Wind Energy, Technical University of Denmark, Risø Campus, 4000 Roskilde, Denmark
[2] Key Laboratory of Advanced Materials (MOE), School of Materials Science and Engineering, Tsinghua University, Beijing 100084, China

E-mail: olmi@dtu.dk

Abstract. The recrystallization behavior of Ni processed by accumulative roll bonding to a total accumulated von Mises strain of 4.8 has been examined, and analyzed with respect to heterogeneity in the deformation microstructure. The regions near the bonding interface are found to be more refined and contain particle deformation zones around fragments of the steel wire brush used to prepare the surface for bonding. Sample-scale gradients are also observed, manifested as differences between the subsurface, intermediate and central layers, where the distributions of texture components are different. These heterogeneities affect the progress of recrystallization. While the subsurface and near-interface regions typically contain lower frequencies of cube-oriented grains than anywhere else in the sample, a strong cube texture forms in the sample during recrystallization, attributed to both a high nucleation rate and fast growth rate of cube-oriented grains. The observations highlight the sensitivity of recrystallization to heterogeneity in the deformation microstructure and demonstrate the importance of characterizing this heterogeneity over several length scales.

1. Introduction

Accumulative roll bonding (ARB) [1,2] is a severe plastic deformation technique, in which a single cycle consists typically of 50% rolling, cutting of the rolled material in half, cleaning and wire brushing of the surface, and then stacking the halves prior to repeating this process in subsequent cycles. This ARB process produces intrinsically heterogeneous microstructures as bonding interfaces are introduced during the multi-pass process. Regions near these interfaces have been found to be more refined than regions in between the interfaces [3–6], termed core regions in this work. In addition to these microscopic heterogeneities, the ARB process introduces severe sample-scale heterogeneities, which are reflected both in the through-thickness distribution of different texture components and also in the microstructure [3–10]. In recent publications on ARB-processed nickel, it has been shown that both microscopic and sample-scale heterogeneities have a significant influence on recrystallization [11,12]. Nevertheless, a strong cube texture has been observed in nickel processed by ARB to high strains [12]. In the present paper, formation of the cube texture is analyzed in nickel processed by ARB to a high von Mises strain, $\varepsilon_{vM} = 4.8$. Both the nucleation and growth of grains with different crystallographic orientations are characterized in this material, and the results are discussed in relation to the microstructural heterogeneities. The results are also compared to those obtained for nickel recrystallized after conventional rolling to similar strains.

2. Experimental

Strips of ~99.97% pure nickel with an average initial grain size of ~20 μm were deformed using a mineral oil as a lubricant [13]. The first 50% rolling pass was conducted on a strip with an initial thickness of 2 mm. The 1-mm thick sample was then cut in half, degreased and wire-brushed, after which the halves were stacked and rolled 50% again, creating a bonding between the two strips. After 6 passes, the total von Mises strain ε_{vM} was 4.8, and the sample consisted of 32 individual layers. The deformed material was then annealed in air at 220 °C for different periods of time.

Specimens from the longitudinal section containing the rolling direction (RD) and the normal direction (ND) were prepared for transmission electron microscopy (TEM) and scanning electron microscopy including electron backscatter diffraction (EBSD). A step size of 30 nm to 50 nm was used for a detailed analysis of deformed and recovered microstructures. Maps covering the entire sample thickness were obtained using a step size of 0.5 μm. For analysis purposes, the data set covering entire sample thickness was divided into two subsurface layers, two intermediate layers and one central layer, each with a thickness of 0.2 mm. The data obtained from the two opposite layers (either subsurface or intermediate) were averaged. Thus, only three distinct layers (subsurface, intermediate and central) are described in the following work.

Low angle boundaries (LABs) and high angle boundaries (HABs) were defined in the EBSD maps as those with misorientations θ between 2° and 15° and θ >15°, respectively. Recrystallized grains were defined as regions having a size of >3 μm with a misorientation angle of at least 1° (excluding twin misorientations). Three texture types were identified in the EBSD data: (i) the rolling texture, described as a combination of the Brass {110}⟨112⟩, S {123}⟨634⟩, Copper {112}⟨111⟩ and Dillamore {4 4 11}⟨11 11 8⟩ orientations; (ii) the shear texture, described as a combination of the {100}⟨011⟩, {111}⟨011⟩, {111}⟨112⟩, and {112}⟨110⟩ orientations; and (iii) the cube {001}⟨100⟩ texture.

3. Results

3.1. Deformed microstructure

The as-deformed microstructure is shown in figure 1(a), where it is apparent that regions within 50 μm to 80 μm from the immediate surface contain an increased frequency of the shear texture components. Orientations of the rolling texture dominate in the intermediate and center layers (figure 1(a) and table 1). A small frequency (~1%, see table 1) of cube-oriented lamellae and subgrains are present even in these layers. Considering the sample-scale heterogeneity, the subsurface contains the smallest fraction of the rolling texture components (f_{Rol}), and the smallest HAB spacing (d_{HAB}) along the ND (table 1). The difference in the average boundary spacing $d_{\theta>2°}$ along the ND is however small, 155–160 nm.

The black elongated features in figure 1 correspond to fragments of the steel wire brush used for cleaning the surface before stacking and roll-bonding. These steel fragments are imprinted along the bonding interfaces (see figure 2). Statistical analysis has shown that the frequency of coarse (> 10 μm along the RD) steel fragments is higher along the interfaces created during the last two (the fifth and the sixth) passes [11]. The presence of these fragments results in formation of characteristic particle deformation zones (figure 2) similar to those previously observed around coarse particles in conventionally rolled materials [14,15]. Therefore, these fragments are also termed steel particles in the present work.

Table 1. Microstructural parameters and texture in the subsurface, intermediate, and central layers of the deformed material (data reproduced from [11] with permission of Springer).

Layer	f_{Rol} (%)	f_{Shear} (%)	f_{Cube} (%)	$d_{\theta>2°}$ (nm)	d_{HAB} (nm)
Subsurface	48	17	1.7	160	285
Intermediate	82	2	1.1	156	310
Center	80	2	1.1	155	302

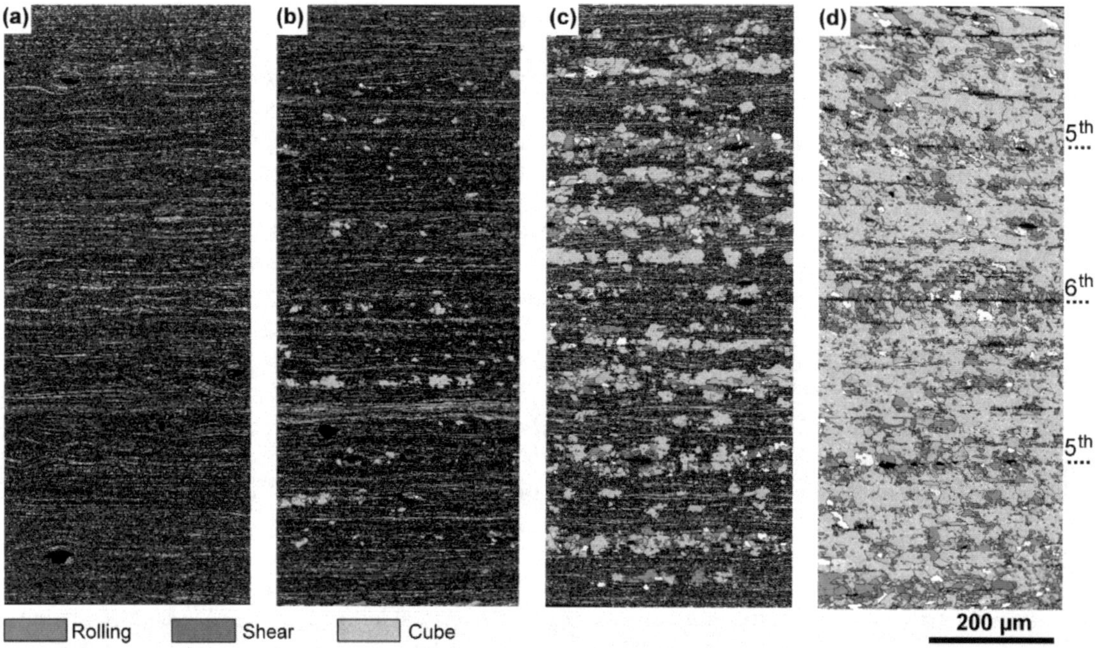

Figure 1. Orientation maps showing the microstructure over the entire thickness of the as-deformed sample (a) and of the samples annealed at 220 °C for 15 min (b); 30 min (c); and 90 min (d). The interfaces during the last two (the fifth and the sixth) rolling passes are indicted by the dashed lines on the right-hand side. LABs, HABs, and twin boundaries are shown as white, black and purple lines, respectively. The RD is parallel to the scale bar.

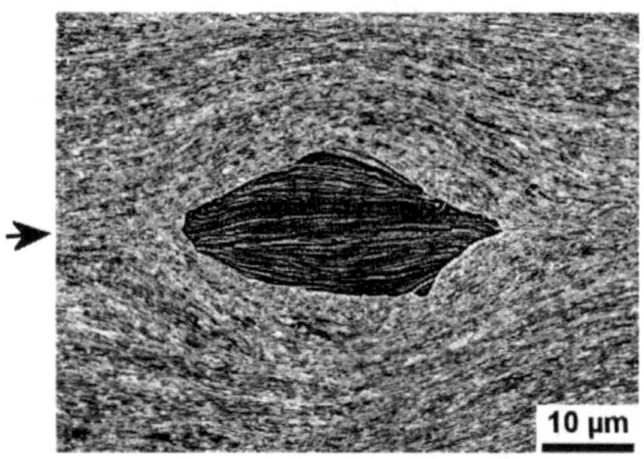

Figure 2. Backscatter electron image showing a particle deformation zone around a fragment of the steel wire brush used for cleaning the surface before stacking and roll-bonding. The arrow marks the position of a bonding interface.

3.2 Annealed microstructure

Annealing at 220 °C for 15 min results in the onset of recrystallization, with a large frequency of recrystallized grains nucleated near the bounding interfaces (typically at a distance of at least 1 μm from the closest interface), in particular, within particle deformation zones (figure 3). Other nucleation sites are shear bands and cube-oriented lamellae within the core regions. Many successful nuclei have orientations of the cube texture as seen in figure 1(b). In contrast, grains nucleated within the particle deformation zones and along the bonding interfaces formed during the last two rolling passes are characterized by a greater variety of orientations (figure 1(b) and 3(b)). The average size of recrystallized cube-oriented grains is 2 to 3 times greater than that for any other orientation class (see table 2). For each orientation class, the difference in the average grain size between the three layers is very small.

Table 2 also presents the nucleation rate, calculated as $N_v = N/(A\ d_{Rex+twins})$, separately for the different texture classes, where N is the number of recrystallized grains belonging to a certain texture class in an area A, and $d_{Rex+twins}$ is the average diameter of those grains, where twins are considered to be individual grains. It is seen that N_v for the cube-oriented grains in the subsurface is about one half of that in the other two layers. In each layer, the highest value of N_v is obtained for the rolling+shear texture, while the lowest value of N_v is recorded for the cube texture (see table 2). However, when the nucleation rate is normalized by the corresponding texture fraction in the as-deformed microstructure, a different situation emerges, where the normalized nucleation rate is the highest for the cube-oriented grains, and is lowest for the rolling+shear texture grains (table 2).

Table 2. Parameters characterizing the recrystallized grains, taking twins into account, in the ARB-processed Ni sample after 15 min at 220 °C. Here "roll+sh" correspond to regions of either the rolling or shear texture.

Layer	$d_{Rex+twins}$ (μm)			$N_v \times 10^{-3}$ (μm^{-3})			Normalized $N_v \times 10^{-3}$ (μm^{-3})		
	cube	roll+sh	other	cube	roll+sh	other	cube	roll+sh	other
Subsurface	5.5	2.1	3.0	0.08	0.57	0.43	5.0	0.8	1.9
Intermediate	5.7	2.1	2.9	0.15	0.75	0.57	13.8	0.9	3.8
Center	5.8	1.9	2.8	0.14	0.57	0.24	13.3	0.7	1.5

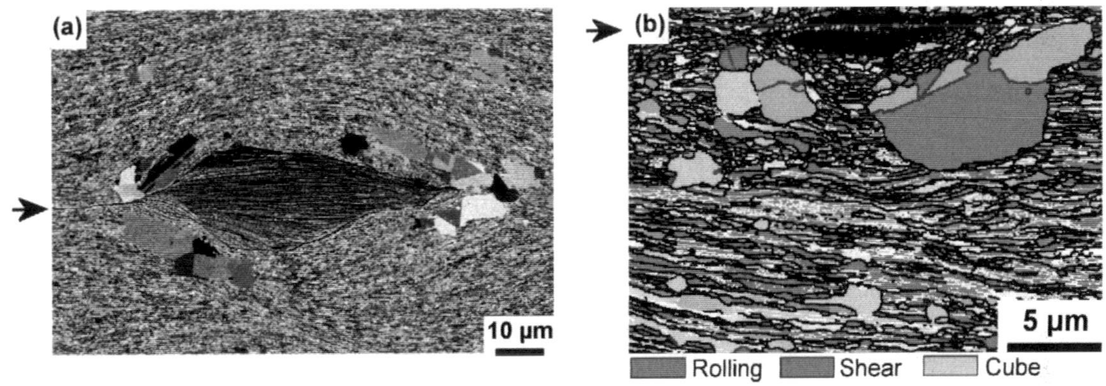

Figure 3. Examples of nucleation within particle deformation zones during 15 min of annealing at 220 °C: (a) backscatter electron image; (b) EBSD map, where LABs, HABs, and twin boundaries are shown as white, black and purple lines, respectively. Arrows mark bonding interfaces.

After 30 min of annealing the fraction recrystallized f_{Rex} is considerably higher in the intermediate layer than in the other layers (figure 4(a)). As f_{Rex} reaches at least 86% after 90 min of annealing, the difference in f_{Rex} between the layers becomes less significant. The average size of cube-oriented grains in each partially recrystallized condition is larger than that for any other orientation class (figure 4(b)), and the difference in $d_{Rex+twins}$ for the cube-oriented grains in the different layers remains small even in the almost fully recrystallized conditions. It is also noteworthy that grains with the rolling or shear texture have the smallest average size. There is also almost no difference in the evolution of texture between the intermediate and central layers. However, the subsurface layer contains consistently lower fractions of the cube texture, ending up with 55 % after 210 min of annealing, as compared to 66-69% in the other two layers (figure 5).

4. Discussion

ARB introduces both sample-scale and microscopic deformation heterogeneities arising directly from the nature of the process. The sample scale heterogeneities are similar to those produced by conventional rolling with large draughts [16,17] where a pronounced shear texture is developed near the surface of a rolled flat product. In ARB, after the cutting, wire brushing, and stacking process, some of this subsurface material is moved to the mid-thickness of the sample. The rolling texture in these regions after subsequent 50% is still slightly weaker than elsewhere, especially along the interfaces formed during the last two passes. Since these former subsurface regions experienced an additional shear deformation, the boundary spacing here is smaller, and hence the stored energy is higher than in the core regions. Additionally, coarse steel particles result in the development of particle deformation zones, which also contribute to a greater variety of orientations, a weaker rolling texture and a greater stored energy in the near-interface regions. The higher stored energy near the interfaces provides a higher driving force for recrystallization, which explains the high frequency of nuclei observed in the near-interface regions (figure 1(b–d)).

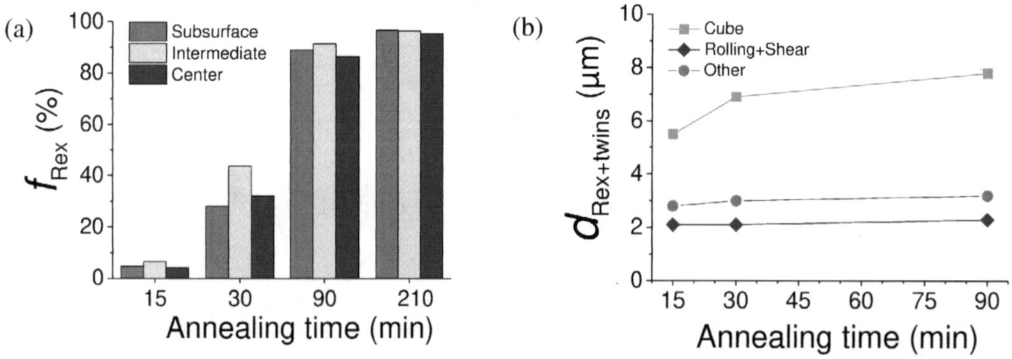

Figure 4. Parameters of recrystallized grains developed in ARB-processed Ni during annealing at 220 °C: (a) area fraction; (b) average sizes of grains calculated taking twin boundaries into account and covering the entire sample thickness.

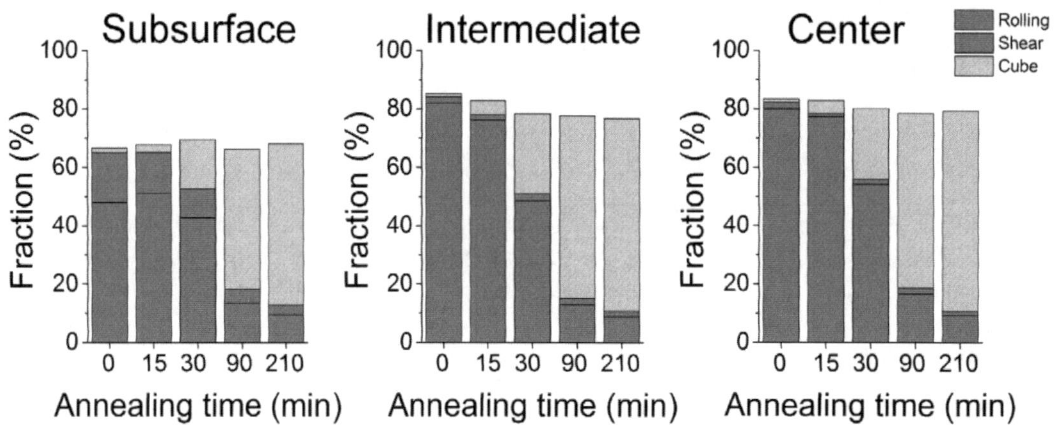

Figure 5. Fractions of different texture components in the three layers as a function of annealing time at 220 °C.

During recrystallization the cube grains quickly outgrow grains of other orientations, similar to the previous observations of conventionally rolled Ni [18, 19]. This growth advantage does not seem to be associated with the influence of impingement, as both cube and non-cube grains appear equally clustered (see figure 1(c)), though it is certainly only a qualitative assessment. A more detailed analysis based on full determination of the Cahn-Hagel growth rates [20] for each orientation class is needed to confirm this.

Although recrystallization in the core regions, resulting in the strong cube texture, appears phenomenologically similar to that in conventionally rolled nickel [18,19], the presence of microscopic deformation heterogeneities in the form of interfaces with imprinted steel particles makes the annealing behavior and the recrystallized microstructure in the ARB-processed material rather unique. The microscopic heterogeneities produced by ARB are inherited in the recrystallized microstructure, where the near-interface regions typically contain lower frequencies of cube-oriented grains. This pattern has also been observed in Ni samples recrystallized after ARB to high and ultrahigh strains [12].

The sample scale deformation heterogeneities also affect the heterogeneity of the recrystallized microstructure. In the subsurface, the cube texture is consistently lower than in the other layers for each annealing duration (figure 5). The lower fraction of the cube texture in the subsurface is attributed to lower nucleation density of cube-oriented grains than in the other layers (see table 2), which is mainly because of the presence of the shear texture in the first 80 μm below the immediate surface, and hence a reduced rolling texture. It seems surprising that despite the reduced rolling texture, and therefore a smaller fraction of favorable misorientations between the cube-oriented grains and the recovered matrix in the subsurface [11], the cube-oriented grains here can grow to an average size similar to that in the other two layers. One possible reason is that the growth of a significant number of cube-oriented grains in each layer is confined by the bonding interface, so that the growth advantage of these grains in the rolling textured matrix, as typically seen in cold rolled nickel [18,19], is restricted along the ND.

5. Summary

The recrystallization behavior of Ni processed by accumulative roll bonding to a total accumulated von Mises strain of 4.8 has been examined, and analyzed with respect to heterogeneity in the deformation microstructure. It is found that the deformation microstructure near bonding interfaces is more refined and contains a higher stored energy compared to other regions. Particle deformation zones are also observed at bonding interfaces, resulting from embedded fragments of the steel wire brush used to prepare surfaces for the bonding process. The spatial arrangement of bonding interfaces produced during the ARB process consequently results in a sample-scale pattern of heterogeneity. In addition, a sample scale variation in local texture is also found, with distinct differences found in the subsurface, intermediate and center layers of the rolled material. The sample-scale and microscopic heterogeneities resulting from the ARB process are inherited in the recrystallized microstructure, where subsurface and near-interface regions typically contain lower frequencies of cube-oriented grains than anywhere else in the sample. Despite the microstructural heterogeneities, a strong cube texture forms in the sample after recrystallization, due to both the nucleation and growth advantage of cube-oriented grains. The complex pattern of heterogeneity developed during the ARB process, and the sensitivity of recrystallization to these variations, opens up the possibility for microstructural control in terms of both grain size and texture during thermomechanical processing, and highlights the importance of a thorough characterization of the deformation microstructure.

References

[1] Saito Y, Tsuji N, Utsunomiya H, Sakai T and Hong R G 1998 *Scr. Mater.* **39** 1221
[2] Saito Y, Utsunomiya H, Tsuji N and Sakai T 1999 *Acta Mater.* **47** 579
[3] Lee S H, Saito Y, Tsuji N, Utsunomiya H and Sakai T 2002 *Scr. Mater.* **46** 281
[4] Su L, Lu C, Li H, Deng G and Tieu K 2014 *Mater. Sci. Eng. A* **614** 148
[5] Najafzadeh N, Quadir M Z and Munroe P 2015 *Metall. Mater. Trans. A* **46** 4772
[6] Duan J Q, Quadir M Z and Ferry M 2016 *Metall. Mater. Trans. A* **47** 471
[7] Kim H W, Kang S B, Tsuji N and Minamino Y 2005 *Metall. Mater. Trans. A* **36** 3151
[8] Kamikawa N, Tsuji N, Huang X and Hansen N 2007 *Mater. Trans.* **48** 1978
[9] Li S, Sun F and Li H 2010 *Acta Mater.* **58** 1317
[10] Zhang Y B, Mishin O V and Godfrey A 2014 *J. Mater. Sci.* **49** 287

[11] Mishin O V, Zhang Y B and Godfrey A 2017 *J. Mater. Sci.* **52** 2730
[12] Zhang Y B and Mishin O V 2017 *Mater. Char.* **129** 323
[13] Zhang Y B, Mishin O V, Kamikawa N, Godfrey A, Liu W and Liu Q 2013 *Mater. Sci. Eng. A* **576** 160
[14] Humphreys F J 1977 *Acta Metall.* **25** 1323
[15] Mishin O V and Godfrey A, Juul Jensen D and Hansen N 2013 *Acta Mater.* **61** 5354
[16] Truszkowski W, Krol J and Major B 1980 *Metal. Trans. A* **11** 749
[17] Schoenfeld S E, Asaro R J 1996 *Int. J. Mech. Sci.* **38** 661
[18] Zhang Y B, Godfrey A, Juul Jensen D 2009 *Proc. 30th Risø Int. Symp. on Materials Science* (Roskilde, Denmark) ed. J C Grivel et al. pp 423
[19] Li X L, Liu W, Godfrey A, Juul Jensen D, Liu Q 2007 *Acta Mater.* **55** 3531
[20] Juul Jensen D 1992 *Scr. Mater.* **27** 533-538.

IOP Conf. Series: Materials Science and Engineering **219** (2017) 012035 doi:10.1088/1757-899X/219/1/012035

Characterization of deformed pearlitic rail steel

Dimitrios Nikas, Knut Andreas Meyer and Johan Ahlström

Department of Industrial and Materials Science, Chalmers University of Technology, SE-412 96 Gothenburg, Sweden

E-mail: nikas@chalmers.se

Abstract. Pearlitic steels are commonly used for railway rails because they combine good strength and wear properties. During service, the passage of trains results in a large accumulation of shear strains in the surface layer of the rail, leading to crack initiation. Knowledge of the material properties in this region is therefore important for fatigue life prediction. As the strain is limited to a thin surface layer, very large strain gradients can be found. This makes it very difficult to quantify changes in material behavior. In this study hardness measurements were performed close to the surface using the Knoop hardness test method. The orientation of the pearlitic lamellas was measured to give an overview of the deformed microstructure in the surface of the rail. Microstructural characterization of the material was done by optical microscopy and scanning electron microscopy to evaluate the changes in the microstructure due to the large deformation. A strong gradient can be observed in the top 50 µm of the rail, while deeper into the rail the microstructure of the base material is preserved.

1. Introduction

Carbon steels with a pearlitic microstructure are the most common materials for the manufacturing of rails today. Pearlite is a lamellar microstructure consisting of colonies of similarly oriented cementite lamellae embedded in a softer ferrite matrix. A large number of randomly oriented pearlitic colonies gives the material an isotropic behavior on the macroscopic level. This type of microstructure provides a good combination of wear and strength properties for railway applications and thus various types of such steel grades have been developed and used in the railways over the years. In this paper we consider the grade R260, which is one of the most common rail steel grades.

The surface layers of rails are subjected to very high rolling contact loads during their service life, which lead to large plastic deformation. As a result, material anisotropy develops, which strongly influences many of the defects that are observed in rails, such as pitting, shelling, spalling and head checks [1]. These defects are often denoted rolling contact fatigue (RCF) and are some of the main problems and sources for increased maintenance costs for the railway industry [2],[3]. The accumulation of strains has been investigated by e.g. [4] and [5] for rails and wheels respectively, where the strains were measured using the shear lines visible after etching. Micro hardness measurements also showed that the material had work hardened close to the surface, and in some cases [4], phase transformation to martensite had occurred (i.e. white etching layer formation).

The main goal of this study is to identify methods for determining microstructural quantities suitable for comparing different samples. To this end, measuring the reorientation of the pearlitic lamellar structure is considered in addition to the previously mentioned methods. By comparing

IOP Conf. Series: Materials Science and Engineering **219** (2017) 012035 doi:10.1088/1757-899X/219/1/012035

two microstructures using these methods, it may be possible to determine if an artificially produced microstructure (e.g. through deformation of test specimens) is representative of the microstructures found in used rails. The benefit of creating a microstructure in the laboratory that resembles the one in the field, is that more extensive mechanical testing can be performed and gradients can be limited, giving more representative local properties.

2. Material and experimental

The material studied in this investigation is the railway rail steel grade R260. The microstructure of this material is almost fully pearlitic with nominal composition as shown in table 1.

Table 1. Chemical composition of R260 rail material, maximum levels, in wt% [6]

C	Si	Mn	P	S	V	Cr	Al	N	O	H
0.80	0.60	1.25	0.03	0.03	0.03	0.15	0.004	0.01	0.002	0.00025

2.1. Sample extraction

Samples for characterization were extracted from a rail section at Jonsered Station in Sweden. The rail was installed in 2005 on the main line between Gothenburg and Stockholm, in a curve with a radius of 590 m. The rail exhibited a highly deformed profile shown in figure 1. Samples were taken from the top of the rail head (T) and gauge corner (G) (see figure 1) and were prepared for hardness measurements and microscopy. They were mechanically ground and polished to 0.04 µm using colloidal silica suspension. Etching was done using Nital (3% HNO_3 in ethanol).

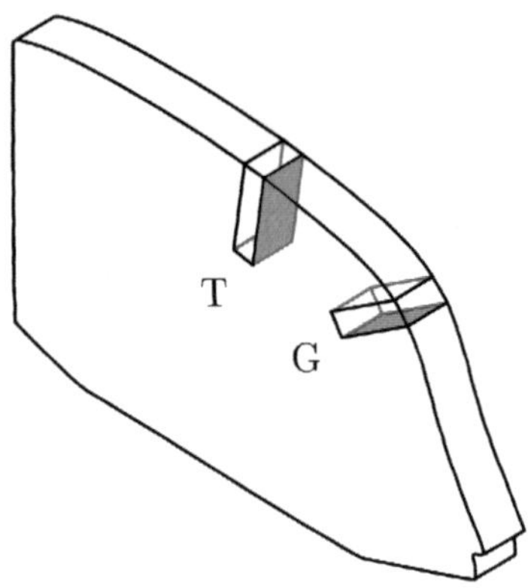

(a)

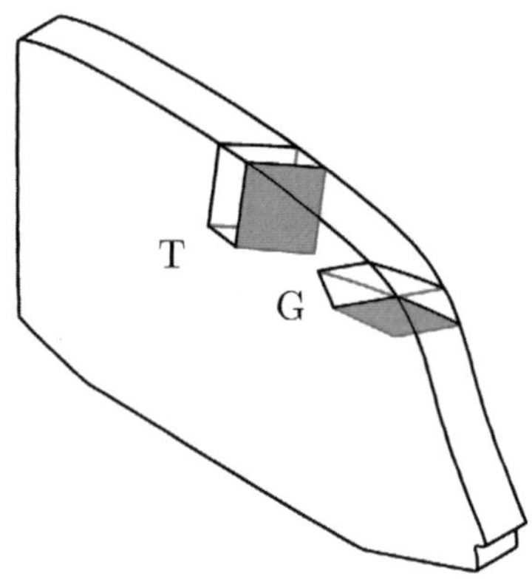

(b)

Figure 1. Samples extracted from the top of the rail (T) and gauge corner (G) with two orientations (90° (a) and 45° (b))with respect to the rail longitudinal direction. The orange colored sides denote the surfaces studied.

2.2. Hardness measurements

Both the Vickers and the Knoop method were used to evaluate the material hardness. Due to the strong gradients close to the surface, the Knoop method was chosen with a 10 g load (HK0.01), as this allowed for closer spaced indents with a sufficient size for optical measurement.

The measurements were conducted on an Struers DuraScan 70 hardness machine, using a 600× magnification. A global hardness map of the rail was obtained using a Vickers hardness indenter with a 10 kg load (HV10). This hardness measurement reduces scatter by averaging over a larger volume.

2.3. Optical flow lines
Three optical micrographs were taken using an optical microscope in the near-surface region to observe the shear deformation. The plastic shear strain was evaluated by measuring the angle α of the shear lines relative to the surface normal, see e.g. [4] and [5]. The angles α are obtained by manually tracing visible lines, and then using image analysis with linear regression to calculate the tangent direction of those lines at different depths. A measure of shear, γ, can then be calculated as $\gamma = \tan(\alpha)$.

2.4. Lamella orientation
The orientations of the cementite lamellae were evaluated from the gray scale gradient in SEM images, using the Sobel-Feldman operator [7]. While more direct methods that identify single lamellae give less noise in the data, they also produce less data per image, and hence more images are required to produce accurate distributions. This work mainly considers the variation with depth, and it was found that the differences in orientation distributions between the depths became more clear when using a gradient based analysis. Between 15 and 25 images per depth were used in this analysis. Fewer images were required close to the surface where the structure is very aligned, while more images were required to analyze the less aligned regions deeper in the rail.

3. Results and Discussion
A hardness map over the deformed railhead is considered in figure 2, to better understand the local variations closer to the surface. The hardness in the deeper parts of the rail show a small spread, and are mostly between 260 and 280 HV10. Closer to the surface a small increase in hardness is identified starting at a depth of about 5 mm.

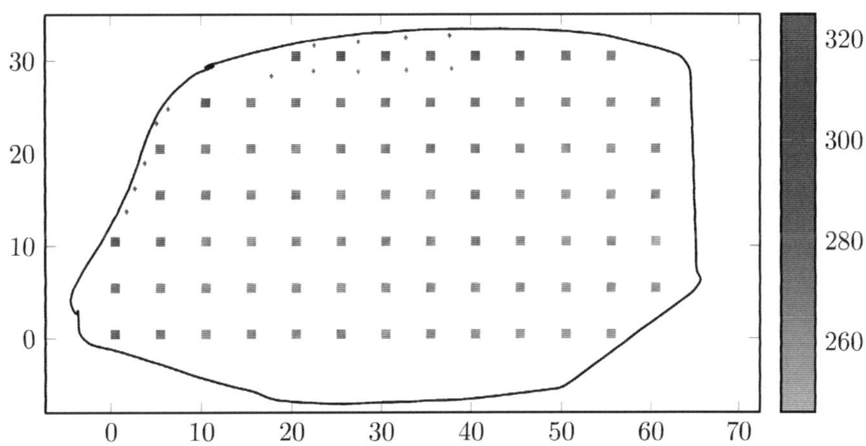

Figure 2. Vickers HV10 hardness overview of the deformed rail head (axis dimensions in mm). The grid points consists of 4 local points in a 1x1 mm grid, represented by the corners of the squares. The smaller diamond markers are additional single measurement points.

The critical area for fatigue in rails is the surface region close to the running band. We therefore consider the hardness variation with depth for the two regions in figure 1, measured on the 90° samples. The hardness results in figure 3 show that the hardness at the very top of the rail reaches around 550 HK0.01 and after about 100 µm it drops to 450. Deeper into the rail the hardness approaches the nominal value for the grade, as is shown by the Vickers overview map in figure 2.

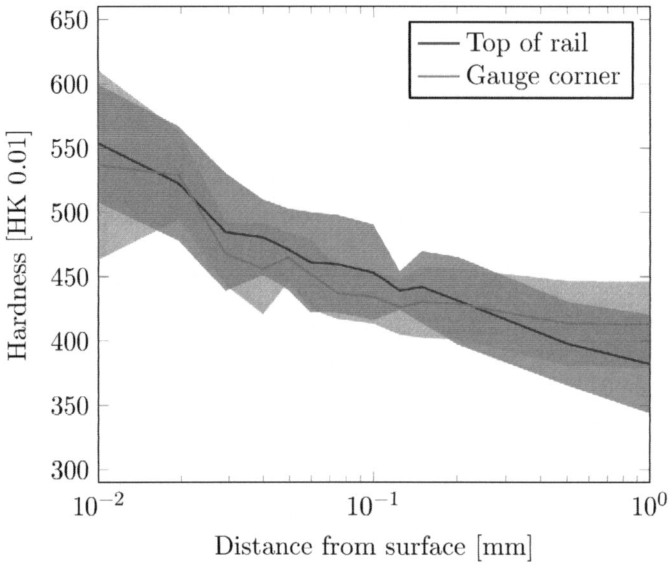

Figure 3. Hardness depth variation, filled region indicate ± one standard deviation.

The initial microstructure of the R260 rail steel is almost fully pearlitic, as shown in figure 4. In figure 4a the different orientations between the colonies are visible, and in figure 4b the grain refinement (but without any visible shear lines) close to the surface after rail rolling is visible.

(a) (b)

Figure 4. Initial microstructure of the R260 rail steel, depicted with an SEM image (a) and an optical micrograph (b)

During service the near-surface region becomes heavily deformed, as can be seen by the shear lines in figure 5a. The measure of shear strain γ shown in figure 5b shows a strong increase close to the surface. At the very surface some local shear effects resulted in a wave-like pattern and hence those values were excluded. The measurements are most accurate between a depth of 50 µm and 200 µm, where most lines could be identified.

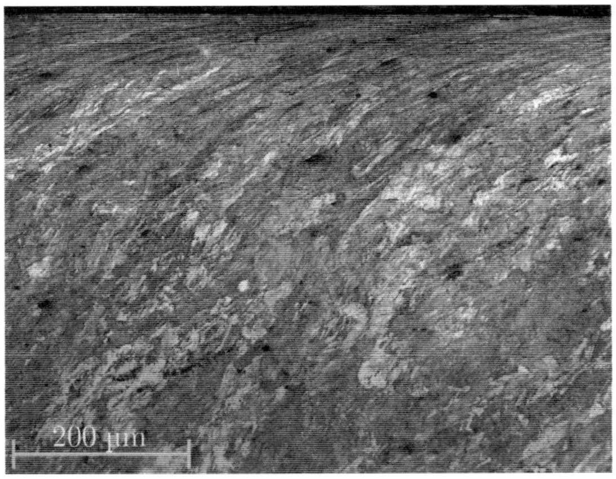

(a)

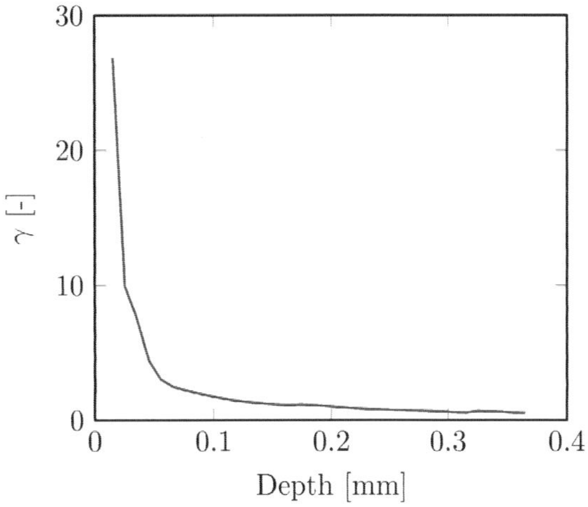

(b)

Figure 5. The deformation of the surface layer observed and measured using an optical microscope. (a) shows the deformed surface layer and (b) the measured shear strain γ as a function of depth

The SEM investigation also shows severe deformation close to the surface, where the lamellar structure becomes distorted such as seen in figure 6.

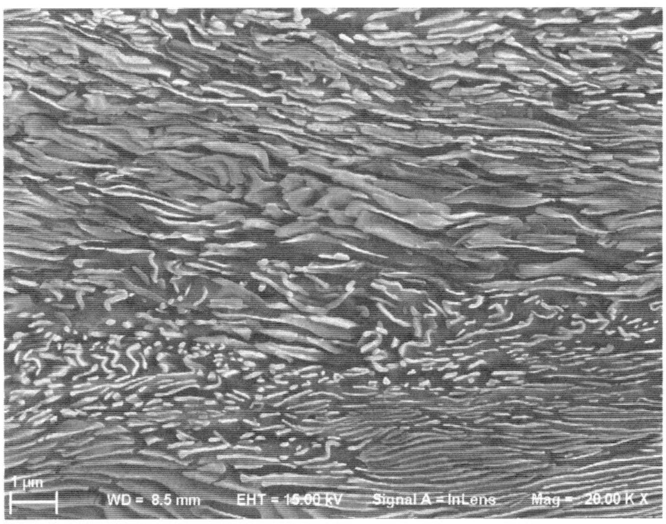

Figure 6. SEM images of the near-surface zone with the heavily sheared lamellas (taken from a depth of 20 µm).

The lamellar orientation distribution is calculated using images such as in figure 6 with a 20000× magnification. The results from this in figure 7 show a clear alignment of the microstructure close to the surface. Deeper in the rail the average angle increases, which implies smaller shear strains. The tails of the distribution also increase and the peaks decrease, indicating a less aligned structure.

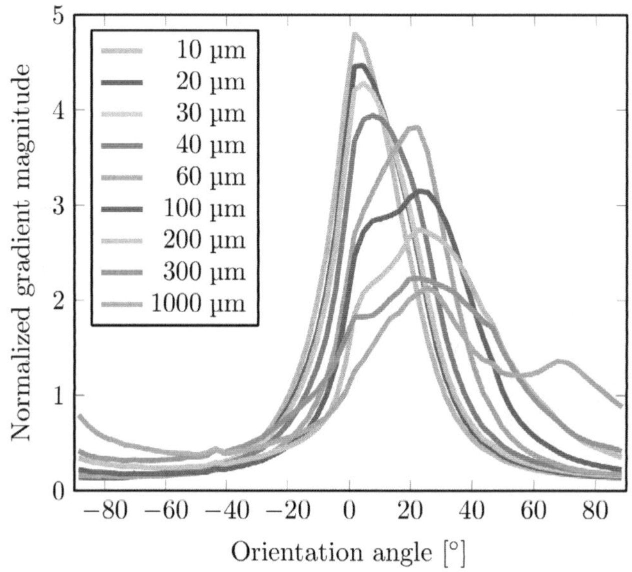

Figure 7. Pearlite orientation distributions obtained from SEM images

The main goal of the characterization in this paper, is to define a method in which some specific measurements can be used to compare the rail material with another microstructure created artificially in the laboratory. A biaxial machine will be used, combining compression and torsion to produce test bars that will also have a sheared microstructure. By using the method described in this paper these two microstructures can be compared. In case the microstructure created in the laboratory is similar, fatigue experiments can be run on those test bars to acquire the fatigue life and mechanical properties of a heavily deformed rail microstructure. Such tests are very important for the railway industry since testing directly on the deformed rail material is very difficult.

4. Conclusion

In this work a rail steel R260 was examined. Rail samples were extracted from an in-service section of railhead with large deformations present close to the surface. Hardness measurements showed a steep gradient with high hardness close to the surface (due to work hardening), reducing to the expected value for the base material deeper in the rail. The flow line method was used to give a measure of the strain close to the surface, showing that large deformations are present in the top layer of the rail. An alternative measurement of the strains has also been proposed, by considering the alignment of the microstructure using SEM images. Three quantifiable measurements of the microstructure have been used to evaluate the rail material. Using these methods to compare field samples with microstructures obtained through laboratory induced deformations, will make it possible to verify that the obtained microstructure is similar to that found in field samples. Hence, mechanical testing on these laboratory deformed specimens can be used to determine mechanical properties, such as constitutive behavior and fatigue life, of the deformed surface layer in rails.

References

[1] Larijani N, Brouzoulis J, Schilke M and Ekh M 2014 *Wear* **314** 57–68 ISSN 00431648
[2] Haidemenopoulos G N, Zervaki A D, Terezakis P, Tzanis J, Giannakopoulos A E and Kotouzas M K 2006 *Fatigue and Fracture of Engineering Materials and Structures* **29** 887–900 ISSN 8756758X
[3] Olofsson U and Nilsson R 2002 *Proceedings of the Institution of Mechanical Engineers, Part F: Journal of Rail and Rapid Transit* **216** 249–264 ISSN 0954-4097
[4] Alwahdi F A M, Kapoor A and Franklin F J 2013 *Wear* **302** 1453–1460 ISSN 00431648
[5] Cvetkovski K and Ahlström J 2013 *Wear* **300** 200–204 ISSN 00431648
[6] EN13674-1 2011 Railway applications - Track - Rail - Part 1: Vignole railway rails 46 kg/m and above. EUROPEAN COMMITTEE FOR STANDARDIZATION
[7] Sobel I and Feldman G 1968 *An isotropic 3 by 3 image gradient operator for Image Processing* (presented at the Stanford Artificial Intelligence Project (SAIL))

IOP Conf. Series: Materials Science and Engineering **219** (2017) 012036 doi:10.1088/1757-899X/219/1/012036

Tailoring hierarchical structures in selective laser melted materials

Jon Olsen[1], Xin Zhou[2], Yuan Zhong[1,2], Leifeng Liu[1], Dianzheng Wang[2], Chenfan Yu[2], Yafei Wang[2], Kailun Li[2], Leilei Xing[2], Jing Ma[2], Daqing Cui[1], Wei Liu[2]*, Zhijian Shen[1,2]*

[1] Department of Materials and Environmental Chemistry, Arrhenius Laboratory, Stockholm University, S-106 91 Stockholm
[2] School of Materials Science and Engineering, Tsinghua University, Beijing 100084, China

* Corresponding author: shen@mmk.su.se (Z. J. Shen)

Abstract. With selective laser melting the potential to manufacture a wide variety of geometries from different materials has presented itself. Interest in this technology keeps growing every year, and with that growth a deeper understanding of the process and resulting materials is urgently needed. In this paper we present a short overview of the structural elements that appear during selective laser melting, and explain how to tailor them to achieve specific structures and material properties. Melt-pools, texture and grains, subgrain cells, and inclusions are the elements discussed herein, and tailoring of these elements can have effects on density, and corrosion resistance, as well as mechanical properties in general.

1. Introduction

Additive manufacturing (AM) has in the last decades become a broad field, covering many different techniques the manufacture of parts [1]. The number of materials that can be used in AM is also rising and today materials such as metals, polymers, ceramics, and wax can all be utilized [2,3]. AM is based on a layer-by-layer processing principle where each layer of solid material is added on top of the previous one. In this way, a three-dimensional part is built up from the bottom to the top, one layer at a time. A computer-aided design (CAD) file is loaded into the controlling computer and is then followed by the machine to get the desired size and geometry of the parts. This design process gives excellent control over the resulting shape in a way that cannot be easily achieved by conventional manufacturing methods. This makes the process ideal for rapid prototyping [4,5]. In the early days of additive manufacturing, fast prototyping was indeed the central focus of the technology, but nowadays many different industries and companies are starting to realize the possibilities of the technique for production of real parts [6]. Such industrial sectors as aerospace [7,8], automotive [9], biomedical [10–12] and energy [13,14] are involved in the research on innovative additive manufacturing methods and applications.

Selective laser melting (SLM) is one of the possible methods in this category. A bed with powder is exposed to a high-energy laser beam that melts the powder granules and forms a solid part. This manufacturing method gives rise to several different structural elements that need to be fully understood. Melt pools, grains, sub-grains, and inclusions are some of the essential characteristics of SLM-manufactured materials. In this article, these elements are given a short explanation together with tailoring possibilities and we discuss how resulting changes in structure can impact different aspects of the performance of the SLM-manufactured material.

2. Melt pools

Melt pools are one of the hierarchically highest levels of the structure observed in SLM-manufactured material. This structural feature is something that is only found when using the powder bed fusion type of additive manufacturing to produce parts. They form around the spot where the laser, or other beam used to melt the material, hits the material and give a distinct appearance when they solidify. The size and form of these melt pools can readily be manipulated by the different parameters that control the laser beam. These parameters include (*i*) the power of the laser, (*ii*) the distance between each scanning track (also called hatch distance), (*iii*) the layer thickness, and (*iv*) the scanning speed of the laser. If the machine is using a discontinuous laser, the scanning speed is replaced by exposure time and the distance between exposure points.

2.1.Density

Melt pool dynamics and size also have a tremendous impact on the density of the manufactured material [15]. When heating up the material surface using a high-energy laser the center point gets much hotter than the surrounding material, which induces complex motions in the melt. Other driving forces for melt motion are laser-induced vapor plumes and plasma. These effect the melt by leading to the formation of an uneven surface after it cools down and solidifies. Parameter optimization to form bigger melt pools will decrease the effect of flow in the melt pool on surface roughness and thus increase the density. Good overlap of the melt pools is also crucial to get a smooth and fully dense material. If the melt tracks do not have enough overlap, un-melted powder granules may be present in the finished part (figure 1).

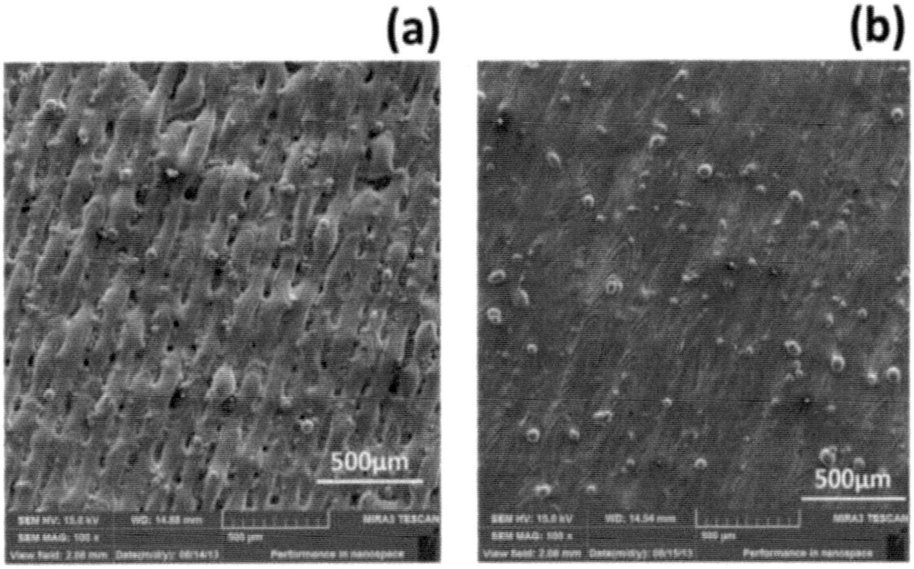

Figure 1. Scanning electron microscopy (SEM) images that show the surface morphologies of the Co-Cr-Mo alloy samples manufactured with different laser powers: (a)130W, showing a lack of overlap between melt tracks and (b) 200W, showing a smooth surface. Other parameters applied: hatch distance 125µm and laser speed 700 mm/s[15].

2.2.Corrosion resistance

Corrosion resistance is one material property that is critical in for example biomedical applications. Tailoring this property is possible for CoCrMo alloys manufactured with SLM. Using three different scanning parameters, where all of them were optimized for full density, and comparing the results to conventionally cast material, it is observed that all the samples manufactured with SLM have higher resistance to corrosion and lower metal release than the cast sample [11]. It can also be seen that changing the parameters, and in effect changing the melt pools forms and sizes, can to some extent alter the amount of metal released and the corrosion resistance. The likely reason for this behavior is that the melt pool boundaries are weaker to corrosion. A more evident finding is the difference in corrosion resistance when using combined potentiodynamic polarization. The difference between the different SLM-manufactured samples are in this case more apparent, and a clear trend can be observed where a higher amount of melt pool boundaries present on the top plane leads to less passivization of the material (figure 2). There are also more spikes in the the curves for samples containing more melt pool boundaries, which also implies lower corrosion resistance.

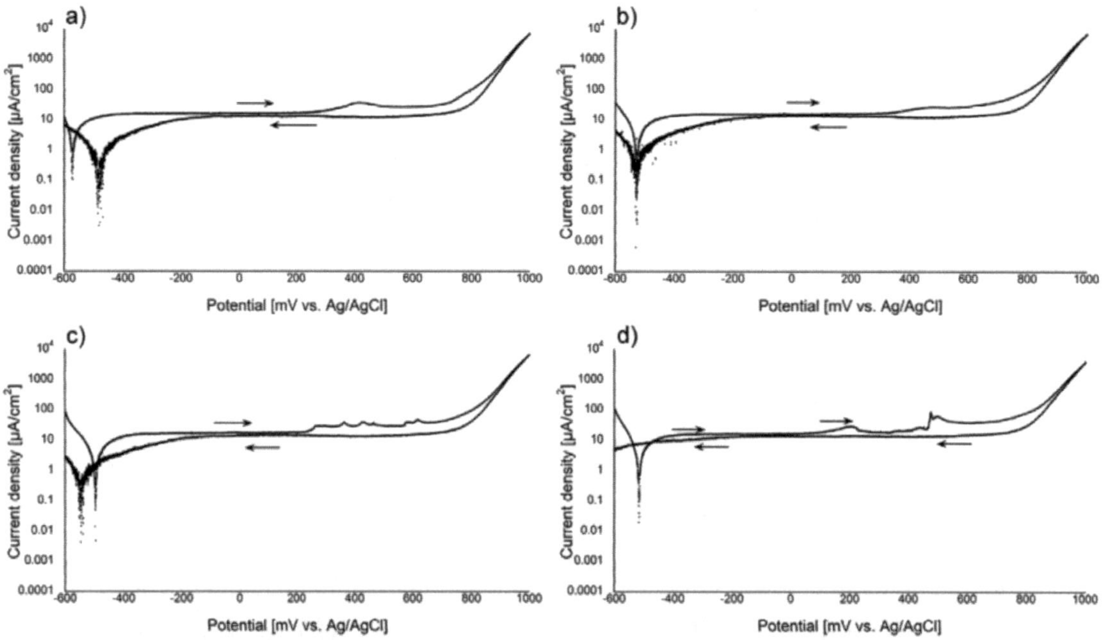

Figure 2. Potentiodynamic polarization (vs. Ag/AgCl) of SLM parts with an increase in melt-pool boundaries on the upper surface (a-c) and a conventionally cast part (d).[11]

3. Texture and Grains

During SLM process, the grains in the previous layer act as the nucleation sites for the further solidification of the melt through heterogeneous nucleation without altering crystallographic orientation, i.e., an epitaxial growth situation. Due to the directional solidification, the grains tend to grow along the temperature gradient direction with a preferred orientation. The morphological texture and crystallographic texture can both be tuned by adjusting processing parameters in CoCr alloy, specifically the scanning strategy, the hatch spacing, and the scanning speed.

The scanning strategy is important because the temperature gradient direction is determined by the laser scanning direction. If the laser is not rotated in adjacent layers, the grains tend to grow not only in the natural growth direction <100> but also the <110> direction on the outermost plane, and they tend to grow much stronger in the <100> direction along the scanning direction in the bulk material. The columnar grains are larger and distinguishable in this case. If the laser is rotated 67° in adjacent layers, the crystallographic texture is much weaker, and the grains are generally much smaller (figure 3). The grain growth direction becomes more random by changing the laser direction after each layer, so the columnar directional microstructure is fragmented under the induced complex thermal field.

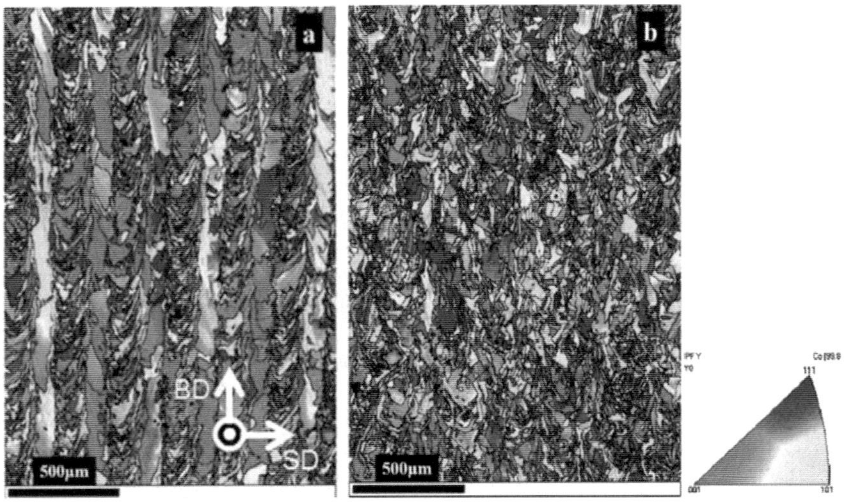

Figure 3. Electron backscatter diffraction (EBSD) orientation maps on the side plane of a CoCr sample consolidated by SLM with laser rotated 0° (a) and 67° between layer (b) [16]. Rotating the direction of laser scanning between layers results in a weaker texture and smaller grains.

The hatch spacing can also be used to tune the grain morphology and texture. Normally there are small equiaxed grains in the center line of melt tracks due to constitutional supercooling. These equiaxed grains can be removed by decreasing the hatch spacing as seen in figure 4. With the small hatch spacing shown in figure 5a, the adjacent melt tracks are firmly bonded, and the epitaxial grain growth diminishes the fusion line. The larger heat input generates lower G×R value (where G is the temperature gradient, and R is the grain growth rate) and thus results in larger grains and a stronger texture, which in turn results from epitaxial grain growth. The new melt track cannot re-melt the previous one, and thus a random distribution structure is retained with larger hatch spacing, as seen in figure 5b.

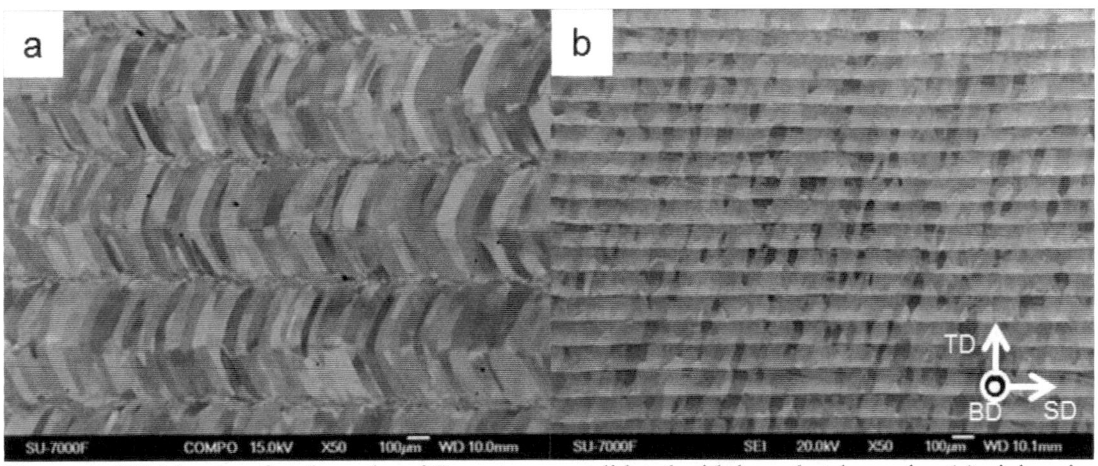

Figure 4. SEM image of melt tracks of Tungsten consolidated with large hatch spacing (a) giving rise to equiaxed grains, and small hatch spacing (b) where no equiaxed grains can be seen.

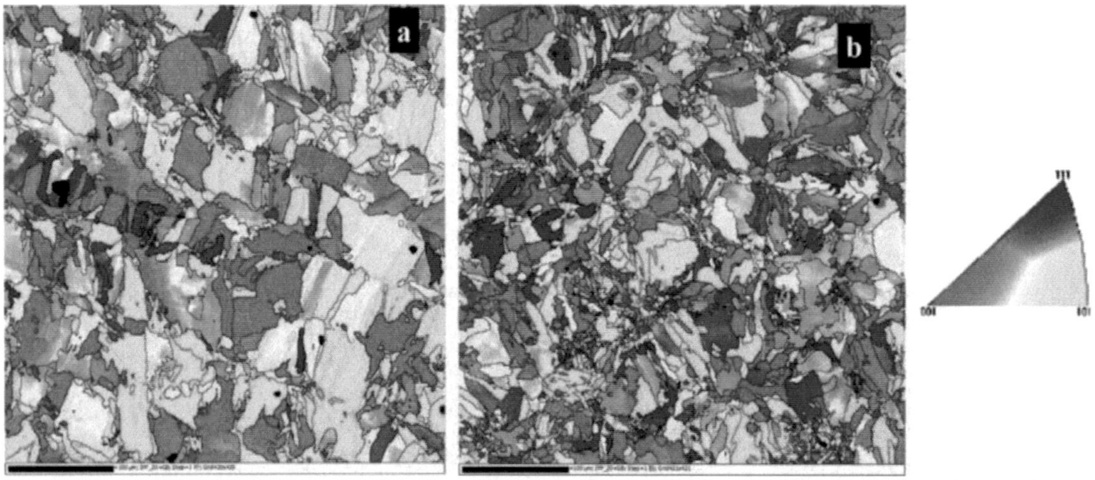

Figure 5. EBSD orientation maps on the top plane of SLM CoCr with hatch spacing 60 μm (a) and 175 μm (b) [16]. In (a) an overlap of the melt pools occurs and thus remelting of the smaller grains, allowing for epitaxial grain growth.

Higher scanning speeds result in a faster solidification rate and a shallow and elongated melt pool shape, which gives rise to a <001> growth preference from the previous layer into the present layer. This can result in a stronger texture [17].

4. Subgrain cellular structure

Subgrain features are found inside the grains of 316LSS manufactured by SLM. Scanning electron microscope (SEM) images of the etched surface of SLM-manufactured 316LSS shows a cellular structure with the cell boundaries less etched and the inside of the cells more etched (figure 6). The cells are around 500 nm in diameter, although the diameter varies in different regions. In most of the cases, spherical cells or elliptical cells are found, but tube like features may also form. This implies that the cells are columnar in shape. The lengths of the columnar cells can be up to hundred micrometers, which is much longer than the size in the other two dimensions.

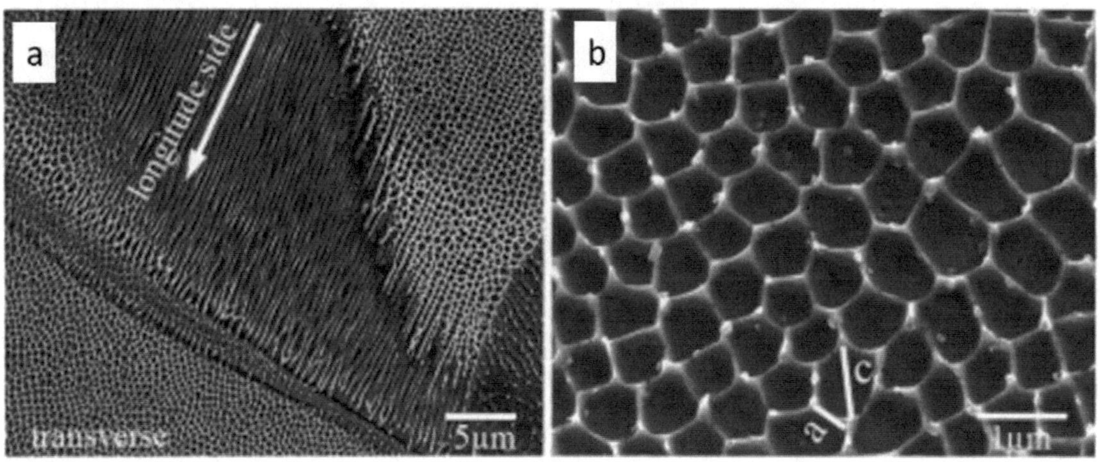

Figure 6. (a) low magnification SEM image of the etched surface of SLM- manufactured 316LSS showing the cellular structure from both the longitudal and radial cross-sections; (b) High magnification SEM image showing the average diameter of the cells of around 500 nm [18].

The cell boundaries are more corrosion resistant than the insides of the cells, resulting in the difference of the contrast in the SEM images. The difference in corrosion resistance is probably due to differences in the compositions. As shown in figure 7, the boundaries of the cells show lighter contrast than the inner part in the dark field scanning transmission electron microscopy (DF-STEM) images, implying the segregation of heavy elements at the boundaries. This is also confirmed by energy-dispersive X-ray spectroscopy (EDS) maps of the Cr distribution.

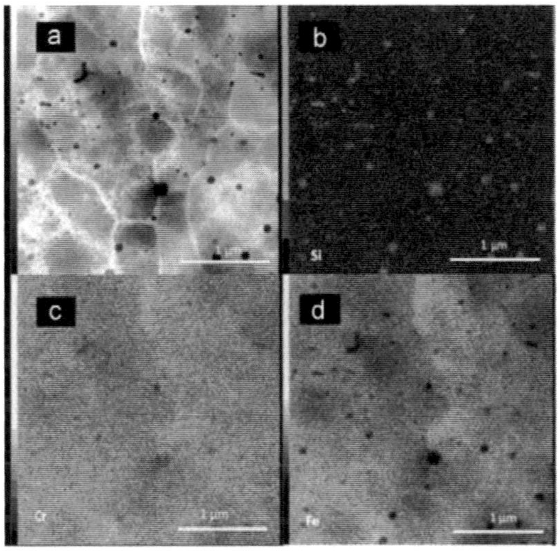

Figure 7. (a) DF-STEM image of the subgrain cellular structure; (b-d) EDS maps of the Si, Cr and Fe elements[19].

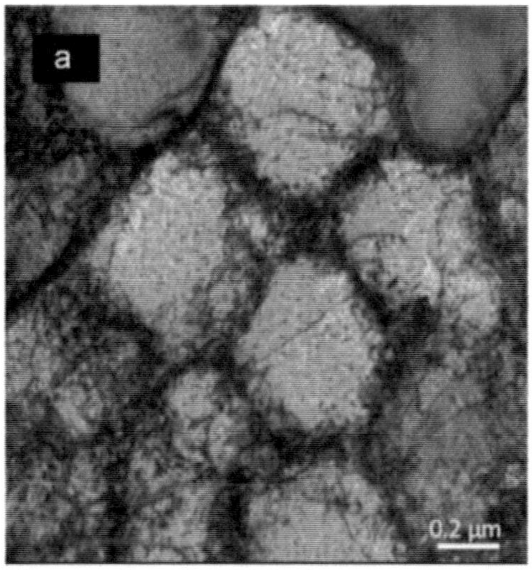

Figure 8. BF-TEM image of SLM-manufactured 316LSS showing the distribution of dislocations at the cell boundaries [19].

Besides the segregation of heavy elements, dislocations are often found to concentrate at the cell boundaries. In the bright field (BF) TEM image in figure 8, the dislocations are shown as dark lines. It is found that there are a few dislocations in the inner part of the cells while a high density of dislocations can be found at the boundaries.

Dislocations are line defects known to have a significant effect on the mechanical properties of metallic materials. The high density of dislocations stored in the structure can effectively increase the critical shear stress for dislocation motion, and hence introducing dislocations is often used for improving the strength of metals in conventional metal processing techniques e.g. rolling and drawing. In the SLM-manufactured 316LSS, the subgrain structure results in a high density of dislocations in the grains. As expected, the yield strength of the SLM-manufactured sample is notably higher than the annealed counterpart.

The formation of the subgrain structure is closely related to the process condition during SLM. Therefore, by changing the process parameters, it should be possible to alter the size and configuration of the subgrain cellular structure and thus the strength of the metal. This means that even in the same sample different regions could have different strengths, by tailoring the local subgrain cellular structure.

5. Oxide dispersion strengthening

The size of *in-situ* formed nano-inclusions can be adjusted by changing the oxygen partial pressure in the processing chamber. Almost no nano-inclusions can be observed in specimens consolidated in an extremely low oxygen chamber environment, as seen in figure 9a,b. By increasing the oxygen level to 300-500 ppm, large particles with size 2 μm are formed. Nevertheless, further increasing oxygen level to 1000-1200 ppm leads to smaller nano-inclusions with size around 50 nm. The higher oxygen level in the laser chamber results in a higher dissolved oxygen in the melt, and thus a larger number of oxide particles that can be generated. At a high level (1000-1200 ppm), the maximum radius of nano-inclusions is controlled by the shorter diffusion radius R_d due to the higher density of the initial reactive cores as seen in figure 10. In the medium level range (300-500 ppm), the nano-inclusions can grow larger due to the smaller number of oxide particles and the longer value of R_d.

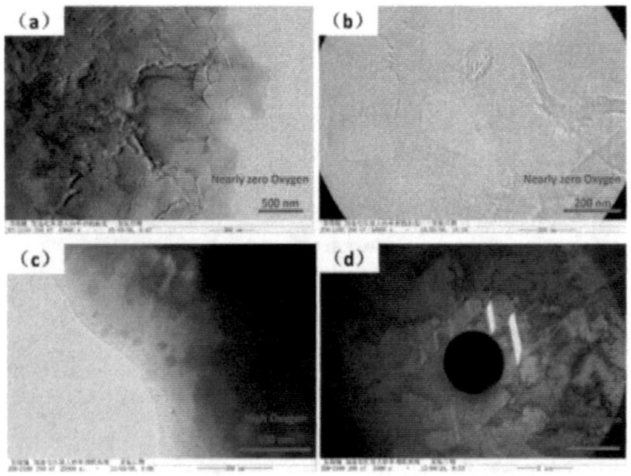

Figure 9. Distribution of oxide particles in stainless steel 316L, chamber oxygen level of less than 50 ppm (a,b), 1000-1200 ppm (c), 300-500 ppm (d) [20]

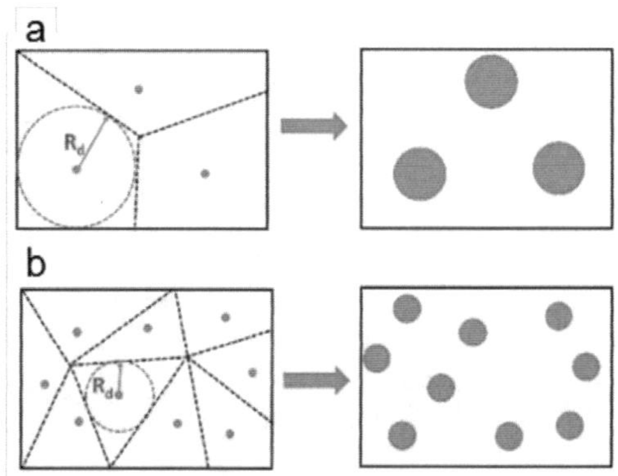

Figure 10. Oxide nano-inclusion growth in medium oxygen level 300-500 ppm (a) and high oxygen level 1000-1200 ppm (b) [20]

6. Conclusions and perspectives

Several different conclusions can be drawn about the possibility to tailor the microstructure to achieve desired properties for materials manufactured by selective laser melting.

1. By manipulating the size and shape of the melt pools it is possible to reach full density in materials, but also to lower the density if that is desirable. The corrosion resistance of the SLM-manufactured materials can also correlate with the melt pools size in such a way that a smaller area of melt-pool boundaries only at the surface exposed to an oxidizing environment will lead to better corrosion resistance and less metal release.
2. Grains tend to grow in the direction of the temperature gradient of the melt pool. By changing the scanning strategy the growth pattern and final size of the grains can be manipulated.
3. Cellular structures may also form in some cases consisting of dislocation networks that can help improve the mechanical properties of the material.
4. By monitoring and controlling the oxygen level in the building chamber, the size of *in-situ* formed oxide inclusions can be tailored.

The broad challenges for future work are twofold. The first is to find other characteristics of both microstructure and properties of the material that can be tailored by control of the SLM manufacturing conditions. The second, is to improve current methods of tailoring the structure and properties to achieve better control and more reliable results. In these ways, materials with better properties and/or improved performance than their conventional counterparts can be developed by selective laser melting.

References

[1] Wong K V and Hernandez A 2012 A Review of Additive Manufacturing *ISRN Mech. Eng.* **2012** 1
[2] Calignano F, Manfredi D, Ambrosio E P, Biamino S, Lombardi M, Atzeni E, Salmi A, Minetola P, Iuliano L and Fino P 2017 Overview on Additive Manufacturing Technologies *Proc. IEEE* **105** 593
[3] Frazier W E 2014 Metal additive manufacturing: A review *J. Mater. Eng. Perform.* **23** 1917

[4] Karunakaran K P, Bernard A, Suryakumar S, Dembinski L and Taillandier G 2012 Rapid manufacturing of metallic objects *Rapid Prototyp. J.* **18** 264

[5] Campbell I, Bourell D and Gibson I 2012 Additive Manufacturing: Rapid Prototyping comes of age *Rapid Prototyp. J.* **18** 255

[6] Guo N and Leu M C 2013 Additive manufacturing: Technology, applications and research needs *Front. Mech. Eng.* **8** 215

[7] Liu R, Wang Z, Sparks T, Liou F and Newkirk J 2017 *13 – Aerospace applications of laser additive manufacturing* (Elsevier Ltd)

[8] Shapiro A A, Borgonia J P, Chen Q N, Dillon R P, McEnerney B, Polit-Casillas R and Soloway L 2016 Additive Manufacturing for Aerospace Flight Applications *J. Spacecr. Rockets* **53** 952

[9] Leal R, Barreiros F M, Alves L, Romeiro F, Vasco J C, Santos M and Marto C 2017 Additive manufacturing tooling for the automotive industry *Int. J. Adv. Manuf. Technol.* 1

[10] Yadroitsev I, Krakhmalev P and Yadroitsava I 2014 Selective laser melting of Ti6Al4V alloy for biomedical applications: Temperature monitoring and microstructural evolution *J. Alloys Compd.* **583** 404

[11] Hedberg Y S, Qian B, Shen Z, Virtanen S and Odnevall Wallinder I 2014 In vitro biocompatibility of CoCrMo dental alloys fabricated by selective laser melting *Dent. Mater.* **30** 525

[12] Qian B, Saeidi K, Kvetková L, Lofaj F, Xiao C and Shen Z 2015 Defects-tolerant Co-Cr-Mo dental alloys prepared by selective laser melting *Dent. Mater.* **31** 1435

[13] Zhong Y, Rännar L-E, Wikman S, Koptyug A, Liu L, Cui D and Shen Z 2017 Additive manufacturing of ITER first wall panel parts by two approaches: Selective laser melting and electron beam melting *Fusion Eng. Des.* **116** 24

[14] Zhong Y, Rännar L-E, Liu L, Koptyug A, Wikman S, Olsen J, Cui D and Shen Z 2017 Additive manufacturing of 316L stainless steel by electron beam melting for nuclear fusion applications *J. Nucl. Mater.* **486** 234

[15] Zhou X, Wang D, Liu X, Zhang D D, Qu S, Ma J, London G, Shen Z and Liu W 2015 3D-imaging of selective laser melting defects in a Co-Cr-Mo alloy by synchrotron radiation micro-CT *Acta Mater.* **98** 1

[16] Zhou X, Li K, Zhang D, Liu X, Ma J, Liu W and Shen Z 2015 Textures formed in a CoCrMo alloy by selective laser melting *J. Alloys Compd.* **631** 153

[17] Dehoff R R, Kirka M M, Sames W J, Bilheux H, Tremsin A S, Lowe L E and Babu S S 2015 Site specific control of crystallographic grain orientation through electron beam additive manufacturing *Mater. Sci. Technol.* **31** 931

[18] Zhong Y, Liu L, Wikman S, Cui D and Shen Z 2016 Intragranular cellular segregation network structure strengthening 316L stainless steel prepared by selective laser melting *J. Nucl. Mater.* **470** 170

[19] Saeidi K, Gao X, Zhong Y and Shen Z J J 2015 Hardened austenite steel with columnar sub-grain structure formed by laser melting *Mater. Sci. Eng. A* **625** 221

[20] Zhou X, An Z, Shen Z, Liu W and Yao C 2017 Particles control in selective laser melting in-situ oxide dispersion strengthened method *IOP Conf. Ser. Mater. Sci. Eng.* **167**

IOP Conf. Series: Materials Science and Engineering **219** (2017) 012037 doi:10.1088/1757-899X/219/1/012037

From an understanding of structural restoration mechanisms towards a selective processing of extreme nanolamellar structures

O Renk[1*], P Ghosh[1] and R Pippan[1]

[1] Erich-Schmid-Institute of Materials Science, Austrian Academy of Sciences, Jahnstrasse 12, 8700 Leoben Austria

* corresponding author: oliver.renk@oeaw.ac.at

Abstract. It has been successfully proven that severe cold deformation can be used to generate bulk nanostructured materials. However, there is a limit to grain refinement, as recovery processes, together with grain boundary and triple junction motion, occur to continuously fragment and remove grains. Although believed to be mechanically driven at low temperatures, it is clear that such movement can be thermally assisted, as reflected in decreased grain aspect ratios at elevated deformation temperatures. Interestingly, for tantalum, a different behaviour can be found. Grain aspect ratios after high pressure torsion increase steadily up to deformation temperatures of 673 K before dropping again. Moreover, extremely large aspect ratios of 10 can be generated. The reasons for this behaviour are discussed and it is suggested from results based on nickel that this phenomenon can be adapted to any metal. This offers the possibility to selectively process materials with extreme grain aspect ratios, to achieve structures which are thought to be the key to the design of materials that combine both, exceptional strength and good damage tolerance.

1. Introduction

Severe cold working of metals significantly enhances their strength as their grain size is decreased efficiently, but along with the increase in strength, generally ductility and fracture toughness are reduced [1–3]. Overcoming this strength-ductility trade-off remains challenging as further strain hardening in such materials is hard to realize. Nevertheless, results for nanostructured materials showing high strength in combination with good ductility have occasionally been reported. These studies rely on the introduction of nanotwins inside grains, or the use of proper annealing treatments to produce bimodal grain size distributions, which were shown to be useful strategies to maintain ductility, despite the observed strength increase [4–7]. Although these results seem promising, especially the concept of nanotwins is not easily transferable to any bulk material system, as synthesis is mainly limited to deposition processes [4,5] or dynamic plastic deformation at cryogenic temperatures [8]. Moreover, generation of suitable bimodal grain size distributions with sufficient volume fractions of coarser grains by annealing procedures is also not easy to achieve. In many cases nanocrystalline (NC) materials remain stable up to certain temperatures, but then coarsen rapidly, often accompanied by abnormal grain growth, while in some cases even though a bimodal grain structure was produced, ductility was not enhanced compared to the as-processed state [9–11].

Results obtained on bulk nanostructured materials, synthesized by high pressure torsion (HPT) suggest, however, that the elongated grain shape, arising from the deformation process, can be effectively used to process materials combining enhanced strength and acceptable toughness values [12]. Similar results were obtained for cold rolled tungsten foils in which elongated grain structures were also developed [13], suggesting that this approach could be applicable work for any kind of material.

These excellent combinations of strength and toughness result from highly anisotropic properties that can be measured in such nanolamellar structures. The pronounced toughness anisotropy arises due to the preferred crack path in sub-micron or nanocrystalline (NC) materials, which in most cases is along the numerous grain boundaries [12]. This leads to low energy consuming crack paths when pre-cracks are introduced parallel to the elongated grain structure, while testing in either of the two directions perpendicular to this results in considerably higher toughness values, as intergranular fracture occurs, leading to crack deflection and delamination, both of which contribute to the measured toughness values [12]. This suggests that one of the essential structural parameters to overcome the strength-toughness conflict in high strength materials is the aspect ratio. Cold working of metals therefore might be an easy and effective approach to produce such highly elongated grain structures. Unfortunately, for most pure single phase materials or alloys, although deformed to severe strains at low homologous temperatures, typically aspect ratios only of 2 – 4 can be obtained [14]. The main restriction for grain elongation and refinement is the occurrence of dynamic recovery mechanisms. These processes, including deformation induced grain boundary (GB) migration [15] and motion of triple junctions (TJs) [16], continuously shorten and fragment grains. Although it is generally recognized that the aspect ratio decreases with an increase in deformation temperature [14,17–19], recent experiments on tantalum have shown that up to a certain deformation temperature, the aspect ratio can even increase [20]. It can be expected that despite the temperature increase, recovery processes are still occurring albeit inefficiently, as otherwise the grains would become more globular.

The present work will summarize the main findings about GB and TJ motion in highly strained structures with a special emphasis on recent results obtained on tantalum. Additional experiments on nickel suggest that the trend of increasing aspect ratios up to specific deformation temperatures is a general one and allows further conclusions how such nanolamellar structures can be generated.

2. Recovery after large and severe strains

Large and severe strains cause significant grain refinement in metal structures as dislocations are continuously produced, stored and rearranged, which subdivides the coarse grains. Details of this grain fragmentation process can be found elsewhere [17,21]. The most important results are that with increasing strain the structural size decreases while the misorientation is increasing, but refinement terminates after a certain strain, when an equilibrium between refinement and recovery is obtained. On the structural scale, this necessitates that grain boundaries have to be removed. Such net removal of grain boundaries already starts at moderate strain levels of $\varepsilon > 1$ [22], with the underlying processes being GB and TJ motion [15,16].

For TJ migration the main driving force stems from the dihedral angles, θ, being far from their equilibrium value, i.e. 120° for a junction consisting of three equal high angle grain boundaries (HAGB) [23]. TJ motion tries to shift θ towards the equilibrium value, thereby lamellae can disappear or reduce their aspect ratio [23,24], as illustrated schematically in figure 1. After severe deformation, the dihedral angles will be far from equilibrium and enlarged driving forces for TJ motion can be expected during annealing. Additionally, during deformation of such structures the thickness of the lamellae will reduce further, which could decrease θ and thereby cause an increased tendency for TJ migration. However, one should be careful to conclude that generally thin lamellae will have smaller dihedral angles, thus higher driving forces for TJ migration. Although, measurements on severely rolled aluminium point in this direction [24], the observed scatter is relatively large and local grain boundary diffusion could also cause an increase of θ [25].

On the other hand, GB migration during deformation can fragment adjacent grains, which may shrink further either by GB migration or subsequent movement of the newly created TJs, see figure 1. Because this kind of GB migration can be observed even at low deformation temperatures it is generally accepted that it is mechanically activated. Nevertheless, the driving forces are still controversially debated. Coupling of GBs with shear stresses was suggested to cause migration of the GBs [26–28], while other studies have provide clear evidence for the role of plastic strain [29]. The latter would be in line with the formation of deformation textures after severe deformation similar to the ones expected for dislocation based plasticity [30]. Moreover, calculations have shown that strain energy differences between grains yield reasonable values for the driving force of several MPa [30]. However, on the atomic scale grain boundary migration has been attributed to the movement of disconnections i.e. GB steps associated with a dislocation [31]. Such GB defects can be formed easily due to the interaction of lattice dislocations with GBs [31,32]. Especially in cases when the mobility of disconnections is limited they presumably pile up, leading to the formation of macro steps [28,31]. The bulging of GBs that leads to fragmentation of neighboring grains, see figure 1, may therefore be directly associated to the formation of such macro steps.

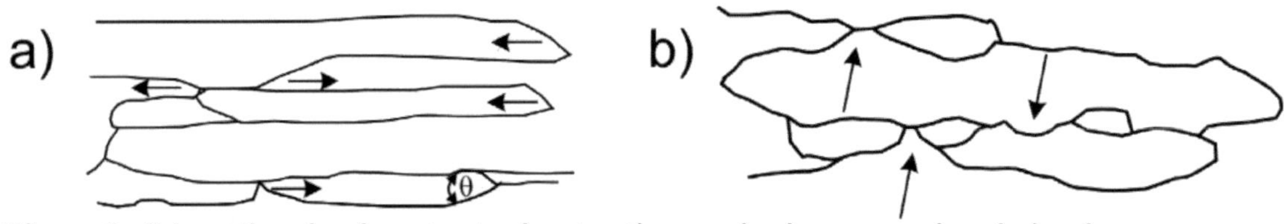

Figure 1. Schematics showing structural restoration mechanisms occurring during large or severe strains; a) movement of triple junctions, and b) movement of grain boundaries. Both processes can fragment or shrink individual lamellae, leading to a structural coarsening and thus limiting refinement and the aspect ratio.

Because both GB and TJ migration counteract grain elongation, the generation of bulk nanostructured materials with enhanced aspect ratios necessitates the development of strategies that subdue them to a great extent. Furthermore, deformation modes or temperatures where shear band formation may occur should also be avoided, as such localized shear can fragment lamellae as well [16,33]. Strain path changes have been suggested as a means to overcome the problem of structural recovery and to suppress these processes, leading to a reinvigoration of boundary generation, offering the possibility for further grain refinement [33]. However, results on structures already deformed to saturation have shown that a strain path change does not necessarily result in further refinement. On the contrary, pure ultrafine-grained copper synthesized by HPT and further deformed by cold rolling revealed, up to strains of $\varepsilon \sim 1$, even a slight coarsening of the grains [15] (see figure 2). At higher rolling strains, the lamellae were refined again, although no significant reduction of the minimum lamella spacing compared to the as-HPT deformed state could be achieved.

As indicated, GB and TJ migration are thought to be mainly mechanically induced, especially when severe deformation is carried out at low homologous temperatures. Nevertheless, it can be expected that they are thermally facilitated, which explains the observation of larger grains with reduced aspect ratios after severe deformation at elevated temperatures [14,17–19].

Accordingly, one would intuitively think that very large aspect ratios could be generated during cryogenic deformation, where the restoration processes can be retarded more efficiently. However, recent results on tantalum have shown a different trend, where the aspect ratio increased with deformation temperature up to 673 K [20], see figure 3. The reason for this distinct increase in grain elongation is related to a much stronger length increase of the long grain axis compared to the minor axis, as shown in figure 3. As the minimum grain dimensions continuously increase with deformation temperature it is clear that the overall restoration processes have to increase. Nevertheless up to 673 K recovery is still clearly insufficient to fragment the lamellae, as the length of the major grain axis increases in this temperature range.

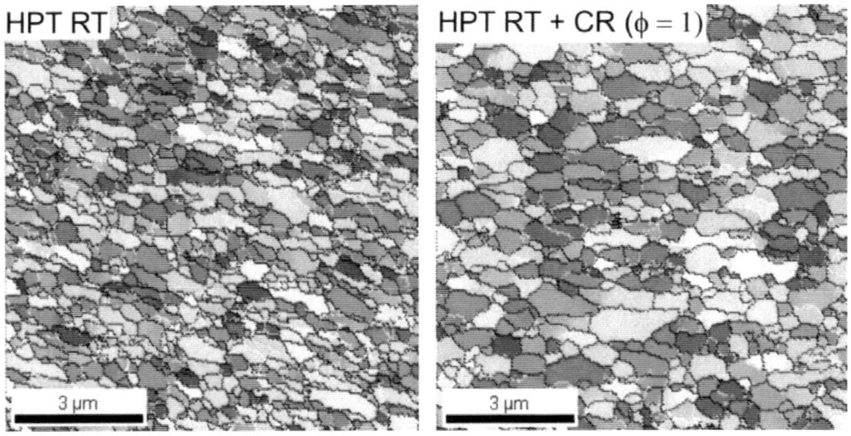

Figure 2. Structural evolution of ultrafine grained Cu processed by HPT and subjected to further cold rolling. Instead of a further refinement of the lamellae, structural coarsening proceeds up to strains of ε ~ 1.

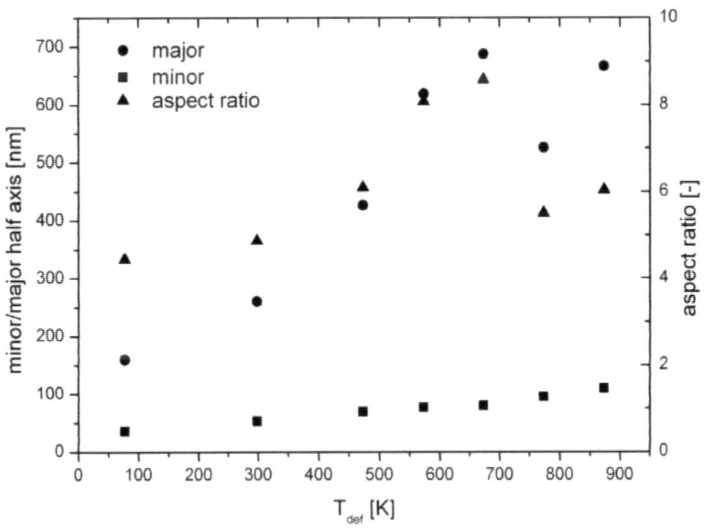

Figure 3. Evolution of the major and minor grain axes, as well as the aspect ratio, of tantalum severely deformed by HPT at different temperatures. Data taken from Ref. [20].

It is not surprising that the structures deformed at 77 K and RT do not differ significantly, as experiments have already shown, that the grain size changes at low homologous temperatures are not pronounced any more, as thermal activation is negligible and mechanically induced motion prevails [14,30]. In fact, as the structures generated after severe deformation at low temperatures consist of rather planar GBs, thermally driven GB migration will not play a role until the structure changes into a curved, more globular one, for instance by thermally induced TJ motion. Such structural changes have already been observed due to thermally induced TJ motion during static annealing of severely deformed structures [34]. However, thermally induced TJ motion will occur only above a certain temperature, accessible from static annealing treatments. For tantalum this temperature was found to lie between 673 K and 773 K, exactly the temperature regime were the aspect ratio starts to drop again [20], (figure 3). The occurrence of a critical temperature T_0, above which frequent TJ motion takes place seems to be related to the presence of lattice defects at temperatures below this, as loose dislocations and interconnecting boundaries within the lamellae act as pinning points for TJ motion [35]. Therefore, fast TJ motion is unlikely to occur during deformation at lower temperatures as the probability of intragranular defect storage, and so the number of pinning points, are enhanced.

Consequently, the continuous increase of the aspect ratio with HPT temperature below T_0 has to be explained by a reduction of grain fragmentation events when the deformation temperature is increased. For tantalum the GBs appeared more flat at elevated processing temperatures, and bulges, as marked by arrows in figure 1, can be rarely observed, although the mobility of the GBs had to increase as larger lamellae thicknesses can be observed for higher deformation temperatures. The explanation for these results might be directly linked to the mobility of disconnections. TEM studies have revealed that on the mesoscale grain boundary migration is caused by the movement of disconnections. These boundary defects can be created at any temperature by the interaction of lattice dislocations with GBs, however their mobility, will differ markedly [31]. At low temperatures, and hence limited mobility, they will pile up frequently, causing the formation of macro steps [31], which appear similar in size to the bulges that can be observed during deformation [15,20] and in turn lead to fragmentation of adjacent grains. During processing at elevated temperatures, when their mobility is increased, macro step formation will occur less frequently but the GB migration rate will be larger. This would explain not only why the minor axis of the tantalum grains slightly increased with temperature but also why fragmentation occurred rarely, enabling the generation of large grain elongations.

For this reason the main conclusion in [20] was that the resulting grain aspect ratio after severe deformation reaches a maximum at a temperature low enough to maintain enough lattice defects for effective TJ pinning but high enough to prevent frequent grain fragmentation events caused by GB migration. Of course, this will only hold true, for homogenous deformation, where the generated structure results from a balance between refinement and the restoration processes and will not be fulfilled for deformation modes, alloys or temperatures where intense shear banding is present. Such localized shear can again lead to fragmentation of the lamellae and thus deviation from the above description [16,33].

3. Generation of extreme nanolamellar structures – a possibility for all metals?

These conclusions on the occurrence of large aspect ratios in tantalum suggest that the generation of such structures should not solely be restricted to high melting point metals like tantalum. Although its high melting point certainly eases such observations, the principle should work for fcc metals or lower melting point metals as well. For this reason, we carried out a similar study on pure nickel with 99.99% purity. Disks with 8 mm diameter and a thickness of 0.8 mm were deformed by quasi-constrained HPT at different deformation temperatures, ranging from 77 K to 673 K. One sample deformed at 298 K was used for static annealing treatments to identify the onset of structural coarsening. The microstructures were analyzed in the radial direction at a radius of $r = 3$ mm using electron backscatter diffraction (EBSD), see inset image in figure 4. For quantification of the structural sizes a standard software package was used. Additional microhardness measurements were taken on the statically annealed samples.

Annealing treatment on the samples HPT deformed at 298 K showed typical recovery up to temperatures of about 423 K, see figure 4. As the grain size remained stable within this temperature range, the decrease in hardness occurs due to dislocation annihilation. Between 423 K and 473 K recrystallization and grain growth occurs, resulting in grain sizes up to 5 µm in size. Moreover, grains with still smaller size appear already equiaxed.

Figure 5 shows inverse pole figure (IPF) maps of the structures deformed at different deformation temperatures. Although the differences are as pronounced as for tantalum, similar trends can be observed. The grains become more elongated with increase of the HPT deformation temperature, with a maximum aspect ratio observed for deformation at 418 K. For higher HPT deformation temperatures, the aspect ratio dropped again, see figure 5.

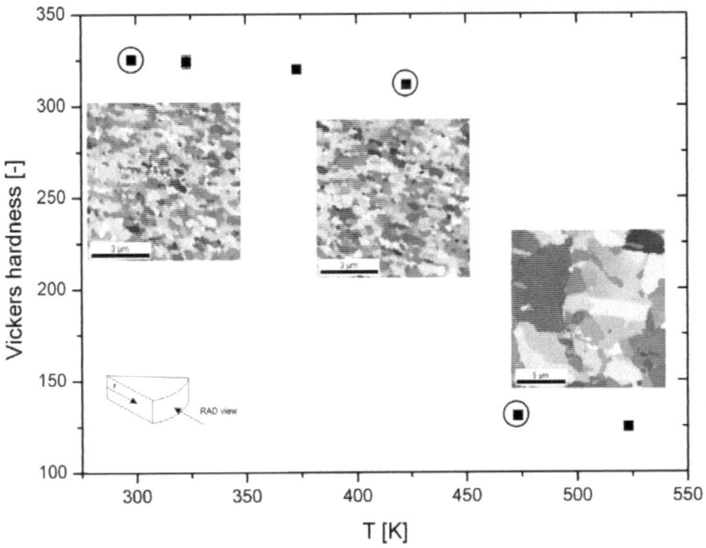

Figure 4. Room temperature microhardness of a nickel sample HPT-deformed at 298 K as a function of the annealing temperature. Microstructures remain stable up to ~ 423 K, as only defect recovery occurs. For temperatures above this fast coarsening of the structure takes place.

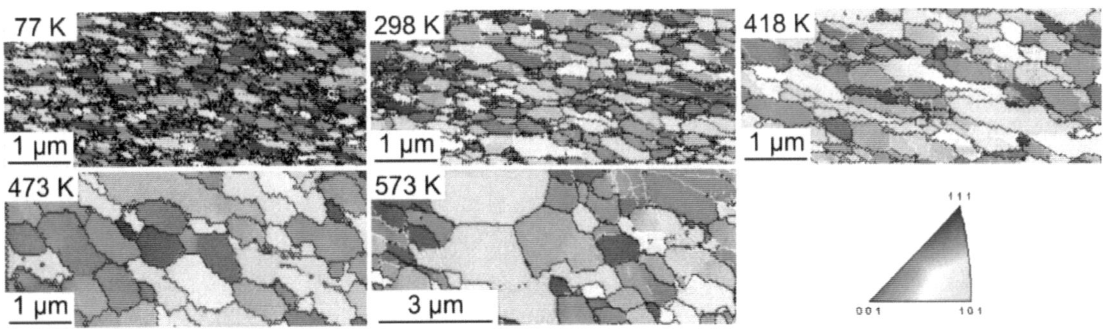

Figure 5. Representative inverse pole figure maps of nickel structures (radial view) deformed at different temperatures. At temperatures up to 418 K the grain aspect ratio clearly increases.

Similar to the observations for tantalum samples, the onset of structural coarsening during static annealing and the deformation temperature where the aspect ratio starts to reduce are again similar. Although grain elongations are not as pronounced as for tantalum, the results on nickel show that elevated deformation temperatures may be the key to obtaining structures with pronounced aspect ratios.

4. Conclusions and Outlook

Experiments on pure tantalum and nickel have shown that processing at elevated temperatures leads to the formation of ultrafine-grained structures with pronounced grain aspect ratios. A maximum of the aspect ratio can be obtained by deformation at temperatures, ~T_0, similar to that leading to first structural coarsening during static annealing. Despite that the recovery mechanisms increase with temperature, these mechanisms may be insufficient to promote fragmentation, thereby enabling the observed increase in grain elongation. The reason for this seems to arise from inefficient movement of triple junctions at temperatures below T_0 due to pinning by intragranular defects or at dislocation boundaries. Moreover, fragmentation events of adjacent grains due to grain boundary migration occur less frequently at elevated temperatures. Although studies on other materials should be performed, it seems clear that the observed trends are general ones and the grain aspect ratio does not necessarily decrease with temperature.

To maximize the aspect ratio for a given material, the deformation temperature has therefore to be sufficiently high to subdue frequent grain fragmentation by boundary migration, but lower than the temperature where defect annihilation, and so a reduction of the pinning sites for triple junctions, takes place.

Future experiments should be carried out to clarify whether alloying elements or impurities can amplify the observed effect, as it is known that they not only stabilize intragranular defects but also to influence the grain boundary itself. In addition, the possibility to generate nanostructures with different aspect ratios, is expected to allow a thorough evaluation of its effect on mechanical properties.

Acknowledgements
Financial support by the European Research Council under ERC Grant Agreement No. 340185 USMS is gratefully acknowledged.

References

[1] Meyers M, Mishra A and Benson D 2006 *Prog. Mater. Sci.* **51** 427–556
[2] Dao M, Lu L, Asaro R J, Hosson J T M De and Ma E 2007 *Acta Mater.* **55** 4041–4065
[3] Estrin Y and Vinogradov A 2013 *Acta Mater.* **61** 782–817
[4] Lu L, Schwaiger R, Shan Z W, Dao M, Lu K and Suresh S 2005 *Acta Mater.* **53** 2169–2179
[5] Shen Y F, Lu L, Lu Q H, Jin Z H and Lu L 2005 *Scr. Mater.* **52** 989–994
[6] Wang Y H and Ma E 2004 *Acta Mater.* **52** 1699–1709
[7] Wang Y, Chen M, Zhou F and Ma E 2002 *Nature* **419** 912–915
[8] Zhao W S, Tao N R, Guo J Y, Lu Q H and Lu K 2005 *Scr. Mater.* **53** 745–749
[9] Thuvander M, Abraham M, Cerezo A and Smith G D W 2001 *Mater. Sci. Technol.* **17** 961–970
[10] Hibbard G D, McCrea J L, Palumbo G, Aust K T and Erb U 2002 *Scr. Mater.* **47** 83–87
[11] Renk O, Hohenwarter A and Pippan R 2012 *Adv. Eng. Mater.* **14** 948–954
[12] Pippan R and Hohenwarter A 2016 *Mater. Res. Lett.* **4** 127–136
[13] Reiser J, Wurster S, Hoffmann J, Bonk S, Bonnekoh C, Kiener D, Pippan R, Hoffmann A and Rieth M 2016 *Int. J. Refract. Met. Hard Mater.* **58** 22–33
[14] Vorhauer A and Pippan R 2008 *Metall. Mater. Trans. A Phys. Metall. Mater. Sci.* **39** 417–429
[15] Renk O, Hohenwarter A, Wurster S and Pippan R 2014 *Acta Mater.* **77** 401–410
[16] Yu T, Hansen N, Huang X and Godfrey A 2014 *Mater. Res. Lett.* **2** 160–165
[17] Hansen N, Mehl R F and Medalist A 2001 *Metall. Mater. Trans.* A **32** 2917–2935
[18] Chen Y C, Huang Y Y, Chang C P and Kao P W 2003 *Acta Mater.* **51** 2005–2015
[19] Wang Y Y, Sun P L, Kao P W and Chang C P 2004 *Scr. Mater.* **50** 613–617
[20] Renk O, Ghosh P and Pippan R 2017 *Scr. Mater.* **137** 60-63
[21] Hughes D A and Hansen N 2000 *Acta Mater.* **48** 2985–3004
[22] Godfrey A and Hughes D A 2000 *Acta Mater.* **48** 1897–1905
[23] Gottstein G and Shvindlerman L S 2009 Grain Boundary Migration in Metals Thermodynamics, Kinetics, Applications 2nd ed. (CRC Press Taylor & Francis Group)
[24] Yu T, Hansen N and Huang X 2011 *Proc. R. Soc. A Math. Phys. Eng. Sci.* **467** 3039–3065
[25] Raj R and Lange F F 1985 *Acta Metall.* **33** 699–703
[26] Legros M, Gianola D S and Hemker K J 2008 *Acta Mater.* **56** 3380–3393
[27] Rupert T J, Gianola D S, Gan Y and Hemker K J 2009 *Science (80)* **326** 1686–1690
[28] Rajabzadeh A, Legros M, Combe N, Mompiou F and Molodov D A 2013 *Philos. Mag.* **93** 1299–1316
[29] Fang T H, Li W L, Tao N R and Lu K 2011 *Science* **331** 1587–90

[30] Ghosh P, Renk O and Pippan R 2017 *Mater. Sci. Eng.* A **684** 101–109
[31] Rajabzadeh A, Mompiou F, Lartigue-Korinek S, Combe N, Legros M and Molodov D A 2014 *Acta Mater.* **77** 223–235
[32] Priester L 2013 Grain Boundaries - From Theory to Engineerinig (Springer)
[33] Huang Y and Prangnell P B 2008 *Acta Mater.* **56** 1619–1632
[34] Yu T, Hansen N and Huang X 2013 *Acta Mater.* **61** 6577–6586
[35] Yu T, Hughes D A, Hansen N and Huang X 2015 *Acta Mater.* **86** 269–278

Characterization of Cu Distribution in an Al-0.3%Cu Alloy Cold Rolled to 98%

L F Shuai[1], T L Huang[1,2,*], G L Wu[1,2], N Hansen[3] and X Huang[3]

[1] College of Materials Science and Engineering, Chongqing University, Chongqing 400045, China

[2] Electron Microscopy Center of Chongqing University, Chongqing University, Chongqing 400045, China

[3] Section for Materials Science and Advanced Characterization, Department of Wind Energy, Technical University of Denmark, DK-4000 Roskilde, Denmark

E-mail: huangtl@cqu.edu.cn

Abstract. In this study, the distribution of Cu element in a Al (99.9996% purity)-0.3%Cu alloy cold rolled to 98% has been characterized in detail by using three-dimensional atom probe (3DAP) and ChemiSTEM techniques. The cold rolled structure is a typical high strain lamellar structure with an average boundary spacing of 200 nm, indicating a strong role of the small amount of Cu element in stabilizing the microstructure to form the fine scale structure. A heavy segregation of Cu element in the lamellar boundaries of high angles has been observed and the Cu concentration in the boundaries can be as high as 20 times of the nominal concentration of the alloy, which is considered as the main reason for a formation of a stable nanoscale lamellar structure.

1. Introduction

Nanostructured and ultrafine-grained metals have in general high strength but low ductility and poor thermal stability [1-5]. Introducing dispersion fine particles can facilitate the generation of nanostructured metals by high strain deformation even in the case when the matrix is of ultrahigh purity [6-14]. Another way to stabilize nanostructured metals is to use alloying elements that may form boundary segregation or precipitates in the nanostructured metals. Different alloying elements may have different efficiency in stabilizing nanostructured metals [15, 16]. In our present studies [17-18], we found that a nanostructured lamellar structure with an average lamellar boundary spacing of 200 nm can be produced after 98% cold rolling in an Al-0.3%Cu alloy with an ultra-high pure Al matrix (99.9996%) and an addition of a small amount (0.3wt%) Cu. In this study, the distribution of the Cu element is characterized in detail to understand the mechanism of Cu in stabilizing the nanostructure.

2. Material and Experimental

An Al-0.3%Cu alloy ingot was produced by using an ultra-high pure (99.9996%) Al base material and addition of 0.3wt.% oxygen free high conductivity (OFHC) Cu. The ingot was cold rolled to a thickness reduction of 98% (corresponding to ε_{vM}=4.5). The microstructure of the cold rolled sample was characterized from the longitudinal section (containing the normal direction, ND, and rolling direction, RD) by means of transmission electron microscopy (TEM) and scanning electron microscopy (SEM). The TEM foils were prepared by twin jet electropolishing and examined in a JOEL JEM2100 electron microscope operating at 200 kV. An online Kikuchi-line analysis system installed in the microscope was used to measure the orientations and misorientations [19]. Dislocation density was calculated based on average dislocation spacing obtained by intersection method and sample thickness determined by CBED method. The TEM foils were also characterized on an FEI Titan G2 60-300 ChemiSTEM with a super X energy dispersive spectrometer (EDS) system. SEM samples were mechanically ground to 4000 SiC paper followed by electropolishing in a 10% perchloric acid and ethanol solution at -20°C. SEM observations were conducted with an FEI Nova400 nanoSEM equipped with a high sensitive backscattering electron (BSE) detector. A CAMECA LEAP4000HR three dimensional atom probe (3DAP) was applied to characterize the element distribution in the cold rolled sample. Needle specimens for 3DAP were prepared by electropolishing using a 5% perchloric acid and ethanol solution. The 3DAP experiments were done with a specimen temperature of 50K, voltage pulse rate of 200 kHz and ion collection rate of 0.5%.

3. Results and Discussion

3.1. Microstructure

Figure 1 shows the microstructure formed in the 98% cold rolled sample. A typical lamellar structure is developed, which is composed of extended lamellar boundaries along the RD and short boundaries interconnecting the lamellar boundaries. The lamellar boundaries have an average boundary spacing of 200 nm. A previous study showed that the ultra-high pure Al (99.9996%) [6] has a very poor thermal stability during cold rolling and dynamic recrystallization may take pace when the rolling strain is above 30%. The addition of 0.3% Cu enabled the generation of the nanoscale lamellar structure [17,18]. The lamellar boundary spacing is comparable with that reported for commercial purity Al (99.2 – 99.5% pure) [20-23] deformed by accumulative roll bonding (ARB) to 6 cycles. Note that the lamellar boundary spacing obtained in the Al-0.3%Cu alloys is finer than that (700-1000 nm) reported for a 99.99% pure Al [24] processed by 6 ARB cycles and that (230 nm) for an Al-1%Si alloys cold rolled to 98% [6]. Clearly the addition of a small amount Cu facilitated the formation of a fine scale lamellar structure. In figure 1a, small particles are seen at a few lamellar boundaries. However, SEM-BSE observations over relatively larger areas showed that most of the lamellar boundaries are free of particles in the 98% cold rolled state, as shown in figure 1b.

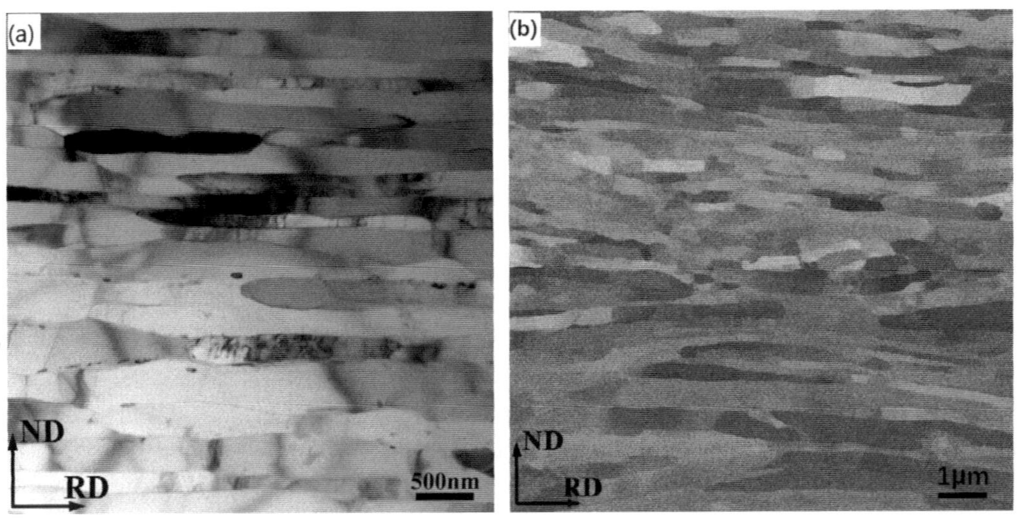

Figure 1. TEM (a) and SEM-BSE (b) images showing the nanoscale lamellar structure of Al-0.3%Cu alloy cold rolled to 98%.

The density of dislocations presented in the volumes between lamellar and interconnecting boundaries was measured to be 1.75×10^{14} m^{-2}, which is a few times higher than those observed in the ARB processed commercial purity Al 1100 [23] and 99.99% purity Al [24], probably due to the pinning effect of solid solute Cu atoms in the matrix.

3.2. Cu distribution

Figure 2a shows a 3DAP map of the 98% cold rolled sample. The red dots in the map represent the Cu atoms. For clarity, the Al atoms are not shown in the map. The volume analyzed is $34\times34\times309$ nm^3. In figure 2a, there is a line with high concentration of Cu element. Orientation analysis using the method proposed by reference [25] showed that the line corresponds to a high angle lamellar boundary. The details of Cu distribution within the marked column in (a) is shown in figure 2b. Evidently, the number of Cu atoms at the boundary is much larger than that in the matrix. The Cu concentration in the column was quantified along a distance from the top to bottom and is shown in figure 2c. It is seen that the Cu concentration at the lamellar boundary can be higher than 3at.% (about 7wt.%), which is higher than 20 times of the nominal concentration of the alloy.

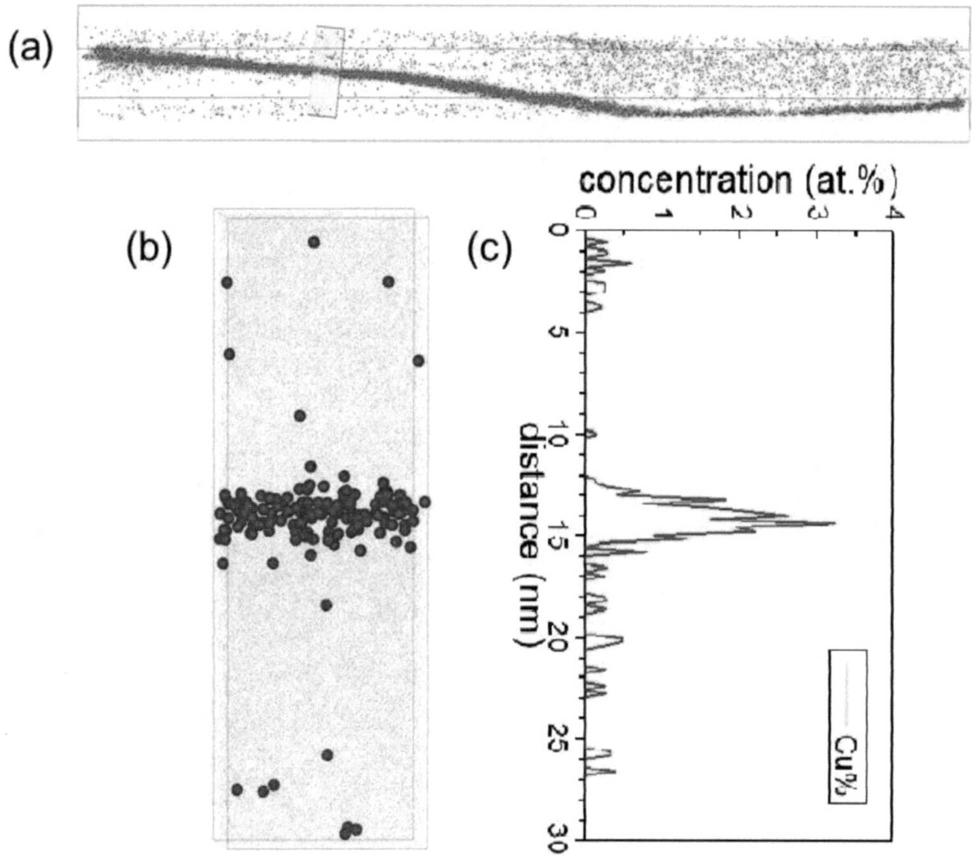

Figure 2. Distribution of Cu atoms (in red color) in the 98% cold rolled Al-0.3Cu alloy. (a) 3D reconstructed map of Cu atoms; (b) distribution of Cu atoms within the column marked in a); (c) distribution of Cu concentration versus distance in (b).

To show the Cu distribution over larger areas, EDS mapping was conducted in the FEI Titan ChemiSTEM. One example of the EDS mapping results is shown in figure 3, in which the corresponding TEM image of the mapped area is shown. The misorientation angles across the lamellar boundaries seen in the area were measured and are indicated in figure 3. From the EDS map, it is seen that there are some traces of high concentration of Cu elements, which are well related to the lamellar boundaries. However, it should be noted that not all the boundaries have Cu segregation. For example, the only low angle lamellar boundary (8.9°) does not show a clear segregation of Cu on the EDS map. Furthermore, the interconnecting boundaries, which are often of low misorientation angles, show no Cu segregation. These observations indicate a possible effect of boundary misorientation on the segregation of Cu.

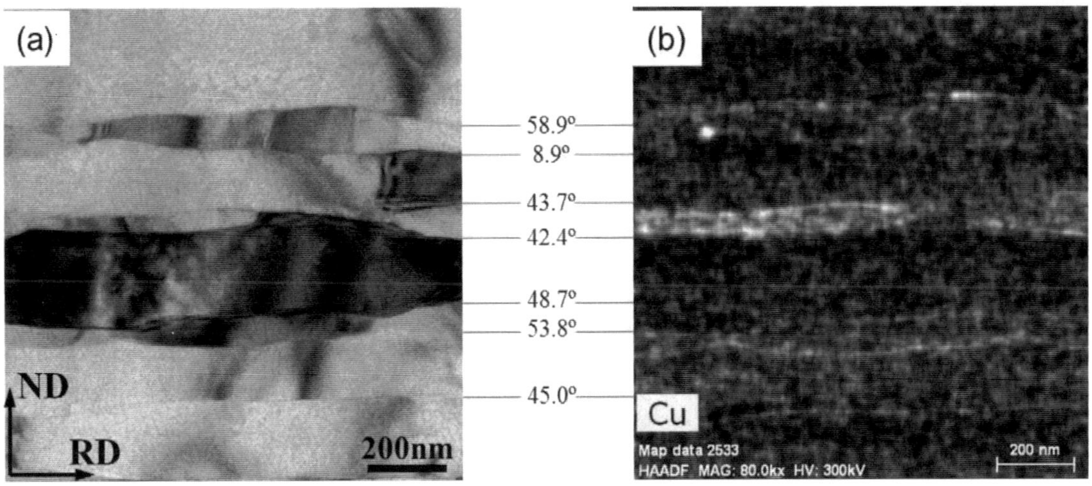

Figure 3. Cu segregation at boundaries in the 98% cold rolled Al-0.3%Cu alloy. (a) TEM bright field image; (b) EDS map of the same area. Misorientation angles of lamellar boundaries are indicated.

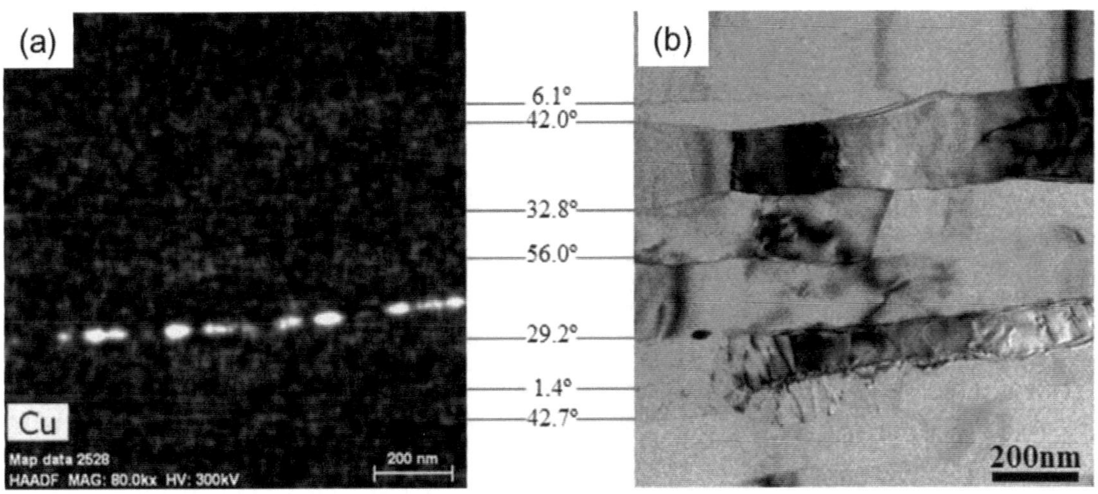

Figure 4. Cu segregation at boundaries in the 98% cold rolled Al-0.3%Cu alloy. (a) EDS map and (b) TEM bright field image of the same area. Misorientation angles of lamellar boundaries are marked.

Figure 4 shows another example of EDS mapping in the 98% cold rolled sample. In this case, most of the lamellar boundaries (both high and low misorientation angles) and all interconnecting boundaries seen in the TEM image cannot be identified on the EDS map. However, along one lamellar boundary (29.2°) clear Cu segregation and even nano-sized precipitates (Al_2Cu) are observed. These observations suggest that the segregation of Cu atoms does not occur at all high angle boundaries, which may be a result of non-uniform distribution of Cu elements at a more global scale in the deformed sample.

To further illustrate the Cu segregation at the lamellar boundaries, the 98% deformed sample was annealed at 200°C for 1 hour. Figure 5 shows the SEM-BSE image of the annealed sample. Compared with the as deformed state (figure 1b), many nano-sized second phase particles can be observed in this annealed sample. Most of the particles are aligned along the direction parallel to the RD and associated with the lamellar boundaries. A few particles are seen within the grains but their alignment with the RD indicates that they were formed originally at the lamellar boundaries but left inside the volume after the boundaries moved during the annealing.

The above 3DAP and EDS mapping results showed clearly that the Cu atoms preferred to segregate at the lamellar boundaries in particular the high angle lamellar boundaries. The absence of segregation at low angle lamellar boundaries and interconnecting boundaries suggests that the high misorientation angles enhance the segregation. The preferred segregation of Cu atoms and the occasional formation of nano-sized particles at the high angle lamellar boundaries will play a role in pinning the boundaries and prevent them being removed by triple junction motion during the cold rolling [26], thus chancing the formation of nanoscale lamellar boundaries. Similar pinning effect on the high angle boundaries may occur during annealing, which will in combination with the effect of relatively low mobility of low angle boundaries (without Cu segregation) lead to an enhanced thermal stability.

Figure 5. SEM-BSE image of Al-0.3%Cu cold rolled 98% and annealed 200°C for 1 hour.

4. Conclusions

The distribution of Cu element in an Al (99.9996% purity)-0.3%Cu alloy cold rolled to 98% has been characterized by means of three-dimensional atom probe and ChemiSTEM techniques. The main observations are summarized as follows:

- Segregation of Cu at lamellar boundaries is revealed by both 3DAP and ChemiSTEM techniques. The Cu concentration at segregated boundaries can be as high as 7wt.%, which is about 20 times of the nominal concentration of the alloy.
- High segregation of Cu element is only found at high angle lamellar boundaries, while low angle lamellar boundaries and interconnecting dislocation boundaries do not show obvious Cu segregation, indicating that boundary misorientation angles play a role in enhancing the segregation of Cu.

Acknowledgments

The authors wish to thank the financial support of the Natural Science Foundation of China (NSFC, Grants No. 51327805, 51471039 and 51421001), State Key Research and Development Program of MOST of China (2016YFB0700401, 2016YFB0700403). NH thanks the support of the "111" Project (B16007) by the Ministry of Education and the State Administration of Foreign Experts Affairs of China.

References

[1] Meyers M, Mishra A and Benson D 2006 *Prog. Mater. Sci.* **51** 427–556
[2] Valiev R Z and Zhu Y 2015 *Research Society of Japan* **40** 309–18
[3] Yu C, Kao P and Chang C 2005 *Acta Mater.* **53** 4019–28
[4] Lu K, Lu L and Suresh S 2009 *Science* **324** 349–52
[5] Huang X, Tsuji N and Pantleon W 2009 *Risoe International Symposium on Materials Science* **30** 59–80
[6] Huang T, Dong Q, Gong X, Hansen N, Liu Q and Huang X 2012 *J. Mater. Sci.* **47** 7914–20
[7] Huang T, Li C, Wu G, Yu T, Liu Q, Hansen N and Huang X 2014 *J. Mater. Sci.* **49** 6667-73
[8] Huang T, Wu G, Liu Q and Huang X 2014 *Mater. Sci. Forum* **783-786** 2629-34
[9] Cheng S, Zhao Y, Zhu Y and Ma E 2007 *Acta Mater.* **55** 5822-32
[10] Song R, Ponge D, Raabe D and Kaspar R 2005 *Acta Mater.* **53** 845-58
[11] Song R, Ponge D and Raabe D 2005 *Scripta Mater.* **52** 1075-80
[12] Liu G, Zhang G, Jiang F, Ding X, Sun Y and Sun J, et al. 2013 *Nat. mater.* **12** 344-50
[13] Zhao Y, Liao X, Cheng S and Ma E 2006 *Adv. Mater.* **18** 2280-3
[14] Huang T, Wu G, Huang X and Hansen N 2015 *IOP Conf. Series: Materials Science and Engineering* **89** 012028
[15] Zhang H, Lu K, Pippan R, Huang X and Hansen N 2011 *Scripta Mater.* **65** 481–4
[16] Jiang L, Li J, Cheng P, Liu G, Wang R, Chen B, Zhang J, Sun J, Yang M and Yang G 2014 *Scientific reports* **4** 3605
[17] Li X, Wakeel A, Huang T, Wu G and Huang X 2015 *IOP Conference Series: Materials Science and Engineering* **89** 012032
[18] Wakeel A, Huang T, Wu G, Mishin O V and Huang X 2015 *IOP Conference Series: Materials Science and Engineering* **82** 012050
[19] Liu Q 1995 *Ultramicroscopy* **60** 81-9
[20] Tsuji N, Ito Y, Saito Y and Minamino Y 2002 *Scripta Mater.* **47** 893-9
[21] Tsuji N, Saito Y, Lee S and Minamino Y 2003 *Adv. Eng. Mater.* **5** 338-44
[22] Huang X, Hansen N and Tsuji N 2006 *Science* **312** 249-51
[23] Kidmose J 2014 PhD Thesis, *Technical University of Denmark*
[24] Kamikawa N, Huang X, Tsuji N and Hansen N 2009 *Acta Mater.* **57** 4198-208
[25] Araullo-Peters V, Gault B, Shrestha S and Yao L, et al. 2012 *Scripta Mater.* **66** 907-10
[26] Yu T, Hansen N, Huang X and Godfrey A 2014 *Mater. Res. Lett.* **2** 160-5

IOP Conf. Series: Materials Science and Engineering **219** (2017) 012039 doi:10.1088/1757-899X/219/1/012039

4D Study of Grain Growth in Armco Iron Using Laboratory X-ray Diffraction Contrast Tomography

J Sun[1,2], A Lyckegaard[1], Y B Zhang[2], S A Catherine[4], B R Patterson[4], F Bachmann[1], N Gueninchault[1], H Bale[3], C Holzner[3], E Lauridsen[1], D Juul Jensen[2]

[1]Xnovo Technology, Køge, Denmark
[2]Department of Wind Energy, Technical University of Denmark, Denmark
[3]Carl Zeiss X-ray Microscopy, Pleasanton, CA, United States
[4]Department of Materials Science and Engineering, University of Florida, FL, United States

Email: jsun@xnovotech.com, alyckegaard@xnovotech.com

Abstract. Using a novel laboratory diffraction contrast tomography (LabDCT) technique, a non-destructive 4D study was conducted to investigate the evolution in 3D of the grain structure during grain growth in an Armco iron sample. The 3D grain morphology and the crystallographic orientations of more than 300 grains were determined at three temporal states during annealing. The correlation between growth of grains and grain orientation is explored. The results demonstrate the capability of the LabDCT technique to allow detailed studies of grain growth, and thereby provide the necessary 4D experimental evidence required for further understanding of grain growth.

1. Introduction

The majority of metallic engineering materials are polycrystalline and their properties are strongly dependent on thermomechanical processing such as plastic deformation and annealing. During annealing grain growth can occur, with some of the grains growing at the expense of others. Grain growth is important since it affects the distribution of boundaries and sizes of the grains in the microstructure, and this is intimately linked to the properties of a material.

The kinetics of grain growth has been a main focus of several studies, investigating factors such as the topology of the grains and the grain boundary properties [e.g. 1-3]. While considerable efforts have been made to model the grain growth process, there have been considerably fewer efforts to provide experimental verification of these models [4, 5]. Essentially, grain growth is a 3D structural evolution process, so that neither 2D examination on section, nor 3D measurements based on serial sectioning are capable of providing direct 4D experimental evidence to validate the theoretical models. With the development during the last two decades of novel X-ray-based diffraction-imaging techniques, non-destructive 3D mapping of grain structures is now possible. Experiments using high-energy synchrotron X-ray beams to study the 3D grain structure evolution during annealing have been reported [5-9], however, the experiments dedicated to grain growth studies are still small in number due to the limited access to international synchrotron facilities.

The novel laboratory-based X-ray diffraction contrast tomography system (LabDCT), operating on a commercially available X-ray microscope (XRM), enables the wider accessibility and routine use of non-destructive, time-evolution experiments, thereby allowing the necessary large number of experiments required for grain growth studies to be carried out [10, 11]. In this work, preliminary results from a 4D study of the grain growth behavior in an Armco iron sample using LabDCT are presented. Essential information regarding the microstructural evolution, including both morphology and crystallographic orientation of the grains, is revealed. Analysis of the grain growth kinetics in correlation with grain orientation is presented to demonstrate the potential of LabDCT in advancing the understanding of grain growth.

2. Experimental method

The material used in the present study is Armco iron with very low carbon content. The initial material was obtained from a hot-rolled bar. A pre-treatment was carried out to refine the grain size, consisting of 75% cold rolling reduction in thickness and subsequent annealing at 880 °C for 5 days. The average starting grain size after pre-treatment was approximately 75 μm. The microstructure consisted of only polygonal ferritic grains without presence of carbides. A cylindrical sample with diameter of 1 mm was prepared by electrical discharge machining for the LabDCT grain growth studies. The axis of the cylinder was taken along the normal direction (ND) of the initial hot-rolled bar.

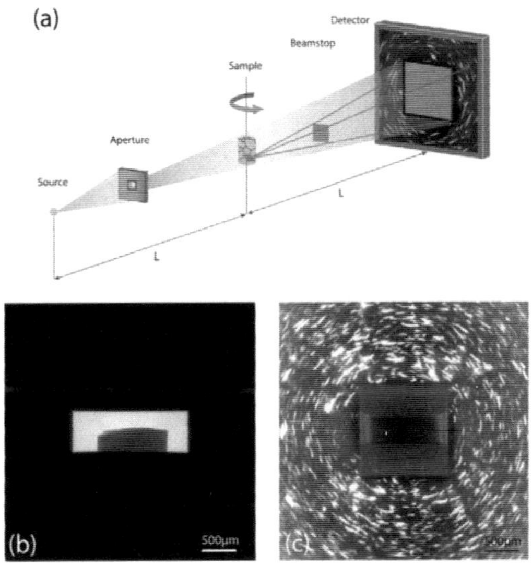

Figure 1. (a) Schematic showing the experimental setup of LabDCT in the laboratory X-ray microscope [10]. (b) Example absorption contrast projection with the direct beam illuminating the top part of the Armco iron sample. (c) Example projection showing diffraction spots from grains within the scanned volume of the sample.

The LabDCT characterization was conducted on a commercially available ZEISS Xradia 520 Versa X-ray Microscope (with LabDCT module). Figure 1(a) schematically illustrates the working principles of LabDCT. The instrument uses a polychromatic, divergent X-ray beam, instead of the parallel monochromatic beam typically used in the synchrotron X-ray DCT technique. An aperture is placed between the source and the sample, to constrain the incoming X-ray beam and to illuminate the sample only in the center region of the detector, as seen in the example shown in figure 1(b). For the DCT scan, a beam-stop is used to block the transmitted X-rays to increase the sensitivity of the diffraction measurements. A high-resolution detector is placed at the Laue focal plane with equal source-detector distance. In this arrangement the crystal grains fulfilling the Bragg condition focus the divergent X-ray beams into a line in the diffraction pattern. Figure 1(c) shows an example of the diffraction spots of the current sample at a single rotation position. As the sample rotates, a specified number of DCT scan projections can be collected through a 360-degree rotation. The collected data were processed and reconstructed with the GrainMapper3D™ analysis package developed by Xnovo Technology ApS. The reconstruction generates a 3D map of the grains in the scanned volume, including both the crystallographic orientation and morphology of the grains. In the present work, the volumes of interest in the as-treated sample were first scanned acquiring 181 DCT projections over a 360-degree rotation. Annealing treatments of 18 hrs at 880 °C and then 21 hrs at 880 °C were subsequently carried out using an external furnace. After each annealing treatment, the volumes of interest in the sample was then scanned with the same experimental settings to track the microstructural evolution. At each state two volumes with a specified overlapping distance in height were scanned and subsequently stitched together. The three temporal states are named as t_0, t_1 and t_2 in the following text.

3. Results and Discussion

Figure 2(a) shows the reconstructed 3D grain maps of the same volume at the three temporal states. It should be noted that these 3D grain maps are preliminary results implementing a new algorithm for shape reconstruction and a full validation is still ongoing. The color chosen is based on the crystallographic orientation of the grains with respect to the sample rotation axis, which is the RD of the initial hot-rolled bar. Each voxel in the reconstructed volume has a side length of 5 μm. The orientation accuracy is this study is estimated to be better than 0.2°. For analysis purposes, only the interior grains are used, with those adjacent to the edge of the volumes removed, as shown in figure 2(b). The remaining numbers of interior grains are 525, 351 and 343 for states t_0, t_1 and t_2.

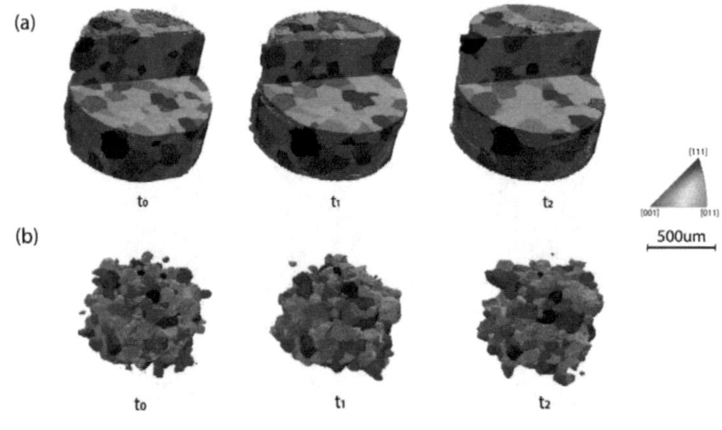

Figure 2. (a) The reconstructed 3D grain maps of the corresponding volume at the three temporal states t_0, t_1 and t_2. (b) The 3D grain maps with the grains adjacent to the edges of the volume removed. The colors of the grains are based on the orientation relative to the rotation axis.

The equivalent sphere diameter (ESD) is used to describe the grain size, as calculated from the volumes of the grains. The distributions of the grain size at the three temporal states are shown in figure 3. The size of the grains is distributed within a similar range after the two annealing treatments, and it is seen that grain growth primarily occurred during the 1st annealing treatment while the 2nd annealing treatment only resulted in minor changes in the grain size distributions. As revealed from the 2D slices at similar position within the reconstructed volumes (figure 4), the three temporal states have similar microstructure on the overall scale with only local variations.

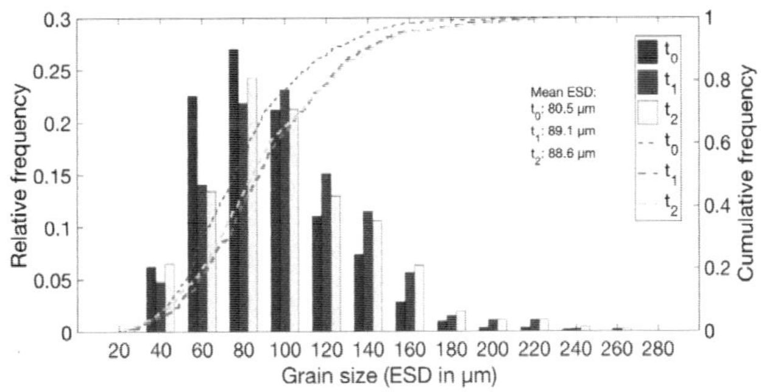

Figure 3. Distributions and cumulative frequency curves for the grain size of the three temporal states.

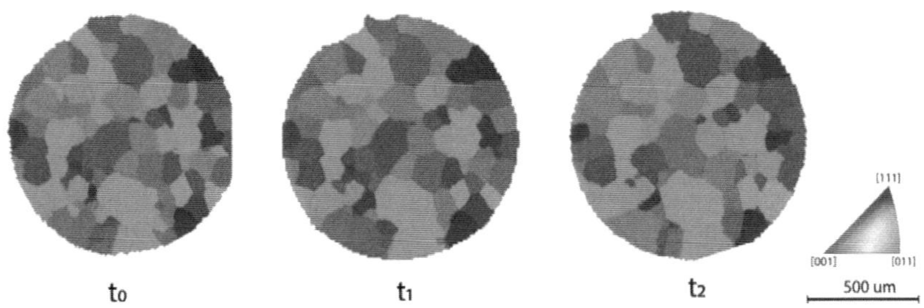

Figure 4. Slice from similar position in the reconstructed volume showing the microstructure of the sample at three temporal states. The colors of the grains are based on the orientation of the grain relative to the rotation axis.

Among the reconstructed 3D grains (excluding those grains adjacent to the volume edge) mapped in the three temporal states, 287 grains were found to have an exact match in orientation and reasonable match in center-of-mass positions. The grain growth behavior for these grains was characterized by the change in equivalent sphere diameter and it was found that only 18 grains had grown more than 10% in their equivalent sphere diameter after the two annealing treatments. The orientations of these growing grains, as well as their grain boundary misorientations to the neighboring grains, were analyzed. As shown in figure 5(a), the orientation of these growing grains (red dots) are almost randomly distributed, and the distribution of misorientation between the growing grains with their neighbors is similar with that of all the grains (figure 5b). No specific orientation correlation with the growing of grains is observed for the current sample.

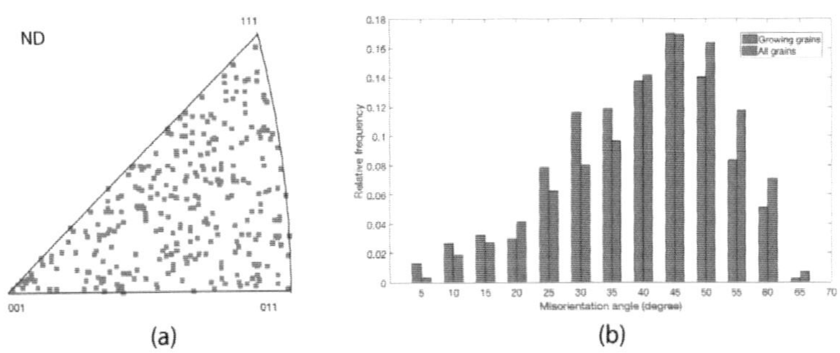

Figure 5. (a) Inverse pole figure of the sample rotation axis (ND for the hot-rolled bar). (b) Distribution of the misorientation angle between grains and their neighbors. The color scheme is the same for (a) and (b): red represents the grains that have grown more than 10% in the ESD and grey represents all the 287 grains.

4. Summary and Outlook

The paper presents a first 4D study of grain growth kinetics enabled by the use of laboratory X-ray diffraction contrast tomography (LabDCT). It is demonstrated that essential information for the study of grain growth, such as the shape, size and crystallographic orientation of the grains, can be obtained non-destructively at multiple temporal states. This methodology provides access to the necessary 4D experimental evidence which previously were not readily available. It is found that for the current Armco iron specimen, no substantial grain growth has occurred with current annealing treatment, and that for the small number of grains that have grown, there is no obvious correlation with their crystallographic orientation.

This work is ongoing and several aspects are suggested for future study:

- New experiments with more time steps should be conducted. The current study only includes two annealing treatments and is limited by the relative small amount of grain growth occurring during the annealing treatments given to the samples.
- Topological parameters such as the number of faces for individual grains should also be analyzed and correlated with the grain growth kinetics, with the results used to validate the existing theoretical models.

Acknowledgement

This work is partially funded by the Innovation Fund Denmark (IFD) under the project 'Multimodal 4D Characterization of Materials in the Laboratory' (case number: 5190-00044B).

References

[1] Weaire D and Glazier J 1993 *Philosophical Magazine Letters* **68** 363–5
[2] Kim B, Hiraga K and Morita K 2003 *Mater Trans* **44** 2239–44
[3] Rollett A, Brahme A and Roberts C 2007 *Mater Sci Forum* **558-559** 33–42
[4] Rowenhorst D, Lewis A and Spanos G 2010 *Acta Mater* **58** 5511–9
[5] McKenna I , Poulsen S, Lauridsen E, Ludwig W and Voorhees P 2014 *Acta Mater* **78**125–34
[6] Schmidt S, Nielsen S, Gundlach C, Margulies L, Huang X and Juul Jensen D 2004 *Science* **305** 229–32
[7] Van Boxel S, Schmidt S, Ludwig W, Zhang Y, Juul Jensen D and Pantleon W 2014 *Mater. Trans.* **55** 128–36
[8] Lauridsen E, Schmidt S, Nielsen S, Margulies L, Poulsen H and Juul Jensen D 2006 *Scripta Mater* **55** 51–6
[9] Schmidt S, Olsen U, Poulsen H, Sørensen H, Lauridsen E, Margulies L, Maurice C and Juul Jensen D 2008 *Scripta Mater* **59** 491–4
[10] Holzner, Lavery L, Bale H, Merkle A, McDonald S, Withers P, Zhang Y, Juul Jensen D, Kimura M, Lyckegaard A, Reischig P and Lauridsen E 2016 *Microsc Today* **24** 34–43
[11] McDonald S, Reischig P, Holzner C, Lauridsen E, Withers P, Merkle A and Feser M 2015 *Sci Reports* **5** 14665

IOP Conf. Series: Materials Science and Engineering **219** (2017) 012040 doi:10.1088/1757-899X/219/1/012040

Effect of strain-path change on the anisotropic mechanical properties of a commercially pure aluminum

P L Sun[1,*] **and S J Huang**[2]

[1]Department of Materials and Optoelectronic Science, National Sun Yat-sen University, Kaohsiung 80424, Taiwan
[2]Department of Materials Science and Engineering, Feng Chia University, Taichung 40724, Taiwan

Email address: plsun@mail.nsysu.edu.tw

Abstract. Samples of commercially pure aluminum were subjected to equal channel angular extrusion (ECAE) using a 90° square die by routes A and C, where the specimens are not rotated and are rotated 180° between extrusion passes, respectively. Qualitatively similar anisotropic responses under compressive loading along the three orthogonal directions of the ECAE billet are seen in both cases. The plastic anisotropy is related to the effect of strain-path change, namely that different slip activities are induced for specimens loaded along different directions with respect to the last ECAE pass. The anisotropic mechanical behavior is more evident in the sample deformed by route C. Considering the shear patterns imposed in each ECAE route, the characteristics of dislocations introduced in ECAE should affect the mechanical response in post-ECAE loading. It is suggested that during the ECAE process, dislocations on fewer slip systems are activated in route C than in route A, and therefore, a stronger plastic anisotropy results in this sample. The as-ECAE specimens were also heat treated to achieve a recovery-annealed state. The plastic anisotropy persists in the annealed specimens to slightly reduced extent, which can be ascribed to partial annihilation of pre-existing dislocations.

1. Introduction

It has been well established in the past two decades that metals with ultrafine-grained (UFG) structures can be generated by the use of equal channel angular extrusion (ECAE) [1-3]. After processing by ECAE at ambient temperature, the microstructures of most alloys are quite complex. The morphology of the subgrains/grains constituting the microstructure may vary from lamellar to polyhedral shape [3-5]. The boundaries separating neighboring subgrains/grains are mixtures of segments with misorientation angles ranging from low to high angles, and the boundary structure can vary from dislocation walls (DW), to partially transformed boundaries (PTB) and to grain boundaries (GB) [4]. It was found that low temperature annealing can be applied to slightly tailor the grain size and boundary structure of UFG metals produced by ECAE, in which annealing is carried out at a temperature below the onset of recrystallization [6, 7].

According to Yu *et al.* [8], an evident transition of tensile behavior appears as the grain size is reduced from the micrometer to submicrometer range in aluminum. It was proposed that as the grain size decreases below a critical value, which may correspond to the cell size, the mean free path of dislocations is no longer determined by the dislocation structure, but is limited by the grain boundaries

[8-11]. This causes an enhancement of dynamic recovery as well as a reduction of work hardening in UFG metals, and consequently leads to reduced tensile elongation [8]. The boundary structure was shown to play an important role in the tensile behavior of UFG aluminum [12]. Quasistatic tests revealed an increase in ductility with an increase in the percentage of low-angle boundaries [12].

In the ECAE process metals are deformed via a nearly simple shear [13], which imparts substantial texture and microstructure evolution. In multiple ECAE passes, the processing route and number of passes influence the microstructure and texture development, and thus affect the mechanical properties of the product. Anisotropy in initial flow stress and work-hardening have been reported in metals deformed by ECAE [14-20]. As the last pass in the process of multiple-pass ECAE treatments is believed to be most influential on properties of the processed materials, a knowledge of the anisotropic properties of materials processed by single-pass ECAE should facilitate the understanding of multi-passed materials. Work done on the anisotropic mechanical properties of aluminum processed by ECAE has mainly been focused on that deformed by single pass ECAE, while relatively little work has been done on aluminum processed by multiple pass ECAE [18-20]. The work of Haouaoui *et al.* [14] reveals that copper subjected to only one ECAE pass can result in both tensile and compressive yield stresses decreasing following the order of along Y-axis (perpendicular to the flow plane) > along Z-axis (normal to the top surface at the point of exit) > along X-axis (the extrusion direction). Strain-path change (SPC) is involved in uniaxial tension or compression tests of materials pre-deformed by ECAE. According to Li [17], SPC is an important factor, in addition to texture, to be considered in interpreting the plastic responses of ECAE processed metals subjected to uniaxial loading.

Compared to single-pass ECAE-processed materials, multiple-pass processed counterparts are expected to possess quite different texture and microstructure, which may be strongly affected by material, die angle, processing route, and number of passes used in the deformation process. Horita *et al.* [18] used AA1100 commercially pure aluminum, deformed by ECAE route Bc for 6 passes, and the yield stress decreased following the order of X-axis > Y-axis > Z-axis. Beyerlein *et al.* [19] employed oxygen-free pure copper and pure aluminum to study the anisotropic mechanical behavior of the as-ECAE samples, and found that the order of flow stresses among the three directions evolves similarly with pass number in the two materials. After the first pass, it is in the order of Y-axis > Z-axis > X-axis and after subsequent passes, Y-axis > X-axis ≈ Z-axis. It should be noted that single-pass processed materials was considered as plastically deformed *coarse-grained materials*, while multi-pass processed materials have *UFG structures*, which may behave quite differently from coarse-grained materials.

This work aims at understanding the anisotropic mechanical behavior of UFG aluminum produced by ECAE. Samples of commercial purity aluminum processed by ECAE routes A and C were investigated, both in the as-ECAE state and in a low temperature annealed state. The microstructures produced by ECAE routes A and C are distinctly different because these two routes provide plastic deformation via different shear patterns [21]. In particular, aluminum processed by ECAE route A possesses higher proportion of high angle boundaries (HABs) than route C (67% versus 35% [22]), which has been shown to play an important role in tensile behavior of UFG aluminum [12]. Therefore, a comparison between aluminum processed by ECAE routes A and C should improve our understanding of the effect of boundary structure on the anisotropic mechanical responses of UFG aluminum. Moreover, most of the studies reported in the literature on the anisotropic properties of ECAE metals are focused on the as-ECAE condition. A direct comparison between the results of a recovery annealed and as-ECAE counterpart can help to elucidate the influence of the residual dislocations on the anisotropic mechanical response of ECAE processed aluminum.

2. Experimental

A sample of commercially pure aluminum, AA1050, was homogenized at 873 K for 12 hours and slowly cooled to room temperature in air. The resulting average grain size was 330 μm. Rectangular billets were used for ECAE process. Billets were then subjected to ECAE route A or route C in a 90° square die. Billets were subjected to eight passes, giving an accumulated von Mises strain of ≈ 8.4. All extrusions were performed at room temperature. A detailed description of the ECAE processing used for these samples can be found in a previous paper [3]. Recovery annealing was conducted at 523 K for 1 hour after ECAE. In the following text, **A** and **C** will refer to samples processed by route A and route C in the as-ECAE state, respectively, while the annealed counterparts will be denoted as AR and CR. The ECAE extrusion direction is defined as the *X*-axis, the flow plane normal is designated as the *Y*-axis, and the *Z*-axis is normal to both the *X*- and *Y*-axes.

The grain size of the specimens was measured on thin foils sectioned from all three orthogonal planes (the *X*-, *Y*, and *Z*-planes) using a transmission electron microscope (TEM) (Philips CM200) operated at 200 kV. The grain size of each grain is defined as the diameter of an equivalent circle with the same area as the grain. Additionally, due to the formation of elongated grains after ECAE, the grain shape was described by its length and width. Ellipses were used to fit every grain. The lengths of the long and short axes of the best fitting ellipse represent the grain length and width, respectively.

Samples for compression tests were sectioned to have dimensions of $6 \times 4 \times 4$ mm^3 from the center of the as-ECAE billets, in which the long axis was parallel to the loading direction (either the *X*-, *Y*- or *Z*-axis). Samples were ground and polished to have smooth surfaces prior to compression testing. Compression tests along all three loading directions were conducted at an initial strain rate of 7.1×10^{-4} s^{-1} at room temperature by using an Instron 5582 universal testing machine. Duplicate tests were conducted for all conditions to assure reproducibility. The sample surfaces were observed in a stereomicroscope and a HITACHI S3000 scanning electron microscope (SEM) at 15 kV after compression testing.

3. Results

3.1. Microstructure

The boundary character and texture of the same material after processing by ECAE route A and C for 8 passes have been reported in previous papers [3, 22]. The microstructural characteristics of the A and C specimens are summarized here. The texture of the as-ECAE specimen was observed to persist after annealing up to grain sizes greater than 10 μm in this material [6]. Therefore, the texture of the annealed specimens (AR and CR) is believed to be similar to that of the as-ECAE counterpart. Aluminum processed by 8 ECAE passes with either route A or route C exhibits similar crystallographic texture [22], which has the same characteristics of single-pass processed aluminum [23]. Texture strengths calculated from orientation distribution functions (ODFs) for A and C are 1.58 and 1.38, respectively, implying moderately weak textures [22, 24].

High angle boundaries are defined as boundaries with misorientation angles greater than 15° while low angle boundaries (LABs) have misorientation angles smaller than 15°. The aluminum processed by ECAE route A possesses higher proportion of HABs than route C (67% versus 35% [3, 22]). The experimental results from our previous work indicate that the ECAE samples annealed at 523K for 4 hours show similar proportion of HABs to that of the as-ECAE state [6]. Therefore, this indicates that the HAB% remains unchanged in both A and C specimens after annealed at 523K for one hour.

The microstructures of A, C, AR and CR were examined on the three orthogonal planes by using TEM and the results are given in figures 1-3. The quantitative results of the grain size and shape are summarized in table 1. Grains in specimen A exhibit mainly elongated shape on the *X*- or *Y*-plane but roughly equiaxed shape on the *Z*-plane. Based on the grain morphology on the three orthogonal planes, the grains in sample A can be considered as pancake (or disc) shape with the face approximately parallel to the *Z*-plane. The majority of grains in sample C are roughly equiaxed (aspect ratio ~ 2) on all three planes (table 1). The grains in sample A were measured to be slightly smaller than those in sample C. In both samples A and C, obvious diffraction contrast variation appears within the grain interior, which indicates the presence of considerable amount of defects and internal stresses in the as-ECAE condition (figure 1).

IOP Conf. Series: Materials Science and Engineering **219** (2017) 012040 doi:10.1088/1757-899X/219/1/012040

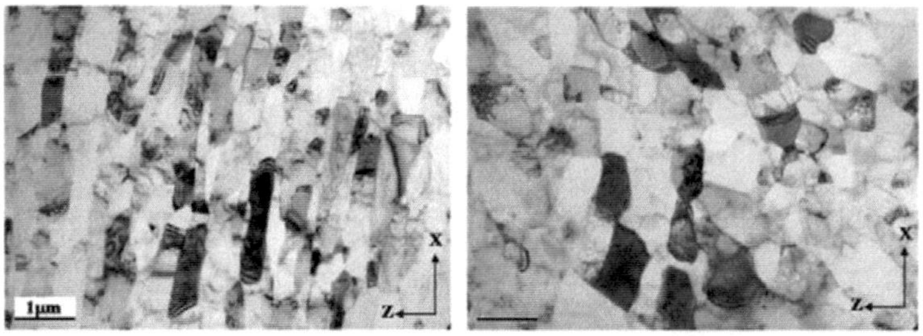

(a) (b)

Figure 1. TEM micrographs of (a) specimen A and (b) specimen C on the Y-plane.

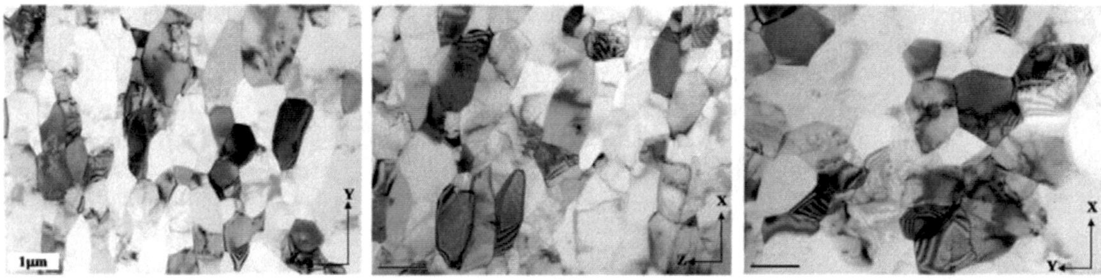

(a) (b) (c)

Figure 2. TEM micrographs on the (a) X, (b) Y and (c) Z planes for the AR specimen.

(a) (b) (c)

Figure 3. TEM micrographs on the (a) X, (b) Y and (c) Z planes for the CR specimen.

Table 1. Quantitative microstructural parameters obtained from measurements on the three *orthogonal* planes for each sample.

material	**X-plane**		**Y-plane**		**Z-plane**		**D (μm)**
	d_X (μm)	R_X	d_Y (μm)	R_Y	d_Z (μm)	R_Z	
A	0.54±0.21 [3]	2.56	0.56±0.26 [3]	2.71	0.68±0.39	1.82	0.59
C	0.68±0.29 [3]	2.08	0.61±0.30 [3]	1.89	0.70±0.40	1.67	0.66
AR	0.64±0.28	1.90	0.64±0.25	1.92	0.94±0.44	1.53	0.73
CR	0.76±0.38	2.00	0.71±0.38	1.68	0.94±0.45	1.56	0.80

*$D = (d_X d_Y d_Z)^{1/3}$, where d_X, d_Y, and d_Z are grain sizes measured on the three orthogonal planes. Grain size is defined as the equivalent circle diameter. Grain aspect ratio, R, is defined as the ratio of grain length over width.

After annealing, the grain structures in the AR and CR are more comparable, in which the majority of grains in both specimens have equiaxed shape (figures 2 and 3). The average grain size in sample CR is slightly larger than that in sample AR (table 1). As compared to samples A and C, the diffraction contrast in samples AR and CR is more uniform, indicating defect density and internal stresses are substantially reduced. Additionally, most of the grains in samples AR and CR are enclosed by boundaries with clear grain boundary fringes (figures 2 and 3).

3.2. Compression tests

The true stress-strain curves obtained from compression tests along three loading directions of the specimens are presented in figure 4. The results indicate that specimen A has higher yield stress and flow stress than specimen C in all three loading directions. Compressive yield stresses along the three loading directions are found to be follow the order of Y-axis > Z-axis > *X*-axis in both the A and C specimens, and the differences in yield stresses are observed to be more significant in sample C than in sample A. The as-ECAE specimens exhibit slight work-softening when testing in both the X- and Y-directions, but slight hardening in the Z-direction.

The annealed samples, AR and CR, show reduced yield and flow stresses as compared to the as-ECAE counterparts (figure 4). Similar to the as-ECAE samples, specimen AR has higher yield stress and flow stress than specimen CR due to the smaller grain size and higher fraction of HABs in the AR specimen. The flow curves of the AR specimen display the "yielding peak" phenomenon in all three loading directions. On the other hand, the yielding peak is less evident in specimen CR, and can only be observed in the Y-direction. Yield stresses along the three loading directions are determined to follow the order of *Y*-axis > *Z*-axis > *X*-axis in specimen AR, while they follow the order of *Y*-axis ≈ *Z*-axis > *X*-axis in specimen CR. The differences in yield/flow stress become less significant in the annealed specimens as compared to their as-deformed states. A "yielding peak" only occurs in the annealed state, and is more evident in the AR sample than in the CR sample. In terms of loading direction, it is most apparent in the Y-direction.

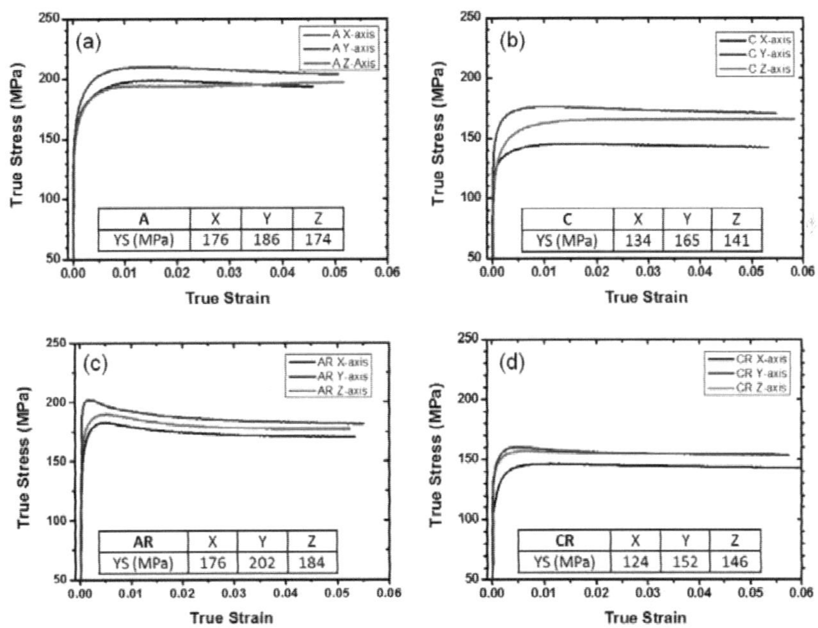

Figure 4. Compressive true stress-strain curves for specimens, (a) A, (b) C, (c) AR and (d) CR, deformed along three orthogonal directions.

3.3. Shear bands

After compression tests, the as-ECAE specimens show a uniform distribution of fine slip traces, while the annealed specimens exhibit distinct bands of flow localization (shear bands). The appearance of shear bands in the annealed specimens also depends on the processing route and the loading direction. Both the AR and CR specimens have similar orientations of shear bands with respect to the loading direction, but the shear bands are fainter in specimen CR. Specimen AR compressed along the *X*-direction only exhibits one set of shear bands on the *Z*-plane, which is parallel to the *Y*-axis, while two sets of shear bands are aligned at angles of approximately ±45° to the *X*-axis on the *Y*-plane (figure 5a). Two sets of shear bands aligned at approximately ±45° to the *Y*-axis were observed on the *X*-plane when the specimens were compressed along *Y*-axis (figure 5b and 5c). However, there is only one set of shear bands observed on the *Z*-plane in specimens compressed along the *Y*-direction, which is approximately parallel to the *X*-axis (figure 5b). For specimens compressed along the *Z*-axis, weak signs of shear bands are observed, which can be better described as a distribution of fine slip lines.

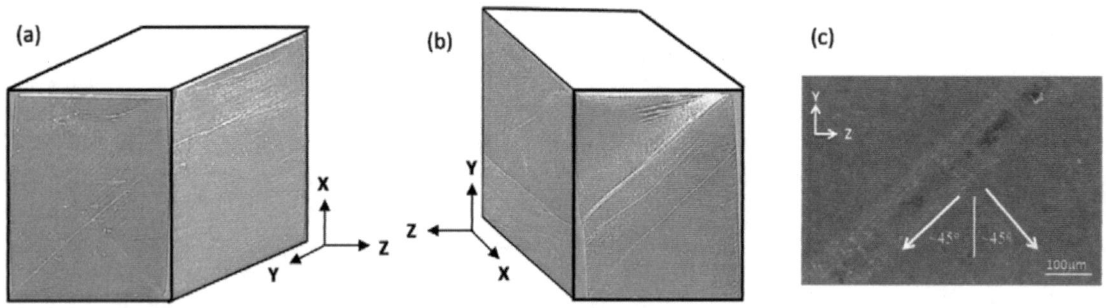

Figure 5. Micrographs showing shear bands on the sample surfaces after compression tests in the AR specimen. (a) Optical micrograph of the Y- and Z- planes of specimen compressed along X-axis, (b) optical micrograph of the X- and Z- planes of the specimen compressed along the Y-axis, and (c) SEM micrograph showing two sets of shear bands on the X-plane of specimen compressed along the Y-axis.

4. Discussion

4.1. Effect of ECAE route (strain-path change) and annealing on the compression behavior

The occurrence of higher yield and flow stresses in specimen A than C is attributed to the smaller grain size and higher proportion of HABs in specimen A. Metals with smaller grain sizes are considered to have higher strength according to the Hall-Petch relationship. However, the grain size effect was found to contribute only 5 MPa to yield stress in A and C specimens from the Hall-Petch slope in this aluminum [8]. The difference in yield stress between samples A and C far surpasses 5 MPa. It is therefore suggested that boundary structure in addition to grain size has a considerable contribution to the strength of this UFG aluminum. Specimen C consists of mainly LABs, which are weaker barriers to dislocation propagation, and consequently result in lower yield stress.

Several factors can affect the anisotropic mechanical response of aluminum, such as texture, grain boundary structure and grain shape. The textures of the two specimens are found to be similar [22] and therefore, the cause of texture can be excluded. The directionality in microstructure might be another possible cause responsible for the anisotropic mechanical behavior. Specimen C has more equiaxed grains surrounded by mainly LABs and some randomly distributed HABs, while specimen A has elongated grains with high proportion of HABs, which are aligned mainly along the X-axis. If the aforementioned factor is the major cause for the anisotropic mechanical response, specimen A should have more pronounced directionality in mechanical behavior. However, the anisotropic behavior is more evident in specimen C. Thus, this proposition could also be ruled out. In mechanical testing, ECAE specimens experience a strain-path change with respect to the previous strain imposed in the ECAE process. It is possible that the anisotropic flow responses could be related to this strain-path change [17, 19]. A more detailed discussion is given in section 4.2.

The occurrence of yielding peak is attributed to the shortage of mobile dislocations, which in turn requires a higher dislocation velocity to satisfy the applied strain rate [25]. It has been shown that UFG aluminum with smaller sizes and low dislocation densities [8] exhibits a significant yielding peak phenomenon. The absence of yielding peaks on the as-ECAE specimens indicates the existence of sufficient mobile dislocations. However, the annealed specimens, AR and CR, exhibit yielding peaks. From the microstructural examinations of the AR and CR specimens, the diffraction contrast is observed to be more uniform than in the as-ECAE state, A and C specimens, indicating defect density and internal stresses are substantially reduced in the annealed state (figures 1 to 3). Therefore, a high dislocation velocity is required to fulfil the applied strain rate and results in yielding peak [8, 25].

4.2. Relation between plastic anisotropy and the shear patterns in ECAE
Shear bands are observed in the annealed specimens in conjunction with the presence of yielding peaks (figure 5). Shear banding is a common feature in UFG metals during mechanical testing, both tensile and compressive, and result from flow localization [8, 26]. The development of shear bands in UFG metals can be related to its inherent low work-hardening rate. Low temperature annealing can only cause partial recovery of the dislocation structure produced by ECAE. The following shear deformation imposed in ECAE can still affect anisotropic flow localization in the annealed specimens.

A shear strain ε_{XZ} is imposed in each ECAE pass to the material. In route A, deformation occurs on two slip planes alternately in consecutive passes, while in route C deformation occurs on one shear plane with opposite shear direction in consecutive passes. Shear bands on specimens compressed along the *X*-direction have the same orientations as the shear planes in ECAE route A. However, compression along the *X*-direction reverses the stress direction applied during ECAE, and can result in a Bauschinger effect [14] to reduce the yield stress along the *X*-axis. Consequently, the yield stress is the lowest among the three orthogonal directions. Compression along the *Y*-direction results in a shear band orientation distinctly different from the shear planes generated in ECAE, and accordingly, new dislocations driven by different stress components need to be activated and a higher flow stress is required. This results in the highest yield/flow stress in the *Y*-axis among the three directions, an effect of strain path-change suggested by Beyerlein *et al.* [19]. For specimens compressed along the *Z*-axis, the maximum shear stress occurs on the same shear plane and direction as those operated in ECAE, and therefore, the yield stress is intermediate among the three orthogonal directions.

The stronger anisotropic response in specimen C compared to A is considered to be related to deformation occurring on single shear plane during the ECAE route C process. The shear direction is continuously reversed on the same shear plane during consecutive extrusions in route C. However, deformation is imposed on two shear planes and four shear directions in the ECAE route A process [21]. It is therefore suggested that during the ECAE process, dislocations on fewer slip systems are activated in route C than in route A. For compression tests along different loading directions of ECAE processed materials, dislocations on different slip systems should be activated to accommodate the applied strain. The characteristics of dislocations introduced in ECAE can affect the mechanical response in post-ECAE loading. Due to the activation of fewer dislocation slip systems in specimen C, a stronger anisotropic mechanical behavior is expected in specimen C than A. The anisotropy in mechanical response is reduced after annealing, which can be ascribed to partial annihilation of the pre-existing dislocations. Specimen CR has fewer activated dislocations left from ECAE and new dislocations are generated during the following compression tests along all three orthogonal directions, resulting in similar yield/flow stresses along the three directions.

5. Conclusions
In this work, the anisotropy of mechanical response of AA1050 processed by ECAE routes A and C has been investigated in the as-processed and recovered conditions. The main results are listed below.

(1) Specimens A and AR exhibit higher strength than specimens C and CR due to the smaller grain size and higher fraction of HABs in specimens A and AR.

(2) Specimens produced by both ECAE routes show qualitatively similar anisotropic responses under compressive loading along the three orthogonal directions of the ECAE billet. The yield stress is the highest in compressive testing along the Y-direction, while the lowest value of yield stress appears in testing along the X-direction. The differences can be related to the effect of strain-path change,

namely that different slip system activities are induced for specimens loaded along different directions with respect to the last ECAE pass.

(3) The plastic anisotropy in specimen A less than that in specimen C. This difference can be explained by the deformation occurring on single shear plane and two shear directions during ECAE route C, while the deformation is imposed on two shear planes and four shear directions in route A. It is then expected that during ECAE process, dislocations on fewer slip systems are activated in route C than in route A. As the characteristics of the dislocations introduced in ECAE may affect the mechanical response in post-ECAE loading then because fewer dislocation slip systems are activated in route C, a stronger anisotropy in mechanical behavior is expected in specimen C than in A.

(4) The plastic anisotropy persists in the annealed specimens, AR and CR, but to a reduced extent. The reduced anisotropy in mechanical response after annealing can be ascribed to the annihilation of pre-existing dislocations resulting from the ECAE process.

(5) Shear bands are observed in annealed specimens in conjunction with the presence of a yielding peak. Both the AR and CR specimens have similar orientations of shear bands with respect to the loading direction, but the shear bands are weaker in the CR specimen.

Acknowledgments
The authors acknowledge the financial support from the National Science Council (NSC) of ROC under contracts NSC 98-2221-E-035-015 and the discussions with Professor Po-We Kao at Department of Materials and Optoelectronic Science, National Sun Yat-sen University.

References

[1] Iwahashi Y, Horita Z, Nemoto M and Langdon TG 1998 *Acta Mater.* **46** 3317
[2] Gholinia A, Prangnell P B and Markushev MV 2000 *Acta Mater.* **48** 1115
[3] Sun P L, Kao P W and Chang C P 2004 *Metall. Mater. Trans.* **35A** 1359
[4] Chang C P, Sun P L and Kao P W 2000 *Acta Mater.* **48** 3377
[5] Humphreys F J, Prangnell P B, Bowen J R, Gholinia A and Harris C 1999 *Phil. Trans. R. Soc. Lond.* **357A** 1663
[6] Yu C Y, Sun P L, Kao P W and Chang C P 2004 *Mater. Sci. Eng.* **A 366** 310
[7] Wang J, Iwahashi Y, Horita Z, Furukawa M, Nemoto M, Valiev R Z and Langdon T G 1996 *Acta Mater.* **44** 2973
[8] Yu C Y, Kao P W and Chang C P 2005 *Acta Mater.* **53** 4019
[9] Hayes R W, Witkin D, Zhou F and Lavernia E J 2004 *Acta Mater.* **52**
[10] Cheng S, Spencer J A and Milligan W W 2003 *Acta Mater.* **51** 4505
[11] Li Y J, Zeng X H and Blum W 2004 *Acta Mater.* **52** 5009
[12] Sun P L, Yu C Y, Kao P W and Chang C P 2005 *Scripta Mater.* **52** 265
[13] Segal V M 1995 *Mater. Sci. Eng.* **A197** 157
[14] Haouaoui M, Karaman I and Maier H J 2006 *Acta Mater.* **54** 5477
[15] Alexander D J and Beyerlein I J 2005 *Mater. Sci. Eng.* **A 410-411** 480
[16] Beyerlein I J, Li S and Alexander D J 2005 *Mater. Sci. Eng.* **A 410-411** 201
[17] Li S 2007 *Scripta Mater.* **56** 445
[18] Horita Z, Fujinami T and Langdon T G 2001 *Mater. Sci. Eng.* **A 318** 34
[19] Beyerlein I J, Alexander D J and Tomé C N 2007 *J. Mater. Sci.* **42** 1733
[20] Xu C, Száraz Z, Trojanová Z, Lukáˇ P and Langdon T G 2008 *Mater. Sci. Eng.* **A 497** 206
[21] Zhu Y T and Lowe T C 2000 *Mater. Sci. Eng.* **A 291** 46
[22] Sun P L, Cerreta E K, Bingert J F, Gray III GT and Hundley M F 2007 *Mater. Sci. Eng.* **A 464** 343
[23] Beyerlein I J and Toth L S 2009 *Prog. Mater. Sci.* **54** 427
[24] Kallend J S, Kocks U F, Rollett A D and Wenk H R 1991 *Mater. Sci. Eng.* **A132** 1
[25] Johnston W G and Gilman J J 1959 *J. Appl. Phys.* **30** 129
[26] Yu C Y, Sun P L, Kao P W and Chang C P 2005 *Scripta Mater.* **52** 359

Synthesis and characterization of a model dual-phase system using the spark plasma sintering technique

M. Teimouri and A. Godfrey[1]

Key Laboratory of Advanced Materials (MOE), Department of Material Science and Engineering, Tsinghua University, Beijing 100084, P. R. China

[1] awgodfrey@mail.tsinghua.edu.cn

Abstract. Samples of a model dual-phase system, consisting of copper and AISI-420 martensitic steel, have been synthesized using spark plasma sintering, with the objective of developing a microstructural analogue for dual-phase steels, in which the volume fraction and size of each phase can be controlled independently. Microstructural investigation of the samples, including fractography of samples deformed in tension until failure, show that densification is strongly temperature dependent. Samples sintered at temperatures of 900 °C or above at a pressure of 60 MPa show a density of more than 98%. The best mechanical properties, in terms of ultimate tensile strength and ductility are found in samples sintered at a temperature of 1000 °C, where a density of nearly 99% is achieved.

1. Introduction

There is much interest at present in the relationship between microstructure and mechanical properties in composite metallic systems consisting of hard and soft phases. One system that has been explored in particular detail is that of dual-phase steels, where a typical microstructure consists of ferrite, as the soft phase, and martensite as the hard phase [1-3]. These materials have been widely explored for use in automotive applications, as they show higher ultimate strength, and higher work hardening rate compared to high-strength low-alloy steels (HLSA), while also demonstrating improved formability and ductility compared to their HSLA counterparts [4].

Some important challenges nevertheless remain with regard to a detailed understanding of the relationship between microstructure and mechanical properties in such dual phase microstructures. One problem concerns the difficulty in exploring the influence of various microstructural parameters as in general it is difficult to control independently using standard thermomechanical processing the size and volume fraction of the martensite phase, and even more so to control the spatial distribution (uniformity or heterogeneity) of the two phases. Moreover, in many cases the microstructures of commercial dual phase steels are too complicated to allow the basic mechanisms controlling deformation to be easily investigated [5].

A related problem is that diffraction-based measurements, which have been used to probe the partitioning of stress and strain between the hard and soft phases, are complicated by the fact that the ferrite and martensite phases have a similar crystal structure with similar lattice parameters. As a consequence, the diffraction peaks from each phase overlap and analysis of the data requires advanced peak fitting methods where certain assumptions about the peak shapes and distribution of each phase are required [5-7].

An alternative approach to develop a model system for fundamental studies of the mechanical behaviour of dual phase systems involves the use of the spark plasma sintering (SPS) technique, sometimes also referred to as field-assisted sintering [8]. This technique is a modified hot powder pressing technique, in which a pulsed electrical current and a uniaxial pressure are applied to a powder compact at low voltage. The current passes through an electrically conductive graphite die and heats the sample. The ability of this technique to produce fully dense bulk material within a very short time and at a relatively low temperature without any secondary processing, offers many advantages compared to other powder metallurgy processes and makes it a cost-effective process, in particular for materials that are highly sensitive to temperature [9]. Although the spark plasma sintering technique was originally developed for ceramics it has also been applied for metals, such as nickel [10,11], copper [12-14], aluminium [15-17], pure iron [18], and a range of steels [3,19].

In this paper, we explore the possibility of using spark-plasma sintering for the preparation of a model dual-phase system by mixing of suitable powders. An advantage of this approach is that it allows independent control of both the size and the volume fraction of the soft and hard phases (and by control of the powder mixing process, their distribution). For this study copper is used as the soft phase, with a martensitic stainless used steel as the hard phase.

2. Experimental

Spherical powders of copper (average size either 3.5, 10.0, or 15μm) and AISI420 stainless steel (average size of 10 or 15μm) were used to prepare samples (for details see table 1). The powders were mixed using a low-energy mechanical milling process. This was carried out using a horizontal ball milling system at room temperature, using stainless steel balls and a polymer vial. For this study samples of copper and AISI420 stainless steel powder were combined in weight ratio of either 50:50 or 70:30. For ball-milling the ball to powder mass ratio was set at 20:1 and filling was performed in an air atmosphere. The powder mixtures were milled for 24h at a constant rotation speed of 120 rpm.

Table 1. Summary of sintering condition and measured density of sintered samples. All samples were sintered with a hold time at the sintering temperature of 5 mins.

Cu (Fe) powder size (μm)	Cu:Fe wt. ratio	Sintering temperature (°C)	Sintering pressure (MPa)	Heating ramp parameters (Q, T_1, t)[a]	Relative density (%)
15 (15)	50:50	850	50	80, 850, 0	88.6
10 (10)	50:50	900	50	100 850, 1	95.0
3.5 (10)	70:30	900	60	100, 850, 1	98.0
3.5 (10)	70:30	950	60	114, 720, 5	98.1
3.5 (10)	70:30	1000	60	114, 720, 8	98.9

[a] Units of Q (°C/min), T_1 (°C), t (min) - see figure 1 for definition of these parameters.

Consolidation of the mixed powders was carried out using the SPS process (SPS-1050, Sumitomo Heavy Industries Ltd., Japan). The powders were first loaded into a graphite die (20 mm diameter). The die was separated from the mould and punches of the SPS system using graphite foils, to prevent welding from powder during the sintering process. Sintering was performed under primary dynamic vacuum and under the application of uniaxial pressure. The temperature, pressure and dwell time were varied over a wide range to investigate the influence of these SPS parameters on the samples density and microstructure. The general heating and loading cycles of the SPS process are shown in figure 1 (table 1 gives the relevant values for each sample studied). The final sintered samples were disc shaped, with a diameter of 20 mm and a height of approximately 4 mm. After sintering the compacted samples were given a heat treatment in order to control the martensitic transformation of AISI420 stainless steel phase. For this the samples were heated to 1000 °C at a heating rate of 10 °C/min and held at this temperature for 15 min before quenching in oil. The samples then were tempered in an air-furnace at 500 °C for 30 min.

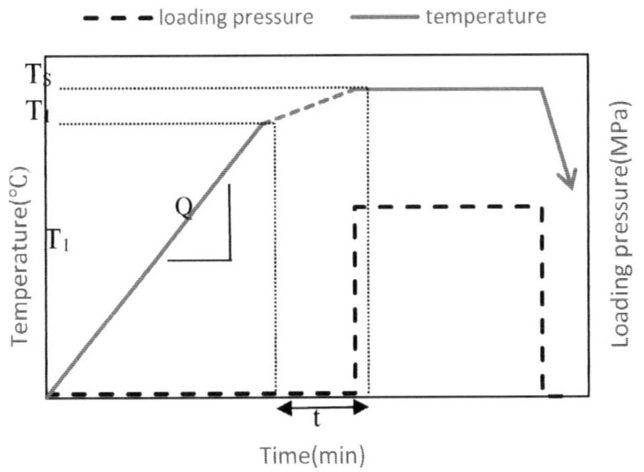

Figure 1. Schematic heating and loading cycles for the SPS process. After an initial heating ramp (heating rate Q °C/min) to temperature T_1, the final sintering temperature was approached over time t (mins). The sintering pressure was applied only during a 5 min hold at the sintering temperature T_s.

The final sample density was determined using Archimedes' method after mechanical grinding and polishing of all sample surfaces. The microstructure of the sintered samples was characterized using a scanning electron microscope (SEM), including elemental mapping using energy dispersive spectroscopy (EDS) to investigate the phase distribution of the samples. Mechanical characterization of the samples was carried out using Vickers hardness measurements (100 g load) and tensile tests of dog-bone shape samples (gauge length of 8 mm, width of 2 mm and thickness of 1 mm) cut from the disks using spark-erosion. Hardness measurements were obtained in the copper and AISI420 phases separately by control of the indent locations. Values are given as the average and standard deviation over at least 7 measurements in each phase. The tensile tests were carried out at room temperature under displacement control using an initial strain rate of 6.3 x 10^{-4} s^{-1}.

3. Results and discussion

The measured densities of samples prepared under a selected range of conditions are given in table 1. In agreement with previous studies [14,16] the results show that the sample density increases both with increasing holding pressure and with increasing sintering temperature. At 850 °C and 50 MPa sintering pressure a resulted in density of only less than 90%. Raising the sintering temperature to 900 °C at the same pressure resulted in improvement of the density to approx. 95%. With additional increase of pressure to 60 MPa at this temperature a density of > 98% was achieved. Further increase of temperature to 1000 °C resulted in a moderate improvement of density to almost 99%.

Figure 2(a-d) shows SEM images of samples sintered under a range of conditions. In these images, the lighter continuous phase is the copper and the darker discrete regions are the AISI420 stainless steel, as illustrated by the elemental EDS maps shown in figure 3 for the sample prepared at 900 °C and 60 MPa. The results of the SEM observations provide microstructural support for the density measurements. As seen in figure 2a, large-scale porosity is present in the sample sintered at 850 °C and under a pressure of 50 MPa. Additionally, extensive fine-scale porosity is present, both between original iron particles, and at the interface with the copper phase. Increasing the temperature to 900 °C at the same sintering pressure (figure 2b) results in a sample without large-scale porosity, although extensive fine scale porosity remains, in particular at the interface region between the AISI420 stainless steel phase and the copper phase. Under these sintering conditions it is found that most of the AISI420 stainless steel particles retain their original round shape after the combined ball-milling and sintering processes. Under SPS conditions of 900 °C and 60 MPa sintering pressure (figure 2c) the interfacial porosity between the copper and AISI420 stainless steel phases is significantly reduced, although a small amount of porosity is still seen in the samples. Furthermore, under these sintering conditions a number of the AISI420 stainless steel particles are deformed and no longer maintain their round shape.

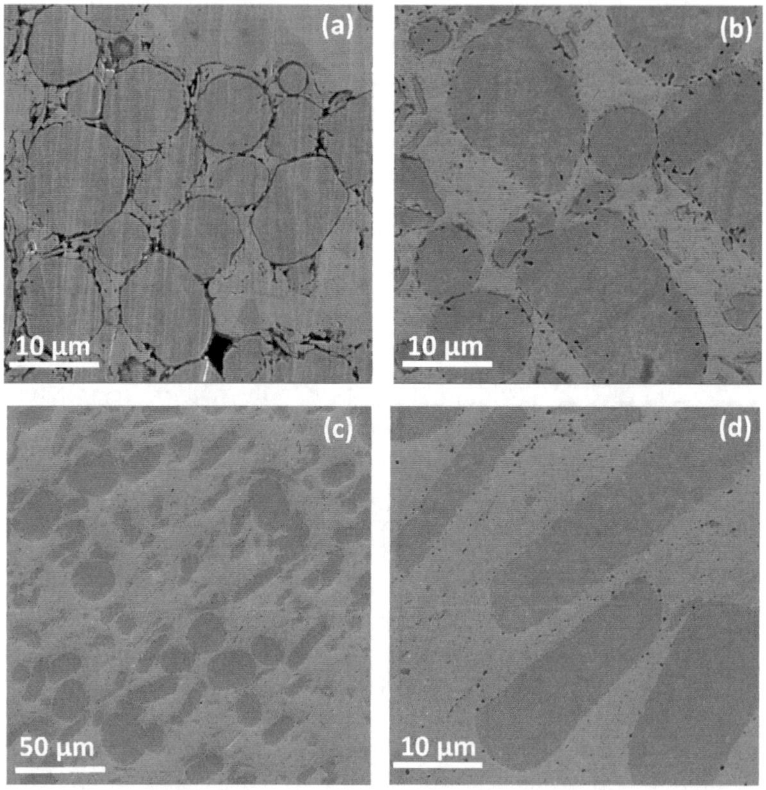

Figure 2. Example SEM images of samples sintered at different temperatures (T) and pressures (P): (a) T = 850 °C, P =50 MPa; (b) T = 900 °C, P = 50 MPa; (c,d) T = 900 °C, P = 60 MPa.

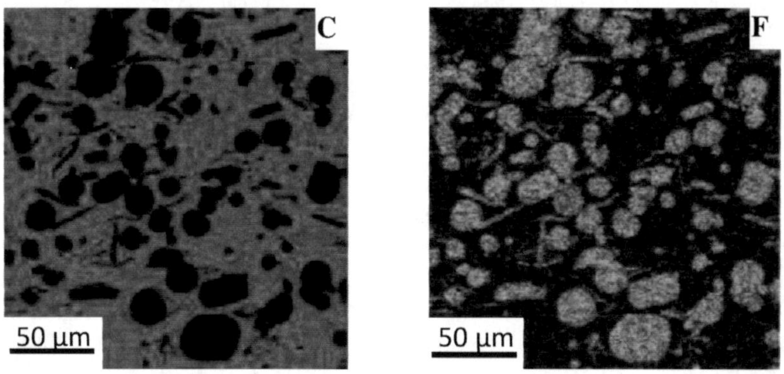

Figure 3. Elemental EDS maps of the sample sintered at 900 °C and 60 MPa showing AISI420 stainless steel particles embedded in a continuous copper phase.

The results of the hardness measurements of the sintered samples are presented in table 2. The data show a non-systematic variation with sintering temperature. It can be seen that with increasing the sintering temperature from 850 °C up to 900 °C, the hardness of the both copper and AISI420 stainless steel phase increase, with a large increase in the hardness of the AISI420 stainless steel phase. Similarly, the specimen sintered at 900 °C and 60 MPa shows a somewhat higher hardness in both phases than for the sample sintered at the same temperature but a lower pressure of 50 MPa. With increasing the sintering temperature up to 1000 °C the hardness of the AISI420 stainless steel phase decreases, although the reason for this decrease is at present unknown (further microstructural investigations, including determination of the grain size within both phases is currently underway).

Table 2. Hardness and ultimate tensile strength of the sintered samples.

SPS temperature (°C)	SPS pressure (MPa)	Hardness (HV) Cu	AISI420	Ultimate tensile strength (MPa)
850	50	138 ± 6	409 ± 56	-
900	50	126 ± 2	528 ± 45	310
900	60	159 ± 13	537 ± 64	486
950	60	154 ± 7	271 ± 45	538
1000	60	159 ± 14	278 ± 36	530

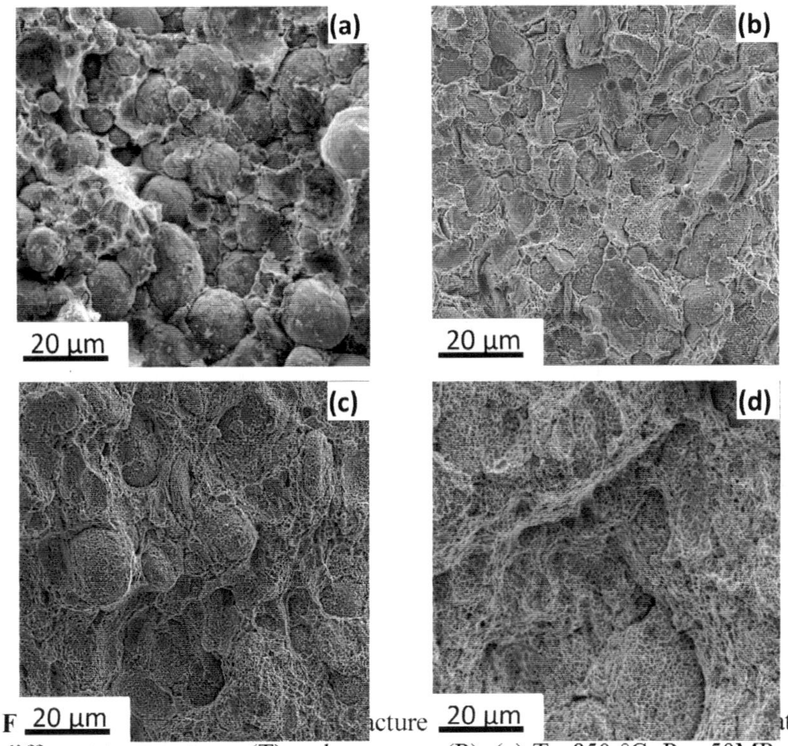

F [illegible] acture [illegible] t different temperatures (T) and pressures (P): (a) T= 850 °C, P = 50MPa, (b) T = 900 °C, P= 50MPa, (c) T = 900 °C, P= 60MPa, and (d) T = 1000 °C, P = 60MPa.

Table 2 also shows the ultimate tensile strength (UTS) achieved during tensile testing of the samples. This is seen to increase both with sintering temperature and with sintering pressure. The increase in UTS seen above 900 °C, despite the lower hardness of the AISI420 phase, is due to the increased ductility of these samples, resulting presumably from the improved density and lack of interfacial porosity. The best ductility obtained is in the sample sintered at 1000 °C, where a total elongation of 15% was achieved. Support for this is seen from the SEM micrographs of fracture surfaces after tensile-deformation until failure. In the sample sintered at 850 °C the shape of individual particles can be clearly discerned, resulting from the lack of interfacial bonding between the copper and stainless steel phases in this sample. In the samples sintered at 900 °C (both at 50 MPa and 60 MPa pressure) extensive areas showing a dimpled morphology are seen, although in some places spherical features that can be identified with the initial powder particles can still be observed in places. In contrast, the fracture surface of the specimen sintered at 1000 °C shows a fully dimpled fracture indicative of an effective consolidation of the material and good bonding at the interphase interfaces.

4. Summary and conclusions

Spark plasma sintering has been used to explore the synthesis of a model dual-phase system using powders of copper and AISI-420 martensitic stainless steel. Densification of the samples, in particular at the interface between the two phases is strongly dependent on temperature, and can also be improved by increased sintering pressure. The best mechanical properties, in terms of ultimate tensile strength and ductility is found in samples sintered at a temperature of 1000 °C, where a density of nearly 99% is achieved. Samples sintered at 900 °C and 60 MPa exhibit a wider difference in hardness between the copper and AISI-420 phases, but although these samples have a density greater than 98%, the tensile fracture surfaces show some evidence of interfacial weakness. Further studies on the post-sintering heat-treatment are underway to allow controlled synthesis of fully dense samples with a range of mechanical strength in each phase.

Acknowledgments

Financial support from the National Natural Science Foundation of China (project number 51671113) is gratefully acknowledged.

References

[1] Ghadbeigi H, Pinna C, Celotto S and Yates JR 2010 *Appl. Mech. Mater.* **24-25** 201
[2] Kamikawa N, Hirohashi M, Sato Y, Chandiran E, Miyamoto G and Furuhara T 2015 *ISIJ Int.* **55** 1781
[3] Ni HW, He H, Li GQ and Liu J 2008 *J. Iron Steel Res. Int.* **15** 73
[4] Senuma T 2001 *ISIJ Int.* **41** 520
[5] Woo W, Em VT, Kim EY, Han SH, Han YS and Choi SH 2012 *Acta Mater.* **60** 6972
[6] Cong ZH, Jia N, Sun X, Ren Y, Almer J and Wang YD 2009 *Metall. Mater. Trans.* A **40** 1383
[7] Jia N, Peng RL, Wang YD, Johansson S and Liaw PK 2008 *Acta Mater.* **56** 782
[8] Saheb N, Iqbal Z, Khalil A, Hakeem AS, Al Aqeeli N, Laoui T, Al-Qutub A and Kirchner R 2012 *J. Nanomater.* **2012** 983470
[9] Dudina DV and Mukherjee AK 2013 *J. Nanomater.* **2013** 625218
[10] Holland TB, Ovid'ko IA, Wang H and Mukherjee AK 2010 *Mater. Sci. Eng.* A **528** 663
[11] Farbaniec L, Dirras G, Krawczynska A, Mompiou F, Couque H, Naimi F, Bernard F and Tingaud D 2014 *Mater. Charac.* **94** 126
[12] Ritasalo R, Cura ME, Liu XW, Söderberg O, Ritvonen T and Hannula SP 2010 *Mater. Sci. Eng.* A **527** 2733
[13] Zhu KN, Godfrey A, Hansen N and Zhang XD 2017 *Mater. Design* **117** 95
[14] Zhu KN, Ruan Q and Godfrey A 2015 *IOP Conf. Series: Mater. Sci. Eng.* **89** 012060
[15] Le GM, Godfrey A, Hansen N, Liu W, Winther G and Huang X 2013 *Acta Mater.* **61** 7027
[16] Le GM, Godfrey A and Hansen N 2013 *Mater. Design* **49** 360
[17] Liu ZF, Zhang ZH, Lu JF, Korznikov AV, Korznikova E and Wang FC 2014 *Mater. Design* **64** 625
[18] Fabrègue D, Piallat J, Maire E, Jorand Y, Massardier-Jourdan V and Bonnefont G 2012 *Powder Metall.* **55** 76
[19] Keller C, Tabalaiev K, Marnier G, Noudem J, Sauvage X and Hug E 2016 *Mater. Sci. Eng.* A **665** 125

Twinning and detwinning mechanisms in a BCC Ti Mo-Fe multilayered alloy

I Gutierrez-Urrutia[1*], C-L Li[1], S Emura[1] and K Tsuchiya[1, 2]

[1] National Institute for Materials Science, Sengen 1-2-1, Tsukuba, Ibaraki, Japan
[2] Graduate School of Pure and Applied Sciences, University of Tsukuba, Ibaraki, Japan

*corresponding author e-mail: gutierrezurrutia.Ivan@nims.go.jp

Abstract. We have investigated {332}<113> twinning and detwinning mechanisms in a multilayered Ti-10Mo-xFe (x =1–3) alloy fabricated by multi-pass hot rolling. A strong influence of the Fe-graded structure on both phenomena is observed. The propagation of twins that are nucleated in Fe-lean regions (~1 wt.% Fe) is interrupted in the grain interiors at about 2 wt% Fe. We ascribe this effect to the role of the Fe content on the stress for twin propagation. The unusual twin structure determines the subsequent detwinning process upon thermal annealing. This process consists of two independent detwinning events that occur at two different microstructural regions, namely at twin tips located at grain interiors and at grain boundaries. Both detwinning modes can be explained from a thermodynamic standpoint whereby the boundary dissociation processes minimize the boundary free energy.

1. Introduction

In the last decade, β-Ti-Mo alloys have received considerable attraction as structural materials in several technological areas including the aerospace, automotive and biomedical industries. These alloys exhibit an outstanding combination of high specific strength, low elastic modulus and excellent bio-compatibility [1-6]. From an engineering standpoint, several processing techniques such as rolling, drawing and forging are typically used to optimize their mechanical behavior. For instance, Min et al. [6] have recently processed a multilayered bcc-Ti-10Mo-xFe (x= 1-3 wt.%) alloy by multi-pass hot rolling that exhibits a superior combination of strength and ductility (yield strength of 800 MPa; total elongation of 28%), compared to those of the individual alloys. This material contains a macroscopic gradient in Fe that tailors the combination of specific deformation mechanisms, namely <111> dislocation slip [7-9] and {332}<113> deformation twinning [10-12], resulting in superior mechanical behavior.

The activation of deformation modes in β-Ti-Mo alloys is associated with the stability of the β-Ti (bcc phase) compared to the β-α′ and β-α″ stress-induced phase transformation [13-15]. In bcc-stable β-Ti-Mo alloys the propagation of {332}<113> twins play an important role on the strain-hardening behaviour. The influence of microstructure on twin propagation is a complex phenomenon due to the different types of interfaces that exist in the microstructure (grain boundaries and twin interfaces), resulting in complex twin-interface interaction phenomena [16-21]. On the other hand, multilayered materials contain another microstructural feature that also influences twin propagation, namely, a chemical graded structure. The propagation of twins through a composition gradient may be different to that occurring in homogeneous alloys due to solute effects on twinning [10, 11, 16, 18-20]. Another relevant feature of twinning is the thermal and mechanical stability of twin interfaces upon thermal annealing, including strain path reversion at room/high deformation temperature or under specific stress states. This phenomenon is associated with the thermal and mechanical stability of twin interfaces, which is controlled by the twin interface structure [16, 21, 22]. In bcc metals, the detwinning mechanism has hardly been analyzed [23] in spite of the large number of reports on {112}<111> and {332}<113> twinning modes [11, 16, 24, 25]. The present work aims at investigating different aspects of {332}<113> twinning and detwinning in a multilayered β-Ti-10Mo-xFe (x: 1-3). We investigate the propagation behaviour of twins through the Fe-graded structure and the detwinning behaviour upon thermal annealing at 900°C. Twin and detwin structures are characterized by electron backscatter diffraction (EBSD) and electron channeling contrast imaging (ECCI).

2. Experimental procedure

The material investigated in this work was a multilayered material consisting of ten alternating layers of Ti-10Mo-1Fe and Ti-10Mo-3Fe (wt.%) alloys. The material was fabricated by multi-pass hot rolling at 1000 °C from individual layers of about 3.5 mm in thickness, and was subsequently solution treated at 900 °C for 1 hour followed by water quenching. Details of the processing procedure are described in [7]. The final thickness of the as-processed 10-layer material was about 2.7 mm. Tensile samples were machined from the multilayered material and from individual alloy sheets, namely, Ti-10Mo-1Fe, Ti-10Mo-2Fe and Ti-10Mo-3Fe. The tensile samples had an 18 mm gage length, a 4 mm gage width and a 2.5 mm gage thickness. The tensile axis was parallel to the rolling direction (RD). The detwinning process was investigated in samples annealed at 900 °C for 60 and 90 seconds in air followed by water quenching, as suggested by a former study [23]. Tensile tests were performed in an Instron 5581 tensile machine at room temperature and at an initial strain rate of $1 \times 10^{-5}\,s^{-1}$. Twin and detwin structures were examined on the longitudinal section defined by the tensile direction (TD) and normal direction (ND). Microstructural characterization was performed using two electron microscopy techniques, namely, electron channeling contrast imaging (ECCI) and electron backscatter diffraction (EBSD) in a Sigma Zeiss field emission gun scanning electron microscope (FEG-SEM) equipped with a backscattering electron detector (BSE) and a TSL orientation imaging microscopy (OIM) EBSD system. ECC images were obtained with optimum channeling contrast following the procedure described in [26]. ECCI was carried out at a working distance of 4 - 5 mm and at 9 kV accelerating voltage using an aperture size of 120 μm. EBSD maps were taken at 20 kV accelerating voltage, a working distance of 15 mm and with step sizes ranging between 100 nm and 2 μm. Elemental mapping was performed using electron probe microanalysis (EPMA) on a JEOL JXA-8900F system operated at 15 kV.

3. Results and discussion

Figure 1 shows the deformation microstructure (a) and the elemental distribution (b) of the multilayered material tensile-deformed to a true strain of 0.04. The elemental map shows that the multilayered alloy contains a macroscopic gradient of Fe, ranging between 1 and 3 wt%, distributed in the form of bands parallel to the RD. The Ti and Mo are homogeneously distributed within the Fe-graded structure [28]. The Fe-graded bands are about 130 μm thick and are situated between the original 1 wt.%Fe and 3 wt.%Fe layers, which are about 50 μm thick.

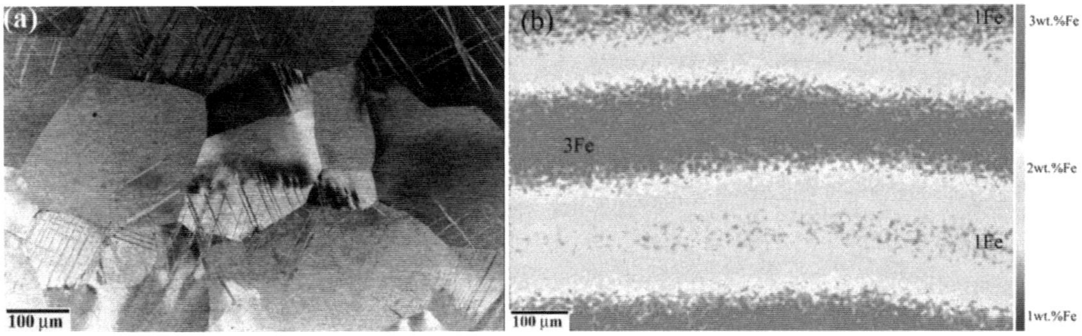

Figure 1. (a) ECC image of the deformation structure of the multilayered Ti-10Mo-xFe (x = 1–3) material tensile deformed to 0.04 true strain. (b) EPMA map showing the Fe distribution [27].

Figure 2(a) shows an example of the deformation structure of a grain oriented close to the [1 15 16] crystallographic direction. The figure reveals that the deformation structure is formed by {332}<113> deformation twins and by <111> slip traces. This structure is similar to that reported in conventional (chemically homogeneous) β-Ti-Mo alloys [9, 10]. However, in the present multilayered Ti-Mo-Fe alloy the propagation of deformation twins is interrupted within the grain interiors. In particular, in the grain shown in figure 2(a) twins are nucleated at the grain boundary GB1 (TN: twin nucleation) and are interrupted in the grain interior at about 1.8 wt% Fe (TT: twin tip). Elemental mapping reveals that the grain contains a Fe gradient of about of about 0.01 wt% Fe/μm over a region 160 μm long. The Fe content varies between 1.2 wt% (at the grain boundary, GB1) and 2.8 wt% (at the low angle grain boundary marked LAGB2). In the same area the Mo content is relatively homogeneous and ranges between 9.8 and 10.2 wt%. The EPMA line analysis in figure 2(b) shows that the Fe variation within the graded structure is smooth, i.e. no sharp homophase interface occurs. Interestingly, under the current thermomechanical conditions, the role of precipitates such as ω-omega phase, on the precipitate-twin interaction can be ruled out due to their small volume fraction (<0.01) and their shearable nature [28]. These observations indicate that the variation in Fe content is the main microstructural parameter controlling twin propagation.

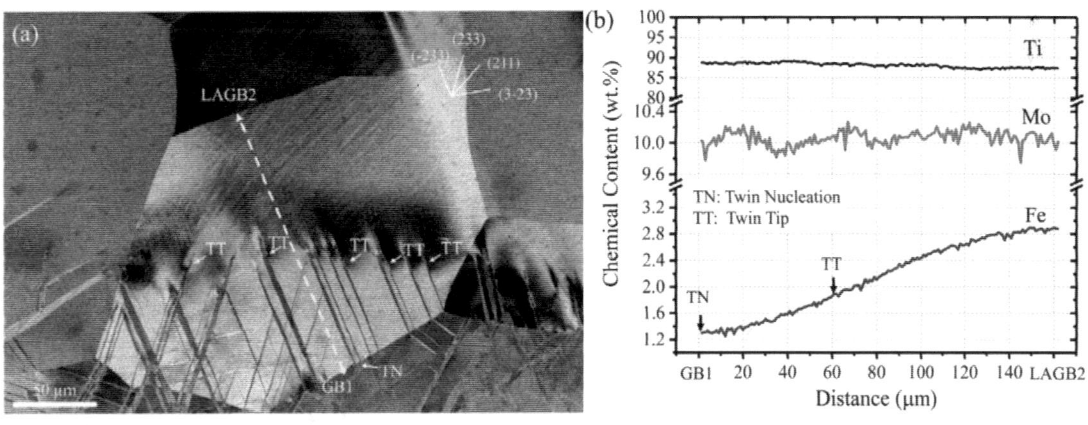

Figure 2. (a) ECC image of the deformation structure of a grain oriented close to [1 15 16]; (b) chemical line analysis performed by EPMA along the dashed line within the grain shown in (a). TN: twin nucleation; TT: twin tip; GB: grain boundary; LAGB: low angle grain boundary [28].

As the processed material does not contain any sharp interface between chemical layers, due to the evolution of the original layer interfaces into migrating grain boundaries, we assume an iso-stress condition for the in-grain propagation of twins. In order to clarify the role of Fe content on twin propagation, we have experimentally determined the twinning stress, σ_{tw}, of the Ti-10Mo-xFe (x = 1–3) system. We define σ_{tw} as the macroscopic stress at which twins are detected. According to dislocation-based models of twinning in bcc metals, the experimental twinning stress corresponds to the stress to grow an existing twin nucleus [16, 19, 20, 25]. Specifically, σ_{tw} for each of the three individual alloys, namely Ti-10Mo-1Fe, Ti-10Mo-2Fe and Ti-10Mo-3Fe (wt.%), was determined by tensile loading experiments. The alloys contained similar average grain sizes of ~250 μm in order to avoid any grain-size effects on the twinning stress. In the alloys with lower Fe content, i.e. Ti-10Mo-1Fe and Ti-10Mo-2Fe, {332}<113> twins readily formed at yielding. For these alloys we thus consider the twin stress to equal the yield stress. In contrast, deformation twins were not visible in the Ti-10Mo-3Fe alloy at any investigated strain level. Accordingly, we can only estimate a lower bound of the twinning stress for this alloy system. The experimental values of the twin stress as a function of Fe content are plotted in figure 3.

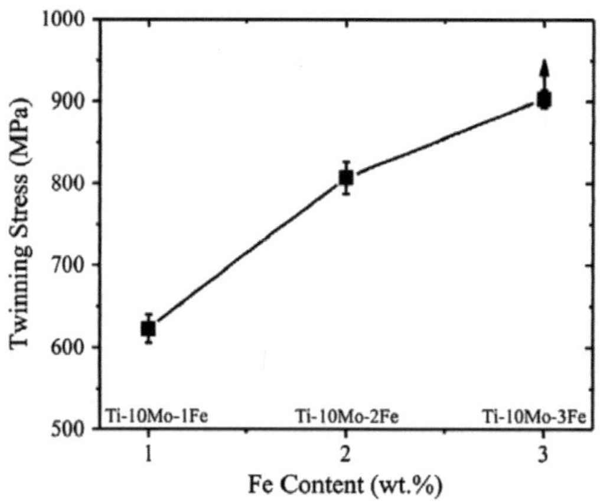

Figure 3. Effect of Fe content on the twinning stress in the Ti-10Mo alloy system. Error bars indicate the experimental deviation [27].

Figure 3 shows that twinning stress increases with Fe content. Specifically, we observe that additions of Fe higher than 1 wt% promote a significant increase of the twinning stress, namely, 2 wt% Fe increases the twinning stress about 205 MPa, and 3 wt% Fe increases the twinning stress at least 330 MPa compared with the 1 wt% addition. These findings indicate that the stress to propagate a twin plate nucleated in a Fe-lean region and propagating through a Fe-graded structure increases significantly with the propagation distance. For the graded alloy in the present study the deformation conditions are such that the macroscopic applied stress is about 830 MPa, which is slightly higher than the twinning stress for the Ti-10Mo-2Fe alloy (σ_{tw} = 780 MPa). This result indicates that twins nucleated in the Fe-lean region (~1 wt% Fe) are able to propagate within the interior of a grain up to an area containing about 2 wt% Fe, which is close to the experimentally observed value of 1.8 wt% Fe.

{332}<113> detwinning was also investigated using EBSD in a grain oriented favorably for deformation twinning, namely the [-13 14 5] crystallographic direction (Schmid factor, SF, for twinning of 0.49), see figure 4. At a true strain of 0.04 (figure 4a) the grain contains two active twin systems, namely (3-32) [1-1-3] and (233) [-311], which correspond to the twin systems with the highest Schmid factors (SF = 0.49 and SF = 0.3, respectively). Due to the negligible volume fraction of secondary twins, we have only analyzed the detwinning mechanism of primary twins, i.e., the (3-32) [1-1-3] system. Figure 4(b,c) shows the further evolution of the twin structure after short annealing treatments at 900°C of either 60 s (figure 4b) or 90 s (figure 4c). The IPF-EBSD maps show that detwinning is characterized by the gradual detachment of the twinned volumes from grain boundaries (GB) and also show the evolution of the boundaries that form the twin tip (TT) into a single twin boundary. These results indicate that the unusual deformation twin structure of the present multilayered material (figure 2) determines the detwinning behaviour, which is characterized by two independent events occurring at different microstructural regions, namely twin tips (TT) located in grain interiors and at grain boundaries (GB).

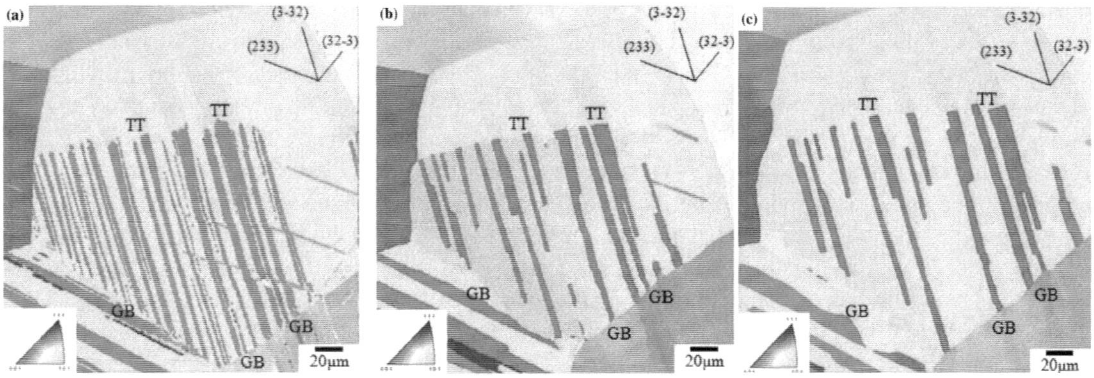

Figure 4. IPF-EBSD maps colored according to the tensile axis direction showing the evolution of the {332}<113>twin structure with annealing time at 900 °C in a grain with [-13 14 5] crystallographic direction; (a) deformation twin structure at 0.04 true strain; (b) detwinned structure at 900 °C/60 s; (c) detwinned structure at 900 °C/90 s. (GB grain boundary; TT twin tip.) [29]

Figures 5(a,b) show the evolution of the detwinning process occurring at twin tips in more detail. The underlying boundary reactions were analyzed by EBSD. Twin crystals are delimited by coherent twin boundaries (CTBs), characterized as twin boundaries with angle/rotation axis pair of 50.5°/<110>. Twin tips are formed by incoherent twin boundaries (ITBs), which are characterized as twin boundaries with angle/rotation axis pair of 60°/<110>, figure 5(a). Upon annealing at 900°C for 60 s, ITBs evolve into a single Σ3 boundary that is characterized by an angle/rotation axis pair of 70.3°/<110>, figure 5(b). The evolution of a high angle grain boundary into a Σ3 boundary has been frequently observed in [110]-misoriented tilt boundaries, such as Σ11 boundaries, in fcc metals [21, 30]. The process can be explained from a thermodynamic standpoint since the reaction process Σ11 → Σ3 minimizes the overall boundary free energy. In the present detwinning event, the driving force for the subsequent mobility of the Σ3 boundaries upon thermal annealing is the stored energy of plastic deformation localized around the twin tips [27].

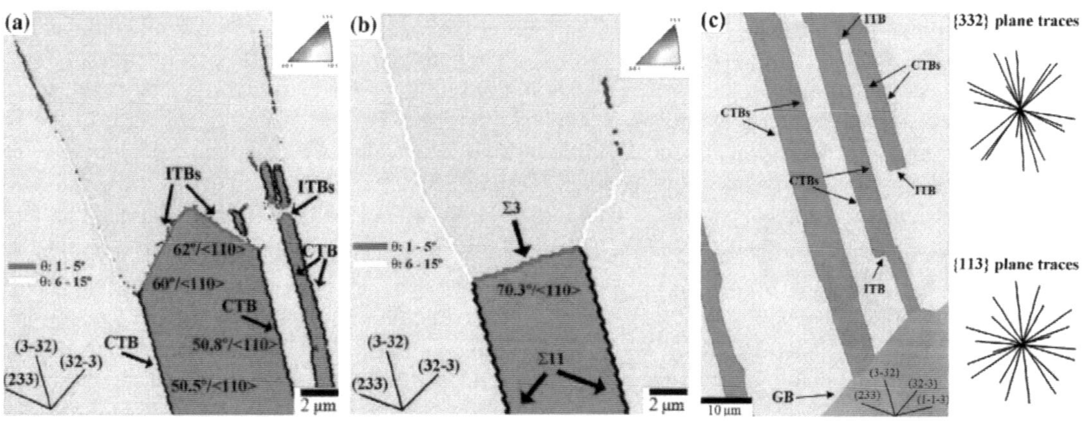

Figure 5. Evolution of the {332}<113> detwinning process at twin tips. (a) Deformation twin structure at 0.04 true strain. (b) Detwinned structure at 900 °C/60 s. (c) Detwinning process at a grain boundary upon annealing treatment of 900 °C/90 s. θ indicates misorientation angle; GB grain boundary; CTB coherent twin boundary; ITB incoherent twin boundary.

Figure 5(c) shows details of the detwinning events occurring at grain boundaries. Specifically, this figure shows the detwinned structure close to a grain boundary after annealing 900°C for 90 s. It can be seen that detwinning process is characterized by the detachment of the twinned volumes from grain boundaries by the formation and migration of incoherent twin boundaries (ITBs) that are perpendicular to the existing Σ11 {332} CTBs, as revealed by plane trace analysis. Crystallographic analysis by EBSD reveals that ITBs correspond to Σ11 {113} incoherent twin boundaries that are characterized by an angle/rotation axis pair of 50.5°/<110>. The detwinning process occurring at grain boundaries can also be explained from a thermodynamic standpoint where the boundary dissociation process minimizes the overall boundary free energy [21,31,32]. Recent calculations of grain boundary energies in bcc metals (Fe and Mo) by the embedded atom method have shown that Σ11 {113} is a low energy grain boundary [31, 32] and hence, its formation reduces the grain boundary free energy. Thermal energy triggers the subsequent migration of ITBs resulting in the formation of large-scaled facets in the twin interface structure upon thermal annealing, as seen in figure 5(c).

The present results indicate that the detwinning process occurring at grain boundaries is controlled by the formation and migration of Σ11 {113} ITBs, which leads to the formation of large facets in the detwinned structure that are perpendicular to the existing CTBs. It is worth commenting that the geometry of the detwinned structure observed in the present bcc-TiMoFe alloy is similar to that commonly observed in fcc metals consisting of Σ3 {111} CTBs containing perpendicular facets formed by Σ3 {112} ITBs [21, 30, 33, 34]. In fcc metals, Σ3 {112} ITBs are represented by a set of Shockley partial dislocations on every {111} plane with a repeatable sequence b_2:b_1:b_3 (where b_i represents the Burgers vector of the Shockley partials) and where the sum of the Burgers vectors in each unit set of three planes equals zero. However, as the crystal defect configurations of Σ11 {332} CTB and Σ11 {113} ITB are still unknown [10–12] we cannot perform a sound analysis of the {332}<113> detwinning structure. Further work is required to elucidate this relevant point. Interestingly, the role of ITBs on the twinning and detwinning processes in fcc metals such as coarse grained, nanocrystalline and sputtered thin films has been recently addressed [33,34]. These works suggest that the formation and migration of ITBs control the kinetics of detwinning. The present results also underline the key role of ITBs on the detwinning of {332}<113> twins in bcc–Ti–Mo alloys and, in particular, on the detachment of twins from grain boundaries. Further work is, however, required to understand the mechanism of formation and migration of {113} ITBs in bcc–Ti–Mo alloys.

4. Conclusions

We have investigated using EBSD and ECCI the propagation and detwinning behavior of {332}<113> twins in a multilayered Ti-10Mo-xFe (x = 1–3) alloy fabricated by multi-pass hot rolling. We observe that the Fe-graded structure has a significant influence on both phenomena. The propagation of twins that are nucleated at grain boundaries in Fe-lean regions (~1 wt% Fe) is interrupted in the grain interior at an Fe concentration of about 2 wt%. We ascribe this effect to the strong influence of the Fe content on the stress for twin propagation. Additionally, it is found that the detwinning process upon thermal annealing consists of two independent detwinning modes that occur in two different microstructural regions, namely twin tips located in grain interiors and at grain boundaries. The first detwinning mode is characterized by the evolution of incoherent twin boundaries (ITBs) into a Σ3 boundary followed by subsequent migration of the Σ3 boundary. The second detwinning process is characterized by the detachment of twin volumes from a grain boundary by the formation and migration of Σ 11 {113} incoherent twin boundaries (ITBs). Both detwinning modes can be explained from a thermodynamic standpoint whereby the observed boundary dissociation processes minimize the boundary free energy.

Acknowledgments

The authors wish to acknowledge financial support from NIMS.

References

[1] Niinomi M 2008 *J. Mech. Beh. Biom. Met.* **1** 30
[2] Al-Zain Y, Sato Y, Kim HY, Hosoda H, Nam TH and Miyazaki S 2012 *Acta Mater.* **60** 2437
[3] Geetha M, Singh AK, Asokamani R and Gogia AK 2009 *Prog. Mater. Sci.* **54** 397
[4] Banerjee D and Williams JC 2013 *Acta Mater.* **61** 844
[5] Clément N, Lenain A and Jacques PJ 2007 *JOM* **59** 50
[6] Min XH, Emura S, Meng FQ, Mi GB and Tsuchiya K 2015 *Scripta Mater.* **102** 79
[7] Christian JW 1983 *Metall. Trans.* **14A** 1237
[8] Taylor G 1992 *Prog. Mater. Sci.* **36** 29
[9] Lee SH, Hagihara K and Nakano T 2012 *Metall. Mater. Trans. A* **43** 1588
[10] Kawabata T, Kawasaki S and Izumi O 1998 *Acta Mater.* **46** 2705
[11] Tobe H, Kim HY, Inamura T, Hosoda H and Miyazaki S 2014 *Acta Mater.* **64** 345
[12] Rusakov GM, Litvinov AV and Litvinov VS 2006 *Met. Sci. Heat. Treat.* **48** 244
[13] Hida M, Sukedai E, Henmi C, Sakaue K and Terauchi H 1982 *Acta Metall.* **30** 1471
[14] Ohyama H and Nishimura T 1995 *ISIJ Int.* **35** 927
[15] Hanada S, Takemura A and Izumi O 1982 *Trans. Jpn. Inst. Met.* **23** 507
[16] Christian JW and Mahajan S 1995 *Prog. Mater. Sci.* **39** 1
[17] Steinmetz DR, Jäpel T, Wietbrock B, Eisenlohr P, Gutierrez-Urrutia I, Saeed-Akbari A, Hickel T, Roters F and Raabe D 2013 *Acta Mater.* **61** 494
[18] Gutierrez-Urrutia I, Zaefferer S and Raabe D 2010 *Mater. Sci. Eng. A* **527** 3552
[19] Mahajan S and Williams DF 1973 *Int. Metall. Rev.* **18** 43
[20] Meyers MA, Vöhringer O and Lubarda VA 2001 *Acta Mater.* **49** 4025
[21] Sutton AP and Balluffi RW 1995 *Interfaces in Crystalline Materials* (Oxford, Oxford University Press)
[22] Wang YM, Sansoz F, LaGrange T, Ott RT, Marian J, Barbee TW and Hamza AV 2013 *Nature Mater.* **12** 697
[23] Qu L, Yang Y, Lu YF, Feng L, Hu JH, Ge P, Zhou W, Han D and Ping DH 2013 *Scripta Mater.* **69** 389
[24] Lai MJ, Tasan CC and Raabe D 2016 *Acta Mater.* **111** 173
[25] Narita N and Takamura JI 1992 Deformation twinning in f.c.c. and b.c.c. metals *Dislocations in Solids* (vol 9 Dislocations and Disclinations) ed Nabarro F R N (Amsterdam, North-Holland) p 135
[26] Gutierrez-Urrutia I, Zaefferer S and Raabe D 2013 *JOM* **65** 1229
[27] Gutierrez-Urrutia I, Li C-L, Min X, Emura S and Tsuchiya K 2016 *STAM* **17** 220
[28] Banerjee S and Naik UM 1996 *Acta Mater.* **44** 3667
[29] Gutierrez-Urrutia I, Li C-L and Tsuchiya K 2017 *J. Mater. Sci.* **52** 7858
[30] Goodhew PJ, Tan TY and Balluffi RW 1978 *Acta Metall.* **26** 557
[31] Tschopp MA, Solanki KN, Gao F, Sun X, Khaleel MA and Horstemeyer MF 2012 *Phys. Rev. B* **85** 064108
[32] Ratanaphan S, Olmsted DL, Bulatov VV, Holm EA, Rollett AD and Rohrer GS 2015 *Acta Mater.* **88** 346
[33] Wang J, Li N, Anderoglu O, Zhang X, Misra A, Huang JY and Hirth JP 2010 *Acta Mater.* **58** 2262
[34] Velasco L and Hodge AM 2016 *Acta Mater.* **109** 142

Observation of simultaneous increase in strength and ductility by grain refinement in a Fe-34.5Mn-0.04C steel

Y H Wang[1,2*], J M Kang[2], Y Peng[1], H W Zhang[1], T S Wang[2] and X Huang[3]

[1] National Engineering Research Center for Equipment and Technology of Cold Strip Rolling, Yanshan University, Qinhuangdao 066004, China
[2] National Key Laboratory of Metastable Materials Science and Technology, Yanshan University, Qinhuangdao 066004, China
[3] Section for Materials Science and Advanced Characterization, Department of Wind Energy, Technical University of Denmark, Risø Campus, DK-4000 Roskilde, Denmark

E-mail: yhwang@ysu.edu.cn

Abstract. Fine grained Fe-34.5Mn-0.04C steel samples with fully recrystallized grain sizes of 3.8 to 2.0 μm were prepared by cold rolling followed by annealing a temperatures of either 650 °C or 800 °C. It is found that a simultaneous increase in both strength and ductility can be obtained by grain refinement, leading to an observation that the best combination of strength and ductility occurs in the sample with the finest recrystallized grain size.

1. Introduction

Grain refinement is an important microstructural design approach for strengthening of metals. However grain refinement induced strengthening is normally associated with decrease in tensile ductility due to reduced work hardening capability. It has been reported, for example, that an abrupt decrease in tensile uniform elongation occurs in both Al and interstitial free steel as the grain size is reduced down to about 1 μm [1]. However, recent studies on high Mn austenitic TWIP steels have shown that the reduction in tensile uniform elongation is less sensitive to grain refinement, and that rather larger tensile uniform elongation is still achievable in fine grained samples with grain sizes in from a few micrometers to the sub-micrometer range [2-5]. For example, a high uniform elongation of 48 % was observed for a Fe-31Mn-3Al-3Si steel sample with a grain size of 1.8 μm [2]. In such a fine grain size deformation twinning was significantly suppressed, leading to a suggestion that the high uniform elongation is related to limited dynamic recovery due to the low stacking fault energy, which enhances work hardening. More interestingly, a simultaneous increase in both yield strength and uniform elongation was observed by Koyama et al. [6] in a binary Fe-33Mn austenitic alloy as the grain size was reduced from 3.8 μm to 1.2 μm. Under these conditions in this alloy the yield strength was increased from 200 MPa to 420 MPa, while the uniform elongation was increased from 35% to 45%. The stacking fault energy for this steel was about 50 mJ/m^2, which is higher than the upper limit of 30-45 mJ/m^2 for the occurrence of deformation twinning [7,8]. No deformation twins or deformation-induced martensite were observed in the tensile deformed ultrafine grains. Instead numerous stacking faults were found, which were considered to be responsible for the observed enhancement in the uniform elongation [6].

In a recent study [9], a fine grained (3.8 μm grain size) Fe-34.5Mn-0.04C steel was produced that showed excellent work hardening and tensile ductility over a wide range of temperature from room temperature down to -180 °C. This steel showed a yield strength of 274 MPa and a uniform elongation of 45%, which are comparable to, and in even slightly higher than, the corresponding values reported for the Fe-33Mn alloy in [6]. Similarly no deformation twinning or martensite formation were observed in the tensile-deformed sample, and the deformation was dominated by dislocation mechanisms. This raises a question as to whether a simultaneous increase in both strength and uniform elongation can be obtained in the Fe-34Mn-0.04C alloy if the grain size is refined to below 3.8 μm. In this study, several samples with grain sizes in the range 3.8-2.0 μm were produced by cold-rolling followed by controlled annealing, and tested. The results show a clear improvement in both strength and tensile ductility with reduction in grain size.

2. Experimental

The steel used in this study had a nominal composition of Fe-34.5Mn-0.04C. An ingot was produced from a vacuum induction furnace, and subsequently forged in the temperature range of 800–1100°C to form a 13 mm thick plate. The plate was consequently cold rolled to a thickness reduction of 90 % using a laboratory rolling mill with a roll diameter of 230 mm. No edge cracking occurred during the course of cold rolling.

It was established in a previous study [9] that after 90 % cold rolling an annealing treatment at 800°C for 1 hour generates a recrystallized grain size of about 3.8 μm. The same heat treatment was applied in this study to prepare a similar grain size. To produce finer grain sizes, the 90 % cold rolled sheet was subjected to annealing treatments at a lower temperature of 650° for either 5, 10, or 30 minutes.

Tensile specimens with gauge dimensions of 10 mm in length, 5 mm in width and 1.3 mm in thickness were machined from the rolled and annealed sheet such that the tensile direction was parallel to the rolling direction (RD). Note that the dimensions of the tensile specimen are different from those used for the low temperature tensile testing in the previous study [9], in particular the cross section area of 5 × 1.3 mm^2 used in this study is much smaller than the value of 12.5 × 1.3 mm^2 used in [9]. Tensile tests were carried out at room temperature at a strain rate of 10^{-3} s^{-1}.

Microstructural characterization was performed by means of optical microscopy, scanning electron microscopy (SEM) coupled with electron backscatter diffraction (EBSD), and transmission electron microscopy (TEM). Thin foil specimens for the TEM observations were prepared by twin-jet electro-polishing in a solution of 100 ml $HClO_4$ + 900 ml CH_3COOH at -15°C. All microstructural observations were carried out on the longitudinal section containing the RD and normal direction (ND).

3. Results and discussion

3.1. Microstructure

Figure 1 shows the appearance in the optical microscope and in the TEM of the 90% cold rolled sample. A banded structure, with the band direction parallel to the rolling direction is seen from the optical image (figure 1a). The band widths vary from about 1 μm to about 10 μm. Some of the bands show somewhat wavy features that are associated with localized shear deformation. TEM observation (figure 1b) shows the formation of a fine scale deformed lamellar structure, with the lamellar boundaries approximately parallel to the RD. The spacings between the lamellar boundaries (figure 1b) are much finer than the band widths seen on the optical image (figure 1a). Measurements of the lamellar boundary spacings along the ND showed a spread from a few nanometers to about 150 nm, with an average value of about 40 nm. In addition to the deformed lamellar structure, nanotwin bundles and localized shear bands were also observed. The nanotwin bundles had an area fraction of about 18 %. Such fine scale heterogeneities in the deformed microstructure may have an effect on the local variation of stored energy, and therefore on the nucleation of recrystallization. However they showed little effect on the homogeneity of the obtained recrystallized microstructure over large areas.

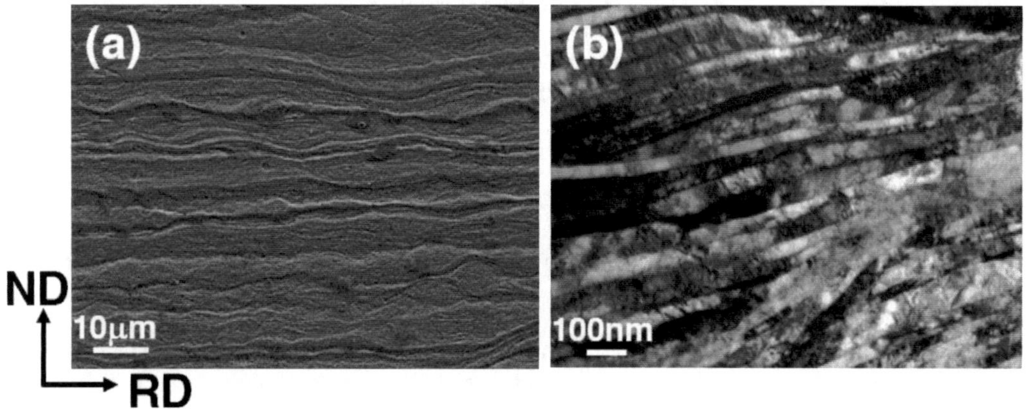

Figure 1. (a) Optical image and (b) TEM image showing the deformed microstructure of the Fe-34.5Mn-0.04C steel cold rolled to 90% thickness reduction.

The 90 % cold-rolled sample was annealed under four different conditions, as described above, to produce four different grain sizes. Figure 2a shows an EBSD orientation map obtained from the sample annealed at 650°C for 5 minutes. It is seen that the grain structure is rather uniform, although there remains about 2% of unrecrystallized deformed material (indicated by DS in figure 2a). All the other three samples, which were annealed at 650°C for 10 minutes, 650°C for 30 minutes and 800°C for 60 minutes, were fully recrystallized. An example EBSD image obtained from the sample annealed at 650°C for 10 minutes is shown in figure 2b to illustrate the formation of a uniform recrystallized grain structure. The mean grain sizes were measured from the EBSD maps for the four annealed samples and the results are given in table 1. The same grain size of 3.8 μm as obtained in the previous study [9] was reproduced after annealing at 800°C for 60 minutes. Annealing at 650°C for 5, 10 and 30 minutes produced three finer, but nevertheless similar, grain sizes of 2.0, 2.1 and 2.2 μm (table 1). Considering the presence of 2% unrecrystallized microstructure after annealing at 650°C for 5 minutes and the similar recrystallized grain sizes after annealing at 650°C for up to 30 minutes, it seems that a grain size of 2.0 μm is the minimum grain size that can be obtained from annealing of the 90 % cold rolled Fe-34.5Mn-0.04C steel.

3.2. Tensile behaviour

Figure 3 shows engineering stress-strain curves for the four samples with different grain sizes. Grain refinement from 3.8 μm to about 2.0 μm resulted in two striking changes. Firstly, a simultaneous increase in both strength and tensile ductility is seen, and secondly, a transition from continuous yielding in the 3.8 μm grain sized sample to a discontinuous yielding in the three finer grain sized samples takes place. These observations are similar to those reported for the Fe-33Mn alloy as the grain size is refined from 3.8 μm to 1.2 μm [6]. Note that the change in the yielding behavior is associated with a significant increase in the yield strength, e.g. from 256 MPa to 405 MPa corresponding to a grain size change from 3.8 to 2.2 μm. Also note that the lower yield strength is very sensitive to the grain size change. A decease of grain size from 2.2 μm to 2.0 μm leads to an increase of the lower yield strength from 405 MPa to 466 MPa. It should be pointed out that the occurrence of discontinuous yielding is in general undesirable for practical applications of the steel. To remove the discontinuous yielding, an annealed sample with a 2.0 μm grain size was further deformed by a slight cold rolling (about 2 % thickness reduction) and tensile tested. It was found that not only the did this results in the flow behavior becoming continuous, but also a further increase in the yield strength, indicating a possibility to further improve the mechanical properties of the steel.

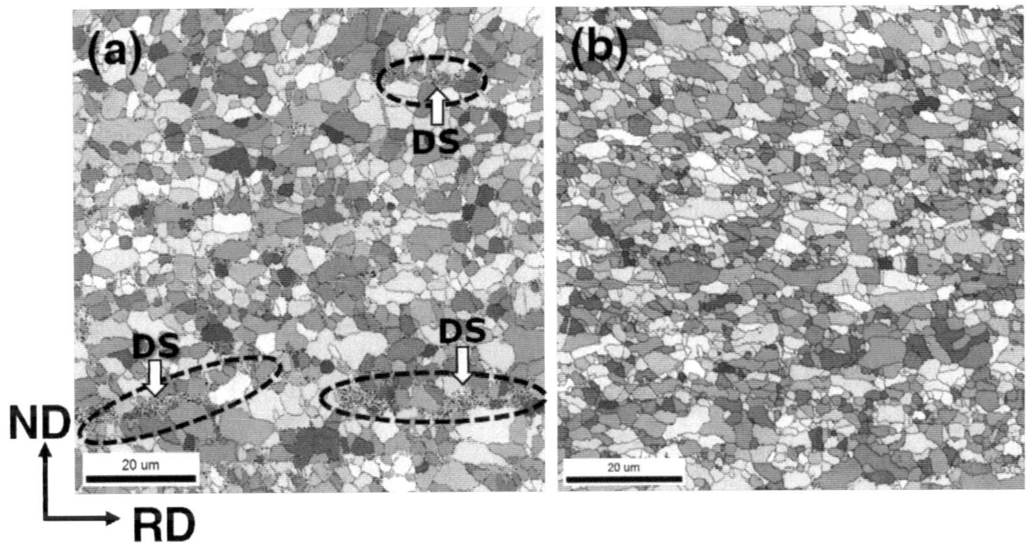

Figure 2. EBSD maps showing the microstructure of the Fe-34.5Mn-0.04C steel cold rolled to 90% and then annealed at 650°C for (a) 5 minutes and (b) 10 minutes.

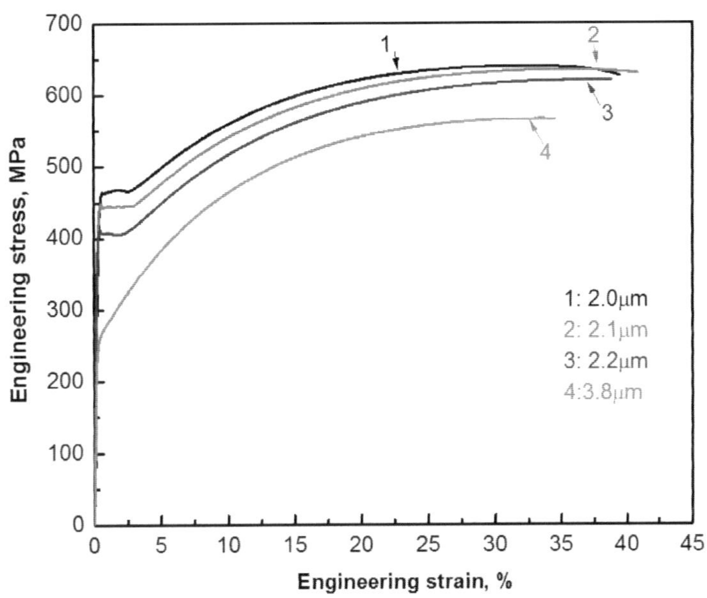

Figure 3. Tensile stress-strain curves for samples with different recrystallized grain sizes.

Table 1. Mean grain sizes and tensile properties of the four samples investigated.

Sample	Mean grain size (μm)	σ (Yield strength) (MPa)	σ (0.2%) (MPa)	UTS (MPa)	Uniform elongation (%)	Total Elongation (%)
650°C, 5 min	2.0	466		640	32.6	39.6
650°C, 10 min	2.1	444		635	35.6	39.4
650°C, 30 min	2.2	405		620	37.5	38.8
800°C, 60 min	3.8		256	550	32.5	34.1

4. Summary

Fe-34.5Mn-0.04C steel samples with fully recrystallized fine grain sizes of 3.8 to 2.0 μm were prepared by cold rolling followed by an annealing treatment either at 650 °C or 800 °C. Investigation of the mechanical properties through tensile testing revealed a simultaneous increase in both strength and ductility occurred with grain refinement in the present steel. Based on the experimental observations it was concluded that the sample with the finest recrystallized grain size exhibits the best combination of strength and ductility.

Acknowledgment

The authors gratefully acknowledge the support from the Project Program of Heavy Machinery Collaborative Innovation Centre, China.

References

[1] Tsuji N, Ito Y, Saito Y and Minamino Y 2002 *Scripta Mater* **47** 893-899

[2] Ueji R, Tsuchida N, Terada D, Tsuji N, Tanaka Y, Takemura A and Kunishige K 2008 *Scripta Mater* **59** 963-66

[3] Gutierrz-Urrutia I, Zaefferer and Raabe D 2010 *Mater Sci Engn A* **527** 3552-3560

[4] Kang S, Jung YS, Jun JH and Lee YK 2010 *Mater Sci Engn* A **527** 745-751

[5] R. Saha, R. Ueji and N Tsuji 2013 *Scripta Mater* **60** 813-16

[6] Koyama M, Sawaguchi T and Tsuzaki K 2011 *Mater Sci Engn A* **530** 659-663

[7] Remy I, Pineau A 1976 *Mater Sci Engn A* **26** 123-132

[8] Allain S, Chateau, Bouaziz O, Migot S, Guelton 2004 *Mater Sci Engn A* **387/389** 158-162

[9] Wang Y, Peng Y, He Y, Wang T, & Huang X 2014 35th *Proceedings of Risø International Symposium on Materials Science* 471-477

IOP Conf. Series: Materials Science and Engineering **219** (2017) 012044 doi:10.1088/1757-899X/219/1/012044

Interaction of C and Mn in a Σ3 grain boundary of bcc iron

A T Wicaksono and M Militzer

Centre for Metallurgical Process Engineering, The University of British Columbia, Vancouver, Canada, V6T 1Z4

E-mail: matthias.militzer@ubc.ca

Abstract. The interaction of alloying elements with migrating interfaces is a key aspect that determines microstructure evolution during thermo-mechanical processing of metals and alloys. Recent advances in atomic scale resolution characterization techniques and atomistic modelling have dramatically increased the potential to generate new knowledge on interfaces thereby enabling paradigm shifts in microstructure design approaches. Computational materials science now offers exciting opportunities to formulate multi-scale process models that bridge the gap between atomistic and continuums approaches. In particular, the development of advanced high strength steels with novel alloying concepts has motivated atomistic scale simulations to predict the interaction of selected alloying elements with grain boundaries and the austenite-ferrite interface in iron. Most of these simulations have so far been carried out for binary systems with one solute species. The extension of this modelling work to multi-component systems is, however, essential in order to account for potential interactions between different alloying elements in industrial steels. Here, we present *ab initio* simulations for the interaction of C and Mn at a Σ3 boundary in bcc iron using density functional theory (DFT). The simulation results confirm the strong co-segregation of C and Mn that has been recently observed in atom probe tomography studies of the austenite-ferrite interface in Fe-Mn-C alloys.

1. Introduction

The development of modern steels such as advanced high strength steels is aided by a better understanding of the role of alloying elements on the kinetics of the austenite-ferrite transformation. The presence of substitutional solutes, e.g. Mn, Nb and Mo, has been shown to dramatically reduce the velocity of the austenite-ferrite interface during phase transformation via a solute drag effect [1–4]. Accurate prediction of the transformation kinetics is challenging because several aspects contribute to the complexity of solute-interface interactions, including the presence of other solute elements in the interface and the difference in their diffusivities. A number of ferrite growth models have been proposed to account for the solute drag effect in ternary Fe-X-C systems, see [5] for a recent survey. These models are frequently based on an assumption in which the solute-interface interaction is solely attributed through a parameter referred to as the binding energy, which captures the affinity of solute X to the interface.

The single binding energy assumption fails to capture the complex interaction among different types of solute elements. For example, recent work that employed atom probe tomography (APT) [6, 7] found that the enrichment level of Mn at the austenite-ferrite interface in the Fe-Mn-C system is twice the bulk Mn content even though the bulk diffusivity and the binding energy of Mn in general grain boundaries of bcc Fe alone are only in the order of 10^{-17}

m^2s^{-1} at 680°C [7] and 5.5 kJ mol^{-1} [8], respectively. On the other hand, no Mn enrichment at the austenite-ferrite interface was reported for the ternary Fe-Mn-N system [6]. This suggests that the high Mn enrichment in the Fe-Mn-C system may be attributed to the presence of carbon at the interface. Such a hypothesis has also been indicated in earlier work [9, 10], although it remains challenging to verify experimentally.

First principle calculations such as density functional theory (DFT) offer a systematic way to verify the interaction between Mn and C in the bulk, at grain boundaries and bcc-fcc interfaces in Fe. The important role of DFT in alloy design has been demonstrated in a number of efforts that employed modelling across different length and time scales, e.g. development of modern steels [11, 12]. Studies that apply DFT as a modelling tool have largely targeted binary systems, e.g. Fe-C [13–19], Fe-Mn [20, 21], Fe-Cr [16, 22], and Fe-N [11, 15], with little attention dedicated to ternary systems. A few examples of DFT calculations for ternary alloys include a study on the Fe-Cr-C system [16] and another on Fe-Nb-X systems (where X = Mo, Ti, Cr, Mn) [23].

This work aims to focus on the Fe-Mn-C system, with the objective of investigating the interaction between Mn and C in the bulk and at the incoherent $\Sigma3(111)$ grain boundary (GB) of bcc Fe using DFT. The $\Sigma3(111)$ boundary is chosen since its structure has a relatively short periodicity, thus providing a reasonable limit on the number of boundary sites that Mn and C can occupy. First, the Mn-C interaction in bulk bcc Fe will be explored as a function of the distance between Mn and C atoms to determine the spatial extent of their interaction. Then, the binding energy of individual Mn and C atoms at a number of boundary sites are computed to identify favourable sites for segregation. Finally, the co-segregation energy of Mn and C atoms for these sites is determined and assessed against recent experiments for the Fe-Mn-C system.

2. Simulation methodology

First-principles spin-polarized calculations are performed using DFT as implemented in the Vienna *ab-initio* Simulation Package (VASP) [24, 25] with the projector-augmented-wave (PAW) method [26]. The exchange correlation functional is modelled using the generalized gradient approximation (GGA) with parameterization by Perdew-Burke-Ernzerhof (PBE) [27, 28]. Targets of 10^{-4} eV and 10^{-2} eV Å^{-1} are aimed for energy and force convergence, respectively. The plane-wave basis kinetic cut-off energy is set at 500 eV for all calculations.

Two types of supercells are constructed, namely a bulk bcc and a bicrystal supercell. The bulk supercell consists of 128 lattice sites (4×4×4 bcc unit cells) with *k*-points distributed on a Monkhorst-Pack mesh of 5×5×5 grids. Periodic boundary conditions are applied in all directions. Bulk calculations show that the lattice parameter of bcc Fe, a, is 2.84 Å and the energy per Fe atom in the bcc crystal, E^0_{Fe}, is −8.31 eV (−802 kJ mol^{-1}). These results are consistent with previous calculations [14, 16]. Solute Mn and C are introduced in the bulk bcc Fe as follows. For Mn, an Fe atom is replaced with a Mn atom (e.g. site U in figure 1(a)) whose initial spin orientation is of the opposite sign from that of Fe atoms (i.e. antiferromagnetic configuration following [21, 23]). For carbon, one C atom with a zero spin is placed on an octahedral interstitial site in bcc Fe. In a bulk supercell that contains a Mn and a C atom, the C atom is placed at an octahedral site and the distance $d_{\text{Mn-C}}$ from the Mn atom is varied, e.g. site V ($d_{\text{Mn-C}} = \frac{1}{2}a$) and site W ($d_{\text{Mn-C}} = \frac{1}{2}\sqrt{2}a$) in figure 1(a), up to a distance $d_{\text{Mn-C}}$ of 2.5a.

A bicrystal supercell is constructed from a set of orthogonal axes that form two $\Sigma3$ incoherent twin boundaries, where the boundary normal is <111> and the remaining orthogonal axes are <112> and <110>, see figures 1(b) and (c). The bicrystal dimensions for subsequent calculations are determined from a size-sensitivity analysis involving supercells of varying dimensions. For a given dimension, the grain boundary energy γ is obtained by taking the difference between the energy of a bicrystal and that of a bulk cell containing an equivalent number of Fe atoms, and normalizing it with the total boundary area of $2L_YL_Z$. The boundary energy varies less than 0.01 eV Å^{-2} for bicrystals with dimensions larger than

$L_{X,0} = 4\sqrt{3}a$ (19.6 Å), $L_{Y,0} = 2\sqrt{6}a$ (13.9 Å) and $L_{Z,0} = 2\sqrt{2}a$ (8.0 Å). The bicrystal with dimensions $L_{X,0} \times L_{Y,0} \times L_{Z,0}$ is chosen for subsequent calculations. It contains 192 lattice sites and has a boundary energy γ of 0.46 eV Å^{-2} (1.58 J m^{-2}), which is consistent with previous work [22].

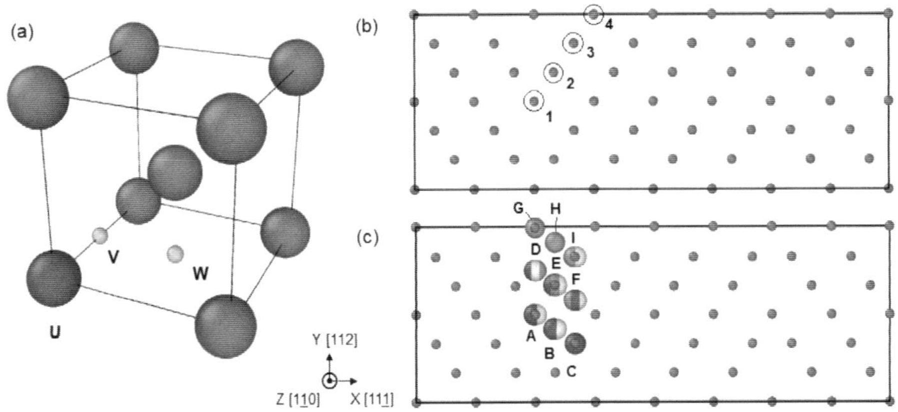

Figure 1. (a) A bcc cell where one atom (U) is highlighted as a substitutional site and two neighbouring octahedral interstitial sites (V and W), (b) A bicrystal that contains two Σ3(111) boundaries with boundary sites labelled as 1, 2, and 3, while site 4 is considered a bulk site, (c) Octahedral sites where blue, yellow and grey sites correspond to sites 1, 2 and 3, respectively. Note, sites G, H and I are equivalent to sites A, B and C, respectively.

In bulk bcc Fe, the interaction between Mn and C atoms separated by a distance $d_{\text{Mn-C}}$ is represented by its solution energy $E^0_{\text{Mn-C}}$, which is defined as

$$E^0_{\text{Mn-C}}(d_{\text{Mn-C}}) = \left(E^0_{\text{Mn-C},total} + E^0_{pure}\right) - \left(E^0_{\text{Mn},total} + E^0_{\text{C},total}\right) \tag{1}$$

where the superscript 0 indicates the calculation belonging to a bulk supercell, $E^0_{\text{Mn-C},total}$ is the total energy of a bulk supercell that contains a Mn and a C atom at a distance $d_{\text{Mn-C}}$ from each other, E^0_{pure} is the energy of a pure bulk Fe supercell with an equivalent number of Fe atoms, $E^0_{\text{Mn},total}$ and $E^0_{\text{C},total}$ are the total energies of a bulk supercell that only contains a Mn or a C atom, respectively. Mn and C atoms favour each other's presence in the bulk Fe if $E^0_{\text{Mn-C}} < 0$.

In order to assess the co-segregation behaviour of Mn-C, the binding energy of a single solute atom X (X = Mn or C) in the Σ3(111) boundary, E^i_X, is first computed via [23]

$$E^i_X = \left(E^i_{X,total} + E^0_{pure}\right) - \left(E^0_{X,total} + E^{gb}_{pure}\right) \tag{2}$$

where the superscript i indicates the boundary site i, $E^i_{X,total}$ is the total energy of a bicrystal that contains a solute X at a boundary site i and E^{gb}_{pure} is the total energy of a bicrystal with only Fe atoms. A negative value of E^i_X indicates that site i is an attractive site for solute X.

The Σ3(111) boundary provides three unique lattice sites for a substitutional atom to reside near the boundary, see figure 1(b). Each substitutional site is associated with six different octahedral sites, consisting of two and four octahedral sites that are $\frac{1}{2}a$ and $\frac{1}{2}a\sqrt{2}$ away from a given boundary site, respectively, as shown in figure 1(c). Octahedral sites further than $\frac{1}{2}a\sqrt{2}$ from a given substitutional site are not considered here. Table 1 summarizes the positions of substitutional and interstitial boundary sites.

Table 1. Positions of substitutional and octahedral sites in the $\Sigma3(111)$ boundary relative to the coordinate of substitutional site 1 (see figures 1(b) and (c)).

Substitutional (S) or octa-hedral (O) site	Site label	Positions relative to site 1 X	Y (in units of a)	Z	Distance* [in units of a] to site 1	2	3
S	1	0	0	0	0	1	1
S	2	$\frac{1}{6}\sqrt{3}$	$\frac{1}{6}\sqrt{6}$	$\frac{1}{2}\sqrt{2}$	1	0	1
S	3	$\frac{1}{3}\sqrt{3}$	$\frac{1}{3}\sqrt{6}$	0	1	1	0
O	A	0	0	$\frac{1}{2}\sqrt{2}$	$\frac{1}{2}\sqrt{2}$	$\frac{1}{2}$	$\frac{1}{2}\sqrt{2}$
O	B	$\frac{1}{6}\sqrt{3}$	$-\frac{1}{12}\sqrt{6}$	$\frac{1}{4}\sqrt{2}$	$\frac{1}{2}$	$\frac{1}{2}\sqrt{2}$	$\frac{1}{2}$
O	C	$\frac{1}{3}\sqrt{3}$	$\frac{1}{12}\sqrt{6}$	$\frac{1}{4}\sqrt{2}$	$\frac{1}{2}\sqrt{2}$	$\frac{1}{2}\sqrt{2}$	$\frac{1}{2}\sqrt{2}$
O	D	0	$\frac{1}{4}\sqrt{6}$	$\frac{1}{4}\sqrt{2}$	$\frac{1}{2}\sqrt{2}$	$\frac{1}{2}$	$\frac{1}{2}\sqrt{2}$
O	E	$\frac{1}{6}\sqrt{3}$	$\frac{1}{6}\sqrt{6}$	0	$\frac{1}{2}$	$\frac{1}{2}\sqrt{2}$	$\frac{1}{2}$
O	F	$\frac{1}{3}\sqrt{3}$	$\frac{1}{6}\sqrt{6}$	0	$\frac{1}{2}\sqrt{2}$	$\frac{1}{2}\sqrt{2}$	$\frac{1}{2}\sqrt{2}$

* the nearest distance between a site i (or its symmetrically equivalent site) and a substitutional site

The segregation energy of a Mn-C pair in the $\Sigma3(111)$ boundary, $E^{ij}_{\text{Mn-C}}$, is defined as the energy change as a result of bringing a single Mn and a single C atom from the bulk bcc Fe to a substitutional site i and an octahedral site j at the boundary, respectively, i.e.

$$E^{ij}_{\text{Mn-C}} = \left(E^{ij}_{\text{Mn-C},total} - E^{gb}_{pure}\right) - \left(\left(E^{0}_{\text{Mn},total} - E^{0}_{pure}\right) + \left(E^{0}_{\text{C},total} - E^{0}_{pure}\right)\right) \quad (3)$$

where $E^{ij}_{\text{Mn-C},total}$ is the total energy of a bicrystal cell that contains an ij pair of Mn-C. The reference states in equation (3) are the binary limits of each solute in bulk bcc Fe, i.e. when Mn and C atoms are sufficiently far apart from each other that they do not interact.

3. Results

3.1. *Mn-C interaction in bulk bcc Fe*

The bulk solution energy of a Mn-C pair, $E^{0}_{\text{Mn-C}}$, is plotted in figure 2 as a function of the Mn-C distance. When Mn and C atoms are separated by a distance $d_{\text{Mn-C}}$ of 1.4 Å (0.5a), the total energy is lowered by 53 kJ mol^{-1} relative to the case of the binary limits (i.e. when Mn and C are sufficiently far apart from each other that they do not interact). $E^{0}_{\text{Mn-C}}$ at a distance of 7.1 Å (2.5a) is -1.1 kJ mol^{-1}, indicating that at this distance the Mn-C pair approaches the interaction limit.

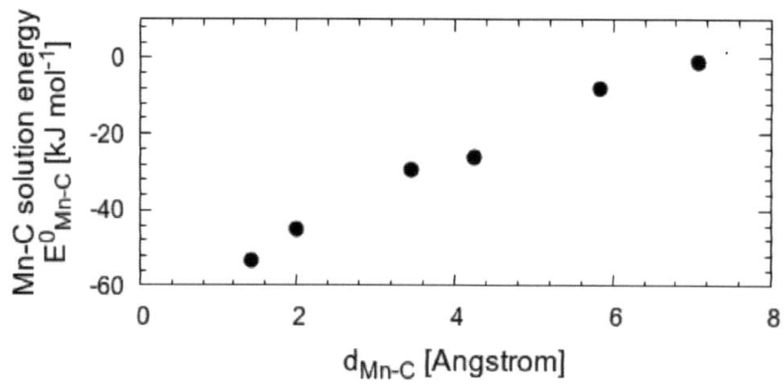

Figure 2. Solution energy of Mn-C pair in bcc Fe as a function of the Mn-C distance.

3.2. *Grain boundary binding energies of single C and Mn atoms*

Figure 3(a) presents the binding energies of a single Mn atom, E^i_{Mn}, on boundary sites 1 to 4 (see figure 1(b)). E^i_{Mn} is the lowest at site 1 (-11 kJ mol^{-1}) and gradually increases to zero at site 4, verifying that site 4 is indeed a bulk site. For comparison, figure 3(a) also presents the binding energy profile for a recently considered Σ5(310) symmetric tilt boundary [23], indicating a number of similarities, including the same order of magnitude of the binding energy. Further, the magnitude of the Mn binding energy at site 1 is about a factor of 2 larger than that obtained from experimental studies on polycrystalline ferrite (-5.5 kJ mol^{-1}) [8].

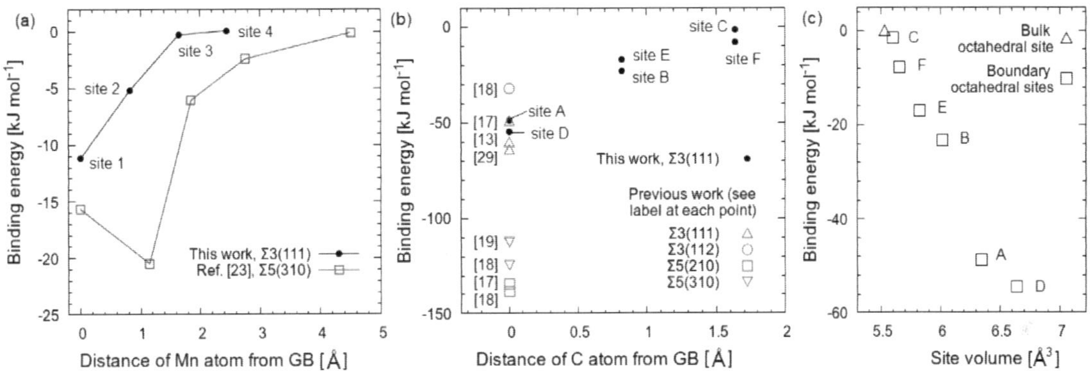

Figure 3. (a) Binding energy profile of a Mn atom from this work and from previous DFT results [23], (b) Binding energy profile of a C atom from this work and previous DFT results [13, 17–19, 29], (c) Relationship between the site volume and the binding energy of a single C atom. See figures 1(b) and (c) for site labels.

Figure 3(b) presents the binding energy profile of a single C atom, E^i_{C}, for octahedral sites A to F (see table 1). DFT results from the literature are also included for comparison. The binding energies of C at zero distance from the Σ3(111) boundary (i.e. -54 and -48 kJ mol^{-1}, see sites A and D in figure 3(b)) are consistent with previous DFT work on the same boundary [13, 17, 29]. The slight discrepancy between our work and Refs. [13, 17, 29] may be attributed to a different cutoff energy for the plane wave basis set and different exchange potential approximations. Finally, the binding energy of C presented here is within the range reported from experiments on general grain boundaries in bcc Fe, i.e. -80 to -30 kJ mol^{-1} [30–33].

The variation of binding energies with different octahedral sites can be rationalized based on the volume of each interstitial site after relaxation, which was obtained using Voronoi tesselation [34]. Figure 3(c) shows the relationship between the site volume and the binding energy of carbon at each site. It is evident that a linear trend emerges, i.e. the magnitude of the binding energy increases with site volume. The correlation between the site volume and the binding energy had also previously been used to rationalize the lower magnitude of binding energies for C in Σ3 boundaries as compared to Σ5 boundaries [18] (see figure 3(b)) since Σ3 boundaries are more compact than Σ5 boundaries [17].

3.3. *Segregation energy of Mn and C*

The segregation energy for each of the 18 ij combinations for the Mn-C pair at the boundary (see table 1) is presented in figures 4(a)-(c). The Mn-C segregation energy ranges from -95 to -51 kJ mol^{-1} and has a profile trend similar to that of the binding energies of single solute atoms shown in figures 3(a) and (b), i.e. $E^{ij}_{\mathrm{Mn\text{-}C}}$ is minimum at the boundary and approaches the Mn-C bulk solution energy when the C atom is further away from the boundary. The most preferred

Mn-C configuration at the boundary is pair 1D followed by pairs 3D and 3A. In the latter two cases Mn is a distance of a away from the boundary occupying a site that had negligible attraction in the binary limit (see figure 3a). It is interesting to note that the magnitude of the Mn-C segregation energy for pairs whose carbon atom occupies site D can be up to 40% larger than that of pairs whose carbon atom is on site A (see e.g. pairs 1A and 1D) even though for both pairs all sites are in the boundary plane and the nominal Mn-C distance is the same, i.e. $\frac{1}{2}\sqrt{2}a$ (2 Å). However, the intra-atomic distance of the pair changes after relaxation. For pair 1D, the distance decreases by 10% to 1.8 Å while it increases for pair 1A by 7.5% to 2.15 Å. Further, the interstitial site volume of A is 10% smaller than for site D. The shorter Mn-C distance and larger site volume for pair 1D is consistent with its lower segregation energy as compared to pair 1A. The effective Mn-C distance for other configurations is found to change by less than ±5%.

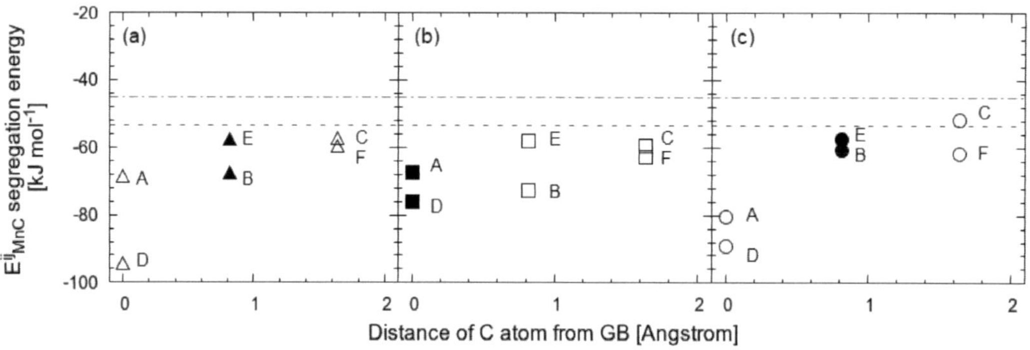

Figure 4. Profile of Mn-C segregation energy for configurations in which the Mn atom is in (a) site 1, (b) site 2, or (c) site 3. The labels (A to F) refer to the site label for the C atom, see table 1. Closed (open) symbols represent Mn-C for $d_{\text{Mn-C}} = \frac{1}{2}a$ ($\frac{1}{2}a\sqrt{2}$). Dashed (dash-dotted) lines indicate $E^0_{\text{Mn-C}}$ when $d_{\text{Mn-C}}$ is $\frac{1}{2}a$ ($\frac{1}{2}a\sqrt{2}$), see figure 2.

4. Discussion

While it is evident from figure 4 that the segregation of Mn and C as a pair to the $\Sigma3(111)$ boundary is favoured, it is also worth analyzing Mn-C segregation as the sum of its contributions from the binding energy of individual Mn and C atoms and their combined interaction, i.e.

$$E^{ij}_{\text{Mn-C}} = E^{i}_{\text{Mn}} + E^{j}_{\text{C}} + \alpha_{ij} \tag{4}$$

where the first two terms on the right hand side correspond to equation(2) and α_{ij} is the Mn-C interaction energy at the boundary. Discussing the Mn-C segregation in terms of α_{ij} is helpful because α_{ij} represents the energy change between the state where Mn and C atoms are at the boundary, within a distance of less than a from each other, and an initial state where they are both at the boundary but sufficiently far apart that they do not interact. As shown in figure 5 the α_{ij} values for all Mn-C configurations range from -60 to -10 kJ mol^{-1}. Except for the pairs 2F and 3F, the magnitude of the Mn-C interaction in the boundary is lower than in the bulk (53 kJ mol^{-1}). In particular for the preferred configurations of Mn-C segregation (i.e. 1D, 3D, 3A), the interaction term is reduced to about 30 kJ mol^{-1} and has its lowest value for the 1A pair (8 kJ mol^{-1}).

The interaction energy α_{ij} can also be used to estimate the co-segregation energy, i.e. the extra binding energy for a given type of solute due to the presence of another type of solute. For example, the co-segregation energy of Mn in site i when there is a C atom in site j is determined from the sum of the binding energy of Mn in site i, E^i_{Mn} in figure 3(a), and α_{ij}. This suggests that, even though Mn itself has a small binding energy in the Σ3(111) boundary, it gains an extra binding energy from -10 to -60 kJ mol^{-1} if there is a C atom around its vicinity at the boundary. For example, when carbon occupies site D the magnitude of the effective Mn binding energy to site 1 is increased from 11 kJ mol^{-1} to 39 kJ mol^{-1} and for site 3 from 0.5 kJ mol^{-1} to 35 kJ mol^{-1}.

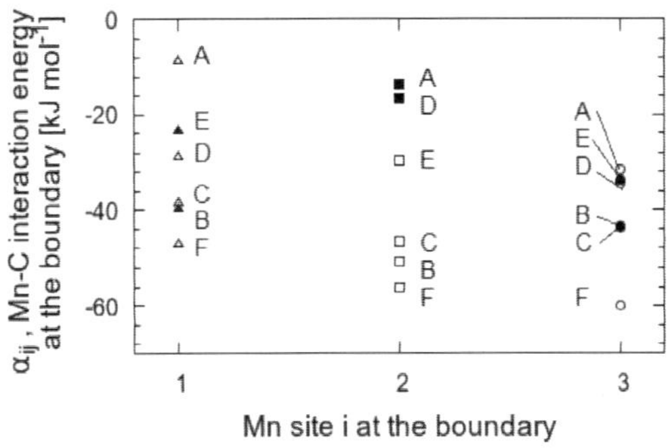

Figure 5. The Mn-C interaction energy at the boundary, α_{ij}, for different Mn-C configurations.

Such an attractive interaction between Mn and C is consistent with recent work using APT [6, 7, 35, 36] which reported high enrichment levels of Mn (a factor of 1.8) at the bcc-fcc interface for the Fe-Mn-C system and no Mn enrichment for the Fe-Mn-N system. The high enrichment level of Mn observed in the Fe-Mn-C system can be rationalized from the fact that C has a relatively high interfacial segregation energy in Fe (about 50 kJ mol^{-1} for the Σ3(111) boundary investigated here) and is a fast diffuser [14]. Both factors contribute to the high segregation level of C at the boundary [18]. The presence of C at the boundary then promotes co-segregation of slower diffusing Mn.

5. Conclusion

We have performed DFT calculations to explore the interaction between Mn and C in bulk bcc Fe and the co-segregation behaviour of Mn and C at a number of boundary sites in a Σ3(111) grain boundary of bcc Fe. Our results demonstrate that Mn and C have an attractive interaction in bulk bcc Fe, with the interaction energy being the strongest (53 kJ mol^{-1}) at a Mn-C distance of 1.4 Å, and gradually vanishing at 7.1 Å. Segregation of single C and Mn atoms to a number of Σ3(111) boundary sites show that the maximum magnitude of the binding energies for C and Mn are 54 kJ mol^{-1} and 11 kJ mol^{-1}, respectively. These results are within the values reported from previous DFT work and experiments.

That the binding energy of C is higher than the binding energy of Mn implies that C segregation is more favoured than Mn segregation in the Fe-Mn-C system. To explain the high enrichment of Mn experimentally observed in the Fe-Mn-C systems, we examined situations where a Mn and a C atom are located 2 Å from each other in the Σ3(111) boundary. Such Mn-C configurations are found to promote a further energy decrease compared to Mn and C individually segregated at the boundary. The energy change associated with the co-segregation of Mn and C ranges from -60 to -10 kJ mol^{-1}, relative to the binding energy of single Mn and C atoms, and depends on the position of the C atom with respect to the boundary and the Mn atom. This finding is consistent with experimental work on Fe-Mn-C alloys which suggest that the high enrichment of Mn at austenite-ferrite interfaces is promoted by carbon segregation.

Acknowledgments
The authors gratefully acknowledge the financial support from the Natural Sciences and Engineering Research Council of Canada and the computing resources provided by WestGrid, a division of Compute Canada.

References
[1] Zurob H S *et al.* 2008 *Acta Mater* **56** 2203–2211
[2] Qiu C *et al.* 2013 *Metall. Mater. Trans. A* **44** 3472–3483
[3] Cahn J W 1962 *Acta Metall.* **10** 789–798
[4] Hillert M 2004 *Acta Mater.* **52** 5289–5293
[5] Gouné M *et al.* 2015 *Mater. Sci. Eng. R.* **92** 1–38
[6] van Landeghem H P *et al.* 2016 *JOM* **68** 1329
[7] Danoix F *et al.* 2016 *Scr. Mater.* **121** 61–65
[8] Guttman M 1975 *Surf. Sci.* **53** 213–227
[9] Mori T 1964 19th committee on reaction in steel making Tech. rep. Department of Metallurgy, Kyoto University
[10] Jarl M 1978 *Scand. J. Metall.* **7** 93–101
[11] Wu R, Freeman A J and Olson G B 1994 *Science* **265** 376–380
[12] Xiong W and Olson G B 2015 *MRS Bulletin* **40** 1035–1044
[13] Wu R, Freeman A J and Olson G B 1996 *Phys. Rev. B* **53** 7504
[14] Jiang D E and Carter E A 2004 *Phys. Rev. B.* **67** 214103
[15] Domain C, Becquart C S and Foct S 2004 *Phys. Rev. B.* **69** 144112
[16] Sandberg N, Henriksson K O E and Wallenius J 2008 *Phys. Rev. B.* **78** 094110
[17] Wachowicz E and Kiejna A 2011 *Modell. Simul. Mater. Sci. Eng.* **19** 025001
[18] Wang J *et al.* 2016 *Acta Mater.* **115** 259–268
[19] Tahir A, Janisch R and Hartmaier A 2014 *Mater. Sci. Eng. A* **612** 462–467
[20] Dick A, Hickel T and Neugebauer J 2009 *Steel Research International* **80** 603
[21] Lintzen S *et al.* 2013 *Journal of Alloys and Compounds* **577** 370–375
[22] Wachowicz E, Ossowski T and Kiejna A 2010 *Phys. Rev. B.* **81** 094104
[23] Jin H, Elfimov I and Militzer M 2014 *J. Appl. Phys.* **115** 093506
[24] Kresse G and Furthmüller J 1996 *Phys. Rev. B* **54** 11169
[25] Kresse G and Furthmüller J 1996 *Comp. Mater. Sci.* **6** 15–50
[26] Kresse G and Joubert D 1999 *Phys. Rev. B* **59** 1758
[27] Perdew J P, Burke K and Ernzerhof M 1996 *Phys. Rev. Lett.* **77** 3865
[28] Perdew J P, Burke K and Ernzerhof M 1997 *Phys. Rev. Lett.* **78** 1396
[29] Yamaguchi M 2011 *Metall. Mater. Trans. A* **42** 319–329
[30] Lejček P and Hofmann S 2016 *J. Phys.: Condens. Matter* **28** 064001
[31] Papazian J M and Besherb D N 1971 *Met. Trans.* **2** 497–503
[32] Hänsel B and Grabke H J 1986 *Scr. Metall.* **20** 1641–1644
[33] Grabke H J 1986 *Steel Research International* **4** 178–185
[34] Rycroft C H 2009 *Chaos* **19** 041111
[35] Guo M *et al.* 2015 *Metall. Mater. Trans. A* **46** 2449–2454
[36] van Landeghem H P *et al.* 2017 *Acta Mater* **124** 536–543

IOP Conf. Series: Materials Science and Engineering **219** (2017) 012045 doi:10.1088/1757-899X/219/1/012045

Structural refinement and property optimization in an Fe-23Cr-8.5Ni duplex stainless steel

L. Xie[1], T.L. Huang[1], Y.H. Wang[2], L. Zhang[1], G.L. Wu[1,*], N. Tsuji[3] and X. Huang[4]

[1] College of Materials Science and Engineering, Chongqing University, Chongqing 400045, China
[2] National Engineering Research Center for Equipment and Technology of Cold Strip Rolling, Yanshan University, Qinhuangdao 066004, China
[3] Department of Materials Science and Engineering, Kyoto University, Kyoto 606-8501, Japan
[4] Section for Materials Science and Advanced Characterization, Department of Wind Energy, Technical University of Denmark, DK-4000 Roskilde, Denmark

E-mail: wugl@cqu.edu.cn

Abstract. An Fe-23Cr-8.5Ni duplex stainless steel was used to prepare samples with different volume-fraction-weighted grain sizes ($d_{\alpha\gamma}$), ranging from the nano-scale to the micrometer-scale by cold rolling and subsequent annealing. The cold rolled sample with $d_{\alpha\gamma}$ of 72 nm showed a high yield strength of about 1.3 GPa but only a small tensile elongation. An abrupt increase of ductility was observed as $d_{\alpha\gamma}$ increased to 375 nm, resulting in a good combination of yield strength of 738 MPa and tensile elongation of 29%. Further increase of $d_{\alpha\gamma}$ up to the micrometer-scale results in continued decreases in yield strength but with only a limited improvement in the ductility.

1. Introduction

Nanostructured single phase metals produced by large strain deformation usually show a high ultimate tensile strength at a relatively small strain followed by a relatively large post elongation [1]. It is difficult to improve the limited ductility, especially the limited uniform elongation, until the grain size is increased up to the micrometer-scale, as widely observed in Al [2], Cu [3], Ti [4] and IF steel [5].

Several microstructural design principles have been put forward recently to improve the strain hardening ability of nanostructured metals, such as gradient structures [6], heterogeneous lamella structures [7] and dual or multi phases structures [8]. In this context dual-phase nanostructures also show a potential ability to achieve both high strength and good ductility. In the present study, we investigate the structural scale effect on the mechanical properties of a duplex stainless steel (DSS) with structural scales from nanometers to micrometers.

2. Materials and methods

The material used in the present study was an Fe-23Cr-8.5Ni DSS, with chemical composition of 0.001% C, <0.01% Si, <0.01% Mn, <0.001% P, <0.001% S, 8.52% Ni, 22.9% Cr, 0.002% N, 0.04% O, and the balance Fe (mass%). The ingot was homogenized at 1150°C for 2 h and hot forged at temperatures above 900°C. The hot forged plate was then cold rolled to a thickness reduction of 90% and subsequently annealed at temperatures from 700-1000°C for 5 min and 30 min in order to tailor the structure into different scales.

Microstructural characterization was carried out using an Axiovert 40 MAT optical microscope (OM) and a JEOL 2100 transmission electron microscope (TEM). The sampling plane was the longitudinal section containing the rolling direction (RD) and normal direction (ND). Samples for OM characterization were mechanically polished followed by electrochemical polishing in an electrolyte of 15 vol.% perchloric acid and 85 vol.% acetic acid at a voltage of 15 V at room temperature. Thin foils for TEM characterization were prepared by mechanical polishing to a thickness of 70 μm and then electropolishing using a twin-jet polisher in a solution of 25 vol.% perchloric acid and 75 vol.% ethanol at a voltage of 20 V at -20 °C. The volume fractions of constituent phases were quantified by X-ray diffraction (XRD) using a Rigaku D/max 2500PC X-ray diffractometer with Cu Kα radiation and a step size of 0.02°. Tensile specimens of gauge dimensions 25 × 5 × 1 mm^3 were machined from the cold rolled and annealed sheets with tensile axis along the RD, and tested at a uniaxial quasi-static strain rate of 6×10^{-4} s^{-1}.

Structural parameters in the DSS samples include grain size, phase size and volume fractions of the two phases. For simplification, we defined the structural size as the volume-fraction-weighted average grain size $d_{\alpha\gamma}$ [9], that is $d_{\alpha\gamma} = d_\alpha f_\alpha + d_\gamma f_\gamma$, where d_α and d_γ are the average grain sizes of the ferrite and austenite phases, respectively (determined by an intercept method along the ND) and where f_α and f_γ are the volume factions of the two phases.

3. Results and Discussion

3.1. Microstructural characterization

Figure 1a shows the XRD spectra of the Fe-23Cr-8.5Ni DSS in the hot forged state and after cold rolling. Both ferrite and austenite diffraction peaks can be observed in the hot forged state. The volume factions of the ferrite and austenite phases were determined to be 44.6% and 55.4%, respectively. After 90% cold rolling, the spectrum exhibits an absence of austenite, indicating the occurrence of a deformation-induced martensitic transformation. After annealing of the cold rolled sheet at 700°C or above, the deformation induced martensite (α′-martensite) transforms back to austenite. As shown in figure 1b, the austenite reversion is dependent on both the annealing temperature and the holding time. In the case of 5 min annealing, the α′-martensite gradually reverts to austenite, resulting in a slow increase of austenite fraction with temperature. The samples annealed for 30 min in the range from 700-1000 °C exhibit a large and almost constant austenite content, close to the amount of austenite in the hot forged state, indicating the completion of austenite reversion.

An example OM image of the 90% cold rolled sample is shown in figure 2a and reveals a microstructure consisting of alternating ferrite and α′-martensite lamellae, with dark and bright contrast, respectively. The samples annealed at 700 and 1000 °C for 5 min (figures 2b and 2c) both exhibit a banded structure similar to the cold rolled sample. Upon further annealing for 30 min, as shown in figures 2d and 2e, the microstructures change into a chain-like morphology.

The TEM image of the cold rolled sample shown in Figure 3a reveals a well-defined lamellar structure, with an average boundary spacing of 72 nm. After annealing at 700 °C for 5 min (figure 3b), recovery in ferrite and the reverse transformation from α′-martensite to austenite are initiated. The constituent phases in the TEM images were identified by electron diffraction. Some austenite grains are indicated by the white arrows and a diffraction pattern from the circled austenite grain is shown in the insert. The transformed austenite grains have an appearance similar to that of recrystallized austenite grains reported in austenitic stainless steels [11]. As the temperature is increased up to

1000°C, as shown in figure 3c, the austenite reversion continues by consumption of the remaining α′-martensite, and some austenite layers are well formed. The austenite layers consist of equiaxed grains, some of which contain annealing twins. In the ferrite phase, on the other hand, a typical recovered structure is seen. As the annealing time is extended to 30 min, as shown in figures 3d and 3e, further structural coarsening takes place. Table 1 lists the microstructural parameters of all samples investigated in the present experiment. It is seen that DSS samples with structural sizes ranging from 72 nm to 2.1 μm were obtained.

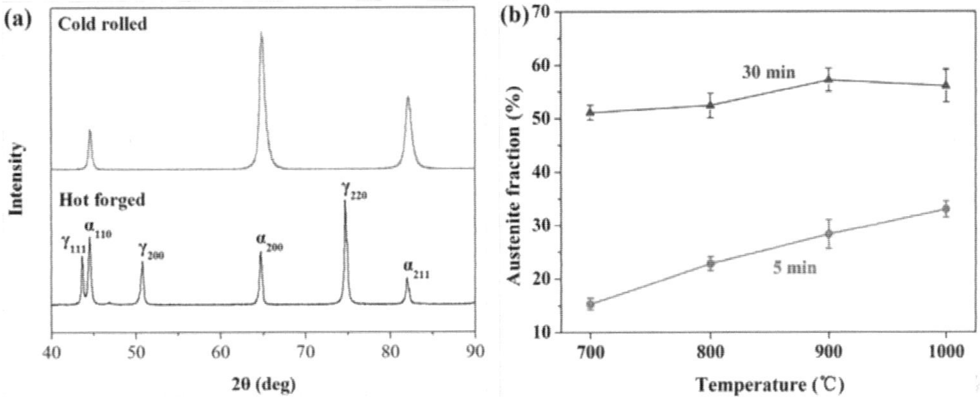

Figure 1. (a) XRD spectra of samples before and after 90% cold rolling. (b) Austenite fraction vs temperature for the cold rolled samples and then annealed at 700-1000°C for 5 and 30 min.

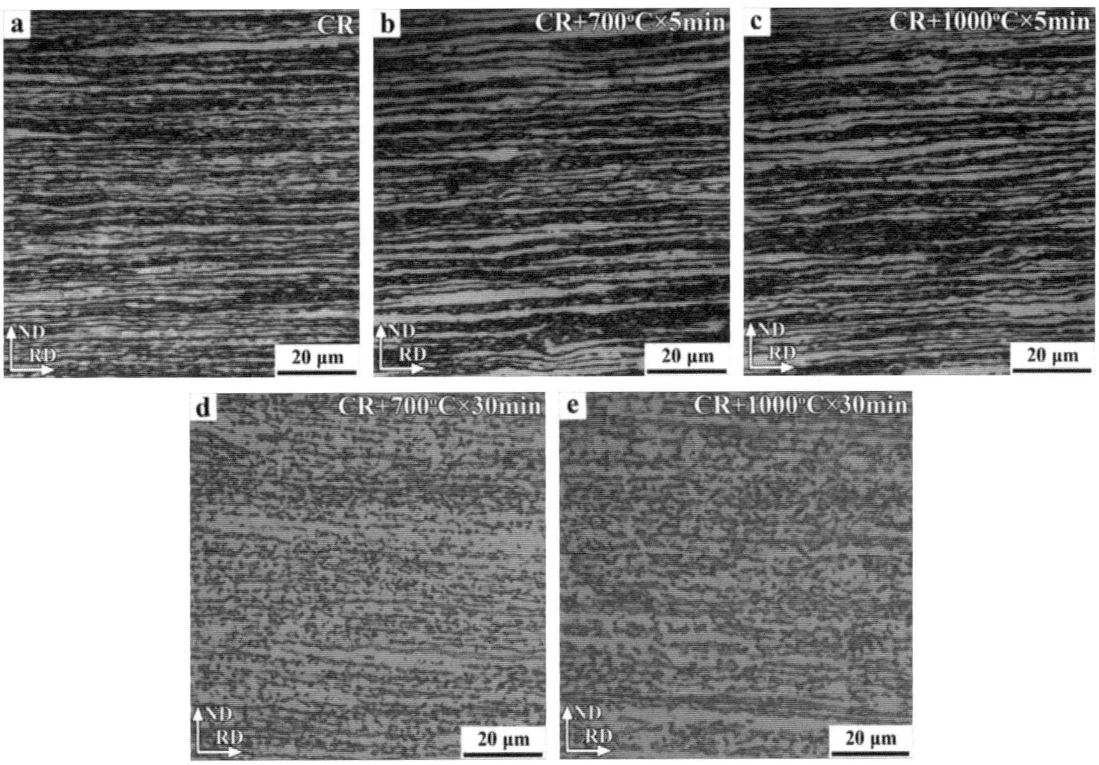

Figure 2. OM images of cold rolled and annealed samples: (a) cold rolled, (b) annealed at 700 °C for 5 min, (c) annealed at 1000 °C for 5 min, (d) annealed at 700 C for 30 min, (e) annealed at 1000 °C for 30 min.

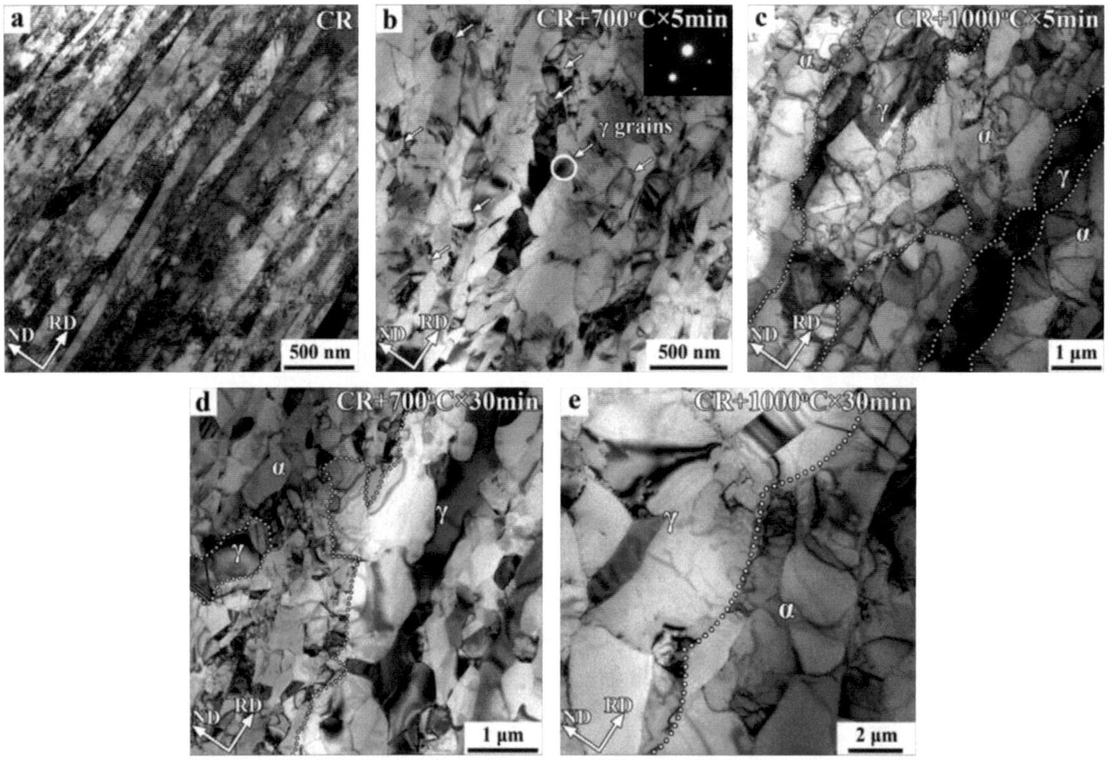

Figure 3. TEM images of cold rolled and annealed samples: (a) cold rolled, (b) annealed at 700°C for 5 min, (c) annealed at 1000°C for 5 min, (d) annealed at 700°C for 30 min, (e) annealed at 1000°C for 30 min.

Table 1. Microstructural parameters of cold rolled and annealed samples.

	$d_{\alpha\gamma}$	d_α	d_γ	f_α	f_γ
CR	72 nm	72 nm	—	100%	0%
CR+700°C×5min	169 nm	168 nm	172 nm	83.3%	16.7%
CR+800°C×5min	301 nm	294 nm	324 nm	77.2%	22.8%
CR+900°C×5min	571 nm	534 nm	665 nm	71.7%	28.3%
CR+1000°C×5min	808 nm	768 nm	890 nm	67.0%	33.0%
CR+700°C×30min	375 nm	347 nm	401 nm	48.9%	51.1%
CR+800°C×30min	723 nm	621 nm	817 nm	47.6%	52.4%
CR+900°C×30min	1.1 μm	1.1 μm	1.1 μm	42.8%	57.2%
CR+1000°C×30min	2.1 μm	1.8 μm	2.3 μm	43.9%	56.1%

3.2. Mechanical properties

Figure 4 shows the engineering stress-strain curves of the 90% cold rolled and annealed samples for volume-fraction-weighted grain sizes ($d_{\alpha\gamma}$) up to 2.1 μm. The cold rolled sample with $d_{\alpha\gamma}$ of 72 nm has a very high yield strength of about 1.3 GPa, but exhibits an ultimate tensile stress followed by necking at a very early stage of tensile deformation, resulting in a limited uniform elongation of less than 2% and a total elongation below 5%. As the volume-fraction-weighted grain size is increased to $d_{\alpha\gamma}$ = 169 nm, the yield strength rapidly decreases to 928 MPa, while the elongation shows almost no increase. An abrupt increase of ductility is observed as $d_{\alpha\gamma}$ is increased to 375 nm. The yield strength under this condition is 738MPa with an elongation of29%. As $d_{\alpha\gamma}$ is further increased the yield strength decreases gradually, but improvements in ductility are limited. A good combination of strength and ductility was therefore achieved in the sample with an average structural size of 375 nm. Under this condition the yield strength is 2.3 times higher than that of the sample with $d_{\alpha\gamma} = 2.1$ μm, while its tensile elongation (29%) is only a slightly smaller than that (33%) of the latter. Note that samples with $d_{\alpha\gamma}$ ranging from 375 to 808 nm exhibit a discontinuous yielding, characterized by a plateau (1~3% Lüders elongation) in the stress-strain curves. This was also observed in ultrafine grained single phase metals such as Al [2], Cu [3], IF steel [5] and austenitic stainless steels [11]. A continuous yielding is seen for structural sizes above 1 μm.

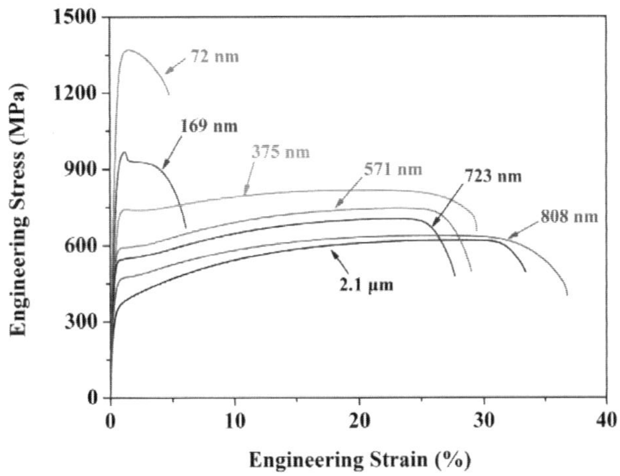

Figure 4. Engineering stress-strain curves of Fe-23Cr-8.5Ni DSS with a wide range of volume-fraction-weighted grain sizes.

It is interesting to observe that a significant increase in uniform elongation occurs at a structural scale of 375 nm. This is much finer than the scale where a similar transition is observed in single phase metals, which is often above a few micrometers. This observation can be attributed to the unique deformation characteristics of dual phase materials. For DSS, the plastic responses of individual phase to the applied strain are different from each other, resulting in a plastic strain gradient across phase interfaces [13]. This requires extra dislocations to be generated near the interfaces due to geometrical necessity, in order to make allow deformation compatibility [14, 15]. The geometrically necessary dislocations and their development with strain can lead to enhanced work hardening and tensile ductility.

4. Conclusions

Fe-23Cr-8.5Ni DSS samples with different structural sizes ranging from the nanometers to micrometers scale have been produced by cold rolling and subsequent annealing. The effect of structural size on the mechanical properties has been studied. The main results are summarized as follows:

- Cold rolling to 90% leads to a significant structural refinement down to 72 nm, which is associated with a high yield strength of about 1.3 GPa but a low tensile elongation of 5%.
- As the structural scale is increased by annealing to 375 nm, a rapid increase in elongation takes place, resulting in a good combination of yield strength and elongation, with values of 738 MPa and 29%, respectively.
- With further increase of the structural scale up to 2.1 μm, the yield strength gradually decreases, but the improvement in tensile elongation is limited.

Acknowledgments

The authors gratefully acknowledge the support from the National Natural Science Foundation of China (NSFC, Grant Nos.51471039, 51327805 and 51671039), and from the Fundamental Research Fund of Central Universities of China (Grant No. 106112017CDJPT130005).

References

[1] Huang X 2009 *Scr. Mater.* **60** 1078-82
[2] Tsuji N, Ito Y, Saito Y and Minamino Y 2002 *Scr. Mater.* **47** 893-9
[3] An H, Wu S, Zhang Z, Figueiredo R B, Gao N and Langdon T G 2012 *Scr. Mater.* **66** 227-30
[4] Li Z, Fu L, Fu B and Shan A 2013 *Mater. Lett.* **96** 1-4
[5] Gao S, Chen M C, Joshi M, Shibata A and Tsuji N 2014 J. *Mater. Sci.* **49** 6536-42
[6] Fang T H, Li W L, Tao N R and Lu K 2011 *Science* **331** 1587-90
[7] Wu X L, Yang M X, Yuan F P, Wu G L, Wei Y J, Huang X and Zhu Y T 2015 *Proc. Natl. Acad. Sci. USA* **112** 14501-5
[8] Koch C C, Scattergood R O and Murty K L 2007 *JOM* **59** 66-70
[9] Fan Z Y, Tsakiropoulos P and Miodownik A P 1992 *Mater. Sci. Technol.* **8** 922-9
[10] Lee S J, Park Y M and Lee Y K 2009 *Mater.Sci. Eng. A* **519** 32-7
[11] Saha R, Ueji R and Tsuji N 2013 *Scr. Mater.* **68** 813-6
[12] Yu C Y, Kao P W and Chang C P 2005 *Acta Mater.* **53** 4019-28
[13] Chen L, Yuan F P, Jiang P, Xie J and Wu X L 2014 *Mater. Sci. Eng. A* **618** 563-71
[14] Ashby M F 1970 *Phil.Mag.* **21** 399-424
[15] Mughrabi H 2001 *Mater. Sci. Eng. A* **317** 171-80

Effect of shot peening on the residual stress and mechanical behaviour of low-temperature and high-temperature annealed martensitic gear steel 18CrNiMo7-6

R Yang[1,2], X Zhang[3], D Mallipeddi[4], N Angelou[5], H L Toftegaard[6], Y Li[7], J Ahlström[4], L Lorentzen [3], G Wu[1,2], X Huang[3]

[1]College of Materials Science and Engineering, Chongqing University, Chongqing 400044, People′ Republic of China
[2]State Key Laboratory of Mechanical Transmission, Chongqing University, Chongqing 400044, PR China
[3]Section for Materials Science and Characterization, Department of Wind Energy, Risø Campus, Technical University of Denmark, DK-4000 Roskilde, Denmark
[4]Industrial and Materials Science, Chalmers University of Technology, Sweden
[5]Section for Test and Measurements, Department of Wind Energy, Risø Campus, Technical University of Denmark, DK-4000 Roskilde, Denmark
[6]Section for Composites and Materials Mechanics, Department of Wind Energy, Risø Campus, Technical University of Denmark, DK-4000 Roskilde, Denmark
[7]Nano Structural Materials Center, Nanjing University of Science and Technology, PR China

Email: xzha@dtu.dk

Abstract. A martensitic gear steel (18CrNiMo7-6) was annealed at 180 °C for 2h and at ~ 750 °C for 1h to design two different starting microstructures for shot peening. One maintains the original as-transformed martensite while the other contains irregular-shaped sorbite together with ferrite. These two materials were shot peened using two different peening conditions. The softer sorbite + ferrite microstructure was shot peened using 0.6 mm conditioned cut steel shots at an average speed of 25 m/s in a conventional shot peening machine, while the harder tempered martensite steel was shot peened using 1.5 mm steel shots at a speed of 50 m/s in an in-house developed shot peening machine. The shot speeds in the conventional shot peening machine were measured using an in-house lidar set-up. The microstructure of each sample was characterized by optical and scanning electron microscopy, and the mechanical properties examined by microhardness and tensile testing. The residual stresses were measured using an Xstress 3000 G2R diffractometer equipped with a Cr Kα x-ray source. The correspondence between the residual stress profile and the gradient structure produced by shot peening, and the relationship between the microstructure and strength, are analyzed and discussed.

1. Introduction

Shot peening is a process in which either spherical shots, made from cast/wrought/drawn high-carbon or stainless steel, or ceramic or glass beads/shots are shot against a target to mechanically treat the surface and to introduce compressive residual stresses parallel to the surface by means of overlapping plastic deformation indents. This process, introducing compressive residual stresses in a surface layer, is well known to improve the fatigue or stress corrosion properties, by preventing or delaying cracking, and is widely used for components and parts such as bearings, shafts and gears, in the aerospace, chemical, petroleum, automotive and power industries. Shot peening has long been applied to the root fillet area of gear teeth to improve their bending strength [1], and from 1982, has also been used to increase the pitting fatigue life of the gear tooth contact surface [2]. In 1992, high-intensity shot peening was reported to improve surface fatigue life of hardened gears [3]. However, little information has been reported about the effect of shot peening process parameters on the microstructure, and residual stress, and on their relationship with mechanical properties. Such information is required to provide a baseline for the design of more complicated and advanced properties through adjusting the parameters, including such aspects as deformation level, (sub)grain size, residual stress, retained austenite, surface morphology and roughness, chemical gradient and hardness.

In the present research, for the understanding of the relationship between microstructure, residual stress, mechanical properties and shot peening, a martensitic gear steel 18CrNiMo7-6 has been annealed at low and high temperatures to design two beginning microstructures, and traditional and high-energy shot peening have been applied.

2. Experimental methods

A 1 mm thick plate made from the martensitic gear steel 18CrNiMo7-6 (composition 0.15-0.21C, 0.17-0.35Si, 0.5-0.9Mn, 1.5-1.8Cr, 1.4-1.7Ni, 0.25-0.35Mo wt.%; balanced Fe [4]) was annealed at 180 °C for 2h and at ~ 750 °C for 1 h to obtain two different starting microstructures for shot peening. The low-temperature annealing treatment maintains the original as-transformed martensite, while the high-temperature annealing treatments results in a microstructure consisting of lath-shaped sorbite with carbides distributed in ferrite matrix and equiaxed ferrite grains. These two materials were shot peened using two different peening conditions. The high-temperature annealed steel, with a sorbite plus ferrite microstructure (HT steel), was shot peened using 0.6 mm conditioned steel shots at an average speed of 25 m/s in a conventional shot peening machine, while the low-temperature annealed steel, with martensite structure (LT steel), was shot peened using 1.5 mm steel shots at a speed of 50 m/s in an in-house developed shot peening machine, using similar shot peening mechanisms at Nanjing University of Science and Technology, China. The shot speeds in the conventional shot peening machine were measured using an in-house lidar set-up through the glass window of the peening chamber [5].

The microstructures of two samples were characterized by optical microscopy (OM) and by scanning electron microscopy (SEM). The mechanical properties were examined by microhardness measurements, using a load of 100 g with 10 s hold at the maximum load, and by tensile testing using dog-bone plate specimens (gauge length of 60 mm, thickness of 1 mm, width of 6 mm; total length of 170 mm) at a strain rate of $4.7 \times 10^{-4}\ s^{-1}$.

The residual stresses were measured using an Xstress 3000 G2R diffractometer equipped with a Cr Kα x-ray source. A standard $\sin^2(\psi)$ technique with five tilts from -45°/+45° was used for determining the stress values. The lattice deformation for the {211} α–Fe peak (diffraction angle 156.4°) was measured. X-rays were irradiated through a circular collimator of diameter 1.5 mm. The diffraction peaks were fitted with the StressTech XTronic software, using a cross-correlation method with a linear background. The stress conversion was carried out assuming an elastic modulus of E = 211 GPa and Poisson's ratio υ = 0.3. A 2.0 mm collimator was also used for one set of angles, and it was verified that collimator size did not influence the result. For measurement of subsurface stresses, an area of 3.5 mm was etched stepwise using a 3M NaCl solution as electrolyte. After every step, a stylus profilometer was used to measure the etch pit depth and geometry.

3. Results

Figure 1 shows example backscatter electron micrographs of the high-temperature and low-temperature annealed martensitic gear steel 18CrNiMo7-6. The LT steel (figure 1b) shows a similar structure to that of the original, with a characteristic martensitic microstructure. The microstructure of HT steel (figure 1a) is heterogeneous, containing both clean ferrite equiaxed grains and a micron-scale lath ferritic structure with (sub)micron-sized carbides (sorbite).

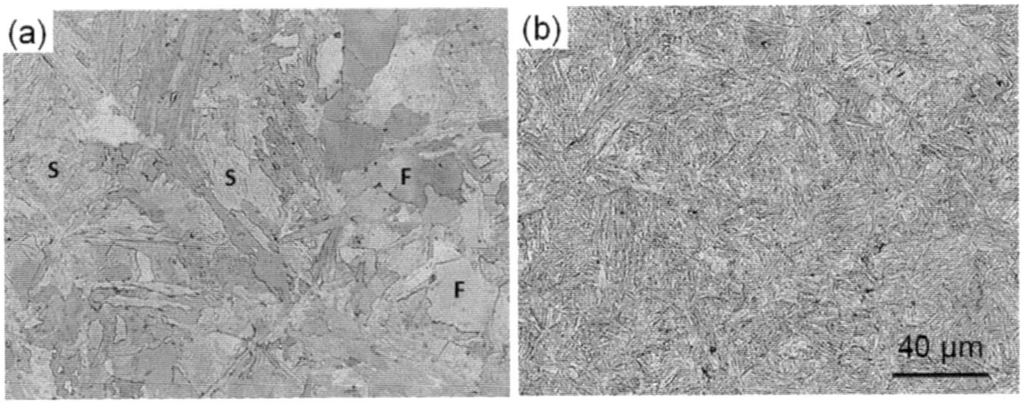

Figure 1. Backscatter electron micrographs showing the high-temperature (a) and low-temperature (b) annealed martensitic gear steel 18CrNiMo7-6. The letters "F" and "S" represent ferrite and sorbite, respectively.

Figure 2 shows the microstructure, in cross section, from the surface regions of the peened specimens. The shot peening introduces a gradient microstructure where the structure gets finer as the surface is approached. In the HT steel a lamellar spacing of ~ 59 nm is seen at the top surface. In the LT steel the gradient structure is not so obvious, due to the fine original martensite structure. A rotation of the martensite lamellar structure to be parallel to the surface is nevertheless seen, with the lamellar structural size reaching a value of 60 nm close to the top surface.

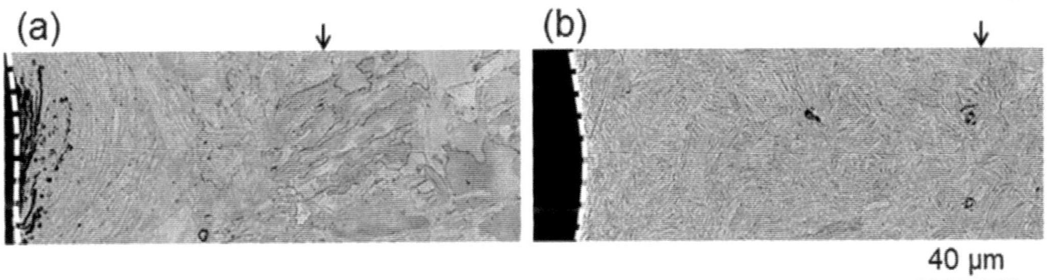

Figure 2. Backscatter electron micrographs showing the gradient structures produced in high-temperature (a) and low-temperature (b) annealed martensitic gear steel 18CrNiMo7-6 by traditional shot peening and high-energy shot peening, respectively. The dashed lines represent the surface positions and the surface direction is vertical. The vertical arrows mark the positions at 120 μm (a) and of 160 μm (b) from the shot-peened surfaces.

Figure 3 shows the microhardness measured from the cross section of the peened specimens. The HT steel has a matrix hardness around 230 kgf/mm^2, while the LT steel has a higher value ~ 480 kgf/mm^2. In the HT steel the shot peening introduces a 56% increase of hardness up to

360 kgf/mm^2 at a distance ~ 10 μm from the peened surface, with a hardened gradient structure to a depth of ~ 120 μm. For the LT steel the high-energy shot peening increases the hardness up to 570 kgf/mm^2 at a distance of 10 μm from the peened surface.

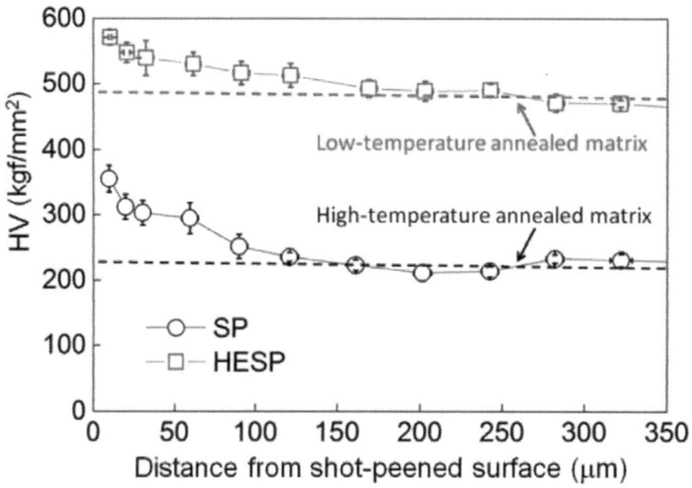

Figure 3. Microhardness as a function of distance from the shot-peened surface of the high-temperature and low-temperature annealed martensitic gear steel. The error bars represent the standard deviation of the hardness data. The dashed lines represent the matrix hardness of both steels. Letters "SP" and "HESP" represent traditional shot peening and high-energy shot peening. Note that the hardness close to the top surface is measured at a distance ~ 10 μm from the actual surfaces taken into account the hardness indent size ~ 6 μm.

Figure 4 presents the compressive residual stress profiles perpendicular to the peened surfaces. The traditional shot peening used for the HT steel introduces a maximum magnitude of compressive residual stress of around 400 MPa, at a distance of 10 μm from the surface. The high-energy shot peening used for the LT steel introduces a compressive stress of maximum magnitude ~ 560 MPa, at a distance of 160 μm from the surface. The residual stress distribution in two samples shows different trends. In the HT steel, the compressive residual stress increases slightly with distance from the surface, and then decreases reaching a small plateau of ~ 320 MPa at a depth of 120 μm from the surface. At further depths the residual stress continues to decrease. In the LT steel, the residual stress increases with increasing depth up to a maximum magnitude at a distance ~ 160 μm from the surface, after which the residual stress decreases sharply, reaching a value of near zero at a depth of around 400 μm. The depth with residual stress of ~320MPa is about 280 μm.

Table 1 shows the tensile properties of four types of specimens. The traditional shot peening, applied to the HT steel, results in an increase of the $\sigma_{0.2\%}$ flow stress by ~ 14%, but to a decrease in the UTS by 7%. The high-energy shot peening, used for the LT steel, results in increases to the $\sigma_{0.2\%}$ flow stress and to the UTS of ~ 11% and ~ 4%, respectively.

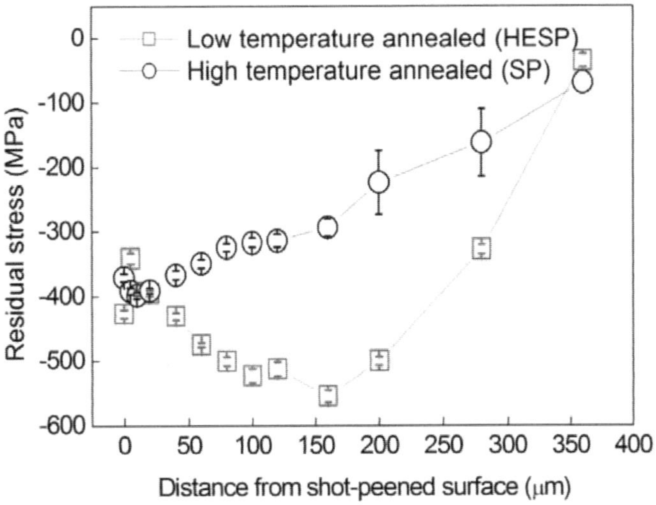

Figure 4. Residual stress as a function of the distance from shot-peened surface for the of high-temperature and low-temperature annealed martensitic gear steel 18CrNiMo7-6.

Table 1. Tensile properties of as heat-treated and shot-peened specimens

Specimen	$\sigma_{0.2\%}$ (MPa)	UTS (MPa)	UTS/$\sigma_{0.2\%}$	Total elongation (%)
HT steel	344.4	582.6	1.69	17.1
LT steel	1124	1410.5	1.25	5.7
HT steel _ SP	392	542	1.38	11.6
LT steel _ HESP	1250	1470	1.18	6.2

4. Discussion

The annealing treatments used in this study result in a sorbite + ferrite microstructure in the HT steel and a tempered martensite in the LT steel, resulting in $\sigma_{0.2\%}$ flow stress almost three times higher for the LT steel. For the HT steel, traditional shot peening results in an increase of the $\sigma_{0.2\%}$ flow stress, but leads to a decrease in the ultimate tensile strength (UTS) and in the total elongation. For the LT steel, high-energy shot peening results in an increase in the $\sigma_{0.2\%}$ flow stress and in the UTS, with a slight reduction in the total elongation. Both traditional and high-energy shot peening work-harden the microstructure and introduce gradient structures and compressive residual stresses in the HT and LT steels. The thickness of gradient structures is in good accordance with the maximum compressive residual stress magnitude (plateau): ~ 120 μm in the HT steel and ~ 160 μm in the LT steel.

It is well known that the effectiveness of shot peening depends both on the intensity and coverage. The intensity dictates the amount of energy transferred from the shot stream into the target parts or components, and is closely related to the shot size, shape, flow, shot material, hardness, velocity and impingement angle. To illustrate the difference between traditional shot peening and high-energy shot peening, we can estimate the energy of a single shot, assuming that in both cases the shots are of similar material and hardness (high carbon steel shot, HRC ~ 60) and that the only differences are the size and velocity of the shot. Such a calculation reveals that the energy of a single shot for high energy shot peening is 62.5 times higher than that for traditional shot peening, on the assumption that the energy (E) of single shot is proportional to the product of the mass (m) and the square of the velocity (v^2):

$$E \sim mv^2 = \frac{4}{3}\pi r^3 \rho \times v^2 \tag{1}$$

where ρ is the density of the shot and r is the radius of the spherical shot. This large difference may be the reason why the high-energy shot peening can still introduce a thicker deformation layer in the hard steel, in comparison with the traditional shot peening on soft steel.

The soft HT steel exhibits a strong work hardening before shot peening (69%, from the data in table 1), but after the peening process, the work hardening capability is reduced in the surface layer and the strain hardening ratio (UTS/$\sigma_{0.2\%}$ flow stress) decreases considerably. During plastic deformation in the tensile test, the residual stress magnitudes decreases [6] as the plastic flow increases but the deformed surface microstructure remains. The reduced strain hardening capability in the surface leads to earlier localization of the fracture, resulting in a lower UTS and lower total elongation. In the hard LT steel exposed to high energy shot peening, the degree of plastic deformation in the surface is much lower, but because of the higher energy and larger indents, plastic deformation extends to larger depth [7]. The maximum shear strain occurring beneath the surface has its correspondence in the maximum in compressive residual stress observed at 160 μm (around 15 μm in the SP). The tensile straining of the high-energy shot peened hard LT steel also reduces the residual stress field on plastic flow, but here the surface layer retains sufficient work hardening capability. Localization of plastic strains leading to necking is delayed, giving higher UTS and less reduced total elongation. The HESP treatment thus improves the performance in monotonic uniaxial straining.

For the further quantitative understanding of the shot peening process on mechanical properties, such as the tensile properties examined in the present experiments, the surface integrity, including surface morphology, roughness and texture needs should be taken into account in addition to the gradient structure and the residual stress profile. More detailed information on the microstructure and (sub)surface texture is being acquired by using electron backscatter diffraction (EBSD) and transmission electron microscopy (TEM), described in our recent studies [8-12] on microscale to nanoscale metallic structures with strength ranging from several hundred MPa up to ~ 5 GPa. These additional studies will constitute a forthcoming paper and will further the understanding of the shot peening process and its effect on the microstructure, texture, surface morphology, roughness, residual stress and mechanical properties including fatigue.

5. Conclusion

A martensitic gear steel 18CrNiMo7-6 was annealed at 180 °C for 2h and at ~ 750 °C for 1h to allow the design of two different starting microstructures for a study of shot peening. One microstructure consists of tempered martensite (LT steel), while the other consists of a mixture of sorbite and equiaxed ferrite (HT steel). Traditional shot peening and high-energy shot peening have been applied to the HT steel and the LT steel, respectively. The conclusions are as follows.

- Both traditional and high-energy shot peening introduce a gradient microstructure in the HT and LT steels.
- The depth of the developed gradient structures and maximum compressive residual stress (plateau) introduced by shot peening are in good accordance: ~ 120 μm in the soft steel and ~ 160 μm in the hard steel.
- The traditional shot peening applied to the softer HT steel results in an increase of the $\sigma_{0.2\%}$ flow stress and a decrease the UTS. In contrast the high-energy shot peening used for the harder LT steel results in increases both in the $\sigma_{0.2\%}$ flow stress and in the UTS.

Acknowledgements

The authors thank Mr. Huilong Lu (Nanjing High Speed Gear Manufacturing Co. Ltd., PR China) for the supply of the raw gear steel 18CrNiMo7-6 used in this investigation. R Yang and G Wu appreciate the financial support from the Natural Science Foundation of China (Grants No. 51471039, 51327805) and from the State Key Research and Development Program (2016YFB0700401, 2016YFB0700403).

References

[1] Daly J J 1977 *Machine Design* May 12
[2] Townsend D P and Zaretsky E V 1982 *NASA technical paper* 2047
[3] Townsend D P 1992 NASA *Technical Memorandum* 105678
[4] Yang R, Wu G L, Zhang X, Fu W T and Huang X 2017 *38th Risø International Symposium on Materials Science: Advanced Metallic Materials by Microstructural Design (IOP Conference Series: Materials Science and Engineering)* DTU Risø campus
[5] Angelou N, Zhang X, Sjöholm M, Lorentzen L and Huang X 2017 *Shot peening speed measurement using lidar technology* DTU Wind Energy
[6] Dalaei B, Karlsson B and Svensson L-E 2011 *Materials Science and Engineering A* **528** 1008
[7] Johnson K L 1985 *Contact Mechanics* Cambridge University Press
[8] Zhang X, Hansen N, Gao Y and Huang X 2012 *Acta Mater.* **60** 5933
[9] Fan G H, Wang Q W, Du Y, Geng L, Hu W, Zhang X and Huang Y D 2014 *Materials Science and Engineering A* **590** 318
[10] Zhang X, Hansen N, Godfrey A and Huang X 2016 *Acta Mater.* **114**, 176
[11] Chen S-L, HU J, Zhang X, Dong H and Cao W 2015 *Journal of Iron and Steel Research International* **22** 1126
[12] Zhu K N, Godfrey A, Hansen N and Zhang X D 2017 *Materials & Design* **117** 95

IOP Conf. Series: Materials Science and Engineering **219** (2017) 012047 doi:10.1088/1757-899X/219/1/012047

Gradient microstructure and microhardness in a nitrided 18CrNiMo7-6 gear steel

R Yang[1, 2], G L Wu[1], X Zhang[3], W T Fu[4] and X Huang[3]
[1] College of Materials Science and Engineering, Chongqing University, Chongqing 400044, China
[2] State Key Laboratory of Mechanical Transmission, Chongqing University, Chongqing 400044, China
[3] Section for Materials Science and Advanced Characterization, Department of Wind Energy, Technical University of Denmark, Risø Campus, DK-4000 Roskilde, Denmark
[4] State Key Laboratory of Metastable Materials Science and Technology, College of Materials Science and Engineering, Yanshan University, Qinhuangdao 066004, China

Email: ranyang@dtu.dk

Abstract. A commercial gear steel (18CrNiMo7-6) containing a tempered martensite structure was nitrided using a pressurized gas nitriding process under a pressure of 5 atm at 530 °C for 5 hours. The mechanical properties and microstructure of the nitrided sample were characterized by Vickers hardness measurements, X-ray diffraction, and backscatter electron imaging in a scanning electron microscope. A micro-hardness gradient was identified over a distance of 500 μm with hardness values of 900 HV at the top surface and 300 HV in the core. This micro-hardness gradient corresponds to a gradient in the microstructure that changes from a nitride compound layer at the top surface (~ 20 μm thick) to a diffusion zone with a decreasing nitrogen concentration and precipitate density with distance from the surface, finally reaching the core matrix layer with a recovered martensite structure.

1. Introduction

Gear steels are required to provide an optimum combination of mechanical properties and the lowest possible cost. Some of the desired mechanical properties for gears are high wear resistance, fatigue strength, and shear strength. These properties demand a hard case on the gear surface and a tough core. A gradient in hardness decreasing from the surface to the center layer is a good option. There are mainly two types of processing to produce such hardness gradients: surface mechanical treatment, such as surface mechanical attrition treatment (SMAT) [1], surface mechanical rolling treatment (SMRT) [2], and shot peening [3], and surface thermo-chemical treatments [4,5], such as carburizing and nitriding.

Because of their low cost, easy operation and large achievable case-hardening depth, surface thermo-chemical treatments are commonly used for high quality gears to form a hard case on the gear surface. Nitriding is often used for components particularly susceptible to dimensional distortion [6,7]. Gaseous nitriding, as compared to plasma and salt-bath process variants, has the eminent advantage of low cost and allows precise control of the microstructure via the nitriding potential [8,9]. Through the use of a nitriding treatment, different layers can be produced from the surface to interior, namely a compound layer at the surface and a diffusion zone beneath, as schematically shown in figure 1. The compound layer can be beneficial for the resistance against wear and corrosion. The diffusion zone brings about a strong increase in fatigue resistance and wear resistance [10].

In this study, a nitriding treatment is applied to a typical gear steel with an aim to investigate the microstructure and hardness gradient in the nitrided sample. To accelerate the nitriding kinetics, a newly developed nitriding process, pressurized gas nitriding (PGN) [5], is used.

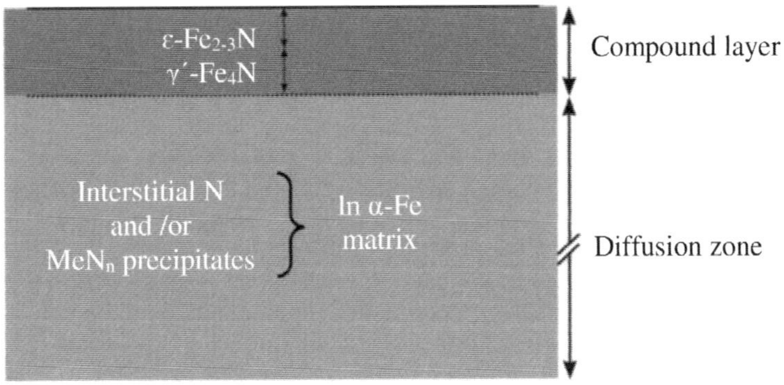

Figure 1. Schematic illustration of the surface region of a nitrided steel.

2. Experimental

A typical gear steel, 18CrNiMo7-6, was used in this investigation. Its chemical composition is listed in table 1. The starting microstructure for nitriding was tempered martensite, which was obtained by the following heat treatment: (i) solid solution treatment at 825 °C for 1 h followed by oil cooling; (ii) tempering at 180 °C for 2 h and then air cooling. The PGN treatment was carried out at 530 °C for 5 hours under flowing NH_3 and H_2 with a nitriding potential of r_N=0.26 atm$^{-½}$ under a gas pressure of 5 atm. The dimensions of the sample for nitriding were 190 mm long, 13 mm wide and 1 mm thick.

Table 1. Chemical composition of 18CrNiMo7-6 steel (wt.%).

	C	Si	Mn	S	Cr	Ni	Mo	H	Fe
18CrNiMo 7-6	0.15-0.21	0.17-0.35	0.50-0.90	≤0.015	1.50-1.80	1.40-1.70	0.25-0.35	≤2.0ppm	Balanced

Optical microscopy (OM) and backscatter electron (BSE) imaging in a scanning electron microscope (SEM) were used to observe the microstructure. Specimens for OM and SEM-BSE observations were etched by 3% nital after polishing.

The phase composition of the compound layer, as well the diffusion zone and the core were examined by X-ray diffraction (XRD) using Cu-K_α radiation (λ = 1.5406 Å) operated at 40 kV and 40 mA. XRD patterns were recorded with a step size of 0.009° and a step duration of 1 s over an angular range of 30°-90°.

Microhardness profiles as a function of distance from the top surface were measured on cross-sections of the sample using a Vickers microhardness tester with a load of 100 g. The case depth was determined according to standard procedures [11] by measuring the distance from the surface to the point where the hardness is 50 HV higher than the hardness of the core substrate.

3. Results

3.1. Microstructure gradient

Figure 2 shows the tempered martensitic microstructure prior to nitriding. The hardness of this tempered martensite is 442 HV.

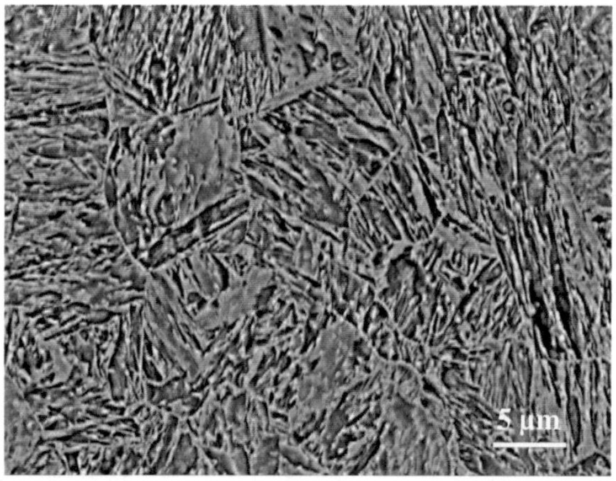

Figure 2. SEM-BSE image showing the tempered martensite prior to nitriding.

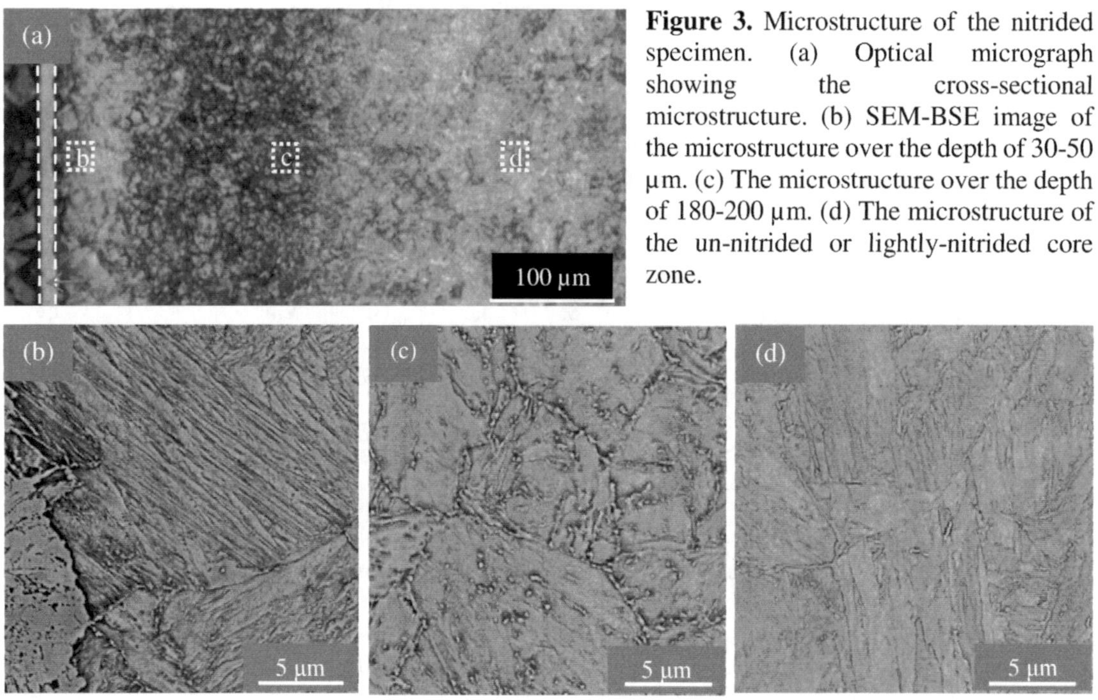

Figure 3. Microstructure of the nitrided specimen. (a) Optical micrograph showing the cross-sectional microstructure. (b) SEM-BSE image of the microstructure over the depth of 30-50 μm. (c) The microstructure over the depth of 180-200 μm. (d) The microstructure of the un-nitrided or lightly-nitrided core zone.

Figure 3 shows the microstructure of a nitrided sample. A thin compound layer of about 20 μm in thickness, indicated by two parallel white dashed lines in the figure, was formed on the top surface (figure 3a). The composition of this compound layer is dependent on the nitriding processing parameters such as temperature and nitriding potential. For the current nitrided sample, as indicated in figure 4, only one series of X-ray diffraction peaks are obtained from the compound layer, which corresponds to the γ′-Fe_4N iron nitride phase. Accordingly the compound layer is single phase γ′-Fe_4N.

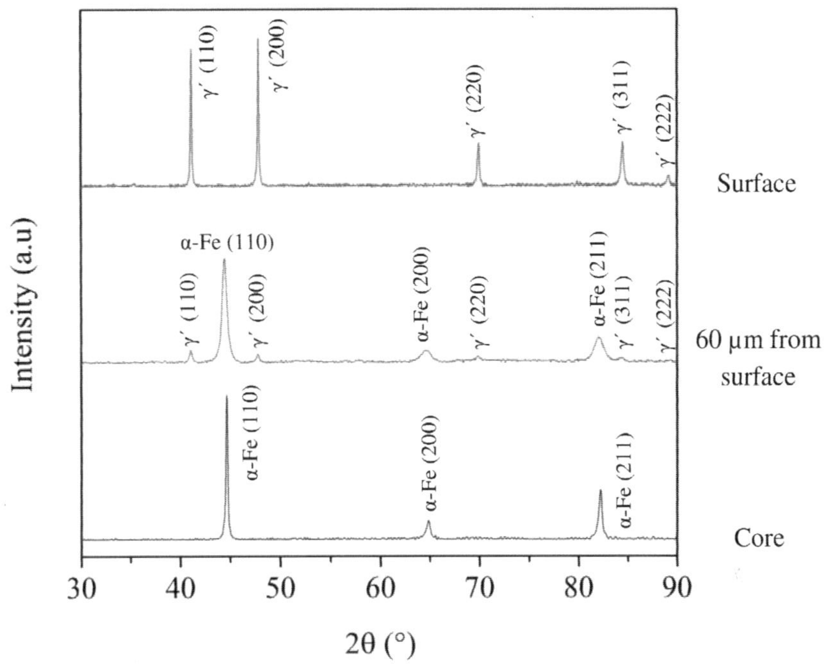

Figure 4. XRD patterns taken from the surface, 60 μm from surface and in the core of the nitrided sample. Only one series of diffraction peaks are seen from the surface data, corresponding to the γ′-Fe_4N iron nitride phase. The diffraction patterns of the core correspond to α-Fe (tempered martensite). At the depth of about 60 μm below the surface, diffraction peaks both from γ′-Fe_4N and α-Fe (tempered martensite) are identified.

Beneath the compound layer there is a diffusion zone (20-350 μm below the surface). Figure 3b gives a representative image of the microstructure over the depth of 30-150 μm below the surface, which is a recovered martensite structure containing dispersed precipitates of γ′-Fe_4N, as verified by the XRD diffraction peaks obtained at a depth of about 60 μm below the surface (see figure 4). In this diffusion zone, the diffraction peaks from the α-Fe (tempered martensite) are wider than those obtained from the core. This means a higher concentration of nitrogen in the interstitial sites of iron in the diffusion zone. At deeper locations (190-350 μm below the surface), in figure 3c, there exists a carbon-rich zone. The presence of nitrogen causes the chromium-containing carbides to dissolve allowing the chromium to react with nitrogen because of the affinity of chromium with nitrogen. The carbon is then free to diffuse to sites with lower chemical potential, for example towards to the surface and the nitriding front, leading to the decarburization of the nitrided case, and to the formation of a carbon-rich zone in the nitriding front as well as to the formation of grain boundary phases [12,13]. The precipitation particles can be seen at the boundaries in figure 3c.

In the un-nitrided or lightly nitrided core zone (350 µm below the surface), the martensitic microstructure was coarsened during the nitriding process, as shown figure 3d.

3.2. Hardness gradient
The Vickers microhardness profile from surface to center direction is shown in figure 5. The hardness shows a maximum (900 HV) near the top surface and decreases gradually toward the sample core. The surface hardness corresponds to a hardness increase of about 600 HV or an enhancement of 200% compared with the core hardness value (300 HV). Note that the core hardness is lower than that prior to nitriding (442 HV). The hardness-depth profile shows a sharp transition from the compound layer to the diffusion zone. The case depth was identified to be about 370 µm as marked by the vertical dashed line in figure 5.

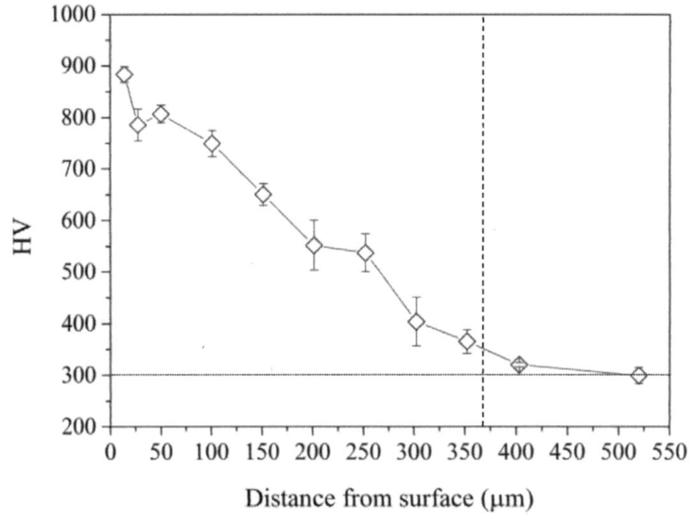

Figure 5. Variation in Vickers microhardness as a function of depth from the surface. The horizontal dotted line indicates the core hardness of 300 HV. The case depth is about 370 µm as marked by the vertical dashed line.

4. Discussion
The nitrided gear steel shows a gradient in microhardness, decreasing from the surface to the center. At the same time there is also a gradient in microstructure along the depth. The microhardness decreases corresponding to the change of microstructure.

Case-hardening of the gear steel 18CrNiMo7-6 is most commonly achieved by carburization [14,15]. In this work, nitriding by the PGN process was applied to create a hardened case. The results show that a typical nitrided layer structure is formed consisting of a compound layer adjacent the surface and diffusion zone beneath this layer.

The case depth produced in the present 18CrNiMo7-6 gear steel is similar to that obtained in a typical nitrided steel (38CrMoAlA) nitrided by PGN using the same nitriding equipment and similar nitriding conditions [5]. It can be concluded that PGN is also feasible method for the generation of case-hardened surfaces for the 18CrNiMo7-6 gear steel with a high efficiency. However for large scale gears carrying high loads, a much thicker case depth is normally desired. In the future efforts to increase the case-hardening depth for this gear steel will be made by varying the nitriding potential, the temperature and the time in the PGN process.

It should be pointed out that the hard and brittle compound layer may spall off in service and therefore is not desirable for gear applications. It is possible to perform the PGN process so that the formation of a compound layer can be eliminated [5] by controlling the nitriding potential. Furthermore, additional plastic deformation after PGN can be used to modify the microstructure of the hardened case and enhance the wear resistance [5]. An appropriate degree of additional plastic deformation is also expected to work harden the core, restoring some of the strength lost as a result of the nitriding process, and as such this possibility also requires further investigation.

5. Conclusions

A gear steel (18CrNiMo7-6) with tempered martensite structure was nitrided using a pressurized gaseous nitriding process. After nitriding, three different layers were observed as expected. The top surface layer with a hardness of about 900 HV is a compound layer consisting of the γ'-Fe_4N phase. Below the compound layer, there is a diffusion zone where the nitrogen concentration and precipitate density decrease with increasing depth, resulting in a gradual decrease of hardness. The core part of the nitrided sample is non-nitrided or lightly nitride, but tempered at the nitriding temperature, causing a decrease in hardness compared with the tempered martensite prior to nitriding. A case depth of about 370 μm was obtained for the conditions used in this study.

Acknowledgments

This work was financially supported by the Natural Science Foundation of China (Grants No. 51471039, 51327805) and the State Key Research and Development Program (2016YFB0700401, 2016YFB0700403). The authors would like to acknowledge Huilong Lu and Haodong Gao (Nanjing High Speed Gear Manufacturing Co. Ltd.) for supplying the gear steel used in the present study.

References

[1] Wang Z B, Lu J, Lu K 2006 *Surf. Coat. Technol.* **201** 2796
[2] Huang H W, Wang Z B, Lu J and Lu K 2015 *Acta Mater.* **87** 150
[3] Zhang X D, Hansen N, Gao Y K, Huang X 2012 *Acta Mater.* **20** 5933
[4] Wang Z B, Lu J, Lu K 2005 *Acta Mater.* **53** 2081
[5] Wang B, Lv Z Q, Zhou Z A, Sun S H, Huang X and Fu W T 2015 *36th Risø int. symp. on materials science* (Risø, Denmark, 7-11 September 2015) vol 89 p 1
[6] Hassani-Gangaraj S M, Moridi A, Guagliano M, Ghidini A and Boniardi M 2014 *Int. J. Fatigue* **62** 67
[7] Basu A, Dutta Majumdar J, Alphonsa J, Mukherjee S and Manna I 2008 *Mater. Lett.* **62** 3117
[8] Hernandez M, Staia M H and Puchi-Cabrera E S 2008 *Surf. Coat. Technol.* **202** 1935
[9] Hosmani S S, Schacherl R E and Mittemeijer E J 2005 *Mater. Sci. Technol.* **21** 113
[10] Hosmani S S, Kuppusami P and Goyal R K 2014 *An Introduction to Surface Alloying of Metals* ed Davim J P (Springer New Delhi Heidelberg New York Dordrecht London) chapter 2 pp 29
[11] JIS G0562 Standard 1995 *Surface Engineer* **11** 57
[12] Mridha S and Jack D H 1982 *Metal Science* **16** 398
[13] Sun Y and Bell T 1991 *Mater. Sci. Eng.*, A **140** 419
[14] Sorsaa A, Leiviskä K, Santa-aho S and Lepistö T 2012 *NDT and E Int.* **46** 100
[15] Feng K, Zhu B Z, Guo Q L, Fang J and Zhang W 2010 *Heat Treat. Met.* **35** 63

IOP Conf. Series: Materials Science and Engineering **219** (2017) 012048 doi:10.1088/1757-899X/219/1/012048

A gradient surface produced by combined electroplating and incremental frictional sliding

T Yu[1], C Hong[1], K Kitamura[2], K Tomatsu[3], A Taniyama[1,2], X Huang[1], N Hansen[1]

[1] Section for Materials Science and Advanced Characterization, Department of Wind Energy, Risø Campus, Technical University of Denmark, DK-4000, Roskilde, Denmark
[2] Advanced Technology Research Laboratories, Nippon Steel & Sumitomo Metal Corporation, 1-8 Fuso-Cho, Amagasaki, Hyogo 660-0891, Japan
[3] Steel Research Laboratories, Nippon Steel & Sumitomo Metal Corporation, 1-8 Fuso-Cho, Amagasaki, Hyogo 660-0891, Japan

Email: tiyu@dtu.dk

Abstract. A Cu plate was first electroplated with a Ni layer, with a thickness controlled to be between 1 and 2 μm. The coated surface was then deformed by incremental frictional sliding with liquid nitrogen cooling. The combined treatment led to a multifunctional surface with a gradient in strain, chemical content, microstructure, and hardness. The chemical profile was measured by glow-discharge optical emission spectroscopy, showing diffusion of Ni into the heavily deformed Cu layer to a depth of about 40 μm. The microstructure and hardness were characterized and compared with a similarly processed Cu plate without Ni coating, showing a strong effect of the coated layer on the deformation. The experimental results are followed by an analysis of strengthening mechanisms and a discussion of the applicability of the new technique for increasing the durability and lifetime of components exposed to friction and wear, e.g. in wind turbines.

1. Introduction

The surface properties of metallic components are critical in determining their durability and lifetime under various conditions, e.g. friction, wear and corrosion. A number of techniques have been developed to improve the surface properties of metallic materials by modifying their chemical compositions and/or microstructures [1-3]. Electroplating is one example widely used to coat one metal by another, for example in order to protect the component against chemical corrosion [4]. Recently, a new technique, namely incremental frictional sliding (IFS), which is able to produce a gradient nanostructured surface layer in a flat sample, has been developed [5]. The resulting microstructural gradient at the surface is known to be effective in enhancing the strength and lifetime of metallic components.

In the present work, we combine electroplating and IFS to produce a multifunctional surface, with a gradient in strain, chemical content, microstructure, and hardness. We choose to coat Ni on a Cu plate since they are completely miscible in the solid state, so a good bonding at the interface is expected. The coated and deformed samples can therefore be used as a model system to investigate the strengthening mechanisms and to evaluate the new combined technique.

2. Experimental details

A phosphorus-deoxidized Cu plate (figure 1), 100 mm in diameter and 5 mm in thickness, was ground on one surface and then annealed at 600 °C for 2 h to obtain a fully recrystallized state (Vickers hardness 46 ± 3). The main impurities are 0.007Zn-0.024P-0.002Fe-0.002Ni (wt%) as measured by spark optical emission spectroscopy (OES). The central part (80 mm in diameter) of the ground surface was etched for 20 s in a solution containing H_2SO_4, HNO_3 and HCl to remove the oxide layer and to enhance bonding, during which the edge and the other side of the surface were protected by a Lacomit varnish. Ni was then electrodeposited on the etched surface of the Cu plate. A duration of 6 minutes with a current of 0.5 A at 55 °C (cathode current density 1 A/dm^2) led to a thickness between 1 and 2 μm (slightly thinner in the center and thicker in the periphery). The water-based solution (pH = 4.5) contained $NiSO_4 \cdot 6H_2O$ (310 g/L), $NiCl_2 \cdot 6H_2O$ (45 g/L) and H_3BO_3 (37 g/L).

The electrodeposited surface of the Cu plate was then deformed by IFS close to liquid nitrogen temperature for 50 minutes at 200 rpm (figure 1c). To maintain a low temperature, the processing was carried out in 25 interrupted sessions. For each session, the plate was first submerged in liquid nitrogen and then processed for a relatively short duration of 2 minutes without further cooling. During each session, the load (about 100 N) along the normal direction (ND) was moved manually back and forth in the radial direction for about 100 times while the plate was rotating [5]. After deformation a bulging of the entire disk was visible, due to the extension of the processed surface along the radial direction, resulting in a bending of the disk.

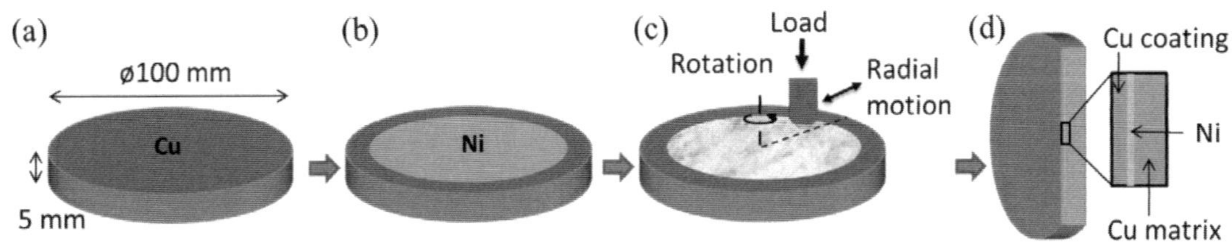

Figure 1. Schematic of the experimental procedure. (a) Geometry of the Cu plate, (b) electroplating of Ni, (c) deformation by IFS, and (d) coating of protective Cu on the processed surface. The marked and magnified area in (d) indicates the area for microstructural and hardness tests.

Microstructural observations and hardness tests of the processed plate were carried out on the tangential section (perpendicular to the radial direction) containing the sliding direction (SD) and the ND, at locations between 20 and 35 mm from the center of the plate (figure 1d). For preparation of the tangential section samples, the processed surface was protected by electroplating of Cu, which was done at room temperature for 25 hours (cathode current density 3 A/dm^2) using a solution containing $CuSO_4 \cdot 5H_2O$ (72.7 g/L), H_2SO_4 (225 g/L) and NaCl (98.7 mg/L). Tangential section samples were prepared by mechanical polishing, with a colloidal silica suspension (0.04 μm) used for the final polishing step. The microstructure was investigated using both electron channeling contrast (ECC) and electron backscatter diffraction (EBSD) in a Zeiss Supra 35 scanning electron microscope equipped with a field emission gun. For EBSD analysis, a step size of 50 nm and a cutoff misorientation angle of 2° were used. Vickers hardness tests were carried out on a Struers DuraScan-70, using a load of 245 mN. The Vickers hardness has a high reliability but as the indent diagonal was between 16 and 26 μm in the current case, nanoindentation tests were therefore also carried out in a Hysitron TI 950 TriboIndenter with a maximum load of 2 mN in order to probe the hardness gradient at the top surface.

In order to determine the chemical gradient along the depth, the processed sample surface (without Cu protective coating) was measured by glow-discharge OES in a Horiba GD-Profiler HR.

3. Results

The combination of electroplating and IFS led to a structural refinement of the processed surface layer. Figure 2 shows examples of ECC observations at the tangential section at different magnifications. The deformation is highly heterogeneous, and the thickness of the heavily deformed zone varies between 5 and 80 μm, below which the initial grain boundaries are discernible. Typically, the top layer of about 40 μm is heavily deformed (figure 2a). This top layer shows a finely spaced lamellar structure, and the extended lamellar boundaries are inclined on average at 30° to the surface, with boundary spacings varying from a few tens of nanometers to a couple of hundred nanometers.

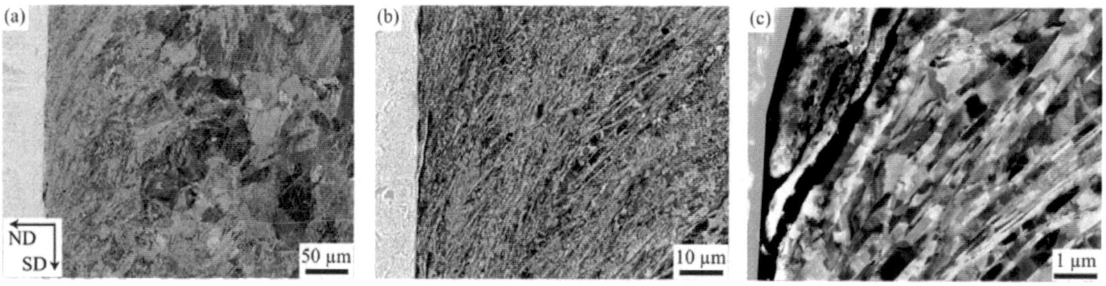

Figure 2. ECC observations at the tangential section of the processed sample at different magnifications. The light shaded area in the left part of each micrograph corresponds to the Cu protective coating.

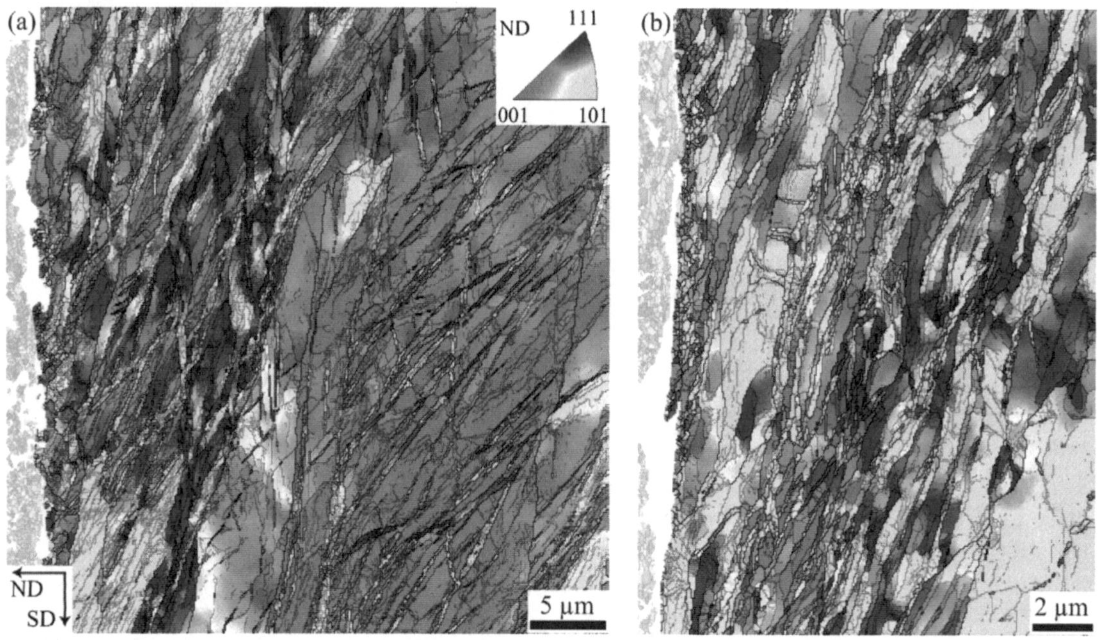

Figure 3. EBSD observations at two different locations at the tangential section of the processed surface layer. The colour is coded according to the orientation of the ND, and the Cu protective layer at the surface is shadowed. High angle boundaries (>15°) are shown by thick lines and low angle boundaries (2-15°) are shown by thin lines.

EBSD investigations were carried out in order to study the details of the nanostructured top surface layer, and two examples are shown in figure 3. The fraction of high angle boundaries (>15°) was found to be around 50%, and their distribution was highly heterogeneous. The deformation structure depended strongly on the initial grain orientation. For some grains, well-defined lamellar structures are developed with a high fraction of high angle boundaries, whereas for others, the structure is relatively coarse and contains nanoscale twins. For the top layer of 40 μm, the average boundary spacing measured along the ND in regions of well-defined lamellar structure is about 180 nm. However, the real lamellar boundary spacing is smaller since they are not perpendicular to the ND and many boundaries may be undetected due to the limited spatial and angular resolution of the EBSD technique.

The strain at the topmost layer of 5 μm is highest. It can be seen from figure 2 that the surface has been ploughed and folded. The ploughed volume has a thickness of 3-5 μm, and this ploughing and folding occurred approximately every 30-40 μm along the SD. The ploughing and folding can be best observed from the distribution of the coated Ni layer by energy dispersive X-ray spectroscopy (EDS). Figure 4 shows the morphology of the Ni layer before and after surface deformation. Before deformation, there is a uniform Ni layer on the top of the Cu plate (figure 4a,b); after deformation, the Ni layer is extended and thus overlapped (as a result of the folding) in some places (figure 4c,d).

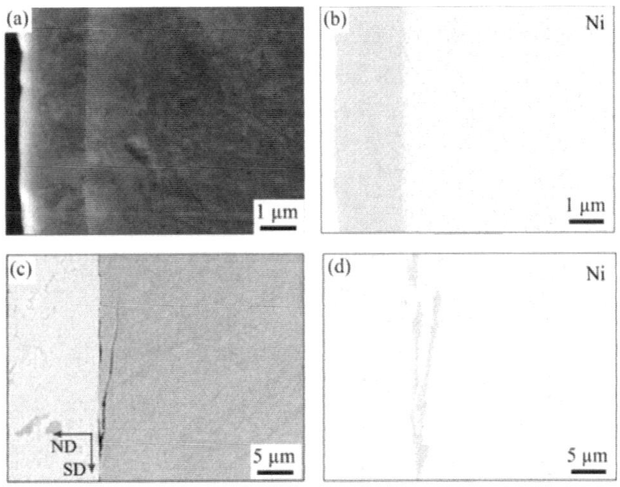

Figure 4. Morphology and Ni distribution observed in the tangential section of the surface layer before IFS (a,b) and after IFS (c,d). Note that there is no Cu protective coating in (a), whereas the Cu protective coating in the left part of (c) is seen in light grey. Note the scale bars are different between (a,b) and (c,d).

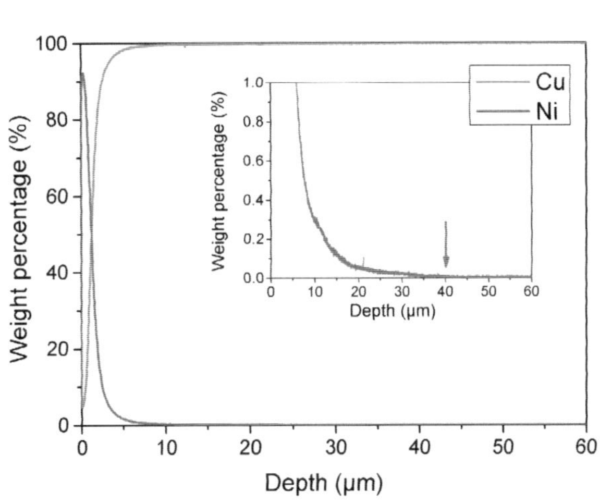

Figure 5. Depth profile revealing diffusion of Ni into the heavily deformed Cu matrix.

EDS is able to differentiate the Ni layer from the Cu matrix, but it cannot accurately detect the amount of Ni diffused into the Cu matrix. The diffusion of Ni was characterized instead by glow-discharge OES, as shown in figure 5. The result shows that on average Ni has diffused to a depth of 40 μm through mechanical alloying, in agreement with the thickness of the heavily deformed surface layer.

The hardness profile of the processed plate has been characterized by Vickers hardness and nanoindentation tests as shown in figure 6 (data in grey), where the Vickers hardness number has been converted to GPa by multiplying 0.0098, to be consistent with a previous study [5]. The hardness decreases gradually from the surface to the bulk interior. Nanoindentation tests resulted in higher hardness values than Vickers hardness tests due to both indentation size effects and different calculation algorithm [6], but the trend is nevertheless similar. The highest hardness (2.8±0.2 GPa) was recorded 1 μm away from the surface by nanoindentation tests, probably due to the very fine microstructure and a mixture of Ni and Cu. The Vickers hardness at the thickness center (0.70±0.02 GPa) has also increased from the fully recrystallized state (0.45±0.03 GPa) as the entire plate was bent during deformation.

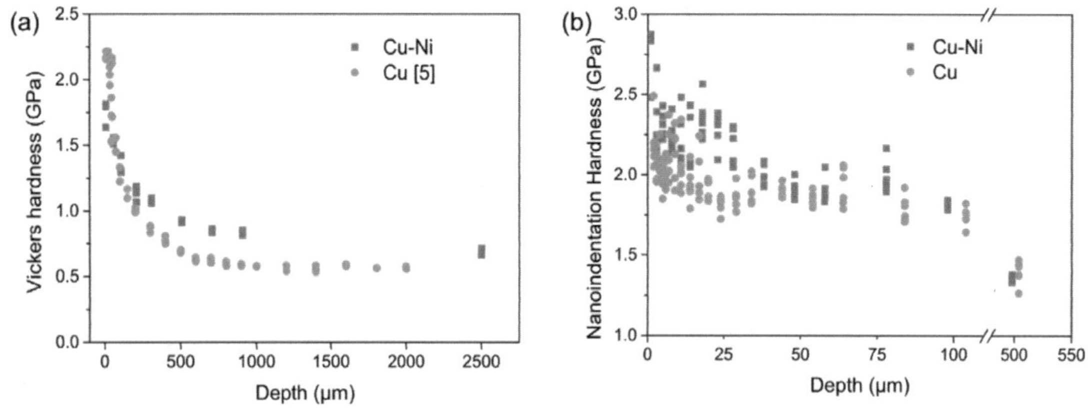

Figure 6. Vickers hardness (a) and nanoindentation hardness (b) profiles of two IFS processed Cu plates, one with Ni coating (Cu-Ni) and the other without Ni coating (Cu).

4. Discussion

In a previous study [5], a similar Cu plate without Ni coating was deformed by IFS under the same experimental conditions. Without coating of a hard Ni layer, the surface layer (about 100 μm thick) underwent a larger strain, the plastic flow was more parallel to the surface, the deformation was more homogeneous, and the average boundary spacing of the surface layer was smaller. For comparison, the Vickers hardness profile of the previous Cu plate [5] is also shown in figure 6a. The Vickers hardness data are consistent with the microstructural observations. A finer microstructure of the surface layer in the previous Cu plate led to higher hardness of the surface layer in that plate, whereas the bending of the current Cu-Ni plate led to higher hardness of the interior volume compared to that in the Cu plate. In order to further compare the surface layer, nanoindentation tests have also been carried out on the previous Cu plate as shown in figure 6b. It is clear that for the top 40 μm, the hardness of the current Cu-Ni plate is higher than that of the previous Cu plate, which might be related to the heterogeneity of the microstructure in the current Cu-Ni plate as nanoindentation tests were carried out in a very small area.

The results show that the Ni coating has a profound effect on the subsequent surface deformation by IFS. The Ni has a nanocrystalline structure and is much stronger than the fully recrystallized Cu. Therefore during deformation, the Cu matrix is subjected to a very different stress condition with a smaller penetration depth of the steel ball of the loading tool, compared to the situation without Ni coating. In the current case, compressive deformation was increased while shear deformation was decreased, leading to a much more inhomogeneous microstructure. The enhanced compression of the surface layer resulted in the significant bending of the plate and a high residual stress in the bent plate; whereas the reduced shear deformation led to less structural refinement in the surface layer. Moreover, the plowing occurred at a much smaller length scale (5 μm thick compared to 50 μm in the case without Ni coating), and therefore the processed surface was smoother.

Ni has diffused into the heavily deformed Cu layer to a depth of about 40 μm (see figure 5). The diffusion is assumed to have occurred during deformation, i.e. through mechanical alloying. During deformation at liquid nitrogen temperature, a high concentration of excess vacancies may also be present, which could enhance diffusion rates. More importantly, a high density of dislocations were created and forced to move, leading to long-distance travel of Ni atoms associated with moving dislocations [7]. On the other hand, thermally activated diffusion through the Cu lattice before deformation and through boundaries and dislocations after deformation are negligible during storage at room temperature.

The diffusion of Ni in to the Cu gives rise to a solute strengthening. However, the direct strengthening contribution is typically considered to be small, and therefore the higher hardness measured at the top surface layer in the Ni-coated plate as shown in figure 6b is not likely to be due to solute strengthening. Compared to the Vickers hardness measurements, nanoindentation tests are more sensitive to residual stresses [6], and thus besides heterogeneity, the presence of higher compressive residual stresses in the current Cu-Ni plate may also contribute to the higher nanoindenation hardness measured in that plate.

5. Concluding remarks

A combined treatment of electroplating and IFS has been demonstrated in a Cu plate, producing a multifunctional surface layer with a gradient in strain, chemical content, microstructure, and hardness. The hard Ni coating layer influences the deformation behavior of the Cu matrix, and also helps to maintain a smooth surface, which may enhance the friction and wear resistance during subsequent application of the processed material. In order to examine the applicability of the new combined technique, fatigue tests will be carried out as a next step. It has been shown already that gradient structures produced by surface mechanical grinding treatment have a high fatigue resistance due to retardation of both fatigue crack initiation on the hard surface and fatigue crack propagation in the work-hardenable interior [3]. In the current case, a further improvement resulting from the chemical gradient is expected.

Acknowledgement

Drs. K. Kawano-Miyata and K. Matsumoto are thanked for coordination of the glow-discharge OES measurements and the nanoindentation tests.

References

[1] Hughes D A and Hansen N 2001 *Phys. Rev. Lett.* **87** 135503
[2] Lu K and Lu J 2004 *Mater. Sci. Eng. A* **375-377** 38–45
[3] Huang H W, Wang Z B, Yong X P and Lu K 2013 *Mater. Sci. Technol.* **29** 1200
[4] Schlesinger M and Paunovic M 2011 *Modern electroplating* (New Jersey: John Wiley & Sons)
[5] Hong C, Huang X and Hansen N 2015 *I O P Conf. Series: Materials Science and Engineering* (*36th Risø International Symposium on Materials Science*), **89**, 012026
[6] Qian L, Li M, Zhou Z, Yang H and Shi X 2005 *Surf. Coat. Technol.* **195** 264
[7] Cohen M 1970 *Transactions of the Japan Institute of Metals* **11** 145

IOP Conf. Series: Materials Science and Engineering **219** (2017) 012049 doi:10.1088/1757-899X/219/1/012049

Hydrogen-induced room-temperature plasticity in TC4 and TC21 alloys

Baoguo Yuan[1,2], Yongyue Jin[1], Chuanshi Hong[2], Xiaoxue Zhang[3], Xiaoxu Huang[2]

[1]School of Materials Science and Engineering, Hefei University of Technology, Hefei 230009, PR China

[2]Section for Materials Science and Advanced Characterization, Department of Wind Energy, Technical University of Denmark, Risø Campus, Roskilde, DK-4000 Denmark

[3]School of Mechanical Engineering, Anhui Sanlian University, Hefei 230601, PR China

E-mail: yuanbaoguo@163.com (Baoguo Yuan)

Abstract. In order to reveal the effect of hydrogen on the room-temperature plasticity of the titanium alloys TC4 and TC21, compression tests have been carried out at room temperature. Results show that an appropriate amount of hydrogen can improve the room-temperature plasticity of both the TC4 and TC21 alloys. The ultimate compression strain of the TC4 alloy containing a hydrogen concentration of 0.5 wt.% increases by 39% compared to the untreated material. For the TC21 alloy the ultimate compression strain is increased by 33% at a hydrogen concentration of 0.6 wt.%. The main reason for the improvement of hydrogen-induced room-temperature plasticity of the TC4 and TC21 alloys is discussed.

1. Introduction

Titanium and its alloys are widely used in the aerospace, chemical, petroleum, automobile and marine industries because of their low density, high specific strength, superior corrosion and erosion resistance, excellent fatigue properties, non-toxicity, and good biochemical properties [1]. However, the plasticity of most titanium alloys at room temperature is low, which restricts their applications. Therefore, how to improve the room-temperature plasticity of titanium alloys has become an important research topic.

The method of using hydrogen to improve the plasticity of titanium alloys was first proposed by Zwicker and Hans in 1959 [2], who found that the hot workability (such as forging, rolling and extrusion) of titanium alloys can be improved after hydrogenation. In recent years, further research has demonstrated that addition of hydrogen can result in improved mechanical properties of titanium alloys [3-5]. The technology of using hydrogen as a temporary alloying element to improve properties of titanium alloys is called thermohydrogen processing (THP) [6-8]. In this technology, hydrogen is introduced into the titanium alloy, which is then processed, and finally hydrogen is removed by vacuum annealing to avoid hydrogen embrittlement during service. THP can not only improve the hot workability of titanium alloys [9, 10], but can also improve its formability by cold working [11-13] and superplastic forming [14]. Hydrogen-induced plasticity of titanium alloys are mainly studied in Russia and China.

Hydrogen-induced room-temperature plasticity of titanium alloy results in an increase in the limiting degree of deformation prior to appearance of the first crack, first noted for the quenched β-titanium alloys VT15 and VT30 [12]. It has been found that the quenched VT30 alloy charged with more than 0.1 wt.% hydrogen (cylindrical specimens of 10 mm in diameter and 15 mm in height) can be flattened into a "cake" with sharp edges and no cracks are found on its lateral surfaces. Kolachev et al report that hydrogen-induced room-temperature plasticity can also be observed in α+β titanium alloys with a large amount of β phase [12, 15]. The limiting degree of deformation prior to appearance of the first crack during jumping-up tests of VT16 alloy at room temperature increases from less than 40% to about 75% as the hydrogen content is increased from 0.005% to 0.1%. Hydrogen-induced room-temperature plasticity has also been observed in the TC4 alloy [11, 16], but the plasticity of hydrogenated TC4 alloy is still not high enough for the cold working of complicated titanium alloy parts. The technology has, however, been applied to produce anchor nuts and large diameter bolts [15]. For wider application further research and development are still required.

2. Hydrogenation method

There are many ways to add hydrogen into a titanium alloy. The first hydrogenation method applied is to melt titanium alloy in a hydrogen atmosphere [2, 17]. The ingots are melted under a gaseous mixture of hydrogen and argon. Hydrogen molecules decompose and diffuse into the melt. The second hydrogenation method is an electrolytic method [18], where a specimen is charged with hydrogen in an electrolyte consisting of phosphoric acid and glycerin for different lengths of charging time. The third hydrogenation method is to diffuse hydrogen into a solid-state titanium alloy at high temperatures, where different hydrogen contents can be obtained by controlling the temperature, hydrogen pressure (or hydrogen flow rate) and the holding time. The actual hydrogen content in a titanium alloy is determined by weighing the specimen before and after hydrogenation using a high precision electronic balance. The third hydrogenation method is preferred for THP studies [19, 20] and is applied to TC4 and TC21 titanium alloys in the present work.

3. Results

In the present work, the effect of hydrogen on the room-temperature plasticity of TC4 and TC21 titanium alloys is studied by room temperature compression tests.

3.1. Compression of TC4 alloy

The nominal composition of the TC4 alloy is Ti-6Al-4V. Compression tests of TC4 alloys with different amounts of hydrogen were carried out at room temperature on a Zwick Z100 machine with a compression speed of 0.05 mm/min. The specimens were cylinders of 4 mm in diameter and 6 mm in height. The specimens were hydrogenated in an atmosphere of hydrogen at 750 °C for 1 h, air cooled to room temperature, and then solution treated at 850 °C for 0.5 h followed by furnace cooling to 700 °C, and finally quenched in water at room temperature. MoS_2 lubricant was coated on both end faces of the specimens. Engineering stress-strain curves of TC4 alloys with different amounts of hydrogen, based on load-displacement measurements are shown in figure 1. From figure 1, it can be seen that compression properties of TC4 alloy are changed obviously after hydrogenation. The plasticity index is defined as the compressive strain when the first macroscopically observable crack appears at the cylindrical surface of the specimen. The strain termed the ultimate compression strain is expressed:

$$\varepsilon_c = \frac{H_k - H_0}{H_0} \times 100\%, \tag{1}$$

where H_0 is the initial height of specimen before compression testing and H_k is the height of the specimen during the compression when the first macroscopic crack appears at the cylindrical surface of the specimen. The compression strength is taken as the maximum stress on the stress-strain curve. The ultimate compression strain and compression strength of TC4 alloys with different amounts of hydrogen are shown in figure 2. From figure 2a, it can be seen that ultimate compression strain of the

TC4 alloy changes slightly at hydrogen contents of lower than 0.4 wt.%. At a hydrogen content of 0.5 wt.%, the ultimate compression strain of the TC4 alloy increases significantly to a value about 39% higher than that of the un-hydrogenated specimen. From figure 2b, it can be seen that the compression strength changes slightly for hydrogen contents lower than 0.4 wt.%, but increases by 18% for a hydrogen charging concentration of 0.5 wt.%. The ultimate compression strain and compression strength of the TC4 alloys with different amounts of hydrogen compressed at 0.5 mm/min show the same trend as those compressed at 0.05 mm/min [21]. For hydrogen contents exceeding 0.5 wt.%, the ultimate compression strain of the TC4 alloy increases first and then decreases with increasing hydrogen content [16]. Hydrogen has also favorable effects on the compressive properties of TC4 alloy at high strain rate [22].

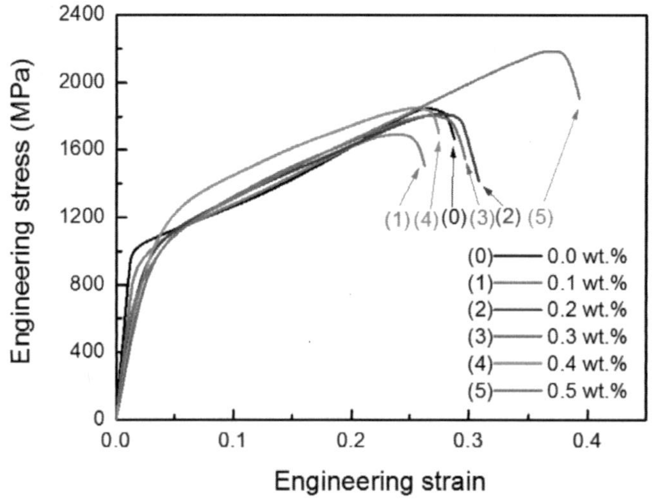

Figure 1. Engineering stress-strain curves of TC4 alloys containing different amounts of hydrogen.

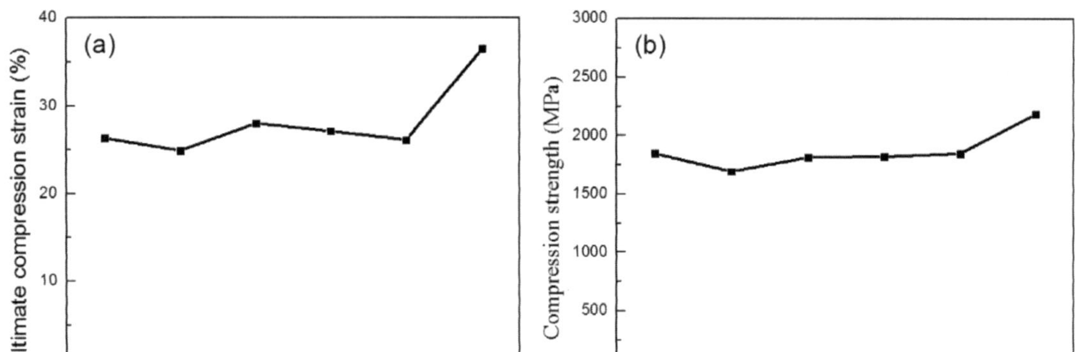

Figure 2. Ultimate compression strain (a) and compression strength (b) of TC4 alloys containing different amounts of hydrogen.

3.2. Compression of TC21 alloy

The alloy TC21 is a kind of α+β damage tolerant titanium alloy, with high strength, high fracture toughness and low crack propagation rate. Its nominal composition is Ti-6Al-2Mo-1.5Cr-2Zr-2Sn-2Nb. Compression tests of TC21 alloys containing different amounts of hydrogen were carried out on a MTS 809 materials testing machine at room temperature. The compression specimens were cylinders of 4 mm in diameter and 6 mm in height. The specimens were first hydrogenated at 750 °C for 2 h, and then air cooled to room temperature. The compression speed was 0.05 mm/min. White Vaseline was used as a lubricant. Engineering stress-strain curves of TC21 alloys with different amounts of hydrogen are shown in figure 3, indicating a marked effect of hydrogenation. The ultimate compression strain and compression strength of TC21 alloys containing different amounts of hydrogen are shown in figure 4. Figure 4a shows that the ultimate compression strain increases with increasing hydrogen content and then decreases. The ultimate compression strain is highest for hydrogen contents in the range of 0.6 wt.% - 1.0 wt.%, with an maximum increase of 33% at 0.6 wt.% hydrogen. The ultimate compression strain decreases by about 56% after introduction of 1.2 wt.% hydrogen into the TC21 alloy. Figure 4b shows that the compression strength decreases slightly with increasing hydrogen concentration, with a decrease of 7% at a hydrogen content of 1.0 wt.%. For hydrogen contents exceeding 1.0 wt.%, the compression strength decreases sharply, showing a decrease of 28% for a hydrogen content of 1.2 wt.%. The present observations are in good agreement with the results for TC21 alloys with different amounts of hydrogen compressed at 0.5 mm/min [13].

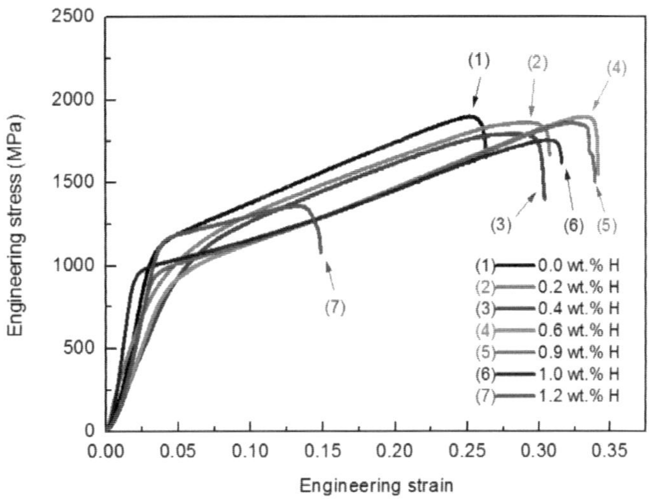

Figure 3. Engineering stress-strain curves of TC21 alloys containing different amounts of hydrogen.

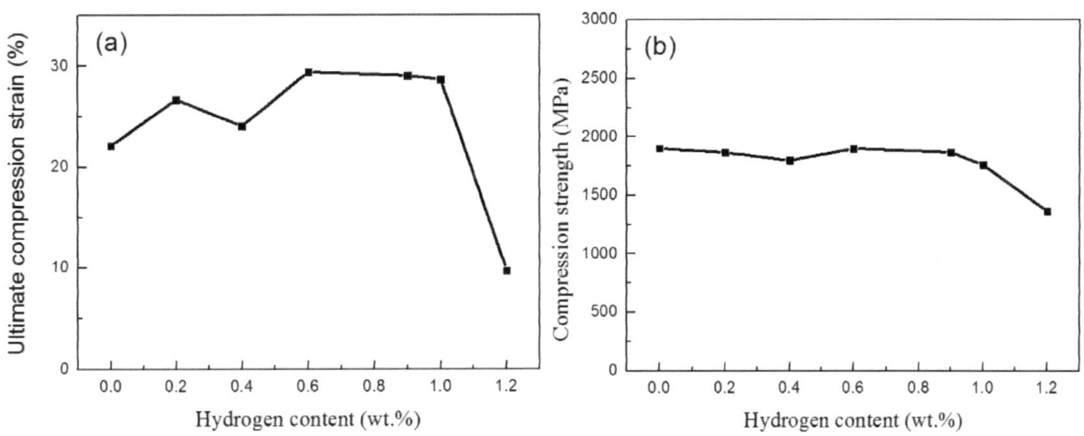

Figure 4. Ultimate compression strain (a) and compression strength (b) of TC21 alloys containing different amounts of hydrogen.

4. Discussion
The above results show that introduction of hydrogen can improve the room-temperature plasticity of both the TC4 and TC21 alloys. Hydrogen has a similar effect on the room-temperature plasticity on both alloys. The main reason for the improvement is that the amount of the ductile β phase increases with hydrogenation, as shown in figure 5. The β phase, being body-centered cubic with 12 slip systems, is more ductile than the α phase, which is hexagonal close packed with only 3 slip systems. Hydrogen is a β-stabilizing element and can lower the β transus temperature in titanium alloys. Therefore, more β phase will be present after hydrogenation. However, hydrides will precipitate widely in the structure if the amount of hydrogen in titanium alloys is high enough, and will cause embrittlement. Therefore, the amount of hydrogen in the TC4 and TC21 alloys should be controlled in order to improve its room-temperature plasticity.

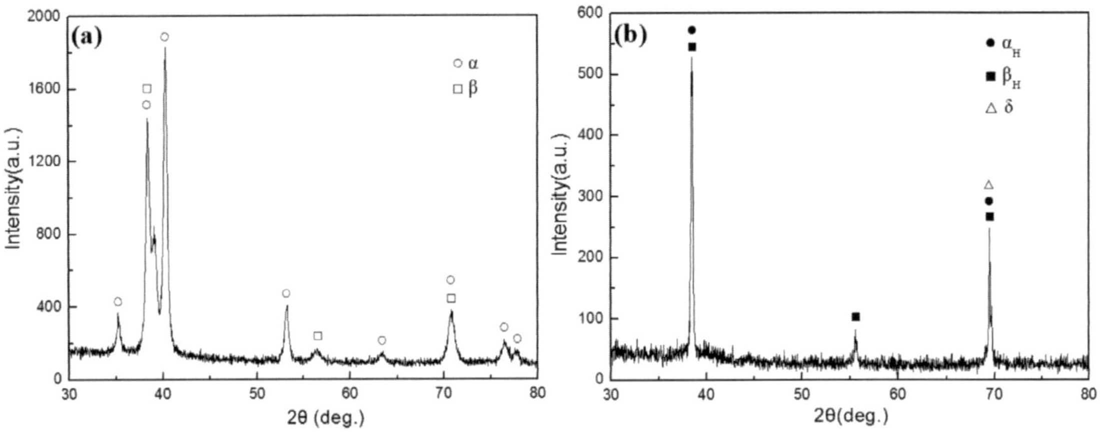

Figure 5. XRD patterns of TC21 alloys: (a) 0.0 wt.% H and (b) 0.9 wt.% H.

5. Conclusion
Two titanium alloys, TC4 and TC21, have been hydrogenated at high temperature and have been tested in compression at room temperature. By taking the compression strain as a measure of formability, the following conclusions have been reached:

1) Hydrogenation up to 0.5 wt.% increases the formability of the TC4 alloy, measured as the strain for the first appearance of macroscopic cracks during compression.

2) Hydrogenation up to 1 wt.% increases the formability of the TC21 alloy, which is related to an increased percentage of the more ductile β phase (bcc) compared to the α phase (hcp).

3) Hydrogenation to 1.2 wt.% of the TC21 alloy reduces the formability due to hydride precipitation.

Acknowledgments
The authors gratefully acknowledge the financial support from the China Scholarship Council, the Natural Science Research Project of Anhui University (Grant No. KJ2016A892) and the National Natural Science Foundation of China (Grant No. 51205102). We also thank Niels Hansen, Xiaodan Zhang and Tianbo Yu for helpful discussions.

References
[1] He J, Liu XY, Sun XX and Li HW 2017 A new idea about active coordination of the gradually-changed microstructure of titanium functionally graded components *J. Netshape Forming Eng.* **9** 17-22

[2] Zwicker U and Schleicher HW Process for improving the workability of titanium alloys 1959 *US Patent* 2892742

[3] Wen DS, Zong YY, Wang YQ, Liu ZY and Shan DB 2016 Positive influence of hydrogen on the hot workability and dynamic recrystallization of a γ-TiAl based alloy *Mater. Sci. Eng.* A

656 151-64

[4] Shen CC and Wang CM 2014 Effects of hydrogen loading and type of titanium hydride on grain refinement and mechanical properties of Ti-6Al-4V *J. Alloys Compd.* **601** 274-9

[5] Chen R, Ma T, Guo J, Ding H, Su Y and Fu H 2016 Deformation behavior and microstructural evolution of hydrogenated Ti44Al6Nb alloy during thermo-compression at 1373–1523 K *Mater. Des.* **108** 259-68

[6] Froes FH, Senkov ON and Qazi JO 2004 Hydrogen as a temporary alloying element in titanium alloys: thermohydrogen processing *Int. Mater. Rev.* **49** 227-45

[7] Senkov ON and Froes FH 1999 Thermohydrogen processing of titanium alloys *Int. J. Hydrogen Energy* **24** 565-76

[8] Eliaz N, Eliezer D and Olson DL 2000 Hydrogen-assisted processing of materials *Mater. Sci. Eng.* A **289** 41-53

[9] Sun DL, Wang Q, Han XL, Xiao JX and Jiang DP 2011 Influence of hydrogen on hot deformation behavior and microstructure of pure titanium *Procedia Eng.* **10** 42-7

[10] Zhao J, Ding H, Zhao W and Jiang Z 2013 Effects of hydrogen on the hot deformation behaviour of Ti–6Al–4V alloy: Experimental and constitutive model studies *J. Alloys Compd.* **574** 407-14

[11] Sun Z, Chen G, Wang Y, Zhou W and Hou H 2010 Investigations on cold deformation mechanisms of the hydrogenated Ti-6Al-4V alloys *Mater. Sci. Eng.* A **527** 1003-7

[12] Kolachev BA 1993 Reversible hydrogen alloying of titanium-alloys *Met. Sci. Heat. Treat.* **35** 586-91

[13] Yuan BG, Zheng YB, Gong LQ, Chen Q and Lv M 2016 Effect of hydrogen content on microstructures and room-temperature compressive properties of TC21 alloy *Mater. Des.* **94** 330-7

[14] Zhang X, Zhao Y, Zeng W 2010 Effect of hydrogen on the superplasticity of Ti40 alloy with large grains *Mater. Sci. Eng.* A **527** 3489-92

[15] Kolachev BA, Malkov AV, Nizkin ID, Mitin AV and Korostylev AF 1991 Feasibility of lowering the hot-heading temperature of alloy VT16 bolts *Izv. Akad. Nauk SSSR, Met.* **3** 67-79

[16] Yuan BG, Li CF, Yu HP and Sun DL 2010 Influence of hydrogen content on tensile and compressive properties of Ti-6Al-4V alloy at room temperature *Mater. Sci. Eng. A* **527** 4185-90

[17] Su Y, Liu X, Luo L, Zhao L, Guo J and Fu H 2010 Hydrogen solubility in molten TiAl alloys *Int. J. Hydrogen Energy* **35** 8008-13

[18] Wen J, Allain N and Fleury E 2016 Hydrogen evolution and its effects on cold rolling behavior in commercial pure titanium *Mater. Charact.* **121** 139-48

[19] Yuan BG, Zheng YB, Wang YJ and Gong LQ 2016 Hydrogen absorption characteristics and microstructural evolution of TC21 titanium alloy *Trans. Nonferrous Met. Soc. China* **26** 599-606

[20] Qazi JI, Senkov ON, Rahim J, Genc A and Froes FH 2001 Phase transformations in Ti-6Al-4V-xH alloys *Metall. Mater. Trans.* A **32** 2453-63

[21] Yuan BG, Li CF, Yu HP and Sun DL 2009 Effect of hydrogen content and stress state on room-temperature mechanical properties of Ti-6Al-4V alloy *Trans. Nonferrous Met. Soc. China* **19** S423-8

[22] Yuan BG, Yu HP and Li CF 2012 Influence of hydrogen content on room temperature compressive properties of Ti–6Al–4V alloy at high strain rate *Trans. Nonferrous Met. Soc. China* **22** 2943-51

Quantification of deformation microstructure at ultra-low tensile strain in pure Al prepared by spark plasma sintering

C L Zhang[1], Y B Zhang[2], G L Wu[3], W Liu[4], D Juul Jensen[2] and A Godfrey[1]

[1] Key Laboratory of Advanced Materials (MOE), School of Materials Science and Engineering, Tsinghua University, Beijing 100084, China
[2] Section for Materials Science and Advanced Characterization, Department of Wind Energy, Technical University of Denmark, Risø Campus, Roskilde 4000, Denmark
[3] College of Materials Science and Engineering, Chongqing University, Chongqing 400045, China
[4] Advanced Photon Source, Argonne National Laboratory, Argonne, IL 60439, USA

E-mail: awgodfrey@mail.tsinghua.edu.cn

Abstract. A sample of Al with grain size of 5.1 μm, prepared by spark plasma sintering, was deformed to a nominal strain of 0.35% under exposure to X-ray synchrotron radiation, allowing spatially resolved orientation measurements to be made during loading by use of a micro-diffraction technique. A significant heterogeneity in the deformation pattern between grains was observed. A statistical analysis shows that grain deformation depends more on crystallographic orientation than on grain size, with grains with tensile axis lying towards the <001>-<101> border of the unit triangle tending to undergo larger deformation. Other possible reasons for the different deformation behaviour between individual grains are briefly discussed.

1. Introduction

One reason for the continued demand for aluminium alloys as structural materials is the high strength-to-weight ratio of these alloys. To achieve further increases in performance without addition of expensive alloying elements, one solution is through the reduction of grain size into the near-micrometre range, where a good balance between strength and ductility may be possible. There are, however, still some open questions regarding the relationship between material properties and grain size in this regime [1-3]. For a better understanding of the underlying deformation mechanisms in the near-micrometre grain-size regime samples with suitable grain sizes and a simple microstructure are needed for controlled experiments.

Traditionally, fine-grained Al materials have been produced using one of several severe plastic deformation (SPD) techniques. In each case, however, the resulting microstructure is inevitably in a deformed state, with a high dislocation density and a large fraction of low angle boundaries (LABs), and the scope for grain size control in the near-micrometre regime by annealing is limited. Recently, an alternative approach for the preparation of fine-grained samples, namely spark plasma sintering (SPS), has been explored. It has been shown that the SPS technique, originally developed for production of ceramic materials, allows the preparation of metallic samples with controlled grain size, where the grains have a random texture, and are in a fully recrystallized condition, containing a low proportion of LABs [4-6]. This technique is, therefore, well suited for preparation of starting materials for research into deformation mechanisms in the near-micrometre grain-size regime.

Over the last decade, the possibility for non-destructive three dimensional (3D) microstructural characterization has also been realized, and it is been reported that in some cases two dimensional (2D) studies are insufficient [7,8] to reveal fully underlying mechanisms. In this work, we present an in situ experimental study on an SPS-processed Al sample with a 5.1 μm grain size, where the differential aperture X-ray microscopy (DAXM) technique [9,10] is employed for collecting spatially resolved diffraction data from the bulk interior with micrometre resolution. The orientation variations within grains after deformation to near yielding are analysed in detail, and furthermore, the relationships between the grain deformation pattern and both grain size and orientation are also discussed. Additionally, a comparison between 2D and 3D orientation characterization is carried out for one grain, thereby demonstrating the advantage of 3D analysis in revealing clearly the deformation behaviour.

2. Experimental

Pure (99.9%) Al powder with an average powder particle size of 5.7 μm was consolidated by SPS. A maximum sintering temperature of 600 °C and a maximum pressure of 50 MPa were used for the sintering process, the details of which can be found elsewhere [6]. The as-sintered samples have a density of 99% of theoretical density and are in the form of disks with a diameter of 20 mm and a height of 4 mm. Dog-bone tensile specimens were cut from the samples by electron discharge machining. The gauge length, width and height of the cross gauge section of the tensile specimens were 12, 1.8 and 0.6 mm, respectively. Prior to loading, the tensile specimens were mechanically polished and then electro-polished to remove machining damage at the surface.

Synchrotron micro-diffraction experiments were conducted at beamline 34-ID-E at the Advanced Photon Source, Argonne National Laboratory [9]. In the experiment, a polychromatic X-ray beam with energies in the range of 7-30 keV was focused using two non-dispersive Kirkpatrick-Baez mirrors to a size of ~ 0.5 μm. The tensile specimen was mounted on a specially designed tensile device offering a nominal strain resolution of ~ 0.05%, then installed at a 45° incident angle towards the X-ray beam. The diffraction patterns from the X-ray illuminated volume were recorded on a flat panel detector (409.6 × 409.6 mm^2, 2048 × 2048 pixels, and 16-bit dynamic range) mounted in a 90° reflection geometry 510.9 mm above the sample. The detector geometry with respect to the incident beam was calibrated using a strain free silicon single crystal. A Pt wire of 100 μm in diameter was used as a differential aperture for depth-resolution of the Laue diffraction patterns. Reconstruction and indexing of the depth-resolved Laue diffraction patterns were conducted using the LaueGo software at 34-ID-E. Reconstructions were performed to a depth of 150 μm into the sample with a depth spacing of 1 μm. The focused beam was scanned in a 51 × 5 grid at 1 μm step size, resulting in a final 3D data set volume of 51 × 5 × 150 μm^3, with 1 μm^3 voxel size.

Data were collected for the same volume of the sample in both the undeformed state and after tensile deformation to near the yield point (a nominal strain of 0.35%) at a strain rate of ~ 1×10^{-4} /s. The stress at this strain, calculated based on the force measured on a load cell attached to the tensile loading frame, was determined as 55 MPa. In this paper, only results for the deformed sample will be presented.

3. Results and discussion

3.1. Microstructure

The microstructure in the characterized volume after deformation to a nominal strain of 0.35% is shown in figure 1a. After deformation, the grains in the volume are still nearly equiaxed and the orientation variations (seen as colour variations in the figure) within individual grains are very small. Individual grains were identified automatically using a clustering algorithm using a critical angle of 0.2°. In total the examined volume contains 605 grains. The grain size distribution, calculated based on inspection of the centre layer of the mapped volume is shown in figure 1b. The average grain size (calculated as the equivalent area diameter of each grain in this layer) is about 5.1 μm, which matches well both the starting powder particle size and the grain size measured using electron backscatter diffraction (EBSD) on a similar sample [11]. The sample has a random texture (see figure 1c).

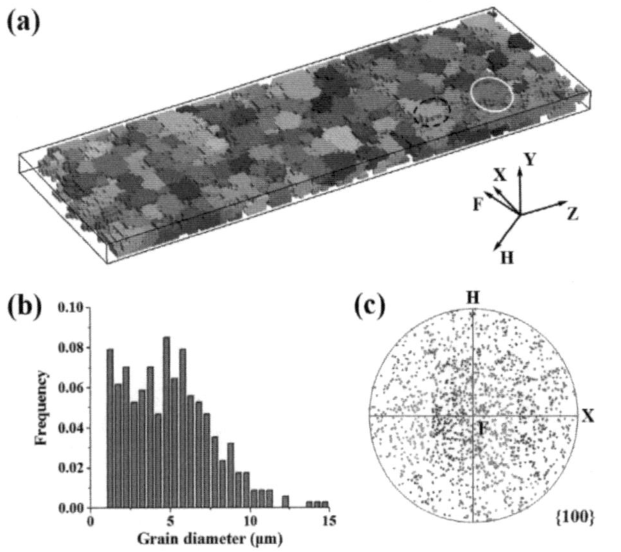

Figure 1. (a) Map (Euler angle colouring) showing the 3D microstructure of the characterized volume (51, 5 and 150 μm along X, Y and Z axis, respectively) after deformation to a nominal strain of 0.35%. The beamline coordinates (X Y Z) and the sample coordinates (X H F) are both defined. The crystallographic orientations are defined in the sample coordinate system. The two grains highlighted by circles are analysed in detail in section 3.2; (b) grain size distribution and (c) {100} pole figure of the characterized volume (colouring corresponding to the grains in (a)).

3.2. Deformation pattern within individual grains

At strain of 0.35%, orientation variations are seen within individual grains. Two grains of similar size chosen for more detailed analysis are indicated by the black-dotted and white-solid circles in figure 1a. As the orientation variation is very small at this strain, typically ~ 0.05°, the orientation data are examined using reference orientation maps based on the grain average orientation after deformation. For these maps the misorientation between each voxel and the average orientation of the grain in which it lies is first calculated in the sample reference frame as an angle:axis pair, and then the results plotted separately for the misorientation angle component (also referred to here as the deviation angle) and the misorientation axis component. It has been shown elsewhere [12] that such maps, in particular the sample-frame misorientation axis map, can be very sensitive to small systematic variations in orientation.

The results are shown in figure 2. For grain G1, an angle deviation of up to 0.2° to the average orientation is seen after deformation (see figure 2a and 3a), and the misorientation rotation axes change gradually and continually from one side of grain G1 to the other side (see figure 2b). In contrast, a different deformation pattern is seen for grain G2. For this grain two distinct regions of different colours are seen in the misorientation rotation axes, implying that grain G2 has subdivided into two parts after deformation.

The distributions of deviation angle to the average orientation for all voxels in grains G1 and G2 are shown in figures 3a and 3c, respectively. An evident difference is seen between the two distributions, which is directly related to the different deformation patterns seen in figure 2. The distributions of neighbour voxel misorientations for the two grains are, however, relatively similar (figure 3b and 3d). The grain orientation spread (defined as the arithmetic mean of the deviation angles between the orientation of each voxels and the average grain orientation) for grains G1 and G2 are 0.08° and 0.07°, respectively, while the grain average misorientation (defined as the arithmetic mean of the misorientation angles between all pairs of nearest neighbour voxels in each grain) is 0.04° for both grains G1 and G2.

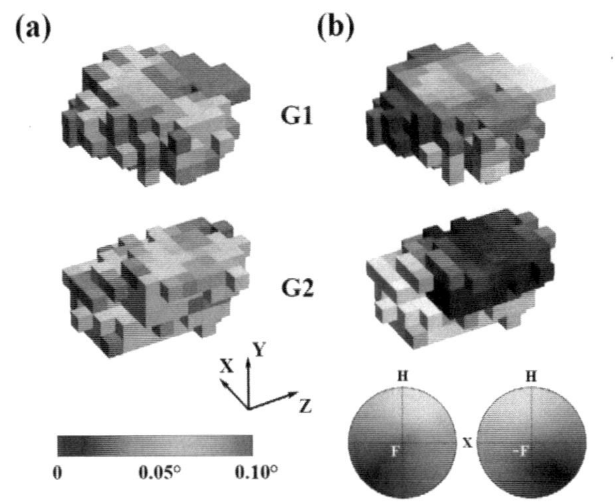

Figure 2. 3D plots of grain G1 and G2 (marked by the white-solid and black-dashed circles, respectively in figure 1a), showing the microstructure after deformation to a nominal strain of 0.35%. The colouring is based on the misorientation between each voxel orientation and the grain average orientation after deformation. (a) Misorientation angle component; (b) misorientation rotation axis component, expressed in the sample coordinate system. The scale bar for the misorientation angle and colour code for the misorientation rotation axis are given.

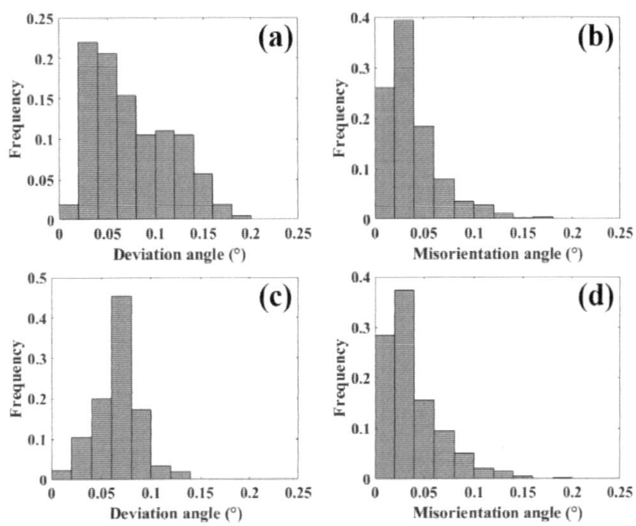

Figure 3. Orientation variation for all voxels in grains G1 and G2. (a) Deviation angle between each voxel orientation and the average orientation for grain G1; (b) misorientation angles between nearest-neighbour voxel pairs in grain G1; (c) - (d) are the corresponding distributions of the two parameters for grain G2.

3.3. Comparison between 2D and 3D measurements

The DAXM technique probes the spatially resolved orientation of the sampled volume, however, most frequently used characterization techniques are performed in 2D. Thus interest arises as to how much difference there is between the 2D and 3D measurements. As mentioned in section 3.1, a layer based analysis from a 3D volume is comparable to an EBSD measurement. In this section, grain G2 is used to demonstrate possible differences between 2D and 3D orientation analysis.

For this demonstration the grain was analysed layer-by-layer. In each layer a calculation was made of the misorientation between the voxel orientations and the average orientation of the grain based only on the measurement in the layer. A quasi-3D data set was then obtained by recombining the data for all layers. For grain G2, the layer based orientation analysis shows smaller deviation angles compared to the volume based result (see figure 4a and 2a), with a mean values of 0.04° and 0.07°, respectively. Moreover, instead of a clearly observed grain subdivision along the Y axis, as seen in figure 2b, the layer based misorientation rotation axis map shows a weaker more gradual change of orientation, predominantly along the Z axis (see figure 4b).

For a better demonstration of the differences between the 2D and 3D results, a locally magnified view of one pole of a {100} pole figure for grain G2 is plotted by the two methods (see figure 4c and 4d), with the size and colour of each point representing the deviation angle and misorientation rotation axis of the corresponding voxel. A clear difference between the patterns of subdivision can be seen, indicated by the clusters of points with similar colouring, as can the larger deviation angles of the 3D results compared to the 2D analysis. The analysis highlights the fact that 2D data, particularly at low plastic strain and in fine grains, can give misleading or incomplete information about the real pattern of grain subdivision.

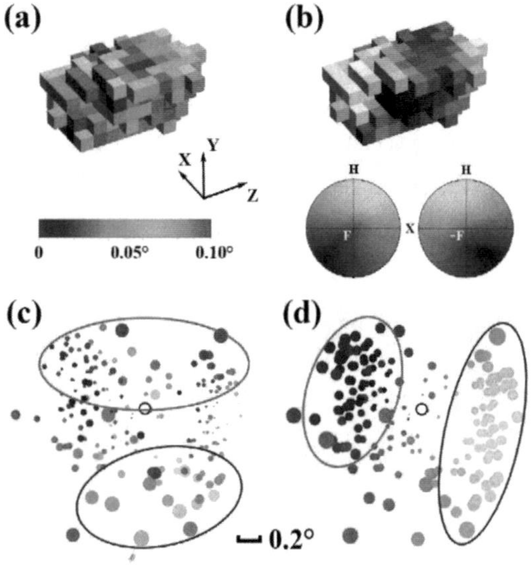

Figure 4. Layer based orientation variation analysis for grain G2 showing (a) misorientation angle component, and (b) misorientation rotation axis (sample coordinate system) component; (c) magnified region of a {100} pole figure, with colouring using the same definition as (b) and where the point size represents the misorientation angle component. The grain average orientation is highlighted by the black circle, and the scale bar corresponds to the point size; (d) corresponding magnified pole figure for the same grain according to the volume based orientation variation result (shown in figure 2). The red and blue circles in (c) and (d) indicate two clusters of voxels with similar colour, i.e. misorientation rotation axis.

3.4. Deformation behaviour of all grains

In order to assess the variation in deformation behaviour for all the grains in the sampled volume, the grain orientation spread and grain average misorientation have been calculated for each grain and are plotted as a function of grain volume in figures 5a and 5b, respectively. A general tendency is seen for larger grains to have a larger grain orientation spread (see figure 5a), although some smaller grains also show a large grain orientation spread. As discussed in [13] this result may arise from a correlation between grain orientation spread and grain size, resulting from the presence of in-grain orientation gradients. In contrast, as shown in figure 5b, the grain average misorientations for large grains are similar to those for small grains, i.e. grain average misorientation does not depend on grain size. The grain average misorientation therefore provides a better estimation of the deformation of individual grains, although it should be noted that the absolute value of this parameter is step-size dependent.

Using this parameter the effect of crystallographic orientation on the deformation has been examined. Figure 5c shows the orientations of the grains having relatively large grain average misorientations (> 0.09°). The figure shows that in general these grains have no preferential orientation. However, it appears that grains with orientations in the lower part of the unit triangle, i.e. in the region towards the <001>-<101> border, have comparatively larger average misorientations than those with orientation close to the <111> corner.

Besides the crystallographic orientation, several other factors may be important for the deformation differences between individual grains, such as the grain shape, the influence of neighbouring grains, and the initial dislocation density in the undeformed state. These factors will be investigated in detail in the future using both experimental and computational (e.g. crystal plasticity finite element modelling) approaches.

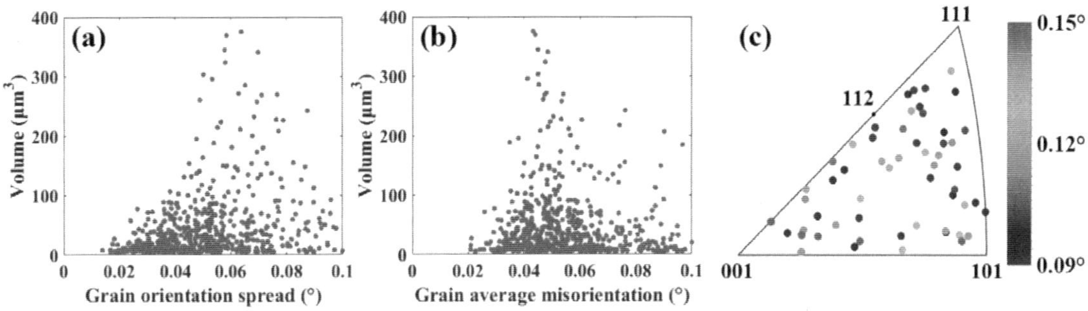

Figure 5. (a) Grain orientation spread and (b) grain average misorientation as a function of grain volume for all 605 grains in the characterized volume. (c) Crystallographic orientations along the tensile direction (H axis in figure 1a) for grains with an average misorientation larger than 0.09°; the colour of each point represents the average misorientation angle of the corresponding grain.

4. Summary

The results show that the 3D X-ray diffraction microscopy technique is a very powerful tool for investigation of deformation microstructure, allowing non-destructive orientation measurements with a spatial resolution of 1 μm and an angular resolution of 0.01°. This combination of characteristics allows for the first time a study of deformation at very low strain in aluminium with near-micrometre grain size. The results show a heterogeneity in the deformation pattern, as quantified by the grain average misorientation. Based on this parameter it is found that grain deformation in this near-micrometre grain-size regime depends more on crystallographic orientation than on grain size, with a weak tendency for grains lying towards the <001>-<101> border of the unit triangle to undergo larger deformation. A comparison between layer based and volume based orientation variation analysis highlights the advantage of 3D measurements for revealing the pattern of grain subdivision.

Acknowledgements

Financial support from the National Natural Science Foundation of China (project numbers 51671113 and 51471095) is gratefully acknowledged. Use of the Advanced Photon Source was supported by the U.S. Department of Energy, Office of Science, Office of Basic Energy Sciences, under Contract No. DE-AC02-06CH11357.

References

[1] Tsuji N, Ito Y, Saito Y and Minamino Y 2002 *Scr. Mater.* **47** 893-9
[2] Hung P C, Sun P L, Yu C Y, Kao P W and Chang C P 2005 *Scr. Mater.* **53** 647-52
[3] Huang X, Hansen N and Tsuji N 2006 *Science* **312** 249-51
[4] Kubota M and Wynne B P 2007 *Scr. Mater.* **57** 719-22
[5] Srinivasarao B, Oh-Ishi K, Ohkubo T and Hono K 2009 *Acta Mater.* **57** 3277-86
[6] Le G M, Godfrey A and Hansen N 2013 *Mater. Des.* **49** 360-7
[7] Lin F X, Godfrey A, Juul Jensen D and Winther G 2010 *Mater. Charact.* **61** 1203-10
[8] Xu C L, Zhang Y B, Godfrey A, Wu G L, Liu W, Tischler J Z, Liu Q and Juul Jensen D 2017 *Sci. Rep.* **7** 42508
[9] Larson B C, Yang W, Ice G E, Budai J D and Tischler J Z 2002 *Nature* **415** 887-90
[10] Larson B C and Levine L E 2013 *J. Appl. Cryst.* **46** 153-64
[11] Le G M, Godfrey A, Hansen N, Liu W, Winther G and Huang X 2013 *Acta Mater.* **61** 7072-86
[12] Hong X, Godfrey A, Zhang C L, Liu W and Chapuis A 2017 *Mater. Sci. Eng.* A **693** 14-21
[13] Delannay L, Mishin O V, Juul Jensen D and Van Houtte P 2001 *Acta Mater.* **49** 2441-51

IOP Conf. Series: Materials Science and Engineering **219** (2017) 012051 doi:10.1088/1757-899X/219/1/012051

Orientation Dependence of the Deformation Microstructure of Ta-4%W after Cold-Rolling

J Zhang[1], G Q Ma[1], A Godfrey[2], D Y Shu[3], Q Chen[3], G L Wu[1,*]

[1] College of Materials Science and Engineering, Chongqing University, Chongqing 400045, China

[2] Key Laboratory of Advanced Materials (MoE), School of Materials Science and Engineering, Tsinghua University, Tsinghua, 100084, China

[3] Southwest Technology and Engineering Research Institute, Chongqing 400039, China

E-mail: wugl@cqu.edu.cn

Abstract: One of the common features of deformed face-centered cubic metals with medium to high stacking fault energy is the formation of geometrically necessary dislocation boundaries. The dislocation boundary arrangements in refractory metals with body-centered cubic crystal structure are, however, less well known. To address this issue a Ta-4%W alloy was cold rolled up to 70% in thickness in the present work. The resulting deformation microstructures were characterized by electron back-scattering diffraction and the dislocation boundary arrangements in each grain were revealed using sample-frame misorientation axis maps calculated using an in-house code. The maps were used to analyze the slip pattern of individual grains after rolling, revealing an orientation dependence of the slip pattern.

1. Introduction

Due to their excellent properties of high density, high corrosion resistance, high melting point, good plasticity even at low temperatures, and a moderately high elastic modulus [1-3], tantalum (Ta) and tantalum–tungsten (W) alloys are regarded as materials with great potential for high temperature applications, as well as in nuclear engineering, aerospace and other areas [4], with extensive studies have conducted on these alloys in various fields [5-7]. However, despite these studies information regarding the evolution of the deformation microstructure in Ta and Ta-W alloys is scarce. Such information regarding the dominant deformation mechanisms and the evolution of microstructure during thermo-mechanical processing is nevertheless required to help with optimization of materials properties during forming and processing.

In the past few decades many studies have been carried out on the deformation mechanism and the evolution of deformation microstructures in face centred cubic (FCC) metals [8, 9], as well as in some ferritic steels [10]. These studies have led to a framework describing the microstructure morphology in terms of dense dislocation walls, geometrically necessary boundaries (GNBs), incidental dislocation boundaries (IDBs) and lamellar boundaries (LBs). However, the evolution of deformation microstructures in refractory body-centred cubic (BCC) metals such as tantalum, where dislocation mobility is expected to be limited at room temperature, is considerably less studied [11, 12]. The aim of this investigation is therefore to examine the deformation microstructure, and especially the GNB structure, of a Ta alloy after cold-rolling to low and medium strains.

2. Experimental

The starting material used in this study was a Ta alloy containing 4 wt.% W. The Ta-4%W alloy was prepared by an electron beam melting method. The cast rod was hot forged and annealed to obtain a recrystallized rod 50 mm in diameter. The resulting alloy had grain sizes ranging from 10 to 125 μm with an average of 26 μm, and an almost random texture. Plates of either 2 mm or 5 mm in thickness were cut from the rod, perpendicular to the hot-forging direction. The plates were then rolled at ambient temperature to reductions of 10%, 30%, 50% (initial 2 mm thickness) and 70% (initial 5 mm thickness) in a rolling mill with 150 mm diameter rollers. The specimens were cooled in cold water after each rolling pass. For electron channelling contrast (ECC) and electron backscattering diffraction (EBSD) observations, the surfaces of the specimens were mechanically ground, followed by electro-polishing in an 90ml H_2SO_4 and 10ml HF solution at room temperature for 9 minutes using a current of 0.1A/cm^2. All the ECC and EBSD experiments were conducted on the RD (rolling direction)-ND (normal direction) section. The microscope used in the present study was a JEOL 7800F scanning electron microscope (SEM) operating at 20 kV, equipped with an Oxford AZtec EBSD system.

Figure 1. SEM ECC micrograph of the initial Ta-4%W alloy.

Standard EBSD orientation maps, such as inverse pole figure (IPF), directly reconstructed from experimental data, were used initially to characterize the microstructure in each sample. However, the deformation is mainly accommodated by slip in the present alloy, and many deformation features associated with dislocation slip, especially low angle dislocation boundaries, cannot be revealed clearly in IPF maps. Therefore, an in-house code was used, in which the misorientation axis in the sample reference frame of each pixel to the grain average orientation was calculated, and the resulting maps used to probe the evolution of dislocation boundaries during cold rolling [13, 14]. To construct these maps individual grains in each EBSD map were first detected and the grain average (g_{av}) orientation determined. Then the misorientation between each map pixel (g_p) and the average orientation of the grain in which it lies was calculated. Finally, an orientation map was reconstructed showing the misorientation axis, expressed in the sample coordinate frame, for each pixel in the EBSD data.

3. Results and discussion

Figure 2 shows the orientation maps of the Ta-4%W alloy after 10% rolling reduction. From the IPF map, it is seen that some grains were subdivided into several domains (figure 2a) with large orientation changes (10-15°) between neighbouring domains, and with a wide transition in orientation, as shown by the misorientation profile (figure 2c) along the line drawn on figure 2a. The local misorientations between neighbouring points are nevertheless very small (~1° or less). Neither the IPF map, nor boundary misorientation angle maps, show a clear structure after 10% rolling. In contrast, a dislocation boundary microstructure could be readily distinguished in many grains in the sample-frame misorientation axis maps (figure 2b). In these maps it can be seen that many grains, or domains within a grain, contain either one set or two sets of well-defined dislocation boundaries inclined ±25~55° to RD. In some grains, however, dislocation boundaries are not seen.

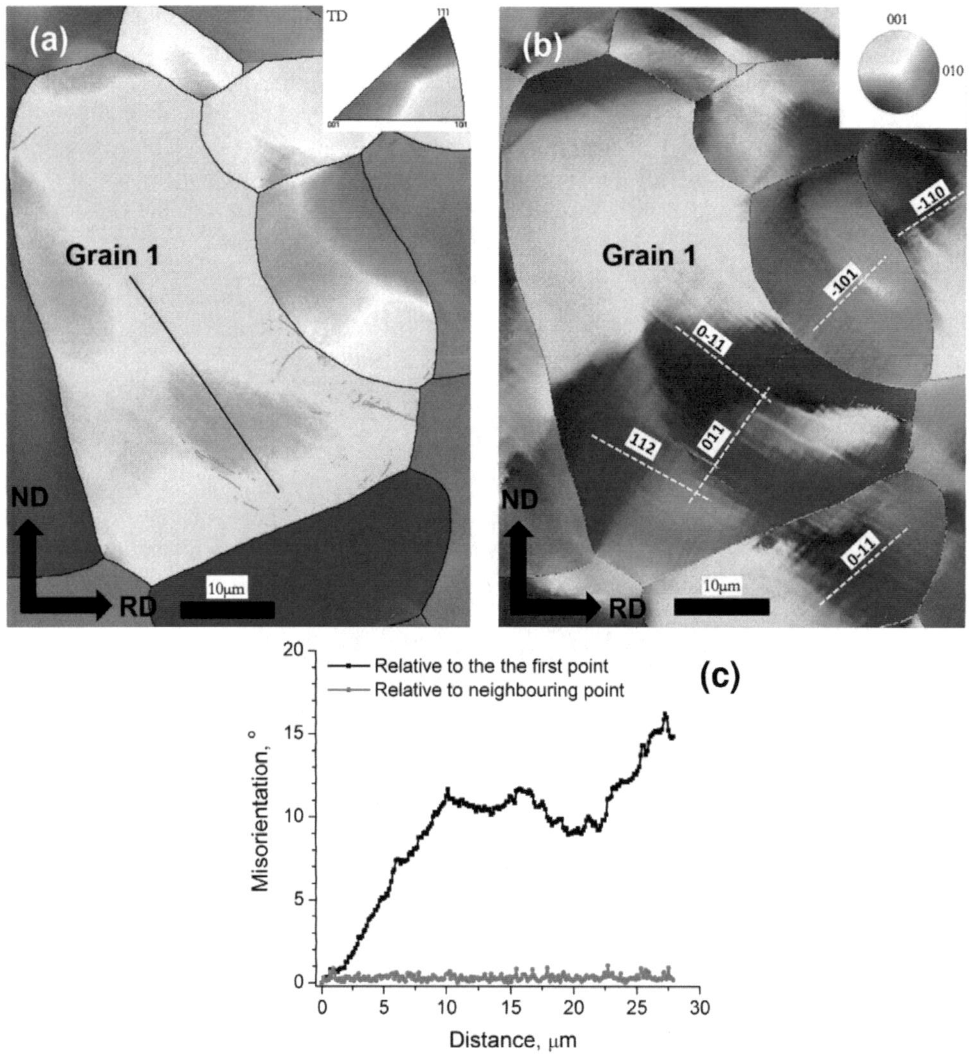

Figure 2. (a) IPF map and (b) misorientation axis map of Ta-4%W after 10% cold rolling; (c) misorientation angle versus distance along the line in (a). The colours in these figure are different, as indicated by the inserted colour scales. Black lines represent misorientation angles ≥ 15°, and grey lines ≥ 2° and < 15°.

For each grain where well defined dislocation boundaries were seen, the traces of {110} and {112} slip planes on the longitudinal section were calculated. If the trace of a slip plane is within 5° to a dislocation boundary on the misorientation axis map, it is assumed that this slip system was strongly activated during cold rolling producing the dislocation boundary. In this manner all the visible dislocation boundaries were examined. Some examples of active slip traces are also shown on the map in figure 2b. For the grain 1, the slip traces of two neighbouring domains are identified as (011) & (0-11) and (011) & (112), respectively. Therefore, the activated slip systems of these two neighbouring domains are different, and can be attributed to the difference in orientation between these regions

Figure 3a and 3b are the IPF and the misorientation axis maps after rolling to 30% reduction. A similar deformation microstructure to that seen at 10% rolling is seen. However, a larger density of low angle misorientation boundaries is now seen in the microstructure. Additionally, more grains or domains are observed containing two sets of dislocation boundaries. These dislocation boundaries are aligned about ±15~50° to RD, and are again found to be mostly parallel to the slip traces of {110} or {112} planes. Domains inside grains can also be found, with the transition regions between domains much narrower than at the lower strain. One example is shown in figure 3c, which gives the misorientation versus distance along the line on figure 3a. The local misorientations vary typically from 2-7°. The misorientation between two neighbouring domains can, however, be as large as 20-25°, with this large orientation change occurring over a distance of ~1.5 μm. The pole figure for the region indicated by a circle in figure 3a, which contains the two domains, shows that these domains are rotated in opposite directions about a common [110] axis (as indicated by the circle in figure 3d), which is also approximately parallel to the TD. The misorientation changes across some transition regions are so abrupt that high angle boundaries can be formed between neighbouring domains, as indicated by arrows in grain 2 on figure 3a.

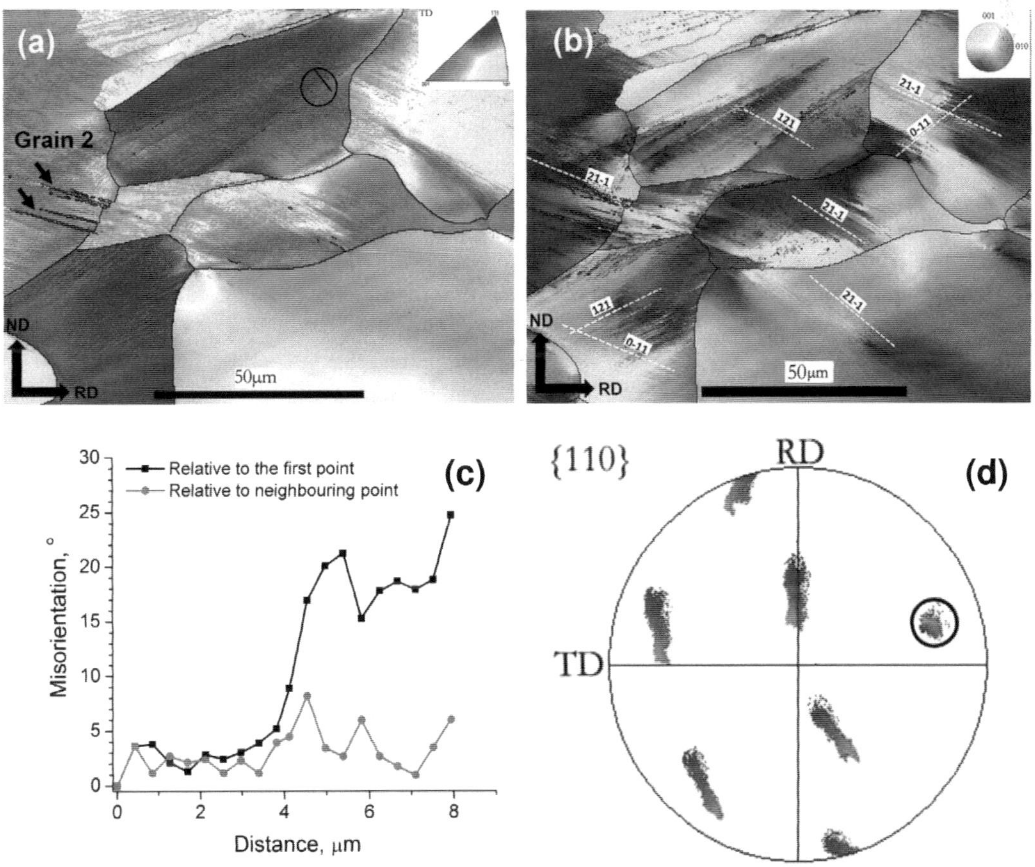

Figure 3. (a) IPF and (b) misorientation axis map of Ta-4%W after 30% cold rolling; (c) misorientation angle versus distance along the line on (a); (d) {110} pole figure of the data inside the circle in (a).

Figures 4a and 4b show the IPF map and rotation axis map of the 50% cold rolled sample. It is seen that many grains are elongated along RD. Many grains or domains contain well-defined dislocation boundaries and some grains or domains with large sizes only have low angle dislocation boundaries. The dislocation boundaries are aligned ±10~50° to RD. Some example slip traces of these boundaries are shown in figure 4b. Fragmented areas containing high angle boundaries to neighbouring regions are also observed in a number of places. These fragmented areas are believed to be created by further development of the rapid orientation changes across neighbouring domains as seen in the sample after 30% rolling.

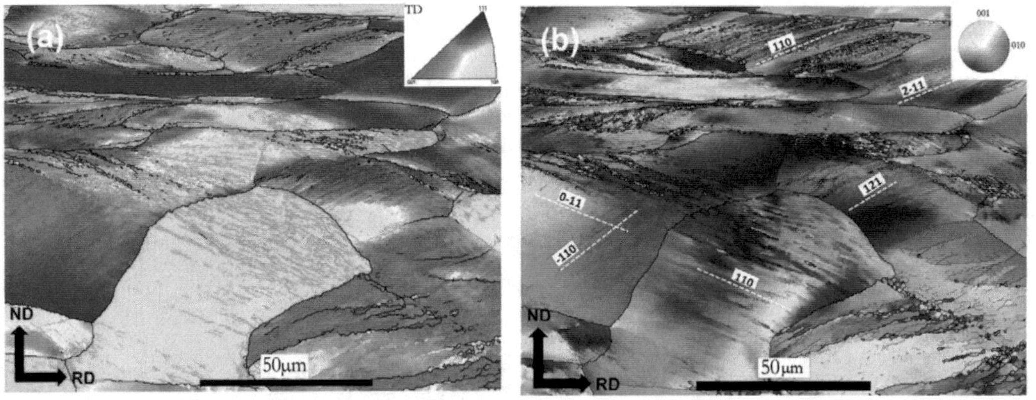

Figure 4. (a) IPF and (b) misorientation axes map of the Ta-4%W after 50% cold rolling.

After 70% cold rolling reduction, the original grains are very clearly elongated and it becomes much more difficult to clearly distinguish features in the more complicated microstructure seen at this strain. Well-defined dislocation boundaries can, nevertheless, still be recognized in many grains and domains. Such dislocation boundaries are inclined ±10~45° to RD, and in many case are close to slip traces (figure 5a). S-shaped bands are also seen in the 70% rolled sample. The formation of S-bands (as shown in figure 5b) is a result of the intersection of two dislocation boundaries, in which localized micro-shearing takes place. The dislocation boundaries in some areas are disturbed, being bent towards original grain boundaries (as shown in figure 5c). Accordingly, in these regions the boundaries are typically no longer parallel to the traces of {110} or {112} planes.

In total, 110 grains were examined in the present experiment. Grains containing either one or two sets of well-defined dislocation boundaries were observed in all examined specimens. The observed dislocation boundaries are believed to be GNBs, as seen previously in FCC (i.e. Al, Ni, Cu) metals and IF steel, where their formation has been extensively studied. The present results for Ta-W alloy (as a BCC refractory metal with a very high melting point and a high stacking fault energy) are similar to those found for IF steel, in which the GNBs were found parallel to either {110} or {112} slip traces.

One interesting point for the GNB structure in the present results is that the local misorientation is quite small after deformation to 10% and 30% so that the GNB structure cannot be revealed very clearly, even in the sample-frame misorientation axis maps. However, the orientation gradients are very large, ~1°/μm and 10°/μm for the two examples in figures 2c and 3c, respectively. Such a large orientation gradient indicates a dislocation density that is very high to accommodate the lattice curvature. However, due to high melting point of this material these dislocations are not able to dynamically recover during plastic deformation by the heat generated by rolling processing. As such it is expected that the dislocation boundaries will not be as narrow (sharp) as seen in deformed Al alloys, but more similar to those seen in Ni. This will be further explored by transmission electron microscopy studies.

In the present results, many of the dislocation boundaries are parallel to the traces of {110} or {112} slip planes, presumed to be active during rolling. For all the grains or domains having identified dislocation boundaries inside, the Schmid factors of all {110} and {112} slip systems were calculated, according to the average orientation of the region where they were found after deformation. It was found that the slip systems corresponding to the observed slip traces do not always have the maximum Schmid factors, but always correspond to one of the slip systems with the largest three Schmid factors. Similar results were also found in IF steel during cold rolling.

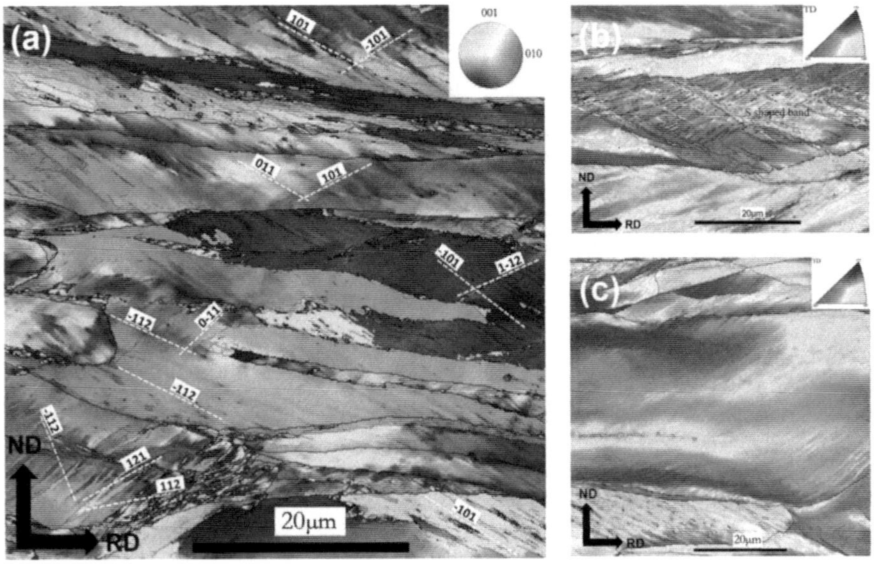

Figure 5. (a) Misorientation axis map of Ta-4%W after 70% cold rolling; (b) IPF map showing the S shaped bands; (c) IPF map showing distorted dislocation boundaries in near-grain boundary regions.

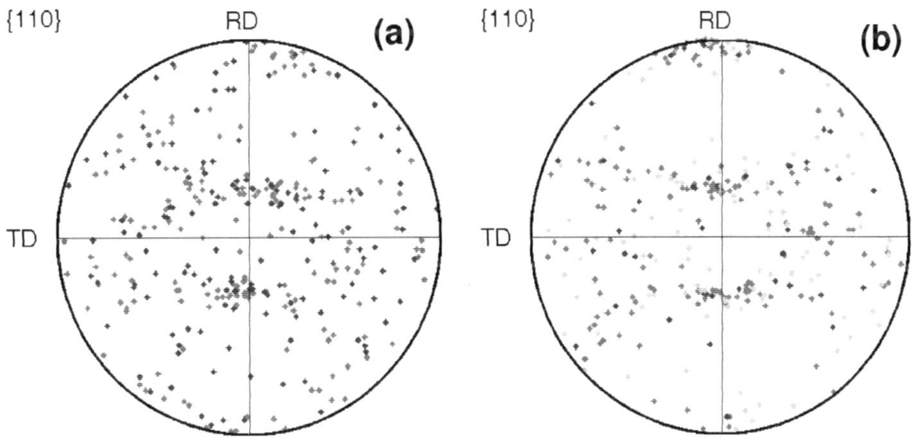

Figure 6. Pole figure showing grains with (a) one set and (b) two sets of dislocation boundaries. In the pole figure, red points represent GNBs close only to {110} traces; blue only to {112} traces, and green for type B grains where one set is close to a {110} trace and the other to a {112} trace.

In earlier studies of rolled Al and Cu a clear relationship between dislocation pattern and grain orientations was found. In the present study the grains or domains where dislocation boundaries are clearly seen are classified into two types: (i) type A (containing one set of GNB), and (ii) type B (containing two sets of GNBs). The orientations associated with each type are shown in figure 6. It is seen that the orientations of grains having type A structure are rather random. In contrast, the grains with type B structure have orientations belonging to the α-fibre (<110> || RD). The results demonstrate therefore an orientation dependence of slip pattern in rolled Ta-4%W. For grains with one set of GNBs, the orientations of the grains where boundaries are close to a {110} slip plane and a {112} slip plane are similar. For type B grains in most cases the GNBs are both close to {110} slip planes. Type B grains with GNBs near to two sets of {112} boundaries are only observed after 70% rolling.

In this investigation only the GNBs have been characterized. IDBs, as observed in deformed FCC metals by using transmission electron microscopy, cannot be determined in the present experimental because of the limited resolution of the EBSD technique. The density and arrangements of these boundaries are however also important and will be studied in future work.

4. Conclusions

A Ta-4%W alloy has been cold rolled to strains of 10% to 70%. The deformation microstructures after each rolling strain were characterized by the EBSD technique, and dislocation boundaries were revealed using an in-house code to plot maps showing the misorientation axis of each pixel to grain average in the sample reference frame. It is found that:

1. Grains develop one set or two sets of dislocation boundaries after rolling and the dislocation boundaries are typically parallel to {110} or {112} slip planes;
2. The grains having one set of dislocation boundary have random orientations, while grains having two sets of dislocation boundaries belong to the α-fibre.

Acknowledgments

This work was supported by the National Natural Science Foundation of China (51471039) and State Key Research and Development Program of China (2016YFB0700403).

References

[1] Kaufmann D, Mönig R, Volkert CA, Kraft O 2011 *Inter. J. Plast.* **27** 470
[2] Hupalo MF, Sandim HRZ 2001 *Mater. Sci. Eng. A* **318** 216
[3] Zhou JQ, Khan AS, Cai R, Chen L 2006 *J. Iron Steel Res.* **13** 68
[4] Kuznietz M, Livne Z, Cotler C, Erez G 1998 *J. Nucl. Mater.* **152** 235
[5] Deng C, Liu SF, Ji JL, et al 2014 *J Mater. Proc. Tech.* **214** 462
[6] Nesterenko VF, Meyers MA, LaSalvia JC, et al 1997 *Mater. Sci. Eng. A* **229** 23
[7] ElAbedin SZ, Welz-Biermann U, Endres F 2005 *Electrochemistry Commun.* **7** 941
[8] Liu Q, Hansen N 1995 *Scripta Metall. Mater.* **32** 1289
[9] Liu Q, Juul Jensen D, Hansen N 1998 *Acta Mater.* **46** 5819
[10] Li BL, Godfrey A, Meng QC, Liu Q, Hansen N 2004 *Acta. Mater.* **52** 1069
[11] Wang S, Chen C, Jia YL, et al 2016 *Int. J. Refract. Metals Hard Mater.* **54** 104
[12] Moussa C, Bernacki M, Besnard R, Bozzolo N 2015 *IOP Conf. Ser.: Mater. Sci. Eng.* in *Proc. of 36th Risø Inter. Sym. on Mater. Sci.* **89** 012038
[13] Hong X, Godfrey A, Zhang CL, Liu W, Chapuis A 2017 *Mater. Sci. Eng. A* **693** 14
[14] Albou A, Driver JH, Maurice C 2010 *Acta Mater.* **58** 3022

Heterogeneous multi-layered IF steel with simultaneous high strength and good ductility

Ling Zhang[1,2]*, Xiaojuan Jiang[1], Yuhui Wang[3], Qiang Chen[4], Zhen Chen[1], Yonghong Zhang[1], Tianlin Huang[1,2], Guilin Wu[1]

[1] College of Materials Science and Engineering, Chongqing University, Shazhengjie 174, Chongqing 400044, China
[2] Electron Microscopy Center of Chongqing University, Chongqing University, Chongqing 400044, P.R. China
[3] National Engineering Research Center for Equipment and Technology of Cold Strip Rolling, Yanshan University, Qinhuangdao 066004, China
[4] Southwest Technology and Engineering Research Institute, Chongqing 400039, China

* Corresponding author: zhangling2014@cqu.edu.cn

Abstract. Multi-layered IF steel samples were designed and fabricated by hot compression followed by cold forging of an alternating stack of cold-rolled and annealed IF steel sheets, with an aim to improve the strength of the material without losing much ductility. A very good combination of strength and ductility was achieved by proper annealing after deformation. Microstructural analysis by electron back-scatter diffraction revealed that the good combination of strength and ductility is related to a characteristic hierarchical structure that is characterized by layered and lamella structures with different length scales.

1. Introduction

In the past several years, achieving simultaneously high strength and good ductility has been realized by utilizing combinations of length scale, such as multi-layered materials, and bimodal (or multi-modal) grain size distributions [1-9]. The results of these studies have shown that it is possible in some cases to do better than just a simple trade-off between strength and ductility (i.e. sacrificing some strength for ductility). Recently, Wu et al. reported a heterogeneous lamella structure (a special case of a bimodal structure, with soft micro-grained lamella embedded in a hard ultrafine-grained lamella matrix) in Ti produced by asymmetric rolling and partial recrystallization that possesses an unprecedented property combination: as strong as ultrafine-grained metal and at the same time as ductile as conventional coarse-grained counterpart material [10]. In the cases above, the metals with bimodal structure have a mechanical incompatibility due to the bimodal grain size distribution, inducing a strain gradient and complex stress state, which generates geometrically necessary dislocations (GNDs) to accommodate the strain gradient and promote the generation and interaction of dislocations [10]. Furthermore, interfaces in the multi-layered materials are believed to be responsible for the observed high strain hardening and ductility, across which there were heterogeneities in

* To whom any correspondence should be addressed.

chemical composition, grain size, hardness and texture [1]. Some researchers have found that gradient distribution of stress near the interface significantly contributes to working hardening in multi-layered metals. Yang et al. [11] proposed that a strain gradient and back stress strengthening caused by the pileup of GNDs played a critical role in the observed high stress and high strain hardening. Therefore, it is promising that multi-layered metals with heterogeneous structure may yield a good combination of strength and ductility by combining both bimodal structure and interface effects.

In the present study, we tried three methods to obtain multi-layered IF steel sheets with an aim to improve the strength of the material without losing much ductility, including accumulative roll bonding (ARB), hot compression followed by hot rolling (HR) and cold forging (CF). The results of the last method will be discussed in detail. Different to other reported multi-layered materials, such as copper with bronze [1], Cu–10Zn with Cu [4], and Nb with Cu [5, 12, 13], single phase IF steel sheets with different initial structures (deformed and annealed) were stacked alternatively into multi-layers in the present experiment. The stacked plates were "welded" together by hot compression, followed by hot/cold deformation. A heterogeneous lamella structure was obtained by this strategy and a very good combination of strength and ductility was achieved by adjusting the heat treatment schedule after hot compression plus cold forging deformation. A characteristic hierarchical structure that is defined by two types of layered or lamella structures with different length scales has been fabricated, referred to hereafter in this work as a L2 structure (layer + lamella), and is discussed in detail.

2. Experimental

The initial material is 1 mm thick Ti-added (0.012 wt%C - 0.061 wt%Ti) commercial cold-rolled (78% reduction in thickness) IF steel sheet. One cold rolled sheet was annealed at 780°C to achieve complete recrystallization with average grain size about 20 μm. For the multi-layered heterogeneous sheets, the cold rolled (CR) and the annealed steel sheets (referred to hereafter by AR) were cut into plates with a diameter of 15 mm. The sample surfaces were ground and then degreased with acetone. Then, the CR and AR steel sheets were stacked in an alternate sequence into 11 layers, along the original rolling direction (RD). To weld the layers together, the stacked multilayer plates were hot compressed at 500°C by using a Gleeble 3800 thermo-mechanical simulator in vacuum. The hot-compressed specimens were air cooled and then cold forged (CF) at room temperature to a thickness of about 1.6 mm using a hydraulic press. After cold forging, the nominal deformation strains of the initial CR and AR layers are 96.8% and 85.5%, corresponding to von Mises strains of 4.0 and 2.2, respectively. Note that unless otherwise specified CR and AR indicate the initial CR and AR layers in the specimens. The cold forged specimens were then annealed at 625 °C for 0.5 hour or 1 hour. Tensile tests were performed along the original RD direction to evaluate the mechanical properties. Dog-bone shaped tensile samples with a gage length of 12 mm and a width of 2.5 mm were used. An Oxford Aztec electron back-scatter diffraction (EBSD) detector attached to a JEOL 7800F scanning electron microscope (SEM) was used to characterize the microstructures of the specimens. All the observations were conducted on the original longitudinal section (RD-ND plane).

3. Results and discussion

The microstructure of the 0.5 hour annealed sample is given in figure 1. Figure 1 (a) shows an inverse pole figure map and (b) shows a boundary map in which black lines represent high angle grain boundaries and red lines indicate low angle grain boundaries. It is seen that the mapping area contains an AR layer and a CR layer. In the AR layer, there are dominantly coarse (recrystallized) grains seen in blue (with ND aligned along the <111> direction). Within this layer, there are also fine-grained (deformed) lamellae (indicated by white arrows), approximately parallel to RD. In contrast in the CR layer, the microstructure is composed of fine-grained (deformed) matrix with mixed red color (with ND aligned along the <001> direction) and blue color. Within this layer, there are also coarse grains (indicated by yellow arrows). The red and blue regions also form heterogeneous lamellae similar to those seen in Ti [10]. The coarse grains are distributed only in the blue region so that the blue regions are the coarse lamellae in the CR layer, while the red regions are the fine lamellae. This characteristic hierarchical layer structure combined with a heterogeneous lamella structure differs from previously reported heterogeneous structures, including a heterogeneous lamella structure [10], a bimodal structure [14], a harmonic structure [15] and a gradient structure [16]. A bimodal distribution of misorientation angle and grain size are shown in figures 1c and 1d. The average grain sizes of the small grains and coarse grains based on the EBSD measurements are 1.6 μm and 11.1 μm, respectively.

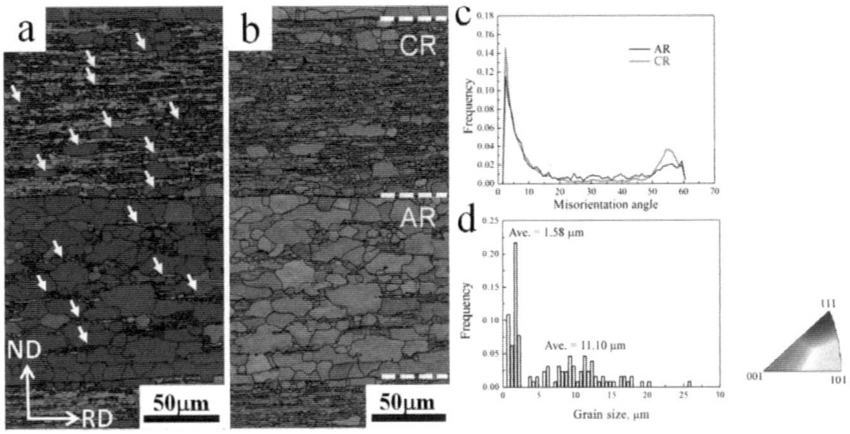

Figure 1. Example EBSD maps of the specimen annealed for 0.5 h: (a) inverse pole figure map, (b) boundary map. Thick black lines represent high angle grain boundaries and thin red lines small angle grain boundaries. The interfaces are indicated by yellow dashed lines in (b). (c) and (d) are misorientation angle distribution and grain size distribution based on the EBSD mapping. In order to show the bimodal grain size distribution, same number of grains were counted for the coarse and fine grains.

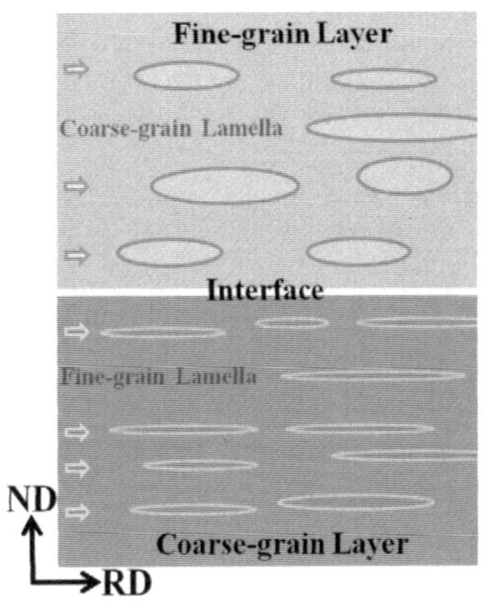

Figure 2. Schematic illustration of the L2 structure.

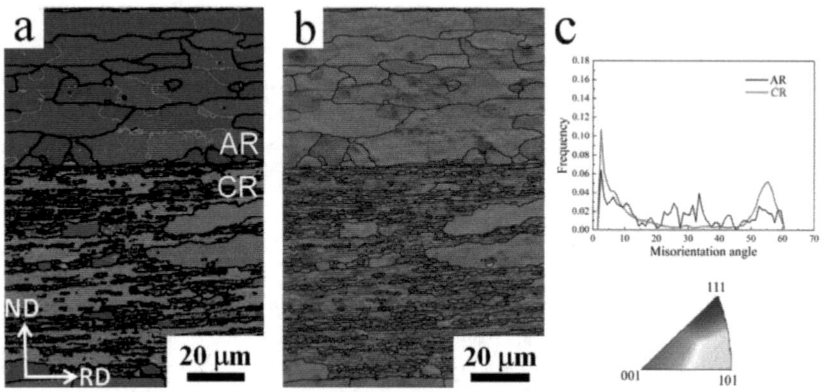

Figure 3. Example EBSD maps of the sample annealed for 1h: (a) inverse pole figure map, (b) boundary map; (c) misorientation angle distribution.

Figure 2 shows schematically the structure of present steel after annealing. The structure contains alternating layers, labelled here as fine-grained layers and coarse-grained layers. In the fine-grained layers, some coarse-grained lamellae are distributed approximately along the RD; meanwhile, fine-grained lamellae decorate the coarse-grained layers. As a result this combination of features is defined as a L2 (layer + lamella) structure.

Figure 3 shows the microstructure of the 1 hour annealed sample. A similar structure can be seen in this sample, especially in the CR layer. However, due to the longer annealing time, the red lamellae in the AR layer are consumed completely, leaving some small angle grain boundaries. The average grain sizes do not change too much. However, the misorientation distribution in the AR layer is clearly more random, as shown in figure 3c.

Figure 4 gives the true stress - true strain curves (S-S curve) for the annealed samples. The S-S curves of the CF, AR and CR samples are also plotted for comparison. The S-S curve of the AR sample has the largest uniform elongation with a yield strength of about 150 MPa, which is typical for annealed IF steel. The heavily deformed samples, such as the CF sample and CR sample, have almost no uniform elongation. The strength of the CF sample is higher than that of the CR sample, due to its deformation strain. After heat treatment at 625°C for 0.5 h, the uniform elongation is about 30%, and the yield strength is twice that of the AR sample. Annealing for 1 h results in a decrease in strength, but still higher than that of the sample fabricated by accumulative roll bonding (ARB) to a similar strain followed by annealing, as shown in the "ARB-1.6 μm" curve [17]. The average grain size of the fine grain layer in the 1 hour annealed sample is about 1.6 μm, comparable to the grain size in Ref. [17], but the present results show that a heterogeneous lamella structure can yield better strength than a homogeneous ultra-fine grain structure. Furthermore, the ARB fabricated homogeneous IF steel shows a yield drop phenomenon, which is not observed in the present study. It is also seen that longer annealing time can result in a less typical L2 structure. This might be the reason why the uniform elongation of the 1 hour annealed sample does not increase, as it seems that it does not follow the same trend as that of the 0.5 hour annealed sample.

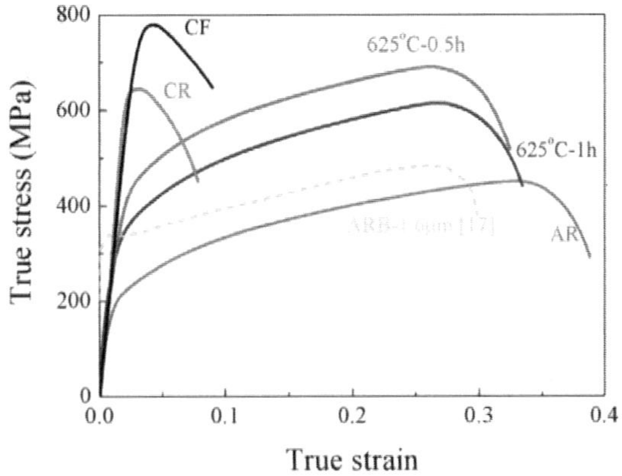

Figure 4. True stress - true strain curves of the cold forged and annealed samples. S-S curves of the AR, CR samples and a 1.6 μm grain size sample (reproduced from Ref. [17]) are also plotted for comparison.

Figure 5 gives the yield strength and uniform elongation combinations obtained in the present study as well as those reported in other references [17-21]. It is clear that the combination of strength and ductility in the present study is superior to those of previously reported data, for example homogeneous ultra-fine grained IF steel prepared by equal-channel angular pressing/extrusion [19, 21] and ARB [17] and heterogeneous structure by asymmetric rolling (AsyR) [20]. It is thus speculated that the ultra-fine grains impart high strength while the large grains stabilize the tensile elongation of the material, accommodating the mechanical incompatibility due to the interfaces between layers and lamellae with different grain sizes and orientations within the layers. The mechanical incompatibility induces a strain gradient and complex stress state, which can generate geometrically necessary dislocations (GNDs) to accommodate the strain gradient and promote the generation and interaction of dislocations. Consequently, a high strength and good ductility can be achieved in the present steel.

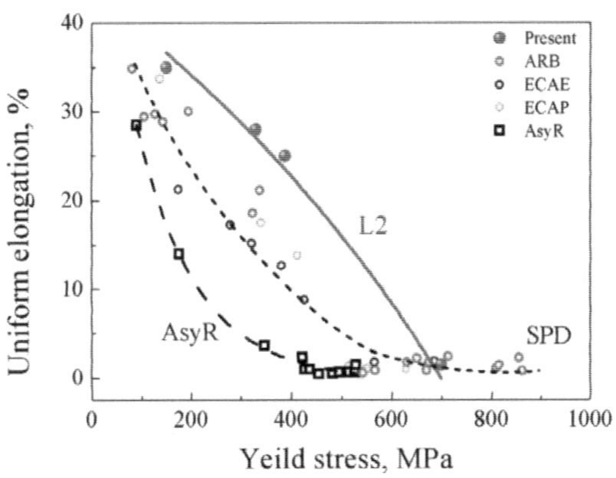

Figure 5. Comparison of yield strength and uniform elongation combinations between the present study and other data obtained from references [17-21] as indicated in the figure.

Important parameters that may affect the L2 structure and mechanical properties are the initial material, layer thickness, sample stacking sequence, welding process, deformation strains, heat treatment parameters, etc. Systematic studies the effects of these afore-mentioned parameters on microstructures and mechanical properties, underlining mechanisms and large-scale production are necessary in the future. It is believed that further improved mechanical properties can be achieved by tailoring the L2 structure. Also, the plausibility of this strategy on other material systems is worth of investigation. The results of the present study nevertheless form a good basis for further studies in this research area.

4. Summary

An IF steel with a characteristic hierarchical layer + lamella structure (L2 structure for short) was produced by interleaving cold rolled and annealed commercial IF steel sheets prior to cold deformation followed by an appropriate post-deformation annealing. Compared to homogeneous fine structured IF steels, it is shown that the L2 structured IF steel exhibits a superior combination of strength and uniform elongation. Both the yield/ultimate strength and ductility in uniaxial tension are improved compared to previous studies, and it is proposed that this enhanced combination of mechanical properties is attributed primarily to the effect of mechanical incompatibility across both layer interfaces and the micro-scale lamella structure with different grain sizes. By tailoring the processing parameters, and thus the L2 structure, still better properties may be achieved.

Acknowledgements

The authors wish to thank the financial support of the Natural Science Foundation of China (NSFC, Grants No. 51541101, 51671039, 51471039), State Key Research and Development Program of MOST of China (2016YFB0700401, 2016YFB0700403).

References

[1] Ma X, Huang C, Moering J, Ruppert M, Höppel H W, Göken M, Narayan J and Zhu Y 2016 *Acta Mater.* **116** 43-52
[2] Wang Y M and Ma E 2004 *Acta Mater.* **52** 1699-1709
[3] Ma E 2006 *JOM* **58** 49-53
[4] Ma X L, Huang C X, Xu W Z, Zhou H, Wu X L and Zhu Y T 2015 *Scr. Mater.* **103** 57-60
[5] Misra A, Hirth J P and Hoagland R G 2005 *Acta Mater.* **53** 4817-4824
[6] Liang F, Tan H F, Zhang B and Zhang G P 2017 *Scr. Mater.* **134** 28-32
[7] Liu H S, Zhang B and Zhang G P 2011 *Scr. Mater.* **65** 891-894
[8] Du Y, Fan G, Yu T, Hansen N, Geng L and Huang X 2016 *Mater. Sci. Eng.* A **673** 572-580
[9] Wang Y, Chen M, Zhou F and EnMa 2002 *Nature* **419** 912-915
[10] Wu X, Yang M, Yuan F, Wu G, Wei Y, Huang X and Zhu Y 2015 *Proc Nat Acad Sci USA* **112** 14501-14505
[11] Yang M, Pan Y, Yuan F, Zhu Y and Wu X 2016 *Materials Research Letters* **4** 145-151.
[12] Zheng S, Beyerlein I J, Carpenter J S, Kang K, Wang J, Han W and Mara N A 2013 *Nature Communications* **4** 1696
[13] Beyerlein I J, Mayeur J R, Zheng S, Mara N A, Wang J and Misra A 2014 *Proc. Nat. Acad. Sci.* USA **111** 4386-4390
[14] Zhao Y, Topping T, Bingert J F, Thornton J J, Dangelewicz A M, Li Y, Liu W, Zhu Y, Zhou Y and Lavernia E J 2008 *Advanced Materials* **20** 3028-3033
[15] Vajpai S K, Sawangrat C, Yamaguchi O, Ciuca O P and Ameyama K 2016 *Mater. Sci. Eng.* C **58** 1008-1015
[16] Wu X L, Jiang P, Chen L, Zhang J F, Yuan F P and Zhu Y T 2014 *Materials Research Letters* **2** 185-191
[17] Tsuji N, Ito Y, Saito Y and Minamino Y 2002 *Scripta Mater.* **47** 893-899
[18] Huang X, Kamikawa N and Hansen N 2008 *Mater. Sci. Eng.* A **493** 184-189
[19] Purcek G, Saray O, Karaman I and Maier H J 2012 *Metall. Mater. Trans.* A **43** 1884-1894
[20] Lapovok R, Orlov D, Timokhina I B, Pougis A, Toth L S, Hodgson P D, Haldar A and Bhattacharjee D 2011 *Metall. Mater. Trans.* A **43** 1328-1340
[21] Hazra S S, Pereloma E V and Gazder A A 2011 *Acta Mater.* **59** 4015-4029

Local microstructure and flow stress in deformed metals

X Zhang[1], N Hansen[1], C V Nielsen[2]

[1]Section for Materials Science and Characterization, Department of Wind Energy, Technical University of Denmark, DK-4000 Roskilde, Denmark
[2]Department of Mechanical Engineering, Technical University of Denmark, Produktionstorvet 425, 2800 Kgs. Lyngby, Denmark

Email: xzha@dtu.dk

Abstract. The microstructure and flow stress of metals are related through many well-known strength-structure relationships based on structural parameters, where grain size and dislocation density are examples. In heterogeneous structures, the local stress and strain are important as they will affect the bulk properties. A microstructural method is presented which allows the local stress in a deformed metal to be estimated based on microstructural parameters determined by an EBSD analysis. These parameters are the average spacing of deformation introduced boundaries and the fraction of high angle boundaries. The method is demonstrated for two heterogeneous structures: (i) a gradient (sub)surface structure in steel deformed by shot peening; (ii) a heterogeneous structure introduced by friction between a tool and a workpiece of aluminum. Flow stress data are calculated based on the microstructural analysis, and validated by hardness measurement and 2D numerical simulations. A good agreement is found over a plastic strain range from ~1 to 5.

1. Introduction

Structure-strength relationships for metallic materials provide guidelines for the design and processing of engineering materials and components. An example of such relationships is the Hall-Petch equation [1,2], which relates the yield (or flow) stress to the inverse square root of the grain size, or to the spacing between boundaries in the structure that act as barriers to dislocation glide [1-3]. Specimens where such a relationship is tested are typically thermo-mechanically processed, for example, by rolling, drawing and torsion, and they are characterized as bulk materials on the assumption that the structure is homogeneous. However, during plastic deformation the stress and strain distribution in the metal may not be homogeneous, for example, where friction and wear are an inherent part of the process. Heterogeneities may also be introduced deliberately, for example by plastic deformation of a surface to increase wear and fatigue resistance of rolling components.

Characteristic heterogeneous structures can be graded or multilayer structures, and the local stress and strain distribution can be characterized by many mechanical and physical methods, such as hardness testing and methods using X-ray diffraction [4,5]. However, as the structural scale may range from the micrometer scale to the nanometer dimension, it has been suggested [4,6] to apply electron microscopy techniques, such as electron backscatter diffraction (EBSD) and transmission electron microscopy (TEM). By using such techniques it is possible to characterize microstructural parameters that can be the basis for calculation of the strength, for example, by applying the Hall-Petch

IOP Conf. Series: Materials Science and Engineering **219** (2017) 012053 doi:10.1088/1757-899X/219/1/012053

relationship. This means that it is possible to estimate the local stress distribution over multiple length scales in a heterogeneous deformation microstructure.

In the present paper we shall demonstrate the application of the microstructural techniques for two heterogeneous microstructures: (i) a gradient lamellar (sub)surface structure in steel deformed by shot peening, and (ii) a heterogeneous structure formed during a metal forming operation due to friction between the tool and a workpiece in aluminum. Calculated stress gradients are validated by microhardness measurements for the fomer case and by 2D numerical simulation for the latter case.

2. Strength-structure relationship

A number of structural parameters from microstructural analysis can be identified as strengthening parameters. These are the spacing (D) between, and the misorientation angle (θ) across, the deformation induced boundaries, where the misorientation angles are separated into low angles (θ < 15°) and high angles (θ ≥ 15°), thereby defining the two main strengthening mechanisms, corresponding to dislocation strengthening and boundary strengthening, respectively.

Dislocation strengthening is typically accounted for by a Taylor strengthening [7] where, for example, the yield stress is taken to be proportional to the square root of the dislocation density stored both in dislocation boundaries and as a small fraction in the volume between boundaries. Boundary strengthening can be accounted for by a Hall-Petch relationship. The flow stress can then be estimated as the sum of these two contributions with the addition of the friction stress and a contribution from dispersed particles (in the case of Al).

However, a modification has been introduced [4] as there appears to exist an angle (β) above which the boundaries change from being penetrable to be impenetrable. This critical angle is estimated to be ~ 6° for Fe [6] and ~ 2° for pure Al [8]. It follows that the structure-strength relationship can be written:

$$\sigma - \sigma_0 - \sigma_{particle} = \left[k_{HP}^{*}\sqrt{f^{\theta>\beta}} + M\alpha G\sqrt{3b\theta^{\theta\le\beta}(1-f^{\theta>\beta})} \right] D_{av}^{-0.5} \quad (1)$$

$$\sigma - \sigma_0 - \sigma_{particle} = k_2 D_{av}^{-0.5} \quad (2)$$

where $f^{\theta>\beta}$ is the fraction of boundaries with a misorientation angle higher than the critical value, σ_0 is the friction stress, $\theta^{\theta\le\beta}$ is the average value of misorientation angles lower than the critical angle, β, and D_{av} is the average spacing between all boundaries measured along random intersection lines.

k_{HP} is typically determined for a polycrystalline structure subdivided by high angle boundaries (≥ 15°). In the present case, boundary strengthening includes contributions from both high angle boundaries (≥ 15°) and low angle boundaries (≥ β). The value of k_{HP}^{*} is larger than k_{HP} and has to be derived experimentally. This is done based on analysis of structures which have been homogeneously deformed in rolling for Fe [4] and in compression for Al. Based on these data and combined with equation (2) the following average values have been obtained: for cold-rolled Fe, $k_2 = 310 \pm 11$ MPa μm$^{0.5}$; for cold compressed Al, $k_2 = 112 \pm 10.5$ MPa μm$^{0.5}$.

3. Experimental results and validation

3.1 Shot peened low carbon steel

A low carbon steel (Fe-0.004 wt. % C) bar of diameter Ø = 12 mm, with a grain size of 40 μm has been shot peened using 0.8 mm high-carbon steel balls (0.91 wt. % C, HRC 62) for 5 min. The shot velocity was ~260 m/s and the coverage was 200 %.

Backscatter imaging and electron backscatter diffraction (EBSD) have been applied to quantify the microstructure [9-12]. The step sizes for EBSD scanning are 1000 nm, 200 nm, 120 nm, 50 nm and 20 nm, chosen according to the fineness of the deformation structure with finer map step sizes used for finer deformation structures.

Figure 1 shows the microstructure from the top surface to a position ~ 460 μm from the surface in the shot peened low carbon steel. The EBSD microstructure is consistent with the microhardness profile as shown in figure 1e, with the structure divided into three zones: Zone A (0-110 μm) has a lamellar structure parallel to the surface, with a sharp hardness increase as the surface is approached; Zone B (110-600 μm) has deformed coarse grains with dislocation boundaries and tangles with a slow hardness increase as the surface is approached; and Zone C (600 μm-) is the undeformed matrix with an almost constant hardness.

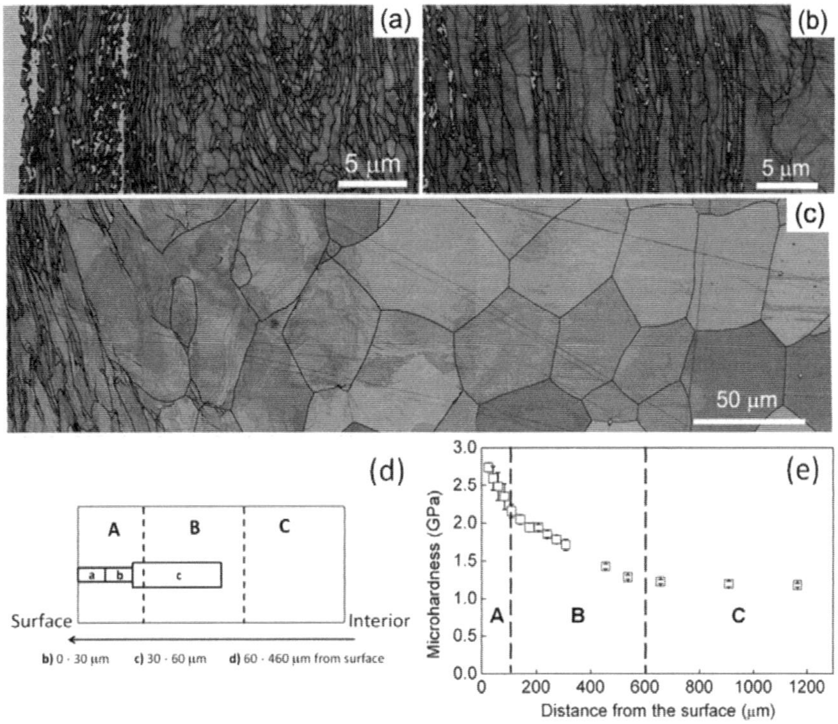

Figure 1. (a-c) EBSD microstructure (greyscale corresponding to band contrast) at the locations shown in the schematic illustration (d). Capital letters A, B and C correspond to areas marked in the microhardness distribution (e) as a function of distance from the surface in a shot-peened low carbon steel [4].

The key structural parameter, D_{av}, (see equation (2)) is given in table 1 at different distances from the surface. The calculated flow stress is shown in figure 2 as a function of the distance from the surface. Note that the maximum stress is ~ 1400 MPa at the surface, which is 6-7 times the flow stress of the undeformed steel in the interior volume. The calculated flow stress is validated by microhardness data included in figure 2, where the flow stress is converted via the relationship HV/3.5 = $\sigma_{(0.2\%)}$, and where good agreement is found. However, the microstructural parameters can be quantified up to the surface, whereas the microhardness data are limited to depths of below 25 μm from the surface. For this region we extrapolate the hardness curve of the top 300 μm area to the surface by $HV = 4830 \times d^{-0.17}$, where d is the distance to the surface and HV is the Vickers hardness determined with a load of 50g with the holding time of 20 s. Based on equation (1), D_{av} at the surface can be estimated to be ~ 70 nm.

Table 1. Average boundary spacing for shot-peened samples (SP) at different distances from the surface (data based on EBSD analysis) [4].

Location	D_{av} (nm)
10-0μm_SP	115
20-10μm_SP	181
40-20μm_SP	260
60-40μm_SP	360

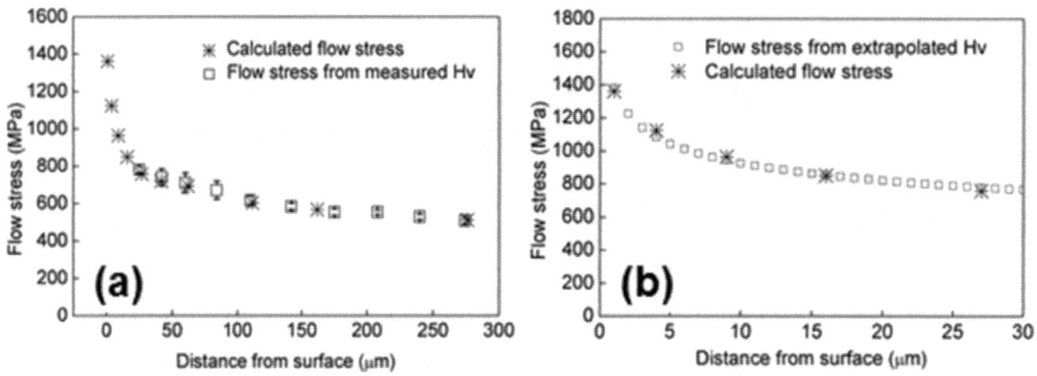

Figure 2. Calculated flow stress and measured flow stress based on microhardness (a) and its extrapolation (b) for the shot-peened low carbon steel. [4]

3.2 Compressed AA 1050 ring

In this experiment the objective is to investigate the stress distribution in a sample deformed heterogeneously. A six–step ring compression test was carried out, where the friction between the tool and the workpiece was maximized by choosing rough tools and poor lubrication, resulting in close to full sticking conditions (a friction factor of $f = 1$). Kerosene, commonly used for cleaning purposes, was used as a lubricant. The initial ring was prepared with an outer diameter of $D_0 = 18$ mm, an inner diameter of $d_0 = 9$ mm, and a height of $h_0 = 6$ mm.

Figure 3 shows the cross section of a ring compressed to a height reduction of 54.7 % (true strain of 0.6) where seven representative areas have been characterized by EBSD. The heterogeneous microstructure is pronounced, including (i) slightly deformed areas where the original grain shape is preserved and with a small amount of dislocation boundaries in the grains where the sticky zone is located (Area 3, figure 3d); (ii) compressed original grains with elongated dislocation boundaries in the grains (Area 7, figure 3h); (iii) equiaxed dislocation cells in the original grains [13, 14] (Area 2, figure 3c); (iv) equiaxed subgrains surrounded by low angle dislocation boundaries (Area 1, figure 3b); (v) inclined dislocation cell structures with one set of dense dislocation walls (DDWs) mixed with equiaxed dislocation cells (Area 5, figure 3f); (vi) inclined dislocation cell structures with two sets of DDWs dividing the original grains (Area 6, figure 3g); and (vii) parallel lamellar dislocation structures with shear bands (indicated by white arrows), which represent the microstructure typical of high strain deformation (Area 4, figure 3e). The average spacing between all boundaries, D_{av}, in each of these areas is shown in table 2. Note that Area 3, is almost undeformed, containing only the original grain boundaries, so the method based on deformation sub-grain structures is not applied here.

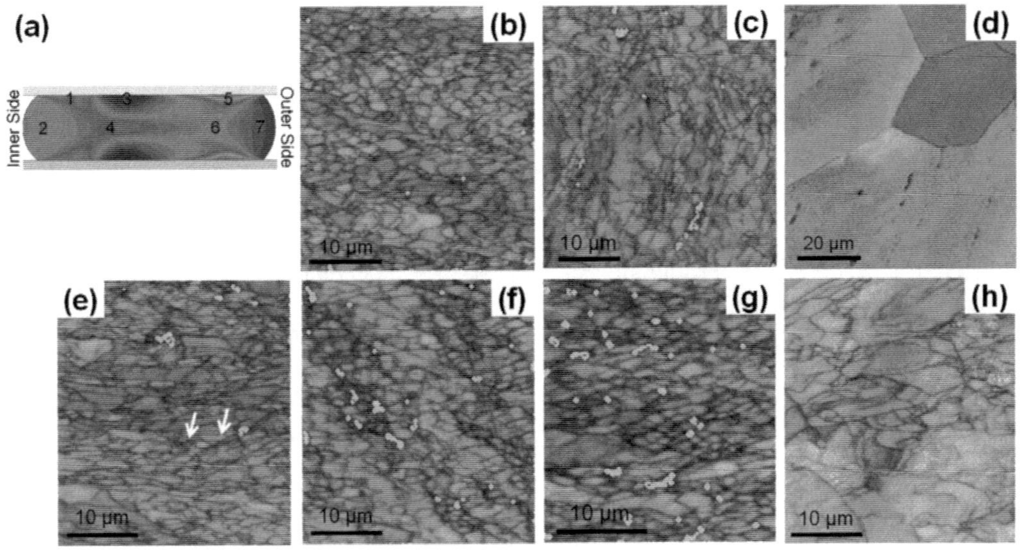

Figure 3. EBSD micrographs showing dislocation structures of the microstructure in the compressed AA 1050 ring. (b)-(h) show the microstructure of Areas 1-7 as marked in the simulated effective strain distribution map of the ring cross section (a).

Table 2. D_{av} of representative areas in the compressed AA 1050 ring

Position	Area 1	Area 2	Area 4	Area 5	Area 6	Area 7
D_{av} (μm)	0.95	1.05	0.74	0.84	0.85	1.21

In this study, the experimental analysis is validated by applying axisymmetric finite element modeling with the underlying calculations being built on the irreducible flow formulation, obtained by energy minimization. The material is assumed to be rigid-plastic and the tool is assumed to be perfectly rigid. For more details on the numerical implementation, see [15]. The friction model employed between the workpiece material and the tool surface is [16],

$$\tau = f\alpha k \tag{3}$$

where f is the friction factor to be calibrated, α is the ratio of real to apparent contact area, and k is the instantaneous shear flow stress obtained by the finite element simulation, taking strain hardening into account. The real contact area, is determined by an analytical expression suitable for finite element implementation [17].

For the seven areas, the calculated and simulated flow stresses are shown in table 3. The numbers are in good agreement taking into account an estimated standard deviation of around 6 %.

Table 3. Simulated strain, calculated flow stress (σ_{cal}) and simulated flow stress (σ_{simul}) of the seven representative areas in the ring (see figure 3).

Area	**2**	**1**	**4**	**3**	**6**	**5**	**7**
ε_{simul}	1.06	1.8	1.84	0.03	1.13	2.55	0.41
σ_{cal} **(MPa)**	127.1 ± 10.3	132.7± 10.8	147.9 ± 12.2	-	139.2 ± 11.4	140.1 ± 11.5	119.7 ± 9.6
σ_{simul} **(MPa)**	127.7	144.2	145	54.9	129.6	156.3	102.6

4. Discussion

This study encompasses four types of deformation microstructure, where two are homogeneous in rolled steel and compressed aluminum, and the other two are heterogeneous in shot-peened steel and friction deformed aluminum. All four structures are typical deformation microstructures subdivided by low and high angle dislocation boundaries. For all specimens except one, the effective strain is in the range from 1 to 5. The structure in each case is a bamboo structure [18] with extended medium and high angle boundaries (geometrically necessary boundaries (GNBs)) and interconnecting low angle boundaries (incidental dislocation boundaries (IDBs)). In previous analyses, the microstructure has been described in terms of the spacing between, and the misorientation across, the GNBs and IDBs as determined by TEM [3]. In the present more simple analysis, the structural parameters are quantified by EBSD where the angular resolution is ~ 1° and where the spacing between IDBs and GNBs, respectively is replaced by the average boundary spacing between all boundaries independent of the structural morphology.

In applying equation (2), it should be noted that k_2 is not constant, and may be expected in fact to increase with strain, due to an increase of θ^{LAB}, which is proportional to the square root of the strain [19]. However, at large strains an increased contribution from low-angle boundaries will be balanced by an increased fraction of high angle boundaries. As a result, k_2 is found not to be strongly dependent on the strain.

The local flow stress in the two heterogeneous structures has been validated by stress profiles obtained by microhardness measurements (for steel) and numerical simulation (for Al). A good agreement has been found supporting the application of equation (2) to a structural scale as low as about 50 nm. A further reduction in the structural scale down to or below ~10 nm requires an optimization of thermomechanical treatments and structural parameters. Example systems where such a scale may be reached include pearlitic steel wires [9, 11, 20-25], Cu alloyed with Fe [26] and Ni alloyed with Mo [27]. An extension of the present research will require a validation of the calculated stress profiles on an extremely fine scale together with mechanical testing at high spatial resolution, for example, in-situ TEM testing [28] and nano-indentation. However, such kinds of fine-scale testing also must take into account test specimen size effects or indentation size effects, to be analyzed in a forthcoming paper.

5. Conclusion

A microstructural method is described that allows the estimation of the local flow stress in a deformation microstructure, and its use has been demonstrated for two heterogeneous microstructures in steel and aluminium deformed to a maximum effective strain of 5 with a local structural scale as low as 50 nm. Based on this work the following conclusions can be made:

- The local flow stresses can be calculated with good accuracy based on microstructural parameters determined by EBSD, as demonstrated through a good agreement with microhardness measurements and with the results of numerical simulations.
- The microstructural method allows the flow stress distribution in a deformed sample to be determined in a precise manner, and can be used to underpin both analytical and numerical modelling, thereby finding applications in the design, processing and performance of metallic metals for engineering components.

Acknowledgements

The authors would like to thank The Danish Council for Independent Research (Grant number: DFF-1335-00230) for financial support.

References

[1] Hall E O 1951 *Proc. R. Soc. Lond. B.* **64** 747
[2] Petch N J 1953 J. Inst Steel Inst. **174** 25

[3] Hansen N 2004 *Scripta Mater.* **51** 801
[4] Zhang X, Hansen N, Gao Y and Huang X 2012 *Acta Mater.* **60** 5933
[5] Larson B C, Yang W, Ice G E, Budai J D and Tischle J Z 2002 *Nature* **415** 887
[6] Heilmann P, Clark W A T and Rigney D A 1983 *Acta Mater.* **31** 1293
[7] Taylor G L 1934 *Proc. R. Soc. Lond.*. **145** 362
[8] Kamikawa N, Huang X, Tsuji N, Hansen N 2009 *Acta Mater.* **57** 4198
[9] Zhang X, Godfrey A, Hansen N, Huang X, Liu W and Liu Q 2010 *Mater. Charac.* **61** 65
[10] Fan G H, Wang Q W, Du Y, Geng L, Hu W, Zhang X and Huang Y D 2014. *Materials Science and Engineering A* **590** 318.
[11] Zhang X, Godfrey A, Hansen N and Huang X 2013 *Acta Mater.* **61** 4898
[12] Zhu K N, Godfrey A, Hansen N and Zhang X D 2017 *Materials & Design* **117** 95
[13] Huang X and Hansen N 1997 *Scripta Mater.* **37** 1
[14] Le G M, Godfrey A, Hong C S, Huang X and Winther G 2012 *Scripta Mater.* **66** 359
[15] Nielsen CV, Zhang W, Alves L M, Bay N and Martins P A F 2013 *Modeling of thermoelectro-mechanical manufacturing processes with applications in metal forming and resistance welding*. Springer-Verlag, London, UK.
[16] Wanheim T and Bay N 1978 *Annals of the CIRP* **27** 189
[17] Bay N 1987 *Journal of Mechanical Working Technology* **14** 203
[18] Kuhlmann-Wilsdorf D and Hansen N 1991 *Scripta Metall. Mater.* **25** 1557.
[19] Pantleon W and Stoyan D 2000 *Acta Mater.* **48** 3005
[20] Zhang X D, Godfrey A, Liu W and Liu Q 2011 *Mater. Sci. Techno.* **27** 562
[21] Zhang X, Hansen N, Godfrey A and Huang X 2016 *Acta Mater.* **114**, 176
[22] Zhang X, Godfrey A, Huang X, Hansen N, Liu W and Liu Q 2014 *30th Risø International Symposium on Materials Science: Nanostructured metals: Fundamentals to applications (Risø, Denmark, 7-11 September 2014)* pp 409
[23] Zhang X, Hansen N, Godfrey A and Huang X 2014 *35th Risø International Symposium on Materials Science: New Frontiers of Nanometals (Risø, Denmark, 1-5 September 2014)* pp 153
[24] Zhang X, Godfrey A, Huang X, Hansen N and Liu Q 2011 *Acta Mater.* **59** 3422
[25] Li Y, Raabe D, Herbig M, Choi P-P, Goto S, Kostka A, Yarita H, Borchers C and Kirchheim R 2014 *Phys. Rev. Lett.* **113** 106104
[26] Hughes D A and Hansen N 2014 *Phys. Rev. Lett.* **112** 135504
[27] Hu J, Shi Y N, Sauvage X, Sha G and Lu K 2017 Science **355** 6331
[28] Zhang X, Godfrey A, Winther G, Hansen N and Huang X 2012 *Mater. Charac.* **70** 21

IOP Conf. Series: Materials Science and Engineering **219** (2017) 012054 doi:10.1088/1757-899X/219/1/012054

Synchrotron measurements of local microstructure and residual strains in ductile cast iron

Y B Zhang [1], T Andriollo [2], S Fæster [1], W Liu [3], A Sturlason [4], R Barabash [5]

[1] Section for Materials Science and Advanced Characterization, Department of Wind Energy, Technical University of Denmark, Risø Campus, Roskilde 4000, Denmark
[2] Department of Mechanical Engineering, Technical University of Denmark, DK-2800 Kgs. Lyngby, Denmark
[3] Advanced Photon Source, Argonne National Laboratory, Argonne, IL, 60439-4800, USA
[4] Vestas Wind Systems A/S, 8200 Aarhus N, Denmark
[5] Materials Science & Technology Division, Oak Ridge National Laboratory, P.O. Box 2008, Oak Ridge, TN, 37831-6118, USA

E-mail: yubz@dtu.dk

Abstract. The local microstructure and distribution of thermally induced residual strains in ferrite matrix grains around an individual spherical graphite nodule in ductile cast iron (DCI) were measured using a synchrotron X-ray micro-diffraction technique. It is found that the matrix grains are deformed, containing dislocations and dislocation boundaries. Each of the residual strain components in the matrix grains exhibits a complex pattern along the circumferential direction of the nodule. Along the radial direction of the nodule, strain gradients from the interface to the grain interior are seen for some strain components, but only in some matrix grains. The observed residual strain patterns have been analysed by finite element modelling, and a comparison between the simulation and experiments is given. The present study of local residual stress by both experimental characterization and simulation provide much needed information for understanding the mechanical properties of DCI, and represent an important contribution for the microstructural design of new DCI materials.

1. Introduction

The mechanical properties of cast iron depend strongly on the microscopic morphology and distribution of the graphite particles present in the iron matrix. It is well known that spherical graphite nodules in ductile cast iron (DCI) result in better impact and fatigue resistance than graphite flakes in grey iron. Even within the DCI family, the mechanical properties are related directly to the nodularity of the graphite particles [1]. It is suggested that whereas sharp graphite flakes create stress concentration points within the metal matrix, rounded nodules reduce stress concentration and inhibit the creation of cracks [2], thereby enhancing the mechanical properties. To understand this effect and to design new DCI with improved fracture toughness and fatigue life, it is important to obtain information about the local stress in the matrix grains around individual graphite nodules.

Recently, the thermal residual strain/stress in a metal mold DCI sample has been measured using a novel synchrotron X-ray technique. It has been found that a compressive residual elastic stress with a maximum of ~180 MPa is present near the graphite nodules, extending into the matrix to a distance of

IOP Conf. Series: Materials Science and Engineering **219** (2017) 012054 doi:10.1088/1757-899X/219/1/012054

about 20 μm, where the elastic stress is near zero [3]. However, only one strain/stress component in one matrix grain around each nodule is reported in [3].

Following this previous research, the local strain tensors for several matrix grains around one graphite nodule in DCI are studied using synchrotron X-ray micro-diffraction. DCI with ferrite matrix and used broadly in modern industry was chosen for the study. The aim is to quantify the pattern of the local residual strain/stress in different matrix grains around a selected graphite nodule, and to use a finite element modelling [4] to predict the residual stresses in the matrix grains.

2. Experimental

A DCI sample cast in a metal mold was chosen for the study. The sample consists of almost spherical graphite nodules and a metal matrix with a relatively homogeneous structure, being mainly ferrite with a small fraction of pearlite (~5%). The graphite nodules, with a volume fraction of 11.5% and an average size of ~30 μm, are distributed relatively uniformly in the matrix. The ferrite matrix has an average grain size of ~30 μm. More information about the sample can be found in [3].

The synchrotron experiments were performed at beamline 34-ID-E at the Advanced Photon Source, Argonne National Laboratory using the white beam Laue micro-diffraction technique [5]. A focused microbeam with Lorentzian profile and a full-width half maximum of ~0.5 μm was used. The X-ray energies were in the range 7-30 keV. A Pt-wire of 100 μm diameter was used as a differential aperture to resolve the white beam Laue diffraction pattern from individual volume elements at different depths. The depth-resolved Laue patterns were then indexed for determining the crystallographic orientations and the deviatoric elastic strain tensor of the diffracting volume elements. This analysis was conducted using the LaueGo software package available at APS beamline 34-ID-E [6].

A graphite nodule with a size (equivalent sphere diameter) of ~50 μm was chosen, based on a tomographic characterization using a Zeiss Xradia 520 Versa μ-CT system carried out in the home laboratory prior to the synchrotron measurements. The detailed scanning position of the synchrotron X-rays relative to the selected nodule is sketched in figure 1.

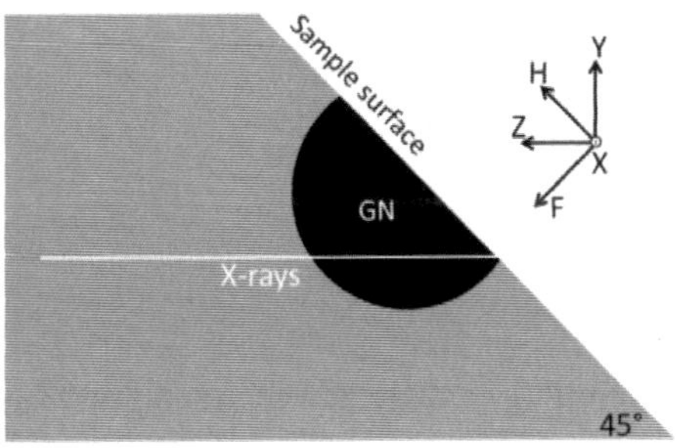

Figure 1. Sketch showing a side view of the X-ray scanning position relative to the selected graphite nodule (GN, shown in black). The white line represents the projection of the mapping plane illuminated by the incoming X-rays, which are along the Z direction. The laboratory coordinate system (X, Y, Z) and sample coordinate system (X, H, F) are defined.

3. Results and discussion

3.1. Microstructure

The microstructure in a slice (i.e. a 2D section) cutting through the selected nodule (GN) is shown in figure 2. In figure 2a, dislocation boundaries with misorientation angle higher than 0.1° are shown together with the crystallographic orientations of the matrix grains. It is found that some of the matrix grains are plastically deformed as they contain dislocation boundaries with misorientation angles up to 1°.

The number of indexed spots in the depth-resolved pattern for each volume element is shown in figure 2b. There is nearly no diffraction signal from the graphite nodule, appearing as "black holes". In the matrix deep bulk volumes appear darker due both to X-ray absorption and plastic deformation. The non-indexed volume in the deep region may be pearlite. The volumes close to sample surface also appear dark. This is an artifact arising from reconstruction of the depth-resolved diffraction images, which creates a positional uncertainty of 2-3 μm at the real sample surface.

Figure 3a is a plot of misorientation angle to the first point measured along each of the 5 lines marked in figure 2a. The angle increases with increasing distance from the interface. The angle increase is almost monotonic, reflecting a strain gradient formed during cooling due to the different thermal contraction of the nodule and matrix. The strain gradient is a result of the strain accommodation process, leading to the formation of geometrically necessary dislocations (GNDs) [7]. The density of these dislocations can be estimated based on the misorientation angle and distance using the following equation [8]:

$$\rho = K/D_c \cdot \theta_c/b, \quad (1)$$

where K is a number typically equal to 3, D_c is the average spacing between dislocation boundaries (taken here as the distance between neighboring measurement points), θ_c is the misorientation angle between two neighboring points, and b is the Burgers vector.

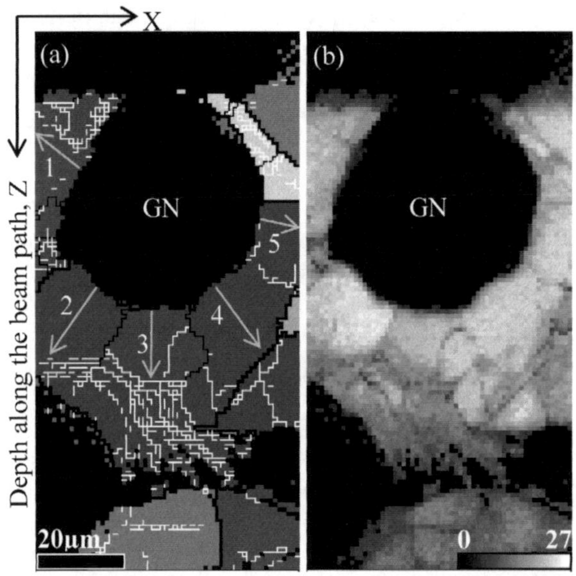

Figure 2. Microstructures of the DCI around the selected nodule GN characterized using white beam micro-diffraction. (a) is colored based on the crystallographic orientations of the matrix grains, while (b) is based on the number of indexed spots from the depth-resolved Laue diffraction images (see the inserted scale bar). In (a), dislocation boundaries with misorientation angles in the range of 0.1 – 1°, 1 – 3°, and > 3° are shown in thin white, thin black and thick black lines, respectively. The colors of the matrix grains correspond to the crystallographic orientation along the sample normal direction in the sample coordinate system (F direction in the XHF system in figure 1). Five numbered arrows mark places and directions of the misorientation analysis and quantification of dislocation density shown in figure 3.

The GND density as a function of distance from nodule/matrix interface along the 5 lines is shown in figure 3b. Despite large variations in the plot, the general tendency is that the GND density decreases within the first 5-7 μm from the interface. The increase of GND density at distances of 10-20 μm from the interface may be related to the influence of other nodules, damage during surface polishing and/or pearlite structures. The average accommodation strain within the first 5-7 μm from the interface corresponds to a dislocation density of GNDs of ~4 × 10^{12} m^{-2}.

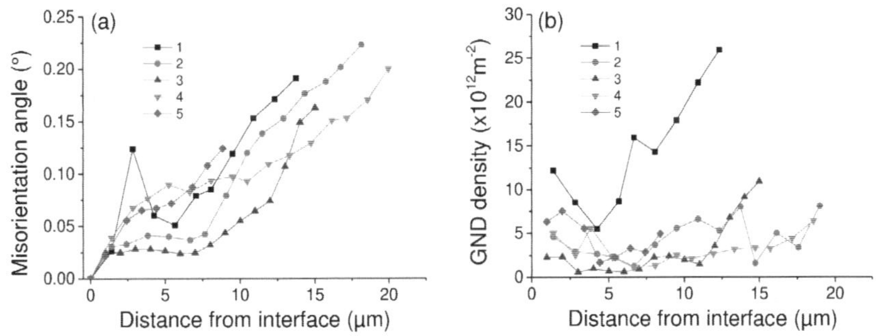

Figure 3. (a) Misorientation angles to the first point and (b) GND density estimated based on the misorientation between neighboring points along the 5 lines marked in figure 2.

3.2. Residual elastic strains

The deviatoric strain components, which represent the shape change of the crystal lattice, were determined based on the indexed depth-resolved Laue diffraction patterns. The resulting strain component maps for the matrix grains around nodule GN are shown in figure 4. The strains were determined in the sample coordinate system (i.e. XHF system in figure 1). As discussed in Refs. [9], dislocation boundaries in the matrix grains result in spot streaking in the Laue diffraction patterns and in turn exaggerate severely the deviatoric strains. Therefore, in the present study, only parts of a few grains (see figure 4) are included in the strain calculation. These parts of the grains have internal misorientation angle < 0.05°, implying that the effect of stored dislocations on spot streaking is limited, and the calculated strains are therefore trustworthy [3].

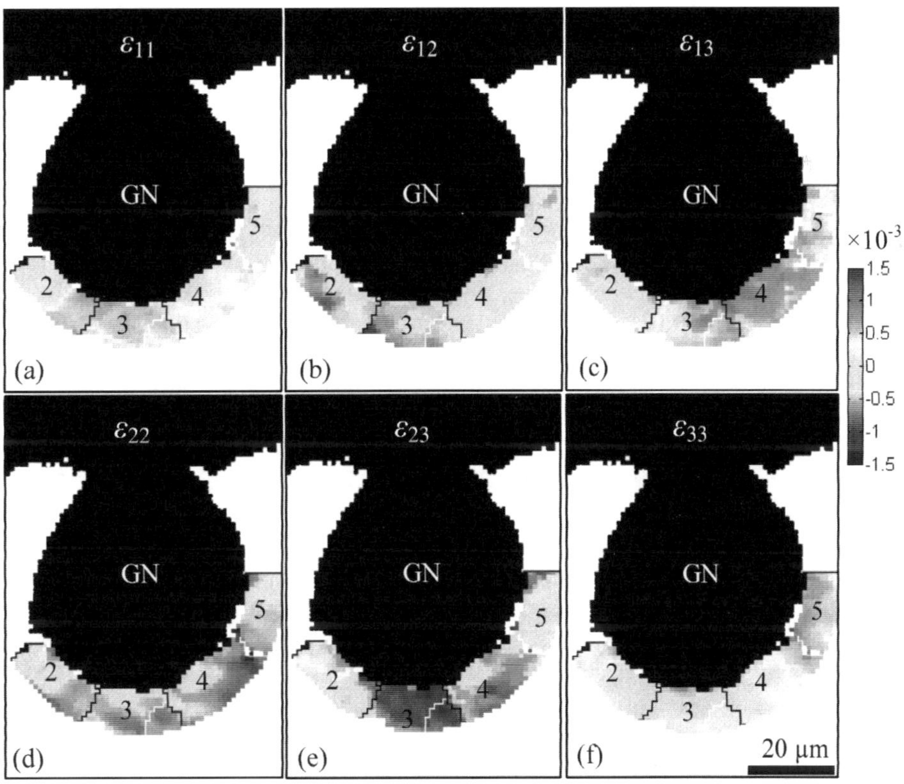

Figure 4. Strain maps colored according to the deviatoric strain components for the grains around nodule GN. Numbers are used to identify individual grains.

The strain patterns for the different strain components are quite different. When inspecting grains circumferentially around the nodule, the ε_{11} strains change gradually from mainly tensile in the bottom grain 3 to mainly compressive in grain 5. The opposite trend is seen for ε_{33}. Mainly compressive ε_{22} strains are seen for all 4 grains. For the shear strain components, the sense of ε_{13} changes from grain 2 to grains 3-5. The opposite trend is seen for ε_{23}. The majority of the ε_{12} strain has negative value, while the ε_{12} strain in grain 4 and 5 are close to zero.

Along the nodule radial directions, the strain pattern are rather complex. Strain gradients from the interface to the grain interior are seen for some strain components within some parts of the grains. For example, the magnitude of ε_{11} and ε_{33} in grains 3 and 5 decreases, while the magnitude of the compressive ε_{22} strain increases in grain 3 and in parts of grains 2, 4 and 5. In grain 4, the strain gradients for most of the strain components are less pronounced.

3.3. Finite element modelling

To understand the complex pattern of the residual strains seen in figure 4, a 3D finite element model of residual elastic strain, developed and presented in a previous publication [5], is used to calculate the residual stress/strain distribution. In this model, a graphite volume fraction of 11.5 % is used. Detailed parameters for the model can be found elsewhere [3,5].

The predicted stress maps on an X-Z section equivalent to that marked by white line in figure 1 are shown in figure 5. The predicted stress patterns are quite different for different stress components. However, symmetrical distributions with respect to the vertical Z axis or horizontal X axis or both axes can be seen for all the stress components. Due to the lack of experimental data, this symmetry cannot be examined for the experimental results.

By comparing the simulated stress pattern with the experimental pattern, some agreements are found. For example, for the ε_{11} strain component, changes from tensile at the bottom of the nodule gradually to compressive at the right side of the nodule along the nodule circumferential direction are seen both in the experimental and simulated strain pattern. Similar agreement is found for the ε_{33} strain component. For the ε_{12} component, negative and close to zero strains is seen both for the grains 4 and 5 in the experimental data and for the simulated results at the same locations. Similarly, negative ε_{23} strains for grains 3 and 4 agree with the predicted strains. However, the predicted ε_{22} strains, mainly tensile at the lower half of the nodule, are completely different from the experimental results. Similar disagreement is observed for ε_{13} component.

If the strain gradients along the radial direction are considered, the model predicts well-defined gradients for all strain components. Compared to the experimental strain gradients for ε_{11} and ε_{33} components, the model captures the gradient direction correctly. In spite of a different strain direction between the simulated and experimental results for ε_{22} component, the gradient direction is the same.

There could be several reasons for the disagreements between the simulated and experimental results. First, in the model the metal matrix is assumed to be elastically isotropic, while multiple grains with different crystallographic orientations (see figure 2a) reveal that the matrix is elastically anisotropic. Additionally, this orientation difference leads to different deformation behavior between grains [10]. Both factors can lead to local strain variations from grain to grain. More importantly, experimentally the plastic deformation is not homogeneous even within individual grains, leading to local orientation variations (see figures 2a and 3). As plastic deformation can relax the residual elastic strain/stress, the inhomogeneous plastic deformation therefore is expected to lead to inhomogeneous strain/stress distribution, which can even change the local strain gradients, along both the radial and circumferential directions. Secondly, an ideal spherical nodule is used in the model, while in reality the nodule is not perfectly spherical, which can lead again to local strain variations from location to location along the circumferential direction of the nodule. Thirdly, experimentally the nodule is exposed to free sample surface, while in the model the nodule is fully embedded in the bulk. As shown in the previous study [3], free surface can lead to reduced magnitude of ε_{33} strain for the grains at the bottom part of the nodule. To fully address how the free surface combined with a non-spherical shape of the nodule affects the strain changes for other strain components, additional simulation work is underway.

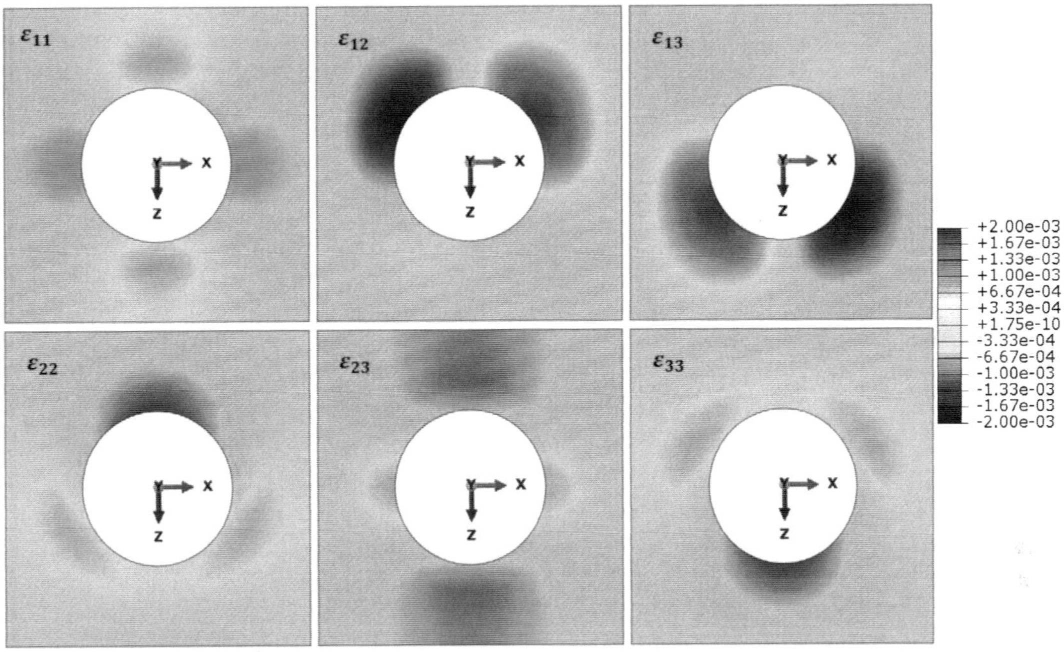

Figure 5. Strain maps colored according to the predicted strain components.

Irrespective of these reasons, a completely different strain direction for the ε_{22} and ε_{13} components between the experimental and simulated results cannot be understood. This suggest that the thermal residual stresses developed due to the difference in thermal expansion coefficients between graphite nodule and metal matrix may not be fully described by a simple thermal contraction process as used in the present model and other models based on Eshelby's equivalent inclusion method [11]. Further studies are required to clarify this aspect.

4. Conclusions

The synchrotron X-ray micro-diffraction method has been used in the present study to characterize the local microstructure and residual strain pattern in matrix grains around a graphite nodule in a ductile cast iron sample. To understand the experimentally measured strain pattern, a finite element model has been used to calculate the residual stress/strain distribution. The main findings are:

i) During cooling of DCI, differences in thermal contraction lead to deformation of the iron matrix both elastically and plastically. Within a distance of 5-7 μm from the surface of a nodule strain gradients are revealed corresponding to storage of geometrically necessary dislocations with an estimated density of $\sim 4 \times 10^{12}$ m^{-2}.

ii) Complex strain patterns are seen for all strain components of the matrix grains around the nodule. Strain gradients are seen along both the circumferential and radial directions for some grains and for some strain components.

iii) Finite element modelling considering simple thermal contraction processes captures some of the experimentally observed strain gradients along the circumferential and radial directions. However, disagreement for the ε_{22} and ε_{13} components between the predicted and experimental results suggests the mechanism for the development of thermal residual stress may be more complex than heretofore assumed.

Acknowledgments
Part of this work has been supported by the Strategic Research Center ''REWIND -Knowledge based engineering for improved reliability of critical wind turbine components,'' Danish Research Council for Strategic Research, grant no. 10-093966. Use of the Advanced Photon Source was supported by the U. S. Department of Energy, Office of Science, Office of Basic Energy Sciences, under Contract No. DE-AC02-06CH11357.

References

[1] Labrecque C and Gagne M 1998 *Can Metall Q.* **37** 343-378
[2] *Ductile iron data.*:http://www.ductile.org/ductile-iron-data-2/
[3] Zhang Y B, Andriollo T, Fæster S, Liu W, Hattel J and Barabash R I 2016 *Acta Mater.* **121** 173-180
[4] Andriollo T, Thorborg J, Tiedje N and Hattel J 2016 *Model Simulations Mater Sci Eng.* **24** 055012
[5] Larson B, Yang W, Ice G, Budai J and Tischler J 2002 *Nature* **415** 887-890
[6] Tischler J Z 2014 *Strain and Dislocation Gradients from Diffraction Ch. 10* Eds. Barabash R and Ice G E (London: Imperial College Press)
[7] Barabash R I, Ice G E and Pang J W L 2005 *Mater Sci Eng A.* **400-401** 125-131
[8] Hansen N 1994 *15th Risø Int. Symp. on Materials Science* (Roskilde: Risø) p 325-334
[9] Larson B C and Levine L E 2013 *J Appl Crystallogr.* **46** 153-164
[10] Huang X and Winther G 2007 *Philos Mag.* **87** 5189-5214
[11] Eshelby J D 1957 *Proc R Soc A Math Phys Eng Sci.* **241** 376-396

Enhanced mechanical properties in fully recrystallized ultrafine grained ZKX600 Mg alloy

Ruixiao Zheng[1], Tilak Bhattacharjee[1,2], Si Gao[1], Wu Gong[2], Akinobu Shibata[1,2], Taisuke Sasaki[3], Kazuhiro Hono[3] and Nobuhiro Tsuji[1,2]

[1] Department of Materials Science and Engineering, Kyoto University, Yoshida Honmachi, Sakyo-ku, Kyoto 606-8501, Japan
[2] Elements Strategy Initiative for Structural Materials (ESISM), Kyoto University
[3] National Institute for Materials Science, 1-2-1 Sengen, Tsukuba 305-0047, Japan

E-mail: zheng.ruixiao.7m@kyoto-u.ac.jp

Abstract. Mg-Zn-Zr-Ca (ZKX600) alloy specimens with various fully recrystallized grain sizes ranging from 0.77 μm to 23.3 μm were fabricated by high pressure torsion (HPT) followed by subsequent annealing. Tensile tests carried out at room temperature revealed that ultrafine grained (UFG) specimens exhibited enhanced combinations of strength and ductility compared to their coarse-grained counterparts. Observation of deformation microstructures suggested that basal slip and {10-12} extension twinning were the dominant mechanisms during deformation of coarse grained specimens, while deformation twinning was significantly inhibited in the UFG specimens. Based on the observations it is proposed that non-basal slip activity is responsible for the enhanced mechanical properties, and should be the focus of further investigations.

1. Introduction

Due to their ultra-low density, and high specific strength, magnesium (Mg) alloys have been increasingly used as structural components in the automotive and aerospace industries to achieve weight reduction and fuel economy improvements [1-3]. However, their anisotropic hexagonal close packed (HCP) crystallographic structure provides only a limited number of independent deformation modes, resulting in relatively low tensile ductility as well as poor formability at room temperature [4].

Grain refinement is considered as one of the promising ways for optimizing the strength of Mg alloys without sacrificing ductility [5]. However, the microstructures of Mg alloys refined by conventional processes (such as rolling and extrusion) are usually inhomogeneous and the mean grain sizes are usually above 10 μm. In contrast, severe plastic deformation (SPD) processes, such as equal channel angular pressing (ECAP), accumulative roll bonding (ARB) and high pressure torsion (HPT), have already been proven as effective approaches for realizing ultra-grain refinement in various metallic materials [6]. It is, however, quite challenging to refine the grain size of Mg alloys by using ECAP and ARB because of cracking and fracture during heavy deformation. For example, Cetlin *et al.* [7] have shown very clear examples of serious segmentation in an annealed ZK60 Mg alloy and an extruded AZ31 Mg alloy processed by ECAP only after one pass at different temperatures.

In contrast, a high hydrostatic pressure is applied to disc-shaped specimens during torsion deformation in HPT, so that undesirable cracking can be avoided even at room temperature [8].

Recently, Meng *et al.* [9] carried out HPT deformation of a Mg-3.4Zn (at.%) alloy at ambient temperature and found that equiaxed grains with an average diameter of 140 nm could be obtained after HPT by 20 revolutions. However, due to the high density of retained lattice defects such as dislocations, the HPT-processed UFG Mg alloys did not show good ductility. Thus, their mechanical properties were characterized only by using simple hardness testing. Such a limited ductility resulting from the characteristics of the deformed microstructure is a common feature of SPD-processed metallic materials.

In order to overcome the above deficiency of HPT-processed Mg alloys, a process including room temperature HPT followed by a subsequent annealing treatment is proposed in the present study. A Mg-Zn-Zr-Ca (ZKX600) alloy is selected for realizing optimal nanostructures, because this alloy contains a high density of thermally stable nano-precipitates even in the as-cast state. It is expected that these nano-precipitates should be effective for controlling the grain size during thermo-mechanical processes.

2. Experimental Procedures

An as-received ZKX600 Mg alloy ingot with a composition of Mg-6.2%Zn-0.5%Zr-0.2%Ca (mass%) was firstly homogenized at 350 °C for 48 h followed by a solid solution treatment at 400 °C for 1 h using a box-type resistance furnace. Disc-shaped samples with a diameter of 10 mm and thickness of 0.80 mm were cut from the ingot for HPT. The HPT deformation was conducted at room temperature under a compressive pressure of 6 GPa at a rotation speed of 0.2 rpm. The total rotation angle applied was 360°. After the HPT process, the specimens were given an annealing treatment at various temperatures for different periods of time, using a salt bath furnace.

The areas at a radial distance of 3.0 mm from the center on transverse sections containing the rotation axis of the HPT-processed discs were observed by transmission electron microscopy (TEM) using a JEOL 2010 microscope operated at 200 kV. The areas at a distance of 3.0 mm from the center on sections perpendicular to the rotation axis of the annealed discs were characterized by field emission scanning electron microscopy (FE-SEM) using a JSM 7100F scanning electron microscope equipped with an electron backscatter diffraction (EBSD) system. Grain sizes were measured using a linear intercept method based on the obtained EBSD maps.

The mechanical properties of the specimens were characterized by uniaxial tensile testing. Small-sized tensile specimens with a gauge length of 2 mm and a cross section of 1 mm × 0.5 mm were cut from the discs such that the center of the gauge length was coincident with a radial distance of 2.5 mm from the center of the HPT disks with the tensile direction taken perpendicular to the radial direction of the discs. Tensile tests were carried out at a quasi-static strain rate of 8.3×10^{-4} s^{-1}, with a direct measurement of the displacement of the gauge section using a CCD video camera extensometer.

3. Results and Discussion

Figure 1 shows microstructures of the as-solution treated (ST) (a,b), 180° HPT-processed (c), and 360° HPT-processed (d) specimens. Figure 1(a) is an EBSD inverse pole figure (IPF) map indicating crystallographic orientations parallel to the normal direction of the as-ST disc specimen. The EBSD-IPF map reveals that the as-ST specimen has a coarse grained structure with a mean grain size of about 57 μm and did not have strong crystallographic texture. TEM observation (figure 1(b)) indicated that a high density of rod-shaped nano-precipitates (β-phase) with an average length of 35 nm are retained, even after the solution treatment [10]. After HPT by 180°, an ultrafine lamellar structure, elongated along the shear direction, was obtained. The mean spacing of the lamellar boundaries was about 100 nm, as indicated by the dark field TEM image shown in figure 1(c). After further deformation by HPT to 360°, the lamellar structure was fragmented and replaced by a nearly equiaxed nanocrystalline structure, as shown in figure 1(d). The mean grain size of the nanocrystals was about 100 nm, which is significantly finer than that of pure Mg (grain size ~1 μm) subjected to the HPT process up to similar rotation angles [11]. The result indicates that the alloying elements in the present Mg alloy play an important role in reducing the grain size that can be obtained by SPD.

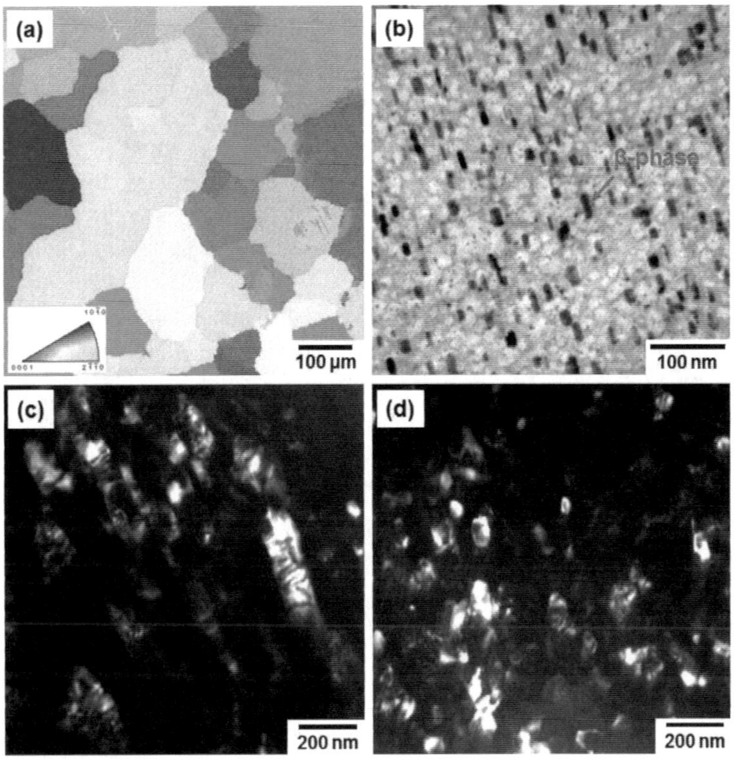

Figure 1. (a) EBSD-IPF map and (b) bright field TEM image of the as-ST specimen showing a coarse grained structure with a mean grain size of 57 μm and a high density of nano-precipitates (β-phase). (c) Dark field TEM image of the 180° HPT processed specimen showing an ultrafine lamellar structure. (d) Dark field TEM image of the 360° HPT processed specimen showing a nearly equiaxed nanocrystalline structure with a mean grain size of about 100 nm.

Figure 2 shows EBSD results of the specimen HPT-processed by 360° and then annealed at 400 °C for 1 min (a-c) and 500 °C for 30 min (d-f). The colors in the IPF maps (a, d) indicate crystallographic orientations parallel to the normal direction of the HPT discs. The blue and green lines in the grain boundary (GB) maps (b, e) correspond to high angle grain boundaries (HAGBs) having misorientations larger than 15° and low angle grain boundaries (LAGBs) with misorientations between 2° and 15°. After rapid annealing at 400 °C for 1 min, a fully recrystallized UFG structure with a mean grain size of 0.98 μm was successfully obtained (figure 2(a,b)). In addition, the peak intensity in the (0001) pole figure (figure 2(c)) is low and the peak intensity regions located far from the center of the pole figure, indicating that the strong basal texture usually observed in wrought Mg alloys was not seen in the present annealed specimen. After the high temperature annealing at 500 °C for 30 min, a coarse grained specimen with a mean grain size of 23.3 μm was obtained (figure 2(d,e)) and the texture was further weakened and randomized (figure 2(f)).

For understanding the mechanical properties of the alloy processed under different conditions, tensile testing was carried out on specimens with various grain sizes. The resulting nominal stress-strain curves are presented in figure 3. The 360° HPT-processed specimen (d = 0.1 μm) showed a very high yield strength (YS, 0.2% offset of nominal stress) of 311 MPa, but the stress-strain curve became unstable immediately after yielding, resulting in a quite limited tensile elongation of 1.2%. This is typical of the mechanical behavior of nanocrystalline or UFG materials having deformed microstructures [12]. It should be noted, on the other hand, that fully recrystallized UFG specimens showed very good combination of strength and ductility. In particular, the YS, ultimate tensile strength (UTS), uniform elongation and total elongation of the specimen having a mean grain size of 0.77 μm were 235 MPa, 328 MPa, 20.5% and 26.1%, respectively. The results clearly demonstrate that by making the microstructure a fully recrystallized UFG, the tensile ductility can be remarkably retained without sacrificing the tensile strength. With increasing grain size, the YS decreases, but the ductility (total elongation) remains almost constant for mean grain sizes smaller than 6.11 μm. Further increasing the grain size results in a simultaneous decrease in strength and ductility. For instance, the YS, UTS, uniform elongation and total elongation of the specimen with mean grain size of 23.3 μm are 90 MPa, 256 MPa, 17.6% and 20.3%, respectively.

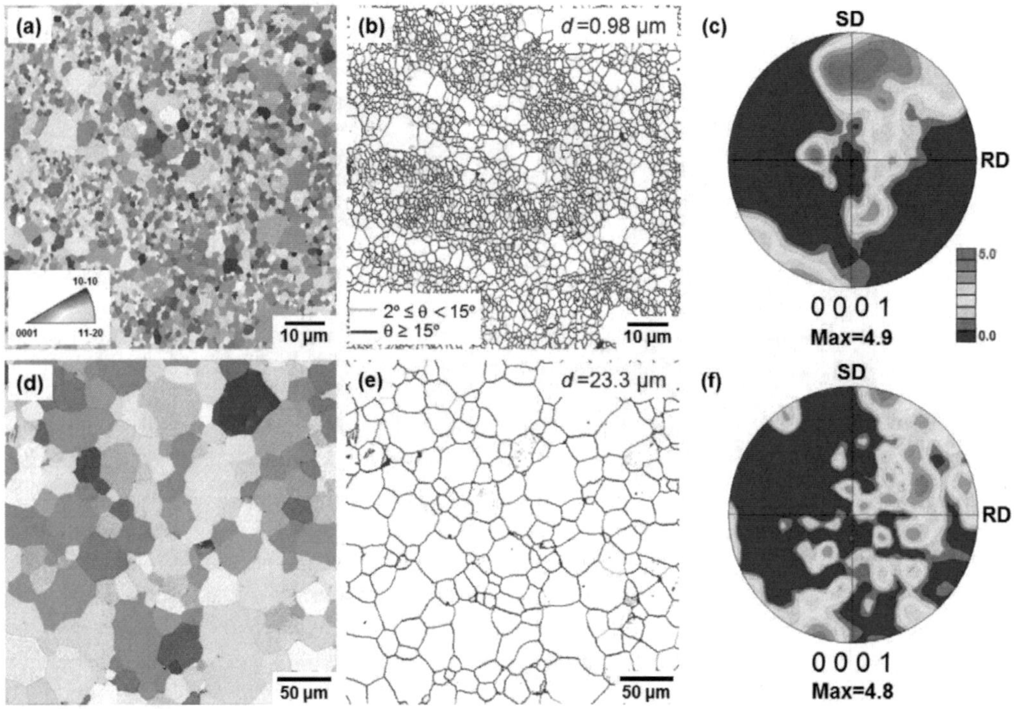

Figure 2. (a, d) IPF maps, (b, e) GB maps and (c, f) (0001) pole figures obtained from EBSD measurements and analysis of the specimens HPT-processed by 360° rotation and then annealed at (a-c) 400 °C for 1 min and (d-f) 500 °C for 30 min. SD and RD in the pole figures represent the shear direction and the radial direction, respectively.

In order to reveal the underlying deformation mechanisms contributing to the enhanced tensile ductility in the fully recrystallized UFG specimens, the deformation microstructures were observed after tensile deformation to specified strains. Figure 4 shows EBSD-IPF maps of a coarse grained specimen (d = 23.3 μm) and an UFG specimen (d = 0.98 μm) after tensile deformation to similar strains. Colors in the IPF maps indicate crystallographic orientations parallel to the normal direction of the sheet-type tensile specimens. As shown in figures 4(a) and (b), a high density of deformation twins, with an area fraction of 18.6%, was generated in the coarse grained specimen after a tensile strain of 15%. Crystallographic analysis indicated that the twins were exclusively {10-12} extension twins. It is known that {10-12} extension twinning has much lower critical resolved shear stress than other twinning systems [13]. In contrast, a quite limited number of deformation twins with an area fraction of only 1.5% was seen in the UFG specimen after a similar tensile strain of 14% (see figure 4 (c) and (d)).

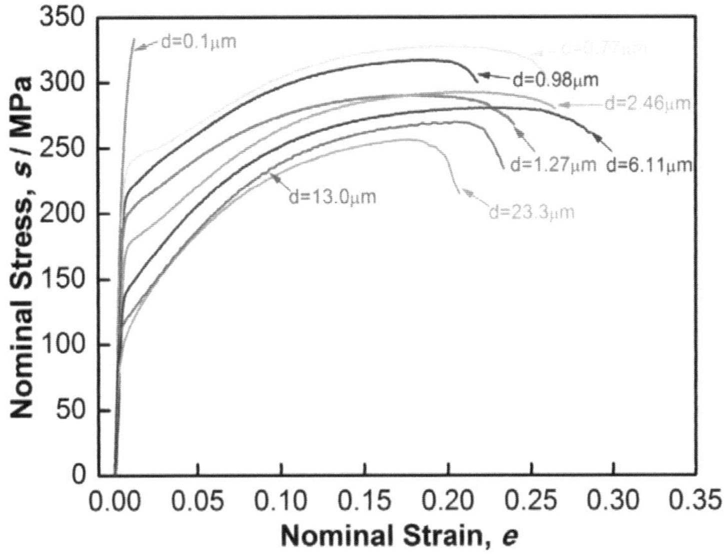

Figure 3. Nominal stress-strain curves of specimens with different mean grain sizes.

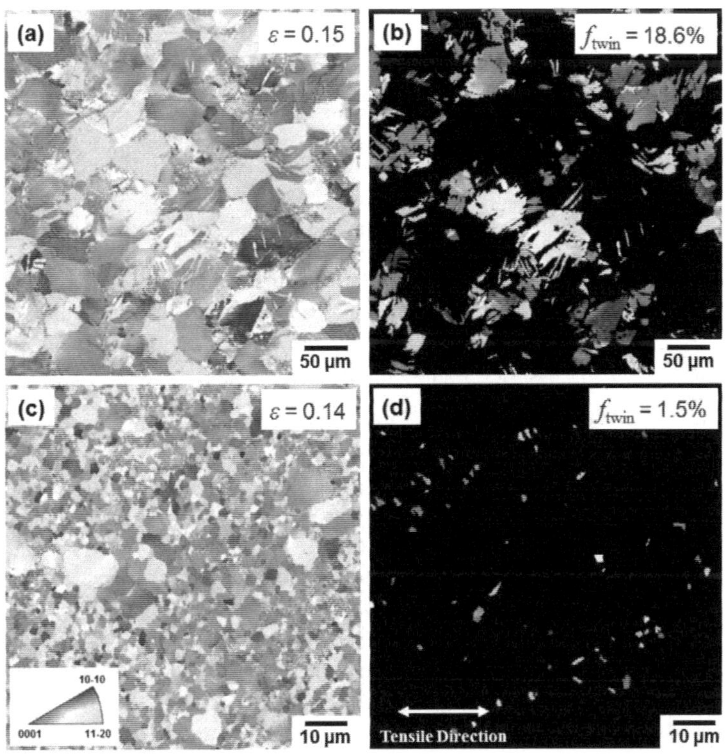

Figure 4. EBSD results of (a, b) a coarse grained specimen (d = 23.3 μm) and (c, d) an UFG specimen (d = 0.98 μm) after tensile deformation to similar strains. (a) IPF map and (b) corresponding twin-partitioned map of the coarse grained specimen after 15% tensile deformation, showing a large area fraction of deformation twins (f_{twin} = 18.6%). (c) IPF map and (d) corresponding twin-partitioned map of the UFG specimen after 14% tensile deformation, showing a very small area fraction of deformation twins (f_{twin} = 1.5%).

Although deformation twinning was significantly inhibited in the UFG specimen, the sample exhibited better ductility than the coarse grained specimen. Thus, there must be some other deformation mechanisms responsible for the enhanced strain-hardening and ductility, for example, the activation of non-basal slip systems. Because the Burgers vector of the primary slip system, i.e., basal slip, of Mg and Mg alloys only contains an <***a***> component, the operation of any deformation mechanism involving a <***c***> deformation component is necessary for large plasticity. Koike *et al.* [14] reported a large tensile elongation up to 47% in a commercial AZ31 alloy having a mean grain size of 6.5 μm. They concluded that the superior tensile ductility was due to the activation of <***c+a***> non-basal slip. Similar activation of unusual slip systems that including a <***c***> component is expected in the present Mg-Zn-Zr-Ca alloy having fully-recrystallized UFG structures. We are planning to make further investigations to clarify this issue in the present material.

4. Conclusion

In summary, fully recrystallized Mg-Zn-Zr-Ca alloy specimens with various mean grain sizes ranging from 0.77 μm to 23.3 μm were successfully fabricated by the combination of HPT and subsequent annealing. Simultaneously enhanced strength and ductility were realized in fully recrystallized UFG specimens. A detailed comparison of the deformation microstructures of an UFG specimen and a coarse-grain specimen clarified that deformation twinning was significantly inhibited in the UFG specimen. It is suggested activation of non-basal slip systems having <***c***> component plays an important role in the enhanced ductility of recrystallized UFG samples, and this will be investigated further in future studies.

Acknowledgments
This work was financially supported by the Elements Strategy Initiative for Structural Materials (ESISM) and the Grant-in-Aid for Scientific Research (S) (No. 15H05767), both through the Ministry of Education, Culture, Sports, Science and Technology (MEXT), Japan.

References

[1] Kim N 2014 *Mater. Sci. Technol.* **30** 1925
[2] Suh B, Shim M, Shin K and Kim N 2014 *Scr. Mater.* **84** 1
[3] Hono K, Mendis C, Sasaki T and Oh-Ishi K 2010 *Scr. Mater.* **63** 710
[4] Agnew S and Nie J 2010 *Scr. Mater.* **63** 671
[5] Kim W, Jeong H and Jeong H 2009 *Scr. Mater.* **61** 1040
[6] Tsuji N, Saito Y, Lee Y and Minamino Y 2003 *Adv. Eng. Mater.* **5** 338
[7] Cetlin P, Aguilar M, Figueiredo R and Longdon T 2010 *J. Mater. Sci.* **45** 4561
[8] Zhilyaev A and Langdon T 2008 *Prog. Mater. Sci.* **53** 893
[9] Meng F, Rosalie J, Singh A, Somekawa H and Tsuchiya K 2014 *Scr. Mater.* **78** 57
[10] Bhattacharjee T, Nakata T, Sasaki T, Kamado S and Hono K 2014 *Scr. Mater.* **90** 37
[11] Edalati K, Yamamoto A, Horita Z and Ishihara T 2011 *Scr. Mater.* **64** 880.
[12] Tsuji N, Ito Y, Saito Y and Minamino Y 2002 *Scr. Mater.* **47** 893
[13] Barnett M 2007 *Mater. Sci. Eng. A* **464** 1
[14] Koike J, Kobayashi T, Mukai T, Watanabe H, Suzuki M, Maruyama K and Higashi K 2003 *Acta Mater.* **51** 2055